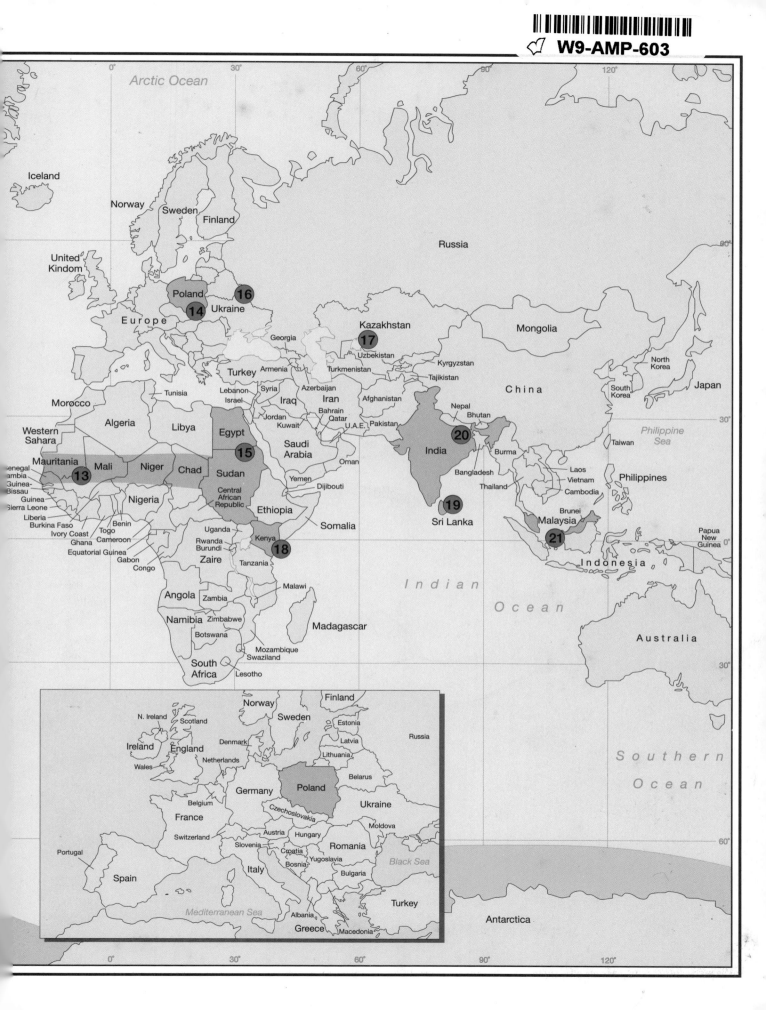

ENVIRONMENT

# Peter H. Raven
Missouri Botanical Garden

# Linda R. Berg
University of Maryland

# George B. Johnson
Washington University

*Saunders College Publishing*
Harcourt Brace College Publishers
Fort Worth   Philadelphia   San Diego   New York
Orlando   Austin   San Antonio   Toronto
Montreal   London   Sydney   Tokyo

ENVIRONMENT

Text Typeface: Goudy Old Style
Compositor: York Graphic Services, Inc.
Acquisitions Editor: Julie Levin Alexander
Development Editors: Jane Tufts, Gabrielle Goodman
Managing Editor: Carol Field
Senior Project Manager: Marc Sherman
Project Editor: Janet Nuciforo
Copy Editors: Mary Patton, Andy Potter
Manager of Art and Design: Carol Bleistine
Art Assistant: Caroline McGowan
Text Designer: Tracy Baldwin
Layout Artist: Dorothy Chattin
Cover Designer: Lawrence R. Didona
Text Artwork: Nadine Sokol; Elizabeth Rohne Rudder; Gary Williamson; Pagecrafters, Inc.; Rolin Graphics, Inc.
Photo Researcher: Robin Bonner
Permissions Editor: Dena Digilio-Betz
Director of EDP: Tim Frelick
Production Manager: Charlene Squibb


Cover Credit: Illustration by Pat Morrison


Frontmatter photo credits: Half-Title and Title page spread, Frans Lanting/Minden Pictures; Foreword, Larry C. Price; Preface, courtesy of U.S. Windpower, Livermore, Calif.; Contents Overview, Chip Clark; Contents, E. R. Degginger.


Printed in the United States of America


ENVIRONMENT

ISBN 0-03-014329-2

Library of Congress Catalog Card Number: 92-056728

3 4 5 6  032  9 8 7 6 5 4 3 2

THE EARTH SUMMIT, held in Rio de Janeiro during the first two weeks of June 1992, stands as a warning about the most critical issue of our time, and as a beacon of hope. Representatives of 178 nations gathered to look for new ways to manage the planet Earth together and in harmony. The issue was the environment, and the challenge was to avoid its destruction. Danger signs were and are everywhere, in a world divided decisively into "haves" and "have-nots," and it has become critical that the two begin to cooperate more effectively. The ways in which we can all work together will determine what kind of a life our children and grandchildren will lead, what kinds of expectations they can reasonably have, and how they will be able to contribute to our common prosperity.

This book is intended to be a contribution to your understanding of the way the world works, and what is happening to it as its human population expands. The environmental sciences have usually been viewed in the past as a series of disconnected subjects with few integrating themes. In this book we attempt to tie these many statements together, for only with such a synthesis can a student understand how the world works, and what we can expect of it. Every educated person needs to know the principles that are involved, and each of us must strive to understand the fundamental issues that are presented here, so that he or she can make informed decisions about appropriate actions to take. Our future way of life will be based, ultimately, on our ability to deal with the Earth intelligently.

Since 1950, the population of the world has grown from 2.5 billion people to over 5.4 billion; a fifth of the topsoil that makes it possible for us to grow the crops that we eat and feed to our domestic animals has been lost; a third to a half of all forests, depending on the region of the world, has been cut over; the characteristics of the atmosphere have been changed drastically, with thin spots in the stratospheric ozone subjecting us to damaging ultraviolet radiation and increases in carbon dioxide and other greenhouse gases inexorably leading to global warming; and thousands of species of plants, animals, fungi, and microorganisms are being lost forever with every passing year. The 23% of the world's population who live in industrialized countries (a rapidly decreasing fraction of the total) are consuming about 80–90% of what the world is capable of producing, while the 77% of people who live in developing countries have to make do with the rest. Poor people constitute more than one fifth of the world's population, with over a billion living on less than $1 a day; half of them are malnourished.

Over the next three decades, about 3 billion more people will be added to the world population, the greater majority of them in cities of the third world. Most of these people are likely to be condemned to a life of poverty and reduced expectations, as they cut over the remaining forests and exhaust the depleted soils of their native lands. We cannot sit aside and watch. The ability of our country to interact trade with other nations, critical to our economic welfare, will be sustained only to the extent that we are able to remain in meaningful contact with other nations and contribute to their stability—while our own resources are being exhausted also.

For all of these reasons, it is necessary for informed citizens everywhere to take effective steps to counteract the global problems that we confront. In dealing with them, we will find new solutions to our own problems, and help to secure our own future. Pessimism is worthless as an attitude or a strategy. Rather, knowledge must be used to provide the key to effective action in a future that may only dimly resemble the familiar past. We offer you this book as a means of learning the basic facts about how the world functions, and hope that it may help to provide the tools that you will need to lead full and complete lives. We hope that it inspires you to seek additional knowledge and to take the kind of meaningful action on which our common future so clearly depends.

PETER H. RAVEN
St. Louis, Mo.
*January 1993*

THE CHALLENGE of creating and maintaining a sustainable environment is probably the single most pressing issue that will confront students throughout their lives. Today, environmental science is not only relevant to students' personal experience, but vital to the future of the entire planet. As humans increasingly alter Earth's land, water, and atmosphere on local, regional, and global levels, the resulting environmental problems can seem insurmountable. Armed with the proper tools, however, students need not find these issues overwhelming. *Environment* equips students with the most essential of these tools: an understanding of the concepts that underlie the problems.

One of our principal goals in preparing this book is to convey to students an appreciation of the marvelous complexity and precise functioning of natural ecosystems. *Environment* begins with an exploration of the basic ecological principles that govern the natural world, and considers the many ways in which humans affect the environment. From the opening pages, we acquaint students with current environmental issues—issues that have many dimensions and that defy easy solutions. Later chapters examine in detail the effects of human activities, including overpopulation, energy production and consumption, depletion of natural resources, and pollution.

Although we do not sugarcoat these problems—many are very serious indeed—we try to avoid the gloomy predictions of disaster so common in environmental science textbooks today. Instead, students are encouraged to take active, positive roles, using the practical and conceptual tools presented in this book, to meet the environmental challenges of today and tomorrow.

*Environment* integrates important information from a number of different fields, such as biology, geology, chemistry, physics, sociology, government and politics, and demographics. Because environmental science is an interdisciplinary field, this book is appropriate for use in environmental science courses offered by a variety of departments, including (but not limited to) biology, geology, geography, and agriculture.

This book is intended as an introductory text for undergraduate students, both science and nonscience majors. Although relevant to all students, *Environment* is particularly appropriate for those majoring in education, journalism, political science/government, and business, as well as the traditional sciences. We assume our students have very little prior knowledge of how ecosystems work, how matter and energy move through ecosystems, and how population dynamics affects and is affected by ecosystems. These important ecological concepts and processes are presented in a straightforward, unambiguous manner.

## INSTRUCTIONAL FEATURES

*Environment* is written in an interesting, conversational style that will help students remember important concepts. The up-to-date coverage of environmental topics includes many unique applications and interesting case studies throughout. Numerous learning aids are used.

1. **Chapter outlines** reflect the main headings within each chapter and provide students with an overview of the material covered.
2. **Tables and graphs** summarize and organize information.
3. Carefully rendered **illustrations** and stunning color **photographs** support concepts covered in the text, elaborate on relevant issues, and add visual detail.
4. **New terms** are in boldface, permitting easy identification and providing emphasis. Occasional boxed **Mini-Glossaries** define closely related terms or potentially confusing new terms and are located within the chapter for handy reference.
5. **Focus On** boxes spark student interest, present applications of concepts discussed, and familiarize students with current issues.
6. **You Can Make a Difference** boxes suggest specific courses of action or lifestyle changes students can make to improve the environment.
7. **Envirobriefs** provide additional topical material about relevant environmental issues, highlighting comparative data across domestic and international regions. Envirobriefs can be used as a starting point for lively classroom discussions. Unless otherwise noted, the source for all Envirobrief data is *EcoSource*, a Canadian environmental research and writing group located in Guelph, Ontario.
8. **Meeting the Challenge** boxes discuss pressing environmental dilemmas that defy easy answers. Examples of how others have responded to challenges allow the student to consider effective problem-solving.
9. **Chapter summaries** are presented as numbered statements at the end of each chapter to provide a quick review of the material presented.
10. **Discussion Questions** encourage critical thinking and focus on important concepts and applications.
11. **Suggested Readings** provide current references for further learning.
12. Eight **interviews** with prominent individuals who are making a positive impact on the environment explore the issues and trends they consider to be most vital, and emphasize the importance of active involvement in environmental affairs.
13. Immediately following Chapter 24, a four-page spread is devoted to the **"Earth at Night,"** an annotated composite of satellite photos dramatically depicting the human presence on Earth.
14. Four **appendices** provide useful supplemental material. "A Review of Basic Chemistry" examines the fundamentals of chemistry for students who did not take chemistry in college or high school and for all other students to use as a handy reference. "How to Make a Difference" is a guide to environmental organizations and government agencies as well as information on internships. "Green Collar Professions" serves as a reference for students interested in pursuing careers in the field. "Units of Measure" provides useful conversions of units of measure used in the text.
15. A separate **glossary** is provided, facilitating rapid location of definitions.
16. In keeping with the global focus of *Environment*, the **index** provides country and U.S. state names for easy reference of regional issues.
17. A **world map** highlighting some of the environmentally sensitive areas discussed in the text appears inside the front cover and is conveniently chapter-referenced.

## THE ORGANIZATION OF *ENVIRONMENT*

### Part 1 Humans in the Environment

Chapter 1, Our Changing Environment, introduces environmental science and highlights a number of serious environmental issues. Chapter 2, Solving Environmental Problems, expands on the nature of science and presents the steps that should be followed to resolve environmental problems. Part 1 includes an interview with Lester Brown, founder of the Worldwatch Institute in Washington, D.C., on the Environmental Revolution.

### Part 2 The World We Live In

Part 2 provides a detailed introduction to the principles of ecology. It is organized around the ecosystem, which is the fundamental unit of ecology. Chapter 3, Ecosystems and Energy, discusses the linear flow of energy through ecosystems, and Chapter 4, Ecosystems and Living Things, examines the living organisms that comprise ecosystems. In Chapter 5, Ecosystems and the Physical Environment, the cycling of materials in ecosystems is discussed. The influence of climate on ecosystems is also examined in Chapter 5. Chapter 6, Major Ecosystems of the World, examines the major biomes of the terrestrial environment as well as major ecosystems of the aquatic environment. Human activities relating to economics and government policies are covered in Chapter 7, Ecosystems, Economics, and Government. Part 2 includes an interview with George Woodwell, Director of the Woods Hole Research Center in Massachusetts, on Global Ecology and the Human Factor.

### Part 3 A Crowded World

The principles of population ecology are discussed in Chapter 8, Understanding Population Growth. Although the human population is the focus of this chapter, the fact that human populations follow the same principles of population ecology as other living things is emphasized. Chapter 9, Facing the Problems of Overpopulation, examines sociological and cultural factors that affect human population growth. Part 3 includes an interview with Stanford University's noted population authorities, Anne and Paul Ehrlich, on A Population Policy for the Super-Consumers.

### Part 4 The Search for Energy

The environmental impact of the human quest for energy is considered in this section. Chapter 10, Fossil Fuels, discusses the problems associated with the use of oil, coal, and natural gas. In Chapter 11, Nuclear Energy, the use of nuclear power as a viable energy source in the future is considered. Chapter 12, Renewable Energy and Conservation, sur-

veys energy alternatives to fossil fuels and nuclear energy. Part 4 includes an interview with L. Hunter Lovins, co-founder of the Rocky Mountain Institute in Colorado, on Efficiency Technology: Less Energy, More Power.

### Part 5 Our Precious Resources

Overusing and abusing our natural resources is considered in detail in Part 5. Chapter 13, Water: A Fragile Resource, describes the problems that can arise from an overabundance or a lack of water resources. In Chapter 14, Soils and Their Preservation, the significance of soil, the least appreciated natural resource, is explained. Chapter 15, Minerals: A Nonrenewable Resource, discusses some of the problems associated with our increasing use of minerals. In Chapter 16, Wildlife: Our Plant and Animal Resources, the importance of biological diversity is developed. Chapter 17, Land Resources and Conservation, examines how we use land and the important ecological contributions made by natural areas. Chapter 18, Food Resources: A Challenge for Agriculture, explains the challenge of providing enough food to the ever-expanding human population. Part 5 includes two interviews, one with Russell Train, Chairman of the World Wildlife Fund in Washington, D.C., on Preserving Biological Diversity in the Developing World, and one with Richard Mahoney, Chairman and CEO of Monsanto Company in St. Louis, on Delivering Biotechnology to Developing Nations.

### Part 6 Environmental Concerns

The effects of pollution are examined in this Part. Chapter 19, Air Pollution, looks at the local effects of air pollution, including indoor air pollution, whereas Chapter 20, Global Changes, considers regional and global effects of air pollution: acid deposition, global climate change, and stratospheric ozone destruction. Chapter 21, Water and Soil Pollution, discusses the closely related issues of water and soil pollution. Pesticides pollute air, water, soil, and food and contaminate so much of the biosphere that they are considered in a separate chapter, Chapter 22, The Pesticide Dilemma. Chapter 23, Solid and Hazardous Wastes, examines the problem of disposing materials we no longer need or want. Part 6 contains an interview with Frank Press, President of the National Academy of Sciences, on Transnational Cooperation in Environmental Science.

## Part 7 Tomorrow's World

Part 7 concludes the book with Chapter 24, Tomorrow's World, which presents the opinions of the authors on social responsibilities, identifying some of the most critical issues that must be grappled with today in order to assure a better tomorrow. Because Chapter 24 provides a comprehensive overview of current global problems and emphasizes the way in which all are interrelated, it can be assigned at any time during the course. Part 7 contains an interview with Jerry Greenfield, co-founder of Ben & Jerry's Ice Cream in Vermont, on The Business of Social Responsibility.

## SUPPLEMENTS

The package accompanying *Environment* is the most comprehensive on the market, and includes several unique items developed specifically to augment students' understanding of environmental issues and concerns. Together, these ancillaries provide instructors and students with interesting and helpful teaching and learning tools.

**1. Regional Environmental Issues Supplements** contain a wealth of information about environmental issues specific to eight different regions: Northeast, MidAtlantic, Southeast, Great Lakes, Midwest, Southwest, Northwest, and Canada. Regional Supplements are available shrinkwrapped free with the text.

These innovative regional supplements provide students with compelling and thought-provoking environmental issues. Each regional packet includes background material such as essays or reports by scientists, activists, businesses, or government agencies; editorials or news articles; policy statements; proposed legislation; and opposition statements. Additionally, discussion exercises address ten environmental issues, encouraging students to analyze issues and use problem-solving skills either in verbal or written form. An outline enables instructors to organize and lead the discussion.

To inspire further student interest and activity, the supplements contain lists of environmental groups in each state; government agencies; and addresses and telephone numbers of key local, state, and federal officials. Pertinent essays and articles suggest ways in which students can contribute positively to energy and environmental conservation.

**2.** Two **videos** feature Peter Raven discussing two critical topics: destruction of the world's rain forests and species extinction.

**3. Instructor's Resource Manual with Test Bank** prepared by Dr. Jacqueline Webb of Friday Harbor Laboratories, University of Washington contains a chapter outline, chapter objectives, a lecture outline, key terms, and teaching tips; suggestions for discussion topics, classroom demonstrations, and field trips; and audiovisual and book resources. The Test Bank consists of short-answer essays, and true/false, multiple-choice, and fill-in questions.

**4. Overhead transparencies** include 50 transparencies and slides with full-color figures from *Environment*. Instructors receive 20 transparency masters. Each figure is labeled with large-print labeling for easy viewing in any classroom.

**5. ExaMaster+ Computerized Test Bank** enables instructors to create or modify tests derived from the Printed Test Bank, and includes the capability to print out tests with answer keys and student answer sheets. Available in $5\frac{1}{4}''$ IBM, $3\frac{1}{2}''$ IBM, and Macintosh.

**6.** The **Laboratory Manual** by Robert Wolff of Trinity Christian College assists in the teaching of laboratory, problem-solving, and thinking skills in environmental science.

## ACKNOWLEDGMENTS

The development and production of *Environment* was a process involving interaction and cooperation among the author team and between the authors and many individuals in our home and professional environments. We appreciate the valuable input and support from family, friends, editors, colleagues, and students. We thank our families and friends for their understanding, support, and encouragement as we struggled through many revisions and deadlines.

### The Editorial Environment

Preparing this book has been hard work, but working with the outstanding editorial and production staff at Saunders College Publishing has been enjoyable. We thank our Editor-in-Chief Elizabeth Widdicombe for her support and help. Julie Alexander (Senior Acquisitions Editor) expertly launched the project, guiding the book through the channels of development and production. We appreciate the willingness of Liz and Julie to stand by us and maintain our good humor throughout the process of intensive work.

Developmental Editor Susan Driscoll gave us valuable input during the beginning phase of the project. Susan was succeeded by Developmental Editor Jane Tufts, whose experience and insight helped us establish the proper tone for *Environment*. Jane's line-by-line editing improved the style and readability of our prose immensely. She did an outstanding job of helping us sort through and make decisions about sometimes conflicting suggestions from numerous reviewers. Jane was ably assisted by Linda Allen, who provided line-by-line editing for many of the chapters. We are indebted to Gabrielle Goodman, who edited the book's interviews, coordinated the final stages of development, and provided us with expert suggestions before the project went into production.

We thank Ed Dolan, co-author of *Economics*, 6th edition (Dryden Press, Hinsdale, Illinois, 1992). Ed provided the inspiration and much of the content for Chapter 7, Ecosystems, Economics, and Government. We thank Tony Leighton, Editor of *EcoSource* magazine, who researched and wrote most of the text's Envirobriefs, and Jennifer Berg, who prepared the glossary.

We thank Photo Researcher Robin Bonner for helping us find the wonderful photographs that enhance the text. Nadine Sokol helped conceptualize and draw much of the art; Liz Rudder of Pagecrafters, Inc., designed most of the figures with graphs and maps. Art Director Carol Bleistine coordinated the art development and ensured that a consistent standard of quality was maintained throughout the illustration program.

We are grateful to Alison Munoz, Lori Weaber, and Gary Porter of Porter Productions for overseeing and coordinating the development of the supplemental package. Photo and Permissions Editor Dena Digilio-Betz obtained permissions for the vast amount of timely information so important to this book.

We greatly appreciate our Project Editor, Marc Sherman, for guiding us through the many deadlines of production. When Marc was promoted to another position within Saunders College Publishing, he was replaced by Janet Nuciforo, who continued to provide the strong leadership needed during the final stages of production.

Our colleagues and students have provided us with valuable input and have played an important role in shaping *Environment*. We thank them and ask for comments and suggestions from instructors and students who use this text. You can reach us through our editors at Saunders College Publishing.

## The Professional Environment

We here express our thanks to the many professors and researchers who have read the manuscript during various stages of its preparation and provided us with valuable suggestions for improving it. Their critiques and comments have contributed greatly to our final product. They are as follows:

Richard Bates
*Rancho Santiago Community College*
Bruce Bennett
*Community College of Rhode Island-Knight Campus*
Kelly Cain
*University of Arizona*
Ann Causey
*Prescott College*
Gary Clambey
*North Dakota State University*
Harold Cones
*Christopher Newport College*
Bruce Congdon
*Seattle Pacific University*
Donald Emmeluth
*Fulton-Montgomery Community College*
Kate Fish
*EarthWays*, St. Louis, Missouri
Neil Harriman
*University of Wisconsin-Oshkosh*
Denny Harris
*University of Kentucky*
John Jahoda
*Bridgewater State College*
Jan Jenner
*Talladega College*
David Johnson
*Michigan State University*
Norm Leeling
*Grand Valley State University*
Joe Lefevre
*SUNY at Oswego*
Gary Miller
*University of North Carolina at Asheville*
Robert Paoletti
*Kings College*

Ervin Poduska
*Kirkwood Community College*

W. Guy Rivers
*Lynchburg College*

Barbra Roller
*Miami Dade Community College-South Campus*

Richard Rosenberg
*The Pennsylvania State University*

Lynda Swander
*Johnson County Community College*

Jack Turner
*University of South Carolina at Spartanburg*

Jeffrey White
*Indiana University*

James Willard
*Cleveland State University*

P.H.R.
L.R.B.
G.B.J.

# Contents Overview

Humans in increasing numbers are encroaching upon the traditional environments of native peoples, such as this Amazonian boy. *(Loren McIntyre)*

**Humans in the Environment**

Part 1

# Lester Brown

## The Environmental Revolution

*In 1974, Lester Brown founded the Worldwatch Institute, a private, nonprofit research institute devoted to the analysis of global environmental issues. Based in Washington, D.C., the Institute is internationally recognized for its bimonthly magazine,* World Watch, *and for its annual series assessing environmental affairs worldwide,* State of the World. *Hailed as the "guru of the global environmental movement," Mr. Brown received his undergraduate degree in agricultural science from Rutgers University, and has an M.S. in agricultural economics from the University of Maryland and an M.P.A. from Harvard. In 1964, he was a foreign agricultural policy advisor to the U.S. Secretary of Agriculture, and in 1966 was appointed Administrator of the International Agricultural Development Service. In 1969, he helped establish the Overseas Development Council. Winner of the United Nations' 1989 environment prize, Mr. Brown is the author of a dozen books, including* Man, Land and Food; World Without Borders; By Bread Alone; *and* Building a Sustainable Society.

**In State of the World 1992, you call for an "Environmental Revolution." What does this mean and how might it occur?**

The health of our planet has deteriorated dangerously during the past twenty years. We have so far failed to stem the tide of continuing rapid human population growth leading to deforestation, loss of plant and animal habitats and ultimately species extinction, overgrazing, soil erosion, and loss of productive cropland, and of water and air pollution with their attendant health costs and potential for altering the entire ecosphere. The decline in living conditions once predicted by some ecologists has become a reality for one sixth of humanity. As a result, our world faces potentially convulsive change. The question is, what sort of change? Will we enact strong worldwide initiatives that reverse the degradation of the planet and restore hope for the future? Or will we suffer the economic decline and social instability that will result from continued environmental deterioration? The policy decisions we make in the years immediately ahead will determine whether our children live in a world of development or decline. Muddling through will not work. Either we turn things around quickly, or the deterioration-and-decline scenario will take over.

**What will it take to reverse these trends?**

Building an environmentally sustainable future depends on restructuring the global economy, major shifts in human reproductive behavior, and dramatic changes in values and lifestyles. These changes add up to a revolution. If this Environmental Revolution succeeds, it will rank with the Agricultural and Industrial Revolutions as one of the great economic and social transformations in human history. But, unlike these past events, the Environmental Revolution must be compressed into a few decades.

**To what extent will the success of this revolution be in the hands of governments?**

Some of the conditions of environmental sustainability can be satisfied by individual choices, such as deciding to have fewer children or to use energy more efficiently. But policy decisions, such as phasing out chlorofluorocarbons, replacing fossil fuels with solar energy, or protecting the planet's biodiversity depend on national governments and international agreements. And without clear policy guidance from governments, corporations—which control a large share of the world's finances—are not likely to change environmentally destructive business practices.

**Internationally, what are some of the measures governments are taking to move their economies onto an environmentally sustainable path?**

Germany has committed itself to a 25% reduction in carbon emissions by the year 2005. Australia, Austria, Denmark, and New Zealand have pledged to cut emissions by 20% in the same period. Some developing countries, particularly China, are committing themselves to stabilizing population size. Bangladesh hopes to decrease the average number of children per woman from 4.9 in 1990 to 2.3 in 2000. Nigeria's goal is to reduce this number from 6.2 in 1990 to 4 in 2000, and Mexico plans to cut its 1990 population growth rate in half by 2000. Denmark has completely banned the use of disposable beverage containers, and the Netherlands has adopted the most comprehensive plan, involving reductions of carbon, sulfur dioxide, and nitrous oxide emissions; nitrogen and phosphorus pollution; and a switch from cars to bicycles and trains. Even the most progressive countries, however, are still in the early stages of the transformation.

**What kind of measures would you like to see governments use more often?**

By far their most effective tool is to tax environmentally destructive activities, such as the generation of hazardous wastes and the use of virgin resources. Taxing carbon emissions, for instance, discourages the use of fossil fuels, while encouraging investment in renewable energy sources. Such taxes also can have a direct economic impact on corporations, which generally have been accustomed to passing on the expense of the environmental disruption they cause to society at large.

**Can we then expect to see increasing resistance from the corporate sector to this kind of governmental action?**

Some corporate leaders will welcome industry-wide regulations and taxes, which would permit them to reduce environmental damage without being put at a competitive disadvantage. Corporations are facing an enormous amount of change in the years ahead. Shifting to a reuse–recycle economy, protecting the ozone layer, reducing air pollution, acid rain, and hazardous wastes are among the environmental influences that will increasingly shape the global economy. The switch from fossil fuels to a solar–hydrogen system alone will affect every sector of the economy, from transportation to food. Corporate leaders will have to respond. California regulations will require 2% of all cars sold there after 1998 to have zero emissions, that is, to be electric. General Motors plans to market its new electric car there by the mid-1990s.

**Certain types of industries will be able to respond to the new environmental priorities more easily than others, won't they?**

For some, the prospective changes are relatively modest. Manufacturers of electrical appliances, for example, can concentrate on designing them to be more energy-efficient and more easily repaired and recycled. Automakers can shift to electric or hydrogen-powered models. Producers of incandescent light bulbs can easily switch to super-efficient compact fluorescents. Many products manufactured on a limited scale today have an enormous market potential in an environmentally sustainable world. Among them are refillable beverage containers, photovoltaic cells, thermally efficient building materials, wind electric generators, high-speed rail cars, rooftop solar water heaters, and water-efficient plumbing appliances.

Other companies will have to make more fundamental decisions, simply because there will

no longer be a place for their products. For example, coal and oil companies can either continue to conduct business as usual and face a bleak future, or they can help develop renewable energy sources. Like everyone else, corporations have a stake in a sustainable future. It is difficult to sustain profits in a declining economy. Those who see the need for change and move to the forefront will fare better than those who attempt to maintain the status quo.

**What can we do, as members of the general public, to help launch the Environmental Revolution?**

Individuals can do many things independently, but the Environmental Revolution depends on systemic change, and that requires an organized means of exerting sustained pressure, like that now coming from the thousands of environmental groups worldwide. These range from large international organizations, such as Greenpeace, to local, single-issue groups, such as the rubber tappers trying to save the Amazonian rain forest, or the women's groups in Kenya planting trees. These groups research issues, educate the public, litigate when necessary, and organize citizens to press local and national governments to abandon environmentally destructive policies.

**Could the news media do a better job at disseminating the information that these groups generate?**

Environmental issues are quite complex, and news editors, like society at large, sometimes have difficulty immediately grasping their importance and sifting through the information. Because it is environmental trends that are increasingly shaping our future, it is time for media organizations to reassess the resources devoted to environmental coverage. Newspapers should consider adding a daily environment section, and featuring at least as many environmental columnists on the editorial page as there are political ones. TV news programs might have a daily environment report, as well as a business one.

**What's the single most important ingredient needed for a successful Environmental Revolution?**

If it is to succeed, the Environmental Revolution will need the support of far more people than it now has. Up until now it has been viewed by society much like a sporting event— one where thousands sit in the stands watching, while only a handful are on the playing field actively attempting to influence the outcome of the contest. Success in this case depends on erasing the imaginary sidelines that separate spectators from participants so we can all get involved. Saving the planet is not a spectator sport.

# Chapter 1

Glacier National Park, Montana. (Visuals Unlimited/John Gertach)

# Our Changing Environment

The world's population is expected to surpass 6 billion by 1998. The support of so many people places a great strain on the Earth's resources and resilience. Environmental science attempts to identify and remedy the many problems that can arise when the environment is so severely stressed. These problems can be as small as the fate of a wildflower, as threatening as the explosion of a nuclear power plant, as final as the extinction of many of the Earth's species of animals and plants. All of these problems and many more face us as the 20th century draws to a close.

## HUMAN NUMBERS

The view in Figure 1–1 is a portrait of 250 million people. It is the United States at night, photographed by satellite in the spring of 1990. All the little specks of light you see, blinking like tiny stars, are cities. The great metropolitan areas are ablaze with light. The northeastern seacoast stands out like a glowing beacon, a great strip of humanity. At its hub, the corner where the seacoast turns from north-south to east-west, is New York City. The light from individual buildings cannot be seen—the scale is far too small for that—and from the satellite it is not obvious that even at night New York City teems with people. At the moment of the snapping of this picture, millions of people at that glowing corner were talking, hundreds of thousands of cars struggled through traffic, hearts were broken, babies born, and promises made. Were our lens but sharp enough, we would see under this one blur of light, frozen in time, all of this and more—15,700,000 people busy at life. All over the country the story is repeated, each light reflecting the same picture on a different scale, all the stories making up a panorama of modern industrial society.

Our futures, and those of all other people on Earth, are linked to each of the unseen 250 million people in this photograph. Each of us is in the picture. The way we lead our lives will have a significant impact upon the environment we share, and our consumption of resources will affect life in many other countries. Indeed, a factor equal to, if not more important than, population *size* is a population's level of *consumption*. Inhabitants of the United States and other developed countries consume many more resources per person than do citizens of developing countries such as Nigeria, India, and Peru. Our high rate of resource consumption affects the environment as much as, or more than, the explosion in population that is occurring in other parts of the world. Thus, as human numbers and consumption increase worldwide, so does humanity's impact on Earth, our common environment, posing new challenges to us all.

## ENVIRONMENTAL SCIENCE

How humankind can best live within Earth's environment is the subject of what is loosely called **environmental science,** the interdisciplinary study of how humanity affects other living organisms and the nonliving physical environment Environmental science encompasses many complex and interconnected problems involving human numbers, Earth's natural resources, and environmental pollution. Environmental science is interdisciplinary because it uses and combines information from many disciplines: biology (particularly ecology), chemistry, geology, physics, economics, sociology (particularly demographics, the study of population dynamics), natural resources management, and politics.

**Figure 1–1**
A satellite view of the United States at night. (Visuals Unlimited/Science Vu)

Humans do not live alone on Earth, nor are we beings above the laws of nature whose actions have no consequences. On the contrary, we have many partners who share Earth with us, and we would not live long without them. Think of the number of animals and plants that had to live in order for you to get through this day: much of the oxygen you breathe was produced by plants, as were the fibers in the paper of this page and in the cloth of your cotton or linen clothing; if you eat meat, it was an animal once, and so was the leather of your shoes; animals produced the wool of your sweaters and socks.

Every human being lives within a complex community of organisms. One of the principal goals of environmental science is to identify ways to avoid upsetting the delicate balance of the biological systems that support us. When imbalances do arise, one of the tasks of environmental science is to suggest how to deal with them in the most constructive way possible. For example, destruction of the Earth's ozone shield by industrial chemicals will have a disastrous impact if allowed to continue; pollution of the environment by industrial wastes and overutilization of particular resources such as fresh water and forest trees are other problems of this kind. Environmental scientists are called on to find solutions to all these problems.

Now look again at Figure 1–1. The central problem of environmental science, the one that links all others together, is that there are many people in this picture—and soon there will be many more. In 1950 only eight cities in the world had populations larger than 5 million; the largest was New York, with 12.3 million. By 1990 the largest city, Tokyo–Yokohama, had 20.5 million inhabitants, and the combined population of the world's ten largest cities was 143.5 million. By the year 2000 there will be 28 megacities with populations greater than 8 million, 22 of them in developing countries such as Brazil, India, and Indonesia. (See Chapter 8 for a discussion of developed and developing countries.)

In 1992 the human population of the Earth as a whole passed a significant milestone: 5.4 billion individuals (Figure 1–2). Since 1650, and probably

**Figure 1–2**
Human population numbers for the last ten thousand years. It took thousands of years for the human population to reach 1 billion, 130 years to reach 2 billion, 30 years to reach 3 billion, 15 years to reach 4 billion, and 12 years to reach 5 billion. At our present growth rate the sixth billion will be added by 1998.

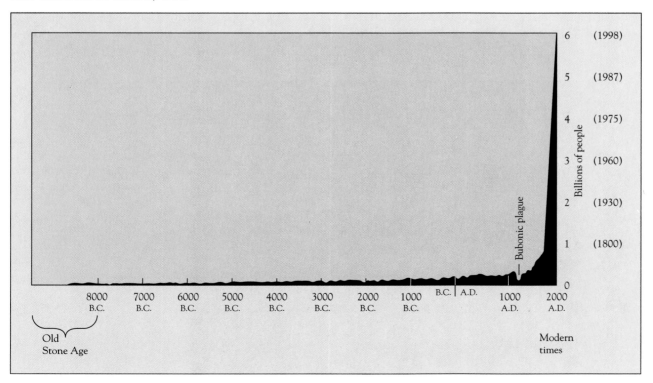

**Figure 1–3**
A slum in Kenya, Central Africa. Many of the world's people live in extreme poverty. (Visuals Unlimited/K. Doyle)

**Choosing Between Poison and Poverty**

Africa has been called "the world's toxic dumpsite." African countries are regularly approached by international dealers looking to dispose of American and European waste. Typical of this trade was the 1988 discovery of 8,000 drums of highly toxic waste behind fisherman Sunday Nano's house in the small village of Koko, Nigeria. An Italian disposal firm left tons of methyl melamine, dimethyl formaldehyde, ethylacetate formaldehyde, and polychlorinated biphenyls (PCBs) in a compound behind Nano's house, where the barrels popped quietly in the sun, oozing liquid poison and acid fumes.

Nano's nightmare represents a fraction of the 24 million tons of hazardous wastes that were dumped in West Africa alone during 1988. In that same year, Sierra Leone was offered $25 million to take American trash and Guinea-Bissau was scheduled to receive 16 million tons of western waste from a consortium of American and European companies in exchange for $120 million per year for five years—roughly the same as its gross national product. "Deals like these," says the environmental group Greenpeace, "force governments to make the unfair choice between poison and poverty."

for much longer, the average global human birth rate has remained nearly constant, at about 30 births per 1000 people per year. The birth rate has declined slightly in recent years (the 1992 value is 26, for example). However, with the spread of better sanitation and improved nutrition and medical techniques, the death rate has fallen steadily, to an estimated 1992 level of about 9 deaths per 1000 people per year. The difference between these two figures amounts to an annual worldwide increase in human population of approximately 1.7 percent. In 1992 this translated to a global increase of 93 million people, a number equivalent to the total population of Northern Europe.[1] In just one year this increase compensates numerically for all the lives lost due to all the wars of the 20th century. More than 254,000 people are added to the world's population each day, about 175 every minute! At this growth rate, the world's population would reach 6.2 billion people by the year 2000—and then increase by almost 2 billion more (the entire world's population in 1930) during the following 20 years.

All these people consume a lot of food and water, use a great deal of energy and raw materials, and produce much waste. As a group, they also

---

[1] The 1992 combined population of Denmark, Estonia, Finland, Iceland, Ireland, Latvia, Lithuania, Norway, Sweden, and the United Kingdom was 93 million.

have the potential to solve many, perhaps most, of the problems that arise in an increasingly crowded world. In this book, we delve into the details of how today's styles of living affect the environment within which all future humans must live, and discuss the efforts being mounted to lessen adverse impacts and increase potential benefits.

Despite the vigorous involvement of most developing countries with family planning, population growth rates cannot be expected to change very soon, owing to age distribution in developing countries. Three billion people will be added to the world in the next three to four decades, so even if we continue to be concerned about the overpopulation problem and even if our solutions are very effective, the coming decades may very well be clouded with tragedy. About 1.2 billion people are now living in extreme poverty (Figure 1–3), half of them getting less than 80 percent of the minimum calories recommended by the United Nations.

Given the efforts that are under way and considering the direction of growth trends, it is estimated that the world population may stabilize by the end of the 21st century. Population experts have made various projections for the population at that time, from almost twice what it is now (about 10.4 billion people) to more than four times our current population (greater than 20 billion people).

No one knows whether the world can support so many people indefinitely. Finding ways for it to do so represents one of the greatest challenges of our times. Among the tasks that we must accomplish is the development of new ways to feed, clothe, and shelter a world population more than twice as large as the present one without destroying the biological communities that sustain us. The quality of life available to our children and grandchildren will depend to a large extent on our ability to achieve this goal.

Success in this effort will require the cooperation of all the world's peoples. It will not be enough for the United States, consuming 20 to 40 times as much per person as developing countries, to preach environmental preservation to those same countries. Until the people of developing countries can meet their urgent needs, preserving tropical rain forests will not be a high priority for them, and if we in the United States want to help preserve the *trees* in the tropics, we must help the *people* in the tropics.

## OUR IMPACT ON THE ENVIRONMENT

One of the best ways to gain a sense of environmental science is to examine some of the problems that today's environmental scientists identify and attempt to solve. Some of the problems are truly global in scope, such as destruction of the Earth's ozone shield; others are more regional, such as acid precipitation; and still others are local, such as the pollution of a river. Some problems are cases of upsetting the delicate balance of nature by introducing foreign animals or plants; the invasions of fire ants and killer bees are problems of this sort. Other problems involve the complete destruction of ecosystems, such as filling in the coastal wetlands of the Atlantic and Gulf of Mexico coasts or clear-cutting the old-growth forests of the Pacific Northwest.

To gain some idea of the diversity of problems dealt with by today's environmental scientists, look at a newspaper. You will see many stories about the environment and our impact on it. Following (in no particular order) are just a few of the stories you might have seen in the last few years. Every one of them affects the fabric of our lives.

### Nuclear Energy Disaster: Chernobyl

The man in Figure 1–4 is testing for radioactivity. A month earlier, at 1:24 in the morning of April 26, 1986, one of the four reactors of the Chernobyl nuclear power plant, which you see in the background, exploded. Located in the former Soviet Union, Chernobyl was one of the largest nuclear power plants in Europe.

Before dawn on April 26, workers at the plant hurried to complete a series of tests, and they took a foolish shortcut: they shut off the safety systems. A power surge occurred during the test, and there was nothing to dampen it. Power zoomed to hundreds of times the maximum safe level, and a white-hot blast with the force of a ton of TNT partially melted the fuel rods and heated a vast head of steam, which blew the reactor apart.

**Figure 1–4**
From a helicopter above the Chernobyl nuclear power plant, the man on the right tests for radioactivity. The blast damage is evident below. (The Bettmann Archive)

The explosion and heat sent up a plume 4.8 km (3 mi) high, carrying some 50 tons of radioactive uranium fuel and fission products—ten times the fallout of Hiroshima. This cloud traveled first northwest, then southeast, spreading radioactivity in a band across Central Europe from Scandinavia to Greece. Within a 32-km (20-mi) radius of the reactor, at least one-fifth of the population, some 24,000 people, received serious radiation doses. In the western parts of the former Soviet Union and in the rest of Europe, the radiation dose was much lower but still significant. Agricultural land was contaminated by radioactive isotopes, particularly downwind from the reactor, and many other serious effects occurred. (See Chapter 11 for further discussion of Chernobyl.)

One lesson the nuclear disaster at Chernobyl teaches us is that, as our technology advances, so does the impact we have on the environment. It is a theme you will encounter repeatedly as you proceed through this text.

## Water Pollution: Fighting to Save the Rhine

The river in Figure 1–5 is the Rhine, a broad ribbon of water running through the soul of Europe. From high in the Alps that separate Italy and Switzerland, it flows north across the heart of industrial Germany before it reaches the Netherlands and the sea. On the first day of November, 1986, the Rhine almost died.

**Figure 1–5**
The Rhine River in Germany. The Rhine flows from Switzerland through Germany and the Netherlands before it empties into the North Sea. (M. L. Dembinsky, Jr./Dembinsky Photo Associates)

The blow that struck the Rhine did not seem so deadly at first. That morning firemen were fighting a blaze in Basel, Switzerland. The fire was gutting a huge warehouse belonging to a giant chemical company, Sandoz, and the firefighters aimed streams of water into the building to dampen the flames. In the rush to contain the fire, no one thought to ask what chemicals were stored there. By the time the fire was out, the water that had quenched the fire had also washed about 30 tons of mercury and pesticide into the Rhine.

Flowing down the river, the deadly wall of poison killed everything it passed. For hundreds of kilometers, the surface of the river was blanketed with dead fish. Even the aquatic plants in the river seemed to die. Many cities that use the waters of the Rhine for drinking had little time to make other arrangements. All across Europe, from Switzerland to the sea, the river reeked of rotting fish, and not one drop was safe to drink or even touch. Nothing could be done except to wait until the poison had washed out of the river ecosystem.

By 1990, Swiss and German environmental scientists monitoring the effects of the accident were able to report that the blow to the Rhine was not mortal. Enough small invertebrate water life and plants had survived to provide the basis for a vigorous return of fish and other aquatic organisms. A lesson difficult to ignore, the spill on the Rhine has caused the governments of Germany and Switzerland to intensify efforts to protect the river from future industrial accidents and to regulate the growth of chemical and industrial plants on its shores.

The Rhine is examined again in Chapter 13, in the context of international management of important water resources. The member governments of the European Economic Community are now cooperating to solve many of Europe's environmental problems. A major challenge looms ahead in Central and Eastern Europe, including Russia, Byelorussia, and the Ukraine, where substantial environmental damage has been done and must soon be addressed.

## Vanishing Species: Our National Parks

In January 1987, William Newmark, a graduate student of biology, published his doctoral thesis results in the journal *Nature*. For several years Newmark had visited the national parks of the United States and Canada to see what species of mammals were found in each. He reviewed sighting records going back decades, and came to a startling conclusion: in all but the very largest of the parks, a high propor-

**Figure 1-6**
Without careful management, mammals such as the grizzly bear often become extinct when isolated in a national park area. (Johnny Johnson © 1993 Animals Animals)

tion of the mammals were becoming extinct (Figure 1–6).

No animals were entirely lost, because those extinct in one park were still found in another. However, the pattern is clear: only the two largest parks (Yellowstone and Banff–Jasper) have not lost from one-fourth to more than one-third of their mammal species since they were founded around the turn of the century.

Environmental scientists were quick to suggest a reason. For over 15 years ecologists had speculated that in natural reserves many animal species would die out, because a national park is in essence an island, isolated from other natural areas by expanding human development. Islands, unless they are very large, cannot maintain very many species. When the parks were established, their animals were members of much larger communities; but then, walled off from these larger areas by human development for more than half a century, the animals became *restricted* to the parks, and in many cases extinction has been the result.

Environmental scientists have learned an important lesson: only in very large parks can we afford to leave our natural heritage untended. The parks were formed with the idea that people should "keep their hands off"—that, left alone, nature would take care of itself. In most of our national parks, however, the continued natural course of events would cause species to be lost on a regular basis, and our heritage would disappear before our children see it. If we want to maintain animal di-

versity for future generations, the parks will have to be managed carefully and lost species reintroduced. Our national parks, like gardens, must be tended. (Wildlife imbalances in our national parks are addressed in greater detail in Chapter 17.)

## Soil Pollution: Salinization of Central California

The San Joaquin Valley runs down the center of California and is one of the most fertile agricultural areas on Earth. Its fresh fruits and vegetables are eaten throughout the United States. The valley was formed about 2 million years ago from a vast, shallow freshwater lake, and its soils are unusually rich. The climate is mild, but not much rain falls. To support intensive farming, the land is irrigated. Before 1900 this valley was a near desert, but irrigation has made it bloom.

The plant you see in Figure 1–7, however, is not blooming, because it has been killed by salt. Irrigation, which is responsible for the rich agricultural harvest of the San Joaquin Valley, is slowly killing the valley. Irrigation water contains dissolved salts, and the continued application of such water, season after season, year after year, leads to the gradual accumulation of salt. When the water evaporates, the salts are left behind, particularly in the upper layers of the soil—the layers that are most important for agriculture. Given enough time, the salt concentration can rise to such a high level that plants are poisoned or their roots dehydrated.

In the San Joaquin Valley the problem is worsened by local geology. A layer of clay 6 to 30 m (20 to 100 ft) below the surface, laid down beneath the valley when it was still a lake, keeps water from draining out. Like an enormous bathtub, the valley slowly fills as irrigation water is added. Beneath some 73,000 hectares (about 180,000 acres) of land, the water level sits within 3 m (10 ft) of the surface. This water has already become very salty because it has been dissolving salts from the soil for

many thousands of years. With continued irrigation, the salty water rises closer and closer to the surface, posing a very serious threat to continued agriculture. There is no easy solution to the problem of soil salinization (discussed further in Chapter 21).

## Acid Precipitation

The smokestacks you see in Figure 1–8 are those of the Four Corners Power Plant in New Mexico. This facility burns coal, sending the smoke high into the atmosphere with the stacks, each of which is more than 60 m (200 ft) tall. The smoke belched out by the stacks contains high concentrations of sulfur, which smells bad (like rotten eggs) and produces acid when it combines with the water vapor in air. The intent of those who designed the plant was to release the sulfur-rich smoke high up in the atmosphere, where the winds would disperse and dilute it. This sort of solution to the problem of burning high-sulfur coal was first introduced in Great Britain in the mid-1950s and rapidly became popular in the United States and Europe. There are now about 800 such stacks in the United States alone.

Environmental scientists first noted in the 1970s that the exporting of industrial smoke into the upper atmosphere was producing acid rain and

**Figure 1–7**
A bush killed by salt at the edge of the Kesterson Reservoir. This reservoir is the end of the San Luis Drain of the underground runoff of central California's irrigation water. (Ed Kashi © 1985 Discover Magazine)

**Figure 1–8**
The Four Corners Power Plant in New Mexico. (Visuals Unlimited/Doug Sokell)

snow. Sulfur introduced into the atmosphere combines with water vapor to produce sulfuric acid, and when the water later falls as rain or snow, the precipitation is acid. Because the Earth spins counterclockwise (from west to east), the atmosphere passes from west to east across the Earth's surface, and the sulfur emissions released in the midwestern states return to the surface in the precipitation that falls on the eastern states. Similarly, the many tall stacks of the Ruhr Valley in Germany are responsible for the often strongly acid precipitation that falls over broad areas of Northern Europe.

Among its other effects, acid precipitation destroys aquatic life. Thousands of the lakes of Sweden and Norway no longer support fish. Many of the lakes in the northeastern United States and in eastern Canada also appear to have been acidified to death. Acid precipitation may also be a factor in the decline of forests. The Black Forest in Germany, for example, has suffered enormous damage.

At first, the solution seems obvious: capture and remove the emissions instead of releasing them into the atmosphere. There are, however, both economic and political blocks to such a solution. The economic problem is that it is expensive: reliable estimates of the cost of installing and maintaining the necessary scrubbers in smokestacks in the United States are on the order of $4 to $5 billion a year. The political problem is that the polluter and the recipient of the pollution are often far from one another, and neither is eager to pay so much for what is seen as someone else's problem. Canada, which is suffering from acid rain produced by the United States, has urged the United States to clean up its sulfur emissions. Encouragingly, the United States in 1990 passed clean-air legislation that should significantly improve the situation. (Acid precipitation is explored in greater detail in Chapter 20.)

## The Introduction of Exotic Species: We Are Being Invaded

In the mid-1980s, Tampa, Florida, had a new kind of visitor—a cockroach that had arrived on a ship from Asia. Unlike its cousin, the common German cockroach, which lives in houses and scurries away from light, this new immigrant lives outdoors, is attracted to light, and flies. German cockroaches have wings but, like chickens, don't often use them. Asian cockroaches come buzzing in through open windows and land on television screens and lampshades. The Asian cockroach is now firmly established in Florida, and it is not known how rapidly it will spread through the Southeast, although it seems almost certain that it will.

In 1992 another unwanted visitor entered our country. Several crates marked "Reptiles" were shipped from Lagos, Nigeria, to animal dealers in the United States. In fact, what they contained was a far more dangerous cargo—1,000 giant African snails (Figure 1–9). This species is considered by biologists to be the most environmentally destructive snail in the world. As large as baseballs, the snails were sold as pets by dealers in 25 states. Weighing up to 0.45 kg (1 lb) and with 80,000 rasping teeth, one giant African snail is able to eat almost anything and, if it escapes into the wild, can give rise to as many as 16 quadrillion descendants in five years. Federal officials are trying to collect the snails before an escape leads to disastrous consequences.

Unfortunately, such invasions are the rule, not the exception. Most of us have read of the recent invasions of fire ants, Africanized killer bees, water hyacinths, kudzu, and others. Each of these incursions seems a unique accident when it happens, but the overall problem is general and very serious: our highly mobile society has facilitated the movement of foreign animals and plants, often with disastrous consequences. (Chapter 16 expands on the problems of introduced species.)

**Figure 1–9**
The giant African snail. (Visuals Unlimited/Joe McDonald)

## Vanishing Species: Destroying the Tropical Forests

The fire you see in Figure 1–10 is in a tropical forest in the Amazon Basin of Brazil. The fire will burn for weeks, turning thousands of acres of forest trees into charred stumps, for no one will try to halt its relentless destruction. This fire was not started by a dropped cigarette or a careless camper. It was set deliberately in order to clear the forest and produce grazing land. All over the world, the tropical forests are being cleared—cut for timber or firewood or burned to make pasture or agricultural land. More than half of all the world's tropical evergreen forests had been destroyed by 1991. What is left, some 6 million square kilometers (about 2.3 million square miles), is less than the area of the United States, and each year another 169,000 km² (65,000 mi²), an area larger than the size of Washington state, is cleared. In the early 1990s, we are destroying tropical forests at the rate of 0.54 hectares (1.3 acres) per second.

Tropical forests are among the richest and most diverse of the Earth's biological communities, but when they are cut down or burned, they are very slow to return. We will not see the forest in Figure 1–10 again in our lifetime. The destruction of the world's tropical forests is a tragedy, for countless species of plants and animals occur in those forests, many of them unique forms only now being described by scientists, and many others awaiting discovery. None of the organisms that have inhabited the forest burning in Figure 1–10 will be discovered, however. Everything that lived there is gone now. And this same story is repeated every day for hundreds of square kilometers. Who knows which of these plants and animals might have been of great use to future generations of humans? (See Focus On: The Environmental Movement in Brazil for a glimpse of Brazilians' efforts to conserve their forests; tropical forest destruction is addressed in greater detail in Chapter 17.)

**Figure 1–10**
Fires are destroying tropical rain forest in Brazil, which is being cleared for agriculture. (Dr. Nigel Smith © Earth Scenes)

## Damage to the Atmosphere: Ozone Depletion

The swirling colors of Figure 1–11 are a view of the South Pole from a satellite. This is not the picture your eye would see, but rather a computer reconstruction, in which the colors represent different concentrations of ozone, a form of oxygen gas. As you can easily see, there is a large ozone "hole" over

**Figure 1–11**
A computer-generated image of the Southern Hemisphere in October 1991 reveals the ozone "hole" over Antarctica. The area surrounded by the lavender and dark purple in the center of the hole represents the lowest amount of ozone measured. (NASA)

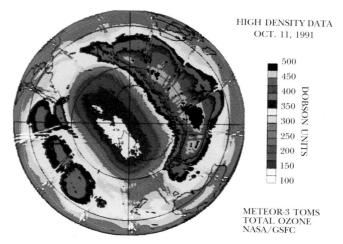

HIGH DENSITY DATA
OCT. 11, 1991

500
450
400
350
300
250
200
150
100

DOBSON UNITS

METEOR-3 TOMS
TOTAL OZONE
NASA/GSFC

## Focus On

### The Environmental Movement in Brazil

The highly visible environmental movement in the United States, which tackles problems both at home and abroad, may make some Americans think that we are the only nation in the world with a strong environmental consciousness. In actuality, people in developing countries are increasingly aware of their environmental problems and are working hard to solve them.

Consider Brazil, the world's fifth largest country and one of the richest in natural resources, particularly fresh water, forests, wildlife, and minerals. Hundreds of Brazilian environmental groups are working to preserve Brazil's forests, wildlife, and coasts. These organizations may not have been in existence for as long as many environmental groups in the United States, but they have accomplished a great deal in a relatively short period of time. Many of them cosponsor projects with environmental organizations in the United States as well as with international corporate sponsors. Some Brazilian environmental organizations are:

*The Pro-Nature Institute,* founded in 1986, which administers programs in several rain forests and has established a wildlife sanctuary

*Funatura,* founded in 1986, which oversees a wildlife sanctuary in the Atlantic coastal forest as well as a conservation program in the Grande Sertao Veredas National Park

*The Blue Wave Foundation,* founded in 1989, which conducts a national television campaign to clean and protect the beaches of Brazil

Although their funds are often limited, several environmental organizations have produced radio and television ads to educate Brazilians about the destruction of their natural heritage. The fact that grass-roots environmental organizations are continually being founded in Brazil attests to the heightened environmental awareness of that society.

---

Antarctica, within which the ozone concentration is much lower than elsewhere; it covers an area about the size of the United States. This ozone hole was first reported in 1985 by British environmental scientists. Looking back at earlier satellite data, we now see that the zone of thinning ozone appeared for the first time in 1975. The hole is not a permanent feature but a seasonal phenomenon, evident only for a few months at the onset of the Antarctic winter in September. Every September from 1975 to the present, the ozone hole has reappeared, and each year the layer of ozone has been thinner. In 1990 the minimum ozone concentration in the hole was 50 percent lower than the minimum ten years earlier. And by 1992 there was clear evidence that the ozone was also being depleted over the Arctic.

Environmental scientists are worried about the Antarctic ozone hole because it seems to portend a thinning of the ozone layer worldwide. In the heavily populated mid-latitudes of the planet, winter ozone levels dropped by as much as 4 to 6 percent during the 1980s. Why is this worrisome? In the upper atmosphere, ozone absorbs harmful ultraviolet radiation from sunlight; without its protection, human skin cancers caused by ultraviolet radiation would become far more common. The increased ultraviolet radiation might also damage other plant and animal species.

Probing the cause of the ozone hole with high-flying aircraft, environmental scientists in 1987 reported significant amounts of chlorine in the upper atmosphere, suggesting that the ozone in the ozone hole had reacted chemically with chlorine. The main source of the chlorine is a human-made group of chemicals called chlorofluorocarbons (CFCs), familiar to many of us as aerosol propellants in spray cans (now banned in the United States) and as the Freon cooling agent in refrigerators and air conditioners.

As a direct result of public awareness of the problem, strong laws were passed restricting use of CFCs in the United States and Canada. In 1990, some 90 other nations agreed to a total ban on CFC production by the year 2000, and in 1992 efforts were made by the United States to move up that deadline. Because CFCs are very stable, they can survive in the atmosphere 120 years or more, so the changes in Earth's atmosphere initiated by the release of CFCs may not be quickly reversible. (Ozone depletion in the upper atmosphere is examined in Chapter 20.)

## Oil Spills: Killing Sea Otters

On March 24, 1989, the oil supertanker *Exxon Valdez* ran aground on a reef in the northern part of Prince William Sound, Alaska, discharging 260,000 barrels (11 million gallons) of crude oil that quickly covered thousands of square kilometers. The spill proved to be one of the most devastating to wildlife, and sea otters were among the most severely hurt. In the first week, 250 dead otters were collected, and 135 live, oil-coated otters were captured and cleaned (Figure 1–12); only 55 of the cleaned otters survived. Post-mortem examinations revealed that the oil had poisoned most of them, severely damaging their livers, kidneys, and lungs. All told, several thousand sea otters and more than 30,000 birds are known to have died. (See Chapter 10 for a discussion of the Alaskan oil spill.)

As a result of the extensive publicity given this natural disaster, new oil tankers are being constructed with double hulls. A reef might tear the

### Figure 1–12

Animal care specialists at Sea World in San Diego, California, treat an Alaskan sea otter for the lingering toxic effects of ingesting crude oil. Other scientists from the Sea World Institute/Hubbs Marine Research Center flew to Alaska to assist in removing oil from the otters' fur coats. Because an oiled coat cannot insulate the animal, most of the sea otters died from hypothermia. (© 1993 Sea World of California, Inc. All rights reserved.)

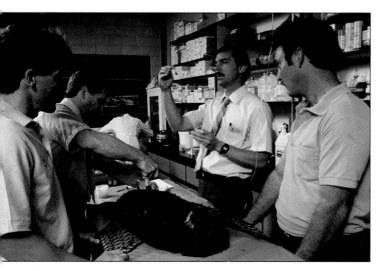

outer hull but would not be likely to damage the inner one. The first such "environmental" tanker was launched in Japan in 1991.

It is important to understand that, although oil spills are spectacular environmental disasters, the major forces driving many marine organisms to extinction are not spills, but fertilizers, pesticides, soil erosion, and other effects of onshore human operations, including the general input of chemicals of all kinds into the open oceans. Less visible than oil spills, this "quiet" pollution is, in the final accounting, far more deadly, for it has a much greater impact on the global marine ecosystem.

## Global Climate Change: Carbon Dioxide Levels

During the past two centuries, as the world's population has grown to ten times its former size, the level of carbon dioxide ($CO_2$) in the Earth's atmosphere has increased dramatically (Figure 1–13). The causes of the increase in atmospheric $CO_2$ are no mystery: the burning of fossil fuels (coal, oil, and natural gas) and the clearing and burning of forests by farmers. Environmental scientists are growing increasingly concerned that the rising levels of $CO_2$ may change the Earth's climate. Carbon dioxide levels rose from 315 parts per million (ppm) in 1958 to 354 ppm in 1990 (latest data available). Just as the panes of glass in a greenhouse let light in but do not allow heat out, so $CO_2$ in the atmosphere allows solar radiation to pass through but does not allow heat to radiate back into space. Instead, the heat is reflected back to the Earth's surface. As $CO_2$ accumulates in the Earth's atmosphere, enough heat may be trapped to gradually warm the Earth.

The decade of the 1980s saw the six warmest years in U.S. weather records, and environmental scientists estimate that if trends are not changed, the Earth's mean temperature could rise 1.5 to 4.5°C (2.5 to 7.5°F) by the middle of the next century, making the atmosphere warmer than it has been at any time in the last 100,000 years. This might produce major shifts in patterns of rainfall, and might initiate melting of the West Antarctic ice sheet (as did the last warm period 120,000 years ago). This melting would cause ocean levels to rise. Such a rise is alarming, as it might put many of the Earth's major cities at least partly underwater. With each new warm year in the 1990s, the possibility seems more real that a significant warming trend has begun. Hard conclusions about long-term trends are difficult to reach, but an increasing number of environmental scientists are concerned.

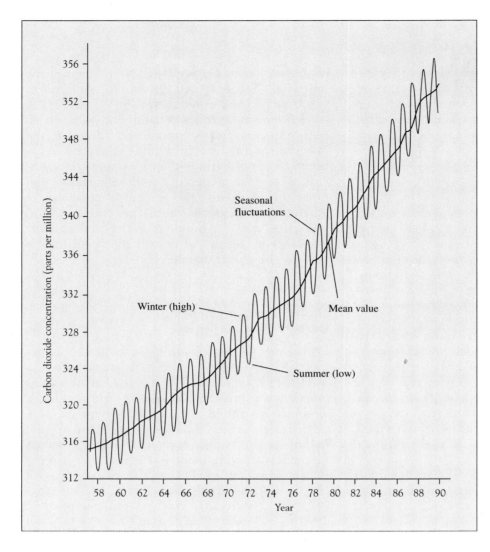

**Figure 1–13**
The concentration of carbon dioxide in the atmosphere has shown a slow but steady increase for many years. These measurements were taken at the Mauna Loa Observatory, far from urban areas where carbon dioxide levels are high because of the large number of factories, power plants, and motor vehicles. The seasonal fluctuation corresponds to winter (a high level of $CO_2$) and summer (a low level of $CO_2$) and is caused by greater photosynthesis in summer.

(Global warming is examined further in Chapter 20.)

## THE GOALS OF ENVIRONMENTAL SCIENCE

Unlike biology, geology, chemistry, and physics—sciences that seek to establish general principles about how the natural world functions—environmental science is, by its very nature, an *applied* science, a form of problem solving: it is the search for constructive alternatives to environmental damage. The science of **ecology,** a discipline of biology that studies the interrelationships between living organisms and their environment, is the basic tool of environmental science, and so we begin our study with a detailed treatment of ecology. Using what we have learned about ecology, we

*(Text continues on page 18.)*

**Envirobrief**

**The Reuseable Shipping Crate**

To reduce solid waste, industries are turning to reuseable packaging. The trend got its first major push in 1989 when General Motors declared it would no longer accept disposable packaging at its automobile plants. The result was the development of rugged steel and plastic containers that make dozens of round trips before being recycled. The most popular of these new-breed containers are large plastic crates with built-in pallet grooves that allow a fork lift to move them around just like wooden pallets. The plastic crates collapse to about one-third their erected size for return trips. They pay for themselves through reuse and avoidance of disposal costs.

# The Earth Summit

In June 1992, representatives from around the world met in Rio de Janeiro, Brazil, for a summit conference officially called the United Nations Conference on Environment and Development. Countries attending the conference examined environmental problems that cross international borders—problems that are truly global in nature: pollution and deterioration of the planet's atmosphere and oceans; destruction of forests; and a decline in the number and kinds of living organisms.

It is easy for the representatives of a country to say that that country supports a cleaner environment, but the actual specifics—what is going to be done, how soon, how much it's going to cost, and who's going to pay—are very difficult to agree upon. The issues that were discussed and debated at the 1992 Earth summit included:

*Climate change:* A treaty that would curb carbon dioxide emissions, thereby reducing the greenhouse effect. Although no timetable was agreed upon, the treaty at least starts the process of stabilizing carbon dioxide emissions.

*Biological diversity:* A treaty that would decrease the rate of extinction of the world's endangered species.

*Deforestation:* A statement of principles regarding the destruction of the world's forests.

*Agenda 21:* A complex action plan for the 21st century in which developed nations would provide money to help developing countries become industrialized without harming the environment.

*Earth Charter (The Rio Declaration):* A statement of philosophy about environment and development.

The statement of principles on deforestation was originally to have been a legally binding treaty that would stop developing nations from burning tropical forests. Developing countries objected that the treaty unfairly focused on tropical forests and did not address the logging of old-growth forests in the United States, Canada, and Europe. When a compromise could not be reached, the treaty was scrapped for a weaker statement that is not legally binding.

The pre-summit negotiations for the remaining topics were equally complex. In general, the poorer developing nations felt that their top priority had to be economic survival rather than saving the environment. They expressed a willingness to follow the environmental mandates of industrialized countries, but only if those countries contributed money to help them protect the environment. Industrialized nations, for their part, acknowledged that they had some responsibility toward developing nations but also stressed that the current rapid population growth and industrialization of developing nations threatens to do more environmental damage than was done by developed countries when *they* were becoming industrialized. Thus, developed countries want developing countries to focus on slowing their rates of population growth and industrialization in order to preserve the environment.

Although the Earth summit did not accomplish all that environmentalists had hoped it would, it was a resounding success in many ways. It was the largest international gathering ever to concentrate on serious environmental issues, and, because it received so much international attention, it increased worldwide awareness of global issues. Also, the Earth summit demonstrated just how far apart developed and developing nations stand on many issues. This awareness will be needed in future negotiations of international environmental issues.

Meeting the Challenge

Environmentalist Jacques Cousteau (left) and other delegates (including Cuban President Fidel Castro, top right) at the Earth summit, June 13, 1992.
(Reuters/Bettmann Archive)

then directly address human population growth and three of the major results of that growth: increasing need for energy, depletion of resources, and rising pollution. Many of the environmental problems we consider in this book are serious ones that must be addressed—if not by us, then unavoidably by our children.

Environmental science is not, however, simply a "doom and gloom" listing of problems coupled with predictions of a bleak future. To the contrary, its focus, and *our* focus as individuals and as world citizens, is on identifying, understanding, and *solv-*

*ing* problems that we and our ancestors have created. A great deal is being done, and more must be done—at individual, country, and worldwide levels—to address the problems of today's world. Most environmental issues are complex, however, and cannot be solved by science alone because they interact with numerous competing interests and goals (see Meeting the Challenge: The Earth Summit). We fill a considerable amount of space in this text examining some successful approaches to environmental problems and exploring other problems that defy easy solutions.

# SUMMARY

1. The world's human population is expected to surpass 6 billion by 1998. The increasing population is placing considerable stress upon the environment, as humans consume ever-increasing quantities of food and water, use a great deal of energy and raw materials, and produce enormous amounts of waste and pollution.

2. Environmental science is the interdisciplinary study of how humanity affects other living organisms and the nonliving physical environment. Environmental science encompasses many complex and interconnected problems involving human numbers, Earth's natural resources, and environmental pollution. Environmental science is interdisciplinary because it uses and combines information from many disciplines: biology, chemistry, geology, physics, economics, sociology, natural resources management, and politics.

3. Among the many challenges to the environment posed by human activities is the release of materials that may harm the environment: the escape of radioactive materials into the atmosphere, the spillage of pollutants such as pesticides and oil into rivers and oceans, the emission into the atmosphere of industrial pollutants that lead to acid precipitation, the release of chemicals

such as CFCs that attack the ozone shield in the upper atmosphere, and the production of carbon dioxide that may alter climate.

4. Other environmental challenges are created by our attempts to manage other life forms with which we share the planet: most of our national parks are too small to protect larger species from extinction without careful monitoring, whereas populations of foreign species sometimes invade our ecosystems in dramatic fashion.

5. Our "progress" in developing natural resources has led to other challenges: extensive irrigation in some cases salts up the soils; commercial development is rapidly destroying entire habitats. On a much greater scale, the cutting down of the tropical forests of the world to make pasture and cropland is producing a wave of extinction among tropical plants and animals.

6. All of these challenges to the environment can and must be addressed. Environmental science is the study of such challenges. Environmental scientists attempt to identify how human activities affect the environment and search for constructive solutions to the problems that arise.

# DISCUSSION QUESTIONS

1. Discuss several ways in which people are making the environment unsuitable for other living organisms.
2. How has recent history demonstrated the root causes of environmental problems?
3. What are the advantages of our industrialized society? What are the disadvantages?

4. Is it possible to have unlimited human population growth indefinitely? Why or why not?
5. Do you think that all or most environmental problems can be solved by technology? Why or why not?

# SUGGESTED READINGS

Jones, P., and T. Wigley. Global warming trends. *Scientific American*, August 1990, 84–91. Analyses of land and marine records confirm that our planet has warmed 0.5°C in the last century, although future trends remain uncertain.

Mohnen, V. The challenge of acid rain. *Scientific American*, August 1988, 30–38. New advances in technology may soon offer environmentally and economically attractive ways to control acid rain.

Repetto, R. Deforestation in the tropics. *Scientific American*, April 1990, 36–42. The argument is made that government policies that encourage exploitation are largely to blame for the accelerating destruction of tropical forests.

Sadik, N. *The State of World Population 1990.* United Nations Population Fund Report, New York, 1990. Summarizes world population issues and our options during the 1990s.

Toon, O., and R. Turco. Polar stratospheric clouds and ozone depletion. *Scientific American*, June 1991, 68–74. Clouds high over the Antarctic conspire with CFCs to create the ozone hole that opens every spring. A very clear account of the chemistry.

White, R. The great climate debate. *Scientific American*, July 1990, 36–43. The prospect for global warming by carbon dioxide- and methane-induced acceleration of the greenhouse effect raises a key question: Should we take steps now?

# Chapter 2

*Scientific assessment is an important part of the process of addressing environmental problems. Here a scientist climbs to an observation station high in a rain forest. (Visuals Unlimited/Jane Thomas)*

# Solving Environmental Problems

I n studying the many environmental problems that face the world today, it is important not to lose sight of the fact that much can be done to improve our situation. The role of environmental science is not only to identify problems but also to suggest and evaluate potential solutions. Although the choice to implement a proposed solution is almost always a matter of public policy, environmental scientists play key roles in educating both officials and the general public. Solving environmental problems successfully involves five steps: scientific assessment, risk analysis, public education, political action, and follow-through.

## SOLVING ENVIRONMENTAL PROBLEMS: AN OVERVIEW

As discussed in Chapter 1, environmental science is the interdisciplinary study of how humanity affects other living organisms and the nonliving physical environment. Its role is to develop the basic information on which wise environmental decisions can be based, and in that sense it is fundamentally a problem-solving science. Before we begin a detailed examination of the environmental problems that face our society today, it is useful to consider the many elements that go into the solving of environmental problems. How is information gathered, and at what point can conclusions be regarded as certain? Who makes the decisions, and what are the trade-offs?

The problems described in Chapter 1 are not insurmountable. A combination of scientific investigation and public action can solve them; pollution of the Earth's atmosphere, land, and water can be halted, and resources can be protected for the future. How can this success be achieved? Viewed simply, there are five components in the solving of any environmental problem:

1. *Scientific Assessment.* The first stage of addressing any environmental problem is scientific assessment, the gathering of information. Data must be collected and experiments performed in order to construct a model that describes the situation. Such a model can be used to make predictions about the future course of events.
2. *Risk Analysis.* Using the results of the scientific investigation as a tool, it is possible to analyze the potential effects of intervention—what could be expected to happen if a particular course of action were followed, including any adverse effects the action might create (see Focus On: An Assessment of Risks).
3. *Public Education.* When a clear choice can be made among alternative courses of action, the public must be informed. This involves explaining the problem, presenting all the available alternatives for action, and revealing the probable costs and results of each choice.
4. *Political Action.* The public, through its elected officials, selects a course of action and implements it (see Focus On: Poisons in the Environment).
5. *Follow-Through.* The results of any action taken should be carefully monitored, both to see if the environmental problem is being solved and, more basically, to evaluate and improve

the initial evaluation and modeling of the problem.

## THE SCIENTIFIC ANALYSIS OF ENVIRONMENTAL PROBLEMS

The key to the successful solution of any environmental problem is rigorous scientific evaluation, and it is important that we understand clearly just what the words "scientific investigation" mean. What is "science"? The word conjures up images of people in white lab coats peering at instruments and shaking test tubes. What are they doing, and why?

Science is a particular way to investigate the world, a systematic attempt to understand the Universe. Science seeks to reduce the apparent complexity of our world to general principles, which can then be used to solve problems or provide new insights.

A number of areas of human endeavor are not scientific. Ethical principles often have a religious foundation, and political principles reflect social systems. Some general principles, however, derive not from religion or politics, but from the physical world around us. If you drop an apple, it will fall whether or not you wish it to, despite any laws you may pass forbidding it to do so. Science is devoted to discovering the general principles that govern the operation of the natural world.

### The Nature of Science

How does a scientist discover such general principles? Where are they written? They are "written" wherever we look in the world around us. A scientist is above all an observer, someone who examines the world in order to understand how it works. Stated briefly, a scientist determines principles from observation.

Discovering general principles by the careful examination of specific cases is called **inductive reasoning.** The scientist begins by organizing data (facts) into manageable categories and asking the question "What do these facts have in common?" He or she continues by seeking a unifying explanation for the facts. Inductive reasoning is the basis of modern experimental science.

As an example of inductive reasoning, consider the following:

*Fact:* Gold is a metal that is heavier than water.
*Fact:* Iron is a metal that is heavier than water.

## An Assessment of Risks

Each of us takes risks every day of our lives. Walking on stairs involves a small risk, but a risk nonetheless, because some people die from falls on stairs. Using household appliances is slightly risky, because some people die from electrocution when they operate appliances with faulty wiring or use appliances in an unsafe manner. Driving in an automobile and flying in a jet offer risks that are easier for most of us to recognize. Yet few of us hesitate to fly in a plane, and even fewer hesitate to drive in a car because of the risk.

Estimating the risks involved in a particular action so that they can be compared and contrasted with other risks is known as **risk assessment.** Risk assessment helps us estimate the probability that an event will occur and enables us to set priorities and manage risks in an appropriate way (Table 2–1).

As an example, consider a person who smokes a pack of cigarettes a day and drinks well water containing traces of the cancer-causing chemical trichloroethylene (in acceptable amounts as established by the Environmental Protection Agency). Without knowledge of risk assessment, this person might buy bottled water in an attempt to reduce his or her chances of getting cancer. Based on risk assessment, the annual risk from smoking a pack of cigarettes per day is $3.6 \times 10^{-3}$, whereas the annual risk from drinking water with EPA-accepted levels of trichloroethylene is $2 \times 10^{-9}$. This means that this person is 180 million times more likely to get cancer from smoking than to get it from drinking such low levels of trichloroethylene. Knowing this, the person in our example would, we hope, be induced to stop smoking.

One of the most perplexing dilemmas of risk assessment is that people often ignore substantial risks but get extremely upset about minor risks. The average life expectancy of smokers is more than eight years shorter than that of nonsmokers; almost one-third of all smokers die from diseases caused or exacerbated by their habit. Yet many people get much more upset over a one-in-a-million chance of getting cancer from pesticide residues on food than they do over the relationship between smoking and

cancer. Perhaps part of the reason for this attitude is that behaviors such as diet, smoking, and exercise are parts of our lives that we can control *if we choose to.* Risks over which most of us have no control, such as pesticide residues, tend to evoke more fearful responses.

One final point needs to be made regarding risk assessment of environmental pollutants. Often the only disease used in chemical risk assessment is cancer, but environmental contaminants are linked to a number of other serious diseases. Smoking is a major contributor to heart disease, for example, whereas certain chemicals cause birth defects, damage the immune response, or attack the nervous system. Cancer, then, is not the only disease that is caused or aggravated by chemicals.

### A Balanced Perspective on Risks

Threats to our health, particularly from toxic chemicals in the environment, make big news. Many of these stories are more sensational than factual. If they were completely accurate, people would be dying left and right, whereas, in fact, human health is better today than at any time in our history, and our life expectancy continues to increase rather than decline.

This does not mean that we should ignore human-made chemicals in the environment. Nor does it mean we should discount the stories that are sometimes sensationalized by the news media. These stories serve an important role in getting the regulatory wheels of the government moving to protect us as much as possible from the dangers of our technological and industrialized world.

People cannot expect no-risk foods, no-risk water, or no-risk anything else. Risk is inherent in all our actions and in everything in our environment. We do, however, have the right to expect the risks to be minimized. For example, given the fact that we consume some natural carcinogenic substances in our food, we should try to avoid consuming uncontrolled amounts of human-made carcinogens in addition. Simply stated, we should not ignore small risks just because larger ones exist.

| Table 2–1 Some Risks of Daily Living* | |
|---|---|
| **Relative Risk**** | **Type of Risk** |
| 0.2 | Disease from PCBs in diet |
| 0.3 | Disease from DDT and DDE in diet |
| 1 | Disease from drinking 1 quart of municipal water per day (contains traces of chloroform) |
| 18 | Dying by electrocution in any given year |
| 60 | Disease from drinking 12 ounces of diet cola per day (contains saccharin) |
| 367 | Falls, fires, poisonings in the home |
| 667 | Respiratory illness caused by air pollution (for those living in eastern U.S.) |
| 800 | Dying in auto accident in any given year |
| 2,800 | Disease from drinking 12 ounces of beer per day (contains ethyl alcohol) |
| 12,000 | Disease from smoking one pack of cigarettes per day |

*These are hypothetical worst case risks, not "real" risks.

**Lower numbers indicate less risk. As an example of how to interpret the information in this chart, consider drinking water versus beer. The risk of disease from drinking 12 ounces of beer per day is 2,800 times greater than the risk from drinking 1 quart of city water per day.

*Fact:* Silver is a metal that is heavier than water.

*Conclusion based on inductive reasoning:* All metals are heavier than water.

Even if inductive reasoning makes use of facts that are all correct, the conclusion may be either true or false. As new facts come to light, they may show that the generalization arrived at inductively is false. Experimental science has shown, for example, that the density of lithium, the lightest of all metals, is about half that of water. When one adds this fact to the preceding list, a different conclusion must be formulated. Inductive reasoning, then, produces new knowledge but is error-prone.

Science also makes use of **deductive reasoning,** which proceeds from generalities to specifics. Deductive reasoning adds nothing new to knowledge,

but it can make relationships among data more apparent. For example:

*General rule:* All birds have wings.
*A specific example:* Robins are birds.
*Conclusion based on deductive reasoning:*
    All robins have wings.

This is a valid argument. The conclusion that robins have wings follows inevitably from the information given. No other conclusion is possible. Deductive reasoning is used by scientists to determine the type of experiment or observations necessary to test a hypothesis.

**Testing Hypotheses**

Scientists learn which general principles are true, among the many that might be true, by attempting systematically to demonstrate that certain proposals are *not* valid—that is, are not consistent with what scientists have learned from experimental observation—then rejecting those invalid proposals. For the time being they retain proposals that they are not yet able to disprove as useful, because they fit the known facts. Later, even these proposals might be rejected if, in the light of new information, they are found to be incorrect.

We call a proposal that might be true a **hypothesis,** and the test of a hypothesis an **experiment.** An experiment evaluates alternative hypotheses. Say, for example, that you face two closed doors. "There is a tiger behind the door on the left" is a hypothesis; an alternative hypothesis is "The door on the right has a tiger behind it"; a third hypothesis might be "There is no tiger behind either door." An experiment works by eliminating one or more of the hypotheses. To test these alternative hypotheses, you might open the door on the right. Let us say that, when you do this, a tiger leaps out at you. Your experiment has disproved the third hypothesis, for it is clearly incorrect to say that there is no tiger behind either door.

Note that a test such as this does not prove that only one alternative is true, but rather demonstrates that one of them is not true. In this instance, the fact that a tiger is behind the door on the right does not rule out the possibility that a tiger also lurks behind the door on the left. A successful experiment is one in which one or more of the alternative hypotheses are demonstrated to be inconsistent with experimental observation and thus rejected. Scientific progress is made in the same way a wooden statue is—by chipping away unwanted bits.

## Focus On

### Poisons in the Environment

The current role of the federal government in assessing human health risks caused by hazardous substances is spelled out in such legislation as the Clean Air Act and the Toxic Substances Control Act. This box discusses how the scientific community determines the health effects of human exposure to hazardous substances. The federal government uses these results to make difficult decisions: if, when, and how human exposures to potentially hazardous substances should be regulated.

**Hazardous Substances**  The science of poisons is known as **toxicology.** It encompasses the effects of toxins on living organisms as well as ways to counteract their toxicity. A number of chemicals classified as toxins will be considered in various chapters. Most toxins are not restricted to the air or the soil or water, but are found in many places in the environment. Many toxins are of concern because they are known to cause cancer.

**Identifying Carcinogenic Substances**
The most common method of determining whether a chemical causes cancer is to expose rats to extremely large doses of that chemical and see whether they develop cancer. There are some problems with this method, however (see figure). For one thing, although humans and rats are both mammals, rats are different organisms and may respond to carcinogens differently. Another problem is that the rats are exposed to massive doses of the suspected carcinogen relative to their body size, whereas humans are usually exposed to only trace amounts. It is assumed that one can work backwards from the huge doses of chemicals and the high rates of cancer they cause in rats to determine the rates of cancer that might be expected in humans exposed to trace amounts of the same chemi-

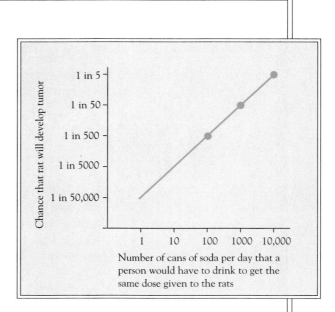

Attempting to extrapolate backward from data on the incidence of cancer in rats exposed to high levels of a substance to predict the incidence of cancer in humans exposed to trace amounts of that substance may not be scientifically sound.

cals. However, there is little evidence to indicate that extrapolating backwards in this area of study is scientifically sound.

Epidemiologic evidence, including studies of human groups who, for one reason or another, have been exposed to high levels of suspected carcinogens, is also used to determine whether or not chemicals are carcinogenic. For example, in 1989 epidemiologists in Germany established a direct link between dioxin and cancer (dioxins are discussed in Chapter 22). They observed the incidence of cancer in workers exposed to high amounts of dioxin during an accident at a chemical plant in 1953, and found unexpectedly high and consistent incidences of cancers of both the digestive and respiratory tracts.

## Controls

Most often, the processes we want to learn about are influenced by many factors. We call each factor that influences a process a **variable.** In order to evaluate alternative hypotheses about one variable, it is necessary to hold all the other variables constant so that we don't get misled or confused by them.

To test a hypothesis about a variable, we carry out two forms of the experiment in parallel. In the experimental test we alter the chosen variable in a known way. In the **control** test we do not alter that variable. We make sure that in all other respects the two tests are the same. We then ask, "What difference is there between the outcomes of the two tests?" Any difference that we see must be due to the influence of the variable that we changed, because all other variables remained the same. Much of the challenge of experimental science lies in designing control tests and in successfully isolating a single variable from all other variables.

## The Importance of Prediction

A successful scientific hypothesis needs to be not only valid but useful—it needs to tell you something you want to know. A hypothesis is most useful when it makes predictions, because the predictions provide a very important way to test the hypothesis' validity: a hypothesis that your experiment does not reject, but which makes a prediction that your experiment *does* reject, must be rejected. The more verifiable predictions a hypothesis makes, the more valid that hypothesis is. There is something very satisfying about a successful prediction, because the prediction being tested to verify the hypothesis is generated by the hypothesis itself, and the result is not known ahead of time.

## Theories

A hypothesis supported by a large body of observations and experiments becomes a **theory.** A good theory relates facts that previously appeared to be unrelated. A good theory grows as additional facts become known. It predicts new facts and suggests new relationships among phenomena.

By demonstrating the relationships among classes of facts, a theory simplifies and clarifies our understanding of the natural world. Theories are the solid ground of science, the concepts of which we are most sure. This definition contrasts sharply with the general public's usage of the word "the-ory," implying *lack* of knowledge, or a guess. In this book, the word "theory" is always used in its scientific sense, to refer to a broadly conceived, logically coherent, and very well supported concept.

Some theories—for example, Newton's theory of gravity, Darwin's theory of evolution, and Einstein's theory of relativity—are so strongly supported that the likelihood of their being rejected in the future is very small. Yet there is no absolute truth in science—only varying degrees of uncertainty. The possibility always remains that future evidence will cause a theory to be revised. A scientist's acceptance of a theory is always provisional.

## A CASE HISTORY: THE RESCUE OF LAKE WASHINGTON

Just as generals study old battles in order to learn how battlefield decisions are made, we can analyze how environmental problems are solved by studying an environmental battle that was successfully waged in the 1960s. We will follow the events in this very real environmental drama step by step. Then, with it as a backdrop, we will discuss (1) how scientific evidence is gathered, (2) how decisions to intervene are made, (3) the sometimes conflicting intents of public policy, and (4) the importance of properly evaluating results. Throughout this book, these same elements will repeatedly play important roles in our analyses of a wide range of environmental issues. They are the working skeleton of environmental science.

The battle was fought over the pollution of Lake Washington, a large (86-square-kilometer or 33.2-square-mile), deep freshwater lake that forms the eastern boundary of the city of Seattle (Figure 2–1). As the Seattle metropolitan area expanded eastward toward the lake from the shores of Puget Sound during the first part of the 20th century, Lake Washington came under increasingly intense environmental pressures. Recreational use of the lake expanded greatly, and so did its use for waste disposal. Sewage arrangements in particular had a major impact upon the lake.

The quality of the water in Lake Washington in the early 1990s is better than at any time in recent memory, despite the explosive expansion of suburban Seattle in the years after World War II. This good news is not an accident, but the result of concerted action by environmental scientists, local politicians, and ordinary citizens to develop and execute a reversal of the lake's disruption by urban

**Figure 2–1**
Lake Washington is a large freshwater lake near Seattle, Washington. (Mark E. Gibson)

pollution. Even though the solution was costly, the citizens of the region supported and implemented it after hearing a clear, no-nonsense statement of the problem, the proposed action, and the stark alternative to action. Environmental scientists defined two futures for Lake Washington, and the voting citizens selected one of them.

## Birth of an Environmental Problem

Seattle began discharging raw sewage into the waters of Lake Washington at the beginning of the 20th century, as the city first began to expand eastward from Puget Sound (Figure 2–2). By 1926 the lakeshore was sufficiently unpleasant that the city of Seattle passed a bond issue for a system of sewer lines to divert the city's sewage from Lake Wash-

**Figure 2–2**
Sewage discharge. (Doug Wechsler)

ington to a treatment plant that discharged directly into Puget Sound. By 1941, the last sewer discharge into Lake Washington from Seattle proper had stopped, and the lake again became a pleasant place where people came to boat, swim, and fish.

Like many cities in the United States, Seattle is ringed by suburbs with individual municipal governments. These suburbs expanded rapidly in the 1940s, creating an enormous waste disposal problem. Between 1940 and 1953, ten suburban sewage treatment plants began operating at points around the lake, with a combined daily discharge of 80 million liters (21 million gallons) into Lake Washington. Each plant treated the raw sewage to break down the organic material within it and release the "harmless" effluent (that is, treated sewage) into the lake.

By the mid-1950s, although raw sewage dumping had ended, a great deal of treated sewage had been dumped into the lake. Try multiplying 80 million liters per day by 365 days per year by 5 to 10 years: enough effluent was dumped to give about 27 to 54 liters (7 to 14 gallons) of it to every man, woman, and child living on Earth.

The effects of this discharge on the lake were first noted by G. Comita and F. Anderson, doctoral students at the University of Washington in Seattle. Their studies of the lake's microscopic single-celled organisms in 1953 and 1954 indicated that

### Envirobrief

#### When Business Listens to Environmentalists

The highly publicized 1991 decision by McDonald's Corp. to abandon polystyrene "clamshell" burger boxes in favor of paper wrappers was reached in consultation with The Environmental Defense Fund (EDF), a Washington-based environmental group. This collaboration between a large American corporation and a group of ecological activists was an important breakthrough in cooperation. EDF continues to work with McDonald's, helping the company reach its ambitious goal to reduce its solid waste by 80%. In another collaborative first, the supermarket chain Safeway worked with the environmental group Earth Island Institute in 1991 to create a dolphin-safe policy regarding all tuna products sold in its stores.

filamentous cyanobacteria (what biologists used to call blue-green algae) were growing in the lake. These are long strings of photosynthetic bacterial cells strung together. Their appearance in Lake Washington was unexpected, because the growth of cyanobacteria requires a plentiful supply of nutrients, and deepwater lakes such as Lake Washington do not usually have enough dissolved nutrients to support cyanobacterial growth.[1] Deepwater lakes are particularly poor in the essential nutrient phosphorus. The presence of filamentous cyanobacteria in Lake Washington's waters hinted that the lake was somehow changing, becoming richer in dissolved nutrients such as phosphorus (nutrient enrichment of lakes is discussed further in Chapter 21).

## Sounding the Alarm

The first public alarm was sounded on July 11, 1955, in a technical report by the Washington Pollution Control Commission. Its author, citing the work of Comita and Anderson, concluded that the treated sewage effluent that was being released into the lake's waters was raising the lake's levels of dissolved nutrients to the point of serious pollution. Whereas primary treatment followed by chlorination of the sewage was ridding it of bacteria (see Chapter 21), it was not eliminating many chemicals, particularly phosphorus (a major component of detergents). In essence, the treated sewage was fertilizing the lake by enriching it with dissolved nutrients.

The process of nutrient enrichment of freshwater lakes is well understood by ecologists, who call it **eutrophication.** Eutrophication is undesirable because, as Comita and Anderson had already begun to observe, high nutrient levels lead to the growth of filamentous cyanobacteria. These photosynthetic organisms need only three things in order to grow: light for photosynthesis (which they get from the sun), carbon atoms (which they get from carbon dioxide dissolved in water), and nutrients such as nitrogen and phosphorus (which were being provided by the treated sewage). Without the nutrients, cyanobacteria cannot grow; supply them,

and soon mats of filamentous cyanobacteria form a green scum over the surface of the water, and the water begins to stink as dead cyanobacteria rot in the sun.

Then the serious problem begins: the bacteria that decompose the masses of dead cyanobacteria multiply explosively, consuming vast quantities of oxygen in the process, until the lake's waters become so depleted that they can no longer support other organisms that require oxygen to live. Fish can no longer extract enough oxygen through their gills, and neither can the myriad of tiny invertebrates that populate freshwater lakes. For all intents and purposes, the eutrophic lake dies (Figure 2–3).

The local newspaper, the *Seattle Times*, mentioned the Pollution Control Commission's technical report in a July 11, 1955, article, "Lake's Play Use Periled by Pollution." The article did not grab the public's attention, but a month later, something else did: the annual Gold Cup yacht races, with their view of a magnificent sailboat's prow slicing cleanly through green scum and the not-so-subtle odor of rotting cyanobacteria. These additions to what had been a popular summer holiday raised protest among spectators and lakeshore residents.

Local authorities discounted the possibility that the cyanobacteria were the result of sewage into the lake, blaming them instead on the unusually sunny weather. But on the very day of the yacht race, F. Anderson collected a water sample from the lake that was to forever banish such sunny

**Figure 2–3**

A eutrophic pond in Texas is covered with the green scum of filamentous cyanobacteria. (E. R. Degginger)

---

[1] *Low* levels of nutrients are desirable in freshwater lakes because they permit the controlled growth of photosynthetic organisms that are the base of the food chain. When a body of water contains a *high* level of nutrients such as nitrogen and phosphorus, the photosynthetic organisms are present in vast numbers, upsetting the natural balance in the lake: When these photosynthetic organisms die, for example, their decomposition removes most of the dissolved oxygen from the water so that many fish suffocate and die.

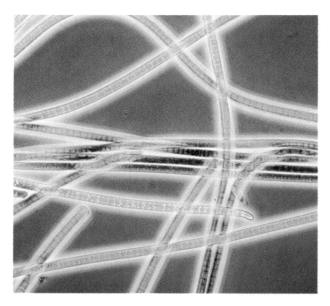

**Figure 2–4**
*Oscillatoria* is a photosynthetic cyanobacterium whose filaments are composed of chains of cells. (J. R. Waaland, University of Washington/Biological Photo Service)

explanations. The sample contained a filamentous cyanobacterium that neither Anderson nor earlier investigators had ever encountered in the lake: *Oscillatoria* (Figure 2–4). The presence of this cyanobacterium proved to be a vital clue. When Anderson's professor at the University of Washington, W. T. Edmondson, reviewed the literature on eutrophication, he came across the name *Oscillatoria* again and again in the lists of organisms found in polluted lakes. In one review of the history of human-induced eutrophication in Europe and North America, he underlined *Oscillatoria* each time the word appeared in the text, and discovered that the organism was a nearly perfect indicator of eutrophication. The destruction of Lake Zurich in Switzerland decades earlier seemed a clear parallel with Lake Washington. Lake Zurich—also a large, deep lake—had been enriched by sewage effluent; cyanobacteria began to be noted; and, soon after *Oscillatoria* appeared, water quality began to decline drastically.

To Edmondson, the appearance of *Oscillatoria* in Lake Washington was a clear warning. On October 13, 1955, the University of Washington *Daily* ran a story, "Edmondson Announces Pollution May Ruin Lake," in which Edmondson announced the appearance of *Oscillatoria* and its likely meaning. From this point on, the scientific case was clear: the eutrophication of Lake Washington was demonstrably at an advanced stage, and unless it was reversed, it would soon destroy the water quality of the lake.

## Scientific Assessment

The purpose of the scientific assessment of an environmental problem is, first, to identify that a problem exists and, second, to build a sound set of observations from which to proceed in seeking a solution. Lake Washington's microscopic life had been the subject of long-term ecological studies by students at the University of Washington since 1933. Thus, when the telltale signs of pollution first appeared in 1952, they were quickly detected by Edmondson's students as changes from previous studies. Without the earlier students' careful analyses of the many forms of microscopic creatures living in the lake, understanding of the changes that were occurring would have been delayed.

Edmondson examined and compared the earlier studies of the lake and confirmed that there had indeed been a great increase in dissolved nutrients in the lake's water. Surmising that the added nutrients were the result of sewage treatment waste discharge into the lake by suburban communities, Edmondson formed the hypothesis that treated sewage was introducing so many nutrients into the lake that its waters were beginning to support the growth of photosynthetic cyanobacteria.

Edmondson's hypothesis made a clear prediction: the continued addition of phosphates and other nutrients to the lake would change its surface into a stinking mat of rotting cyanobacteria, unfit for swimming or drinking, and the beauty of the lake would be only a memory. Bolstering his prediction was the fact that lakes near other cities, such as Madison, Wisconsin, had deteriorated after receiving sewage discharges.

The appearance of *Oscillatoria* in 1955 confirmed Edmondson's prediction: pollution was progressing in a classic pattern, its seriousness signaled by this almost-universal indicator of future trouble.

**Making a Model**    By 1955 the ecology of the lake had been extensively studied, and a great deal was known about it. Edmondson used this data base to construct a hypothetical model of the lake, which traced the general quantitative relationships between nutrient additions from sewage treatment plants and the growth of cyanobacteria in the lake's waters. By 1957 his model was sufficiently detailed to be used for quantitative predictions about the lake's future. It predicted a serious and rapid decline in water quality. Importantly, Edmondson's model also predicted that the decline could be reversed: if the pollution was stopped, the lake would clean itself at a predictable rate, reverting to its previous, unpolluted state within five years.[2]

Could anything be done to reverse the process? In April 1956, Edmondson outlined three steps that would be necessary in any serious attempt to save the lake: (1) comprehensive regional planning by the many suburbs that ringed the lake, (2) complete elimination of sewage discharge into the lake, and (3) research to identify the key nutrients that were causing the cyanobacteria to grow. His proposals received widespread publicity in the Seattle area, and the stage was set to bring scientists and civic leaders together.

## Risk Analysis

It is one thing to suggest that the addition of treated sewage to Lake Washington stop, and quite another to devise an acceptable alternative. Further, treatment of sewage can remove some nutrients, but it is not practical to remove all of them. The alternative is to dump the sewage somewhere else—but where? In this case, officials chose to discharge the treated sewage into the Pacific Ocean. In their plan, a ring of sewers to be built around the lake would collect sewage treatment discharges, treat them further, and then transport them to be discharged at great depth into Puget Sound.

Why go to all the trouble and expense of treating the discharges further, if you are just going to dump them? And why bother discharging them deep under water? Because it is important that the solution to one problem not create another. The plans to further treat the discharge and release it at great depth were formulated in an attempt to minimize the environmental impact of diverting Lake Washington's discharge into Puget Sound. It was assumed that sewage effluent would have less of an impact on the great quantity of water in the ocean than on the much smaller amount of water in Lake Washington.

Practically any course of action that can be taken to reverse an environmental problem has its own impacts on the environment, which must be assessed when evaluating potential solutions (see Focus On: Environmental Impact Statements). Environmental impact analyses often involve studies by geographers, chemists, and engineers as well as ecologists and other biologists. Furthermore the decision whether or not to implement a plan to restore or protect the environment is almost always affected by nonscientific factors and concerns. Any proposal is inevitably and rightly constrained by existing laws and by the citizens who will be affected by the decision.

## Public Education

The scientific studies indicating the progressive pollution of Lake Washington first received public notice from the Washington Pollution Control Commission, which used the studies as the basis of its 1955 technical bulletin (already mentioned). Local sanitation authorities, however, were not convinced that urgent action was necessary. Public action required further education, and it was at this stage that scientists played a key role. Edmondson and other scientists wrote articles for the general public that contained concise explanations of what nutrient enrichment was and where it would lead. As these articles were picked up by the local newspapers, the general public's awareness of the problem increased.

In December 1956, Edmondson, concerned about the delay of action, wrote a letter in an effort to alert the chairman of a committee (formed by the mayor of Seattle) on regional problems affecting Seattle and its suburbs. He explained that even well-treated sewage would soon destroy the lake, and that Lake Washington was already showing signs of deterioration. Edmondson received an encouraging response and prepared for the committee a nine-page report, in nontechnical language, of his scientific findings. After presenting his data showing that the mass of cyanobacteria varied in strict proportion to the amounts of nutrients being added to the lake, Edmondson posed a series of questions: "How has Lake Washington changed?" "What will happen if nothing is done to halt nutrient accumulation?" "Why not poison the cyanobacteria and then continue to discharge the effluent?" He then answered the questions, outlined two alternative courses of public action—do nothing or stop adding nutrients to the lake—and made a clear prediction about the consequences of each.

## Political Action

Edmondson's report was widely circulated among local governments, but implementing its proposals presented serious political problems, because there was no governmental mechanism that would permit the many local suburbs to act together on re-

---

[2] In freshwater lakes, iron reacts with phosphorus (an important nutrient for cyanobacteria) to form an insoluble complex that sinks to the bottom of the lake and is buried in the sediments. Thus, if additional phosphorus were not introduced into the lake from sewage effluent, the lake would slowly recover.

## Environmental Impact Statements

The **National Environmental Policy Act** (NEPA), signed into law in 1970, is the cornerstone of U.S. environmental policy. It requires that the federal government consider the environmental impact of any proposed federal action (for example, financing highway or dam construction) before deciding whether to implement that action. The NEPA provides the basis for commissioning a detailed **environmental impact statement** (EIS) to accompany every federal recommendation or proposal for legislation. EISs are meant to help federal officials and the public make informed decisions. Each EIS must include:

1. The nature of the proposal and why it is needed
2. The environmental impact of the proposal, including short-term and long-term effects and any adverse environmental effects
3. Alternatives to the proposed course of action, including a no-action alternative, that will lessen the adverse effects

A draft of the EIS must be available for public scrutiny and review by other federal agencies for at least 45 days before the final EIS is written. Individuals, citizen groups, and environmental groups can send letters commenting on the proposal, and these letters must be evaluated when the final EIS is prepared.

The final decision on the proposal cannot be made until at least 30 days after the final EIS is available or 90 days after the draft EIS is available, whichever is later. After the final decision is made, any challenges to the decision are made in court.

As a result of the NEPA, individuals, citizen groups, and environmental groups have filed hundreds of lawsuits against agencies of the federal government. These suits have forced the federal government and individuals doing business with the federal government to focus on the environmental impacts of their projects.

The NEPA has influenced environmental legislation in many states and a number of other countries. Thirty-six states have passed similar legislation requiring EISs for state-funded projects. Australia, Canada, France, New Zealand, and Sweden are some of the countries that now require EISs for government-sponsored projects.

Although almost everyone agrees that the NEPA has helped federal agencies reduce adverse environmental impacts of their activities and projects, it is not without its critics. Environmentalists complain that some EISs are incomplete or are ignored when decisions are made. Other critics think the EISs delay important projects because they are too detailed, take too long to prepare, and are often the targets of lawsuits.

---

gional matters such as sewage disposal. In late 1957 the state legislature passed a bill permitting a public referendum in the Seattle area on the formation of a regional government with six functions: water supply, sewage disposal, garbage disposal, transportation, parks, and planning. The referendum was defeated in March 1958, apparently because suburban voters felt that the plan was an attempt to tax them for the city's expenses.

Understanding the urgency of Edmondson's proposals, an advisory committee immediately submitted to the voters a revised bill limited to sewage disposal. Over the summer there was widespread discussion of the lake's future, and when the votes were counted on September 9, 1958, the revised bill had passed by a wide margin.

At the time it was passed, the Lake Washington plan was the most ambitious and most expensive pollution control project in the United States. Every household in the area had to pay $2 a month in additional taxes for construction of a massive trunk sewer to ring the lake, collecting all the effluent, treating it, and discharging it into Puget Sound, where the tides would carry it to sea.

The role of environmental science in addressing problems such as the pollution of Lake Washington is limited to assessing the problem, evaluating alternative solutions, and educating the public about these matters. Events then pass into the public arena, as they should. The proposed solution of Lake Washington's problems involved considerable expense for every citizen in the area, as well as a radical reorganization of sewage treatment that transferred it from local to regional control. Recall that the first ballot proposal, to establish a regional district to deal with this and other problems, was defeated by voters because many other issues in addition to the pollution of Lake Washington affected the vote. Only by limiting the regional effort to sewage disposal and thus separating it from other concerns was the proposal eventually accepted by the voters.

The lesson is that the public is always concerned about many other matters in addition to environmental ones. Although the statement "The citizens of the Seattle area were presented with two clearly defined futures and were asked to choose between them" is true, it does not begin to take into account the many other factors, such as increased taxation and conflicting commercial and political interests, that influenced the public decision.

**Implementing the Plan of Action**  Ground-breaking ceremonies for the new project were held in July 1961. As Edmondson had predicted, the lake had deteriorated further. Visibility in lake water declined from 4 meters (12.3 ft) in 1950 to less than 1 meter (3.1 ft) in 1962, the water being clouded with cyanobacteria. On October 5, 1963, a suburban newspaper dubbed Lake Washington "Lake Stinko." In 1963 the first of the waste treatment plants around the lake began to divert its effluent into the new trunk sewer; one by one, the others diverted theirs, until the last effluent was diverted in 1968. The lake's deterioration stopped by 1964, and then its condition began to improve (Figure 2–5).

## Following Through

By carefully analyzing what was happening in the lake, Edmondson could predict that the lake would recover fully. Not all environmental scientists agreed with him, many arguing that dissolved phosphorus, the key nutrient regulating cyanobacterial growth, would not dissipate for decades, if ever. A lot depended on assumptions about the chemical makeup of the sediment at the bottom of the lake.

**Figure 2–5**

Relationship between nutrient additions to Lake Washington from sewage treatment plants and the growth of cyanobacteria. (a) Dissolved phosphorus in Lake Washington from 1955 to 1975. The amount of phosphorus contributed by sewage effluent is indicated in the shaded area. (b) Cyanobacterial growth during Lake Washington's recovery, 1965 to 1975. As the level of phosphorus dropped in the lake, the number of cyanobacteria also declined (as measured indirectly by the amount of chlorophyll).

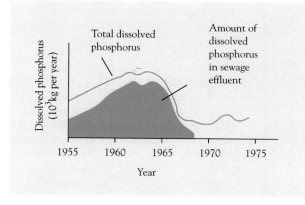

(a)

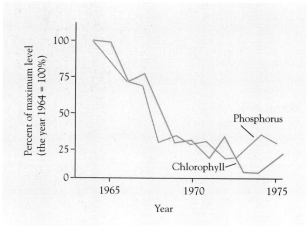

(b)

Edmondson was right. Water transparency returned to normal within a few years (Figure 2–6). *Oscillatoria* persisted until 1970, but eventually it too disappeared. By 1975 the lake was back to normal. Indeed, by 1980 the lake was clearer than at any time in recent memory, with visibility exceeding 12 meters (39.4 ft) at times. Before the recovery, the presence of filamentous cyanobacteria such as *Oscillatoria* had restrained growth of the lake's population of a microscopic organism called *Daphnia* (cyanobacterial filaments clog *Daphnia*'s feeding apparatus). The disappearance of *Oscillatoria* and other filamentous cyanobacteria allowed the lake's *Daphnia* population to flourish and become dominant among the many invertebrate species that live there. Because *Daphnia* are very efficient eaters of nonfilamentous algae, levels of these algae in the water fell, too, so that the water became even clearer. A dozen years later, in 1992, the lake remains clear.

Every environmental intervention is an experiment. The diversion of sewage discharge from Lake Washington was nothing more or less than a large-scale experiment in nutrient cycling (see Chapter 5) in a freshwater lake, and Edmondson's model made clear predictions about the results of the experiment. By carefully monitoring the outcome of the sewage diversion, Edmondson was able to confirm that it matched his model's predictions.

Monitoring is necessary because environmental scientists work with imperfect tools. There is a great deal we don't know, and every added bit of information increases our ability to deal with future problems. The knowledge that Edmondson's approach to modeling the Lake Washington situation worked provides helpful information to today's environmental scientists. They would not have this information if the lake's recovery had not been monitored. Knowledge about the effects of environmental interventions is almost always valuable.

It is a mistake, however, to assume that we always know just what is going to happen. Edmondson's model did not predict, for example, that the lake would become even clearer than before, because the role of *Daphnia* in keeping down the levels of nonfilamentous algae was not anticipated. The unanticipated always lurks just beneath the surface of any experiment carried out in nature, because our knowledge is limited. There is much to be learned from careful observation of the results of environmental "experiments."

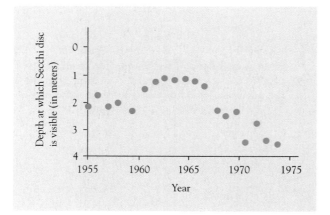

**Figure 2–6**

Effect of pollution and recovery of water transparency in Lake Washington. Measurements were taken in July or August from 1955 to 1975 using a round disc called a Secchi disc attached to a rope. The greater the water transparency, the deeper the Secchi disc can be lowered and still be visible. If the Secchi disc disappears at one meter, it means the water is very cloudy from the growth of cyanobacteria. If the Secchi disc can be lowered over three meters before disappearing from sight, it means there are few cyanobacteria present.

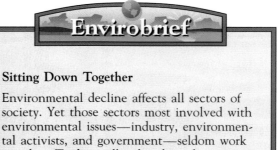

**Envirobrief**

**Sitting Down Together**

Environmental decline affects all sectors of society. Yet those sectors most involved with environmental issues—industry, environmental activists, and government—seldom work together. Traditionally, they have been antagonists. Recognizing the need for cooperation, each of the 10 Canadian provinces, as well as the Canadian federal government, formed "roundtables" on environment and the economy in 1989. These "multi-stakeholder" groups include representatives from industry, government, the environmental community, and Native American groups. They meet regularly to discuss environmental issues and offer recommendations to government. While the roundtables have generally bogged down in rhetoric and process, demonstrating how difficult it is to reconcile divergent points of view, they have at least brought together influential groups that were formerly not even speaking to each other—an important first step.

## WORKING TOGETHER

The reversal of the pollution of Lake Washington is a particularly clear example of how environmental science can work to identify, address, and solve environmental problems. One element of the Lake Washington story is of particular importance as we try to address far more complex global environmental problems today: the lake's pollution problem was solved only because the many small towns involved in the problem cooperated in seeking a solution. Environmental science provided the information that identified the source of the problem and suggested a solution, but implementation of that solution depended critically on a political decision. In the final analysis, it is people, not scientists, who must solve environmental problems (Figure 2–7).

In the same way in which these small towns needed to cooperate to solve their environmental problem, counties, states, and nations need to cooperate to solve regional and global problems, including the cleanup of the Mediterranean Sea, the

(a)

(b)

**Figure 2–7**
People create environmental problems, but they also can solve them. On February 7, 1990, for example, an oil spill occurred near Newport Beach, California. (a) The spill coated the beach with oil. (b) Employees and volunteers assisted in the cleanup of the beach. (Wallace Kleck/ Terraphotographics)

### Envirobrief

**Getting Tough With Trash**

Germany has the toughest trash laws of any country in the world. Since January 1, 1993, it has been mandatory in Germany that all forms of packaging be reused or recycled. No longer can they be buried in landfill sites or burned. The legislation also stipulates that consumers can return packaging to the place where it was purchased. This has led a coalition of more than 600 German companies—including packaging companies, consumer goods companies, and large food retailers—to form an independent collection system called Duales System Deutschland (DSD). Any company that markets a packaged product pays DSD a fee that earns them the right to display a "Green Dot" logo on their packaging. All products with a Green Dot are collected and recycled by DSD at huge sorting plants throughout the country. By 1995, it is estimated that 80% of all German packaging will be recycled, drastically reducing the amount of garbage the country must bury or burn. In the United States, by comparison, only the most readily recycled packaging, such as aluminum cans, glass, newsprint, and plastic bottles, is being recycled in meaningful amounts. Most forms of packaging are simply buried or burned.

## The Tragedy of the Commons

Garret Hardin is a professor of human ecology who writes about human environmental dilemmas. In 1968 he published a short essay, "The Tragedy of the Commons," in which he contended that our inability to solve complex environmental problems is the result of a struggle between short-term individual welfare and long-term societal welfare.

Hardin used the commons to illustrate this struggle. In medieval Europe, the inhabitants of a village shared pasture land, called the commons, and each herder could bring animals onto the commons to graze. The more animals a herder brought onto the commons, the greater the advantage to that individual. When every herder in the village brought as many animals onto the commons as possible, however, the plants were killed from overgrazing, and the entire village suffered (see photo). One of the outcomes of the eventual destruction of the commons was private ownership of land, because when each individual owned a parcel of land, it was in that individual's best interest to protect the land from overgrazing.

Hardin's parable has relevance today. The "commons" are those parts of our environment that are available to everyone but for which no single individual has responsibility—the atmosphere and the oceans, for example. These modern-day commons, sometimes collectively called the **global commons,** are experiencing increasing environmental stress. Because they are owned by no individual, jurisdiction, or country, they are susceptible to overuse. Although their exploitation may benefit only a few, everyone on Earth must pay the environmental costs of exploitation.

Clearly, the world needs legal and economic policies to prevent the short-term degradation of our global commons and insure the long-term well-being of our natural resources. There are no quick fixes, because solutions to global environmental problems are not as simple or short-term as solutions to some local problems are, but instead are complex and long-term. Most environmental ills are inextricably linked to other persistent problems such as poverty, overpopulation, and social injustice, problems that are beyond the ability of a single nation to resolve.

Cooperation and international commitment are essential if we are to alleviate poverty, stabilize the human population, and preserve our environment for future generations.

The effects of overgrazing in northern Ghana. When too many animals graze on a given piece of land, the grasses die and the land cannot support any cattle. (Robert E. Ford/ Terraphotographics)

**Figure 2–8**
The cleanup of pollution in the Mediterranean Sea will require the combined efforts of all the nations that border it.
(Visuals Unlimited/Frank M. Hanna)

elimination of acid rain, and the reduction of chemicals that destroy ozone and contribute to global warming (Figure 2–8). Unfortunately, humans haven't yet learned to cooperate well at any level. We desperately need to develop mechanisms, including laws, to induce us to do so (see Focus On: The Tragedy of the Commons).

Environmental education will play an important role in engendering this cooperation. The more environmental education people have, the more they understand risks and the more capable they are of making good decisions. This sort of education will be particularly important in developing countries, which have rapidly growing populations but relatively few of the world's scientists and engineers and therefore are less equipped to work out clear, scientifically based alternatives for themselves. With assistance from countries such as the United States, they may be able to avoid repeating some of the mistakes developed countries have made. International cooperation will be increasingly needed as ecological problems become more global in scope.

## SUMMARY

**1.** Science is a particular way of investigating the world, a systematic attempt to understand the Universe. Science seeks to reduce the apparent complexity of our world to general principles, which can then be used to solve problems or provide new insights.

**2.** Scientists make use of two kinds of reasoning, inductive and deductive. Inductive reasoning begins with specific examples and seeks to draw a conclusion or discover a unifying rule or general principle on the basis of those examples. Science also makes use of deductive reasoning, which operates from generalities to specifics. Deductive reasoning adds nothing new to knowledge, but it can make relationships among data more apparent.

**3.** A hypothesis is a suggestion or idea about how the world might work. Hypotheses are most useful when they make predictions that can be tested.

**4.** An experiment is a test of a hypothesis. A factor that can affect the outcome of an experiment is called a variable. Experiments are usually conducted in pairs, with all but one of the variables held constant. By testing variables one at a time, the single variable responsible for an experimental observation can be identified.

**5.** A hypothesis supported by a large body of observations and experiments becomes a theory. A good theory relates facts that previously appeared to be unrelated, grows as additional facts become known, predicts new facts, and suggests new relationships among phenomena.

**6.** Damage to the environment can often be reversed, as demonstrated by the rescue of Seattle's polluted Lake Washington. The dumping of treated sewage into the lake had raised its level of nutrients to the point where the lake supported the growth of filamentous cyanobacteria. Disposal of the sewage in another way solved the lake's pollution problem.

**7.** A successful approach to solving an environmental problem usually involves five components: scientific assessment, risk analysis, public education, political action, and follow-through. Scientific assessment of environmental issues involves identifying potential environmental problems and suggesting possible solutions.

**8.** Risk analysis is the evaluation of the potential consequences of solutions to environmental problems. Some of these consequences may be new environmental problems; others may raise political or social issues.

**9.** Public education involves placing into the public arena the results of scientific assessment and risk analysis. The public is thus made aware of the problem and of the consequences of alternative actions.

**10.** Political action is the implementation of a particular plan of action by elected or appointed officials; ultimately, the decisions are made by the voting public.

Follow-through is the assessment of the effects of an action that was taken in an attempt to correct a perceived environmental problem.

**11.** An element of risk is inherent in everything we do. Risk assessment, the estimation of risks for comparative purposes, helps us set priorities and manage risks.

**12.** The National Environmental Policy Act requires that environmental impact statements be prepared to aid decision making about federally funded projects. EISs ensure that environmental values are given appropriate consideration in planning and decision making.

# DISCUSSION QUESTIONS

**1.** When Sherlock Holmes amazed his friend Watson by determining the general habits of a stranger based on isolated observations, what kind of reasoning was he using? Explain.

**2.** The Seattle area was presented with two alternatives for dealing with the pollution of Lake Washington: (1) do nothing or (2) divert the sewage. What other alternatives might have been proposed? Why weren't they?

**3.** The Lake Washington case is presented as a model in this chapter because it clearly demonstrates the five components of solving any environmental problem. The final outcome in the Lake Washington case—dumping treated sewage into the ocean—is not an ideal solution, however. Explain why.

**4.** Can you name an environmental problem that appears to have received adequate scientific assessment and to have been extensively reported by the press, and yet remains unsolved? What do you think has made the difference between the fate of this issue and the fate of the pollution of Lake Washington?

**5.** Can you think of several examples of the global commons other than those mentioned in the chapter?

**6.** Determining whether a substance causes cancer currently requires extensive testing. Do you think there will ever be a quick and accurate test for carcinogenicity in humans? Why or why not?

**7.** Is it possible to reduce risk to zero? Why or why not?

# SUGGESTED READINGS

Ames, B. N., R. Magaw, and L. S. Gold. Ranking possible carcinogenic hazards. *Science*, vol. 236, 17 April 1987. The authors challenge the assumption that one can extrapolate from experiments on rodents to predict risks to human health of long-term exposure to low doses of chemicals.

Burgis, M. J., and P. Morris. *The Natural History of Lakes.* Cambridge University Press, New York, 1987. An excellent background for understanding what happened to Lake Washington.

Edmondson, W. T. *The Uses of Ecology: Lake Washington and Beyond.* University of Washington Press, Seattle, 1991. After a detailed account of the saving of Lake Washington, the author presents a broad perspective on the critical role played by scientific research in solving several other environmental problems.

French, H. F. "Strengthening Global Environmental Governance," in *State of the World 1992, a Worldwatch Institute Report on Progress Toward a Sustainable Society.* W. W. Norton & Company, New York, 1992. Using international treaties and other forms of environmental governance to guard the global commons.

Lehman, J. "Control of Eutrophication in Lake Washington," in *Ecological Knowledge and Environmental Problem Solving: Concepts and Case Studies.* National Academy Press, Washington, D.C., 1986. A fine overview of the Lake Washington story is presented on pages 302–312.

Wilson R., and E. A. Crouch. Risk assessment and comparison: An introduction. *Science*, vol. 236, 17 April 1987. The element of uncertainty makes risk assessment hard to grasp, but comparing risks with commonplace situations helps.

**Mt. Sanford in Wrangell–St. Elias
National Park, Alaska.**
(David Muench)

The World We Live In

Part 2

# George Woodwell

## Global Ecology and the Human Factor

George M. Woodwell is the director and founder of the Woods Hole Research Center in Woods Hole, Massachusetts. Through research, education and the application of science in public policy-making, the Center addresses global environmental problems generated by the expansion of human activities over our finite Earth. Dr. Woodwell received his undergraduate degree in biology from Dartmouth College, and his Masters and Doctoral degrees in botany from Duke University. He has worked extensively on issues such as global warming, ecological effects of ionizing radiation and nuclear war. Conducting major studies on pesticides and other toxins in North America's forests and estuaries, he led the campaign in 1972 to ban the insecticide DDT.

**What led you to found the Woods Hole Research Center in 1985? How is its mission, with its special focus on the ecologist's perspective, different from that of other research institutions?**

As an ecologist, I've worked continuously with natural communities. The central question is: How does the world work? Ecologists recognize that plant and animal communities are the basis of all life, including human life. We're interested in how those communities work to maintain a human environment, in how they are changed as a result of human activities, and in how to keep them working. We look at these biotic resources locally, regionally, and globally. There are very few institutions equipped to deal with those topics.

**So, there was an institutional vacuum that needed to be filled?**

Yes. I've been a member of the faculties of various universities. In the 1960s, I went to Brookhaven National Laboratory in New York to establish a research program on the effects of ionizing radiation on forests. Brookhaven was run then by the Atomic Energy Commission, later by the Department of Energy. Brookhaven's interests were in high energy physics, not ecology. In 1975, I moved to the Marine Biological Laboratory in Woods Hole where the major interest is in the cellular biology of human health—again, not ecology. Experience showed me that the vigor of basic research in ecology hinges on having a free-standing institution whose purpose is not in question. The Woods Hole Research Center has a board of trustees committed to advancing basic knowledge of ecology.

**The interaction between our planet's many ecosystems is so complex that it is not yet fully understood, even by ecologists. What is the best way to understand these global relationships?**

If we try to look at the Earth in great detail, the picture is quite complicated. But, if we view it as a whole, we can certainly see the net effect of human activity. We know that the scale of these activities has now reached a point where it rivals or exceeds the normal fluxes of energy, and certain elements in the natural environment. We introduce nitrogen into the world in quantities that approach the total nitrogen flux in nature. We introduce carbon dioxide into the atmosphere at a rate that is changing the carbon dioxide burden of the atmosphere. At the same time, we are changing water flows over the surface of the Earth so significantly that we change the volume of water in rivers, streams, and estuaries, and their ability to support life.

**We are also releasing toxins into the global environment. Are we any closer now than we were 25 years ago to understanding how toxins behave in the ecosphere?**

The assumption used to be that there is an assimilative capacity in nature for virtually any toxin or human activity, and that if we stay below or well within that assimilative capacity then the activity is acceptable. Experience with toxic substances, however, has shown this assumption to be misleading.

The DDT story is a good example. DDT was sprayed, often from airplanes, to control various insects, including the mosquitoes that carry malaria. DDT is highly effective used in this way, and not very toxic to people. But some of the droplets from the spray dry out before they reach their target and form small crystals of DDT that are carried by wind around the world. We had, in a very few years, a global pollution with DDT and concentrations that were accumulating in nature. If there was an "assimilative capacity," it was much lower than the amount used globally.

**And if it landed in an aquatic area, for example, what were the results?**

DDT is not very soluble in water; it's most soluble in fats, and it breaks down very slowly. That means that if you put it in the general environment it will be absorbed into anything that has fat in it, algal cells for instance, and will stay there. DDT is then passed up the food chain from the algae to multicellular animals to fish, and then into birds, people, and other top carnivores, accumulating in higher quantities each step of the way. If it's released anywhere in a salt marsh, it will show up in the birds, and continue to accumulate in them for years. One of its effects is to interfere with the reproduction of birds. The osprey of the eastern United States didn't reproduce for 10 years; the peregrine falcon became extinct over a large portion of its range in eastern North America.

So, with a substance like this, the assimilative capacity of nature is elusive, if it exists at all. The only practical policy was zero release, the policy adopted by the U.S. in 1972.

**Although DDT was banned in this country, it is still being used elsewhere. Who is still using DDT?**

DDT is still being used in other countries, although it's generally recognized around the world as an undesirable toxin. But it also turns up in certain pesticides in the United States today, listed as an inert ingredient. There is a peculiar loophole in the laws governing toxins: The primary toxin is listed as the effective ingredient, but other substances can be present in smaller quantities, and called "inert."

**You've said that you prefer to focus on long-term trends, and you identify the main trend as "biotic impoverishment." What do you mean by this term?**

All of the Earth's species evolved on a planet that, over millions of years, was essentially stable. Now, in the relatively short period of a century, or even a few decades, humans are changing the physical, chemical, and biotic conditions of the Earth. Forests are reduced to shrublands and then to eroded soil or even bare rock. The result is that the tough plants and animals on the Earth—the insects, the hardy roadside herbs—these may survive. They are the small-bodied organisms that reproduce rapidly. But we lose the larger, longer-lived organisms—the trees, the elephants, the whales—these are the vulnerable species. I call this chronic assault on the structure of nature and its effects "biotic impoverishment." Furthermore, as the environment becomes impoverished, so do people. We lose opportunities for using resources, our economy loses potential, and people become financially poorer. So, biotic impoverishment leads to human impoverishment.

**Yet the world views held by economists and industry are often diametrically opposed to those of ecologists. Is there any prospect for a meeting of minds here?**

Economists have been very slow to recognize the need for a full-cost accounting of economic activities. They have been able to think about national and global economies, ignoring the environment, which they consider to be an externality. And they are aided in this unrealistic thinking by industrial and commercial interests who focus on profits, and allow their costs to diffuse over the world's population in the form of a degraded environment. Now economists are gradually realizing that this process does affect the economy, and that they must incorporate those costs into their calculations. Companies are increasingly being held accountable for what they're doing. People are realizing that they pay these costs, in shorter lives, less intelligent children, a less healthful habitat.

These are signs that the entire population of the world is becoming concerned about this trend of biotic impoverishment. The most serious evidence of this was the 1992 Earth Summit in Rio de Janeiro, where the nations of the world gathered to address major environmental issues and issues of economic development. These topics are high on the political agenda now, and will only become more important as human populations continue to expand and human influences extend further around the world.

# Chapter 3

*Lupines and other plants absorb solar energy for photosynthesis. (David Cavagnaro)*

# Ecosystems and Energy

An ecosystem encompasses all the interactions among organisms living together in a particular area and between those organisms and their physical environment. The interactions can be self-sustaining—that is, the ecosystem can continue to operate—only as long as energy is provided to the system. Energy flows through an ecosystem in one direction, and once it has been used by a living organism, it is not available for reuse by any organism in the ecosystem. Living organisms in an ecosystem assume the roles of producer, consumer, and decomposer, depending on their sources of energy. Energy flows from producers to consumers to decomposers.

## THE HOUSE WE LIVE IN

The concept of ecology was first developed in the 19th century by Ernst Haeckel, who also created its name—*eco* from the Greek word for "house" and *logy* from the Greek word for "study." Thus, ecology literally means the study of one's house. Viewed from the standpoint of ecology, nature is something like a great estate in which the physical environment and living organisms interact in an immense and complicated web of relationships. **Ecology,** then, is the study of the interactions among organisms and between organisms and their physical environment.

### What Ecologists Study

The focus of ecology can be local and very specific or global and quite generalized, depending on the scientist's view. One ecologist might determine the temperature or light requirements of a single species of oak, another might study all the organisms that live in a forest where the oak is found, and another might examine how matter flows between the forest and surrounding communities.

#### Mini-Glossary of Ecology Terms

**population:** A group of organisms of the same species that live together.

**community:** All the living organisms found in a particular environment. Includes all the populations of different species that are living together.

**ecosystem:** A community and its environment. Includes all the interactions between living things and their physical environment.

**biosphere:** All of the Earth's living organisms. Includes all the communities on Earth.

**ecosphere:** The largest, worldwide ecosystem. It encompasses all the living things on Earth and their interactions with each other, the land, the water, and the atmosphere.

How does the field of ecology fit into the organization of the biological world? As you may know, one of the characteristics of life is its high degree of organization (Figure 3–1). Atoms are organized into molecules, which are organized into cells. In multicellular organisms, cells are organized into tissues, tissues into organs (such as the brain and stomach), organs into organ systems (such as the nervous system and digestive system), and organ systems into individual organisms (dogs, humans, cacti, ferns, and so on).

The levels of biological organization that interest ecologists are those above the level of the individual organism. Organisms are arranged into **populations,** members of the same species[1] that live together in the same area at the same time. A population ecologist might study a population of polar bears or a population of marsh grass. (Populations are discussed in Chapters 8 and 9.)

Populations are organized into communities. A **community** consists of all the populations of different species that live and interact within an area. A community ecologist might study how organisms interact with one another—including who eats whom—in a coral reef community (Figure 3–2) or in an alpine meadow community. "Ecosystem" is a more inclusive term than "community" because an **ecosystem** is a community together with its environment. Thus, an ecosystem includes not only all the interactions among the living organisms of a community but also the interactions between organisms and their physical environment. An ecosystem ecologist might examine how temperature, light, precipitation, and soil affect the organisms living in a desert community or a coastal bay community (Figure 3–3).

All the communities of living things on the Earth are organized into the **biosphere.** The organisms of the biosphere depend on one another and on the other divisions of the Earth's physical environment: the atmosphere, hydrosphere, and lithosphere. The **atmosphere** is the gaseous envelope surrounding the Earth; the **hydrosphere** is the Earth's supply of water (liquid and frozen, fresh and salty); and the **lithosphere** is the soil and rock of Earth's crust. The term **"ecosphere"** encompasses the biosphere and its interactions with the atmosphere, hydrosphere, and lithosphere. Ecologists who study the biosphere or ecosphere examine the complex interrelationships among the Earth's atmosphere, land, water, and living things.

Ecology is the broadest field within the biological sciences, and it is linked to every other biological discipline. The universality of ecology also brings subjects into view that are not traditionally part of biology. Geology and earth science are extremely important to ecologists. Because humans are biological organisms, *all* of our activities have a bearing on ecology—even economics and politics have profound ecological implications (see Chapter 7).

---

[1] A species is a group of similar organisms whose members freely interbreed with one another in the wild and do not interbreed with other sorts of organisms.

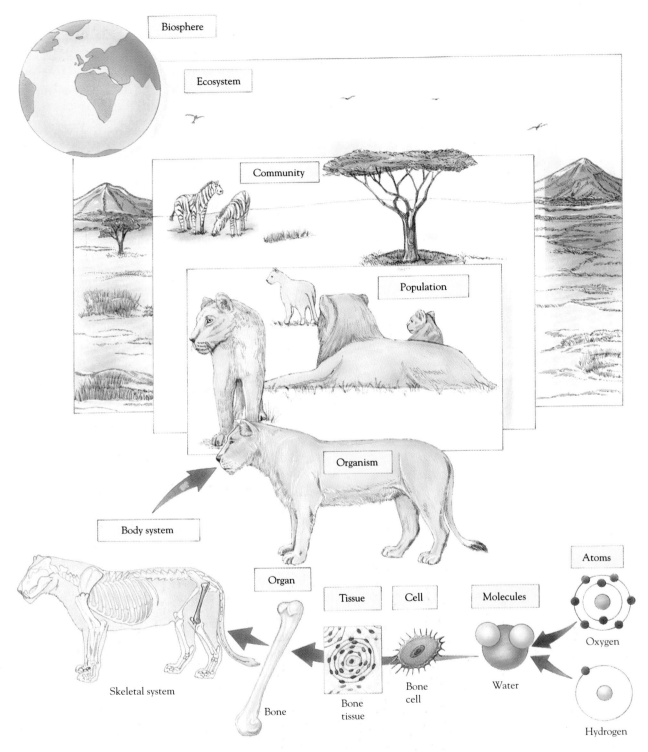

**Figure 3–1**

Levels of biological organization. Starting at the simplest level, atoms are organized into molecules, which are organized into cells. Cells are organized into tissues, tissues into organs, organs into organ (or body) systems, and organ systems into individual multicellular organisms. A group of individuals of the same species is a population. Populations of different species interact to form communities. A community and its physical environment is an ecosystem. The highest level of organization is the biosphere, which consists of all communities of living things on Earth. Ecologists study the highest levels of biological organization: populations, communities/ecosystems, and the biosphere.

**Figure 3–2**
A coral reef community in the Red Sea. Coral reef communities have the greatest species diversity and are the most complex kind of aquatic community. (Robert Shupak)

**Figure 3–3**
Many ecosystem studies require elaborate equipment. The floating rings, called mesocosms, extend downward into the water, enclosing part of the water column. (Courtesy of Dr. M. R. Reeve, National Science Foundation)

## THE INHABITANTS OF ECOSYSTEMS

Ecosystems often contain an astonishing assortment of organisms that interact with each other and are interdependent in a variety of ways. Consider for a moment a salt marsh in the Chesapeake Bay on the east coast of the United States. This bay is the world's richest estuary (a semi-enclosed body of water where fresh water drains into the ocean). Biological diversity and productivity abound wherever fresh water and salt water form a salinity gradient (a gradual change from unsalty fresh water to salty ocean water), as they do in the Chesapeake Bay. The salinity gradient in the bay results in three distinct marsh communities: freshwater marshes at the head of the bay, brackish (moderately salty) marshes in the middle bay region, and salt marshes on the ocean side of the bay. Each community has its own characteristic organisms.

Sail to one of the salt marsh islands in the Chesapeake Bay, such as South Marsh Island, and you can explore a salt marsh community that is relatively unaffected by humans (Figure 3–4). A salt marsh presents a monotonous view—miles and miles of flooded meadows of cordgrass (*Spartina*). High salinity (although not as high as that of ocean

water) and twice-daily tidal inundations create a challenging environment to which only a few plants have adapted.

Nutrients such as nitrates and phosphates, which drain into the marsh from the land, promote rapid growth of both cordgrass and microscopic algae. These organisms are eaten directly by some animals, and when they die, their remains (called detritus) provide food for many inhabitants of both the salt marsh and the bay.

A casual visitor to a salt marsh would observe two different types of animal life, insects and birds. Insect pests such as salt marsh mosquitoes and horseflies number in the millions. Birds nesting in the salt marsh include seaside sparrows, laughing gulls, and clapper rails. Migratory birds spend time in the salt marsh as well.

Study the salt marsh carefully and you'll find it has numerous other species. Large numbers of invertebrates seek refuge in the water surrounding the cordgrass. Here they eat, hide to avoid being eaten, and reproduce. Many of them gather in the intertidal zone because food (detritus, algae, protozoa, and worms) is abundant there. A variety of crusta-

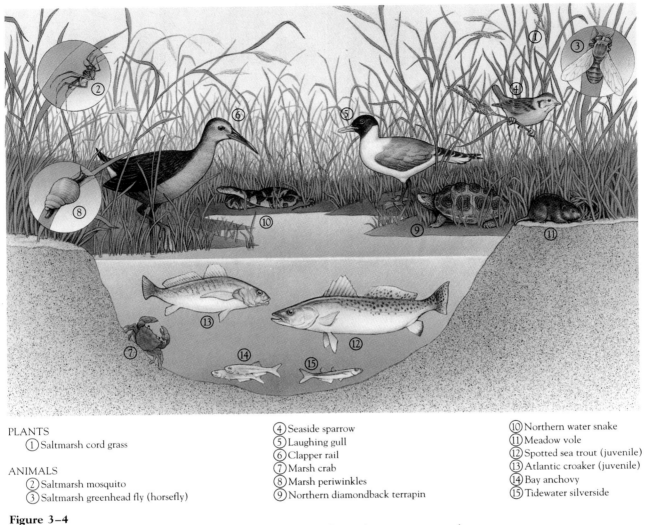

PLANTS
    ① Saltmarsh cord grass

ANIMALS
    ② Saltmarsh mosquito
    ③ Saltmarsh greenhead fly (horsefly)

④ Seaside sparrow
⑤ Laughing gull
⑥ Clapper rail
⑦ Marsh crab
⑧ Marsh periwinkles
⑨ Northern diamondback terrapin

⑩ Northern water snake
⑪ Meadow vole
⑫ Spotted sea trout (juvenile)
⑬ Atlantic croaker (juvenile)
⑭ Bay anchovy
⑮ Tidewater silverside

**Figure 3–4**
Salt marshes, which teem with life, are found in the transitional areas between ocean and land.

ceans live in the salt marsh. The marsh crab, for example, is a common inhabitant that eats cordgrass and small animals as well as detritus. Mollusks include the marsh periwinkle, a snail that moves along the cordgrass skimming off attached algae for its food. Marsh periwinkles climb up the cordgrass to avoid becoming prey for larger marsh animals such as terrapins.

Almost no amphibians inhabit salt marshes, because the salty water dries out their skin, but a few reptiles have adapted—the northern diamondback terrapin, for example. It spends its time basking in the sun or swimming in the water searching for food—snails, crabs, worms, insects, and fish. Although a variety of snakes abound in the dry areas adjacent to salt marshes, only the northern water snake (which preys on fish) is adapted to brackish water.

Mammals are represented in the salt marsh by the meadow vole, a small rodent that constructs its nest of cordgrass on the ground above the high-tide zone. Meadow voles are excellent swimmers and scamper about the salt marsh day and night. Their diet consists mainly of insects and cordgrass.

The Chesapeake Bay marshes are an important nursery for numerous marine fish—spotted sea trout, Atlantic croaker, striped bass, and bluefish, to name just a few. Other fish, such as bay anchovies, bull minnows, and tidewater silversides, never leave the estuary, spending their summers in the salt marsh shallows and their winters burrowed in the mud or swimming in the deeper waters of the bay.

Add to all these visible plant and animal organisms the unseen microscopic world of the salt marsh, which contains uncountable numbers of protozoa, fungi, and bacteria, and you can begin to appreciate the complexity of a salt marsh community.

Ecosystems such as the Chesapeake Bay salt marsh teem with life. Where do these organisms get the energy to live? And how do they harness this energy?

## THE ENERGY OF LIFE

**Energy** is the capacity or ability to do work. In living things, the biological work that requires energy includes processes such as growing, moving, reproducing, and repairing damaged tissues.

Energy exists in several forms: heat, radiant energy (electromagnetic radiation from the sun), chemical energy (stored in chemical bonds of mole-cules), mechanical energy, and electrical energy. Energy can exist as stored energy—called **potential energy**—or as **kinetic energy,** the energy of motion (Figure 3–5). You can think of potential energy as an arrow on a drawn bow. When the string is released and the arrow shoots through the air, the potential energy is converted to kinetic energy. Thus, energy can change from one form to another. The study of energy and its transformations is called **thermodynamics.** There are two laws about energy that apply to all things in the universe: the first and second laws of thermodynamics.

### The First Law of Thermodynamics

According to the **first law of thermodynamics,** energy cannot be created or destroyed, although it can be transformed from one form to another. As far as we know, the energy present in the universe

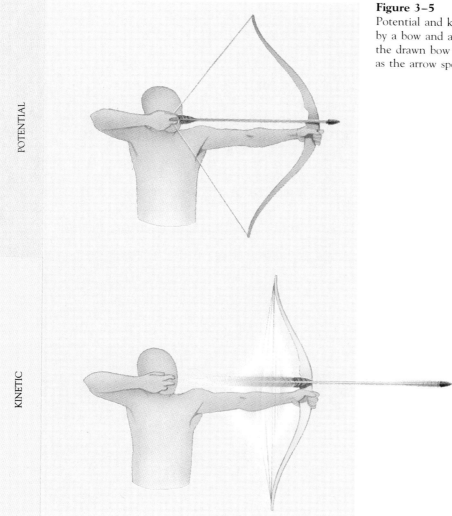

POTENTIAL

KINETIC

**Figure 3–5**
Potential and kinetic energy can be represented by a bow and arrow. Potential energy is stored in the drawn bow and is converted to kinetic energy as the arrow speeds toward its target.

at its formation, approximately 15 billion years ago, equals the amount of energy present in the universe today. This is all the energy that can ever be present in the universe. Similarly, the energy of any object and its surroundings is constant. An object may absorb energy from its surroundings, or it may give up some energy into its surroundings, but the total energy content of that object and its surroundings is always the same.

As stipulated by the first law of thermodynamics, then, living organisms cannot create the energy they require to live. Instead, they capture energy from the environment and use it to do biological work. This process involves transforming energy from one form to another. Through the process of photosynthesis, for example, plants absorb the radiant energy of the sun and convert it into the chemical energy contained in the bonds of food molecules. Similarly, the chemical energy of food can be transformed into the mechanical energy of walking, running, slithering, flying, or swimming.

## The Second Law of Thermodynamics

As each energy transformation occurs, some of the energy is changed to heat energy that is then given off into the surroundings. This energy can never again be used by the living organism for biological work, but because of the first law of thermodynamics, it is not "gone;" it still exists in the surroundings.

The **second law of thermodynamics** can be stated most simply as follows: when energy is converted from one form to another, some useful energy (that is, energy available to do work) is degraded into a lower-quality, less useful form—usually heat that disperses into the surroundings. As a result, the amount of useful energy available to do work in the universe decreases over time.

Low-quality energy is more dilute, or disorganized. **Entropy** is a measure of this disorder, or randomness; organized, useful energy has a low entropy, whereas disorganized, low-quality energy has a high entropy (Figure 3–6). Entropy is continuously increasing in the universe, and at some time billions of years from now, all energy will exist as low-quality heat that is uniformly distributed throughout the universe. When that happens, the universe will cease to operate because no work will be possible; everything will be at the same temperature, so there will be no way to convert the thermal energy of the universe into useful mechanical energy. Another way to explain the second law of thermodynamics, then, is that entropy, or disorder, in a system tends to increase over time.

**Figure 3–6**
Entropy can be represented by beakers filled with marbles of two different colors. The beaker on the left, in which all marbles of the same color are located together, represents a system with low entropy. The beaker on the right, in which the marbles are randomly arranged regardless of color, represents a system with greater entropy. (Dennis Drenner)

Living things have a high degree of organization and at first glance appear to refute the second law of thermodynamics; that is, as living things grow and develop, they maintain a high level of order and do not appear to become more disorganized. However, living things are able to maintain their degree of order over time only with the constant input of energy. That is why plants must photosynthesize and animals must eat.

## Photosynthesis and Cell Respiration

**Photosynthesis** is the biological process in which light energy from the sun is captured and transformed into the chemical energy of food. Photosynthetic pigments such as **chlorophyll** (which is green and gives plants their green color) absorb radiant energy. This energy is used to manufacture a sugar called glucose ($C_6H_{12}O_6$) from carbon dioxide ($CO_2$) and water ($H_2O$), with the liberation of oxygen ($O_2$) as a waste product:[2]

$$6CO_2 + 12H_2O + \text{radiant energy} \longrightarrow$$
$$C_6H_{12}O_6 + 6H_2O + 6O_2$$

[2] The chemical equation for photosynthesis is read as follows: 6 molecules of carbon dioxide plus 12 molecules of water plus light energy are used to produce 1 molecule of glucose plus 6 molecules of water plus 6 molecules of oxygen. See Appendix I for a review of basic chemistry.

**Figure 3-7**
The chemical energy in plant tissues is transferred to the bighorn ram as he eats. (Carolina Biological Supply Company)

Photosynthesis, which is essential for life on Earth, is performed by plants, algae,[3] and a few bacteria. Photosynthesis provides these organisms with a ready supply of energy (in glucose molecules) that they can use as the need arises. The energy can also be transferred from one organism to another—for instance, from plants to the organisms that eat plants (Figure 3-7). Photosynthesis also produces oxygen, which is required by living things when they break down food.

The chemical energy that plants store in food molecules is released within cells of plants, animals, or other living organisms through **cell respiration.** In this process, food molecules such as glucose are broken down in the presence of oxygen (and water) into carbon dioxide and water:

$$C_6H_{12}O_6 + 6O_2 + 6H_2O \longrightarrow$$
$$6CO_2 + 12H_2O + energy$$

Cell respiration makes the chemical energy stored in food molecules available to the cell for biological work. All living organisms respire to obtain energy.

[3] Algae, photosynthetic aquatic organisms that range from single cells to seaweeds well over 50 m in length, were originally classified as plants. Most biologists currently classify algae as protists (simple eukaryotic organisms) rather than as plants, because algae lack many of the structural features of plants.

# THE FLOW OF ENERGY THROUGH ECOSYSTEMS

The passage of energy in one direction through an ecosystem is known as **energy flow.** Energy enters an ecosystem in the form of the radiant energy of sunlight. Some of it is trapped by plants during the process of photosynthesis. Now in chemical form, it is stored in the bonds of organic (carbon-containing) molecules such as glucose. When these molecules are broken apart by cell respiration, the energy becomes available to do work such as repairing tissues, producing body heat, or reproducing. As the work is accomplished, the energy escapes the living organism and dissipates into the environment as low-quality heat. Ultimately, this heat energy radiates into space. Thus, once energy has been used by living things, it becomes unavailable for reuse.

## Producers, Consumers, and Decomposers

The organisms of a community can be divided into three categories based on how they get nourishment: producers, consumers, and decomposers (Figure 3-8). Most communities contain representatives of all three groups, which interact extensively with one another.

Sunlight is the source of energy that powers almost all life processes on the face of the Earth. **Producers,** also called **autotrophs** (Greek *auto,* "self," and *tropho,* "nourishment"), manufacture complex organic molecules from simple inorganic substances (carbon dioxide and water), usually using the energy of sunlight to do so. In other words, producers perform the process of photosynthesis. By incorporating the chemicals that they manufacture into their own bodies, producers make their bodies or body parts a potential food resource for other organisms. Whereas plants are the most significant producers on land, algae and certain types of bacteria are important producers in aquatic environments. In the salt marsh community, cordgrass, algae, and photosynthetic bacteria are all important producers. In abyssal hot spring communities deep in the ocean, nonphotosynthetic bacteria are the producers (see Focus On: Life Without the Sun).

Animals are **consumers;** that is, they use the bodies of other plant and animal organisms as sources of food energy and body-building materials. Consumers are also called **heterotrophs** (Greek *heter,* "different," and *tropho,* "nourishment"). Consumers that eat producers are called **primary**

(a)    (b)

**Figure 3–8**
Producers, consumers, and decomposers.
(a) *Heliconia* and other plants are producers.
This *Heliconia* is growing in a tropical rain
forest in Panama. (b) Macaws and other animals are consumers in the tropical rain forest
community. (c) Decomposers such as this cup
fungus (*Cookeina tricholoma*) in a Peruvian
rain forest reduce dead organisms to their
mineral constituents, carbon dioxide, and
water. Most of the body of the cup fungus,
not visible in this photograph, consists of
threadlike hyphae that penetrate widely
through the substrate on which it grows.
(a, David Cavagnaro; b, Richard H. Gross; c, James L.
Castner)

(c)

**consumers,** which usually means that they are exclusively **herbivores** (plant eaters). Cattle and deer
are examples of primary consumers, as is the marsh
periwinkle in the salt marsh community. **Secondary consumers,** which consume primary consumers, include **carnivores,** animals that consume
other animals exclusively. Lions and tigers are examples of carnivores, as are the northern diamondback terrapin and the northern water snake in the
salt marsh community. Other consumers, called
**omnivores,** eat a variety of plant and animal organisms. Bears, pigs, and humans are examples of omnivores; the meadow vole, which eats both insects
and cordgrass in the salt marsh community, is also
an omnivore. Many animals do not fit readily into
one of these three categories because they modify
their food preferences to some degree when the
need arises.

Some consumers, called **detritus feeders** or
**detritovores,** consume the organic matter originat-
ing in plant and animal remains. Detritivores are
especially abundant in aquatic habitats, where they

burrow in the bottom muck and consume the organic matter that collects there. For example,
marsh crabs are detritus feeders in the salt marsh
community. Earthworms are terrestrial (land-dwell-
ing) detritus feeders, as are termites and maggots
(the larvae of flies). Detritus feeders work together
with decomposers to destroy dead organisms and
waste products. An earthworm, for example, actually eats its way through the soil, digesting much of
the organic matter contained there. Earthworms
also aerate the soil and redistribute its minerals and
organic matter with their extensive tunneling.

**Decomposers** (also called **saprophytes**) are
heterotrophs that break down organic material and
use the decomposition products to supply themselves with energy. They typically release simple
inorganic molecules, such as carbon dioxide and
mineral salts, that can then be reused by producers.
Bacteria and fungi are important examples of decomposers. Dead wood, for example, is invaded first
by sugar-metabolizing fungi that consume the
wood's simple carbohydrates, such as glucose.

**Focus On**

## Life Without the Sun

In 1977 an oceanographic expedition aboard the submersible research craft *Alvin* studied the Galapagos Rift, a deep cleft in the ocean floor off the coast of Ecuador. The expedition revealed a series of hot springs at the bottom of the ocean where seawater had apparently penetrated the ocean floor and been heated by the hot rocks below. During its time within the Earth, the water had also been charged with mineral compounds, including hydrogen sulfide, $H_2S$.

At the tremendous depths (greater than 2,500 meters) of the Galapagos Rift, there is no light for photosynthesis. But the hot springs support a rich and bizarre web of life that contrasts with the surrounding lightless "desert" of the deep ocean floor. Many of the species found in these oases of life were new to science. For example, giant blood-red tube worms almost 3 meters (10 feet) in length cluster in great numbers around the vents. Other animals around the hot springs include clams, crabs, barnacles, and mussels.

The mystery is, what do these species live on? Most deep-sea communities depend on the little organic material that drifts down from surface waters; that is, they depend on energy derived from photosynthesis. But the Galapagos Rift community is too densely clustered and too productive to be dependent on chance encounters with organic material from surface waters.

The base of the food chain in such an aquatic oasis consists of certain bacteria which can survive and multiply in water so hot (exceeding 200°C) that it would not even remain in liquid form were it not under such extreme pressure. These bacteria function as producers, but they do not photosynthesize. Instead, they cause hydrogen sulfide to react with oxygen, producing water and sulfur or sulfate. The chemical reactions provide the energy required to support life in deep ocean hot springs: many of the Galapagos Rift animals consume the bacteria directly, whereas others, such as the giant tube worm, get their energy from bacteria that live inside their bodies.

Thus, it is accurate to say that *almost* all organisms on Earth depend on the sun for energy. The organisms in deep-sea hot springs are an interesting and unique exception.

The inhabitants of the Galapagos Rift. (a) A scanning electron micrograph (×5200) of bacteria that are the base of the food chain in hydrothermal vents. These bacteria obtain their energy from chemicals issuing from the vents rather than from sunlight, which is absent in their blackened realm. (b) Some of the bacteria live in the tissue of these tube worms. Tube worms, lacking digestive systems, depend on the organic compounds provided by the bacteria and on certain materials filtered from the surrounding water. Also visible in the photograph are some filter-feeding mollusks. (a, Courtesy of Carl O. Wirsen/Woods Hole Oceanographic Institution; b, J. Frederick Grassle/Woods Hole Oceanographic Institution)

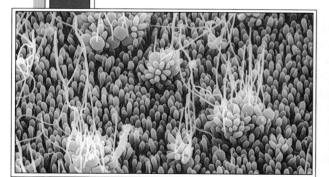

(a)

(b)

## Changes in Antarctic Food Webs

Although the icy waters around Antarctica may seem to be a very inhospitable environment, a rich variety of life is found there. The base of the food chain is microscopic algae, which are present in vast numbers. The algae are eaten by a huge population of tiny shrimplike animals called **krill,** which in turn support a variety of larger animals. Among the main consumers of krill are the baleen whales (such as blue whales, humpback whales, and right whales), which take in great mouthfuls of frigid water and then filter the water through special bristle-like "teeth" (the baleen), leaving the krill to be consumed. Krill are also consumed in great quantities by squid and fish, which in turn are eaten by other animals: toothed whales such as the sperm whale, elephant seals and leopard seals, king penguins and emperor penguins, and birds such as the albatross and petrel.

Humans have had an impact on the Antarctic food web, as they have on most other ecosystems. Before the advent of whaling, baleen whales consumed huge quantities of krill. For the past 150 years, however, whaling has steadily reduced the number of large baleen whales in Antarctic waters; many whale populations are so decimated that they are in danger of becoming extinct (see Chapter 16). As a result of fewer whales eating krill, more krill has been available for other krill-eating animals, whose populations have consequently increased: seals, penguins, and smaller baleen whales have replaced the large baleen whales as the main eaters of krill.

Now that commercial whaling is regulated (see Chapter 16), it is hoped that the number of large baleen whales will slowly increase. It is not known whether the whales will return to or be excluded from their former position of dominance (in terms of krill consumption) in the food web. Marine ecologists will monitor changes in the entire Antarctic food web as the whale populations recover.

More recently, a human-related development has taken place in the atmosphere over Antarctica that has the potential for much greater effects on the entire Antarctic food web: the depletion of the ozone layer in the stratospheric region of the atmosphere. This ozone "hole" allows more of the sun's ultraviolet radiation to penetrate to the Earth's surface. Ultraviolet radiation contains more energy than visible light; it is so energetic that it can break the chemical bonds of some biologically important molecules. It is not yet known how increased ultraviolet radiation will affect the organisms in the Antarctic food web, but environmental scientists are most concerned about its effects on the lowest levels of the food chain. If the algae and krill are harmed by the greater ultraviolet radiation, negative repercussions will extend throughout the entire Antarctic food web. The problem of stratospheric ozone depletion is discussed in detail in Chapter 20.

When these carbohydrates are exhausted, fungi, often aided by termites and bacteria, complete the digestion of the wood by breaking down cellulose, a complex carbohydrate that is the main constituent of wood.

Communities such as the Chesapeake Bay salt marsh contain balanced representations of all three ecological categories of organisms—producers, consumers, and decomposers. Producers and decomposers have indispensable roles in ecosystems. Producers provide both food and oxygen for all life. Decomposers are also necessary for the long-term survival of any community because without them, dead organisms and waste products would accumulate indefinitely. Without decomposers, important elements such as potassium, nitrogen, and phosphorus would permanently remain in dead organisms and therefore be unavailable for use by new generations of living things. Although consumers are an important part of most ecosystems, they are not essential to the long-term survival of producers or decomposers.

### The Path of Energy Flow: Who Eats Whom in Ecosystems

Energy flow in an ecosystem occurs in **food chains,** in which energy from food passes from one organism to the next in a sequence (Figure 3–9). Producers start the food chain by capturing the

**Figure 3–9**
A terrestrial and an aquatic food chain.

sun's energy through photosynthesis. Herbivores (and omnivores) eat the plants, obtaining both building materials—from which they construct their own tissues—and the chemical energy of the producers' molecules. Herbivores are in turn consumed by carnivores (and omnivores), who reap the energy stored in the herbivores' molecules. At the end of a food chain are decomposers, which

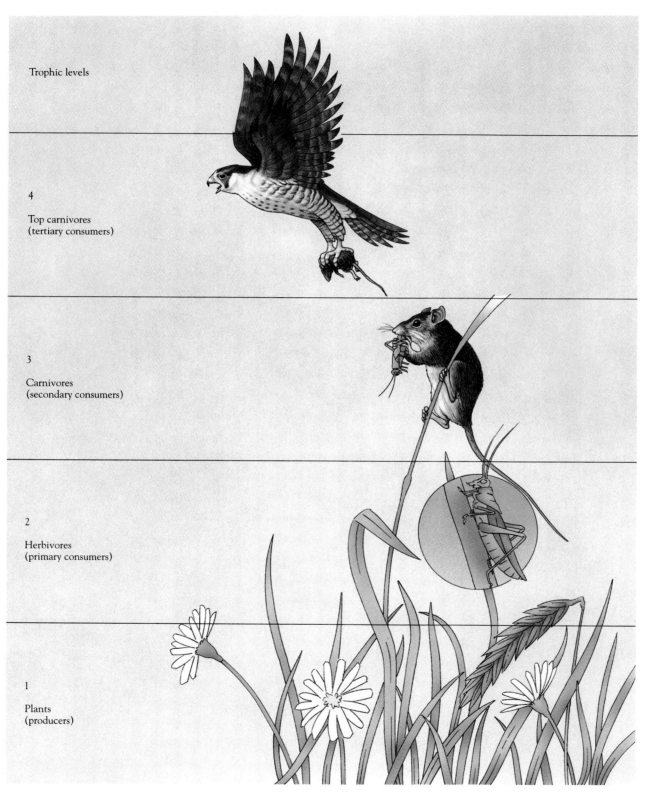

Trophic levels

4

Top carnivores
(tertiary consumers)

3

Carnivores
(secondary consumers)

2

Herbivores
(primary consumers)

1

Plants
(producers)

**Figure 3–10**
Trophic levels in an ecosystem.

respire organic molecules in the remains (the carcasses and body wastes) of all other members of the food chain.

Each level in a food chain is called a **trophic level** (Figure 3–10). The first trophic level is formed by producers (photosynthesizers), the second trophic level by primary consumers (herbivores), the third trophic level by secondary consumers (carnivores), and so on.

Simple food chains such as the one just described rarely occur in nature, because few organisms eat just one other kind of organism. More typically, the flow of energy and materials through an ecosystem takes place in accordance with a range of choices of food on the part of each organism involved. In an ecosystem of average complexity, numerous alternative pathways are possible. Thus, a **food web,** a complex of interconnected food chains in an ecosystem, is a more realistic model of the flow of energy and materials through ecosystems. (See Focus On: Changes in Antarctic Food Webs, page 50, for an examination of how humans have affected the complex food web in Antarctic waters.)

The most important thing to remember about energy flow in ecosystems is that it is *linear*, or one-way. That is, energy can move along a food chain or food web from one organism to the next as long as it is not used. When energy is used, it becomes unavailable for use by any other organism in the ecosystem.

## Ecological Pyramids

An important feature of energy flow is that most of the energy going from one trophic level to another in a food chain or food web dissipates into the environment. The relative energy values of trophic levels are often graphically represented by **ecological pyramids.** There are three main types of pyramids—a pyramid of numbers, a pyramid of biomass, and a pyramid of energy.

A **pyramid of numbers** shows the number of organisms at each trophic level in a given ecosystem, with greater numbers illustrated by wider sections of the pyramid (Figure 3–11). In most pyramids of numbers, each successive trophic level is occupied by fewer organisms. Thus, in a typical grassland the number of zebras and wildebeests (herbivores) is greater than the number of lions (carnivores). Reverse pyramids of numbers, in which higher trophic levels have *more* organisms than lower trophic levels, are often observed among decomposers, parasites, tree-dwelling herbivorous insects, and similar organisms. One tree

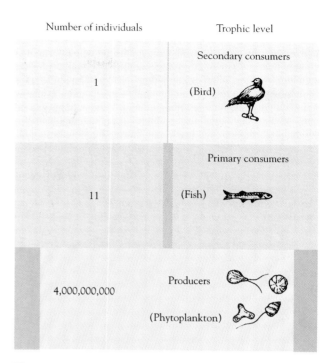

**Figure 3–11**
A pyramid of numbers is based on the number of organisms at each trophic level.

can provide food for hundreds of leaf-eating insects, for example.

A **pyramid of biomass** illustrates the total biomass at each successive trophic level. **Biomass** is a quantitative estimate of the total mass, or amount, of living material. Its units of measure vary: biomass may be represented as total volume, as dry weight, or as live weight. Typically, pyramids of biomass illustrate a progressive reduction of biomass in successive trophic levels (Figure 3–12). On the assumption that there is, on the average, about a 90 percent reduction of biomass for each trophic level,[4] 10,000 kg of grass should be able to support 1,000 kg of crickets, which in turn support 100 kg of frogs. By this logic, the biomass of frog eaters (such as herons) could be, at the most, only about 10 kg. From this brief exercise you can see that, although carnivores may eat no vegetation, a great deal of vegetation is still required to support them.

A **pyramid of energy** illustrates the energy relationships of an ecosystem by indicating the energy content (usually expressed in calories) of the biomass of each trophic level (Figure 3–13). On the whole, energy pyramids resemble biomass pyramids in shape, but they help to make another consequence of the nature of trophic levels clearer: *most food chains are short* because of the dramatic reduction in energy content that occurs at each trophic

[4]The 90 percent reduction in biomass is an approximation; actual biomass reduction in nature varies widely.

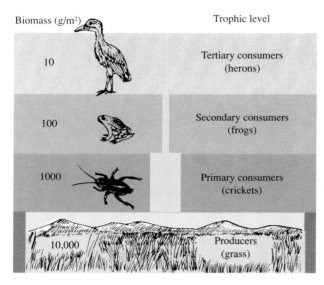

Figure 3–12
A pyramid of biomass for a hypothetical area of a temperate grassland. Pyramids of biomass are based on the biomass at each trophic level.

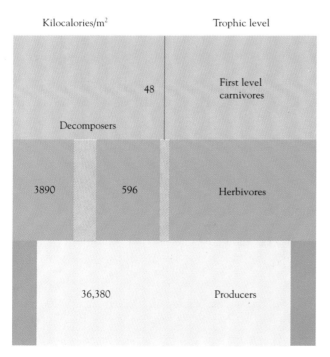

Figure 3–13
A pyramid of energy. Note the relatively large role played by decomposing bacteria. These decomposers actually operate on all trophic levels, not just the first (as depicted).

level. (In Chapter 18, the eating habits of humans as they relate to food chains and trophic levels are discussed; see You Can Make a Difference: Vegetarian Diets.)

**A Chain of Poison**

Toxic substances that resist biodegradation can accumulate in the tissues of living organisms. This is called biological magnification. The poison becomes more concentrated as it moves up the food chain, harming organisms and causing other unforeseen and often disastrous effects. American biologists Paul and Anne Ehrlich observed just such a catastrophe in Borneo: It began when the World Health Organization sprayed large quantities of the insecticide DDT to control mosquitoes carrying deadly malaria. The mosquitoes died, but so did a predatory wasp that naturally controlled caterpillars. Without the wasp around, the caterpillars bred rapidly and proceeded to devour the thatched roofs of local homes, causing many to collapse. More spraying was done indoors to kill house flies, and the DDT entered the food chain. Gekko lizards, which had controlled the flies, now ate their bodies, ingesting high concentrations of the poison. The dying lizards were caught and eaten by house cats, who received massive doses of DDT and also died. The lack of cats led to an abundance of rats, which ate the people's food and threatened to cause the worst plague of all—bubonic plague. Eventually, the government of Borneo grew so anxious it dropped healthy cats into villages by parachute to hold back the growing rat population.

A secondary consumer requires an enormous home range—the area needed to obtain enough food—to encompass all the necessary producers, especially if it is a large animal. Thus, a large solitary predator such as a tiger may have a home range exceeding 250 square kilometers (96.5 square miles), whereas a cottontail rabbit, which occupies a lower trophic level, can live confortably on 6 hectares (14.8 acres). These area estimates reflect not only diet, but also other factors such as size of the organism, so the best comparison of the home ranges of primary consumers and secondary consumers is between animals similar in every way ex-

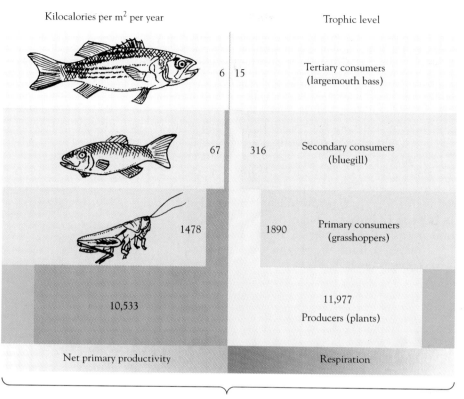

Kilocalories per m² per year                    Trophic level

| | | |
|---|---|---|
| 6 | 15 | Tertiary consumers (largemouth bass) |
| 67 | 316 | Secondary consumers (bluegill) |
| 1478 | 1890 | Primary consumers (grasshoppers) |
| 10,533 | 11,977 | Producers (plants) |

Net primary productivity                    Respiration

Gross primary productivity

**Figure 3–14**
A pyramid of energy for a river ecosystem, illustrating gross primary productivity and net primary productivity. Measurements are in kilocalories per square meter per year.

cept eating habits. Such a comparison is possible, for example, between the home ranges of two species of mice. Living in the same community, the seed-eating white-footed mouse requires about 1.5 hectares (3.7 acres), but the carnivorous grasshopper mouse requires 5 hectares (12.4 acres).

## Variation in Productivity of Ecosystems

The **gross primary productivity** of an ecosystem is the *rate* at which energy accumulates (as biomass) during photosynthesis—that is, the total amount of photosynthesis in a given period of time. Of course, plants must respire to provide energy for their life processes, and cell respiration acts as a drain on photosynthesis. Energy that remains (as biomass) after cell respiration has occurred is called **net primary productivity.** In other words, net primary productivity is the amount of biomass found in excess of that broken down by a plant's cell respiration. Net primary productivity represents the *rate* at

which organic matter is actually incorporated into plant bodies so as to produce growth (Figure 3–14).

$$\underbrace{net\ primary\ productivity}_{plant\ growth} =$$

$$\underbrace{gross\ primary\ productivity}_{total\ photosynthesis} - plant\ respiration$$

Only the energy represented by net primary productivity is available for the nutrition of consumers, and of this energy only a portion is actually utilized by them. Both gross primary productivity and net primary productivity can be expressed in terms of kilocalories (of energy fixed by photosynthesis) per square meter per year, or in terms of dry weight (grams of carbon incorporated into tissue) per square meter per year.

What determines productivity? A number of factors may interact. Some plants are more efficient photosynthesizers than others. Environmental factors are also important. The influx of solar energy, availability of mineral nutrients, availability of water, and other climatic factors are important, as are the degree of maturity of the community, the severity of human modification, and other factors that are difficult to assess. For example, the high productivity of intertidal communities along an ocean shoreline is largely due to wave action. Many intertidal organisms are sedentary detritus filter feeders (such as mussels) that have their food carried to them by wave action and so need to expend less of their own energy to obtain food.

Ecosystems differ strikingly in their productivity. Terrestrial communities are generally more productive than aquatic ones, partly because of the greater availability of light for photosynthesis[5] and partly because of higher concentrations of available mineral nutrients. However, adverse temperatures and lack of water limit the productivity of certain terrestrial ecosystems. Aquatic ecosystems have an abundance of water, which moderates tempera-

tures, but are usually limited by the scarcity of mineral nutrients (especially in the open sea) and the low light intensity.

The net primary productivity of an ecosystem tells us little about how much biomass is present at any given time. Despite plant growth, under natural conditions there is about as much grass in a section of prairie this year as there was last year. The reason is that the **turnover** of plant biomass from its consumption by animals and decomposers is usually about the same as the net primary productivity of the ecosystem. It is this balance that determines the current plant biomass, or **standing crop.**

[5] Aquatic environments receive less light than terrestrial environments. When sunlight hits the surface of water, much of it is scattered or reflected. Also, as light penetrates water, some of it is absorbed by water molecules, reducing the amount of light at greater depths.

# SUMMARY

1. The study of the relationships between organisms and their physical environment is called ecology. Ecologists study populations, communities/ecosystems, and the biosphere/ecosphere. A population is all the members of the same species that live together. A community is all the populations of different species living in the same area; an ecosystem is a community and its environment. The biosphere is all the communities on Earth—in other words, all of the Earth's living organisms. The ecosphere, the largest ecosystem on Earth, comprises the interactions among the biosphere, atmosphere, lithosphere, and hydrosphere.

2. The first law of thermodynamics states that energy can be converted from one form to another, but it can be neither created nor destroyed. The second law of thermodynamics states that entropy continually increases in the universe as useful energy is converted to a lower-quality, less useful form—usually heat.

3. Living organisms, like everything else in the universe, obey the two laws of thermodynamics. Because organisms cannot create energy (as stipulated by the first law), producers must photosynthesize, and consumers and decomposers must eat, in order to obtain the energy that is required for biological work.

4. Almost all ecosystems obtain their energy from the sun. Photosynthesis, performed by producers, converts radiant energy from the sun into the chemical energy stored in the bonds that hold food molecules together. The chemical energy of food is released and made available for biological work by the process of cell respiration. All living organisms respire.

5. Energy flows through an ecosystem linearly, from the sun to producer to consumer to decomposer. Much of this energy is converted to less useful heat as the energy moves from one organism to another, as stipulated in the second law of thermodynamics.

6. Living organisms in ecosystems assume the roles of producer, consumer, and decomposer. Producers, or autotrophs, are the photosynthetic organisms that are at the base of almost all food chains; they include plants, algae, and some bacteria. Consumers, which feed upon other organisms, are almost exclusively animals. Decomposers feed on the components of dead organisms and organic wastes, degrading them into simple inorganic materials that can then be used by producers to manufacture more organic material. Both consumers and decomposers are heterotrophs.

7. Trophic relationships may be expressed as food chains or, more realistically, food webs, which show the multitude of alternative pathways that energy can take among the producers, consumers, and decomposers of an ecosystem. Ecological pyramids express the progressive reduction in numbers of organisms, biomass, and energy found in successively higher trophic levels.

8. Gross primary productivity of an ecosystem is the rate at which organic matter is produced by photosynthesis. Net primary productivity expresses the rate at which some of this matter is incorporated into plant bodies; net primary productivity is less than gross primary productivity because of the losses resulting from plant respiration.

# DISCUSSION QUESTIONS

**1.** What is the simplest stable ecosystem that you can imagine?

**2.** Could a balanced ecosystem be constructed that contained only producers and consumers? Only consumers and decomposers? Only producers and decomposers? Explain the reason for your answer in each case.

**3.** How are the following forms of energy significant to living things in ecosystems? (a) radiant energy (b) mechanical energy (c) chemical energy (d) heat

**4.** Why is the concept of a food web generally preferred over that of a food chain?

**5.** Draw a food web containing organisms found in a Chesapeake Bay salt marsh.

**6.** Suggest a food chain that might have an inverted pyramid of numbers (that is, greater numbers of living organisms at higher trophic levels than at lower trophic levels).

**7.** Is it possible to have an inverted pyramid of energy? Why or why not?

**8.** Since photosynthesis is the conversion of $CO_2$ and $H_2O$ to sugar and oxygen, and cell respiration is the conversion of sugar and oxygen to $CO_2$ and $H_2O$, why doesn't each process cancel out the effect of the other in a plant that is both photosynthesizing and respiring?

# SUGGESTED READINGS

Brewer, R. *The Science of Ecology,* 2d ed. Saunders College Publishing, Philadelphia, 1993. A good general textbook on the principles of ecology, including ecosystem ecology.

Cousins, S. Ecologists build pyramids again. *New Scientist* 106, 4 July 1985. An analysis of the value of ecological pyramids in understanding energy flow through ecosystems.

Pimm, S. L., J. H. Lawton, and J. E. Cohen. Food web patterns and their consequences. *Nature* 350, 25 April 1991. A well-written review article on current ecological knowledge of food webs.

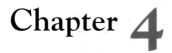

# Chapter 4

*Dall sheep are adapted to survive the cold of Alaska. (M. Kazmers/Shark Song/ Dembinsky Photo Associates)*

# Ecosystems and Living Things

**E**ach species confronts the challenge of survival in its own unique fashion. The way in which an organism interacts with other living things and with its physical environment defines that organism's niche, or role in an ecosystem. Different species in an ecosystem often interact with one another in such a way that over a very long period of time they develop intimate associations and influence each other's evolution. The species composition of an ecosystem in a given area undergoes orderly changes over time.

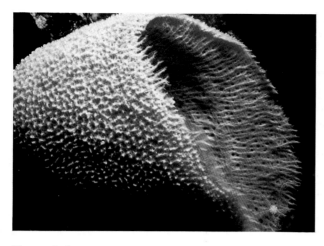

**Figure 4–1**
Sponges such as this pink vase sponge, photographed off Little Cayman Island in the Caribbean, are part of a larger seabed community. Each sponge is also a community in itself, often harboring a variety of organisms such as little crabs. (M. Kazmers/Shark Song/Dembinsky Photo Associates)

## LIVING ORGANISMS INTERACT

As you may recall from Chapter 3, the term "community" has a far broader sense in ecology than in everyday speech. For the biologist, a **community** is an association of organisms of different species living and interacting together (see Focus On: The Five Kingdoms of Life for an overview of the categories of organisms found in communities). Thus, you, your dog, and the fleas on your dog are all members of the same community! You could also add to the list cockroaches, silverfish, dandelions, grasses, maple trees, and much more.

Communities vary greatly in size, lack precise boundaries, and are rarely completely isolated. They interact with and influence one another in countless ways that are not always apparent. Furthermore, communities are nested within one another like Chinese boxes; that is, there are communities within communities (Figure 4–1). A forest is a community, but so is a rotting log in that forest. The log contains bacteria, fungi, slime molds, worms, insects, and perhaps even mice. The microorganisms living within the gut of a termite in the rotting log also form a community. On the other end of the scale, the entire living world can be considered a community.

Living organisms exist in a nonliving environment that is as essential to their lives as are their interactions with one another. Minerals, air, water, and sunlight are just as much a part of a honeybee's environment, for example, as the flowers that it pollinates and from which it takes nectar. Together the nonliving environment and the living communities it contains make up an **ecosystem.** Although this chapter emphasizes the living community, communities and their physical environments are inseparably linked (see Focus On: Microcosms).

## THE NICHE

Every organism has its own role within the structure and functions of an ecosystem; we call this role its ecological **niche** (Figure 4–2). An organism's

(a) Cape May warbler

(b) Bay-breasted warbler

(c) Blackburnian warbler

(d) Black-throated green warbler

(e) Yellow-rumped warbler

**Figure 4–2**
Some songbird ecological niches. Each of these species of *Dendroica,* common warblers, spends most of its feeding time in a distinct portion of the trees it frequents and also consumes somewhat different insect food from the others. The colored regions indicate where each species spends at least half of its foraging time. (After MacArthur)

## The Five Kingdoms of Life

For hundreds of years, biologists regarded living things as falling into two broad categories—plants and animals. With the development of microscopes, it became increasingly obvious that many organisms did not fit very well into either the plant or animal kingdom. For example, bacteria have a prokaryotic cell structure: they lack nuclear envelopes and other internal cell membranes. This feature, which separates bacteria from all other organisms, is far more fundamental than the differences between plants and animals, which have similar cell structures. Hence, it became difficult to regard bacteria as plants. Furthermore, certain microorganisms such as *Euglena*, which is both motile and photosynthetic, seem to possess characteristics of both plants and animals.

These and other considerations have led to the system of classification used by most biologists today, which has five kingdoms: Prokaryotae, Protista, Fungi, Plantae, and Animalia. Bacteria, with a prokaryotic cell structure, have their own kingdom—Prokaryotae. The remaining four kingdoms are composed of organisms with a eukaryotic cell structure. Eukaryotic cells have a high degree of internal organization, containing such organelles as nuclei, chloroplasts (in photosynthetic cells), and mitochondria. The kingdoms Fungi, Plantae, and Animalia differ from one another in—among other features—their types of nutrition: fungi secrete

digestive enzymes into their food and then absorb the predigested nutrients; plants use radiant energy to manufacture food molecules by photosynthesis; and animals ingest their food, then digest it inside their bodies. That leaves us with the kingdom Protista, which is composed of eukaryotic organisms that are either single-celled or relatively simple multicellular organisms: algae, protozoa, slime molds, and water molds.

Although the five-kingdom system is a definite improvement over the two-kingdom system, it is not perfect. Most of its problems concern the kingdom Protista, which includes some organisms that may be more closely related to members of other kingdoms than to certain other protists. For example, green algae are protists that are clearly similar to plants but do not appear to be closely related to other protists such as slime molds and red algae.

In addition, all members of a kingdom should ideally have a common ancestor. There seems to be no common ancestor for the kingdom Protista, because it appears that eukaryotic cells evolved several separate times. Some biologists want to resolve this problem by dividing the kingdom Protista into several kingdoms along more natural groupings. However, most biologists would rather deal with the limitations of the five-kingdom system than with additional kingdoms.

ecological niche takes into account all aspects of the organism's existence—all the physical, chemical, and biological factors that the organism needs to survive, to remain healthy, and to reproduce. Among other things, the niche includes the physical surroundings in which an organism lives (its **habitat**) and how it interacts with and is influenced by the nonliving components of its environment (for example, light, temperature, and moisture). An organism's niche also encompasses the organisms it eats, the organisms that eat it, and the living organisms with which it competes. The niche, then, represents the totality of an organism's adaptations, its use of resources, and the life style to which it is

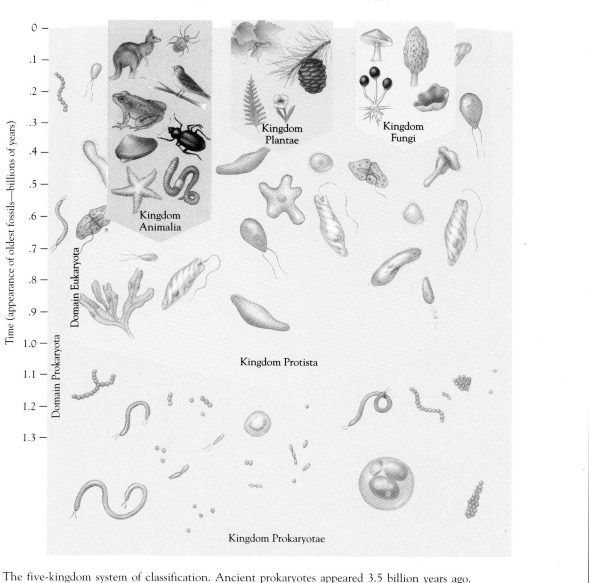

The five-kingdom system of classification. Ancient prokaryotes appeared 3.5 billion years ago. Protists, which are simple eukaryotic organisms, evolved from ancient prokaryotes (kingdom Prokaryotae); the oldest protist fossils are 1.2 billion years old. Animals (which appeared 700 million years ago), plants (which appeared 420 million years ago), and fungi (which appeared 390 million years ago) arose from different ancestral protists.

fitted. Obviously, a complete description of an organism's ecological niche has numerous dimensions.

There are two aspects to an organism's ecological niche: the role the organism *could* play in the community and the role it actually *fulfills*. The niche may be far broader potentially than it is in actuality. As an analogy, a person might be capable of becoming a doctor *and* a lawyer, but few people manage to be both. A person's actual life style, including his or her career, is chosen from among many possibilities. Put differently, an organism is usually capable of utilizing much more of its environment's resources or of living in a wider assort-

### Microcosms

A balanced aquarium—that is, an aquarium containing fish and plants along with decomposer bacteria—has always been popular as an illustration of ecosystems. Theoretically, if the aquarium is properly set up, it should be possible for its inhabitants to survive indefinitely even if it is totally sealed off from the outside world. Unfortunately, when this experiment has actually been tried, the organisms inside the aquarium have usually died in a very short period of time.

This outcome is of more than academic interest because of the development of space flight. Spacecraft sent on journeys lasting years cannot be expected to carry all the food and oxygen they will need. It is obvious that they must be balanced ecosystems, growing their own food, recycling their own wastes, and producing their own oxygen by photosynthesis. In the Soviet Bios experiments, in which humans were sealed inside completely closed systems (on Earth) to simulate spacecraft, the walls became covered with green slime and the humans contracted severe diarrhea. Obviously, this situation could not be tolerated in a ship on a long-term mission to Mars.

In 1977, Joe Hanson of NASA developed the first stable, sealed ecosystems; they contained shrimp and algae. Engineering Research Associates of Tucson, Arizona, produced versions of Hanson's systems that are now available commercially. By extensive experimentation, the researchers developed a controlled mixture of as many as 100 species of organisms (mostly microorganisms) that worked well together. For example, it was necessary to include bacteria in addition to shrimp and algae, to convert the toxic ammonia excreted by the shrimp into nitrite. Nitrite, also toxic, is in turn converted to nitrate by a different species of bacteria. The nitrate is then utilized as a nitrogen source by the algae. The entire system, called a **microcosm,** is set up inside a sealed glass sphere resembling a paperweight. These little ecosystems may provide us with clues about the management of spacecraft, including our own spaceship, Earth.

An aquarium-like microcosm. (Courtesy of Ecosphere Associates, Ltd.)

(a)

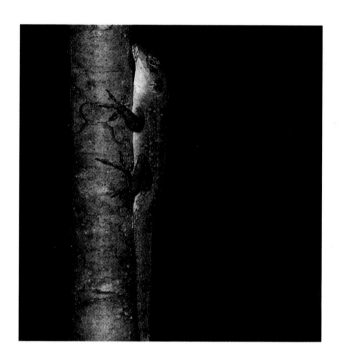

(b)

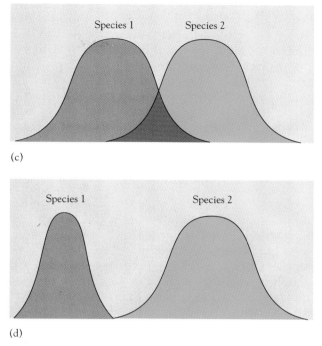

(c)

(d)

**Figure 4–3**
Competition can restrict an organism's realized niche.
(a) The green Carolina anole is native to Florida. (b) The brown Cuban anole was introduced in Florida. (c) The fundamental niches of the two lizards overlap. Species 1 is the Carolina anole, and species 2 is the Cuban anole. (d) The Cuban anole was able to outcompete the Florida anole, restricting its niche. (a, Runk/Schoenberger, from Grant Heilman; b, Connie Toops)

ment of habitats than it actually does. The potential ecological niche of an organism is its **fundamental niche,** but various factors such as competition with other species may exclude it from part of its fundamental niche. Thus, the life style that an organism actually pursues and the resources that it actually utilizes comprise its **realized niche.**

An example may help clarify this distinction. The little Carolina anole (Figure 4–3a), native to Florida, perches on tree trunks or bushes during the day and waits for insect prey. In past years these lizards were widespread in Florida. Several years ago, however, a related species, the Cuban anole, was introduced in Florida and quickly became com-

mon, especially in urban areas (Figure 4–3b).[1] Suddenly the Carolina anole became rare—apparently driven out of its habitat by competition from the larger Cuban lizard. Careful investigation disclosed, however, that Carolina anoles were still around but were now confined largely to the foliated crowns of trees, where they were less easily seen.

The habitat portion of the Carolina anole's fundamental niche includes the trunks and crowns of trees, exterior walls of houses, and many other locations. The Cuban anoles were able to drive Carolina anoles out from all but the tree crowns, and the latter's realized niche became much smaller as a result of this environmental competition (Figure 4–3c, d). Because all natural communities consist of numerous species, many of which compete to some extent, the complex interactions among them produce the realized niche of each.

[1] See Chapter 16 for a discussion of the introduction of nonnative species into a new habitat.

## Competitive Exclusion

When two species are very similar, as are the Carolina and Cuban anoles, their fundamental niches may overlap. However, no two species can occupy the same niche in the same community indefinitely, because **competitive exclusion** eventually occurs. In this process, one species is excluded from a niche by another as a result of competition between species (interspecific competition). Although it is possible for two different species to compete for a single resource without being total competitors, two species with absolutely identical ecological niches cannot coexist. Coexistence *can* occur, however, if the overlap in the two species' niches is reduced. In the lizard example, direct competition between the two species was reduced as the Cuban anole competitively excluded the Carolina anole from most of its former physical habitat until the only place that remained open to it was the tree.

Competition between different species, then, determines a species' realized niche. The initial evidence for this came from a series of experiments

**Figure 4–4**

Competition among paramecia. The top graph shows how a population of each species of *Paramecium* grows in separate cultures (i.e., each in a single-species environment). The bottom graph shows how they grow together in a single culture (in competition with each other). *P. aurelia* outcompetes *P. caudatum* and drives it to extinction.

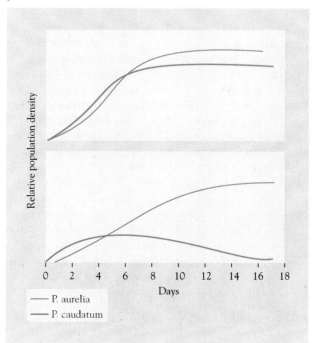

conducted by the Russian biologist A. F. Gause in 1934. In one study Gause grew two species of *Paramecium* (a type of protozoa), *P. aurelia* and the larger *P. caudatum* (Figure 4–4). When the two were grown in separate test tubes, each species quickly increased its population to a high level, which it maintained for some time thereafter. When the two were grown together, however, only *P. aurelia* thrived; *P. caudatum* dwindled and eventually died out. Under different sets of culture conditions, *P. caudatum* prevailed over *P. aurelia*. Gause interpreted this to mean that one set of conditions favored one species, and a different set favored the other. Because the two species were similar, given time one or the other would eventually triumph at the other's complete cost.

Competitive exclusion of a wild mouse (*Mus musculus*) population by voles (small rodents with short tails) apparently occurred in California during the 1960s. The aggressive voles ate much of the mice's food supply, and the voles' continual proximity may have also disturbed the mouse population in other ways. The wild mice exhibited less vigor and a lower reproductive rate that eventually resulted in their local extinction.

Apparent contradictions to the competitive exclusion principle sometimes occur. In Florida, for instance, native fish and introduced (non-native) cichlid fish seem to coexist in identical niches. Similarly, in the same area botanists have observed closely competitive plant species. Although such situations seem to contradict the concept of competitive exclusion, the realized niches of these organisms may differ significantly in some way that scientists do not yet understand.

## Limiting Factors

The factors that actually determine an organism's realized niche can be extremely difficult to identify. For this reason the concept of the ecological niche is largely abstract, although some of its dimensions can be experimentally determined. Whatever environmental variable tends to restrict the realized niche of an organism is called a **limiting factor.**

What factors actually determine the realized niche of a creature? An organism's niche is basically determined by the sum of its structural, physiological, and behavioral adaptations. Such adaptations determine, for example, the tolerance an organism has for environmental extremes. If any feature of its environment lies outside the bounds of its tolerance, then the organism cannot live there. That is why you would not find a cactus living in a pond or water lilies growing in the desert.

## The Zebra Mussel: How a Niche Became a Nightmare

As killer bees, gypsy moths, and the noxious weed kudzu have demonstrated, accidental introduction of non-native species can create havoc in an ecosystem. One of North America's latest threats is the zebra mussel, a native of the Caspian Sea that was probably introduced through ballast water flushed into the Great Lakes by a foreign ship in 1986. Since securing a niche, the tiny freshwater mussel, which clusters in extraordinary densities, has massed on hulls of boats, piers, buoys, and, most damaging of all, on water intake systems. One power utility has estimated it will cost $50 million to $100 million a year just to keep Great Lakes intake pipes open. The zebra mussel's strong appetite for algae, phytoplankton, and zooplankton is also cutting into the food supply of native mussels and clams, threatening their survival. By 1991, the zebra mussel had advanced from the Great Lakes into the Mississippi, Ohio, and Susquehanna rivers. Some scientists think that by the year 2000 it will have adapted to the warmer waters of Florida and Texas, and the cost of controlling it will run into the billions.

Most of the limiting factors that it has been possible to investigate are simple variables such as the mineral content of soil, temperature extremes, amount of precipitation, and the like. Such investigations have disclosed that any factor that exceeds an organism's tolerance for it or is present in quantities smaller than the minimum required by the organism limits the occurrence of that organism in a community. By their interaction, such factors help to define an organism's realized niche.

The concept of limiting factors was originated in the 19th century by the agricultural chemist J. von Liebig, who propounded what is now called the **law of the minimum.** As amended in 1913 by V. E. Shelford, the law of the minimum holds that the growth of each organism is limited by whatever essential factor is in shortest supply or is present in harmful excess.

This consideration applies throughout the life cycle of an organism. For instance, although adult blue crabs can live in almost fresh water, they cannot become permanently established there because their larvae cannot tolerate fresh water. Similarly, the ring-necked pheasant, a popular game bird, has been introduced widely in North America but does not survive in the southern United States. The adult birds do well, but the eggs cannot develop properly in the high southern temperatures.

As a result of more recent studies of limiting factors, ecologists now understand that von Liebig viewed limiting factors much too narrowly. He understood, rightly, that an excess of one limiting factor cannot make up for the deficiency of another. But what von Liebig didn't realize is that when an organism is near the limits of its tolerance for *several* factors, their interactions collectively restrict its realized niche more severely than would be expected from simple addition of the effects of the individual limiting factors.

We have seen that an organism's ecological niche takes into account all aspects of that organism's existence. Now we examine coevolution and symbiosis, two biological factors that strongly influence an organism's niche.

## COEVOLUTION

Sometimes two different species develop an intimate association so that, over time, the course of each species' evolution is affected. **Coevolution** is the interdependent evolution of two or more species that occurs as a result of their interactions. Flowering plants and their animal pollinators provide an excellent example of coevolution. Because plants are rooted in the ground, they lack the mobility that animals have when mating. Many flowering plants rely on animals to help them mate. Bees, beetles, hummingbirds, bats, and other animals transport the male reproductive structures, called pollen, from one plant to another, in effect giving plants mobility (Figure 4–5). How has this come about?

During the millions of years over which these associations developed, flowering plants evolved a number of ways to attract animal pollinators. One of the rewards for the pollinator is food—nectar (a sugary solution) and pollen. Plants often produce food that is precisely correct for one type of pollinator. The nectar of flowers that are pollinated by bees, for example, usually contains between 30 percent and 35 percent sugar, the concentration that

(a)

(b)

(c)

**Figure 4–5**
The coevolution of flowering plants and their animal pollinators. (a) A bumblebee worker feeds on nectar. Bumblebees and other insects are valuable pollinators of flowering plants. (b) A ruby-throated hummingbird obtains nectar from and pollinates a trumpet creeper flower. (c) A lesser long-nosed bat visits flowers to obtain nectar. The pollen grains on the bat's fur will be carried to the next flower.
(a, Dwight R. Kuhn; b, Dan Dempster/Dembinsky Photo Associates; c, Merlin D. Tuttle/Bat Conservation International)

bees need in order to make honey. Bees will not visit flowers with lower sugar concentrations in their nectar. Pollen also attracts many pollinators. Bees, for example, use pollen to make bee bread, a nutritious mixture of nectar and pollen that is eaten by their larvae.

Plants have also evolved a variety of ways to get the pollinator's attention, most of which involve colors and scents. Showy petals visually attract the pollinator as a neon sign or golden arches attract a hungry person to a restaurant. Different animal pollinators perceive colors differently. Insects, for example, see the blue and yellow range of the visible spectrum but don't perceive red as a distinct color. Thus, plants that are pollinated by insects often have blue or yellow petals. Insects can

also see the ultraviolet range of the electromagnetic spectrum, which is invisible to the human eye; insects see ultraviolet as a color called "bee's purple." Many insect-pollinated flowers have parts that reflect ultraviolet (making them appear purple to insects) and parts that absorb ultraviolet (making them appear as other colors, such as yellow). This creates patterns on the flower, which direct the insect to the center of the flower where the pollen and nectar are located (Figure 4–6).

Scents are also an effective way to attract pollinators. Insects have a well-developed sense of smell, and many insect-pollinated flowers have a strong scent, which may be pleasant but is not always. The carrion plant, for example, is pollinated by carrion flies and smells like rotting flesh. Its petals are dappled with a reddish-brown color that

(a)

(b)

**Figure 4–7**
The carrion plant, *Stapelia variegata*. This East African desert plant gets its name because its coloration and bad smell are reminiscent of decaying flesh. It is pollinated by carrion flies, several of which have alighted on the flower.
(Gary J. James, Biological Photo Service)

**Figure 4–6**
Many insect-pollinated flowers have ultraviolet markings that are invisible to humans but very conspicuous to insects. (a) A flower as seen by the human eye is solid yellow. (b) The same flower viewed under ultraviolet radiation provides clues about how the insect eye perceives it. The outer, light-appearing portions of the petals reflect both yellow and ultraviolet, which makes them appear purple to the bee's eyes. The inner parts of the flower absorb ultraviolet and reflect yellow, which makes them appear yellow to the bee. These differences in coloration draw attention to the center of the flower, where the pollen and nectar are located. (Thomas Eisner)

looks like dried blood (Figure 4–7). Flies move from one flower to another, looking for a place to deposit their eggs, and in the process pollen is transferred from one flower to another.

During the time plants were evolving specialized features to attract pollinators, the animal pollinators coevolved specialized body parts and behaviors that enabled them to both aid pollination and obtain nectar and pollen as a reward. Many insects have mouthparts that fit into certain flowers much as a lock fits into a key. Thus, even though other insect species may be attracted to the flowers, they cannot obtain the rewards because they lack these specialized mouthparts. The behavior of animals has also coevolved. For example, male wasps, which mature before female wasps, attempt to copulate with certain orchid flowers that resemble female wasps in coloring and shape. As a result, pollination is achieved as the males move from flower to flower. When female wasps emerge at a later time, the males finally get to mate for real.

## SYMBIOSIS

**Symbiosis** is any intimate relationship or association between members of two or more different species. The partners of a symbiotic relationship, called **symbionts,** may benefit from, be unaffected by, or be harmed by the relationship. The thousands, or even millions, of symbiotic associations in nature are all products of coevolution, and they fall into several categories.

## Mutualism: Sharing Benefits

**Mutualism** is a symbiotic relationship in which both partners benefit. One example of mutualism is the association between reef-building coral animals and microscopic algae. These symbiotic algae, which are called zooxanthellae, live inside cells of the coral, where they photosynthesize and provide the animal with carbon and nitrogen compounds as well as oxygen. Zooxanthellae have a stimulatory effect on the growth of corals, which deposit calcium carbonate skeletons around their bodies much faster when the algae are present. The coral, in turn, supplies its zooxanthellae with waste products such as ammonia, which the algae use to make nitrogen compounds for both partners.

Mycorrhizae are mutualistic associations that take place between fungi and the roots of almost all plants (see Chapter 14). The fungus absorbs essential minerals from the soil and provides them to the plant, and the plant provides the fungus with food produced by photosynthesis. Plants grow more vigorously in the presence of mycorrhizae, and they are better able to tolerate environmental stresses such as drought and high soil temperatures.

Frequently, mutualistic partners are completely dependent on one another. For example, an obligatory relationship exists between the yucca, a plant with stiff leaves found in the southwestern United States, and the yucca moth. The moth transfers pollen between plants, and the plants provide food and a safe habitat for the moth larvae, which hatch from eggs laid inside the flower. Neither species could exist without the other. Without the yucca moth, pollination—and therefore successful reproduction—would not occur in the yucca, and it would die out. Likewise, without the yucca, the yucca moth would be unable to reproduce successfully because it lays its eggs only inside yucca flowers.

## Commensalism: Taking Without Harming

**Commensalism** is a type of symbiosis in which one organism benefits and the other one is neither harmed nor helped. One example of commensalism is the relationship between two kinds of insects, silverfish and army ants. Certain kinds of silverfish live with army ants and share the food caught by

### Life on a Sloth

The concept of symbiosis has evolved to extremes in some species. In his 1986 book, *Life Above the Jungle Floor,* Donald Perry noted that the fur of a three-toed sloth is often occupied by green algae, as seen here, as well as pyralid moths, house mites, a number of beetle species, and several kinds of arthropods. In fact, a single sloth can be home to over 900 beetles.

A three-toed sloth. (Norbert Wu)

the ants. The army ants derive no apparent benefit (or harm) from the silverfish. Another example of commensalism is the relationship between a tropical tree and its **epiphytes,** smaller plants that live attached to the bark of the tree's branches. The epiphyte anchors itself to the tree but does not obtain nutrients or water directly from the tree. Its position on the tree enables it to obtain adequate light, water (rainfall dripping down the branches), and minerals (washed out of the tree's leaves by rainfall). Thus, the epiphyte benefits from the association, whereas the tree remains largely unaffected.

## Parasitism: Taking at Another's Expense

**Parasitism** is a symbiotic relationship in which one member, the **parasite,** benefits and the other, the **host,** is adversely affected. The parasite obtains nourishment from its host, and although it may weaken the host, it rarely kills it. (A parasite would have a rough life if it kept killing off its hosts!) Some parasites, such as ticks, live outside the host's body; others, such as tapeworms, live within the host.

When a parasite causes disease and sometimes the death of a host, it is called a **pathogen.** For example, humans sometimes get histoplasmosis, a serious and often fatal disease caused by a fungus. Humans are infected when they breathe the spores of the fungus into their lungs. The spores grow and invade the lung tissue, causing chronic coughing and fever. Eventually the disease progresses to other organs of the body. The spores of the fungus are common in soils that have high concentrations of bird droppings, and the disease is more prevalent in warm tropical regions of the world.

Crown gall disease, which is caused by a bacterium, occurs in many different kinds of plants and results in millions of dollars of damage to ornamental and agricultural plants each year. The bacterium, which lives in the soil, enters plants through small wounds such as those caused by insects. It causes galls, or tumorlike growths, often at the crown (between the stem and the roots) of a plant. Although plants seldom die from crown gall disease, they are weakened, grow more slowly, and often succumb to other pathogens.

Many parasites do not cause disease. For example, humans can acquire the pork tapeworm by eating poorly cooked pork that is infested with immature tapeworms. Once the tapeworm is inside the human digestive system, it attaches itself to the wall of the small intestine and grows rapidly by absorbing nutrients that pass through. Pork tapeworms that live in the human digestive tract do not cause any noticeable symptoms.

We have seen that each organism has its own niche within its community and that organisms form intimate relationships among themselves. Now we examine how the types of organisms found within a community change over time.

### Mini-Glossary of Symbiosis

**symbiosis:** Any intimate relationship between two or more different species.

**mutualism:** A symbiotic relationship in which both partners benefit.

**commensalism:** A symbiotic relationship in which one partner benefits and the other partner is unaffected.

**parasitism:** A symbiotic relationship in which one partner (the parasite) obtains nutrients at the expense of the other (the host).

# HOW COMMUNITIES CHANGE OVER TIME

A community of organisms does not spring into existence full-blown but develops gradually through a series of stages until it reaches maturity. The process of community development over time, which involves species in one stage being replaced by different species, is called **succession.** An area is initially colonized by certain organisms that are replaced over time by other organisms, which are themselves replaced, until a more or less stable community that is in equilibrium with existing environmental conditions develops. The relatively stable stage in a community's development is called a **climax community** or simply a **climax.** Climax communities represent the dominant vegetation of an area, but they are not permanent; they change as environmental conditions change.

Succession is usually described in terms of the changes in the species composition of the vegetation of an area, although each successional stage also has its own characteristic animal life. The time involved in ecological succession is on the order of hundreds or thousands of years, not the millions of years involved in the evolutionary time scale.

## Primary Succession

**Primary succession** is the change in species composition over time in a habitat that has not previously been inhabited by organisms. No soil exists when primary succession begins. A bare rock surface, such as recently formed volcanic lava (Figure 4–8) or rock scraped clean by glacial action, is a potential site for primary succession. Although the details vary from one site to another, one might first observe a community of lichens—dual organisms usually composed of a fungus and an alga (Fig-

**Figure 4–8**
Primary succession. This view shows small lehua plants growing on recently formed volcanic lava in Hawaii Volcanoes National Park. (David Muench)

**Figure 4–9**
Some lichens, such as *Psora decipens*, grow as thin mats encrusted on rock. (Courtesy of James D. Mauseth)

ure 4–9). Because they are the first organisms to colonize bare rock, lichens are called the **pioneer** community. Lichens secrete acids that help to break the rock apart, beginning the process of soil formation. Over time, the lichen community may be replaced by mosses and drought-resistant ferns, followed in turn by tough grasses and herbs. Once sufficient soil has accumulated, grasses and herbs may be replaced by low shrubs, which in turn are replaced by forest trees in several distinct stages. Primary succession from a pioneer community on bare rock to a climax forest community may take hundreds or thousands of years.

**Primary Succession on Rock: A Closer Look**  Few habitats are less hospitable than bare rock. Its temperature may approach 90°C in the sunlight, and unless it is raining, the rock may be totally devoid of moisture. Whatever minerals are present are locked up in its hard crystalline structure, unavailable to living things. Animals would find nothing there to eat; few of them could do more than briefly rest on such a surface.

Lichens are able to live on the surfaces of many rocks and beneath the surface of porous rock in a sheltered and somewhat moister habitat. Lichens are very resistant to desiccation (drying out). They cease to grow when water is unavailable but quickly resume active growth when moisture returns; they can absorb their own weight in water within moments of moistening.

As generations of lichens pass on a rock surface, several important cumulative changes occur. First, the biomass (amount of living material) of the lichen community increases, so more and more solubilized minerals are stored in the living tissue of the community. Second, fine particles of rock become detached from the rock's surface or even within the rock itself (see Chapter 14 for a description of how rock is pulverized during soil formation). Third, as lichens die, their decomposing remains mix with the rock particles to form a rudimentary soil. Fourth, whenever water is available, it is absorbed by the lichens and retained in their tissues and in the new, thin soil layer for longer periods of time than ever before. As all of these changes occur, an increasing number of tiny animals move into the area and make their homes in the lichens and soil.

All of these changes—increased biomass, soil development, water retention, and an increased number of life forms—work together to moderate the harsh conditions under which the pioneer community has lived, making it possible for mosses to grow there. In fact, because mosses can grow faster than lichens, they tend to replace any lichens that die. The higher productivity of mosses results in a greater accumulation of biomass and, ultimately, of soil. This leads to further habitat change, and ferns, grasses, and herbs move in.

**Primary Succession on Sand Dunes** Lake and ocean shores often have extensive sand dunes that have been deposited by wind and water. These dunes are not permanent; they move before the wind. The sand-dune environment is severe, with high temperatures during the day and low temperatures during the night; the sand may also be deficient in certain mineral nutrients needed by plants. As a result, few plants can tolerate the environmental conditions of a sand dune.

Grasses are a common pioneer plant on sand dunes. As the grasses extend over the surface of a dune, their roots help to hold the dune in place and stabilize it. At this point, mat-forming shrubs can invade the dune, further stabilizing it. Much later the shrubs are replaced by pines, which in turn are replaced by oaks. (Sometimes the pine stage is skipped.) Because the soil fertility remains low, oaks are rarely replaced by other forest trees; they are thus the climax community in primary succession of sand dunes.

**Mini-Glossary of Succession**

**succession:** A process of community development that involves a changing sequence of species.

**pioneer community:** The first organisms to colonize (or recolonize) an area.

**climax community:** A relatively stable community that is in equilibrium with current environmental conditions.

**primary succession:** Ecological succession in a habitat that has not previously been inhabited.

**secondary succession:** Ecological succession in a habitat that has previously been inhabited.

## Secondary Succession

**Secondary succession** is the change in species composition over time in a habitat already substantially modified by a pre-existing community; soil is already present. An area opened up by a forest fire (Figure 4–10) and an abandoned field are common examples of sites where secondary succession occurs.

Secondary succession on abandoned farmland has been studied extensively. Although it takes

(a)

(b)

**Figure 4–10**

Secondary succession after the Yellowstone fires of 1988. (a) Gray ash covers the forest floor after the fire. The trees, although dead, remain standing. (b) Less than one year later, in spring 1989, young plants mark the beginning of secondary succession. Many of the dead trees have fallen over. The dominant plant at this stage is trout lily (*Erythronium*), which sprouts up rapidly after a fire because its underground parts are not killed by the fire. (a, Ted and Jean Reuther/Dembinsky Photo Associates; b, Stan Osolinski/Dembinsky Photo Associates)

| Years after cultivation | Dominant vegetation | |
|---|---|---|
| 1 | Crabgrass | |
| 2 | Horseweed | |
| 3 | Broomsedge | |
| 5-15 | Pine seedlings | |
| 25-50 | Pine forest (with developing understory of deciduous hardwoods— not shown) | |
| 150 | Oak-hickory climax forest | |

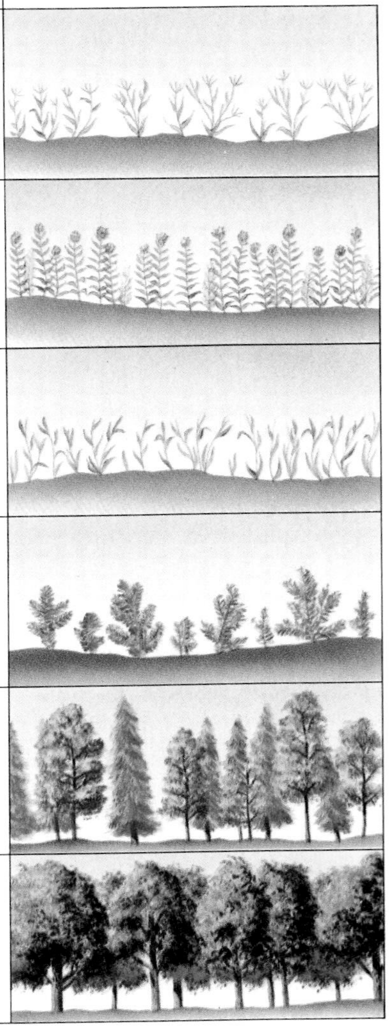

**Figure 4–11**
Secondary succession on an abandoned field in North Carolina.

more than 100 years for secondary succession to occur at a given site, it is possible for a single researcher to study a case of old field succession in its entirety by observing multiple sites in the same area. The scientist examines court records to determine when each field was abandoned.

Abandoned farmland in North Carolina is colonized by a predictable succession of plant communities (Figure 4–11). The first year after cultivation ceases, the field is dominated by crabgrass. The following year the dominant species is horseweed. It does not dominate more than one year, however,

because its decaying roots inhibit the growth of young horseweed seedlings. In addition, horseweed does not compete well with other plants which become established during the third year after the last cultivation, including broom sedge, ragweed, and aster. Typically, broom sedge, which is drought-tolerant, outcompetes aster, which is not.

In years 5 through 15, the dominant plants in the abandoned farmland are pines such as shortleaf pine and loblolly pine. Through the buildup of litter (pine needles and branches) on the soil, pines produce conditions that cause the earlier dominant

## Envirobrief

**Why Forests Need Squirrels**

Until recently, the interdependency of plant and animal species was largely ignored by foresters, who endorsed clearcutting vast tracts of forest land as the only way to manage our timber resources. In doing so, of course, forest companies alter complex forest ecosystems forever by wiping out all tree species and habitats. In his 1990 book, *The Redesigned Forest,* author and biologist Chris Maser demonstrated the close connection between the forest and its inhabitants in a description of the northern flying squirrel. A native of the Pacific Northwest, the squirrel feeds on a certain fungus found in the forest floor. The fungus is digested by the squirrel and excreted in the form of droppings or "pellets." Each pellet, says Maser, is like a "symbiotic pill." It contains four components of great value to the forest: spores of the fungi, yeast, nitrogen-fixing bacteria, and the complete nutrient component for the nitrogen-fixing bacteria. Wherever the pellets fall on the ground, they create fertile conditions. More of the valuable fungi are propagated and the overall fertility of the forest is increased. In a clearcut area, the squirrel disappears; without the squirrel, soil fertility is decreased. The northern flying squirrel illustrates only one of thousands of such relationships in any given forest.

plants to decline in importance. Over time, pines give up their dominance to hardwoods such as oaks. This climax stage of secondary succession depends primarily on the environmental changes produced by the pines. The pine litter causes soil changes, such as an increase in water-holding capacity, that are necessary in order for young oak seedlings to become established.

**Animal Life in Secondary Succession** As secondary succession proceeds, a progression of wildlife follows the changes in vegetation. Although a few species—the short-tailed shrew, for example—are found in all stages of abandoned farmland succession, most animals appear with certain stages and disappear with others. During the crabgrass and weed stages of secondary succession, the habitat is characterized by open fields that support grasshoppers, meadow mice, cottontail rabbits, and birds such as grasshopper sparrows and meadowlarks. As young pine seedlings become established, animals of open fields give way to animals common in mixed herbaceous and shrubby habitats. Now white-tailed deer, white-footed mice, ruffled grouse, robins, and song sparrows are common, whereas grasshoppers, meadow mice, grasshopper sparrows, and meadowlarks disappear. As the pine seedlings grow into trees, animals of the forest replace those common in mixed herbaceous and shrubby habitats. Cottontail rabbits give way to red squirrels, and ruffled grouse, robins, and song sparrows are replaced by warblers and veeries. Thus, each stage of succession supports its own characteristic wildlife.

# SUMMARY

1. A biological community consists of a group of organisms that interact and live together. A living community and its physical environment constitute an ecosystem.

2. The distinctive life style and role of an organism in a community are its ecological niche. The niche takes into account all aspects of the organism's existence—that is, all of the physical, chemical, and biological factors that the organism needs to survive, to remain healthy, and to reproduce.

3. Organisms are potentially able to exploit more resources and play a broader role in the life of their community than they actually do. The potential ecological

niche of an organism is its fundamental niche, whereas the niche an organism actually occupies is its realized niche.

4. It is thought that no two species can occupy the same niche in the same community for an indefinite period of time because competitive exclusion occurs. In this process, one species is excluded by another as a result of competition for resources that are in limited supply. An organism's limiting factors (such as the mineral content of soil, temperature extremes, and amount of precipitation) tend to restrict its realized niche.

5. In coevolution, two different species develop an intimate association that, over time, affects the evolu-

tion of both. Flowering plants and the animals that pollinate them, for example, have coevolved specialized adaptations that enable them to interact and survive.

**6.** Symbiosis is any intimate association between two or more different species. Both partners benefit from a mutualistic association. In commensalism, one organism benefits and the other is unaffected. In parasitism, one organism (the parasite) benefits and the other (the host) is harmed.

**7.** Succession is the orderly replacement of one community by another. Primary succession begins in a habitat that has not previously been inhabited. Secondary succession begins in an area where there was a preexisting community and a well-formed soil.

## DISCUSSION QUESTIONS

**1.** Why is a realized niche usually narrower, or more restricted, than a fundamental niche?

**2.** What portion of the human's fundamental niche are we occupying today? Do you think our realized niche is changing? Why or why not?

**3.** Who was A. F. Gause, and what important ecological concept did he originate?

**4.** Can you think of any coevolutionary relationships in which humans are involved?

## SUGGESTED READINGS

Ahmadjian, V., and S. Paracer. *Symbiosis: An Introduction to Biological Associations.* University Press of New England, Hanover and London, 1986. Covers all types of symbiosis, from viruses in bacteria to plants and their pollinators.

Beardsley, T. Recovery drill. *Scientific American,* November 1990. Some conventional ideas about how communities respond to environmental catastrophes are being challenged by the recovery of Mount St. Helens.

Boucher, D. H. Growing back after hurricanes. *BioScience* 40:3, March 1990. The significance for communities of periodic environmental catastrophes such as hurricanes is causing ecologists to reconsider the idea of a climax community.

Conniff, R. Yellowstone's "rebirth" amid the ashes is not neat or simple, but it's real. *Smithsonian,* September 1989, 36. Secondary succession of the forests that were burned during the September 1988 fire in Yellowstone.

Mohlenbrock, R. H. Mount St. Helens, Washington. *Natural History,* June 1990. Secondary succession of areas devastated by the eruption of Mount St. Helens.

Moore, P. D. Vegetation's place in history. *Nature* 347, 25 October 1990. A brief discussion of the differing views on the nature of communities.

# Chapter 5

*Sunset along the Apache Trail, Arizona. Deserts are determined primarily by a lack of precipitation. (David Muench).*

# Ecosystems and the Physical Environment

Almost completely isolated from everything in the Universe but sunlight, our planet Earth has often been compared to a vast spaceship whose life-support system consists of the living things that inhabit it. These living things produce oxygen, cleanse the air, adjust gases, transfer energy, and recycle waste products with great efficiency. Yet none of those processes would be possible without the nonliving physical environment of our spaceship Earth. Much of the climate to which living things have adapted is determined by the sun, which warms the planet, powers the hydrologic cycle (causes precipitation), and drives ocean currents and circulation patterns in the atmosphere.

## THE CYCLING OF MATERIALS WITHIN ECOSYSTEMS

In Chapter 3 we learned that energy flows in one direction through an ecosystem. In contrast, matter, the material of which living things are composed, moves in numerous cycles from the living world to the nonliving physical environment and back again (Figure 5–1); we call these **biogeochemical cycles.**

The Earth and biosphere are essentially a closed system—that is, a system from which matter cannot escape. The materials utilized by organisms cannot be "lost," although they can end up outside the organisms' reach. Usually, however, materials are reused and recycled both within and among ecosystems.

Four biogeochemical cycles of matter—carbon, nitrogen, phosphorus, and water—are representative of all biogeochemical cycles and are particularly important to living things. Carbon, nitrogen, and water have gaseous components and so cycle over large distances with relative ease. The element phosphorus, however, is completely nongaseous and, as a result, cycles only locally with ease.

## The Carbon Cycle

Proteins, carbohydrates, and other molecules essential to life contain carbon, so living organisms must have carbon available to them. Carbon makes up approximately 0.03 percent of the atmosphere as a gas, carbon dioxide ($CO_2$). It is also present in the ocean as dissolved carbon dioxide—that is, carbonate ($CO_3^{2-}$) and bicarbonate ($HCO_3^{-}$)—and in rocks such as limestone. Carbon cycles between the nonliving (abiotic) environment, including the atmosphere, and living organisms.

During photosynthesis (see Chapter 3), plants remove carbon dioxide from the air and **fix,** or incorporate, it into complex chemical compounds such as sugar (Figure 5–2). The overall equation for photosynthesis is

$$6CO_2 + 12H_2O \xrightarrow{\text{light}} \underset{\text{sugar (glucose)}}{C_6H_{12}O_6} + 6O_2 + 6H_2O$$

Thus, photosynthesis incorporates carbon from the abiotic environment into the biological compounds of producers. Those compounds are usually used as fuel for cell respiration (see Chapter 3) by the producer that made them, by a consumer that eats the

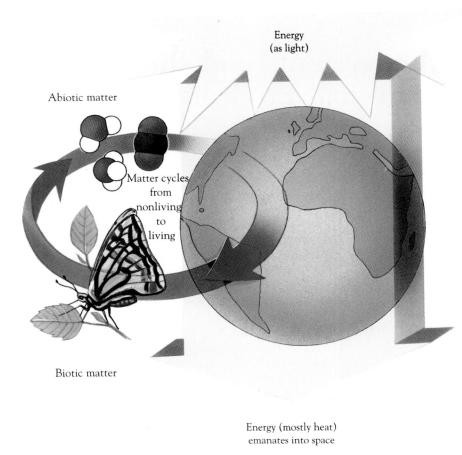

Energy (as light)

Abiotic matter

Matter cycles from nonliving to living

Biotic matter

Energy (mostly heat) emanates into space

**Figure 5–1**
Although energy flows one way through ecosystems, matter continually cycles from the abiotic to the biotic components of ecosystems and back again.

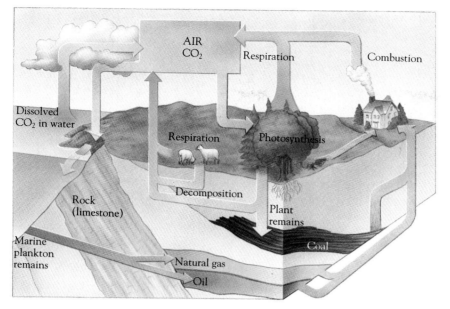

**Figure 5–2**
A simplified diagram of the carbon cycle. Carbon, in the form of carbon dioxide, enters living things from the nonliving environment when plants and other producers photosynthesize. Carbon returns to the environment when living things respire, are decomposed, or are burned (combustion). Fossil fuels, which are carbon-containing compounds formed from the remains of ancient organisms, and the carbon of limestone rock and marine animal shells, may take millions of years to cycle back to the biotic world.

producer, or by a decomposer that breaks down the remains of the producer or consumer. The overall equation for cell respiration is

$$C_6H_{12}O_6 + 6O_2 + 6H_2O \longrightarrow$$
$$6CO_2 + 12H_2O + \text{energy for biological work}$$

Thus, carbon dioxide is returned to the atmosphere by the process of cell respiration. A similar carbon cycle occurs in aquatic ecosystems, between aquatic organisms that photosynthesize (aquatic plants, algae, and cyanobacteria) and dissolved carbon dioxide in the water.

Sometimes the carbon in biological molecules isn't recycled back to the abiotic environment for some time. For example, a lot of carbon is stored in the wood of trees, where it may stay for several hundred years.

Millions of years ago, vast coal beds formed from the bodies of ancient trees that did not decay fully before they were buried. Similarly, accumulations of the oils of unicellular marine organisms in the geological past probably gave rise to today's underground deposits of oil and natural gas. Coal, oil, and natural gas, called **fossil fuels** because they formed from the remains of ancient organisms, are vast deposits of carbon compounds, the end products of photosynthesis that occurred millions of years ago (see Chapter 10).

The carbon in coal, oil, natural gas, and wood can be returned to the atmosphere by the process of burning, or **combustion.** In combustion, organic molecules are rapidly oxidized (combined with oxygen) and thus converted into carbon dioxide and water, with an accompanying release of heat and light.

Scientists think that most of the carbon that leaves the carbon cycle for millions of years is incorporated into the shells of marine organisms.

When these organisms die, their shells sink to the ocean floor and are covered by sediments, forming seabed deposits thousands of feet thick. The deposits are eventually cemented together to form a sedimentary rock called limestone. The Earth's crust is dynamic, and over millions of years, sedimentary rock on the bottom of the sea floor may lift to form land surfaces (the summit of Mount Everest, for example, is composed of sedimentary rock). After limestone is exposed by the process of geologic uplift, it is slowly worn away, or disintegrated, by chemical and physical weathering processes. This returns the carbon to the water and atmosphere, where it is available to participate in the carbon cycle once again.

Thus, one process (photosynthesis) removes carbon from the abiotic environment[1] and incorporates it into biological molecules, and three processes (cell respiration, combustion, and erosion) return carbon to the water and atmosphere of the abiotic environment.

**The Carbon Cycle and Global Warming** Human activities have disturbed the balance of the carbon cycle. From the advent of the Industrial Revolution to the present, humans have burned increasing amounts of fossil fuels—coal, oil, and natural gas. This trend, along with a greater combustion of wood as a fuel and the burning of large sections of tropical forest, has released carbon dioxide into the

[1] Carbon is also removed from the abiotic environment for the shells of marine organisms, which are formed in two ways. Many shells are formed from the products of the organism's metabolism; the carbon in these shells can be traced back to photosynthesis. Other shells are formed by a direct deposit of carbonate from seawater; the carbon in these shells actually represents a second biological process (other than photosynthesis) for shuttling carbon from the abiotic environment to living organisms.

atmosphere at a rate greater than the natural carbon cycle can handle.

The slow and steady rise of $CO_2$ in the atmosphere may be causing changes in climate called global warming. Global warming could result in a rise in sea level, changes in precipitation patterns, death of forests, extinction of animals and plants, and problems for agriculture. It could force the displacement of thousands or even millions of people, particularly from coastal areas. A more thorough discussion of increasing atmospheric $CO_2$ and global warming is found in Chapter 20.

## The Nitrogen Cycle

Nitrogen is crucial for all living things because it is an essential part of proteins, which are important structural components of cells and serve as enzymes and hormones, and nucleic acids, which store genetic information about an organism's traits.

At first glance it would appear that a shortage of nitrogen for living organisms is impossible: the Earth's atmosphere is about 80 percent nitrogen gas ($N_2$), a 2-atom (diatomic) molecule. But molecular nitrogen is so stable that it does not readily combine with other elements; therefore, living things cannot take nitrogen gas directly from the atmosphere and combine it with other elements to manufacture their proteins and nucleic acids. The molecular nitrogen must first be broken apart. The overall reaction that breaks up molecular nitrogen and combines its atoms with such elements as oxygen and hydrogen requires a great deal of energy.

There are five steps in the nitrogen cycle (Figure 5–3): (1) nitrogen fixation, (2) nitrification, (3) assimilation, (4) ammonification, and (5) denitrification. All of the steps except assimilation are performed by bacteria.

**(1) Nitrogen Fixation**  The first step in the nitrogen cycle, **nitrogen fixation,** involves the conversion of gaseous nitrogen ($N_2$) to ammonia ($NH_3$). The process gets its name from the fact that nitrogen is *fixed* into a form that living things can use. Although considerable nitrogen is also fixed by combustion, volcanic action, and lightning discharges and by industrial processes (all of which supply enough energy to break up molecular nitrogen), most nitrogen fixation is biological. It is carried out by nitrogen-fixing bacteria, including cyanobacteria (a type of photosynthetic bacterium), in soil and aquatic environments. Nitrogen-fixing bacteria employ an enzyme called **nitrogenase** to break up molecular nitrogen and combine it with hydrogen.

Because nitrogenase functions only in the absence of oxygen, the bacteria that use nitrogenase must insulate the enzyme from oxygen by some means. Some nitrogen-fixing bacteria live beneath layers of oxygen-excluding slime on the roots of a number of plants. But the most important nitrogen-fixing bacterium, *Rhizobium*, lives in special swellings, or **nodules,** on the roots of legumes such as beans or peas (Figure 5–4) and some woody plants. The relationship between *Rhizobium* and its host

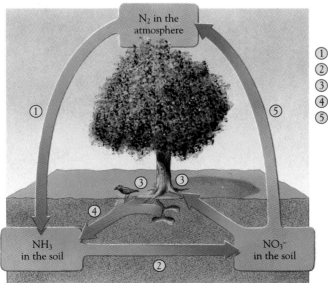

(1) Nitrogen-fixation
(2) Nitrification
(3) Assimilation
(4) Ammonification
(5) Denitrification

**Figure 5–3**
The nitrogen cycle has five steps.
(1) Nitrogen-fixing bacteria, including cyanobacteria, convert atmospheric nitrogen ($N_2$) into ammonia ($NH_3$).
(2) Ammonia is converted to nitrate ($NO_3^-$) by nitrifying bacteria in the soil. Nitrate is the main form of nitrogen absorbed by plants. (3) Plants assimilate nitrate when they produce proteins and nucleic acids; then animals eat plant proteins and produce animal proteins. (4) When plants and animals die, their nitrogen compounds are broken down by ammonifying bacteria. A product of this decomposition is ammonia. (5) Nitrogen is returned to the atmosphere by denitrifying bacteria, which convert nitrate to molecular nitrogen.

plants is mutualistic: the bacteria receive carbohydrates from the plant, and the plant receives nitrogen in a form that it can use.

In aquatic habitats most of the nitrogen fixation is done by cyanobacteria. Filamentous cyanobacteria have special oxygen-excluding cells called **heterocysts** that fix nitrogen (Figure 5–5). Some water ferns have cavities in which cyanobacteria live, somewhat as *Rhizobium* lives in the root nodules of legumes. Other cyanobacteria fix nitrogen in symbiotic association with certain plants or as the photosynthetic partners of certain lichens (see Chapter 4).

The reduction of nitrogen gas to ammonia by nitrogenase is a remarkable accomplishment of living organisms that is achieved without the tremendous heat, pressure, and energy required to manufacture commercial fertilizers. Even so, nitrogen-fixing bacteria must consume the energy equivalent of 12 grams of glucose in order to biologically fix a single gram of nitrogen.

**(2) Nitrification**    The conversion of ammonia ($NH_3$) to nitrate ($NO_3^-$), called **nitrification,** is accomplished by soil bacteria. Nitrification is a two-step process. First the soil bacteria *Nitrosomonas* and *Nitrococcus* convert ammonia to nitrite ($NO_2^-$). Then the soil bacterium *Nitrobacter* oxidizes nitrite to nitrate. The process of nitrification furnishes these bacteria, called nitrifying bacteria, with energy.

**(3) Assimilation**    In **assimilation,** plant roots absorb nitrate ($NO_3^-$) and/or ammonia ($NH_3$) that have been formed by nitrogen fixation and nitrification, and incorporate the nitrogen of these molecules into plant proteins and nucleic acids. When animals consume plant tissues, they also assimilate nitrogen by taking in plant nitrogen compounds and converting them to animal compounds.

**(4) Ammonification**    Living organisms produce nitrogen-containing waste products such as urea (in urine) and uric acid (in the wastes of birds). These substances, plus the nitrogen compounds that occur in dead organisms, are decomposed, releasing the nitrogen into the abiotic environment as ammonia ($NH_3$). The conversion of biological nitrogen compounds into ammonia is known as **ammonification,**

**Figure 5–4**

Root nodules of a clover plant (a legume). Mutualistic *Rhizobium* bacteria live in these nodules, subsisting on energy derived from sugars provided by their legume host. The bacteria fix nitrogen, some of which is utilized by the host plant. The ultimate death and decay of both partners enriches the soil with the nitrogen they have brought into chemical combination. (Carolina Biological Supply Company)

**Figure 5–5**

Many cyanobacteria fix nitrogen, often in association with plants. Shown is *Anabaena,* a cyanobacterium that has distinctive specialized cells, called heterocysts, where nitrogen fixation occurs. (Visuals Unlimited/S. Thomson)

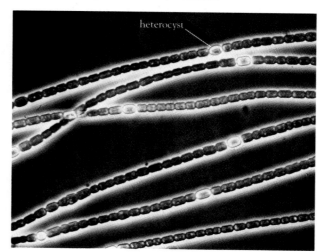

and the bacteria that perform this process both in the soil and in aquatic environments are called ammonifying bacteria. The ammonia produced by ammonification enters the nitrogen cycle and is once again available for the processes of nitrification and assimilation.

**(5) Denitrification** The reduction of nitrate ($NO_3^-$) to gaseous nitrogen ($N_2$) is called **denitrification.** Denitrifying bacteria reverse the action of nitrogen-fixing and nitrifying bacteria; that is, they return nitrogen to the atmosphere as nitrogen gas. Denitrifying bacteria are anaerobic, which means they prefer to live and grow where there is little or no free oxygen. For example, they are found deep in the soil near the water table, an environment that is nearly oxygen-free.

### Mini-Glossary of the Nitrogen Cycle

**nitrogen fixation:** The conversion of atmospheric nitrogen to ammonia, performed by nitrogen-fixing bacteria including cyanobacteria.

**nitrification:** The conversion of ammonia to nitrate, performed by nitrifying bacteria.

**assimilation:** The conversion of inorganic nitrogen (nitrate or ammonium) to the organic molecules of living things.

**ammonification:** The conversion of organic nitrogen (biological molecules containing nitrogen) to ammonia, performed by ammonifying bacteria.

**denitrification:** The conversion of nitrate to nitrogen gas, performed by denitrifying bacteria.

**The Nitrogen Cycle and Water Pollution** Humans affect the nitrogen cycle by producing large quantities of nitrogen fertilizer (both ammonia and nitrate) from nitrogen gas. Although this process in in itself is not harmful, the overuse of commercial fertilizers on the land can cause water quality problems. Rain washes nitrate fertilizer into rivers and lakes, where it stimulates the growth of algae. As these algae die, their decomposition robs the water of dissolved oxygen, which in turn causes other aquatic organisms, including many fish, to die of suffocation. Nitrates from fertilizers can also leach (filter) down through the soil and contaminate groundwater. Many people who live in rural areas drink groundwater, and groundwater contaminated by nitrates is dangerous, particularly for infants and small children. The effects of nitrate contamination on the environment and on human health are discussed in Chapters 14 and 18.

## The Phosphorus Cycle

Phosphorus, which does not exist in a gaseous state and therefore does not enter the atmosphere, cycles from the land to sediments in the oceans and back to the land (Figure 5–6). As water runs over rocks containing phosphorus, it gradually wears away the surface and carries off inorganic phosphate ($PO_4^{3-}$) molecules.

The erosion of phosphorus rocks releases phosphorus into the soil, where it is taken up by plant roots in the form of inorganic phosphates. Once in the plant's cells, phosphates are used in a variety of biological molecules including nucleic acids. Animals obtain most of their required phosphate from the food they eat, although in some localities drinking water may contain a substantial amount of inorganic phosphate. Thus, like carbon and nitrogen, phosphorus moves through the food chain as one organism consumes another. Phosphorus released by decomposers becomes part of the soil's pool of inorganic phosphate that can be reused by plants.

Phosphorus cycles through aquatic communities in much the same way it does through terrestrial communities. Dissolved phosphorus enters aquatic communities via absorption by algae and plants, which are then consumed by plankton and larger organisms. These, in turn, are eaten by a variety of fin fish and shellfish. Ultimately, decomposers that break down wastes and dead organisms release inorganic phosphorus into the water, where it is available to be used by aquatic producers again.

Phosphate can be lost from biological cycles. Some of it is carried from the land by streams and rivers to the ocean, where it can be deposited on the sea floor and remain for millions of years. The geologic process of uplift may someday expose these sea floor sediments as new land surfaces, from which phosphates will once again be eroded.

Some phosphate in the aquatic food chain finds its way back to the land. A small portion of the fish and aquatic invertebrates are eaten by sea birds, which may defecate where they roost on the land. Their manure, called guano, contains large amounts of phosphate and nitrate; on land these minerals may be absorbed by the roots of plants. The phosphate contained in guano may enter terrestrial food chains in this way, although the amounts involved are quite small.

**Humans and the Phosphorus Cycle** Humans affect the natural cycling of phosphorus by accelerating its long-term loss from the land. Corn grown in

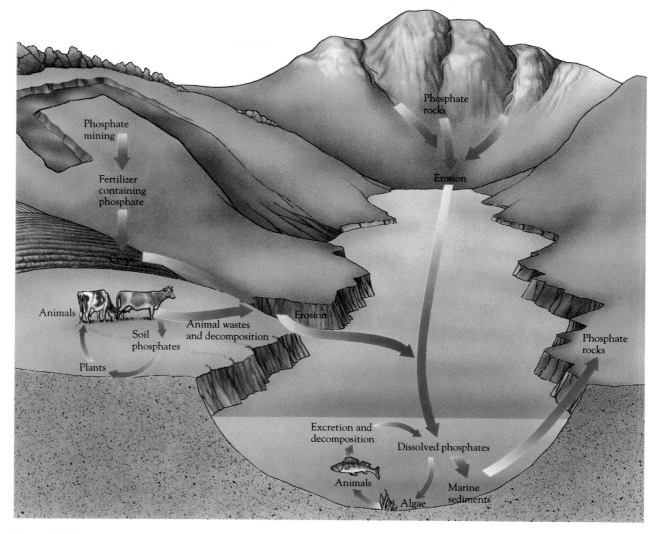

**Figure 5–6**
The phosphorus cycle in terrestrial and aquatic environments. Recycling of phosphorus (as phosphate, $PO_4^{3-}$) is slow because no biologically important form of phosphorus is gaseous. Phosphates that become part of marine sediments may take millions of years to solidify into rock, uplift as mountains, and erode to again become available to living things.

Iowa (which contains phosphate absorbed from the soil) may be used to fatten cattle in an Illinois feedlot. Part of the phosphate absorbed by the roots of the corn plants thus ends up in the feedlot wastes, which probably eventually wash into the Mississippi River. Beef from the Illinois cattle may be consumed by people living far away—in New York City, for instance. Hence, more of the phosphate ends up in human wastes and is flushed down toilets into the New York City sewer system. Sewage treatment rarely removes phosphates, and so they cause water quality problems in rivers and lakes (see Chapter 21). To compensate for the steady loss of phosphate from their land, farmers must add phosphate fertilizer to their fields. More than likely, that fertilizer is produced in Florida from the large deposits of phosphate rock that are mined there.

In natural communities, very little phosphorus is lost from the cycle, but few communities today are in a "natural" state. Phosphorus loss from the soil is accelerated by land-denuding practices such as the clearcutting of timber and by erosion of agricultural and residential land. For practical purposes, phosphorus that washes from the land into the sea is permanently lost from the terrestrial phosphorus cycle, for it remains in the sea for millions of years.

## The Hydrologic Cycle

Water continuously circulates from the oceans to the atmosphere to the land and back to the oceans, providing us with a renewable supply of purified water on land. This complex cycle, known as the **hydrologic cycle,** results in a balance among water in the oceans, water on the land, and water in the atmosphere (Figure 5–7). When water evaporates from the ocean's surface, it forms clouds in the atmosphere. Water also evaporates from soil, streams, rivers, and lakes. **Transpiration,** the loss of water vapor from land plants, also adds water to the atmosphere. Roughly 97 percent of the water absorbed from the soil by a plant is transported to the leaves, where it is transpired back to the atmosphere.

Water moves from the atmosphere to the land and oceans in the form of precipitation (rain, snow,

**Figure 5–7**
The hydrologic cycle. Water cycles from the oceans to the atmosphere to the land and back to the oceans. Although some water molecules are unavailable for thousands of years (locked up in polar ice, for example), all water molecules eventually travel through the hydrologic cycle.

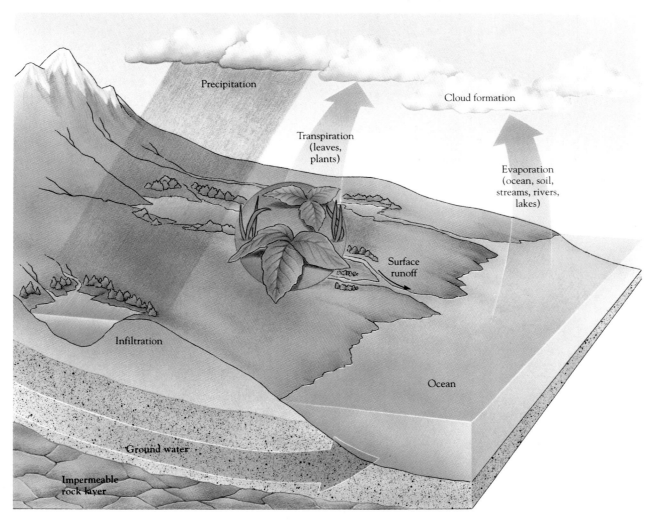

sleet, or hail). Once on land, the water can move in several ways through the hydrologic cycle:

1. It may evaporate from land and re-enter the atmosphere directly.
2. It may flow in rivers and streams to coastal **estuaries** (where fresh water meets the oceans) and into the ocean. The movement of water from land to oceans is called **runoff.**
3. The water may percolate (seep) downward in the soil to become **groundwater.** Groundwater supplies water to the soil, to streams and rivers, and to plants.

Ultimately, the water that falls on land from the atmosphere makes its way back to the oceans.

Regardless of its physical form (solid, liquid, or vapor) and location, every molecule of water eventually moves through the hydrologic cycle. Tremendous quantities of water are cycled annually between the Earth and its atmosphere. The amount of water entering the atmosphere each year is estimated at about 400,000 cubic kilometers (95,000

cubic miles). Approximately three-fourths of this water re-enters the ocean directly as precipitation over water; the remainder falls on land.

## THE PHYSICAL ENVIRONMENT

We have seen how living things depend on the physical environment to supply essential materials for biogeochemical cycles. Physical factors such as climate and soil also affect living things. (See Focus On: The Gaia Hypothesis for an intriguing view of living organisms and their abiotic environment.)

**Climate** comprises the average weather conditions that occur in a place over a period of years. Factors that determine an area's climate include temperature, precipitation, wind, humidity, fog, and cloud cover. Day-to-day variations, day-to-night variations, and seasonal variations in these factors are also important aspects of climate which we discuss in the remainder of this chapter. (Chapter 14 discusses soil, the surface layer of Earth that supports plants and is home to countless numbers of bacteria, fungi, protists, and animals.)

### The Sun Warms the Earth

The sun makes all life on Earth possible. It warms the planet to habitable temperatures. Without the sun's energy, the temperature on planet Earth would approach absolute zero ($-273°C$) and all water would be frozen, even in the oceans. The hydrologic cycle, carbon cycle, and other biogeochemical cycles are powered by the sun, and it is the primary determinant of Earth's climate. The sun's energy is captured by photosynthetic organisms, which use it to make the food molecules required by almost all forms of life. Most of our fuels—wood, oil, coal, and natural gas, for example—represent solar energy captured by photosynthetic organisms. Without the sun, life on planet Earth would cease.

The sun's energy is the product of a massive nuclear fusion reaction (see Chapter 11) and is emitted into space in the form of electromagnetic radiation—especially visible light and infrared and ultraviolet radiation (which are not visible to the human eye) An infinitesimal portion of this energy—one-billionth of the sun's total production—strikes the Earth's atmosphere, and of this tiny trickle of energy a minute part operates the ecosphere.

**Envirobrief**

#### Milking Mountains for Moisture

In many parts of the world, moisture-laden clouds (fog) pass tantalizingly close to extremely arid regions without releasing rainfall. In the Middle Eastern Sultanate of Oman, people have "harvested" water for centuries from such clouds using the leaves of olive trees that grow near mountain summits. Small tanks built at the foot of the trees collect droplets that form on the leaves. The idea has been taken one step further in Chile's Atacama Desert. Fifty low-cost "captors" that resemble volleyball nets have been built along a ridge of the Andes Mountains. Moist clouds from the Pacific Ocean pass through the captors, releasing 7,200 liters of fresh water each day. The water is channeled by an aqueduct to the coastal village of Caleta Chungungo, which formerly depended on weekly truck shipments for its drinking water. Gardens are now being grown with this new source of water. When more captor nets are built the village plans to develop a fish processing plant.

### The Gaia Hypothesis

One of the most unusual and controversial hypotheses to be advanced in recent years is the Gaia hypothesis, which states that the entire Earth can be viewed as a single living organism. According to this model, the planet Earth is alive in the sense that it is capable of self-maintenance. Living organisms on Earth interact with the nonliving environment to produce and maintain Earth's chemical composition, temperature, and other characteristics. Thus, the environment and living organisms of Earth depend on one another and work together as a homeostatic mechanism. (Biological systems have homeostatic mechanisms to help maintain a steady state or constant environment.)

As an example of the Gaia mechanism, consider the Earth's temperature. It is generally accepted that the temperature of the Earth has remained relatively constant at a temperature suitable for life over the past 3.5 to 4 billion years in which life has existed. Yet there is evidence that the sun has been heating up during that time. Why hasn't the Earth increased in temperature? Gaia proponents attribute the constant temperature to a drop in the level of atmosphere-warming $CO_2$ in the atmosphere, which they say happened because the living Earth compensated for increased sunlight by "fixing" $CO_2$ into calcium carbonate shells of countless billions of marine phytoplankton (photosynthetic plankton). As the phytoplankton died, their shells sank to the ocean floor, thus removing $CO_2$ from the system. This Gaia planetary temperature mechanism is an example of a feedback loop between the abiotic environment and the living organisms on Earth, which mutually interact to regulate Earth's temperature.

Another example of interactions between the nonliving and living components of the Earth, according to the Gaia hypothesis, involves the salinity of the oceans. As terres-

The largest organism? According to the Gaia hypothesis, the Earth is a huge "organism" capable of self-maintenance. (NASA)

trial rocks are weathered, oceans tend to get saltier and saltier, in time becoming too saline to support life. However, geological evidence indicates that the salinity of the oceans has remained constant for millions of years. Gaia proponents suggest that there is a feedback loop in which bacteria remove excess salt from the ocean in salt flats, which are shallow bays along tropical and subtropical oceans where bacteria grow in such numbers that they form great mats.

Many scientists are reluctant to accept the Gaia hypothesis. Almost everyone agrees that the environment modifies living organisms and that living organisms modify the environment to some extent, especially on a local scale. However, the idea that Earth's living things *adjust* the physical environment to meet their needs has few backers, in part because it is difficult to test.

---

In the daytime, 30 percent of the solar radiation that falls upon Earth is immediately reflected away by clouds and surfaces, especially snow, ice, and oceans (Figure 5–8). The remaining 70 percent is absorbed by the Earth, where it runs the water cycle, drives winds and ocean currents, pow-

ers photosynthesis, and warms the planet. Ultimately, however, all of this energy is lost by the continual radiation of long-wave infrared (heat) energy into space.

The foregoing values are averages for the entire Earth and vary substantially at different places be-

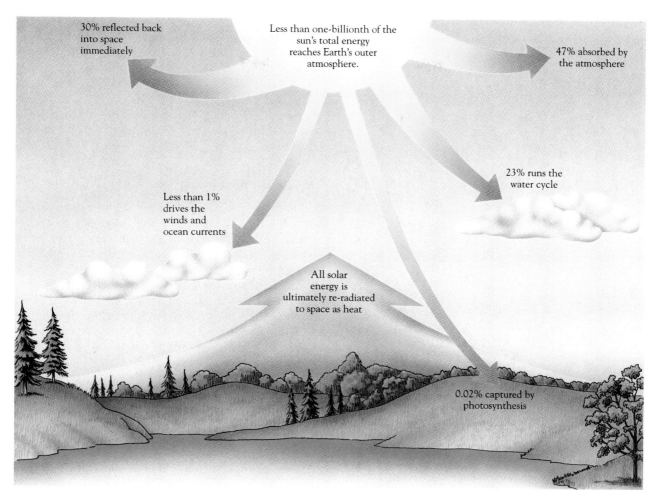

**Figure 5–8**

Most of the energy produced by the sun never reaches the Earth. The solar energy that does reach the Earth warms the planet's surface, powers the water cycle and other biogeochemical cycles, produces our climate, and powers almost all life on Earth through the process of photosynthesis, which converts solar energy into the chemical energy of organic molecules.

cause of local conditions. For example, high clouds increase energy reflection, whereas low clouds increase energy absorption.

**Solar Energy at the Equator and the Poles**  The most significant local variation in Earth's temperature is produced because the sun's energy doesn't reach all places on Earth uniformly. A combination of the Earth's roughly spherical shape and the tilt of its axis produces a great deal of variation in the exposure of the Earth's surface to the energy delivered by sunlight.

The principal effect of the tilt is on the angles at which the sun's rays strike different areas of the Earth at any one time (Figure 5–9a). On the average, the sun's rays hit the Earth vertically near the

equator, making the energy more concentrated and producing higher temperatures. Near the poles the sun's rays hit more obliquely, and as a result their energy is spread over a larger surface area. Also, rays of light entering the atmosphere obliquely near the poles must pass through a deeper envelope of air than those entering near the equator. This causes more of the sun's energy to be scattered and reflected back to space, which in turn further lowers temperatures near the poles. Thus, because the solar energy that reaches polar regions is less concentrated, temperatures are lower.

**Seasonal Variations in Solar Energy**  Seasons are determined by two main factors: the inclination of the Earth's axis (the more important factor) and

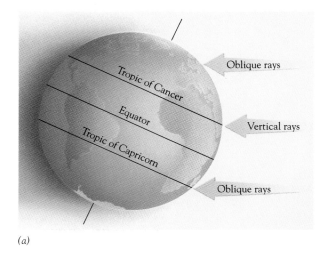

*(a)*

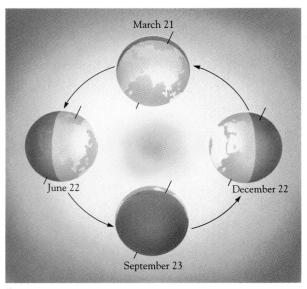

*(b)*

**Figure 5–9**

Variation in solar intensity on Earth. (a) The angle at which the sun's rays strike the Earth varies from one geographical location to another due to the Earth's spherical shape and its inclination on its axis. The month of June is represented here. (b) The inclination of the Earth's axis remains the same as it travels around the sun. Thus, the sun's rays hit the Northern Hemisphere obliquely during the winter months and more directly during the summer. In the Southern Hemisphere, the sun's rays are oblique during their winter, which corresponds to our summer. At the equator, the sun's rays are approximately vertical at all times of the year.

the distance of the Earth from the sun, which varies during the year. Since the Earth's inclination on its axis is always the same (23.5°), during half of the year (March 21 to September 22) the Northern Hemisphere tilts *toward* the sun, and during the other half (September 22 to March 21) it tilts *away* from the sun (Figure 5–9b). (The orientation of the Southern Hemisphere is just the opposite at these times.)

## Atmospheric Circulation

In large measure, differences in temperature caused by variations in the amount of solar energy reaching the Earth at different locations drive the circulation of the atmosphere. The very warm surface of the Earth near the equator heats the air that is in contact with it, causing this air to expand and rise. As the warm air rises it cools, and then it sinks again. Much of it recirculates almost immediately to the same areas it has left, but the remainder of the heated air flows toward the poles, where eventually it is chilled. Similar upward movements of warm air and its subsequent flow toward the poles occur at higher latitudes (farther from the equator) as well (Figure 5–10). As air cools by contact with the polar ground and ocean, it sinks and flows toward the equator, generally beneath the sheets of warm air that simultaneously flow toward the poles. The constant motion of air transfers heat from the equator toward the poles, and as the air returns, it cools the land over which it passes. This continuous turnover does not equalize temperatures over the surface of the Earth, but it does moderate them.

**Surface Winds**    In addition to global circulation patterns, the Earth's atmosphere exhibits complex horizontal movements that are commonly referred to as **winds**. The nature of wind, with its turbulent gusts, eddies, and lulls, is difficult to understand or predict. It results in part from differences in atmospheric pressure and from the rotation of the Earth.

The gases that constitute the atmosphere have weight and exert a pressure that is, at sea level, about 1,013 millibars (14.7 pounds per square inch).[2] Air pressure is variable, however, changing with altitude, temperature, and humidity. Winds tend to blow from areas of high atmospheric pressure to areas of low pressure, and the greater the difference between the high- and low-pressure areas, the stronger the wind.

[2] Atmospheric pressure is expressed in millibars. One millibar equals 0.03 pounds per square inch.

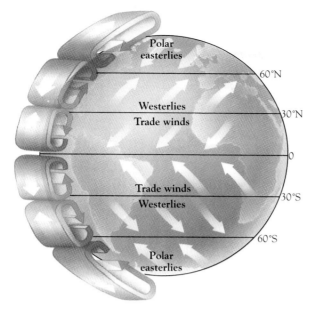

**Figure 5–10**
Atmospheric circulation transports heat from the equator to the poles. The greatest solar energy input occurs at the equator, heating air most strongly in that area. The air rises and travels poleward, but is cooled in the process so that much of it descends again at around 30 degrees latitude in both hemispheres. At higher latitudes the patterns of movement are more complex.

The Earth's rotation also influences the direction of wind. The Earth's rotation from west to east causes moving air to be deflected from its path and swerve to the right in the Northern Hemisphere and to the left in the Southern Hemisphere. This tendency is known as the **Coriolis effect.**

The Coriolis effect can be visualized by imagining that you and a friend are standing about 10 feet apart on a merry-go-round that is turning clockwise. Suppose you throw a ball directly at your friend. By the time the ball reaches the place where your friend was, he or she is no longer in that spot. The ball will have swerved far to the left of your friend. This is how the Coriolis effect works in the Southern Hemisphere.

To visualize how the Coriolis effect works in the Northern Hemisphere, imagine you and your friend are standing on the same merry-go-round, only this time it is moving counterclockwise. Now when you throw the ball, it will swerve far to the right of your friend.

The Earth's atmosphere has three **prevailing winds**—major surface winds that blow more or less continually (Figure 5–10). Prevailing winds that blow from the northeast near the North Pole or from the southeast near the South Pole are called **polar easterlies.** Winds that blow in the mid-latitudes from the southwest (in the Northern Hemisphere) or from the northwest (in the Southern Hemisphere) are called **westerlies.** Tropical winds that blow from the northeast (Northern Hemisphere) or the southeast (Southern Hemisphere) are called **trade winds.**

**Patterns of Circulation in the Oceans**

The persistent prevailing winds blowing over the ocean produce mass movements of surface ocean water known as **currents.** The prevailing winds generate *circular* ocean currents called **gyres.** For example, in the North Atlantic, the tropical trade winds tend to blow toward the west, whereas the westerlies in the mid-latitudes blow toward the east. This helps establish a clockwise gyre in the North Atlantic. Thus, surface ocean currents and winds tend to move in the same direction, although there are many variations on this general rule. Other factors that contribute to ocean currents include the Coriolis effect, the varying density of water, and the positions of land masses.

The paths traveled by surface ocean currents are partly caused by the Coriolis effect (Figure 5–11). The Earth's rotation from west to east causes surface ocean currents to swerve to the right in the Northern Hemisphere, creating a circular, clockwise pattern of water currents. In the Southern Hemisphere, ocean currents swerve to the left, thereby moving in a circular, counterclockwise pattern.

The varying **density** (mass per unit volume) of seawater affects deep ocean currents. Water that is colder is denser than warmer water.[3] Thus, colder ocean water sinks and flows under warmer water, creating currents far below the surface. Deep ocean currents often travel in different directions and at different speeds than do surface currents, in part because the Coriolis effect is more pronounced at greater depths.

The positions of land masses also affect oceanic circulation. As you can see in Figure 5–12, the oceans are not distributed uniformly over the globe: there is clearly more water in the Southern Hemisphere than in the Northern Hemisphere. Therefore, the circumpolar (around the pole) flow of water in the Southern Hemisphere is almost unimpeded by land masses.

[3]The density of water increases with decreasing temperature down to 4°C.

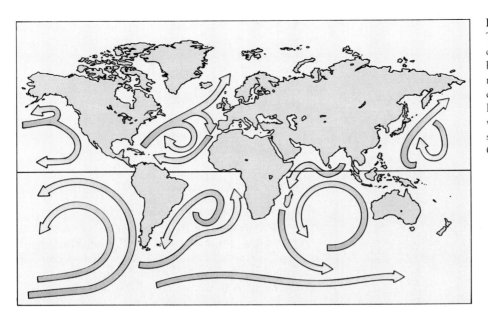

**Figure 5–11**
The basic pattern of surface ocean currents is caused largely by the action of winds. The main ocean current flow—clockwise in the Northern Hemisphere and counterclockwise in the Southern Hemisphere—results partly from the Coriolis effect.

Ocean currents also influence atmospheric patterns (see Focus On: The El Niño–Southern Oscillation Event and the World's Climate).

## What Causes Climate?

Average temperature, temperature extremes, precipitation, the seasonal distribution of precipitation, day length, and season length are the most important dimensions of climate that affect living organisms. Variations in these climatic factors produce the Earth's major ecosystems, including tundra, desert, rain forest, and grassland (see Chapter

**Figure 5–12**
The Northern and Southern Hemispheres have greatly differing proportions of land and water, with far more water occurring in the Southern Hemisphere. (a) The Southern Hemisphere as viewed from the South Pole. (b) The Northern Hemisphere as viewed from the North Pole. Ocean currents are freer to flow in a circumpolar manner in the Southern Hemisphere.

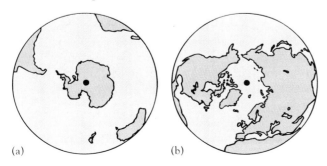

(a)                              (b)

6). Latitude (distance north or south of the equator) and the inclination of the Earth on its axis determine day length, season length, and, to a large degree, temperature.

Differences in precipitation depend upon several factors. The heavy rainfall of some areas of the tropics results mainly from the equatorial upwelling of moisture-laden air. High surface-water temperatures cause the evaporation of vast quantities of water from tropical oceans, and prevailing winds blow the resulting moist air over land masses. Heating of the air by land surface that has been warmed by the sun causes moist air to rise. As it rises, the air cools, and moisture condenses from water vapor to a liquid, then falls as precipitation. The air eventually returns to Earth on both sides of the equator between the Tropics of Cancer and Capricorn (latitudes 23.5° north and 23.5° south). By then most of its moisture has precipitated, and the dry air returns to the equator. This air makes little biological difference over the ocean, but its lack of moisture produces some of the great tropical deserts, such as the Sahara Desert.

Air is also dried by long journeys over land masses. Near the windward (the side from which the wind blows) coasts of continents, rainfall may be heavy. However, in the temperate zones—the areas between the tropics and the polar zones—continental interiors are usually dry, because they are far from oceans that replenish water in the air passing over them.

Moisture is also removed from air by mountains when they cause humid air masses to rise and thus

Focus On

## The El Niño–Southern Oscillation Event and the World's Climate

Seasonal weather forecasting requires an understanding of not only how the atmosphere operates but also how the oceans interact with the atmosphere. Consider the **El Niño–Southern Oscillation** event, a periodic warming of surface waters of the tropical East Pacific that alters both oceanic and atmospheric circulation patterns and results in unusual weather in remote areas of the Earth. Every three to seven years, a warm mass of water that is normally restricted to the western Pacific (near Australia) expands eastward, increasing surface temperatures in the East Pacific to 3 to 4 degrees over normal. Ocean currents, which normally flow westward in this area, slow down, stop altogether, or even reverse and go eastward. The phenomenon is called *El Niño* (Spanish for "the child") because the warming usually reaches the fishing grounds off Peru just before Christmas.

The El Niño–Southern Oscillation has a devastating effect on the fisheries off South America. The higher temperatures and accompanying changes in ocean circulation patterns prevent nutrient-laden deeper waters from upwelling (coming to the surface). This severely decreases the populations of anchovies and other marine organisms. During the 1972 El Niño, for example, the anchovy population decreased by 90 percent.

The El Niño–Southern Oscillation also alters air currents, directing unusual weather to areas far from the tropical Pacific. The 1991 El Niño, for example, resulted in a much warmer-than-usual winter across much of Alaska, western Canada, and the northern United States. El Niño was responsible for the torrential rains that hit Texas and southern California during the 1991–1992 winter season. In addition, the effects of El Niño have been linked to droughts in Africa, Australia, and Hawaii.

Climate patterns associated with an El Niño, which can drastically alter the climate in many areas remote from the Pacific Ocean.

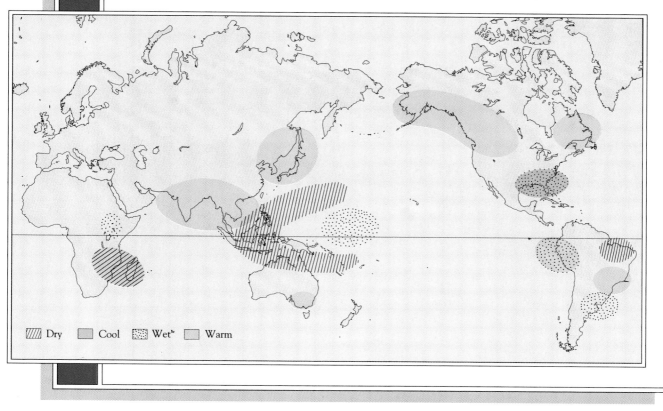

///// Dry ☐ Cool ⬚ Wet* ☐ Warm

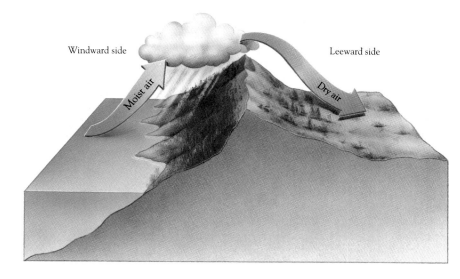

**Figure 5–13**
When wind blows moist air over a mountain range, precipitation occurs on the windward side of the mountain, causing a dry "rain shadow" on the leeward side. Such a rain shadow exists east of the Cascade Range in Washington state.

release their water as precipitation. If prevailing winds blow onto a mountain range, precipitation occurs primarily on the windward slopes of the mountains. This situation exists on the west coast of North America where precipitation falls on the western slopes of the mountains. Downwind (in this case, east of the mountain range), a low-precipitation **rain shadow** (Figure 5–13) develops, often creating a desert. Thus, some of the regional differences in worldwide precipitation (Figure 5–14) result from the drying of air as it is returned to more equatorial areas; some result from long travel over

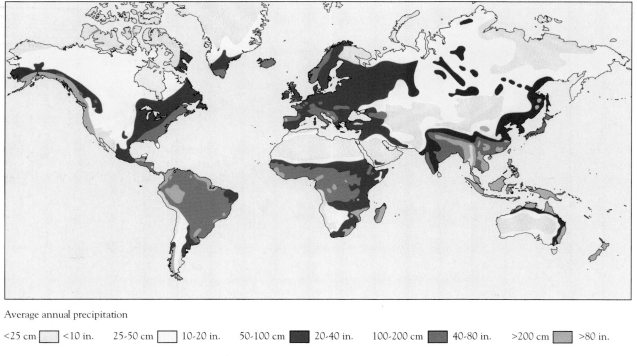

Average annual precipitation

<25 cm  <10 in.    25-50 cm  10-20 in.    50-100 cm  20-40 in.    100-200 cm  40-80 in.    >200 cm  >80 in.

**Figure 5–14**
Average annual precipitation. Notice the light precipitation at extreme northern latitudes and in the centers of many continents.

continents; and some result from cooling produced by mountainous regions.

**Variations in Overall Climatic Conditions** Differences in elevation, in the steepness of slopes and the directions they face, or in exposure to prevailing winds may produce local variations in climate known as **microclimates,** which can be quite different from the overall climate surrounding them. The microclimate of an organism's habitat is the climate it actually experiences and with which it must cope.

Sometimes it is possible for a population of organisms to substantially modify its own microclimate, making it more favorable. For example, trees modify the local climate within a forest so that the temperature is usually lower, and the relative humidity higher, than outside the forest. Beneath the litter of the forest floor, temperature and humidity differ still more; the bottom of the litter is cooler and moister than the surrounding forest. As another example, desert-dwelling organisms burrow in the sand to evade surface climatic conditions that would kill them in minutes. The cooler daytime microclimate in their burrows permits them to survive until night, when the surface cools off and they can leave their retreats to forage or hunt.

# SUMMARY

**1.** In contrast to energy, which moves in one direction through ecosystems, matter is cyclic. All materials vital to life are continually recycled through ecosystems and so become available to new generations of organisms. Biogeochemical cycles are cycles of matter such as carbon, nitrogen, phosphorus, and water from the environment to living things and back to the environment.

**2.** Carbon enters plants, algae, and cyanobacteria as carbon dioxide ($CO_2$), which is incorporated into organic molecules by photosynthesis. Cell respiration by plants, by animals that eat plants, and by decomposers returns $CO_2$ to the atmosphere, making it available for producers again.

**3.** There are five steps in the nitrogen cycle. (1) Nitrogen fixation is the conversion of nitrogen gas to ammonia. (2) Nitrification is the conversion of ammonia to nitrate, one of the main forms of nitrogen used by plants. (3) Assimilation is the biological conversion of nitrates or ammonia into proteins and other nitrogen-containing compounds by plants; the conversion of plant proteins into animal proteins is also part of assimilation. (4) Ammonification is the conversion of organic nitrogen to ammonia. (5) Denitrification converts nitrate to nitrogen gas.

**4.** The phosphorus cycle has no biologically important gaseous compounds. Phosphorus erodes from rock in the form of inorganic phosphates, which are absorbed from the soil by plant roots. Phosphorus enters other living things through the food chain and is released back into the environment as inorganic phosphate by decomposers. When phosphorus washes into the ocean and is deposited in seabeds, it can be lost from biological cycles for millions of years.

**5.** The hydrologic cycle, which continuously renews the supply of water that is so essential to life, involves an exchange of water among the land, the atmosphere, and living things. Water enters the atmosphere by evaporation and transpiration, and leaves the atmosphere as precipitation. On land, water filters through the ground or runs off to lakes, rivers, and oceans.

**6.** Sunlight is the primary (almost the sole) source of energy available to the biosphere. Of the solar energy that reaches the Earth, 30 percent is immediately reflected away and the remaining 70 percent is absorbed, including 0.02 percent that is absorbed by plants. Ultimately, all absorbed solar energy is radiated into space as infrared (heat) radiation.

**7.** A combination of the Earth's roughly spherical shape and the tilt of its axis concentrates solar energy at the equator and dilutes solar energy at the poles. The tropics are therefore hotter and less variable in climate than the temperate and polar areas. Seasons are determined by two main factors: the inclination of the Earth's axis (the more important factor) and the distance of the Earth from the sun, which varies during the year.

**8.** Atmospheric heat transfer from the equator to the poles produces a movement of warm air toward the poles and a movement of cool air toward the equator, thus moderating the climate. In addition to these global circulation patterns, the Earth's atmosphere exhibits complex horizontal movements called winds that result in part from differences in atmospheric pressure and from the rotation of the Earth (the Coriolis effect).

**9.** Surface ocean currents result largely from prevailing winds. Other factors that contribute to ocean currents include the Coriolis effect, the varying density of water, and the positions of land masses.

**10.** Local climate includes average temperature, temperature extremes, precipitation, seasonal distribution of precipitation, day length, and season length. Precipitation is greatest where warm air passes over the ocean, absorbing moisture, and is then cooled, such as when humid air is forced upward by mountains. Deserts develop in the rain shadows of mountain ranges or in continental interiors.

# DISCUSSION QUESTIONS

**1.** Why is the cycling of matter essential to the continuance of life on the Earth?

**2.** How might humans disturb the temperature balance of the Earth?

**3.** What basic forces determine the circulation of the Earth's atmosphere? Describe the general directions of atmospheric circulation.

**4.** How do ocean currents affect the climate on land?

**5.** What are some of the factors that produce areas of precipitation extremes, such as rain forests and deserts?

**6.** Why might industrial polluters think that the Gaia hypothesis gives them permission to pollute the air, water, and soil indefinitely?

# SUGGESTED READINGS

Brewer, R. *The Science of Ecology*, 2d ed. Saunders College Publishing, Philadelphia, 1993. A good general textbook on the principles of ecology, including ecosystem ecology.

Gilliland, M. W. A study of nitrogen-fixing biotechnologies for corn in Mexico. *Environment* 30:3, April 1988. Explains the potential benefit of engineering nitrogen-fixing bacteria to provide nitrogen for crops such as corn.

Schneider, S. H. Debating Gaia. *Environment* 32:4, May 1990. Reviews the Gaia view of a self-regulating Earth.

Watson, A. Gaia. *New Scientist* 131, 6 July 1991. Presents arguments for and against the Gaia hypothesis.

# Chapter 6

*Monteverde Cloud Forest, a tropical rain forest in Costa Rica. (James L. Castner)*

# Major Ecosystems of the World

limate, particularly temperature and precipitation, influences the distribution of the Earth's organisms. In each major kind of climate a distinctive type of vegetation develops. For example, desert plants are associated with arid climates, grasses with semiarid climates, and forests with moist climates. Certain animals and other kinds of organisms are associated with each major type of vegetation. The major terrestrial (land) ecosystems, called biomes, extend over large geographical areas. In like manner, certain aquatic organisms are characteristically assembled in each of the Earth's major aquatic ecosystems.

## THE GEOGRAPHY OF LIFE

One would hardly expect to find a polar bear in Florida or a palm tree in Alaska—at least, outside of a zoo or botanical garden. Yet we are not surprised to find white-tailed deer in both places, along with black bears, song sparrows, honeybees, dandelions, and daisies. It is obvious that organisms are not uniformly distributed throughout the Earth. But what is it that governs their distribution? Basically, living things are restricted to areas in which available habitats (local environments) and potential life styles fit their adaptations. The greater the physical differences among habitats, the greater the differences among the groups of creatures that inhabit them.

## MAJOR TERRESTRIAL BIOMES

A **biome** is a large, relatively distinct ecosystem that is characterized by particular climate, soil, plants, and animals, regardless of where it occurs on Earth (Figure 6–1). Examples of biomes include deserts, tropical rain forests, and tundra. A biome's boundaries are determined by climate more than any other factor. Because the northernmost biome, the tundra, is colder and has shorter growing seasons, for example, it has fewer kinds of vegetation than warmer biomes; few plants can tolerate its extreme conditions. Moving from the poles toward the equator, precipitation becomes a very important climatic factor, producing the temperate communities of forest, grassland, and desert, in decreas-

**Figure 6–1**
The world's major terrestrial life zones, or biomes, are distributed primarily in accordance with two factors, temperature and precipitation. In the higher latitudes temperature is the more important of the two. In temperate and tropical zones, precipitation is a significant determinant of community composition.

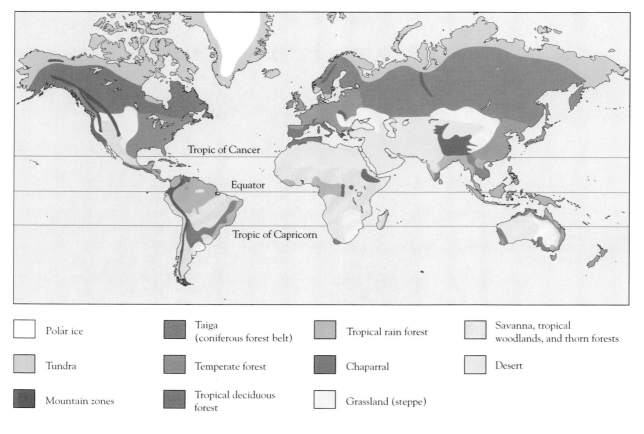

| | | | |
|---|---|---|---|
| ☐ Polar ice | ◼ Taiga (coniferous forest belt) | ◼ Tropical rain forest | ☐ Savanna, tropical woodlands, and thorn forests |
| ☐ Tundra | ◼ Temperate forest | ◼ Chaparral | ☐ Desert |
| ◼ Mountain zones | ◼ Tropical deciduous forest | ☐ Grassland (steppe) | |

ing order of precipitation (Figure 6–2). Thus, a characteristic biome develops in each major kind of climate.

Tropical and subtropical biomes, which occur in the lower latitudes near the equator, lack pronounced temperature differences throughout the year. They are at least as varied as temperate biomes, and like temperate biomes they are determined mainly by the amount and seasonality of precipitation they receive. Thus, there are not only tropical forests, but also tropical grasslands and tropical deserts. In the tropics the seasonal distribution of rainfall is especially important. Some tropical grasslands would be rain forests (in terms of the *amount* of precipitation they receive) except that almost all of their rainfall occurs during two months of the year. Lush rain forest vegetation could scarcely persist for ten months without water!

Altitude also affects ecosystems: changes in vegetation with increasing altitude resemble the changes in vegetation observed with movement from warmer to colder climates (see Focus On: The Distribution of Vegetation on Mountains).

### Tundra: Cold Boggy Plains of the Far North

The **tundra** occurs in the extreme northern latitudes wherever the snow melts seasonally (Figure 6–3). (The Southern Hemisphere has no equiva-

**Figure 6–3**
Alaskan tundra. The yellow flowers are marsh marigolds, which are common in moister sections of Alaskan tundra.
(Visuals Unlimited/Steve McCutcheon)

lent of the arctic tundra because it has no land in the corresponding latitudes.) Tundra is exposed to long, harsh winters and very short summers. Although its growing season, with warmer temperatures, is short (from 50 to 160 days depending on location), the days are long. In many places the sun

**Figure 6–2**
Average monthly temperature (black) and precipitation (blue) for three temperate biomes. Although temperature is approximately the same in all three locations, precipitation varies a great deal, resulting in deciduous forest where precipitation is plentiful, grassland where precipitation is less plentiful and more seasonal, and desert where precipitation is quite low.

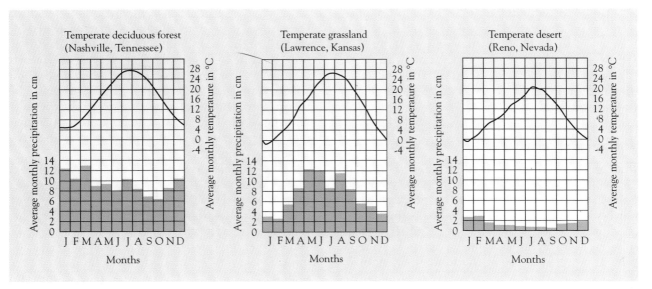

does not set at all for a considerable number of days in midsummer, although the amount of light at midnight is one-tenth that at noon. Over much of the tundra, little precipitation (10 to 25 cm [4 to 10 inches] per year) occurs, with most of it falling during the summer months.

Tundra soils tend to be geologically young, most of them having been formed since the last ice age. They are usually nutrient-poor and have little organic litter. Although the soil's surface melts during the summer, tundra has a layer of permanently frozen ground called **permafrost,** varying in depth and thickness, that interferes with drainage and prevents the roots of larger plants from becoming established. The limited precipitation in combination with low temperatures, flat topography (surface features), and the permafrost layer produces a landscape of broad shallow lakes, sluggish streams, and bogs.

The few species found in the tundra tend to exist in great numbers. Tundra is dominated by mosses, lichens (such as reindeer moss), grasses, and grasslike sedges. There are no readily recognizable trees or shrubs except in very sheltered localities, although dwarf willows and other dwarf trees are common—tundra plants seldom grow taller than 30 cm (12 inches).

The year-round animal life of the tundra includes weasels, arctic foxes, snowshoe hares, ptarmigan, snowy owls, and hawks. In the summer, large herbivores such as musk-oxen and caribou migrate north to the tundra to graze on sedges, grasses, and dwarf willow. There are no reptiles or amphibians. Insects such as mosquitoes, blackflies, and deerflies survive the winter as eggs or pupae and appear in great numbers during summer weeks.

Tundra regenerates very slowly after it has been disturbed. Even casual use by hikers can injure it. Damage that is likely to persist for hundreds of years has been done to large portions of the arctic tundra by oil exploration and military use (see the discussion on the Arctic National Wildlife Refuge in Chapter 10).

## Taiga: Evergreen Forests of the North

Just south of the tundra is the **taiga,** or **boreal forest,** which stretches across North America and Eurasia, covering approximately 11 percent of the Earth's land (Figure 6–4). (A biome comparable to the taiga is not found in the Southern Hemisphere.) Winters are extremely cold and severe, although not as harsh as in the tundra. The growing season of the boreal forest is somewhat longer than

**Figure 6–4**
Taiga, or boreal forest. (Carolina Biological Supply Company)

that of the tundra. Taiga receives little precipitation—perhaps 50 cm (20 inches) per year—and its soil is acidic, mineral-poor, and characterized by a deep layer of partly decomposed pine and spruce needles at the surface. Permafrost is either deep under the soil or absent. Taiga has numerous ponds and lakes—depressions in the Earth's surface created by the grinding ice sheets that covered this area during the last ice age.

Deciduous trees such as aspen and birch, which shed their leaves in autumn, may form striking stands in the taiga, but overall, spruce, fir, and other conifers (cone-bearing evergreens) clearly dominate. Conifers have many drought-resistant adaptations, such as needlelike leaves with minimal surface area for water loss, that enable them to withstand the "drought" of the northern winter months (plant roots cannot absorb water when the ground is frozen).

The animal life of the boreal forest consists of some larger species such as caribou (which migrate from the tundra to the taiga in winter), wolves, bears, and moose. However, most of the animal life is medium-sized to small, including rodents, rabbits, and fur-bearing predators such as lynx, sable, and mink. Most species of birds are abundant in the summer but migrate to warmer climates in the win-

## The Distribution of Vegetation on Mountains

Hiking up a mountain is similar to traveling toward the North Pole with respect to the major life zones encountered. This is because, as one climbs a mountain, the temperature drops, just as it does when one travels north, and the types of plants growing on the mountain change with the temperature.

The base of a mountain in Colorado, for example, might be covered by deciduous trees, which shed their leaves every autumn. Above that altitude, where the climate is colder and more severe, one might find a coniferous forest called subalpine forest, which resembles the northern taiga. Higher still, where the climate is very cold, a kind of tundra occurs, with vegetation composed of grasses, sedges, and small tufted plants; it is called alpine tundra to distinguish it from arctic tundra. At the very top of the mountain, a permanent ice or snow cap might be found, similar to the nearly lifeless polar land areas.

There are important environmental differences between high altitudes and high latitudes, however, that affect the types of organisms found in each place. Alpine tundra typically lacks permafrost and has more precipitation than arctic tundra. Also, high elevations of temperate mountains do not have the extremes in day length, associated with the changing seasons, that occur in high-latitude biomes. Furthermore, the intensity of solar radiation is greater at high elevations than at high latitudes. For example, at high elevations the sun's rays pass through less atmosphere, which results in a greater amount of ultraviolet (UV) radiation (less UV is filtered out by the atmosphere) than at high latitudes.

The cooler temperatures at higher elevations of a mountain produce a series of biomes similar to those encountered when going from the equator toward the North Pole.

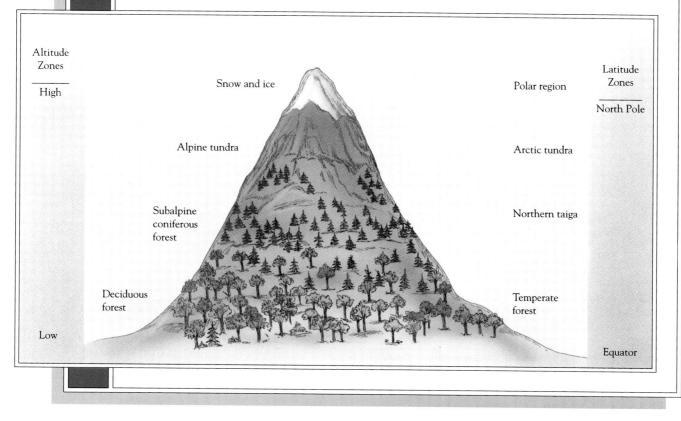

Altitude Zones

High

Low

Snow and ice

Alpine tundra

Subalpine coniferous forest

Deciduous forest

Latitude Zones

North Pole

Polar region

Arctic tundra

Northern taiga

Temperate forest

Equator

ter. Insects are abundant, but there are few amphibians and reptiles except in the southern taiga.

Most of the taiga is not well suited to agriculture because of its short growing season and mineral-poor soil. However, the boreal forest yields vast quantities of lumber and pulpwood (for making paper products), plus furs and other forest products.

## Temperate Forests: Various Kinds in Temperate Areas

In temperate latitudes, precipitation varies greatly with location (see Chapter 5). Continental interiors tend to be dry for a variety of reasons. Permanent high-pressure areas, such as those over the Sahara Desert, may nudge moist air masses aside. Air passing over a large land mass also may dry out without having the opportunity to be recharged with fresh moisture.

The climate of the North American continent, especially the West, is dominated by rain shadows cast by mountain ranges. As prevailing westerly winds push against the bases of the Cascade Range in the Pacific Northwest, masses of moist air from the Pacific Ocean are forced upward to higher altitudes, where they cool and precipitate much of their moisture. Thus, the western slopes of the mountains are so well watered that a temperate rain forest has developed. Considerable precipitation also falls in the upper reaches of the eastern slopes, but by the time the air has sunk back to lower altitudes, most of its available moisture has already been released.

**Temperate Rain Forest**  A coniferous **temperate rain forest** occurs on the northwest coast of North America; similar vegetation exists in southeastern Australia and in southern South America. Annual precipitation in this biome is high, from 200 to 380 cm (80 to 152 inches), and is augmented by condensation of water from dense coastal fogs. The proximity of temperate rain forest to the coastline moderates the temperature so that it has only a narrow seasonal fluctuation—winters are mild and summers are cool. Temperate rain forest has relatively nutrient-poor soil, although its organic content may be high.

The dominant vegetation in the North American temperate rain forest is large evergreen trees such as western hemlock, Douglas fir, Sitka spruce, and western arborvitae. Like tropical rain forest, temperate rain forest is rich in epiphytic vegetation—smaller plants that grow nonparasitically on large trees (Figure 6–5) (see Chapter 4). The epiphytes in temperate rain forest are mainly mosses,

**Figure 6–5**
Temperate rain forest in Washington. (Phil Degginger)

club mosses, lichens, and ferns. Squirrels, deer, and numerous bird species are among the animals found in the temperate rain forest.

Temperate rain forest is one of the richest wood producers in the world, supplying us with lumber and pulpwood. It is also one of the most complex ecosystems on Earth. Care must be taken to avoid overharvesting the original (never logged) old-growth forest, however, because such an ecosystem takes hundreds of years to develop. The logging industry typically harvests old-growth forest and replants the area with trees of a single species that will be harvested in 100 years or less. Thus, the old-growth forest ecosystem never has a chance to redevelop.

**Temperate Deciduous Forest**  Where temperate zone precipitation ranges from about 75 to 125 cm (30 to 50 inches) annually, **temperate deciduous forests** occur. Hot summers and pronounced winters are characteristic of these forests. Typically, the soil of a temperate deciduous forest consists of a topsoil, which is rich in organic material, and a deep, clay-rich lower layer. As organic materials

**Amazon of the North**

When we think of rampant forest destruction, the Amazon region of Brazil comes to mind. But the abundant northern forests of the world are also being cut rapidly. Consider the following:

Size of Canada's boreal forest region: 1.5 million square miles.
Size of Brazil's Amazon rain forest: 1.59 million square miles.

Amount of Canadian forests cleared annually: 5,500 square miles.
Amount of Brazilian Amazon cleared or burned each year: 15,900 square miles.

Portion of total productive Canadian forest now barren or insufficiently restocked with timber-producing species: 10.3%
Portion of Brazil's Amazon region that has disappeared: 12%

Portion of Canada's boreal forest protected from development and resource extraction: 2.6%
Portion of the Brazilian Amazon forest protected: 9.4%

(a)

(b)

**Figure 6–6**
Temperate deciduous forests exhibit dramatic seasonal changes. (a) The branches are clothed in a profusion of greenery in summer. (b) The onset of autumn turns the leaves into a variety of golds, oranges, reds, and browns. (Dennis Drenner)

decay, mineral ions are released. If the ions are not immediately absorbed by the roots of the living trees, they leach into the clay, where they may be retained.

The temperate deciduous forests of the northeastern and mideastern United States are dominated by broad-leaved hardwood trees, such as oak, hickory, and beech, that lose their foliage annually (Figure 6–6). In the southern reaches of the temperate deciduous forest, the number of broad-leaved evergreen trees, such as magnolia, increases.

Temperate deciduous forest originally contained a variety of large mammals including puma, wolves, deer, bison, bears, and other species now extinct, plus many small mammals and birds. Both reptiles and amphibians abounded, together with a denser and more varied insect life than exists today. Much of the original temperate deciduous forest was removed by logging and land clearing. Where it has been allowed to regenerate, temperate decid-

uous forest is now commonly in a seminatural state—
that is, highly modified by humans.

Worldwide, deciduous forests were among the
first biomes to be converted to agricultural use. In
Europe and Asia, for example, many soils that originally supported deciduous forests have been cultivated by traditional agricultural methods for thousands of years without a substantial loss in fertility.
However, American farmers of the 18th and 19th
centuries perceived the vast land as a limitless resource. They typically abandoned the wise soil conservation practices of their ancestors, allowing erosion and other forms of soil depletion to damage
their land.

## Grasslands: Temperate Seas of Grass

Summers are hot, winters are cold, and rainfall is
often uncertain in **temperate grasslands.** Annual
precipitation averages 25 to 75 cm (10 to 30
inches). In grasslands with less precipitation, there
is a greater tendency for minerals to accumulate in
a well-defined layer just below the topsoil instead of
washing out of the soil. Grassland soil has considerable organic material because the aboveground portions of many grasses die off each winter and contribute to the organic content of the soil, while the
roots and rhizomes (underground stems) survive
underground. Also, many grasses are sod formers;
that is, their roots and rhizomes form a thick, continuous underground mat.

The North American Midwest is an excellent
example of a temperate grassland. Here there are
few trees except for those that grow near rivers and
streams, and grasses grow in great profusion in the
thick, rich soil. Formerly, certain species of grasses
grew as tall as a person on horseback, and the land
was covered with herds of grazing animals, particularly bison. The principal predators were wolves,
although in more sparsely vegetated, drier areas
their place was taken by coyotes. Smaller animals
included prairie dogs and their predators (foxes,
black-footed ferrets, and various birds of prey),
grouse, reptiles (such as snakes and lizards), and
great numbers of insects.

**Steppes,** or shortgrass prairies (Figure 6–7),
such as those in South Dakota, are temperate grassland habitats with less precipitation than the
moister grasslands just described but with greater
precipitation than deserts. They have less grass
than the moister grasslands, and occasionally some
bare soil is exposed. Native grasses of the steppes
are drought-resistant.

The development in postcolonial times of the
steel plow and later the tractor spelled the doom of

**Figure 6–7**
A grassland of the North American plains. (© David Muench
1992)

the original North American grassland. This biome
was so well suited to agriculture that little of it was
spared. Almost nowhere can we see even an approximation of what our ancestors saw as they settled the Midwest. It is not surprising that the
American Midwest, the Ukraine, and other moist
temperate grasslands became the breadbaskets of
the world. These habitats provide ideal growing
conditions for grass crops such as corn and wheat.

## Chaparral: Thickets of Evergreen Shrubs and Small Trees

Some temperate habitats have mild winters with
abundant rainfall combined with very dry summers.
Such **Mediterranean climates,** as they are called,
occur not only in the area around the Mediterranean Sea but also in California, western Australia,
portions of Chile, and South Africa. In the North
American Southwest this Mediterranean-type
community is known as **chaparral.** Chaparral soil is
thin and not very fertile. Fires occur naturally and
frequently in this habitat, particularly in late summer and in autumn.

Chaparral vegetation looks strikingly similar
around the world, even though the individual species of different areas are quite distinct. Chaparral
is usually dominated by a dense growth of evergreen
shrubs but may contain drought-resistant pine or
scrub oak trees (Figure 6–8). During the rainy season the habitat may be lush and green, but during
the hot, dry summer the plants lie dormant. Chaparral trees and shrubs often have **sclerophyllous**

**Figure 6–8**
Chaparral in the Santa Monica Mountains, California. Drought-resistant evergreen shrubs are the main vegetation in the chaparral biome. Chaparral develops where hot, dry summers alternate with mild, rainy winters. (Visuals Unlimited/ John Cunningham)

**leaves**—hard, small, leathery leaves that resist water loss. Many plants are also specifically fire-adapted and actually grow best in the months following a fire. When the aboveground parts of these plants burn, their minerals are released. The underground parts are not destroyed by fire, and during the winter rains, with the new availability of essential minerals, the plants sprout vigorously. Mule deer, wood rats, chipmunks, lizards, and many species of birds are the common animals of the chaparral.

The fires that occur at irregular intervals in California chaparral are often quite costly to humans because they consume expensive homes built on the hilly landscape. Unfortunately, efforts to control the naturally occurring fires sometimes create new problems. When periodic fires are prevented, denser, thicker vegetation tends to accumulate; then, when a fire does occur, it is much more severe.

### Deserts: Arid Life Zones

**Deserts** are very dry areas found in both temperate and tropical regions. The low water content of the desert atmosphere leads to a wide daily temperature range. Deserts vary greatly depending upon the amount of precipitation they receive, which is generally less than 25 cm (10 inches) per year. A few deserts are so dry that virtually no plant life occurs in them, as is the case in portions of the African Namib Desert and the Atacama-Sechura Desert of Chile and Peru. As a result, desert soil is low in organic material but typically has a high mineral content. In some regions, the concentration of certain soil minerals reaches toxic levels.

Plant cover is sparse in deserts, so much of the soil is exposed. Both perennials (plants that live for more than two years) and annuals (plants that complete their life cycles in one growing season) are present. Plants in North American deserts include cacti, yuccas, Joshua trees, and widely scattered bunchgrass (Figure 6–9a). Perennial desert plants tend to have reduced leaves or no leaves, an adaptation that enables them to conserve water. Other desert plants shed their leaves for most of the year, growing only in the brief moist season. Desert plants are noted for **allelopathy,** an adaptation in which toxic substances secreted by roots or shed leaves inhibit the establishment of competing plants nearby. Many desert plants are provided with spines, thorns, or toxins to defend them against the heavy grazing pressure they experience in this food- and water-deficient environment.

Desert animals tend to be small. During the heat of the day, they remain under cover or return to shelter periodically; at night they come out to forage or hunt. In addition to desert-adapted insects, there are many specialized desert reptiles— lizards, tortoises, and snakes. Desert mammals include rodents such as the American kangaroo rat, which does not have to drink water but can subsist solely on the water content of its food plus metabolically generated water. In American deserts

(a)

(b)

**Figure 6–9**
Inhabitants of deserts are strikingly adapted to the demands of their environment. (a) Saguaro cacti in the Saguaro National Monument, Arizona. The warmer deserts of North America that are characterized by summer rainfall frequently contain large cacti such as the giant saguaro. Photosynthesis is carried out in the stem, which also serves to store water. Leaves are modified into spines, which discourage herbivores from eating cactus tissue. (b) An antelope jackrabbit in the Saguaro National Monument, Arizona. The jackrabbit got its name because its huge ears resemble those of a jackass. Jackrabbits never drink water, but obtain the liquid they need from the plants on which they feed. (a, David Muench; b, Stan Osolinski/ Dembinsky Photo Associates)

there are also jackrabbits (Figure 6–9b), and in Australian deserts, ecologically equivalent kangaroos. Carnivores such as the African fennec (a fox) and some birds of prey, especially owls, live on the rodents and rabbits.

## Savanna: Tropical Grasslands

The **savanna** biome is a tropical grassland with widely scattered trees (Figure 6–10). It is found in areas of low rainfall or seasonal rainfall with prolonged dry periods. The temperatures in tropical savannas vary little throughout the year; thus, seasons are regulated by precipitation rather than by temperature as in temperate grasslands. Annual precipitation is 85 to 150 cm (34 to 60 inches).

Savanna soil is low in essential mineral nutrients, in part because the parent rock from which it is formed is infertile. Although the African savanna is best known, there are also tracts of savanna in South America and northern Australia.

Savanna is characterized by wide stretches of grasses interrupted by occasional trees such as *Acacia*, which bristles with thorns that protect it against herbivores. Both trees and grasses have fire-adapted features, such as extensive underground root systems, that enable them to survive the periodic fires that sweep through savanna.

The greatest assemblage of hoofed mammals in the modern world occurs in the African savanna in the form of great herds of herbivores—wildebeest, antelope, giraffe, zebra, and the like. Large preda-

**Figure 6–10**
The savanna biome of eastern Africa. These grasslands with scattered acacia trees formerly supported large herds of grazing animals and their predators, both of which are swiftly vanishing under pressure from pastoral and agricultural land use. (Visuals Unlimited/Steve McCutcheon)

tors, such as lions and hyenas, kill and scavenge the herds. In areas of seasonally varying rainfall, the herds and their predators migrate annually.

Tropical grasslands are rapidly being converted to rangeland for cattle and other animals that are replacing the big herds of game animals. In places, severe overgrazing has converted marginal savanna to desert (see Chapter 17).

## Tropical Rain Forests: Lush Equatorial Forests

**Tropical rain forests** occur where temperatures are high throughout the year and precipitation occurs almost daily. The annual precipitation of tropical rain forest is 200 to 450 cm (80 to 180 inches). Much of this precipitation comes from locally recycled water that enters the atmosphere by transpiration (loss of water vapor from plants) of the forest's own trees.

Tropical rain forests commonly occur in areas with ancient, mineral-poor soil that has been extensively leached by high precipitation, producing a soil poor in both nutrients and organic matter. Because the temperature is high year-round, decay organisms and detritus-feeding ants and termites decompose organic litter quite rapidly. Nutrients from the decomposing material are quickly absorbed by plant roots. Thus, the mineral nutrients of tropical rain forests are tied up in the vegetation, not the soil.

**Envirobrief**

### Medicines from Living Things

Biological diversity is one of the Earth's most precious resources. We depend on complex ecosystems containing a wide variety of species to clean and retain water, recycle carbon dioxide, and harbor thousands of insect, animal, and plant species, many of which have medicinal uses. In fact, over 40% of the prescription medicines sold in the U.S. today are derived from wild plant species. Two powerful anti-cancer agents—vincristine and vinblastine—come from one small, delicate plant known as the rosy periwinkle that grows wild on the island of Madagascar off the coast of Africa. Taxol, another cancer-fighting agent, is extracted from the yew tree, found in forests of the U.S. Pacific Northwest. Endod, a drug derived from an Ethiopian plant, may help control the spread of schistosomiasis, a debilitating disease that affects more than 300 million people in tropical countries. It is likely that thousands of other naturally occurring pharmaceuticals lie untapped in the world's forests, yet less than 1% of all plant species have been screened for medicinal use.

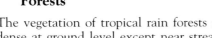

## The Plants in Tropical Rain Forests

The vegetation of tropical rain forests is not dense at ground level except near stream banks or where a fallen tree has opened the canopy. The continuous canopy of leaves overhead produces a dark habitat with an extremely moist microclimate. A fully developed rain forest has at least three distinct stories of vegetation (see part a of figure). The topmost story consists of the crowns of occasional very tall trees, some 80 m (260 ft) or more in height. This story is entirely exposed to direct sunlight. The middle story, which reaches an average height of 50 m (160 ft), forms a continuous canopy of leaves that lets in very little sunlight for the support of the sparse understory, which consists of both the seedlings of taller trees and smaller plants that are specialized for life in the shade.

The trees in a tropical rain forest also support extensive communities of epiphytic plants such as orchids and bromeliads. Although epiphytes grow in crotches of branches, on bark, or even on the leaves of

Tropical rain forest vegetation. (a) Profile of the layers, or stories, of trees in a tropical rain forest. There are at least three stories, with the middle one being so continuous that very little light penetrates to the forest floor. Although shrubs and herbaceous plants are uncommon on the rain forest floor, young trees are found there. Vines and epiphytes are not shown. (David A. Steingraeber)

(a)

Despite the scarcity of mineral nutrients in its soil, tropical rain forest is very productive—that is, its plants capture a lot of energy by photosynthesis—owing to the significant solar energy input and precipitation.

Of all the life zones on land, the tropical rain forest is unexcelled in species diversity. No one species dominates this biome; one could travel for 0.4 km (0.25 mi) without encountering two members of the same species of tree.

The trees of tropical rain forests are usually evergreen flowering plants. Their roots are often shallow and form a mat almost 1 m (about 3 ft) thick on the surface of the soil; the mat catches and absorbs almost all mineral nutrients released from leaves and litter by decay processes. Braces called their hosts, they use their host trees only for physical support, not for nourishment. Many epiphytes have special adaptations to allow them to obtain water and minerals high above the forest floor where they spend their lives. The flowerpot plant, for example, has specialized leaves that form a "flowerpot" (see figure, parts b and c). The pot collects rainwater and various minerals that leach out of the leaves above. A special root runs into the flowerpot leaves to absorb water and dissolved minerals.

Because little light penetrates to the understory, many of the plants living there are adapted to climb upon already established host trees rather than invest their meager photosynthetic resources in the dead cellulose tissues of their own trunks. Tropical vines, some as thick as a human thigh, twist up through the branches of the rain forest trees.

(b)

(c)

(b, c) The flowerpot plant, *Dischidia rafflesiana*. (b) The leaves of the flowerpot plant are modified to collect rainwater. (c) A cutaway view of the "pot" reveals a special root produced by the plant to utilize the collected rainwater. (David A. Steingraeber)

buttresses or swollen bases hold the trees upright and aid in the extensive distribution of the shallow roots (see Focus On: The Plants in Tropical Rain Forests).

Rain forest animals include the most abundant and varied insect, reptile, and amphibian fauna on Earth (Figure 6–11). The birds, often brilliantly colored, are also varied. Most rain forest mammals, such as sloths and monkeys, live only in the trees, although some large ground-dwelling mammals, including elephants, are found in rain forests.

Human population growth and industrialization in tropical countries may spell the end of most or all tropical rain forests by the end of this century. Scientists think many rain forest organisms will become extinct before they have even been discov-

**Figure 6–11**
Some animals of the South American tropical rain forest: (1) howler monkey, (2) scarlet
macaw, (3) three-toed sloth, (4) red spider monkey and baby, (5) silky anteater and baby,
(6) porcupine, (7) emerald tree boa, (8) tapirs, (9) tawny ocelot, and (10) green leaf frog.
Except for the tapirs and ocelot, these animals spend most of their lives in the trees. (Robert
Hynes, artist, © National Geographic Society)

ered and scientifically described. The ecological impacts of tropical rain forest destruction are discussed extensively throughout this text.

## AQUATIC LIFE ZONES

As you might expect, aquatic life zones are different in almost all respects from terrestrial life zones. For example, recall that in terrestrial biomes, temperature and precipitation are the major determinants of plant and animal inhabitants, and light is relatively plentiful (except in certain habitats such as the floor of the rain forest). Temperature is less important in aquatic life zones than it is on land, and water itself is obviously not an important limiting factor.

The most fundamental division in aquatic ecology is probably between freshwater and saltwater habitats. **Salinity** (the concentration of dissolved salts, such as sodium chloride, in a body of water) affects the kinds of organisms present in aquatic ecosystems, as does the amount of dissolved oxygen. Water greatly interferes with the penetration of light, so floating aquatic organisms that photosynthesize must remain near the water's surface, and vegetation attached to the bottom can grow only in the shallowest water. In addition, low levels of essential mineral nutrients limit the number and distribution of living things in certain aquatic environments.

Aquatic habitats contain three main ecological categories of organisms: free-floating plankton, strongly swimming nekton, and bottom-dwelling benthos. **Plankton** are small or microscopic organisms that are relatively feeble swimmers and thus, for the most part, are carried about at the mercy of currents and waves (Figure 6–12). They are unable to swim far horizontally, but some species are capable of large daily vertical migrations and are found at different depths of water at different times of the day or sometimes at different seasons. Plankton are generally subdivided into two major categories. **Phytoplankton** — photosynthetic cyanobacteria and free-floating algae of several types—are producers and form the bases of most aquatic food chains. **Zooplankton** are nonphotosynthetic organisms that include protozoa (animal-like protists) and small animals, including the larval stages of many animals that are large as adults. **Nekton** are larger, more strongly swimming organisms such as fish, whales, and turtles. **Benthos** are bottom-dwelling creatures that fix themselves to one spot (such as oysters and barnacles), burrow into the

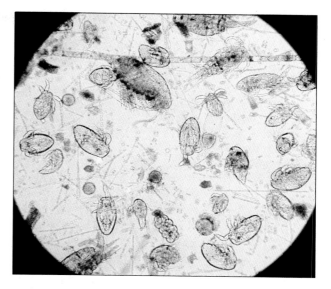

**Figure 6–12**

A variety of marine plankton. Most of the organisms shown here are nauplius larvae, tiny immature crustaceans (animals such as lobsters, shrimp, and crabs) that hatch from eggs. Nauplius larvae eventually mature into several types of small crustaceans, including fairy shrimp, tadpole shrimp, and copepods, all of which graze on diatoms in the plankton and are in turn the food for larger aquatic organisms. (Runk/Schoenberger, from Grant Heilman)

sand (such as many worms and echinoderms), or simply walk about on the bottom (such as lobsters and brittle stars).

### Freshwater Ecosystems

Freshwater ecosystems include rivers and streams (flowing water), lakes and ponds (standing water), and freshwater wetlands. Freshwater wetlands, lands that are transitional between freshwater and terrestrial ecosystems, are usually covered by shallow water and have characteristic soils and vegetation. They include marshes, in which grass-like plants dominate, and swamps, in which woody plants (trees or shrubs) dominate. Freshwater wetlands are discussed in Chapter 17.

**Rivers and Streams**   The nature of a flowing-water ecosystem changes greatly between its source (where it begins) and its mouth (where it empties into another body of water). For example, headwater streams (small streams that are the sources of a river) are usually shallow, swiftly flowing, highly oxygenated, and cold. In contrast, downstream

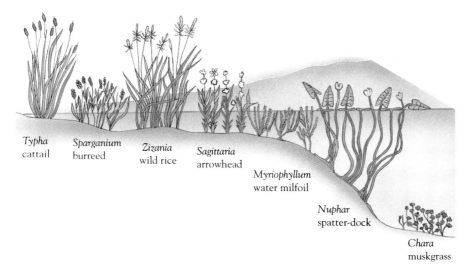

**Figure 6–13**
The littoral zone of lakes and large ponds. As the depth of water increases, the vegetation changes. Plants such as cattails and burreeds grow in the shallower water, whereas muskgrass (a type of alga) and spatter-dock grow in the deeper water.

*Typha* cattail

*Sparganium* burreed

*Zizania* wild rice

*Sagittaria* arrowhead

*Myriophyllum* water milfoil

*Nuphar* spatter-dock

*Chara* muskgrass

from its headwaters a river is wider and deeper, slower-flowing, less oxygenated, and not as cold.

The kinds of organisms found in flowing-water ecosystems vary greatly from one stream to another, depending primarily on the strength of the current. In streams with fast currents, the inhabitants may have adaptations such as suckers to attach themselves to rocks so that they are not swept away, or they may have flattened bodies to enable them to slip under or between rocks. Organisms in large, slow-moving streams and rivers do not need such adaptations, although they are typically streamlined (like most aquatic organisms) to lessen resistance when moving through water. Where the current is slow, plants and animals of the headwaters are replaced by those characteristic of ponds and lakes.

In addition to having currents, flowing-water ecosystems differ from other freshwater ecosystems in their dependence on the land for much of their energy. In headwater streams, for example, up to 99 percent of the energy input comes from detritus, dead organic material (such as leaves) that is carried from the land into the stream by wind or surface drainage after precipitation. Downstream, rivers have more producers and therefore a slightly lower dependence on detritus as a source of energy.

Human activities have several adverse impacts on rivers and streams, including water pollution (see Chapter 21) and the effects of dams (see Chapters 12 and 13), which are built to contain the water of rivers or streams. Both pollution and dams change the nature of flowing-water ecosystems downstream.

**Figure 6–14**
Temperature differences (black line) in a temperate lake during the summer. There is an abrupt transition, called the thermocline, between the upper warm layer and the bottom cold layer.

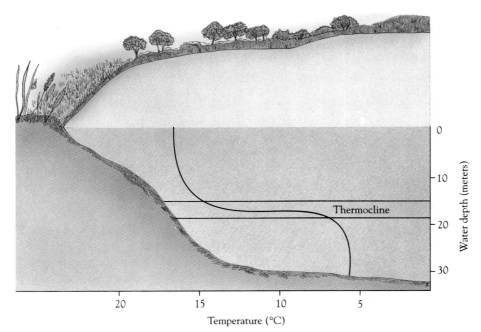

Thermocline

Water depth (meters)

0

10

20

30

20    15    10    5

Temperature (°C)

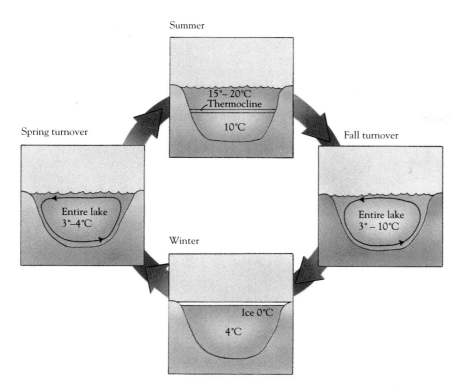

Summer

15°– 20°C
Thermocline
10°C

Spring turnover

Entire lake
3°–4°C

Fall turnover

Entire lake
3° – 10°C

Winter

Ice 0°C
4°C

**Figure 6–15**
Fall and spring turnover, or mixing of upper and lower layers of water in a temperate lake, brings oxygen to the oxygen-depleted depths of the lake and minerals to the mineral-deficient surface waters.

**Lakes and Ponds**  A large lake has three basic life zones: littoral, limnetic, and profundal; smaller lakes and ponds typically lack a profundal zone. The **littoral zone** is the shallow-water area along the shore of a lake or pond. It includes lakeshore vegetation such as cattails and bur-reeds, plus several deeper-dwelling aquatic plants and algae (Figure 6–13). The littoral zone is the most productive zone of the lake (that is, photosynthesis is greatest here), in part because it receives nutrients, which stimulate the growth of plants and algae, from the surrounding land. Animals of the littoral zone include frogs and their tadpoles, turtles, worms, crayfish and other crustaceans, insect larvae, and many fish, such as perch, carp, and bass. Here, too, at least in the quieter areas, one finds surface dwellers such as water striders and whirligig beetles.

The **limnetic zone** is the open-water area away from the shore; it extends down as far as sunlight penetrates. The main organisms of the limnetic zone are microscopic phytoplankton and zooplankton. Larger fish also spend most of their time in the limnetic zone, although they may visit the littoral zone to feed and breed. Owing to the depth of this zone, less vegetation grows here.

The deepest zone of a large lake, the **profundal zone,** is below the limnetic zone. Because of the lack of light, producers do not live in the profundal zone. Much food drifts into the profundal zone from the littoral and limnetic zones. When dead plants and animals reach the profundal zone, decay bacteria decompose them, liberating the minerals contained in their bodies. These minerals are not effectively recycled, because there are no producers to absorb them and incorporate them into the food chain. As a result, the profundal habitat tends to be both mineral-rich and anaerobic (without oxygen), and hence occupied by few forms of higher life.

**Thermal Stratification and Turnover in Temperate Lakes**  The marked layering of lakes caused by how far light penetrates is accentuated by **thermal stratification,** which is characteristic of large lakes in temperate areas. Thermal stratification occurs because the summer sunlight penetrates and warms surface waters, making them less dense.[1] In the summer, cool (and therefore denser) water remains at the lake bottom, separated from the warm (and therefore less dense) water above by an abrupt temperature transition called the **thermocline** (Figure 6–14).

In temperate lakes, falling temperatures in autumn cause a mixing of the lake waters called the **fall turnover** (Figure 6–15). (Because there is little seasonal temperature variation in the tropics, turnovers are not common there.) Fall turnover occurs because, as the surface water cools, its density increases and it displaces the less dense, warmer, mineral-rich water beneath. The warmer water then rises to the surface, where it cools and sinks. This

[1]Recall that the density of water is greatest at 4°C; both above and below this temperature, water is less dense.

process of cooling and sinking continues until the lake reaches a uniform temperature throughout.

When winter comes, the surface water cools below 4°C, and if it is cold enough (0°C), ice forms. Ice is less dense than cold water and thus forms on the surface, and so the water on the lake bottom is warmer than on the surface.

In the spring, a **spring turnover** occurs as ice melts and the surface water reaches 4°C, its temperature of greatest density. Surface water again sinks to the bottom and bottom water returns to the surface. As summer arrives, thermal stratification occurs once again.

The mixing of deeper, nutrient-rich water with nutrient-poor surface water during fall and spring turnovers brings essential minerals to the surface. The sudden presence of large amounts of essential minerals in surface waters encourages the development of large algal populations, which form temporary **blooms** (population explosions) in the fall and spring.

## Estuaries: Where Fresh Water and Salt Water Meet

Where the sea meets the land, there may be one of several kinds of ecosystems: a rocky shore, a sandy beach, an intertidal mud flat, or a tidal estuary. An **estuary** is a coastal body of water, partly surrounded by land, with access to the open sea and a large supply of fresh water from rivers. Estuaries usually contain **salt marshes,** areas dominated by grasses (Figure 6–16) (see Chapter 3), and their salinity fluctuates between that of seawater and that of fresh water. During the course of a year many estuaries undergo significant variations in temperature, salinity, depth of light penetration, and other physical properties. To survive there, estuarine organisms must have tolerance for this wide range of conditions.

The waters of estuaries are among the most fertile in the world, often having much greater productivity than either the adjacent sea or the fresh water upriver. This high productivity is brought about by (1) the action of the ocean's tides, which promote rapid circulation of nutrients and help remove waste products; (2) the transport of nutrients from the land into rivers and creeks that empty into the estuary; and (3) the presence of many plants, which provide an extensive photosynthetic carpet and whose roots and stems also mechanically trap much potential food material. As leaves and plants die, they decay, forming the bases of detritus food chains. Most commercially important fin fish and shellfish spend their larval stages in estuaries among the protective tangle of decaying stems.

Salt marshes have often appeared to uninformed people to be worthless, empty stretches of land. As a result, they have been used as dumps and become severely polluted or have been filled with dredged bottom material to form artificial land for residential and industrial development. A large part of the estuarine environment has been lost in this way, along with many of its benefits: wildlife habitat, sediment trapping, flood control, and groundwater supply (see Chapter 17).

## Marine Life Zones

Although freshwater and marine (ocean) life zones are comparable in many ways, there are also many dramatic differences. The depths of even the deepest lakes, for example, do not approach those of the oceanic abysses (Figure 6–17), which extend more than 6 km (3.6 mi) below the sunlit surface. Oceans are profoundly influenced by tides and currents. The gravitational pulls of both sun and moon produce two tides a day throughout the oceans, but the heights of those tides vary with the phases of the moon (a full moon causes the highest tides), season, and local topography.

The area of shoreline between low and high tides is called the **intertidal zone.** Although the high levels of light and nutrients, together with an abundance of oxygen, make the intertidal zone a biologically productive habitat, it is also a very stressful one. If an intertidal beach is sandy, the inhabitants must contend with a constantly shifting environment that threatens to engulf them and gives them scant protection against wave action (Figure 6–18). Consequently, most sand-dwelling

**Figure 6–16**
A salt marsh at Assateague Island National Seashore in Maryland. The plants are cordgrass (*Spartina*). (Connie Toops)

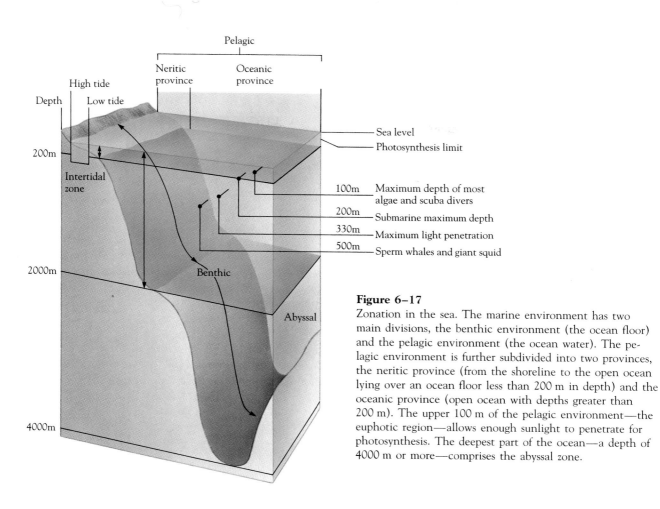

**Figure 6–17**
Zonation in the sea. The marine environment has two main divisions, the benthic environment (the ocean floor) and the pelagic environment (the ocean water). The pelagic environment is further subdivided into two provinces, the neritic province (from the shoreline to the open ocean lying over an ocean floor less than 200 m in depth) and the oceanic province (open ocean with depths greater than 200 m). The upper 100 m of the pelagic environment—the euphotic region—allows enough sunlight to penetrate for photosynthesis. The deepest part of the ocean—a depth of 4000 m or more—comprises the abyssal zone.

**Figure 6–18**
Selected organisms from the life zones of a sandy beach. I. Supratidal zone (above the high-tide zone): ghost crabs and tiger beetles. II. Flat beach zone: bristle worms and clams (not shown). III. Intertidal zone: lugworms, mole crabs, and sand crabs (which follow the retreating or advancing waters; not shown). IV. Subtidal zone (always under water): heart clams, olives, and blue crabs.

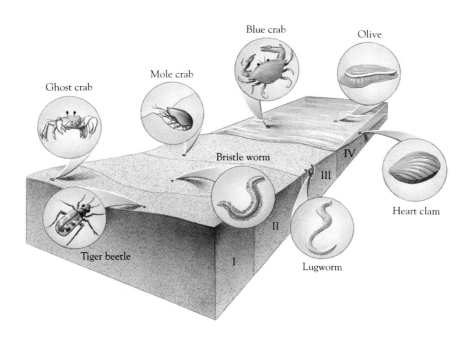

organisms, such as mole crabs, are continuous and active burrowers. Because they are able to follow the tides up and down the beach, they usually do not have any notable adaptations to survive desiccation (drying out) or exposure.

A rocky shore provides a fine anchorage for seaweeds and animals, but is exposed to wave action when immersed during high tides and to drying and temperature changes when exposed to the air during low tides (Figure 6–19). A typical rocky shore inhabitant has some way of sealing in moisture, perhaps by closing its shell if it has one, plus a powerful means of anchorage to the rocks—mussels, for example, have horny, threadlike anchors, and barnacles have a special cement gland. Rocky shore intertidal algae (seaweeds) usually have thick, gummy coats, which dry out slowly when exposed to air, and flexible bodies not easily broken by wave action. Some rocky shore community inhabitants hide in burrows or crevices at low tide, and some small semiterrestrial crabs run about the splash line, following it up and down the beach.

The marine environment has two main divisions, the benthic environment (the ocean floor) and the pelagic environment (the ocean water). The upper reaches of the pelagic environment comprise the **euphotic** region, which extends from the surface to a depth of approximately 100 m (325 ft). Light penetrates the euphotic zone in sufficient amounts to support photosynthesis. The pelagic environment is divided into two provinces, the neritic province and the oceanic province.

The **neritic province** is open ocean from the shoreline to a depth of 200 m (650 ft). Nekton (such as sharks, tunas, and porpoises) and larger benthic organisms (such as corals, spiny lobsters, and starfish) are mostly confined to the shallower neritic waters (less than 60 m, or 195 ft, deep) because that is where their food is. Not only are there seaweeds on the bottoms of shallower areas, but there are also large numbers of phytoplankton in the water itself.

The **oceanic province** is the part of the open ocean that is deeper than 200 m (650 ft). This is most of the ocean; in fact, about 88 percent of the ocean is more than 1.5 km (0.9 mi) deep. Because light cannot penetrate to such depths, the oceanic

**Figure 6–19**
Adaptations to life in a rocky intertidal zone. (a) The sea palm (*Postelsia*) is a sturdy seaweed that is common on the rocky Pacific coast from Vancouver Island to California. Its base is firmly attached to the rocky substrate, enabling it to withstand the pounding wave action. (b) Limpets (*Acmaea digitalis*) adhere to rock in the intertidal zone of Point Lobos State Park, California. Limpets are mollusks whose shells help them resist the pounding waves of their habitat. (a and b, William E. Ferguson)

(b)

(a)

## The Importance of Coral

Coral formations are important ecosystems, as rich in species as a tropical rain forest. Most reefs are between 5,000 and 10,000 years old; some have existed for several million years. A single reef can contain more than 3,000 species of corals, fish, and shellfish. Nearly one-third of all the world's fish live on coral reefs and many more depend on them at some stage in their life cycle. Reefs are abundant fishing grounds, critical to the fishing industries of countries such as the Philippines and Indonesia. They also reduce the energy of waves, thereby protecting shorelines against storms. The destruction of one reef off the coast of Sri Lanka pushed the shoreline back some 350 yards. Yet the world's coral reefs are being degraded and destroyed. Of 109 countries with large reef formations, 90 are damaging them. Silt washing downstream from clearcut inland forests has been smothering the world's reefs. Pollution, land reclamation, tourism, and the mining of corals for building material are also taking a heavy toll. Regeneration cannot keep pace: A new coral colony requires 20 years to grow to the size of a human head.

**Figure 6–20**

A female anglerfish that was discovered at a depth greater than 300 m (1,000 ft). Like many deepsea fish, this one has weak or vestigial eyes and a luminous lure that may help it to locate a mate or attract prey. (© Norbert Wu, 1991)

province supports few organisms. Most of the life that exists under the tremendous pressure and darkness of the abysses depends upon whatever food drifts down into its habitat from the upper, lighted regions. The principal exceptions are found at the deep-sea thermal vents (see Chapter 3, Focus On: Life Without the Sun).

Animals of the abysses are strikingly adapted to darkness and scarcity of food (Figure 6–20). Abyssal fish, for example, have huge jaws that enable them to swallow large food particles they might encounter. (If an organism does not chance upon food very often, it needs to eat as much as possible when food is present.) Many abyssal animals have illuminated organs, enabling them to see one another for mating or food capture. A great many are predators or scavengers (there is no other choice) and live in dispersed populations.

## INTERACTION OF LIFE ZONES

Although we have discussed terrestrial and aquatic life zones as discrete entities, none of them exists in isolation. When parts of the Amazon rain forest flood annually, for example, fish leave the stream beds and swim all over the forest floor, where they play a role in dispersing the seeds of many species of plants. And in the Antarctic, whose waters are much more productive than its land areas, there is hardly any terrestrial community of organisms (there is no "polar biome"), but the many seabirds and seals form a link between the two environments. Although these animals are supported exclusively by the ocean, their waste products, cast-off feathers, and the like, when deposited on land, support whatever lichens and insects occur there.

Some inhabitants of terrestrial and aquatic life zones cover great distances—in the case of migratory fish and birds, even global distances. For example, many young albacore tuna migrate from the California coast across the Pacific Ocean to Japan! Flycatchers spend their summers in Canada and the United States and their winters in Central and South America. Like the flycatchers, many other migratory birds commonly spend critical parts of their life cycles in entirely different countries, which can make their conservation difficult (see Chapter 16, Focus On: Vanishing Tropical Forests). It does little good, for instance, to protect a songbird in one country if the inhabitants of the next put it in the cooking pot as soon as it lands. Such large-scale interaction makes ecological concepts difficult for many people to grasp and apply.

# SUMMARY

1. A biome is a large, relatively distinct ecosystem that is characterized by certain types of climate, soil, plants, and animals, regardless of where it occurs on Earth. A biome's boundaries are determined by climate more than any other factor.

2. Tundra is the northernmost biome. It is characterized by a permanently frozen layer of subsoil called the permafrost, and low-growing vegetation adapted to extreme cold and a very short growing season. The taiga, or boreal forest, lies south of the tundra and is dominated by coniferous trees.

3. Temperate forests occur where precipitation is relatively high. Temperate deciduous forests are dominated by broad-leaved trees that for the most part lose all their leaves seasonally. Temperate rain forest occurs on the northwest coast of North America and is dominated by conifers.

4. Temperate grasslands typically possess deep, mineral-rich soil, have moderate precipitation, and are well suited to growing grain crops. Many tropical grasslands, called savannas, are similar to open woodland, with scattered trees interspersed with grassy areas. The chaparral biome is characterized by thickets of small-leaved shrubs and trees and a climate of wet, mild winters and very dry summers.

5. Deserts, found where there is little precipitation, are communities whose organisms have specialized water-conserving adaptations. Deserts occur in both temperate and tropical areas.

6. Tropical rain forests are characterized by very high rainfall that is evenly distributed through the year. Rain forests have high species diversity, with at least three stories of forest foliage, many epiphytes, and mineral-poor soil.

7. In aquatic life zones, important environmental factors include salinity, amount of dissolved oxygen, and availability of light for photosynthesis. Aquatic life is ecologically divided into plankton (free-floating), nekton (strongly swimming), and benthos (bottom-dwelling). The microscopic phytoplankton are photosynthetic and are the base of the food chain in most aquatic communities.

8. Flowing-water ecosystems (streams and rivers) differ from other freshwater ecosystems in that their water flows in a current. In addition, streams and rivers depend on detritus from the land for much of their energy. The kinds of organisms found in flowing-water ecosystems vary greatly, mostly due to the strength (speed) of the current, which is greater in headwaters than downstream.

9. Freshwater lakes are differentiated into zones on the basis of water depth. The marginal littoral zone contains emergent vegetation and heavy growths of algae. The limnetic zone is the open water away from the shore. The deep, dark profundal zone holds little life other than decomposers. In summer, temperate zone lakes exhibit thermal stratification in which the thermocline separates the warmer water above from the deep, cold water. Annual spring and fall turnovers remix these layers.

10. The main marine habitats include estuaries, the intertidal zone, the neritic province, and the oceanic province. The very productive estuary community receives a high input of nutrients from the adjacent land and serves as an important nursery area for the young stages of many aquatic organisms.

11. Organisms of the intertidal zone possess adaptations that enable them to resist wave action and the extremes of being covered by water (high tide) and exposed to air (low tide). Sandy shore inhabitants usually burrow; rocky shore organisms cling to rocks. The neritic province is characterized by rich plankton and benthic organisms. The oceanic province has no producers (except in the case of abyssal hot spring communities); its animal inhabitants are exclusively predators or scavengers, subsisting on detritus that drifts in from the areas of the ocean where photosynthesis takes place.

# DISCUSSION QUESTIONS

**1.** What climate and soil factors produce each of the major terrestrial biomes?

**2.** In which biome do you live? If your biome does not match the description given in this book, how do you explain the discrepancy?

**3.** Which biomes are best suited for agriculture? Explain why each of the biomes you did not specify are unsuitable for agriculture.

**4.** What environmental factors are most important in determining the adaptations of the organisms found in aquatic habitats?

**5.** How and why do the inhabitants of a rocky beach differ from those of a sandy beach?

**6.** Explain the roles of phytoplankton, zooplankton, and nekton in aquatic food webs.

# SUGGESTED READINGS

Gore, R. Between Monterey tides. *National Geographic* 177:2, February 1990. An enthralling description of the rich species diversity and ecosystem complexity of Monterey Bay, off the coast of California.

Odum, E. P. *Ecology and Our Endangered Life-Support Systems,* Sinauer Associates, Sunderland, Mass., 1989. Contains a chapter on the major ecosystem types of the world.

Sokolov, V. E., and B. V. Vinogradov. Man and the biosphere: The view from above. *Nature and Resources* 22:1–2, January–June 1986. How ecologists use aerial and space imagery to study communities, major life zones, and the biosphere. This article was written primarily about Soviet scientific research.

*A smokestack pollutes the air in River Rouge, a small town near Detroit, Michigan. (John Mielcarek/Dembinsky Photo Associates)*

# Ecosystems, Economics, and Government

As humans strive to live in a world increasingly stressed by our activities, decisions are made every day that either add to or lessen the stress. In large measure the decisions are economic ones, driven not by a concern for the environment but by the demands of individuals, businesses, and governments in the economic marketplace. If we are to preserve our environment, we must seek new approaches that take into account the future costs of the harm being done to the world ecosystem in the name of economic development. At present in the United States, a framework of laws governs the degree to which industry is permitted to affect the environment. These laws and their enforcement, although imperfect, have had a generally positive effect in the direction of protecting and improving the condition of the environment.

# AN ECONOMIST'S VIEW OF POLLUTION

As the human population has grown, its appetite for natural resources—plants, animals, water, minerals, air, land, and so on—has seriously stressed the Earth's environment. Pressures for continued economic and industrial development have proven difficult to resist, and such development has almost always entailed the disruption of natural ecosystems.

In this chapter we examine one form of such disruption, industrial pollution. Lessons about the economics of industrial pollution also apply to other environmental issues where disruption of ecosystems is a consequence of economic activity. For now we focus on the two factors that most critically influence decisions in which development and the environment come into conflict: economic pressures and government policy.

**Economics** is the study of how people use their limited resources to try to satisfy their unlimited wants. Using theories, models, and testing of observations and data, economists try to understand the consequences of the ways in which people, businesses, and governments allocate their limited resources. In a free market system such as that of the United States (Figure 7–1), economists study the prices of goods and services and how those prices influence the amount of a given good or service that is produced and consumed. Like any scientists conducting experiments, economists try to predict

the consequences of particular economic actions. When the actions involve economic development, their predictions may lead to policy decisions that have significant environmental consequences. If, as citizens, we wish to affect these policy decisions, we need to understand how economists view the world.

Seen through an economist's eyes, the world is one large marketplace where resources are allocated to a variety of uses and where goods (a car, a pair of shoes, a hog) and services (a haircut, a tour of a museum, an education) are consumed and paid for. In a free market, the price of a good is determined

**Figure 7–1**
Different countries have different systems of economics. A free market economy (such as those of Canada and the United States) is characterized by private ownership and operation for profit under competitive conditions. A command economy (such as those of Ethiopia and North Korea) has community rather than private ownership.

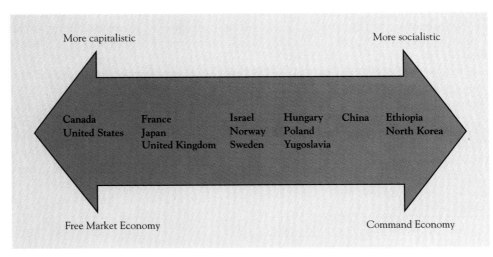

by its supply and by the demand for it. If something in great demand is in short supply, its price will be high. High prices encourage suppliers to produce more of a good or service (as long as the selling price is higher than the cost of producing the good or service). This interaction of consumer demand, producer's supply, prices, and costs underlies much of what happens in our country's economy, from the price of a hamburger to the salary of a corporate executive to the cycles of economic expansion (increasing economic activity) and recession (slowdown in economic activity).

An important aspect of the operation of a free market system is that the person consuming a product should be the one to pay for all the costs of producing that product. When consumption or production of a product has a harmful side effect that is borne by people not directly involved in the market exchange for that product, the side effect is called an **external cost**. Because external costs are usually not reflected in a product's price, the market system does not operate in the most efficient way. For example, if an industry makes a product

and, in so doing, also releases a pollutant into the environment, the product is bought at a price that reflects the cost of making it, but *not* the cost of the damage to the ecosystem by the pollutant. Because this damage is not included in the product's price and because the consumer may not be aware that the pollution exists or that it harms the environment, the cost of the pollution has no impact on the consumer's decision to buy the product. As a result, consumers of the product demand more of it than they would if its true cost (including the cost of pollution) were known. The failure to add the price of environmental damage to the cost of products creates a market force that increases pollution. From the perspective of economics, then, one of the root causes of the world's pollution problem is the failure to consider external costs in the pricing of goods.

## HOW MUCH POLLUTION IS ACCEPTABLE?

In order to come to grips with the problem of assigning a proper price to pollution, economists first attempt to answer the basic question "How much pollution should be allowed?" One can imagine two extremes: an uninhabited wilderness in which no pollution is produced and an uninhabitable sewer that is completely polluted from excess production of goods. In an uninhabited wilderness, the environmental quality would be the highest possible, but many goods that are desirable to humans would be scarce or nonexistent; in an uninhabitable sewer, millions and millions of goods could be produced, but environmental quality would be extremely low. In our world, a move toward a better environment almost always entails a cost in terms of goods.

How do we, as a country and as part of the larger international community, decide where we want to be on the environment–material prosperity continuum? Economists answer such questions by analyzing the marginal costs of environmental quality and of other goods. A **marginal cost** is the additional cost associated with one more unit of something. The tradeoff between environmental quality on the one hand and more goods on the other involves balancing two kinds of marginal costs: (1) the cost, in terms of environmental quality, of enduring more pollution (the marginal cost of pollution) and (2) the cost, in terms of other goods given up, of eliminating pollution (the marginal cost of pollution abatement).

### Mini-Glossary of the Economics of Pollution

**marginal cost of pollution:** The cost, in environmental quality, of a unit of pollution that is emitted into the environment.

**marginal cost of pollution abatement:** The cost to dispose of a unit of pollution in a nonpolluting way.

**optimum amount of pollution:** The amount of pollution that is economically most desirable. It is determined by plotting two curves, the marginal cost of pollution and the marginal cost of pollution abatement. The point where the two curves meet is the optimum amount of pollution from an economic standpoint.

**emission charge policy:** A government policy that controls pollution by charging the polluter for each unit of emissions, that is, by establishing a tax on pollution.

**waste-discharge permit policy:** A government policy that controls pollution by issuing permits allowing the holder to pollute a given amount. Holders are not allowed to produce more emissions than are sanctioned by their permits.

**emission reduction credit (ERC):** A waste-discharge permit that can be bought and sold by companies producing emissions. Companies have a financial incentive to reduce emissions, because they can recover some or all of their cost of pollution abatement by the sale of the ERCs that they no longer need.

**command and control:** Pollution control laws that work by setting pollution ceilings. Examples include the Clean Water Acts and Clean Air Acts.

## The Marginal Cost of Pollution

The **marginal cost of pollution** is the added cost to all present and future members of society of an additional unit of pollution. For each type of pollution (for example, sulfur dioxide, which causes acid rain), economists add up the harm done by each additional unit of pollution (for example, another ton of sulfur dioxide). As the total amount of pollution increases, the harm done by each additional unit usually also increases, so that the curve showing the marginal cost of pollution slopes upward (Figure 7–2). At low pollution levels, the environment may be able to absorb the damage so that the marginal cost of one added unit of pollution is near zero. As the quantity of pollution increases, the marginal cost rises, and at very high levels of pollution the cost soars.

## The Marginal Cost of Pollution Abatement

The **marginal cost of pollution abatement** is the added cost to all present and future members of society of reducing a given type of pollution by one unit. This cost tends to rise as the level of pollution falls (Figure 7–3). It is relatively inexpensive, for example, to reduce automobile exhaust emissions by half, but costly devices are required to reduce

### Envirobrief

**How Much Does a Clean Environment Cost?**

In 1990 the United States spent $115 billion to comply with environmental regulations and clean up pollution. That is about 2 percent of the gross national product.

Most of this amount was spent by industry and consumers rather than by the federal government. The Environmental Protection Agency's expenses represented only about $5.5 billion of the 1990 total.

Pollution control is big business. The pollution abatement industry in the United States employs almost 3 million people and generates nearly $100 billion each year in revenues.

Source: The Environmental Protection Agency

the remaining emissions by half again. For this reason, the curve showing the marginal cost of pollution abatement slopes downward. At high pollution levels the marginal cost of eliminating one

**Figure 7–2**
The marginal cost of pollution may be represented by an upward sloping curve, which shows that as the level of pollution rises, the social cost (in terms of human health and a damaged environment) increases sharply.

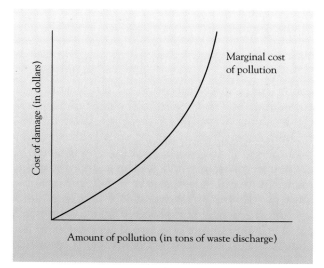

**Figure 7–3**
The marginal cost of pollution abatement is represented by a downward sloping curve, which shows that as more and more pollution is eliminated from the environment, the cost of removing each additional (marginal) unit of pollution increases.

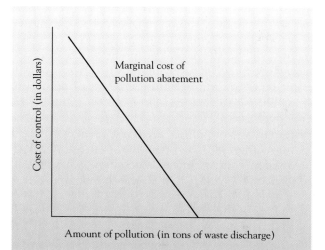

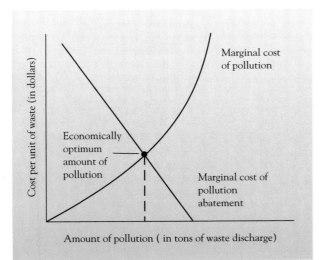

**Figure 7–4**
Economists identify the optimum amount of pollution as the amount whose marginal cost of pollution equals the marginal cost of pollution abatement (the point at which the two curves intersect). If more pollution than the optimum is allowed, the social cost will be unacceptably high. If less than the optimum amount of pollution is allowed, the pollution abatement cost will be unacceptably high.

unit of pollution is low, but as more and more pollution is eliminated, the cost rises.

## How Economists Estimate the Optimum Amount of Pollution

In Figure 7–4, the two marginal-cost curves from Figures 7–2 and 7–3 are plotted together on one graph, called a cost-benefit diagram. Economists use this to identify the point at which the marginal cost of pollution equals the marginal cost of abatement—in the diagram, the point where the two curves cross. As far as economics is concerned, this point represents an **optimum amount of pollution.** If pollution exceeds this amount (is right of the crossover point), the harm done (measured on the marginal-cost-of-pollution curve) exceeds the cost of reducing it (the marginal-cost-of-pollution-abatement curve), and it is economically efficient to reduce pollution. On the other hand, if pollution is less than this amount (is left of the crossover point), then the marginal cost incurred to reduce it exceeds the marginal cost of the pollution. In the latter situation the gain in environmental quality is more than offset monetarily by the decrease in re-

sources available for other uses. In such a situation, an economist would argue, pollution should be increased.

## Flaws in the Optimum Pollution Concept

There are three major flows in the economist's concept of optimum pollution. First, it is difficult to actually measure the monetary cost of pollution. Second, pollution is seen by some as violating basic human rights. Third, the risk of ecosystem disruption is rarely taken into account.

**The Difficulty of Measuring Pollution Cost**  Economists usually measure the cost of pollution in terms of damage to property, damage to health (in the form of medical expenses or time lost from work), and the monetary value of animals and plants killed. It is difficult, however, to place a value on damage to natural beauty—how much is a scenic river worth, or the sound of a bird singing? And how does one assign a value to the extinction of a species? Also, when pollution covers a large area and involves millions of people—as, for example, acid rain does—assessing pollution cost is extremely complex (see Focus On: Natural Resources, the Environment, and the National Income Accounts).

**The Right of Everyone to a Clean Environment**  Many people think that every person has a basic right to clean air and water, and that industries that pollute are nothing less than criminals, slowly killing us without our consent. By destroying our natural heritage, these people say, polluters are stealing from us, our children, and grandchildren, and thus there is no optimum permissible amount of pollution, any more than there is a permissible amount of rape or murder.

**The Risk of Disrupting the Ecosystem**  In adding up pollution costs, economists do not take into account the possible disruption or destruction of an ecosystem. As you've seen in the last few chapters, the web of relationships within an ecosystem is extremely complex and may be vulnerable to pollution damage, often with disastrous results. For an economist to simply add up the costs of lost elements in a polluted ecosystem is like sitting in an airplane adding up the costs of damaged items as someone repeatedly shoots a gun in the cockpit. The sum of costs does not reflect the very real danger that the shooting will cause the plane to crash.

## Natural Resources, the Environment, and the National Income Accounts

Much of our economic well-being flows from natural, rather than human-made, assets—our land, our rivers and oceans, our natural resources (such as oil and timber), and indeed the air that we breathe. Ideally, for the purposes of economic and environmental planning, the use and misuse of natural resources and the environment should be appropriately measured in the national income accounts. Unfortunately, they are not. There are at least two important conceptual problems with the way the national income accounts currently handle the economic use of natural resources and the environment.

**1. Natural resource depletion.** If a firm produces some output but in the process wears out a portion of its plant and equipment, the firm's output is counted as part of gross national product (GNP), but the depreciation of capital is subtracted in the calculation of net national product (NNP). Thus NNP is a measure of the net production of the economy, after a deduction for used-up capital. In contrast, when an oil driller drains oil from an underground field, the value of the oil produced is counted as part of the nation's GNP; but no offsetting deduction to NNP is made to account for the fact that nonrenewable resources have been used up. In principle, the draining of the oil field should be thought of as a type of depreciation, and the net product of the oil company should be accordingly reduced. The same point applies to any other natural resource that is depleted in the process of production.

**2. The costs and benefits of pollution control.** Imagine that a company has the following choices: It can produce $100 million worth of output and in the process pollute the local river by dumping its wastes. Alternatively, by using 10% of its workers to dispose properly of its wastes, it can avoid polluting but will only get $90 million of output. Under current national income ac-

counting rules, if the firm chooses to pollute rather than not to pollute, its contribution to GNP will be larger ($100 million rather than $90 million), because the national income accounts attach no explicit value to a clean river. In an ideal accounting system the economic costs of environmental degradation would be subtracted in the calculation of a firm's contribution to GNP, and activities that improve the environment—because they provide real economic benefits—would be added to GNP.

Discussing the national income accounting implications of resource depletion and pollution may seem to trivialize these important problems; but in fact, since GNP and related statistics are used continually in policy analyses, abstract questions of measurement may often turn out to have significant real effects. For example, economic development experts have expressed concern that some poor countries, in attempting to raise measured GNP as quickly as possible, have done so in part by overexploiting their natural resources and impairing the environment. Conceivably, if "hidden" resource and environmental costs were explicitly incorporated into official measures of economic growth, these policies might be modified. Similarly, in industrialized countries political debates about the environment have sometimes emphasized the impact on conventionally measured GNP of proposed pollution control measures, rather than the impact on overall economic welfare. Better accounting for environmental quality might serve to refocus these debates to the more relevant question of whether, for any given environmental proposal, the benefits (economic and noneconomic) exceed the costs.

Sources: Jonathan Levin, "The Economy and the Environment: Revising the National Accounts," *IMF Survey*, 4 June 1990; Washington: International Monetary Fund, 1990.

# ECONOMIC STRATEGIES FOR POLLUTION CONTROL

Economists have traditionally approached the problem of pollution control in terms of supply and demand. Because their suggested solutions have a significant impact on government policy, it is important for any student of environmental science to understand the nature of their arguments and proposed solutions.

## The Demand Curve

To understand the economist's analysis of the problem of pollution control, examine Figure 7–5. You will recognize it as the diagram presented in Figure 7–4, but we now look at the curve we called "marginal cost of pollution abatement" from a different point of view—that of the industrial polluter. Labeled "Demand for pollution opportunities," the curve now expresses how much a firm would be willing to pay for the opportunity to dump an additional unit of pollution into the environment. From the point of view of cost effectiveness, the firm would pay any amount smaller than the marginal cost of pollution abatement. That is, it would pay for the opportunity to pollute as long as polluting costs less than not polluting. In this sense, an industry's demand to pollute is determined by its desire to avoid abatement costs.

## The Supply Curve

In Figure 7–5, the horizontal axis is in effect a supply curve of pollution opportunities. When the cost of pollution (measured on the vertical axis) is zero, the opportunities for pollution are unlimited, and events are determined simply by the demand curve. As the cost of pollution rises, the horizontal line rises, and its intersection with the demand curve moves to the left (that is, the demand for pollution falls). For any cost of pollution, supply and demand are equal at the point at which their curves intersect.

## The Intersection of Supply and Demand

From an economist's point of view, the problem of excessive pollution can be looked at as an error in cost estimation: because the polluter is not charged for dumping wastes into the environment and therefore the cost of pollution is not included in the product's price, the supply curve is too low, and so the intersection of supply and demand curves is too low. As a result, it is less expensive to pollute than to dispose of wastes properly. Given this view, economists propose a simple and straightforward solution: raise the supply curve by raising the cost of polluting. This approach is often referred to by economists as "adopting a market-oriented strategy" because it seeks to use the economic forces of a free market to alleviate the pollution problem.

A very popular market-oriented strategy for controlling pollution involves raising the supply curve of Figure 7–5 up to the optimum pollution value by imposing an **emission charge** on polluters. In effect, this charge is a tax on pollution. Several European countries, for example, have encouraged drivers to switch to unleaded gasoline by imposing extra taxes on leaded gasoline, which releases more pollutants into the air when it burns. Sweden taxes the active ingredients in pesticides. Finland and Norway impose charges on nonreturnable containers, as do some communities in the United States. Such charges increase the cost of polluting and shift the supply curve of pollution opportunities upward. In theory, polluters react to the increase in cost caused by the emission charge by decreasing pollution, because the cost of abatement is now less than the cost of polluting. In actual practice, this

**Figure 7–5**

Supply and demand of pollution opportunities. The supply curve of pollution opportunities is a horizontal line, whereas the demand curve of pollution opportunities is equivalent to the marginal cost of pollution abatement. It is possible to adjust the supply curve so that supply and demand curves intersect at the optimum amount of pollution.

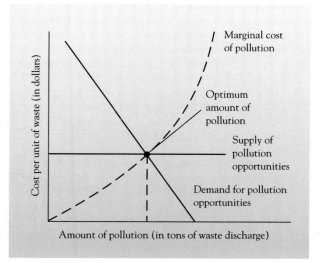

approach has not been notably successful because taxes are almost always set too low to have much effect on the behavior of people or companies.

Many economists argue for a different market-oriented strategy: raising the supply curve to the optimum pollution level by issuing a fixed number of **waste-discharge permits,** each of which allows the holder to emit a certain amount of a given pollutant. The permits (called **emission reduction credits,** or ERCs) can then be freely bought and sold. Economists believe that this approach is an efficient way to move the market toward the point of optimum pollution: potential polluters desiring to add more pollution to the environment will increase their costs by buying additional ERC permits, whereas those polluting less than the optimum level will sell their excess ERC permits and so lower their costs. Because using marketable permits ensures that pollution does not fall below the optimum amount and allows firms with higher abatement costs to control their pollution less, it prevents pollution abatement from impeding economic development and achieves pollution reduction at low costs.

This approach was used by the Environmental Protection Agency (EPA) in mandating lower levels of lead in gasoline. Refineries were assigned quotas of lead, which they could trade with each other. Since 1974 the EPA has also issued air pollution permits as marketable ERCs. When a company wishes to move into a city that fails to meet the standards of the 1970 Clean Air Act, it buys ERC polluting rights from established firms that have cut their own emissions. The Federal Clean Air Act of 1990 includes a similar plan to cut sulfur dioxide emissions with tradable permits for coal-burning electricity utilities. The EPA is also considering using marketable permits to phase out the manufacture of ozone-destroying chlorofluorocarbons.

### The Problem with Free Market Economic Strategies

Although treating pollution opportunities as a commodity on the open market is an efficient *economic* strategy, it has been unpopular with environmentalists. Emission charges and marketable waste-discharge permits are thought to be poor methods of maintaining a healthy environment. Not only is pollution cost difficult to measure with any accuracy, but the true effects of the pollution often do not enter into the calculation because they affect a third party or, frighteningly, because they are simply not known.

From the point of view of sound environmental science, no economic policy that has an adverse impact on the environment should be encouraged if that impact is not completely understood. It is one thing to say that the unavoidable price of development is some lowering in environmental quality, but quite another thing to say that we cannot predict how much environmental quality will be lost. Risk—the possibility of catastrophic consequences—is the element left out of the free market economic calculations. The interrelationships of organisms within ecosystems are not incorporated into economic theories, models, and predictions. Because such calculations produce policies that affect ecosystems in direct and potentially disastrous ways, it is essential that a more serious attempt be made to incorporate potential ecological risk factors into economic calculations.

## GOVERNMENT AND ENVIRONMENTAL POLICY

Given that it is difficult to assess how much, if any, pollution should be allowed, how should we govern levels of pollution in our society? Historically, acceptable levels of pollution have been determined by rough guess, using the guiding principle that cleaner is safer. While this might seem imprecise, setting such guidelines has generally proven effective. To enforce the determined limit on pollution, legislation is usually enacted (Table 7–1).

Historically, most pollution control efforts have involved what economists called **command and control**—the passage of laws that impose rules and regulations and set limits on levels of pollution. Sometimes such laws state that a specific pollution control method must be used (such as catalytic converters in cars to decrease polluting emissions in exhaust). In other cases, a quantitative goal is set. For example, the 1990 Clean Air Act (see Chapter 19) set a goal of a 60 percent reduction in nitrogen oxide emissions in passenger cars by the year 2003. Usually, all polluters must comply with the same rules and regulations regardless of their particular circumstances.

Overall, legislative approaches to pollution control have been successful. For example, the air in many of the world's cities, although still polluted, is far cleaner than it was 15 years ago. Levels of sulfur dioxide emissions from coal-fired power plants have been cut in several large industrial countries, including the United States, the former

## Table 7–1
## Some Important Environmental Legislation

**General**
National Environmental Policy Act of 1970 (NEPA)
International Environmental Protection Act of 1983

**Conservation of Energy**
National Energy Acts of 1978, 1980
Northwest Power Act of 1980
National Appliance Energy Conservation Act of 1987

**Conservation of Wildlife**
Anadromous Fish Conservation Act of 1965
Fur Seal Act of 1966
National Wildlife Refuge System Acts of 1966, 1976, 1978
Species Conservation Acts of 1966, 1969
Marine Mammal Protection Act of 1972
Marine Protection, Research, and Sanctuaries Act of 1972
Endangered Species Acts of 1973, 1982, 1985, 1988
Fishery Conservation and Management Acts of 1976, 1978, 1982
Whale Conservation and Protection Study Act of 1976
Fish and Wildlife Improvement Act of 1978
Fish and Wildlife Conservation Act of 1980

**Conservation of Land**
General Revision Act of 1891
Taylor Grazing Act of 1934
Soil Conservation Act of 1935
Wilderness Act of 1964
Land and Water Conservation Fund Act of 1965
Multiple Use Sustained Yield Act of 1968
Wild and Scenic Rivers Act of 1968
National Trails System Act of 1968
National Coastal Zone Management Acts of 1972, 1980
National Reserves Management Acts of 1974, 1976
Forest and Rangeland Renewable Resources Acts of 1974, 1978
Federal Land Policy and Management Act of 1976
National Forest Management Act of 1976
Public Rangelands Improvement Act of 1978
Soil and Water Conservation Act of 1977
Surface Mining Control and Reclamation Act of 1977
Antarctic Conservation Act of 1978

Endangered American Wilderness Act of 1978
Alaskan National Interests Lands Conservation Act of 1980
Coastal Barrier Resources Act of 1982
Food Security Act of 1985 (amended and renamed the Food, Agricultural, Conservation and Trade Act of 1990)
Emergency Wetlands Resources Act of 1986

**Air Quality and Noise Control**
Clean Air Acts of 1963, 1965, 1970, 1977, 1990
Noise Control Act of 1965
Quiet Communities Act of 1978
Asbestos Hazard and Emergency Response Act of 1986

**Water Quality and Management**
Refuse Act of 1899
Water Resources Research Act of 1964
Water Quality Act of 1965
Water Resources Planning Act of 1965
Federal Water Pollution Control Act of 1972 (amendment to Water Quality Act of 1965)
Ocean Dumping Act of 1972
Ocean Dumping Ban Act of 1988
Safe Drinking Waters Acts of 1974, 1986
Water Resources Development Act of 1986
Clean Water Acts of 1977, 1981, 1987 (up for renewal in 1992)
Water Quality Act of 1987 (amendment of Clean Water Acts)

**Control of Pesticides**
Food, Drug, and Cosmetics Acts of 1938, 1954, 1958
Federal Insecticide, Fungicide, and Rodenticide Acts of 1947, 1972, 1988

**Management of Solid and Hazardous Wastes**
Solid Waste Disposal Act of 1965
Resource Recovery Act of 1970
Hazardous Materials Transportation Act of 1975
Toxic Substances Control Act of 1976
Resource Conservation and Recovery Acts of 1976, 1984
Comprehensive Environmental Response, Compensation, and Liability (Superfund) Acts of 1980, 1986
Nuclear Waste Policy Act of 1982
Marine Plastic Pollution Research and Control Act of 1987

---

West Germany, and Japan (Table 7–2). Lake Erie, which borders the United States and Canada, is no longer a dying ecosystem (Figure 7–6). The level of lead emissions in the United States is one-tenth the 1975 level; this decline is primarily the result of a switch from leaded to unleaded gasoline. There is still plenty of room for progress—for example, in the United States, nitrogen dioxide pollution of city air from automobile exhaust is worse than it

was 15 years ago. But the preceding examples demonstrate that it is possible to reduce pollution by setting legal limits.

In countries where laws have not been passed to control pollution, the situation is far worse. Cities in developing countries (which typically do not regulate air emissions) have far filthier air than those in developed countries. Water and air pollution in Eastern Europe, which never attempted to

| Table 7–2 1988 Per-Capita Sulfur Dioxide Emissions in Selected Countries | |
|---|---|
| **Country** | **Emissions per Capita (kilograms)** |
| Former East Germany | 317 |
| Czechoslovakia | 179 |
| Hungary | 115 |
| Bulgaria | 114 |
| Poland | 110 |
| United States | 84 |
| United Kingdom | 64 |
| Sweden | 25 |
| France | 22 |
| Former West Germany | 21 |

**Figure 7–6**
Lake Erie in the 1990s. After cities and industries embarked on a reduction in pollution, the Lake Erie ecosystem experienced a dramatic comeback from the poor water quality of the 1960s. Most of this recovery was the result of diminished phosphorus levels in the lake. (Mark E. Gibson)

control pollution while under Communist rule,[1] are particularly bad. Approximately half of Poland's water is too polluted even for industrial use, 80 percent of its deep wells are polluted, and one fourth of its soil is too contaminated for safe farming. One fourth of Czechoslovakia's rivers support no fish, and one third of Bulgaria's forests are damaged or dying. (See Focus On: Economics and the Environment in Eastern Europe for a detailed discussion of environmental problems in Eastern Europe.)

Seeking to limit pollution by legislation is only one of many ways in which governments act to protect the environment. In order to better understand the key role governmental policy plays in environmental protection, let's look briefly at the history of environmental legislation in this country.

## National Forests

From the establishment of the first permanent English colony at Jamestown, Virginia, in 1607, the first two centuries of our country's history were a time of widespread environmental destruction. The great forests of the Northeast were leveled within a few generations, and soon after the Civil War in the 1860s, loggers began deforesting the Midwest at an appalling rate. Within 40 years they deforested an area the size of Europe, stripping Minnesota, Michigan, and Wisconsin of virgin forest. By 1897 the sawmills of Michigan had processed 160 billion board feet of white pine, leaving less than 6 billion board feet standing in the whole state (Figure 7–7).

[1] Many laws and regulations *existed* but were not enforced.

There has been nothing like this unbridled environmental destruction since.

In 1875 a group of public-minded citizens formed the American Forestry Association, with the intent of influencing public opinion against the wholesale destruction of America's forests. Sixteen

*(text continues p. 128)*

**Figure 7–7**
Loading up logs in a white pine forest near Lake Huron. By 1897, most of the forests in Michigan had been cut. (The Bettmann Archive)

## Economics and the Environment in Eastern Europe

The fall of Communist governments in Eastern Europe during the late 1980s revealed a grim legacy of environmental destruction. Water in Eastern Europe is so poisoned from raw sewage and chemicals that it cannot be used for industrial purposes, let alone for drinking. Unidentified chemicals leak out of dump sites into the surrounding soil and water, while nearby, fruits and vegetables are grown in chemically laced soil. Power plants pour soot and sulfur dioxide into the air, cre-

ating a persistent chemical haze. Buildings and statues are eroded and entire forests are dead because of air pollution and acid rain. Crop yields are falling despite intensive use of chemical pesticides and fertilizers.

How does this massive pollution affect human health? Many Eastern Europeans suffer from asthma, chronic bronchitis, and other respiratory diseases as a result of breathing the filthy, acrid air. By the time most Polish children are ten years old, for

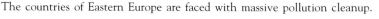

The countries of Eastern Europe are faced with massive pollution cleanup.

example, they suffer from chronic respiratory diseases or heart problems. The levels of cancer, miscarriages, and birth defects are also extremely high. Life expectancies are lower than in other industrialized nations; the average Czech lives four to five years less than the average Western European, for example.

The economic theory behind Communism was one of high production and economic self-sufficiency—regardless of costs or damages—and so pollution in Communist-controlled Eastern Europe went unchecked. Meeting industrial production quotas always took precedence over environmental concerns, even though production was not carried out for profit. The Communist regimes supported heavy industry—power plants, chemicals, metallurgy, large machinery—at the expense of the more environmentally benign service industries. As a result, Eastern Europe is overindustrialized, and because most of its plants were built during the post–World War II period, it lacks the pollution abatement equipment now required in factories in most industrialized countries.

In addition, Communism did little to encourage nature resource conservation, a very effective way to curb pollution. For example, high energy subsidies and lack of competition allowed power plants to provide energy at prices far below its actual cost. Thus, neither industries nor individuals had a strong incentive to conserve energy. On the other hand, the conversion of Eastern European countries to a free market economy and the removal of government-sponsored energy subsidies have driven the price of energy up to more realistic costs, thereby encouraging energy conservation.

Communism also took a toll on the environment because a repressive government run by a single political party cannot be held accountable. Political opposition to any aspect of government operations, including environmental damage, was unthinkable, and people who wanted to scrutinize pollution information found that it was unavailable, having been classified as top secret. Only the collapse of Communism allowed the citizens of Eastern Europe to begin to assess the full extent of damage to their environments.

Clearly the new Eastern European governments face an intimidating task. While switching from Communism with a command economy to democracy with a free market economy, they also face the overwhelming responsibility of improving the environment. They are trying to formulate environmental policies based on the experiences of the United States and Western Europe in the past several decades. What they need most urgently is information and technical assistance. Eastern European governments want to clean up their countries, but do not possess or even know about the latest pollution abatement technologies. They need to deter-

The Nowa Huta Steelworks in Cracow, Poland, which emits large quantities of air pollutants. One of the costs of this plant's pollution is that 80 percent of its workers retire early because of disabilities. (The Image Works)

## Economics and the Environment in Eastern Europe (Continued)

mine which environmental policies and strategies are most effective in the United States and Western Europe so they can adopt them.

The United States and Western Europe can also help by providing money for environmental reconstruction. Water and air pollution move readily across political borders, so Western European countries in particular have a real incentive to help Eastern European nations reduce pollution. In doing so, they help to protect their own environments.

How long will it take? Optimistically, experts predict that years, even decades, will be required to clean up the pollution legacy of Communism. How much will it cost? The figures are staggering. It is estimated that improving the environment in what was formerly East Germany alone will cost up to $300 billion.

Eastern European nations seem committed to making their lands habitable once again. Although the scope of the pollution is daunting, particularly when considered in conjunction with the costly economic and social reforms currently under way, Eastern Europeans seem to have little choice but to focus on environmental issues. The cost of environmental inaction—in terms of human illness and damage to the environment—will continue to block economic recovery.

Soil and water are dangerously polluted in many areas of Eastern Europe. Bulgarian farmlands are irrigated with this water, which is polluted by factory wastes. (© Chervenkv/The Image Works)

years later, the General Revision Act gave the president the authority to establish "forest reserves on the public domain." Benjamin Harrison, Grover Cleveland, and Theodore Roosevelt used this law to put 43 million acres of forest, primarily in the West, out of the reach of loggers. In 1907 angry Northwest congressmen pushed through a bill rescinding the president's powers to establish forest reserves. Theodore Roosevelt responded by designating 21 new national forests that totaled 16 million acres, *then* signing the bill that would have stopped him into law. Today, national forests have multiple uses, from wildlife habitat to recreation to timber harvest (see Chapter 17). (See Meeting the Challenge: Jobs and the Environment.)

### National Parks and Monuments

The world's first national park was created in 1872 after a party of Montana explorers reported on the natural beauty of the canyon and falls of the Yel-

lowstone River; Yellowstone National Park now includes parts of Idaho, Montana, and Wyoming. In 1890 the Yosemite and Sequoia parks in California were created by the Yosemite National Park Bill, largely in response to the efforts of a single man, John Muir, and the organization he founded, the Sierra Club (Figure 7–8). In 1906 Congress passed the Antiquities Act, which authorized the president to set aside as national monuments sites (such as the Badlands in South Dakota) that had scientific, historic, or prehistoric importance. Many of the national parks and monuments that exist today were set aside over the following ten years. By 1916 there were 13 national parks and 20 national monuments, under the loose management of the U.S. Army.

Some environmental battles were lost. John Muir's Sierra Club fought such a battle with the city of San Francisco over its efforts to dam a river and form a reservoir in the Hetch Hetchy Valley

**Figure 7–8**
President Theodore Roosevelt and John Muir on Glacier Point above Yosemite Valley, California. (The Bettmann Archive)

(Figure 7–9), which lay within Yosemite National Park and was as beautiful as Yosemite Valley. In 1913 Congress voted to approve the dam. But the controversy created a strong sentiment that the nation's national parks should be better protected, and in 1916 Congress created the National Park Service to manage the national parks and monuments (there are more than 300 different sites today) for the enjoyment of the public, "without impairment." It was this clause that gave a different outcome to another battle, fought in the 1950s between environmentalists and dam builders over the construction of a dam within Dinosaur National Monument. Nobody could deny that to drown the canyon under 400 feet of water would "impair" it. This victory for conservation established the "use

**Envirobrief**

**Pricing the Rain Forest**

The economic value of a tropical rain forest is normally calculated only by the value of its tree-related products: sawlogs for lumber and other wood that can be converted into pulp for paper. Botanical researchers have recently demonstrated that the non-wood resources of rain forests such as edible fruit, nuts, latex for rubber, and medicinal extractions can be nearly twice as valuable per acre. The challenge is one of changing public policy to reflect these economic opportunities. While timber brings much-needed foreign exchange to poor tropical nations, non-wood products can be harvested manually and traded at local markets by thousands of individual collectors, farmers, and middlemen, without the loss of a single tree. While these vast decentralized trading networks are critical to the people who use them, they are difficult for governments to monitor and thus are often ignored in national accounting models.

**Figure 7–9**
Hetch Hetchy Valley in Yosemite. Before (a) and after (b) Congress approved a dam to supply water to San Francisco. (*a*, National Archives; *b*, Wallace Kleck/Terraphotographics)

(a)

(b)

without impairment" clause as the firm backbone of the legal protection afforded our national parks and monuments.

## The Power of Public Awareness

Until 1970 the voice of environmentalists in the United States was heard primarily through societies such as the Sierra Club and the National Wildlife Federation. There was no generally perceived mass "environmental movement" until the spring of 1970, when the first Earth Day (Figure 7–10) transformed the specialized interests of a few societies into a pervasive popular movement. On the first Earth Day, an estimated 20 million people in the United States demonstrated to demand improvements in resource conservation and environmental quality. By the 20th Earth Day in 1990, the move-

**Figure 7–10**
A crowd of New Yorkers observes the first Earth Day, April 22, 1970. (Freda Leinwand)

ment had spread to all seven continents; approximately 200 million people in 141 nations demonstrated to increase public awareness of the importance of individual efforts in sustaining the Earth ("Think globally, act locally").

Galvanized by ecological disasters such as the 1969 Santa Barbara oil spill and by overwhelming public support for the Earth Day movement, in 1970 the National Environmental Policy Act (NEPA) was signed into law. A key provision stated that before beginning any development project involving federal lands or funds, public agencies and private individuals and corporations would have to examine the likely environmental consequences. The required environmental impact statements (EISs) would be monitored by a newly created federal board, the Council on Environmental Quality, which reports directly to the president.

Because this council had no enforcement powers, the NEPA was thought to be innocuous, generally more a statement of good intentions than a regulatory policy. During the next few years, however, environmental activists took people, corporations, and the government to court to challenge their environmental impact statements or to use the statements to block proposed development. The courts decreed that EISs had to be substantial documents that thoroughly analyzed the environmental consequences of anticipated projects on soil, water, and wildlife and that EISs must be made available to the public for its scrutiny. These rulings put very sharp teeth into the law—particularly the provision for public scrutiny, which placed intense pressure on federal agencies to respect EIS findings (see Chapter 2).

The National Environmental Policy Act revolutionized environmental protection in this country. In addition to overseeing highway construction, flood and erosion controls, military projects, and many other public works, federal agencies own nearly one third of the land in the United States. Their holdings include extensive fossil fuel and mineral reserves as well as millions of acres of public grazing land and public forests. Since 1970 very little has been done to any of them without some sort of environmental review.

In the mid-1970s, following the passage of the NEPA, Congress passed a number of other substantial pieces of pro-environment legislation, notably the Endangered Species Act (see Chapter 16), the Clean Air Act (see Chapter 19), and the Clean Water Act (see Chapter 21). These laws greatly increased federal regulation of pollution, creating a tough interlocking mesh of laws to improve environmental quality.

However, the laws are not perfect. Economists and industries have argued that they make pollution abatement unduly complex and expensive. Nor have the laws always worked as intended. The Clean Air Act of 1977, for example, required coal-burning power plants to outfit their smokestacks with expensive "scrubbers" to remove sulfur dioxide from their emissions, but made an exception for tall smokestacks (Figure 7–11). This loophole led directly to the proliferation of tall stacks that have since produced acid rain throughout the Northeast. The Clean Air Act of 1990, described in Chapter 19, goes a long way toward closing such loopholes.

On balance, despite its imperfections, environmental command and control legislation has had positive and substantial effects. Since 1970, eight national parks have been established and 80 million acres have been declared wilderness. Some 34 million acres of farmland that are particularly vulnerable to erosion have been withdrawn from production. Many previously endangered species are better off than they were in 1970, including alligators, antelope, badgers, bald eagles, bighorn sheep, pelicans, falcons, and wild turkeys. (However, dozens of other species have suffered further decline or extinction since 1970.)

Although we still have a long way to go, pollution control efforts have been particularly successful. Emissions of sulfur dioxide, carbon monoxide, and soot, totaling 181 million metric tons (200 million tons) in 1970, have been reduced by more than 30 percent. The number of secondary sewage treatment facilities (in which bacteria break down organic wastes before the water is discharged into rivers and streams) has increased by 72 percent in the last ten years. The release of certain toxic chemicals into the environment (notably DDT, asbestos, and dioxin) has been banned. The EPA estimates that public agencies and private firms are currently spending about $100 billion a year on pollution control, double the amount ten years ago and five times the amount in 1970. In the last ten years, the U.S. population has increased by some 40 million and the number of cars by more than 60 million. Had pollution control legislation not been in place, environmental quality would almost certainly have declined dramatically.

**Figure 7–11**
Tall smokestacks, which emit sulfur dioxide from coal-burning power plants, were permitted by the Clean Air Act of 1977. (Visuals Unlimited/Albert Copley)

# Jobs and the Environment

The issues in this chapter have deliberately been simplified to help you understand the complex interactions among economic practices, government policies, and environmental concerns. We have created the impression that pollution is bad and that cleaning up pollution should be one of the nation's top priorities. You probably have also concluded that economists don't consider everything that is at stake in environmental issues and that government policies can force industries to solve environmental problems.

Although all of this is true, environmental problems are usually much more complex than we have implied. Most environmental problems have economic, scientific, health, social, cultural, personal, political, community, and international considerations that compete with one another and create situations that defy simplistic solutions. In the real world, there are no easy answers; priorities have to be determined, and tradeoffs often have to be made.

Consider western Oregon and Washington, where a real-life drama involving people's jobs and the environment unfolded during the late 1980s and early 1990s. The confrontation appeared to be between the timber industry and a small owl. At stake were thousands of jobs, the future of 3 million acres of old-growth (virgin) forest, and the continued existence of the spotted owl and other forest creatures.

Biologists view the few remaining old-growth forests of Douglas fir and spruce in the Pacific Northwest as a living laboratory demonstrating the complexity of one of the few unaltered natural ecosystems. To environmentalists, these stable forest ecosystems, with trees ranging in age from 150 to 2,000 years, are a national treasure to be protected and cherished. They provide habitat for many species of wildlife, including the endangered northern spotted owl. Like other, less well known species, the spotted owl is found only in old-growth forests. Provisions of the Endangered Species Act require the government to protect the habitat of endangered species so that their numbers can increase. Enforcement of this law in the Pacific Northwest would require the suspension of

The northern spotted owl, an endangered species found only in old-growth forests in the Pacific Northwest. (Jack Wilburn © 1993 Animals Animals)

logging where the owl lives—that is, in about 3 million acres of federal forest.

The timber industry bitterly opposed enforcement of the act, stating that thousands of jobs would be lost if the spotted owl habitat were set aside. Rural communities in the mountainous areas of the Pacific Northwest do not have diversified economies; timber is their main source of revenue. Thus, a major confrontation ensued between the timber industry and environmentalists over the future of the old-growth forest. Strong feelings were expressed on both sides—witness the bumper sticker reading, "Save a logger, kill an owl."

The situation was more complex than simply jobs versus the owl or even jobs versus the environment, however. The timber industry in the Pacific Northwest was already in a decline in terms of its ability to support people. During the decade between 1977 and 1987, logging in Oregon's national forests increased by more than 15 percent, while employment dropped by 15 percent—an estimated 12,000 jobs. The main cause of the decline in employment was automation of the industry. Even if the timber industry were allowed to continue logging old-growth forests, automation would continue to eat away at the number of logging jobs.

In addition, the timber industry was not operating *sustainably*. Due to harvesting and selling practices now viewed by many as alarmingly shortsighted, trees were being removed faster than the forest could regenerate. If the industry continued to log at its 1980s rates, all of the old-growth forest would disappear within 20 years. Thus, the timber industry in the Pacific Northwest was doomed anyway.

Furthermore, timber is not as important to the economy of the Pacific Northwest as it used to be, and this change started long before the spotted owl controversy. Northwestern states are slowly becoming more diversified; by the late 1980s the timber industry's share of the economy in Oregon and Washington was less than 4 percent.

The availability of jobs in other fields is not very comforting to people who are losing the only jobs they have known. In today's technological world, a person cannot shift careers without assistance that includes job training and income support for the interim period between the old job and the new. Many solutions to the workers' dilemma have been proposed—some by conservation organizations such as The Wilderness Society. One proposal came from the Oil, Chemical, and Atomic Workers Union. They suggested that the United States government establish a "Superfund for workers" to provide financial support for people who lose jobs in environmentally destructive fields. The Superfund money would provide for up to four years of education so that a displaced worker could shift to a new occupation. This sounds like a great solution to the problem of displaced loggers, but it would be expensive. Although some of the funds for the Superfund could come from unemployment compensation and other assistance programs already in existence, the government would have to provide new revenues as well, which translate into tax dollars.

A broad new agenda is needed in the dwindling forests of the Pacific Northwest. Multiple issues and interests are involved, and any attempt at resolution will have to be complex—more complex than a trade of jobs for owls.

Meeting the Challenge

# SUMMARY

1. Economists view pollution in a market economy as a failure in pricing. In addition, the impact of pollution on third parties is not taken into account in the marketplace.

2. From an economic point of view, the appropriate amount of pollution is a tradeoff between harm to the environment and inhibition of development. The cost in environmental quality of a unit of pollution that is emitted into the environment is known as the marginal cost of pollution. The cost to dispose of a unit of pollution in a nonpolluting way is called the marginal cost of pollution abatement.

3. According to economists, the use of resources for pollution abatement should be increased only until the cost of abatement equals the cost of the damage done by the pollution. This results in the optimum amount of pollution—the amount that is economically most desirable.

4. There are three main flaws in the economic approach to pollution: the true cost of environmental damage by pollution is difficult to determine; many people regard any pollution as a violation of fundamental human rights; and the risks of unanticipated environmental catastrophe are not taken into account in assessing the potential environmental damage of pollution.

5. Market-oriented strategies to control pollution include emission charges and waste-discharge permits. An emission charge policy controls pollution by charging the polluter for each given unit of emissions (that is, by establishing a tax on pollution); as the imposed cost (that is, the emission charge) on the supply of permissible pollution rises, the "demand" for pollution falls. Pollution can also be controlled by waste-discharge permits, which allow the holder to pollute a given amount; holders are not allowed to produce more emissions than are sanctioned by their permits, although they can buy and sell permits as needed.

6. Government controls pollution by imposing legal limits on amounts of permissible pollution (that is, by command and control). Such laws had their genesis in this country in 1970, with the passage of the National Environmental Policy Act. By requiring environmental impact statements that are open to public scrutiny, this law initiated serious environmental protection in the United States. Later laws (such as the Endangered Species Act, the Clean Air Acts, and the Clean Water Acts) have added to the environmental safety net, with some success.

# DISCUSSION QUESTIONS

1. How would your life be different if society were run under an environmental ethic rather than under an economic ethic?

2. Does your life style reflect an environmental ethic or an economic ethic? Elaborate.

3. If you were a member of Congress, what legislation would you introduce to deal with each of the following problems?

(a) Poisons from a major sanitary landfill are polluting your state's groundwater.

(b) Acid rain from a coal-burning power plant in a nearby state is harming the trees in your state. Loggers and foresters are upset.

(c) There is a high incidence of cancer in the area of your state where heavy industry is concentrated.

4. How would an economist approach each of the problems listed in question 3? How would an environmentalist?

5. How might pollution abatement legislation actually cause a net increase in the number of jobs in an area?

# SUGGESTED READINGS

Cairncross, F. The environment: An enemy and yet a friend. *The Economist*, 8 September 1990. The author, while admitting that industry pollutes that environment, develops a persuasive argument that environmentalists need industry to help solve pollution problems.

Dolan, E. G., and D. E. Lindsey. *Economics*, 6th ed. Dryden Press, Hinsdale, Ill., 1991. Chapter 33, "Externalities and Environmental Policy," presents the economists' view of pollution in an understandable fashion. Numerous examples help clarify technical economic concepts.

French, H. F. Eastern Europe's clean break with the past. *World Watch* 4:2, March/April 1991. An excellent overview of the environmental damage in Eastern Europe and what can be done about it.

Heaton, G., R. Repetto, and R. Sobin. *Transforming Technology: An Agenda for Environmentally Sustainable Growth in the 21st Century.* World Resources Institute, April 1991. This report, which was financially supported by the National Science Foundation, identifies technologies (both existing and emerging) and policies (such as fiscal incentives and new regulatory approaches) that reconcile economic growth and a cleaner, healthier environment.

Postel, S. Accounting for nature. *World Watch* 4:2, April 1991. Examines how government fiscal policy, including taxes and subsidies, can be used to solve many environmental problems.

Renner, M. G. Saving the Earth, creating jobs. *World Watch*, January/February 1992. Protecting the environment will produce more jobs than it eliminates. The author builds a case for environmental protection despite the sometimes painful local problems it creates.

**Camel market, Luxor, Egypt.** *(Shark Song/M. Kazmers/Dembinsky Photo Associates)*

# Anne and Paul Ehrlich

## A Population Policy for the Super-Consumers

*Anne Ehrlich is a senior research associate in biology and Associate Director of the Center for Conservation Biology at Stanford University, California. She received her undergraduate degree in biology from the University of Kansas. She has written extensively on population issues, environmental protection, and the environmental consequences of nuclear war, and has taught a course in environmental policy at Stanford since 1981. Paul Ehrlich is Bing Professor of Population Studies at Stanford University. He received his B.A. in biology from the University of Pennsylvania, and his M.A. and Ph.D. in entomology from the University of Kansas. A noted expert in population biology, ecology, evolution, and behavior, he has published over 600 articles, scientific papers, and books, including* The Population Bomb *(1968). Together, the Ehrlichs have authored six popular books, including* The Population Explosion *(1990) and, most recently,* Healing the Planet *(1991).*

**You have said that our number one concern should be "natural ecosystem services."**

**Explain what you mean by this phrase, and how these services relate to population issues. What is the relationship between population and other ecological issues?**

Natural ecosystems supply civilization with an array of essential free services that include maintaining the gaseous quality of the atmosphere (and thus helping to stabilize climate); running the hydrologic cycle that supplies humanity with fresh water; generating and maintaining soils and making them fertile; disposing of wastes and recycling the nutrients essential to agriculture; controlling the vast majority of potential pests of crops and carriers of diseases; supplying food from the sea; and maintaining a huge genetic "library" from which humanity has already withdrawn the very basis of civilization in the form of crops, domestic animals, medicines, and industrial products (and which still has the potential for supplying many more benefits).

These services are necessary to support the human economy, but ironically the excessive scale of the human enterprise and its continued expansion are rapidly degrading the capacity of ecosystems to deliver them. A major element in the scale of that enterprise is, of course, vast overpopulation and continuing population growth. Human beings, for example, are already consuming, co-opting, or destroying some 40 percent of the entire food supply of terrestrial animals, resulting in a disastrous and accelerating decline in the diversity of animal populations and species.

The impact of the expanding human enterprise on plant diversity is indicated by the level at which the Earth's land surface has been occupied or converted to human-controlled systems. While only about 2 percent of the surface has been taken over for human habitation and intensive development (highways, airports, industrial areas), some 11 percent is used for growing crops, another 25 percent is used for pasture, and much of the shrinking area of forest—now about 30 percent of the land—has been harvested at least once. Virtually all of the remaining third of the world's land is too cold, too dry, too rugged, or too inaccessible to be useful. And an increasing portion has been too degraded—desertified—by human activities to be productive any longer.

The conversion and degradation processes result in an impoverishment of the plant communities that once thrived in those places, with a concomitant loss of the animal populations that depended on the plants. Yet the animals, plants, and microorganisms ("biodiversity") that are destroyed as land areas are taken over or degraded are working parts of the ecosystems that provide those critical services described above. In short, the very numbers of people are a major factor in reducing the capacity of Earth to support people in the future.

**Environmental activitists often present their concerns as separate and apart from other social problems, such as racism, sexism, poverty, international political unrest. You, however, take a different approach. What is the best way to describe the connection between environmental and social, economic, and political issues?**

If the burgeoning environmental problems facing humanity are to be overcome, they must be viewed in the broadest possible context. The massive poverty of a Bombay slum, people forced to live on the streets of New York, smog rotting the lungs of Los Angelenos, and Arab-Israeli tensions over limited water supplies are all examples of environmental problems. And solving those problems will require a degree of cooperation within and between societies that is incompatible with the second-place status of women in places as different as Kenya and Japan, with the notion that people with dark skins are inferior to those with lighter ones (or vice versa), with the notion that only one's own political views are correct, or that one's own religion is the only true one (and adherents to others are misguided), and so on. Unless we go a lot further toward solving humanity's traditional social problems, civilization is very unlikely to solve its environmental problems.

**If you could change one policy of our government in order to improve the environment, what policy would you select? Are any desirable current policies now being threatened?**

There is no question that the most needed policy in the United States is one to halt American population growth and begin a slow decline. Population growth is the most ignored of the major ingredients of the environmental crisis, and the United States has virtually had a policy of *promoting* population growth at home and abroad since 1984. There are many non-coercive steps that the government could take, starting with the promotion of a simple slogan: "Patriotic Americans stop at two." Beyond that, a restoration of the former U.S. leadership in family planning assistance to developing countries would greatly improve the chances of controlling worldwide population growth before it is too late.

The desirable American policy that is most threatened at the moment is that em-

bodied in the Endangered Species Act: a commitment to stem the loss of biodiversity. Instead, pressure to protect outdated resource-exploiting industries and for short-term economic growth is overwhelming the need for preserving the long-term health of both our economy and our life-support systems.

**You have often pointed out that population growth in the already-developed world poses a greater environmental threat than growth in developing countries. What role does excessive consumption by industrialized nations play in this observation? What are the most effective ways of illustrating this to the American public?**

The most serious population problem in the world is in the United States. Its population is gigantic, the third largest in the world. Each American places, on average, as much destructive pressure on Earth's life-support systems as a dozen citizens of most poor nations. That is because Americans are super-consumers and use technologies that are environmentally destructive to supply that overconsumption. It should be pointed out to Americans at every turn that population problems are not just problems of brute numbers of people—one must always consider what those people *do*. The world can support many more vegetarians who walk to work than steak- and hamburger-eaters who commute daily two hours each way in two-ton automobiles.

**What concrete advice would you offer students of the environment about what *they* can do to help find the solutions to the global issues we all face?**

The most important thing that students can do is put serious effort into learning about the human predicament and its dimensions—the population-resource-environment crisis. If they do not learn how to analyze problems in this area, how to detect the errors and outright lies that often characterize statements on the environment, they will not be equipped to make sensible decisions as citizens. Students shouldn't trust us or any other "experts;" they should learn how to find out for themselves (this book and the library are good places to start). Then they should plan to tithe to society for the rest of their lives—that is, spend at least one tenth of their time helping to solve the problems society faces. The environmental crisis is so pervasive that anyone can make an important contribution regardless of personal life goals—and future generations will bless ours for doing it.

# Chapter 8

*A large family in Mwokora, Burundi. (Dr. Nigel Smith © 1993 Earth Scenes)*

# Understanding Population Growth

ll living organisms tend to reproduce in greater numbers than can survive, but an organism's environment (which includes food, water, and living space) limits the size of its population. The human population is swelling rapidly; if it continues to increase at its current rate, we may reach the limits imposed by our world environment. If that occurs, the human species could be significantly reduced in numbers or even destroyed. In this chapter, we focus on the dynamics of population growth and describe the state of the current human population. Chapter 9 examines the consequences of continued population growth and explores ways to limit the expanding world population.

## THE BIOLOGICAL SUCCESS OF HUMANS

If you are a typical college student, you were probably born in the early 1970s. At that time the human population was slightly less than 4 billion. Today there are well over 5 billion humans, and it is likely that our numbers will increase to more than 8 billion *during your lifetime*. This tremendous increase in population is a measure of our biological success as a species. Humans have been able to provide more food and better nutrition by increasing the productivity of agriculturally important crops and animals through selective breeding. We have made great strides in the fight against diseases with life-saving sanitation practices, ever-advancing medical techniques, and newly developed medicines. All of these factors have not only increased our numbers, but also increased the likelihood that we will live longer.

Unfortunately, our biological success has created innumerable problems for us and the other plant and animal species on our planet; we are in danger of overwhelming the Earth with too many people. The Earth has limited resources, and the human population is using up, encroaching upon, fouling, and wasting them. The pollution, extinction of wildlife, degradation and loss of natural resources, and depletion of energy reserves in today's world are all related to human population growth.

## PRINCIPLES OF POPULATION ECOLOGY

Because the human population is central to so many environmental problems and their solutions, it is important that we understand how populations increase or decrease. The biological principles that affect the sizes of all animal and plant populations also apply to human populations.

### What Causes Populations to Change in Size?

Populations of organisms, whether they are sunflowers, eagles, or humans, change over time. On a *global* scale, this change is due to two factors: the number of births and the number of deaths in the population. For humans, the **birth rate** is usually expressed as the number of births per 1,000 people per year, and the **death rate** is the number of deaths per 1,000 people per year.

The rate of change, or **growth rate** ($r$), of a population is the birth rate ($b$) minus the death rate ($d$).

$$r = b - d$$

As an example, consider a hypothetical population of 10,000 in which there are 2,000 births per year (or 200 births per 1,000 people) and 1,000 deaths per year (or 100 deaths per 1,000 people).

$$r = \underbrace{\frac{2,000}{10,000}}_{b} - \underbrace{\frac{1,000}{10,000}}_{d}$$

$$r = 0.2 - 0.1 = 0.1$$

A value of 0.1 for $r$ means that the population has an annual percentage growth rate of 10 percent.

Another way to express the growth rate of a population is to determine the **doubling time**—the amount of time it would take for the population to double in size, assuming that its rate of increase doesn't change. Doubling time ($t_d$) is calculated as approximately 0.7 divided by the growth rate.[1]

$$t_d = \frac{0.7}{r}$$

In our example ($r = 0.1$), the doubling time would be 0.7/0.1 = 7 years.

In addition to the birth and death rates, migration (movement from one region or country to another) must be considered when changes in populations on a *local* scale are examined. There are two types of migration: **immigration,** by which individuals enter a population and thus increase the size of the population, and **emigration,** by which individuals leave a population and thus decrease its size. The growth rate of a local population of organisms must take into account birth rate ($b$), death rate ($d$), immigration ($i$), and emigration ($e$). The growth rate is equal to the value of birth rate minus death rate, plus the value of immigration minus emigration:

$$r = (b - d) + (i - e)$$

For example, the growth rate of a population of 10,000 that has 1,000 births, 500 deaths, 10 immigrants, and 100 emigrants in a given year would be calculated as follows:

[1]This is a simplified formula for doubling time. The actual formula involves calculus and is beyond the scope of this text.

$$r = \left( \underbrace{\frac{1,000}{10,000}}_{b} - \underbrace{\frac{500}{10,000}}_{d} \right) + \left( \underbrace{\frac{10}{10,000}}_{i} - \underbrace{\frac{100}{10,000}}_{e} \right)$$

$$r = (0.1 - 0.05) + (0.001 - 0.01)$$

$$r = 0.05 - 0.009 = 0.041$$

Note that, although emigration exceeds immigration in this example, the growth rate is still positive due to the very high birth rate. Can you calculate the doubling time for this population?[2]

## Maximum Population Growth

The maximum rate at which a population could increase under ideal conditions is known as its **biotic potential.** Different species have different biotic potentials. A particular species' biotic potential is influenced by several factors, including the age at which reproduction begins, the percentage of the life span during which the organism is capable of reproducing, and the number of offspring produced during each period of reproduction.

Generally, larger organisms, such as blue whales and elephants, have lower biotic potentials, whereas microorganisms have the greatest biotic potentials. Under ideal conditions, certain bacteria can reproduce by splitting in half every 20 to 30 minutes. At this rate of growth, a single bacterium would increase to a population of more than 1,000,000 in just 10 hours (Figure 8–1a), and the population from a single individual would exceed 1 billion in 15 hours!

If one were to plot this increase versus time, the graph would have the "J" shape that is characteristic of **exponential growth,** the constant reproductive rate that occurs under optimal conditions (Figure 8–1b). Regardless of the organism being considered, whenever its biotic potential is plotted versus time, the shape of the curve is the same. The only variable is time; that is, it may take longer for a sea lion population than for a bacterial population to reach a certain size, but its population will always increase exponentially under ideal conditions.

## How Nature Limits Population Growth

Certain populations may exhibit exponential growth for a short period of time. However, organisms cannot reproduce indefinitely at their biotic potentials, because the environment sets limits, which are collectively called **environmental resist-**

| Time (hours) | Number of bacteria |
|---|---|
| 0 | 1 |
| 0.5 | 2 |
| 1.0 | 4 |
| 1.5 | 8 |
| 2.0 | 16 |
| 2.5 | 32 |
| 3.0 | 64 |
| 3.5 | 128 |
| 4.0 | 256 |
| 4.5 | 512 |
| 5.0 | 1,024 |
| 5.5 | 2,048 |
| 6.0 | 4,096 |
| 6.5 | 8,192 |
| 7.0 | 16,384 |
| 7.5 | 32,768 |
| 8.0 | 65,536 |
| 8.5 | 131,072 |
| 9.0 | 262,144 |
| 9.5 | 524,288 |
| 10.0 | 1,048,576 |

(a)

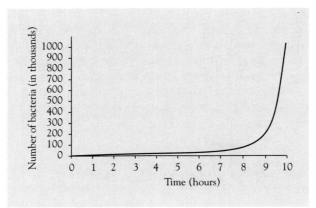

(b)

**Figure 8–1**

Exponential growth. (a) When bacteria divide every 30 minutes, their numbers increase exponentially. This set of figures assumes a zero death rate, but even if a certain percentage of each generation of bacteria died, exponential growth would still occur; it would just take longer to reach the very high numbers. (b) When these data are graphed, the curve of exponential growth has a characteristic "J" shape.

**ance.** Using the earlier example, bacteria would never be able to reproduce unchecked for an extended period of time, because they would run out of food and living space, and poisonous body wastes would accumulate in their vicinity. With crowding, they would also become more susceptible to para-

[2]Answer: doubling time = 0.7/0.041 = 17 years.

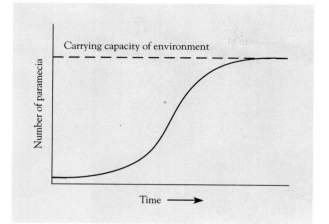

**Figure 8–2**

In many laboratory studies, including Gause's investigation with *Paramecium caudatum*, exponential population growth slows as the carrying capacity of the environment is approached. This produces a curve with a characteristic "S" shape.

**Figure 8–3**

The population of reindeer on one of the islands in the Bering Sea, off the coast of Alaska, experienced rapid growth followed by a sharp decline when the environment was damaged by an excess of reindeer. (Visuals Unlimited/Steve McCutcheon)

sites and predators. As their environment changed, their birth rate (*b*) would decline and their death rate (*d*) would increase due to shortages of food, increased predation, increased competition, and stress. The environmental conditions might worsen to a point where *d* would exceed *b* and the population would decrease. The number of organisms in a population, then, is controlled by the ability of the environment to support it.

Over longer periods of time, the rate of population growth for most organisms decreases to around zero. This leveling out occurs at or near the limit of the environment's ability to support a population. The **carrying capacity** represents the highest population that can be maintained for an indefinite period of time by a particular environment.

When population over a longer period of time is graphed (Figure 8–2), the curve has a characteristic "S" shape that shows the population's initial exponential increase (note the curve's "J" shape at the start), followed by a leveling out as the carrying capacity of the environment is approached. Although the S-curve is a simplification of actual population changes over time, it does appear to fit the population growth observed in many populations that have been studied in the laboratory and a few that have been studied in nature. For example, G. F. Gause (see Chapter 4) grew a single species, *Paramecium caudatum*, in a test tube. He supplied a constant but limited amount of food daily and replenished the media occasionally to eliminate the buildup of metabolic wastes. Under these conditions, as shown in Figure 8–2, the population of *P. caudatum* increased exponentially at first. The paramecia became so numerous that the water was cloudy with them. But then their rate of increase declined and their population leveled off.

Sometimes, when a population exceeds the carrying capacity and environmental degradation results, a population crash occurs. In 1910 a small herd of 26 reindeer was introduced on one of the Pribilof Islands of Alaska. The herd's population increased exponentially for about 25 years until there were approximately 2,000 reindeer, many more than the island could support. The reindeer overgrazed the vegetation until the plant life was almost wiped out. Then, in slightly over one decade, as reindeer died from starvation, the number of reindeer plunged to eight, one-third the size of the original introduced population (Figure 8–3).

## THE HUMAN POPULATION

Now that we have examined some of the basic concepts of population biology, we can apply those concepts to the human population. Examine Figure 8–4, which shows the worldwide increase in the human population since the New Stone Age, approximately 10,000 years ago. Now look back at Figure 8–1 and observe how the human population is increasing exponentially. The characteristic J-curve of exponential growth reflects the decreas-

**Figure 8–4**
The human population has been increasing exponentially from the New Stone Age to the present. (Mike Andrews © 1993 Earth Scenes)

ing amount of time it has taken to add each additional billion people to our numbers. It took thousands of years for the human population to reach 1 billion, 130 years to reach 2 billion, 30 years to reach 3 billion, 15 years to reach 4 billion, and 12 years to reach 5 billion.

One of the first people to recognize that the human population cannot continue to increase indefinitely was Thomas Malthus, a British economist who lived in the 18th century. He pointed out that human population growth was not always desirable (a view contrary to the beliefs of his day) and that the human population was capable of increasing faster than the food supply. He maintained that the inevitable consequences of population growth were famine, disease, and war.

As of 1992, our world population was over 5.4 billion and was increasing by approximately 93 million humans each year (about 175 people per minute).[3] This increase is not due to a rise in the birth rate (*b*). In fact, the worldwide birth rate has actually declined slightly during the past 200 years. The population increase is due instead to a *decrease in*

the death rate (*d*), which has occurred primarily because of greater food production, better medical care, and improved sanitation practices. For example, from about 1920 to 1990, the death rate in

**Figure 8–5**
In Mexico, the birth and death rates have generally declined in this century. Because the death rate has declined much more than the birth rate, however, Mexico has experienced a high growth rate.

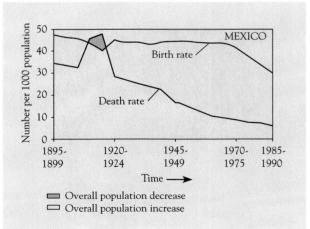

[3] Unless otherwise noted, all population data in this chapter were obtained from the Population Reference Bureau, a private, nonprofit educational organization that disseminates demographic and population information.

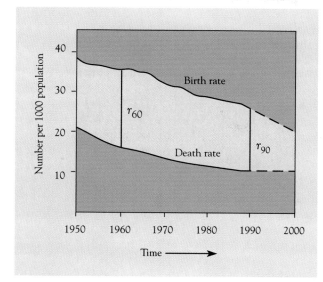

**Figure 8–6**
Worldwide birth and death rates. The difference between the birth rate and the death rate equals the annual rate of population increase (shaded area). Since 1950, both birth and death rates have been declining. Because the number of births has declined slightly more rapidly than the number of deaths, the worldwide rate of increase has also declined; that is, the growth rate in 1990 ($r_{90}$) was slightly less than the growth rate in 1960 ($r_{60}$).

Mexico fell from approximately 40 to 6, whereas the birth rate dropped from approximately 40 to 30 (Figure 8–5).

The human population has reached a turning point. Although our numbers continue to increase, the global *rate* of population growth ($r$) has declined over the past several years (Figure 8–6). Despite this declining growth rate, it will take many years for the world population to stabilize ($r = 0$), primarily because of the momentum provided by our current age structure, with a preponderance of young people (to be discussed shortly).

Population experts at the United Nations and the World Bank have projected that the worldwide rate of population growth will continue to slowly decrease until **zero population growth** is attained. Zero population growth—when the birth rate equals the death rate—is projected to occur around 2089 A.D., when it is anticipated that the human population will level off at approximately 10.4 billion (Figure 8–7). This number is almost twice the 1992 population of the world.

Population projections are "what if" exercises: given certain assumptions about future tendencies in the birth rate, death rate, and migration, an area's population can be calculated for a given number of years into the future. Population projections indicate the changes that may be upcoming, but they must be interpreted with care because they vary depending on the assumptions made. For example, in projecting that the world population will stabilize at 10.4 billion by the end of the 21st century, demographers assume that the average number of children born to each woman in all countries will have declined to just about 2 by 2040 A.D. (in 1992, the average number of children born to each woman in all countries was 3.3). If that decline does not occur by 2040 A.D., our population will not stabilize at 10.4 billion people by the end of the

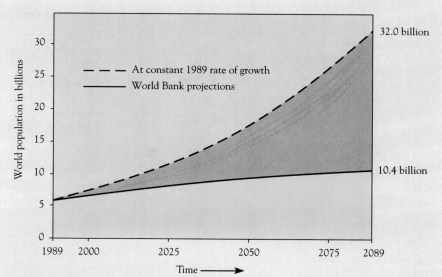

**Figure 8–7**
A comparison of world population projections. Demographers at the World Bank and the United Nations have projected that our growth rate will decrease, resulting in a 2089 population of 10.4 billion. However, if the population continues to increase at its 1989 rate, there will be 32 billion humans by 2089.

21st century, but will stabilize later and at a greater number. For example, if the population were to continue to increase at its 1992 growth rate, there would be more than 30 billion humans toward the end of the 21st century.

The main unknown factor in this population growth scenario is the carrying capacity of the environment. No one knows how many humans can be supported by Earth, and projections and estimates vary widely, depending on what assumptions are made. It is also not clear what will happen to the human population when the carrying capacity is approached. Optimists suggest that the human population will stabilize because of a decrease in the birth rate and an increase in the death rate (more people will die because the Earth cannot support them). Some experts take a more pessimistic view and predict that the widespread degradation of our environment caused by our ever-expanding numbers will make the Earth uninhabitable for humans and that a massive wave of human deaths will occur.

## Demographics of Countries

Whereas worldwide population figures illustrate overall trends, they do not describe other important aspects of the human population story, such as population differences from country to country. **Demographics,** the branch of sociology that deals with population statistics, provides interesting information on the populations of countries. As you probably know, not all countries have the same rates of population increase. Countries can be clas-

sified into two groups—developed and developing—depending upon rate of population growth and other factors, such as degree of industrialization and relative prosperity (Table 8–1).

**Developed countries** (also called **highly developed countries**), such as the United States, Canada, France, Germany, Sweden, Australia, and Japan, have low rates of population growth, are highly industrialized, and have high per-capita incomes relative to the rest of the world. Developed countries have the lowest birth rates in the world. Indeed, some developed countries (such as Germany) have birth rates just below those needed to sustain their populations and are thus declining slightly in numbers. Highly developed countries also have a very low **infant mortality rate** (the number of infant deaths per 1,000 live births). The infant mortality rate of the United States was 9.0 in 1992, for example, compared with a worldwide rate of 68. Highly developed countries also have longer life expectancy (74.5 years versus 62.5 years worldwide in 1992) and a high average per-capita income ($17,900 versus $3,790 worldwide in 1992).

**Developing countries** fall into two subcategories, moderately developed and less developed. Mexico, Turkey, Thailand, and most South American nations are examples of **moderately developed countries.** Their birth rates and infant mortality rates are higher than those of highly developed countries. Moderately developed countries have a medium level of industrialization, and their average per-capita incomes are lower than those of highly developed countries. **Less developed countries** include Bangladesh, Niger, Ethiopia, and Laos. These countries have the highest birth rates, the highest infant mortality rates, the shortest life expectan-

---

**Table 8–1**
**A Comparison of 1992 Population Data in Developed and Developing Countries**

| | Developed | Developing | |
| --- | --- | --- | --- |
| | Ex-United States (highly developed) | Ex-Brazil (moderately developed) | Ex-Kenya (less developed) |
| Fertility rate | 2.0 | 3.1 | 6.7 |
| Doubling time at current rate | 89 | 37 | 19 |
| Infant mortality rate | 9.0 | 69 | 62 |
| Life expectancy at birth | 75.5 | 65 | 61 |
| Per-capita income (U.S. $) | $21,700 | $2,680 | $370 |

cies, and the lowest average per-capita incomes in the world.

A country's doubling time can place it as a highly, moderately, or less developed country: the shorter the doubling time, the less developed the country. At current rates of growth, for example, the doubling times are 19 years for Togo, 25 years for Ethiopia, 30 years for Mexico, 34 years for India, 89 years for the United States, and 347 years for Belgium.

It is also instructive to examine **replacement-level fertility,** that is, the number of children a couple must produce in order to "replace" themselves. Replacement-level fertility is usually given as 2.1 children in developed countries and 2.7 children in developing countries. The number is always greater than 2.0 because some children die before they reach reproductive age. Thus, higher infant mortality rates are the main reason that replacement levels are greater in developing countries than in developed countries. Worldwide, the **total fertility rate**—the average total number of children born to each woman—is currently 3.3, well above replacement levels in developed *and* developing countries. (See Focus On: Fertility Rates in Sub-Saharan Africa for an examination of social and cultural effects on fertility.)

**Mini-Glossary of Population Terms**

**birth rate:** The number of births per 1,000 people.

**death rate:** The number of deaths per 1,000 people.

**growth rate:** The natural increase of a population per year.

**doubling time:** The number of years it will take a population to double in size, given its current rate of increase.

**immigration:** The migration of individuals into a population from another area or country.

**emigration:** The migration of individuals from a population, bound for another area or country to live.

**zero population growth:** The condition when a population is no longer increasing because the birth rate equals the death rate.

**infant mortality rate:** The number of infant deaths per 1,000 live births.

**replacement-level fertility:** The number of children a couple must have to "replace" themselves.

**total fertility rate:** The average number of children born to each woman during her lifetime.

**Demographic Stages** Demographers recognize four demographic stages, based on their observations of Europe as it became industrialized and urbanized (Figure 8–8). These stages converted Europe from

**Figure 8–8**

The demographic transition consists of four demographic stages through which a population progresses as its society becomes industrialized. Note that the death rate declines first, followed by a decline in the birth rate.

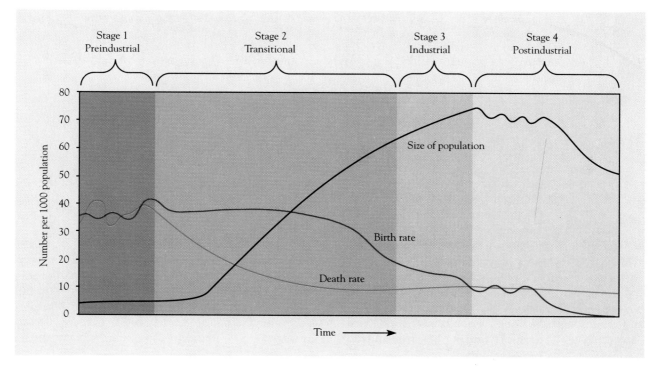

## Fertility Rates in Sub-Saharan Africa

At first glance, reducing fertility rates in any society might seem to be simple: educate people about the economic and health benefits of lower fertility rates, and supply them with contraception. In many areas of the world, fertility rates have begun to decline because of planned *and* unplanned circumstances. The decline has occurred because the structure and belief systems of these societies can accommodate the idea of reduced fertility for the greater benefit of society. The one exception to this general rule occurs in sub-Saharan Africa, the arid regions around the edge of the Sahara Desert.

While fertility rates are declining in most other parts of the world, they are increasing in sub-Saharan Africa. At the current rate of growth, Africans will probably make up more than one-fourth of the human race by the late 2000s. Why? The answer most likely lies in ancient social and family patterns that resist change.

Traditional African society is based on ancestry and descent. From a religious perspective, this means that most Africans (even those who have converted to Islam or Christianity) believe in the intervention of ancestral spirits in their lives. From a social perspective, this means that the preservation of lineage (not the establishment of a family for its own sake) is most important. Although African society is patrilineal (lineage is through the father), the significant family unit is a woman and her children, rather than two parents and their children. And *any* children borne by a woman—not just those sired by a single husband—are accepted as part of the family and of its lineage through the husband. Indeed, more children born to a woman mean a stronger clan and lineage, regardless of who the biological fathers are.

Thus, large families are viewed as positive in sub-Saharan African culture. Indeed, small families are considered bad, and the women of such families are often believed to be tainted by evil. Infertility, or barrenness, is a traditional sub-Saharan woman's greatest fear, followed by the fear that all her children will die, which would mean the end of her husband's line of descent. In such a society, contraception becomes almost unthinkable—let alone abortion or even chastity. Even those women who practice contraception usually do so in order to lengthen the intervals between pregnancies rather than to limit the number of children they bear.

Women and their children are also important to African society because they do most of the agricultural work. Land is usually held communally by a clan rather than by an individual family, and the more children a woman has, the more food the clan can produce.

It might seem that African governments could require family planning. However, African leaders generally fear that instituting family planning programs would bring on accusations that they are trying to undermine African society. That society's beliefs have persisted for thousands of years, and they will not change easily or rapidly.

relatively high birth and death rates to relatively low birth and death rates. Because all highly developed and moderately developed countries with more advanced economies have gone through this demographic transition, demographers generally assume that the same progression will occur in less developed countries as they become industrialized.

In the first stage, called the **preindustrial stage,** birth and death rates are high and population grows at a modest rate; although women have many children, the infant mortality rate is high. Intermittent famines, plagues, and wars also increase the death rate, so the population grows slowly. Finland in the late 1700s is an example of the first demographic stage.

As a result of the improved health care and more reliable food and water supplies that accompany the beginning of an industrial society, the second demographic stage, called the **transitional stage,** is characterized by a lowered death rate. Because the birth rate is still high, the population grows rapidly. Finland in the mid-1800s was in the second stage, and today, much of Latin America, Asia, and Africa are in the second demographic stage.

The third demographic stage, the **industrial stage,** is characterized by a decline in the birth rate and takes place at some point during the industrialization process. The decline in the birth rate, along with the relatively low death rate, slows population growth. For Finland, this occurred in the early 1900s.

The fourth demographic stage, sometimes called the **postindustrial stage,** is characterized by low birth and death rates. In countries that are heavily industrialized, people are better educated and more affluent; they tend to desire smaller families and take steps to limit family size. The population grows very slowly or not at all in the fourth demographic stage. This is the situation in such developed countries as the United States, Canada, Australia, the former U.S.S.R., Japan, and most of Western Europe, including Finland.

Once a country reaches the fourth demographic stage, is it correct to assume it will continue to have a low birth rate indefinitely? The answer is that we don't know. Low birth rates may be a permanent response to the socioeconomic factors that are a part of an industrialized, urbanized society. On the other hand, low birth rates may be a temporary response to socioeconomic factors such as the changing roles of women in developed countries. No one knows for sure.

### Envirobrief

**A Look at World Population**

The demographics of world population are complex and constantly changing. Here are some facts about the world's population in 1992:

Rwanda, in eastern Africa, has the highest total fertility rate in the world (8.0). Hong Kong has the lowest total fertility rate (1.2), with Spain, San Marino, and Italy being close seconds (1.3 each). The total fertility rate in the United States is 2.0.

Afghanistan has the highest infant mortality rate in the world (172), and Liechtenstein, in Western Europe, has the lowest (2.7). The infant mortality rate in the United States is 9.0.

Three countries in Europe—Bulgaria, Hungary and Germany—have declining populations. Italy and Croatia have the highest population doubling time (1,386 years). Gaza has the lowest doubling time (15 years), and three nations—Syria, Marshall Islands, and Zambia—are tied for the second lowest doubling time (18 years). The doubling time for the United States is 89 years.

Switzerland boasts the highest per-capita income (based on U.S. dollars), $32,790; Mozambique ($80), Ethiopia ($120), and Tanzania ($120) have the three lowest. Per-capita income in the United States is $21,700.

Source: Population Reference Bureau

### Envirobrief

**The Polarizing of Wealth**

As the world's economy has grown over the past four decades, wealthy countries have become wealthier and poor countries have slid deeper into poverty. Studies at the University of Pennsylvania show that while *world* average per capita income has doubled since 1950, a closer look reveals that incomes in *developed* nations account for most of this increase. People in industrialized countries have gotten three times richer, while those in developing countries earn the same or less.

Why has the population stabilized in many developed countries? The reasons are complex. The decline in birth rate has been associated with an improvement in living standards. However, it is difficult to say whether improved socioeconomic conditions have resulted in a decrease in birth rate, or a decrease in birth rate has resulted in improved socioeconomic conditions. Perhaps *both* are true. Another reason for the decline in birth rate in developed countries is the increased availability of family planning services. Still other factors influence birth rate, including education, particularly of women, and urbanization of our society; they are considered in greater detail in Chapter 9.

| Table 8–2 Fertility Change in Selected Developing Countries | | |
|---|---|---|
| | **Total Fertility Rate\*†** | |
| **Country** | *1960–65* | *1992* |
| Afghanistan | 7.0 | 6.9 |
| Bangladesh | 6.7 | 4.9 |
| Brazil | 6.2 | 3.1 |
| China | 5.9 | 2.2 |
| Egypt | 7.1 | 4.4 |
| Guatemala | 6.9 | 5.2 |
| India | 5.8 | 3.9 |
| Kenya | 8.1 | 6.7 |
| Mexico | 6.8 | 3.8 |
| Nepal | 5.9 | 6.1 |
| Nigeria | 6.9 | 6.5 |
| Thailand | 6.4 | 2.4 |

\*Total fertility rate = average number of children born to each woman during her lifetime.

†Source: Population Reference Bureau, Washington, D.C.

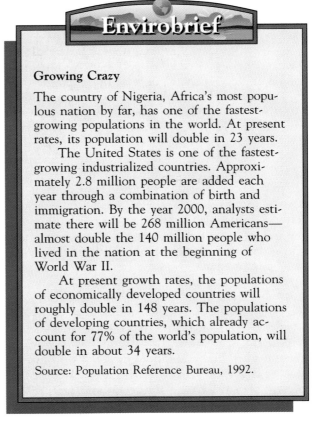

## Envirobrief

### Growing Crazy

The country of Nigeria, Africa's most populous nation by far, has one of the fastest-growing populations in the world. At present rates, its population will double in 23 years.

The United States is one of the fastest-growing industrialized countries. Approximately 2.8 million people are added each year through a combination of birth and immigration. By the year 2000, analysts estimate there will be 268 million Americans—almost double the 140 million people who lived in the nation at the beginning of World War II.

At present growth rates, the populations of economically developed countries will roughly double in 148 years. The populations of developing countries, which already account for 77% of the world's population, will double in about 34 years.

Source: Population Reference Bureau, 1992.

The populations in many developing countries are beginning to approach stabilization (the fertility rate must decline in order for a population to stabilize; see Table 8–2 and note the general decline in total fertility rate in selected developing countries from the 1960s to 1992). Worldwide, the total fertility rate in developing countries has decreased from an average of 6.1 children per woman in 1970 to 3.8 in 1992. In the past decade, fertility rates have declined by at least 25 percent in countries such as Brazil, Indonesia, and Mexico.[4] Fertility rates continue to increase in some African countries—Ethiopia and Cameroon, for example.

## Age Structure of Countries

In order to predict the future growth of a population, it is important to know its **age structure,** the distribution of the population by age. The number

of males and number of females at each age, from birth to death, can be represented in an **age structure diagram** (Figure 8–9). The diagram is divided vertically in half, one side representing the males in a population and the other side the females. The bottom third of the diagram represents pre-reproductive humans (from 0 to 14 years of age); the middle third, reproductive humans (15 to 44 years); and the top third, post-reproductive humans (45 years and older). The widths of these segments are proportional to the population sizes—a greater width implies a larger population.

**Predicting Population Using Age Structure Diagrams** The overall shape of an age structure diagram indicates whether the population is increasing, stable, or shrinking. The age structure diagram of a country with a very high growth rate—for example, Nigeria or Venezuela—is shaped like a pyramid (Figure 8–10a). Because the largest percentage of the population is in the pre-reproductive age group, the probability of future population growth is great. When all these children mature, they will become the parents of the next generation. Thus, *even if the fertility rate of such a country is at replace-*

[4]Although fertility rates in these countries have declined, it should be remembered that they are still greater than replacement-level fertility. Consequently, the populations of these countries are still increasing.

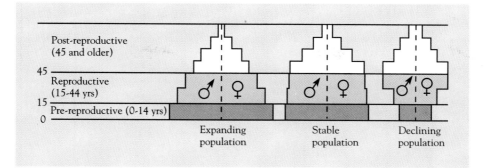

**Figure 8–9**
Generalized age structure diagrams for an expanding population, a stable population, and a population that is decreasing in size. In each, the left half of the diagram represents the males in the population, and the right half represents the females. Each diagram is divided horizontally into age groups, and the width of each segment represents the population size of that group.

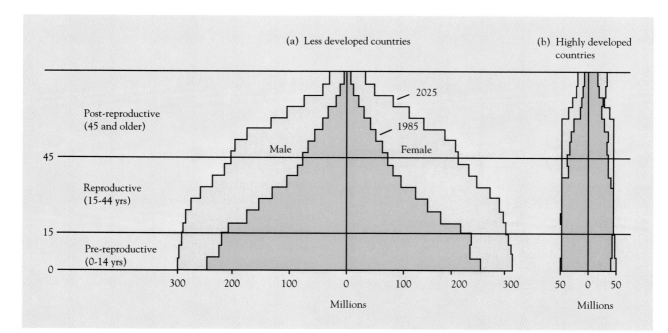

**Figure 8–10**
Age structure diagrams for (a) less developed and (b) highly developed countries. The dark blue region represents actual age distribution in 1985. The light blue region represents projected age distribution in 2025. These age structure diagrams indicate that less developed countries have a much higher percentage of young people than highly developed countries. As a result, less developed countries are projected to have greater population growth than highly developed countries.

ment level, the population will continue to grow. In contrast, the more tapered bases of the age structure diagrams of countries with stable or declining populations indicate a smaller proportion of children to become the parents of the next generation.

The age structure diagram of a stable population, one that is neither growing nor shrinking, demonstrates that the numbers of people at pre-reproductive and reproductive ages are approximately the same (Figure 8–10b). Also, a larger per-

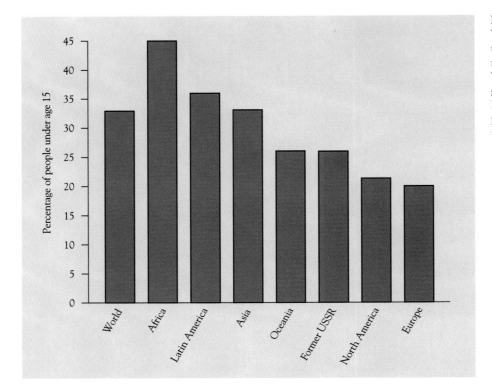

**Figure 8–11**
The percentage of the population under age 15 for various regions of the world in 1992. The higher this percentage, the greater the potential for population growth when those people reach their reproductive years.

---

### A Look at U.S. Population

Here are some facts about the changing and complex population of the United States.

Between 1980 and 1990, Nevada had the highest rate of increase in population (50.4 percent), with Alaska second (37.4 percent). West Virginia and the District of Columbia had population declines (−8.0 percent and −4.8 percent, respectively).

California and Florida had the highest rates of immigration (1987–1988 data), with a whopping 364,243 and 266,353 people added, respectively, as compared to Texas, which lost 122,667 people due to emigration.

California had 533,148 births in 1988, whereas Wyoming had only 7,162.

In 1988 Alaska had the highest percentage of its population below the age of 18 (31.9 percent), whereas Florida had the highest percentage of its population over age 65 (17.8 percent).

Source: Population Reference Bureau

centage of the population is older (post-reproductive) than in a rapidly increasing population. Many countries in Europe have stable populations.

In a population that is shrinking in size, the pre-reproductive age group is *smaller* than either the reproductive or post-reproductive group. Germany, Bulgaria, and Hungary are examples of countries with slowly shrinking populations.

Worldwide, it is estimated that one-third of the population is under age 15 (Figure 8–11). When these people enter their reproductive years, they have the potential to cause a large increase in the population growth rate. Even if the birth rate does not increase, the population growth rate will increase simply because there are more females reproducing.

Most of the worldwide population increase that has occurred since 1950 has taken place in developing countries (as a result of the younger age structure and the higher-than-replacement-level fertility rates of their populations) (Figure 8–12). In 1950, 66.8 percent of the world's population was in developing countries in Africa, Asia (minus Japan), and Latin America; the remaining 33.2 percent of the population was in developed countries in Europe, the U.S.S.R., Japan, Australia, and North America. Between 1950 and 1992, the world's population more than doubled in size, most of that

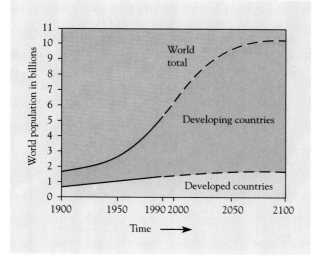

**Figure 8–12**

Most of the worldwide increase in population since 1950 has occurred in developing countries.

will occur during the next century will take place in developing countries.

## Demographics of the United States

The United States has one of the highest rates of population increase of all the developed countries. For example, the U.S. population increased by about 2.8 million from 1991 to 1992. This translates to a 1992 percent annual increase of 0.8, which is high compared to many developed countries—for example, Europe's 1992 percent annual increase was 0.2. (These figures take only birth and death rates into account; migration is not considered.)

Although the U.S. birth rate has been decreasing for several years, we are still experiencing an increase in population growth. There are two reasons for this. First, our population growth has a built-in momentum because of the Baby Boom, the large wave of births that followed World War II (Figure 8–13). The babies born then are now in their reproductive years. Thus, although the number of children born per female has declined (in 1992 the total fertility rate was 2.0), there is an

growth occurring in developing countries. As a result, in 1992 the number of people in developing countries had risen to 77.4 percent of the world's population. Most of the population increase that

**Figure 8–13**

The total fertility rate in the United States from 1917 to 1992. Note the high average number of children born per woman during the Baby Boom years (1945 to 1962).

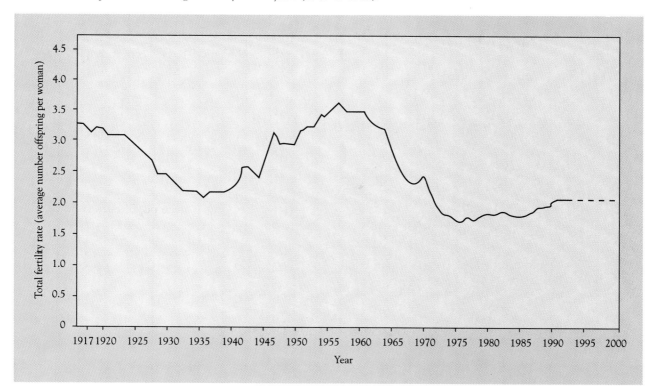

increase in births because of the greater number of females who are bearing children.

A second reason for the large growth rate in U.S. population is immigration, which has a greater effect on population size in the United States than in most other nations (see Focus On: A History of U.S. Immigration). Approximately 30 percent (5 million people) of our population increase of 16 million from 1980 to 1987 was due to immigration. This number represents the immigrants who were *legally* admitted. The number of *illegal* immigrants who gained access to the United States and were not deported is not known.

## Focus On

### A History of U.S. Immigration

Birth control and improved health care services are not the only ways to affect or control population in a certain area. Another way is through immigration laws. Prior to 1875, there were no immigration laws in the United States, and thus no such thing as illegal immigration. In 1875, however, Congress passed a law denying convicts and prostitutes entrance to the United States. In 1882 the Chinese Exclusion Act was passed, and in 1891 the Bureau of Immigration was established. Hence, by the late 1800s a policy of selective exclusion was officially established in the United States and began to shape the population of this country.

During the early 20th century, Congress set numerical restrictions on immigration, including quotas allowing only a certain number of people from each foreign country to immigrate. This severely restricted entrance to the United States by people from Asia, southern and Eastern Europe, and Mexico. With these stronger laws, illegal immigration began to increase. U.S. immigration policy relaxed during World War II, when China became an ally and labor shortages made it possible for workers from Mexico, Barbados, Jamaica, and the British Honduras to gain temporary residence in the United States.

In 1952 the Immigration and Nationality Act was passed, and although it has been revised since then (it is now called the Immigration Reform and Control Act, or IRCA), it is still the basic immigration law in effect. The act applies penalties to U.S. citizens who harbor illegal aliens; in 1986 it was expanded to include employers who knowingly hire undocumented immigrants. Its intent is to control and eventually eliminate the flow of illegal aliens into the United States. Because the IRCA is relatively new, it is not yet known what its overall effect will be on the shape and size of the population of the United States.

Ellis Island in New York harbor was the arrival station for nearly all legal immigrants to the United States. Between 1892, when it opened, and 1954, when it closed, about 20 million immigrants (90 percent of the immigrants entering the United States during that time) passed through Ellis Island. (National Park Service, Ellis Island Immigration Museum)

# SUMMARY

**1.** The principles of population biology that are used to understand populations of living organisms apply to humans.

**2.** On a global scale, the rate of change in a population is due to two factors, the number of births and the number of deaths: populations increase in size as long as the birth rate is greater than the death rate.

**3.** In addition to birth rate and death rate, migration must be considered when examining changes in local populations. The number of people emigrating from a country and the number immigrating into a country affect its population size and growth rate.

**4.** Although certain populations exhibit exponential growth for limited periods of time, eventually the rate of population growth decreases to around zero. This occurs when the carrying capacity of the environment is reached.

**5.** Currently, human population is increasing exponentially. The rate of population increase has declined slightly over the past several years, however, leading demographers to project that the world population will stabilize at approximately 10.4 billion by the end of the 21st century.

**6.** Highly developed countries have the lowest birth rates, the lowest infant mortality rates, the longest life expectancies, and the highest per-capita incomes. Developing countries have the highest birth rates, the highest infant mortality rates, the shortest life expectancies, and the lowest per-capita incomes.

**7.** The age structure of a population greatly influences its population changes. It is possible for a country to have replacement-level fertility and still experience population growth if the largest percentage of the population is in the pre-reproductive years.

**8.** The United States is experiencing population growth caused by immigration and the post–World War II Baby Boom (which has increased our birth rate as these postwar babies have matured and reproduced).

# DISCUSSION QUESTIONS

**1.** If a population of 100,000 has 1,000 births per year and 250 deaths per year, what is its growth rate? What is its doubling time?

**2.** Should the rapid increase in world population be of concern to the average citizen in the United States? Why or why not?

**3.** Explain two different ways a population could have an increase in its growth rate when its birth rate is declining.

**4.** Explain the effects of medical advances on population size; on age structure.

**5.** What is population momentum? Draw an age structure diagram to represent population momentum.

**6.** If all the women in the world suddenly started bearing children at replacement-level fertility rates, would the population stop increasing immediately? Why or why not?

**7.** Should the United States increase or decrease the number of legal immigrants? Present arguments in favor of both sides.

**8.** What is replacement-level fertility? Why is it higher in developing countries?

# SUGGESTED READINGS

Bureau of the Census. *United States Population Estimates by Age, Sex, and Race: 1980–1987*. U.S. Department of Commerce, Washington, D.C. Detailed tables and charts show the trends in population change for the United States during the 1980s.

Caldwell, J. C., and P. Caldwell. High fertility in sub-Saharan Africa. *Scientific American*, May 1990, 118–125. Examines why birth rates and population growth have not declined in sub-Saharan Africa, when they have everywhere else in the developing world.

Espenshade, T. J. A short history of U.S. policy toward illegal immigration. *Population Today* 18:2, February 1990, 6–9. A discussion of the population growth of illegal aliens in the United States and of political efforts to control it.

Haub, C. Understanding population projections. *Population Bulletin* 42:4, December 1987. A comprehensive evaluation of the challenges facing demographers as they calculate population projections.

Population Reference Bureau. *United States Population Data Sheet*. Washington, D.C. Provides an annual demographic picture of the United States. Includes population changes by region and state.

Population Reference Bureau. *World Population Data Sheet*. Washington, D.C. An annually produced chart that provides population data for all countries. Includes birth rates, death rates, infant mortality rates, total fertility rates, and life expectancies as well as other pertinent information.

# Chapter 9

*New York City urbanization (Bruce F. Molnia/Terraphotographics)*

# Facing the Problems of Overpopulation

Even though overpopulation is not the sole reason for world hunger, environmental problems, underdevelopment, poverty, and urban problems, all of those issues are aggravated by rapid population growth. Some of the factors that influence the fertility rate include cultural traditions, age at marriage, availability of family planning services, government policies, and educational opportunities, particularly for women. If population growth is slowed and resource consumption per person is decreased, the world will be in a better position to tackle many of its most serious environmental problems.

## THE HUMAN POPULATION CRISIS

Most people would agree that all people in all countries should have access to the basic requirements of life: food, shelter, and clothing. Over the next century, however, it will become increasingly difficult to meet these basic needs, especially in countries that have not achieved population stabilization. Moreover, it is likely that the social, political, and economic problems resulting from continued population growth in these countries will affect other countries that have already achieved stabilized populations and high standards of living. For these reasons, population growth should be of concern to the entire world community, regardless of where it is occurring.

As our numbers increase during the next 100 years, environmental deterioration, hunger, persistent poverty, and health issues will continue to challenge us. Already, for example, the need for food for the increasing numbers of people living in environmentally fragile dryland areas of the world, such as sub-Saharan Africa,[1] has led to overuse of the land for grazing and crop production (Figure 9–1). As a result of such overuse, these formerly productive lands have been degraded into deserts at the rate of approximately 20.2 million hectares (50 million acres) per year (see Chapter 17). Although it is possible to reclaim such drylands, efforts to do so are made difficult, if not completely impeded, by the large numbers of people and their animal herds trying to live off the land.

You may recall from Chapter 8 that when the carrying capacity of the environment is reached, the population will stabilize or crash due to a decrease in the birth rate, an increase in the death rate, or a combination of both. No one knows whether the Earth can support 10 billion or 6 billion humans. It may be that we have already reached Earth's carrying capacity and that the numerous environmental problems we are experiencing will cause the worldwide population increase to come to a halt.

On a national level, developing countries have the largest rates of population increase and often have the fewest resources to support their growing numbers. In order for a country to support a certain number of humans over an extended period of time, it must have either the agricultural land to raise enough food for those people or enough of other natural resources (for example, minerals or oil) to provide buying power to purchase food.

[1] Arid regions around the edge of the Sahara Desert.

**Figure 9–1**
Overgrazing of grasslands brought on by overpopulation contributes to desertification. This area in the African country of Burkina Faso now offers meager support for wildlife and the domestic animals of humans. (Robert E. Ford/Terraphotographics)

### Population and World Hunger

In 1985, 30 million people in sub-Saharan Africa starved. Although sub-Saharan Africa has the greatest concentration of suffering, many of the world's people do not get enough food to thrive, and in certain areas of the world people, especially children, still starve to death (Figure 9–2). The *cause* of world hunger, however, is anything but clear.[2] Experts agree that complex relationships exist among population, world hunger, poverty, and environmental problems, but they do not agree about the most effective way to stop world hunger.

Those who think that population growth is the root cause of the world's food problems point out that countries with some of the highest fertility rates[3] are also the ones with the greatest food shortages. They argue that it is imperative to reduce population growth, even through drastic measures such as the establishment of world population quotas. Under such a system, a country that exceeded its assigned population size would not be eligible for relief from the international community during times of food shortages.

Some people think the way to tackle world food problems is not by controlling population but by promoting the economic development of countries that are unable to produce adequate food for their people. They presume that development

[2] Famines are a notable exception. The cause of famine can usually be attributed to bad weather, insects, war, or some other disaster.

[3] Recall that total fertility rate is the average number of children born to each woman of a population.

**Figure 9–2**
As the population continues to increase, world hunger will become an even greater problem. Overpopulation is not the only reason that there is not enough food for all the world's people, however. Other underlying causes include poverty and environmental mismanagement. (Thomas S. England/Photo Researchers, Inc.)

would provide the appropriate technology for the people living in those countries to increase their food production. Also, once a country becomes more developed, its fertility rate should decline (see Chapter 8), helping to lessen the population problem.

A third group of people thinks that neither controlling population growth nor enhancing economic development alone will solve world food problems. They argue that the inequitable distribution of resources is the primary cause of world hunger. According to this view, there are enough resources, land, and technologies to produce food for all humans, but people on the lower end of the socioeconomic scale in many developing countries do not have *access* to the resources they need to support themselves.

These differing viewpoints indicate that the relationship between world hunger and population is complex and may be affected by economic development as well as by poverty and the uneven distribution of resources. Regardless of whether or not population growth is the primary cause of world hunger, however, it is clear that world food problems are *exacerbated* by population pressures.

## Population, Resources, and the Environment

The relationships among population growth, utilization of natural resources, and environmental degradation are also complex. We address the details of resource management and environmental problems in later chapters, but for now, we can make two useful generalizations. (1) The resources that are essential to an individual's survival are small, but a rapidly increasing *number* of people (as we see in developing countries) tends to overwhelm and deplete a country's soils, forests, and other natural resources (Figure 9–3a). (2) In developed nations, individual resource demands are large, far above requirements for survival. In order to satisfy their desires rather than their basic needs, people in more affluent nations exhaust resources and degrade the global environment through extravagant *consumption* and "throwaway" life styles (Figure 9–3b).

**Types of Resources**  When examining the effects of population on the environment, it is important to distinguish between the two types of natural resources: nonrenewable and renewable. **Nonrenewable resources,** which include minerals (such as aluminum, tin, and copper) and fossil fuels (coal, oil, and natural gas), are present in limited supplies and are depleted by use. They are not replenished by natural processes within a reasonable period of time. Fossil fuels, for example, take millions of years to form.

In addition to a nation's population, several other factors affect how nonrenewable resources are used—including how efficiently the resource is extracted and processed, and how much of it is required or consumed by different groups of people. Nonetheless, the inescapable fact is that Earth has a finite supply of nonrenewable resources that sooner or later will be exhausted. In time, technological advances may enable us to find or develop substitutes for nonrenewable resources. And slowing the rate of population growth will help us buy time to develop such alternatives.

**Renewable resources** include trees in forests; fish in lakes, rivers, and the ocean; fertile agricul-

(a)

(b)

**Figure 9–3**
People and natural resources. (a) The rapidly increasing number of people in developing countries overwhelms their natural resources, even though individual resource requirements may be low. (b) People in developed countries consume a disproportionate share of the world's natural resources. (*a*, Robert Caputo, © 1988 National Geographic Society; *b*, Pamela Spaulding, © 1988 National Geographic Society)

tural soil; and fresh water in lakes and rivers. Nature replaces these resources, and they can be used forever as long as they are not overexploited in the short term. In developing countries, forests, fisheries, and agricultural land are particularly important renewable resources because they provide food. Indeed, many people in developing countries are subsistence farmers, able to harvest just enough food so that they and their families can survive.

Rapid population growth can cause renewable resources to be overexploited. For example, when fisheries are overharvested to the point where fish populations are so low that they cannot recover, then there will be too few fish to serve as a food source (see Chapter 18). A similar problem arises when land that is inappropriate for farming (such as mountain slopes or tropical rain forests) is used to grow crops (see Chapter 17). Although this practice may provide a short-term solution to the need for food, it does not work in the long run, because when these lands are cleared for farming, their agricultural productivity declines rapidly and severe environmental deterioration occurs. Renewable resources, then, are not always renewable. They must be used in a sustainable way—that is, in a

manner that gives them time to replace or replenish themselves.

Whereas developed countries waste their resources by overconsumption, in developing countries the effects of population growth on natural resources are particularly critical. The economic growth of developing countries is often tied to the exploitation of natural resources. These countries are faced with the difficult choice of exploiting natural resources to provide for their expanding populations in the short term or conserving those resources for future generations. (It is instructive to note that the economic growth and development of the United States and of other highly developed nations came about through the exploitation—and in some cases the destruction—of their resources.)

**Population: Numbers Versus Resource Consumption** Whereas it is true that resource issues are clearly related to population size (more people use more resources), an equally if not more important factor is a population's *resource consumption*. People in developed countries are conspicuous consumers; their use of resources is greatly out of proportion to their numbers. A single child born in a developed

### The American Baby as Consumer

Every minute, 7.78 babies are born in the United States; that's 4,089,600 per year. As soon as each new baby is born, it becomes a consumer, using the resources of Earth. Among other things, in his or her lifetime, this new baby will consume:

Calories equal to 100,000 hamburgers, 2 million french fries, 50,000 chocolate shakes, 50,000 apples, and 4,000 gumdrops

One-half million gallons of water just to bathe or shower, and 1 ton of soap and detergent

Enough energy to drive around the world 1,500 times

One hundred fifty trees for wood and paper

Tons of metals, textiles, plastics, and glass

American babies consume far more in resources than do those from other parts of the world. Although people in developing countries work hard to improve their standards of living, if they all began to consume at the rate of Americans, the world would need 20 times the resources it now consumes.

Source: U.S.A. exhibit, Expo '74.

country such as the United States, for example, causes a greater impact on the environment and on resource utilization than do a dozen or more children born in a developing country. Many natural resources are needed to provide the air conditioners, disposable diapers, cars, video cassette recorders, and other "comforts" of life in developed nations. Thus, the disproportionately large consumption of resources by developed countries affects natural resources and the environment as much as does the population explosion in the developing world.

**Population and Environmental Impact: A Simple Model** Although human impact on the environment is complex, we have identified the three factors that are most important in determining this impact: (1) the number of people in a particular

area, (2) the effect on the environment of obtaining and using the resources in that area, and (3) the resource utilization (amount of resources used) per person. A country is *overpopulated* if it has more people than its resource base can support without damage to the environment. If we combine our three factors in order to compare human impact on the environment in developing and developed countries, we see that a country can be overpopulated in two ways. **People overpopulation** occurs when the environment is worsening from *too many people*, even if those people consume few resources per person. People overpopulation is the current problem in many developing nations. In contrast, **consumption overpopulation** occurs when each individual in a population consumes too large a share of resources. The effect of consumption overpopulation on the environment is the same as that of people overpopulation—pollution and degradation of the environment. Many affluent developed nations suffer from consumption overpopulation: developed nations represent only 20 percent of the world's population, yet they consume about 80 percent of its resources.

### Economic Effects of Continued Population Growth

The relationship between economic development and population growth is complex and difficult to evaluate. Some economists have argued that population growth stimulates economic development and technological innovation. Others hold that developmental efforts are hampered by a rapidly expanding population. At the present time, most major technological advances are occurring in countries where population growth is low to moderate, an observation that seems to support the latter point of view.

The National Research Council of the U.S. National Academy of Sciences published a report in 1986 that examined whether large increases in population were a deterrent to economic development.[4] Their panel of experts took into account the complex interactions among world problems such as underdevelopment, hunger, poverty, environmental problems, and population growth. While concluding that population stabilization alone would not eliminate other world problems, the panel determined that for most of the developing world economic development would profit from

[4] National Research Council, *Population Growth and Economic Development: Policy Questions*, National Academy Press, Washington, D.C., 1986.

slower population growth. Thus, population stabilization would not guarantee higher living standards but would most likely promote economic development, which in turn would raise the standard of living.

If a country's standard of living is to be raised, its economic growth must be greater than its population growth; if a population doubles every 40 years, then its economic goods and services must more than double during that time. Until recently, many developing nations were able to realize economic growth despite increases in population, largely because of financial assistance (usually in the form of loans) from developed nations. However, it may become increasingly difficult for many developing countries to continue raising their standards of living, because the tremendous debts they have accumulated while funding past development preclude future loans. For example, countries in Latin America are overwhelmed by massive foreign debts equivalent to each man, woman, and child owing about $1,000. In 1985, for every $1 that developed nations donated to Africa for famine relief, the Africans returned $2 in debt repayments; if they had not had such a massive debt to repay, Africans would have been able to easily purchase needed food. Today developing nations owe more than $1 trillion to developed nations and foreign banks, and many of their loans are in default.

## Population and Urbanization

The social, environmental, and economic aspects of population growth are influenced not only by an excess of people but also by the geographical *distribution* of people in rural areas, cities, and towns. Throughout recent history, people have increasingly migrated to cities. When Europeans first settled in North America, the majority of the population was farmers in rural areas. Today approximately 5 percent of the people in the United States are involved in farming, and three-fourths of our population live in cities. The increasing convergence of a population on cities is known as **urbanization.**

How many people does it take to make an urban area or a city? The answer varies from country to country; it can be anything from 100 homes clustered in one place to a population of 50,000 residents. The important distinction between rural and urban areas is not how many people live there but how people make a living. Most people residing in a rural area have occupations that involve harvesting natural resources—such as fishing, logging, and farming. In urban areas, most people have jobs that are not directly connected with natural resources.

Cities have grown at the expense of rural populations for a number of reasons. With advances in agriculture, including the increased mechanization of farms, more and more people can be supported by fewer and fewer farmers. Consequently, there are fewer employment opportunities for people in rural settings. Cities have traditionally provided more jobs because cities are the sites of industry, economic development, and educational and cultural opportunities.

The advantages of urban life notwithstanding, today many problems are faced by cities in developed and developing countries. Consider homelessness. Every country, even a highly developed country such as the United States, has people who lack shelter living in cities (see Focus On: Counting the Homeless). Urban problems are usually more pronounced in the cities of developing nations, however (Figure 9–4). In many cities in India and Mexico, thousands of homeless people sleep in the streets each night.

Urbanization is a worldwide phenomenon, but the percentage of people living in cities compared to rural settings is higher in developed countries than in developing countries. In 1991, urban inhabitants made up 73 percent of the total popula-

**Figure 9–4**

Homelessness, a problem in the cities of developed countries, is an even greater problem in cities of developing nations, such as Oaxaca, Mexico. (Steve Skloot/Photo Researchers, Inc.)

### Counting the Homeless

"S-Night" sounds like the code name for a secret operation, but it actually stands for Shelter and Street Night, or the night U.S. census takers conducted the first formal survey of homeless people in the United States (the night of March 20–21, 1990). Although the U.S. census had previously tossed a casual count of people in transient living situations into its pool of information, it had never before considered the homeless as a discrete entity.

S-night was controversial among some groups. Advocates for the homeless, who believed that the count would be so low as to allow officials to consider homelessness only a minor problem, dumped sand in front of the Department of Commerce in Washington, D.C., with a placard reading, "Like grains of sand, the homeless can't be counted" stuck in the sand. In New York, two census takers were shot at. Journalists trailed many of the census takers, possibly compromising the confidentiality of the interviews between census takers and homeless individuals.

But in the end, the staff of 15,000 census workers, each being paid $7.50 per hour (at a total cost of $2.7 million), did its best to interview and tally the numbers of homeless individuals across the country. The 1990 S-night numbers totalled 178,828 people in shelters for the homeless and 49,793 people visible at pre-identified street locations. It is hoped by many that these numbers will help secure funding for programs to ensure that in the future, S-night will become unnecessary.

tion of developed countries, but only 34 percent of the total population of developing countries.

**Urban Growth** Although proportionately more people still live in rural settings in developing countries, urbanization has been increasing there. Urbanization is increasing in developed nations, too, but at a much slower rate.

Consider the United States as representative of developed nations. Here, most of the migration to cities occurred during the past 150 years, when an increased need for industrial labor coincided with a decreased need for agricultural labor. The growth of U.S. cities over such a long period of time was slow enough to allow important city services such as water, sewage, education, and adequate housing to keep pace.

In contrast, the faster urban growth in developing nations has outstripped the capacity of many cities to provide basic services. It has also outstripped their economic growth. Consequently, cities in developing nations are faced with graver challenges than are cities in developed countries. These challenges include substandard housing for most residents, exceptionally high unemployment, and inadequate or nonexistent water, sewage, and waste disposal (Figure 9–5). Rapid urban growth also strains school, medical, and transportation systems.

Illustrating the greater urban growth of developing nations is the fact that most of the world's largest cities today are in developing countries. In 1950, three of the world's ten largest cities were in developing countries: Shanghai (China), Buenos Aires (Argentina), and Calcutta (India). In 1990,

**Figure 9–5**
Rapid urban growth in developing nations, together with overpopulation, poverty, and lack of economic development, has caused severe problems in many cities. The substandard housing seen in this section of Rio de Janeiro, Brazil, is common in cities of developing nations. (Tim Holt/Photo Researchers, Inc.)

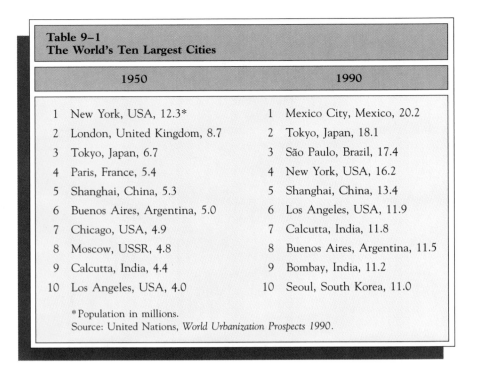

| Table 9–1 The World's Ten Largest Cities | |
|---|---|
| **1950** | **1990** |
| 1 New York, USA, 12.3* | 1 Mexico City, Mexico, 20.2 |
| 2 London, United Kingdom, 8.7 | 2 Tokyo, Japan, 18.1 |
| 3 Tokyo, Japan, 6.7 | 3 São Paulo, Brazil, 17.4 |
| 4 Paris, France, 5.4 | 4 New York, USA, 16.2 |
| 5 Shanghai, China, 5.3 | 5 Shanghai, China, 13.4 |
| 6 Buenos Aires, Argentina, 5.0 | 6 Los Angeles, USA, 11.9 |
| 7 Chicago, USA, 4.9 | 7 Calcutta, India, 11.8 |
| 8 Moscow, USSR, 4.8 | 8 Buenos Aires, Argentina, 11.5 |
| 9 Calcutta, India, 4.4 | 9 Bombay, India, 11.2 |
| 10 Los Angeles, USA, 4.0 | 10 Seoul, South Korea, 11.0 |

*Population in millions.
Source: United Nations, *World Urbanization Prospects 1990*.

seven of the world's ten largest cities were in developing countries: Mexico City (Mexico), São Paulo (Brazil), Shanghai, Calcutta, Buenos Aires, Bombay (India), and Seoul (South Korea) (Table 9–1).

### Rural Exodus

In 1940, only one of every 100 people in the world lived in a city with a population of over one million. By 1980, the proportion had grown to one in every ten people. Today, the trend continues. Two cities—Mexico City and São Paulo—have grown phenomenally, due to rural migrations. By the year 2000, at current rates of growth, Mexico City will have almost 26 million people, roughly equivalent to the present population of Canada. São Paulo will have 22 million people. Twenty of the world's 25 largest cities will be in developing nations. Up to half the people in these cities will live in overcrowded slums.

Urban growth contributes to the deterioration of the nonurban environment, which must provide the large quantities of food, water, energy, building supplies, minerals, and other materials required to maintain a concentrated population. In addition, cities and towns are the sources of water and air pollution that contaminates the surrounding rural areas as well as the cities themselves.

Does urbanization affect the rate at which population grows? Urbanization appears to be a factor in *decreasing* fertility rates, perhaps because family planning services, including access to contraceptives, are more readily available in urban settings.

## REDUCING THE FERTILITY RATE

Migration (moving from one place to another) used to be a solution for overpopulation, but not today. As a species, we have expanded our range throughout Earth; there is no habitable[5] location on Earth to which we can migrate that does not already sup-

[5] "Habitable" is defined as "fit to be lived in." Although humans can exist in Antarctica, it is not considered habitable because people cannot survive there without importing almost all of the necessities of life.

port humans. Nor is increasing the death rate an acceptable means of regulating population size. Clearly, the way to control our expanding population is by reducing the number of births. Because the total fertility rate (the average number of children born to a woman during her lifetime) is influenced by cultural traditions, marriage age, education, family planning, and government policies (which affect fertility by providing economic rewards and penalties), we examine each of these factors.

## Culture and Fertility

The values and norms of a society—what is considered right and important, and what is expected of a person—constitute that society's culture. With respect to fertility and culture, a woman has the number of children that are determined by the traditions of her society.

High fertility rates are traditional in many cultures. The motivations for having lots of babies vary from culture to culture, but overall a major reason for the high fertility is that infant mortality rates are high. (In order for a society to endure, it must continue to produce enough children who survive to reproductive age; thus, if infant mortality rates are high, fertility rates must also be high to compensate.) Although the worldwide infant mortality rate has been decreasing, it will take time for fertility levels to decline. Part of the reason for this slowness is cultural: changing anything that has been traditional, including large family size, takes a long time.

Higher fertility rates in some developing countries are also due to the societal roles of children. In some countries, children usually work in family enterprises (such as farming or commerce), contributing to the family's livelihood (Figure 9–6). When these children become adults, they provide support for their aging parents. In contrast, children in developed countries have less value as a source of labor, because they attend school and because human labor is less important in a mechanized society. Further, developed countries provide a number of social services for the elderly, so the burden of their care does not fall entirely on their offspring.

In addition, many cultures place a higher value on male children than on female children. In these societies, a woman who bears many sons achieves a high status; thus, there is a social pressure that keeps the fertility rate high.

Religious values are another aspect of culture that affects fertility rates. For example, a number of

**Figure 9–6**
In developing countries, fertility rates are high partly because children contribute to the family by working. More children mean a greater family income. (Gary J. James/Biological Photo Service)

studies done in the United States point to differences in fertility rates among Catholics, Protestants, and Jews. In general, Catholic women have a higher fertility rate than either Protestant or Jewish women, and women who do not follow any religion have the lowest fertility rate of all. However, it is difficult to conclude that the observed differences in fertility rates are the result of religious differences alone, because other variables, such as race (certain religions are associated with particular races) and residence (certain religions are associated with urban or with rural living), complicate the situation.

## Marriage Age and Fertility

The fertility rate is affected by the average age at which women marry, which in turn is determined by the laws and customs of the society in which she lives. Women who marry are more likely to bear children than women who do not marry, and the earlier a woman marries, the more children she is likely to have.

The percentage of women who marry and the average age at marriage vary widely among all countries. There is always a correlation between marriage age and fertility rate, however. Consider Sri Lanka and Bangladesh, two developing countries in Asia. In Sri Lanka, the average age at marriage is 25, and the average number of children born per woman is 2.5. In contrast, in Bangladesh, the average age at marriage is 16, and the average number of children born per woman is 4.9.

## Educational Opportunities and Fertility

Women with more education tend to marry later and have fewer children. Figure 9–7 shows the fertility rates of women in the United States with different education levels. Providing women with educational opportunities delays their first childbirth, thereby reducing the number of childbearing years and increasing the amount of time between generations. Education also opens the door to greater career opportunities and changes women's lifetime aspirations. In the United States, it is not uncommon for a woman in her thirties or forties to give birth to her first child, after establishing a career.

In developing countries there is also a strong correlation between the average amount of education a woman receives and fertility rate. For example, women with a secondary (high school) education have fewer children than women with only a primary (elementary school) education. Education increases the likelihood that women will know how to control their fertility, and it provides them with knowledge to improve the health of their families, which results in a decrease in infant mortality. Education also increases women's options, opening doors to other careers and ways of achieving status besides having babies.

Education may have an indirect effect on fertility rate, as well. Children who are educated have a greater chance of improving their living standards, partly because they have more employment opportunities. Parents who recognize this may be more willing to invest in the education of a few children than in the birth of many children whom they cannot afford to educate. The ability of better-

**Figure 9–7**
The number of children a woman has varies with the amount of education she has received. Shown are the 1987 fertility rates for 35- to 44-year-old women in the United States with differing levels of education.

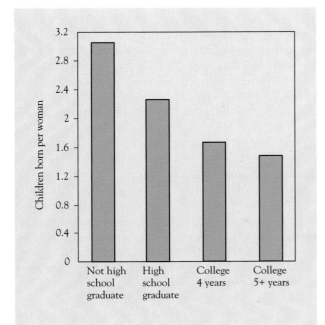

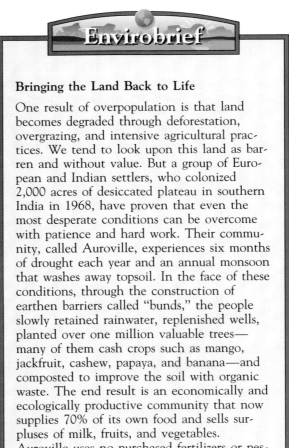

**Bringing the Land Back to Life**

One result of overpopulation is that land becomes degraded through deforestation, overgrazing, and intensive agricultural practices. We tend to look upon this land as barren and without value. But a group of European and Indian settlers, who colonized 2,000 acres of desiccated plateau in southern India in 1968, have proven that even the most desperate conditions can be overcome with patience and hard work. Their community, called Auroville, experiences six months of drought each year and an annual monsoon that washes away topsoil. In the face of these conditions, through the construction of earthen barriers called "bunds," the people slowly retained rainwater, replenished wells, planted over one million valuable trees—many of them cash crops such as mango, jackfruit, cashew, papaya, and banana—and composted to improve the soil with organic waste. The end result is an economically and ecologically productive community that now supplies 70% of its own food and sells surpluses of milk, fruits, and vegetables. Auroville uses no purchased fertilizers or pesticides and has accepted no outside funding.

educated people to earn more money may be one of the reasons why smaller family size is associated with increased family income, although another obvious reason is that fewer children are fewer mouths to feed and, thus, are less of a drain on the family income.

## Family Planning Services and Fertility

Socioeconomic factors may encourage people to want smaller families, but reduction in fertility won't become a reality without the availability of health and family planning services. The governments of most countries recognize the importance of educating people about basic maternal and child health care. Family planning services provide information on reproductive physiology and contraceptives, as well as the actual contraceptive devices, to those who wish to control the number of children they produce or to space their children's births. Family planning does not try to force people to limit their family sizes, but rather attempts to convince people that small families (and contraceptives that promote small families) are acceptable and desirable. The major birth control methods in use today are shown on the next page in Table 9–2.

Contraceptive use is strongly linked to lower fertility rates. Research has shown that 90 percent of the decrease in fertility in 31 countries was a direct result of the increased use of contraceptives. In developed countries, where fertility rates are at replacement levels or lower, the percentage of married women of reproductive age who use contraceptives is often greater than 75 percent. Fertility declines have been noted in developing countries where contraceptives are readily available. During the 1970s and early 1980s, use of contraceptives in East Asia and many areas of Latin America increased significantly, and these regions experienced a corresponding decline in birth rate. In areas where contraceptive use remained low, such as sub-Saharan Africa, there was little or no decline in birth rate.

Family planning centers provide information and services primarily to women. As a result, in the male-dominated societies of many developing countries, such services may not be as effective as they could otherwise be. Polls of women in developing countries reveal that many who say they don't want additional children still do not practice any form of birth control. When asked why they don't use birth control, these women often respond that their husbands want more children.

Reduction in fertility will not result from family planning alone, especially where cultural traditions and religious beliefs prohibit birth control. Only when people are convinced that having smaller families will somehow benefit them (in terms of a better standard of living, better health, and so on) will they embrace family planning.

## Government Policies and Fertility

The involvement of governments in childbearing is well established. Laws determine the minimum age at which people may marry and the amount of education that is compulsory. Governments may allot portions of their budgets to family planning services, education, primary health care, old-age security, or incentives for smaller or larger family size. The tax structure, including additional charges or allowances based on family size, also affects fertility.

In recent years, the governments of many developing countries have recognized the need to limit population growth and have formulated policies (such as economic rewards and penalties) designed to achieve this goal. Most countries sponsor family planning projects, many of which are integrated with primary health care, education, economic development, and efforts to improve women's status. A number of these projects are supported by the United Nations Fund for Population Activities.

Population control measures have been instituted in many developing countries. Here we examine those in China, India (the world's two most populous nations), Nigeria, and Mexico.

**China**  China, with a mid-1992 estimated population of 1.17 billion people, has the largest population in the world. Recognizing that its rate of population growth had to decrease or the quality of life for everyone in China would be compromised, the Chinese government in 1979 instigated an aggressive plan to push China into the third demographic stage. (Recall from Chapter 8 that the third demographic stage is characterized by a decline in the birth rate along with a relatively low death rate.) Announcements were made of incentives to promote later marriages and one-child families. Local jurisdictions were assigned the task of reaching this goal. A couple who signed a pledge to limit themselves to a single child might be eligible for such incentives as medical care and schooling for that child, cash bonuses, preferential housing, and old-age security. Penalties were also instituted, including fines and the surrender of all privileges if a second child was born.

China's aggressive plan brought about an immediate and drastic reduction in fertility, from 5.8

**Table 9–2**
**Methods of Birth Control**

| Method | Failure Rate* | Mode of Action | Advantages | Disadvantages |
|---|---|---|---|---|
| Norplant (arm implant) | 0.2 | Prevents ovulation; affects cervical mucus, preventing entry of sperm into cervical canal | Effective for five years; sexual freedom; fertility returns immediately after removal | Menstrual irregularities and/or minor discomfort in some women; minor surgery required to implant and remove |
| Oral contraceptives | 0.3; 5 | Prevent ovulation; may also affect endometrium and cervical mucus and prevent implantation | Highly effective; sexual freedom; regular menstrual cycle | Minor discomfort in some women; possible thromboembolism, hypertension, heart disease in some users; possible increased risk of infertility; should not be used by women who smoke |
| Intrauterine device (IUD) | 1; 5 | Not known; probably stimulates inflammatory response | Provides continuous protection; highly effective | Cramps; increased menstrual flow; spontaneous expulsion; since mid-1980s, prescribed less frequently due to increased risk of pelvic inflammatory disease and infertility; no longer manufactured in United States |
| Spermicides (sponges, foams, jellies, creams) | 3; 20 | Chemically kill sperm | No side effects (?); vaginal sponges are effective in vagina for up to 24 hours after insertion; sponges also act as physical barriers to sperm cells | Some evidence linking spermicides to birth defects |
| Cervical cap | 3; 10 | Cervical cap mechanically blocks entrance to cervix | No major health risks; can wear for weeks at a time | May be difficult to insert; may irritate cervix |
| Contraceptive diaphragm (with jelly) | 3; 14† | Diaphragm mechanically blocks entrance to cervix; jelly is spermicidal | No side effects | Must be prescribed (and fitted) by physician; must be inserted prior to coitus and left in place for several hours after intercourse |

births per woman in 1970 to 2.1 births per woman in 1981. However, it compromised individual freedom of choice. In some instances, social pressures from the community induced women who were pregnant with a second child to get abortions. Moreover, based on the disproportionate number of male versus female babies reported born in 1981, it is suspected that thousands of newborn baby girls were killed by their parents who, if required to conform to the one-baby policy, wanted it to be a boy. In China, sons traditionally provide old-age security for their parents and—for this and other cultural reasons—are valued more highly than daughters.

## Methods of Birth Control (Continued)

| Method | Failure Rate* | Mode of Action | Advantages | Disadvantages |
|---|---|---|---|---|
| Condom | 2.6; 10 | Mechanical; prevents sperm from entering vagina | No side effects; some protection against STDs, including AIDS | Interruption of foreplay to fit; slightly decreased sensation for male; could break |
| Rhythm§ | 13; 21 | Abstinence during fertile period | No side effects (?) | Not very reliable |
| Douche | 40 | Flushes semen from vagina | No side effects | Not reliable; sperm beyond reach of douche in seconds |
| Withdrawal (coitus interruptus) | 9; 22 | Male withdraws penis from vagina prior to ejaculation | No side effects | Not reliable; is contrary to powerful drives present when an orgasm is approached. Sperm present in fluid secretion before ejaculation may be sufficient for conception |
| Sterilization Tubal ligation | 0.04 | Prevents ovum from leaving uterine tube | Most reliable method | Usually not reversible |
| Vasectomy | 0.15 | Prevents sperm from leaving scrotum | Most reliable method | Usually not reversible |
| Chance (no contraception) | About 90 | | | |

*The lower figure is the failure rate of the method; the higher figure is the rate of method failure plus failure of the user to utilize the method correctly. Based on number of failures per 100 women who use the method per year in the United States.

†Failure rate is lower when diaphragm is used together with spermicidal foam.

§There are several variations of the rhythm method. For those who use the calendar method alone, the failure rate is about 35. However, by taking the body temperature daily and keeping careful records (temperature rises after ovulation), the failure rate can be reduced. Also, by keeping a daily record of the type of vaginal secretion, changes in cervical mucus can be noted and used to determine time of ovulation. This type of rhythm contraception is also slightly more effective. When women use the temperature or mucus method and have intercourse *only* more than 48 hours *after* ovulation, the failure rate can be reduced to about 7.

Source: Solomon, Schmidt, and Adragna, *Human Anatomy and Physiology*, 2d ed., Saunders College Publishing, Philadelphia, 1990.

Recently, China's population control program has used education and publicity campaigns, more than penalties, to achieve its goals (Figure 9–8). China trains population specialists at institutions such as the Nanjing College for Family Planning Administrators. In addition, thousands of secondary school teachers have been taught how to integrate population education into the curriculum. However, as the one-child policy has been deemphasized, the fertility rate has increased slightly, to 2.2 in 1992.

**India** India is the second most populous nation in the world, with an estimated mid-1992 population

**Figure 9–8**
A billboard campaign in China promotes the one-child family.
(Visuals Unlimited/E. F. Anderson)

of 883 million. It was the first country to establish government-sponsored family planning, in the 1950s. Unlike China, India did not experience immediate results from its efforts to control population growth, in part because of the diverse cultures, religions, and customs in different regions of the country. For example, Indians speak hundreds of different dialects, which makes a broad program of family planning education difficult.

In 1976 the Indian government became more aggressive. It introduced not only incentives to control population growth, but also controversial programs of compulsory sterilization in a number of states. If a man had three or more living children, he was compelled to obtain a vasectomy. Compulsory sterilization was a failure; it had little effect on the birth rate and was exceedingly unpopular. It may have been partly responsible for Indira Gandhi being voted out of office in 1977.

More recently, India has integrated development and family planning projects. For example, adult literacy and population education programs have been combined. Multi-media advertisements and education have been used to promote voluntary birth control. India has also emphasized lowering the infant mortality rate, improving women's status, and increasing birth spacing. These changes have had an effect: India's total fertility rate has declined from 5.3 in 1980 to 3.9 in 1992.

**Mexico** Mexico, with a mid-1992 estimated population of 87.7 million, is the second most populous nation in Latin America. (Brazil, with a population of 151 million, is the most populous.) Mexico has a tremendous potential for growth because 38 per-

cent of its population is under 15 years of age. Even with a low birth rate, the population would still increase in the future because of the numbers of people having babies.

In 1973 the Mexican government instigated a number of measures to reduce population growth, such as educational reform, family planning, and primary health care. Mexico has had great success in reducing its fertility level, from 6.7 births per woman in 1970 to 3.8 births per woman in 1992.

Mexico's goal includes not only population stabilization, but balanced regional development. Its urban population is 71 percent of its total population, and most of these people live in Mexico City. Although Mexico is largely urbanized compared to other developing countries, its urban-based industrial economy has not been able to absorb the great number of people in the work force, and unemployment is very high. Consequently, many Mexicans have migrated, both legally and illegally, to the United States.

Mexico's recent efforts at population control include multimedia campaigns. For example, popular television and radio soap operas carry family planning messages. Booklets on family planning are distributed, and population education is being integrated into the public school curriculum. Social workers receive training in family planning as part of their education.

**Nigeria** Nigeria is part of the sub-Saharan region that currently has the most rapid population growth on Earth. Nigeria has the highest population of any African country: in mid-1992 its popu-

### Sponsor a Child in a Developing Country

It is easy to think that there is nothing you can do to help improve the lives of people who live in developing countries. But you can make a difference in one child's life and, in doing so, help the child's family and community. A number of nonprofit organizations offer the opportunity for an individual or group to sponsor a child by donating a fixed amount each month, quarter, or year to assist in the child's education, nutrition, and health care. In addition, your donations help these organizations carry out all-important family and community projects, including the improvement of health facilities, clean water projects, teacher training, community gardens, and family planning. The average sponsorship fee is less than $25 per month, which is roughly equivalent to the monthly cost of cable TV.

Education, improved health care, and family planning can contribute to a lower infant mortality rate and, in many cases, a reduced fertility rate. All of these help control population and thus work toward improving the lives of both adults and children in a community.

When you sponsor a child through one of these organizations, you typically correspond with the child and receive reports from the case workers and community workers about the child's progress in school, the family's economic status, and any community projects being undertaken by the field staff in conjunction with community and family members.

If you think you might want to sponsor a child, contact one of the nonprofit organizations that run these programs. Two of them are PLAN International USA, 155 Plan Way, Warwick, RI 02886, 1-800-556-7918; and Save the Children, 54 Wilton Road, Westport, CT 06881, 1-800-243-5075. You can sponsor a child as an individual, or your instructor might want to organize a sponsorship by the entire class that is carried on by each subsequent class taking this course. Remember that once you begin sponsorship of a child, that child and his or her family will probably count on your assistance for a number of years (although you are not legally bound to continue). Thus, you should be ready to maintain your commitment long after you finish this course.

---

lation was estimated at 90.1 million people. Its total fertility rate is 6.5 births per woman, and it has great reproductive potential because 45 percent of the population is under 15 years of age. The average life expectancy in Nigeria is 48.5 years, which is low in part because of the high infant mortality rate.

The Nigerian government has recognized that its economic goals are more likely to be attained if its rate of population growth decreases. In 1986 Nigeria developed a national population policy that is an integration of population and development projects. Part of the plan involves improving primary health care, including training nurses and other health care professionals. Population education is being used to encourage later marriages and birth spacing.

## A GLOBAL PLAN TO REDUCE POPULATION GROWTH

In this chapter we have considered world problems—including hunger, poverty, underdevelopment, and environmental issues—that are exacerbated by an increase in population. Population stabilization is critically important if we are to effectively tackle these serious problems. All countries must develop policies to bring about an immediate reduction in the rate of population growth. (See Focus On: Population Concerns in Europe on page 172 for viewpoints that oppose population stabilization.) But what kinds of policies will help achieve population stabilization?

Developing countries should increase the amount of money that they allot to family planning

## Population Concerns in Europe

Population has stabilized in Europe, where most countries have lower-than-replacement-level fertility rates, and several countries have experienced a slight decline in population. As a result of fewer births, the proportion of elderly people in the European population is increasing. Population control has thus become a controversial issue in Europe, with two opposing viewpoints emerging.

Those who favor population growth are called pronatalists (see figure). Pronatalists believe that the vitality of their region is at risk because of declining birth rates and that the decrease in population might result in a loss of economic growth. They believe their countries' positions in the world will weaken and that their cultural identity will be diluted by immigrants from non-European countries (the next wave of immigrants to Europe is expected to be primarily from North Africa and the Middle East). Pronatalists are also concerned about the possibility that pension and old-age security systems will be overwhelmed by the large numbers of elderly in Europe unless a larger work force is available to contribute to those systems. The most outspoken pronatalists believe that women should marry young and have many children for the good of society. Pronatalists favor government policies that provide incentives for larger families and penalties for smaller families.

The opponents of pronatalists do not view European society as declining; rather, they think European influence is increasing, because they judge "power" in economic terms rather than in terms of population. Generally, they are not as opposed to immigration, and they believe that Europe should *not* be making a concerted effort to increase its fertility rates when overpopulation is such a serious problem in much of the world. Further, they point out that technological innovations have eliminated many jobs in Europe, and the consequential unemployment would only be made worse by an increase in the labor force caused by a rise in birth rate. Opponents of pronatalists also question whether the elderly are a burden to society. Moreover, they argue that the elderly are not the only ones the work force supports. Their view is that the costs of providing for the young, especially for many years of education, cancel out any perceived benefit from an increased birth rate.

---

services, particularly the dissemination of information on birth control. Governments should take steps to increase the average level of education, especially of women, and women must be given more employment opportunities. Religious and traditional leaders should be involved in discussions *before* new population policies are implemented. (Muslim leaders in Indonesia, who initially opposed efforts to control population growth, became supporters of family planning when they were consulted by the government on the matter.) Likewise, national population policies will not work without local community involvement and acceptance, and community leaders can provide feedback on the effectiveness of programs.

Developed countries can help developing nations achieve a decrease in population growth by providing financial support for the United Nations Fund for Population Activities. Developed countries can also support research and development of new birth control methods. On a personal level, individuals in developed countries can help developing nations by sponsoring children (see You Can Make a Difference: Sponsor a Child in a Developing Country on page 171).

Most important, developed nations need to face their own population problems, particularly consumption overpopulation. Policies to support recycling of resources, eliminate needless production, and discourage overconsumption and the throwaway mentality should be formulated. These policies will also show developing nations that developed countries are serious about facing the issues of overpopulation at home as well as abroad.

Advertisement showing desirability of population growth in Ireland. [Courtesy of Industrial Development Authority of Ireland (IDA Ireland)]

## SUMMARY

**1.** Many human problems, such as hunger, resource depletion, environmental problems, underdevelopment, poverty, and urban problems, are exacerbated by the rapid increase in population.

**2.** Those countries with the greatest food shortages are also the ones with the highest fertility rates. The relationship between world hunger and population growth is complex, however, and may be influenced by the degrees of economic development, poverty, and inequitable distribution of resources.

**3.** Nonrenewable resources are present in a limited supply and cannot be replenished in a reasonable period on the human time scale. Slowing the rate of population growth will give us more time to find substitutes for nonrenewable resources as they are depleted.

**4.** Renewable resources are replaced by natural processes and can be used forever, provided they are used in a sustainable way. Overpopulation causes renewable resources to be overexploited. When this happens, renewable resources become nonrenewable.

**5.** One model of the effect of population on the environment has three factors: the number of people, the amount of resources used per person, and the effect on the environment of using those resources. This model shows that there are two kinds of overpopulation, people overpopulation and consumption overpopulation.

**6.** Developing countries have people overpopulation, in which the population increase degrades the environment even though each individual uses few resources. In contrast, developed countries have consumption overpopulation, in which each individual in a stable population consumes a large share of resources and the result is environmental degradation.

7. Underdevelopment and poverty are associated with high fertility rates. Most economists believe that slowing population growth helps promote economic development.

8. As a nation develops economically, the proportion of the population living in cities increases. In developing nations, most people live in rural settings, but their rates of urbanization are rapidly increasing. This makes it difficult to provide city dwellers with services such as housing, water, sewage, and transportation systems. Urbanization appears to be a factor in decreasing population growth, however.

9. The relationships between fertility rate and cultural, social, governmental, and economic factors are intricate. It appears that a combination of factors is responsible for the overall decline in fertility. These include urbanization, greater educational opportunities (especially for women), higher marriage age, improved status of women, and availability of family planning services. Government policies, including economic rewards and penalties, can also influence population growth.

## DISCUSSION QUESTIONS

1. Keep a list of the natural resources you use in a single day. How would this list compare to a similar list made by a poor person in a developing country?

2. How does a rapidly expanding population affect the utilization of nonrenewable resources? Of renewable resources?

3. Discuss this statement: The current human population crisis causes or exacerbates all environmental problems.

4. Explain how a single child born in the United States can have a greater effect on the environment and the world's resources than a dozen or more children born in Kenya.

5. Discuss some of the ways in which population and economic growth are related.

6. What are some of the problems brought on by rapid urban growth in developing countries?

7. What is the relationship between fertility and marriage age? Between fertility and educational opportunities for women?

8. Explain the rationale behind this statement: It is better for developed countries to spend millions of dollars on family planning in developing countries now than to have to spend billions of dollars on relief efforts later.

9. Discuss some of the difficulties that might be encountered should the United Nations try to develop and enforce population quotas for countries.

## SUGGESTED READINGS

Ehrlich, P. R., and A. H. Ehrlich. Population, plenty, and poverty. *National Geographic*, December 1988, 914–945. Highlights representative families in India, Kenya, China, Hungary, Brazil, and the United States.

Ehrlich, P. R., and A. H. Ehrlich. *The Population Explosion*, Simon & Schuster, New York, 1990. Continues the Ehrlichs' crusade against population growth by connecting an increase in human population to many serious world problems. Concludes with a chapter on what individuals can do to help.

Frey, W. H. Metropolitan America: Beyond the transition. *Population Bulletin* 45:2, July 1990. Discusses demographic trends in U.S. cities.

Jacobson, J. L. India's misconceived family plan. *World Watch* 4:6, November–December 1991. Argues that reducing the birth rate alone will not solve India's many social, economic, and environmental problems.

Keyfitz, N. "Population Growth Can Prevent the Development That Would Slow Population Growth," in *Preserving the Global Environment*, ed. J. T. Mathews. W. W. Norton & Company, New York, 1991. An in-depth examination of the complex interactions among population growth, economics, development, and poverty.

Mellor, J. W. The intertwining of environmental problems and poverty. *Environment* 30:9, 1988. Considers the inseparable problems of environmental problems and poverty in developing countries.

Merrick, T. W. World population in transition. *Population Bulletin* 41:2, 1988. Population Reference Bureau, Washington, D.C. Highlights a number of world population trends.

Sadik, N. *The State of World Population 1990*. United Nations Population Fund, New York, 1990. Summarizes world population issues and our options during the 1990s.

Vibrator trucks in Oman send out vibrations that bounce off underlying layers of rock and provide clues on the location of oil. *(George Steinmetz)*

# Part 4

# L. Hunter Lovins

## Efficiency Technology: Less Energy, More Power

L. Hunter Lovins is President and Executive Director of Rocky Mountain Institute in Old Snowmass, Colorado, which she co-founded with her husband and colleague, Amory Lovins, in 1982. RMI's interrelated programs in Energy, Transportation, Water, Agriculture, Economic Renewal, and Security are designed to foster the efficient, sustainable use of resources as a path to global security. Ms. Lovins holds B.A.s in Political Science and Sociology from Pitzer College, California, and a J.D. from Loyola University. Internationally recognized for their expertise in energy policy, the Lovinses have been consulted by heads of state, government agencies, and multinational corporations around the world.

**What are the primary objectives of Rocky Mountain Institute's Energy Program? Explain what you call the "end-use/least-cost" approach, and how it relates to energy issues.**

Our Energy Program grew from Amory's and my longstanding efforts to find solutions to the energy problem that make sense economically and environmentally. To that end, we use the end-use/least-cost approach to analyze first, how much and what kinds of energy we need for each task, and second, how to get that energy at the lowest price.

**What has this analysis shown?**

Many people believe that we are running out of energy, thus the logical solution is to obtain more of it. In response to the 1973 Arab oil embargo, President Nixon proposed to build, by the year 2000, 450–800 new nuclear reactors and 500–800 new coal fire plants, primarily to produce more electricity. During 1976–1985 alone, this would have cost over $1 trillion, three quarters of all domestic investment dollars. The plan was discarded for obvious reasons.

The real question, however, is, are we truly running out of energy? Analysis of actual energy usage shows that about 58% of America's delivered energy needs are for *heat*, over half of that at low temperatures, and 34% of our energy needs are for liquid fuels, primarily for *transportation*. Only 8% of our energy needs are for *electricity*.

**What are the alternatives?**

Most of our heating needs can be met through efficiency (using the energy we already have more productively) and through various forms of renewable energy sources. An end-use/least-cost analysis shows that the best buys are in the efficiency technologies, which generally cost between half a cent and 2 cents per kilowatt-hour. Compare that to running an existing coal or nuclear plant, which costs about 2–5 cents per kilowatt-hour, or building a new nuclear plant, which costs 15–20 cents per kilowatt-hour.

**Do we need to combine the new technologies with energy conservation?**

People have been led to believe that "conservation" means being colder in the winter and hotter in the summer. Our analysis does not assume conservation in the sense of "doing without," reducing your standard of living, or changing your lifestyle. Our focus is on technological efficiency to do the same job as before, using less energy and less money. We investigate opportunities in energy policy and provide people with information on technological choices. We firmly believe that if people have incentive and opportunity, they will make wise energy decisions for themselves, and in the process, preserve resources for the future.

**Part of your Energy Program is the Competitek Service, which provides consultation on new developments in the use and misuse of electricity. Tell us about that.**

*Competitek* has more than 230 members from over 35 countries, both developed and developing. Most members are utilities, industries, and governments. Saving energy can mean the difference between profit and loss. For example, consider Southwire Corporation, the largest independent U.S. producer of cable, rod and wire. Its ten plants in six states are quite energy-intensive. In the early 1980s, faced with falling prices for its products and rising costs, Southwire cut its energy use per pound of product in half over eight years. That savings *equaled* the company's profit for that period. Without an energy saving program, they would have operated at a loss and perhaps even followed other such firms into bankruptcy. Their efficiency program may well have saved over 4,000 jobs. That's why so many companies are interested in *Competitek*. It's not just a public relations scheme to give themselves a "clean, green" image. We provide hard numbers to improve their financial bottom lines.

**Your prescription for improving the utility industry's bottom line centers around the idea of the "negawatt" or saved watt of electricity. How can a utility make money by selling less energy?**

Utility companies have many costs associated with providing energy to their customers: building new power plants, buying fuel, running the plant, dispatching the energy, etc. It was once true that the more electricity consumers used and the more utilities sold, the cheaper it was for everyone. For complex reasons, in the early 1970s this began to change. Today, it is more financially advantageous for utilities to produce *less* electricity, avoiding the costs of production, and saving money and energy. That is the negawatt, a saved watt. Consumers are not interested in buying raw energy, but in getting access to energy services—heat, light, hot water, etc. If those services can be provided more cheaply through efficiency than by generating more energy, then it is just good business for utilities to supply those services, and to enable their consumers to use energy more efficiently.

**What does the future look like for utilities?**

Five West Coast utilities now project that they will be able to meet 75–100% of their new power needs in the 1990s from efficiency, and the rest from renewables, such as wind power, solar photovoltaics, or small hydroelectric plants. They are not planning to build any new thermal power plants. Many of the renewable energy facilities will be built by private entrepreneurs. Utilities will still exist, but ultimately may come to resemble phone companies, providing energy services to customers without necessarily owning or operating power plants.

**Could the negawatt idea be used by other industries? For example, when super fuel–efficient cars became available, could oil companies make money by selling less gas, rather than by charging us more per gallon?**

Yes. The big oil companies today are essentially like banks, pools of capital in search of something to finance. They believe the business they're in is extracting and selling oil. We argue that the business they're really in is providing energy services to their customers, and what they ought to be selling are *negabarrels*. They should be financing the conversion to super-efficient cars, taking their return as a bank does. Few oil companies have yet begun to redefine their business as service-oriented in this way, but they will eventually. As new efficiency technologies are brought onto the market, the profitability of selling oil will likely decline.

**To remedy our most dire energy problems, what are some of the newest technologies developing now? What technologies are in need of more research?**

At the moment, almost a dozen manufacturers have made prototype cars that get 63–138 mpg, and that isn't even close to state-of-the-art. Using recent advances in aerodynamics, ultralightweight materials, new engine and hybrid electric drive technology, and computer-aided design, you could get 150 mpg in a zippy, comfortable, affordable station wagon. Oil efficiency has not received nearly enough attention, so RMI will be working in that area.

There have been dramatic developments in renewable energy sources. Among the most important to pursue aggressively are: further refinements in new solar absorbing surfaces that can yield high industrial temperatures even in cloudy weather; sustainably producing biofuels from farm and forestry wastes; making photovoltaics even cheaper; and using sunlight to break down water directly into hydrogen. It's already so much cheaper to save fuel than to burn it that we could now abate virtually all of the greenhouse gases associated with global warming, not at a cost but at a profit. Unless government and industry foster the transition to an efficient, sustainable energy path, we will forego these opportunities, and commit ourselves to a future of economic and environmental degradation.

Interview

# Chapter 10

Coal cars on railroad. (Louie Psihoyos, © National Geographic Society)

# Fossil Fuels

Coal, oil, and natural gas are called fossil fuels because they are composed of the remnants of organisms that lived millions of years ago. All fossil fuels are present in the Earth's crust in finite amounts and are therefore considered nonrenewable natural resources. Environmental problems associated with the production and use of fossil fuels include damage to the environment at the site of production, accidental spills during transport or storage, and emission of carbon dioxide and other pollutants during combustion. Synfuels (tar sands, oil shales, gas hydrates, alcohol fuels, coal liquid, and coal gas) are generally more expensive than traditional fossil fuels and have many of the same undesirable environmental effects. Nevertheless, as coal, oil, and natural reserves decline, synfuels may become more important sources of energy.

## ENERGY CRISES

Human society depends on energy. We use it to warm our homes in winter and cool them in summer; to grow and cook our food; to extract and process natural resources, and to manufacture items we use daily; and to power various forms of transportation. Many of the conveniences of modern living depend on a ready supply of energy.

The United States' dependence on energy resources from other countries has been dramatically demonstrated several times in recent years. In 1973, for example, the Organization of Petroleum Exporting Countries (OPEC) restricted oil shipments to the United States, creating an energy crisis. Although total oil consumption was cut by only about 5 percent in response to the restrictions, the United States was thrown into a panic. Because Americans own a lot of automobiles and because automobiles require petroleum fuels, any factor that influences oil availability or cost has tremendous repercussions in our society. The 1973 OPEC oil embargo resulted in escalating prices for gasoline and home heating oil. Long lines at filling stations were commonplace (Figure 10–1), and in some states motorists were restricted to buying gasoline every other day (based on the last digits of their license plate numbers). Car sales dropped, and people who did buy cars generally purchased the more fuel-efficient foreign makes. Because cars get better mileage at moderate speeds, the freeway speed limit was reduced by federal law to 55 miles per hour as an energy conservation measure.

One positive effect of the 1973 oil embargo was the development of automobiles, both foreign and domestic, with greater fuel economy. In 1975 the United States government imposed fuel efficiency standards on the automobile industry; they began to go into effect in 1978 with a corporate average fuel economy of 18 miles per gallon. (The 1992 standard was 27.5 miles per gallon.)

The OPEC oil embargo of 1973 was not the only oil crisis Americans have faced in recent years. In 1979 oil prices skyrocketed from $13 to $34 a barrel due to an oil shortage touched off by the Iranian revolution, although the effects of this oil crisis were not as crippling as those of 1973. By the 1980s, the oil scares of the seventies were largely forgotten as oil prices declined. Americans purchased more cars, both domestic and foreign, than ever before. Gasoline was so cheap and abundant that consumption of gasoline increased during the 1980s. Between 1985 and 1989, oil imports increased from 4.9 million barrels a day to 8 million barrels. At the same time, domestic oil production in the United States declined.

In the 1990s, a greater proportion of our total oil supply is being used to provide gasoline for automobiles than was used in 1973. In 1989 (the most recent data available at this writing), 46 percent of all refined crude oil in the United States was converted to gasoline. After Iraq invaded Kuwait in the summer of 1990, fears of a war in the Persian Gulf caused jitters among oil companies in the United States. The prices of heating oil and gasoline rose dramatically, but they stabilized once it was clear that a war in the Persian Gulf initially

**Figure 10–1**
The oil embargo imposed by OPEC in 1973 resulted in gas rationing and subsequent long lines at filling stations. The effects of the embargo dramatically illustrated that America's dependency on automobiles requires a reliable source of fuel. (Tom McHugh/Photo Researchers, Inc.)

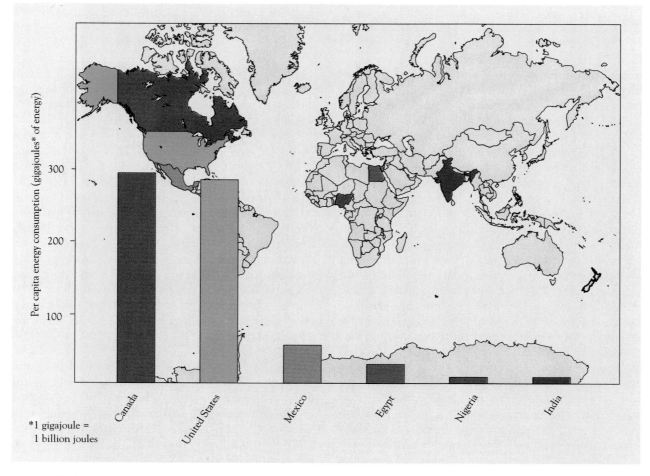

**Figure 10–2**
Annual per capita commercial energy consumption in developed nations is much higher than
it is in developing countries.

would not restrict our domestic oil supply. Nevertheless, the war with Iraq demonstrated that the
United States is still vulnerable when it comes to
energy supplies, and energy conservation is not an
issue of the past.

## ENERGY CONSUMPTION IN DEVELOPED AND DEVELOPING COUNTRIES

A conspicuous difference in per-capita energy consumption exists between developed and developing
nations (Figure 10–2). As you might expect, developed nations consume much more energy per person than developing nations. Although only 22.6
percent of the world's population lives in developed

countries, in 1989 (the most recent data available)
these people used approximately 74 percent of the
commercial energy consumed worldwide.[1] That
means that each person in developed countries uses
approximately ten times as much energy as each
person in developing countries.

A comparison of energy requirements for food
production clearly illustrates the energy consumption differences between developing and developed
countries. Farmers in developing nations rely on
their own physical energy or the energy of animals
to plow and tend fields. In contrast, agriculture in
developed countries involves many machines (such
as tractors, automatic loaders, and combines) that

[1] Worldwide commercial energy use does not take into account the use of firewood, which meets the energy needs of a substantial portion of the population in many developing countries.

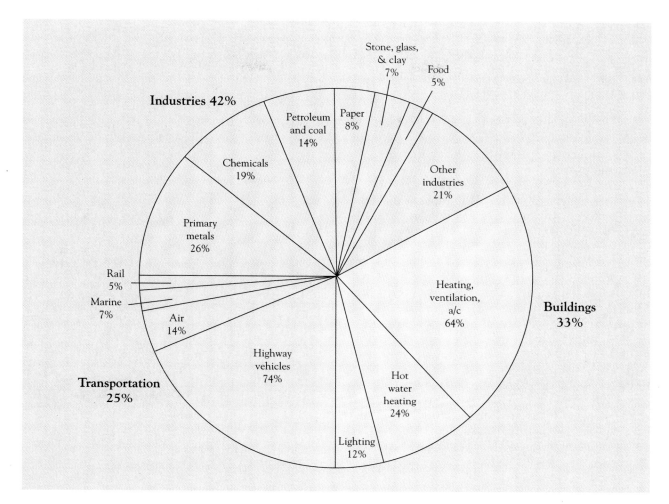

**Figure 10–3**
How energy is consumed in the United States.

require fuel. Additional energy is required to produce the fertilizers and pesticides widely used in industrialized agriculture. A higher energy input is one of the reasons that the agricultural productivity of developed countries is greater than that of developing countries.

Currently, energy consumption is increasing worldwide, with most of the increase occurring in developing countries. One of the goals of developing countries is to improve their standard of living. One way to achieve this is through economic development, a process usually accompanied by a rise in per-capita energy consumption. Furthermore, the world's energy requirements will also increase during the 21st century, as its population continues to climb (see Chapter 8). Most of the population growth will be in developing countries.

In contrast, the population in developed nations is more stable, and those nations' per-capita energy consumption may be at or near saturation. Also, it is possible that additional energy demands may be more than compensated for by increased energy efficiency of such items as appliances, automobiles, and home insulation.

Figure 10–3 illustrates how energy is used in the United States. Approximately 40 percent of the energy we consume is used by industry, which encompasses the production of chemicals, minerals, food, and additional energy resources. Another third of our consumed energy makes buildings comfortable through heating, air conditioning, lighting, and hot water. The remainder of the energy we consume provides for transportation, with the automobile being the major user.

# FOSSIL FUELS

Energy is obtained from a variety of sources, including fossil fuels, nuclear reactors (see Chapter 11), and solar and alternative energy sources (see Chapter 12). Today, most of the energy required in North America is supplied by fossil fuels: oil, natural gas, and coal. A **fossil fuel** is composed of the remnants of organisms that lived millions of years ago. Coal is composed of the remains of prehistoric plants, as evidenced by the fossil imprints of countless species of plants that are found in it. When coal is burned, the organic molecules formed hundreds of millions of years ago by photosynthesis are broken down, and heat is released. Oil and natural gas, also fossil fuels, are composed of the remains of microscopic algae and animals. Unlike coal, which formed from terrestrial plants, oil and natural gas formed from the remains of marine organisms.

Fossil fuels are nonrenewable resources; that is, the Earth has a finite, or limited, supply of them. Although coal and other fossil fuels are still being formed by natural processes today, they are forming too slowly to replace the fossil fuel reserves we are using. Because fossil fuel formation does not keep pace with use, when the Earth's supply of fossil fuels has been used up, we will have to make a transition to other, sustainable forms of energy (see Chapter 12).

## How Fossil Fuels Were Formed

Three hundred million years ago, the climate of much of the Earth was mild and warm, and plants grew year round. Vast swamps were filled with plant species that have long since become extinct. Many of these plants—horsetails, ferns, and club mosses—were large trees (Figure 10–4).

Plants in most environments decay rapidly after death, due to the activities of decomposers such as bacteria and fungi. As the ancient swamp plants died, either from old age or from storm damage, they fell into the swamp, where they were covered by water. Their watery grave prevented the plants from decomposing much; wood-rotting fungi cannot act on plant material where oxygen is absent, and anaerobic bacteria, which thrive in oxygen-deficient environments, don't decompose wood very rapidly. Over time, more and more dead plants piled up. As a result of periodic changes in sea level, layers of sediment (materials deposited by gravity) accumulated, covering the plant material. Aeons passed, and the heat and pressure that accompanied burial converted the plant material into

## Envirobrief

### Cars and the Atmosphere

Each time you burn one quart of gasoline, you are creating about 4.5 pounds of carbon dioxide, adding to the 22 billion tons of carbon dioxide that humans produce each year by burning fossil fuels.

An average air-conditioned automobile contains between 4.5 and 5.5 pounds of chlorofluorocarbon (CFC) coolant, which is the equivalent of about six refrigerators. About one pound of CFCs leak out of each car annually from vibration, *whether the air-conditioning is in use or not.*

Bits of airborne lead, cadmium, and asbestos are constantly emitted by automobiles due to the wear of tires, brakes, and clutch linings. These metals and minerals collect on roads and the landscape that surrounds them. The heavy metals are toxic to human beings.

a carbon-rich rock called coal, and the layers of sediment into sedimentary rock. Much later, geological upheavals raised these layers so that they were nearer the Earth's surface.

Oil was formed when large numbers of microscopic aquatic organisms died and settled in the sediments. As these organisms accumulated, their decomposition depleted the small amount of oxygen that was present in the sediments. The resultant oxygen-deficient environment prevented further decomposition; over time, the dead remains were covered and buried deeper in the sediments. The heat and pressure caused by burial aided in the conversion of these remains to the mixture of **hydrocarbons** (molecules containing carbon and hydrogen) known as oil.

Natural gas, composed primarily of the simplest hydrocarbon, **methane,** was formed in essentially the same way as oil, only at higher temperatures. Over millions of years, as the organisms were converted to oil or natural gas, the sediments covering them were transformed to sedimentary rock.

Deposits of oil and natural gas are often found together. Because oil and natural gas are less dense than rock, they tend to move upward through porous rock layers and accumulate in pools beneath nonporous or impermeable rock layers. Most oil

1. Giant horsetails (two different species)
2. Seed fern
3. Lycopod trees (two different species)
4. Early seed plant
5. Amphibians (three different species, one an immature tadpole)
6. Reptiles (two different species)
7. Shark
8. Fish (two different species)
9. Horseshoe crab
10. Clam-like mollusk
11. Scorpion
12. Cockroach

**Figure 10–4**
Reconstruction of a Carboniferous swamp. Fossils of these organisms have been identified in coal deposits.

and natural gas deposits are found in rocks that are less than 200 million years old.

## COAL

Although coal had been used as a fuel for centuries, it was not until the 18th century that it began to replace wood as the dominant fuel in the Western world, and since then it has had a significant impact on human history. It was coal, for example, that powered the steam engine and supplied the energy needed for the Industrial Revolution. Today

coal is used primarily by utility companies to produce electricity and, to a lesser extent, by heavy industries such as steelmaking.

Coal occurs in different grades, largely as a result of the varying amounts of heat and pressure to which it was exposed during formation. Coal that was exposed to high heat and pressure during its formation is drier, is more compact (and therefore harder), and has a higher heating value. Lignite, bituminous coal, and anthracite are the three most common grades of coal (Table 10–1).

**Lignite** is a soft coal, brown or brown-black in color, with a soft, woody texture. It is moist and produces little heat compared to other coal types.

**Table 10–1**
**A Comparison of Different Kinds of Coal***

| Type of Coal | Color | Water Content (%) | Noncombustible Compounds (%) | Carbon Content (%) | Heat Value (Btu/pound) |
|---|---|---|---|---|---|
| Lignite | Dark brown | 45 | 20 | 35 | 7,000 |
| Bituminous coal | Black | 5–15 | 20–30 | 55–75 | 12,000 |
| Anthracite | Black | 4 | 1 | 95 | 14,000 |

*Courtesy of Thompson and Turk, *Modern Physical Geology,* © 1991, Philadelphia: Saunders College Publishing.

Lignite is often used to fuel power plants. Sizable deposits of it are found in the western states, and the largest producer of lignite in the United States is North Dakota.

**Bituminous coal,** the most common type, is also called **soft coal** even though it is harder than lignite. It is dull to bright black with dull bands. Many bituminous coals contain sulfur, a chemical element that causes severe environmental problems when the coal is burned. Bituminous coal is nevertheless used extensively by electric power plants because it produces a lot of heat.

The highest grade of coal, **anthracite** or **hard coal,** was exposed to extremely high temperatures during its formation. It is a dark, brilliant black and burns most cleanly (produces the fewest pollutants per unit of heat released) of all the types of coal because it is not contaminated by large amounts of sulfur. Anthracite also has the highest heat-producing capacity of any grade of coal. Most of the anthracite in the United States is located east of the Mississippi River, particularly in Pennsylvania.

Coal is usually found in underground layers, called seams, that vary from 2.5 cm (1 inch) to more than 30 m (100 ft) in thickness. Because they are easily located, geologists think that most or all of Earth's major coal deposits have probably been identified. Scientists working with coal are therefore concerned less about finding new deposits than about the safety and environmental problems associated with coal, to be discussed shortly.

## Coal Mining

The two basic types of coal mines are surface, or open-pit, mines and subsurface (underground) mines. The type of mine chosen depends on the location of the coal bed relative to the surface as well as on surface contours (Figure 10–5). If the coal bed is within 30 m (100 ft) or so of the surface, **open-pit,** or **strip, mining** is usually done. This process involves using bulldozers, giant power shovels, and wheel excavators to remove the ground covering the coal seam. The coal is then scraped out of the ground and loaded into railroad cars or trucks. Approximately 60 percent of the coal mined in the United States is obtained by strip mining.

When the coal is deeper in the ground or runs deep into the Earth from an outcrop on a hillside, it is mined underground. **Subsurface mining** accounts for approximately 40 percent of the coal mined in the United States. There are several types of underground mining: drift mining, slope mining, and shaft mining. When a coal outcrop occurs on a hillside, miners simply tunnel into the hill. Such a mine is called a **drift mine.** A coal deposit located a little too deep in the ground for strip mining is usually reached with a **slope mine,** in which the coal is hauled out of the ground through a sloping shaft. Coal deposits located very deep underground are mined by digging a vertical shaft down as much as 150 m (500 ft) or even deeper; this is known as **shaft mining.**

## Coal Reserves

Coal, the most abundant fossil fuel in the world, is found primarily in the Northern Hemisphere. The largest coal deposits are in North America, Russia, and China (Figure 10–6), but deposits are also found in the Arctic islands, Western Europe, India, South Africa, Australia, and eastern South America. The United States has 25.6 percent of the world's coal supply in its massive deposits.

The dearth of coal deposits in the Southern Hemisphere is a result of **plate tectonics,** the movement of huge portions of the Earth's crust (called plates) on which the continents are located. Dur-

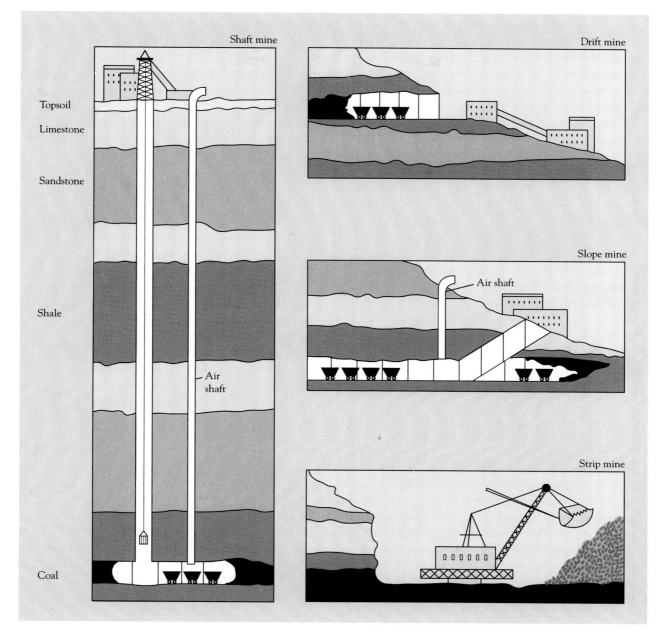

**Figure 10–5**
The types of coal mines. Strip mines are used when the coal is near the Earth's surface. Drift mines, slope mines, and shaft mines are underground mines.

ing the Carboniferous Period, approximately 300 million years ago, the land masses that were to become our present-day South America, Africa, and Australia were joined, making up a massive continent. Known as Gondwanaland, this continent was located near the South Pole (Figure 10–7). Although the climate in the tropical and temperate regions of the Earth was mild, much of Gondwanaland was covered by ice sheets, and because few plants grew there, coal never formed. At this same time, however, much of Europe, North America, and Asia were located closer to the equator, where the warm climate promoted lush vegetation. Since the Carboniferous Period, the continents have separated and migrated to their present locations.

According to the World Resources Institute, known world coal reserves could last for several hundred years at the present rate of consumption.

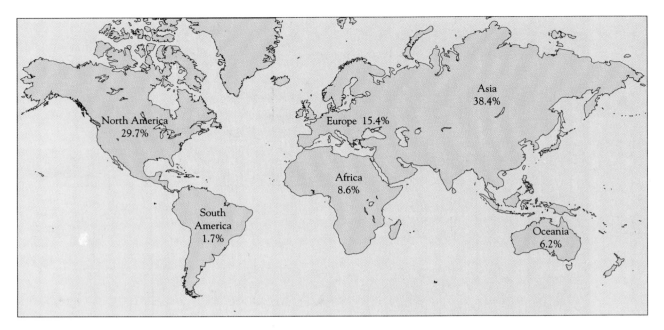

**Figure 10–6**
Distribution of coal deposits. The majority of the world's coal deposits are located in the
Northern Hemisphere.

**Figure 10–7**
Most coal was formed approximately 300 million years ago when the continents were arranged
differently than today. Much of South America and Africa was near the South Pole and was
covered by ice sheets. Consequently, there was little coal formation in those areas.

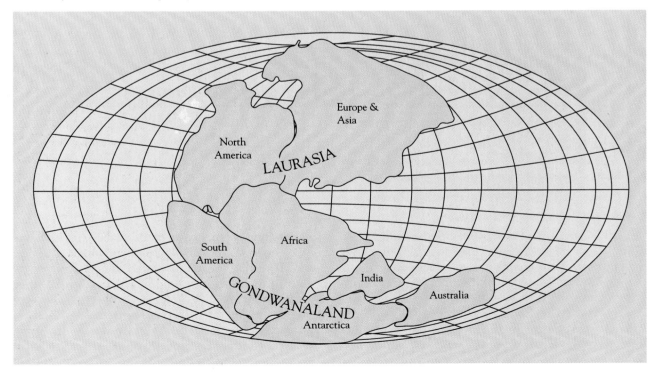

**Figure 10–8**
An underground coal mine in Illinois. The dangers to workers from underground coal mining, which produces about 40% of the coal in the United States, include deaths and injuries due to accidents and respiratory diseases. (Department of Energy)

**Figure 10–9**
Surface mining in eastern Texas. (Chuck Meyers, Office of Surface Mining, Department of the Interior)

Additional coal resources that are currently too expensive to develop[2] have the potential to provide enough coal to last for a thousand or more years (at current consumption rates).

## Safety and Environmental Problems Associated with Coal

Although we usually focus on the environmental problems caused by mining and burning coal, there are also significant human safety and health risks in the mining process itself. Underground mining is an extremely dangerous occupation (Figure 10–8). According to the Department of Energy, during the 20th century more than 90,000 American coal miners have died in mining accidents. Although the number of deaths per year has declined significantly since the earlier part of the century, those not killed or maimed in accidents have an increased risk of cancer and **black lung disease,** a condition in which the lungs are coated with inhaled coal dust so that the exchange of oxygen between the lungs and the blood is severely restricted. It is estimated that these diseases are responsible for the deaths of at least 2,000 miners in the United States each year.

**Environmental Impacts of the Mining Process**
Coal mining, especially strip mining, has substantial effects on the environment. In strip mining,

[2] For example, some coal deposits are buried more than 5,000 feet inside the Earth's crust. Drilling a shaft that deep would cost considerably more than the current price of coal would justify.

vegetation and topsoil are completely removed, causing a loss of habitat for plants and animals and increasing soil erosion and water pollution. Over time, as the coal is removed, strip mining lowers the surface of the ground (Figure 10–9). Prior to passage of the 1977 Surface Mining Control and Reclamation Act, the mines were usually left as large open pits. Acid and toxic mineral drainage from such mines and the removal of topsoil prevent most plants from naturally recolonizing the land (Figure 10–10). Few tree species will grow on land that is badly disturbed as a result of coal mining, called **coal spoils.** Some sites are so severely damaged that only a few types of herbs will grow there. Coal spoils can be restored to prevent further degradation and to make the land productive for other purposes, but restoration is extremely expensive (see Chapter 15, page 330).

**Environmental Impacts of Burning Coal** Burning any fossil fuel releases carbon dioxide, $CO_2$, into the atmosphere. You may recall from the discussion of the carbon cycle in Chapter 5 that a natural equilibrium exists between the $CO_2$ in the atmosphere and the $CO_2$ dissolved in the oceans. Currently we are releasing so much $CO_2$ into the atmosphere through our consumption of fossil fuels that the Earth's $CO_2$ equilibrium has been disrupted. Because the concentration of $CO_2$ in the atmosphere is increasing and $CO_2$ prevents heat from escaping from the planet (the greenhouse effect—see Chapter 20), the temperature of Earth may be affected. An increase of a few degrees in global temperature caused by higher levels of $CO_2$ and other greenhouse gases may not seem very serious at first glance, but a closer look reveals that such an

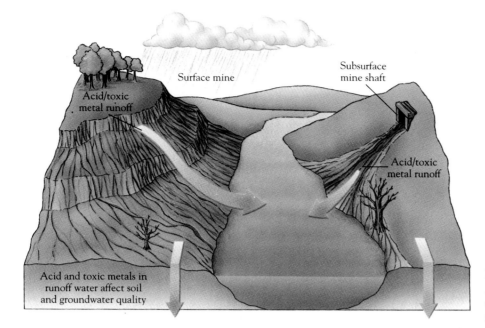

**Figure 10–10**
Acid and toxic mineral runoff from coal mines can cause soil and water pollution.

increase would be catastrophic. For example, as polar ice began to melt, sea levels would rise, flooding coastal areas and placing them at higher risk for storm damage. Other serious environmental consequences of global climate change are considered in Chapter 20. The $CO_2$ problem is made more severe by the burning of coal than by the burning of other fossil fuels, because coal burning releases more $CO_2$ per unit of heat produced.

Coal burning also contributes more of other air pollutants than does the combustion of either oil or natural gas. Bituminous coal contains sulfur and nitrogen that, when burned, are released into the atmosphere as sulfur oxides ($SO_2$ and $SO_3$) and nitrogen oxides ($NO$, $NO_2$, and $N_2O$). Both sulfur and nitrogen oxides form acids when they react with water, and when these reactions occur in the Earth's atmosphere, **acid precipitation** (including **acid rain**) results. The combustion of coal is partly responsible for acid precipitation, which seems to be particularly prevalent downwind from coal-burning power plants. Normal rain is slightly acidic (pH 5.6), but in some areas the pH of acid rain has been measured at 2.1, equivalent to that of lemon juice (see Appendix I for a review of pH). Acid precipitation is a factor in some of the forest decline that has been documented worldwide (Figure 10–11).

Although it is relatively easy to identify and measure pollutants in the atmosphere, it is difficult to trace their exact origins. They are dispersed by air currents and are often altered as they react chemically with other pollutants in the air. Even so, it is clear that some nations suffer the damage of acid rain caused by pollutants produced in other countries, and as a result acid precipitation has become an international issue. The severe environmental repercussions of acid rain are considered in detail in Chapter 20.

It is possible to reduce sulfur emissions associated with the combustion of coal by installing desulfurization systems, or **scrubbers,** in smokestacks; these decrease the amount of sulfur released in the air by 90 percent or more. By federal law, all coal-burning power plants built in the United States since 1978 must have scrubbers. Installing and maintaining scrubbers is extremely expensive; for example, the estimated cost for scrubber installation in older coal-burning power plants in the United States alone is about $10 billion.

Furthermore, the scrubbers themselves give rise to another environmental problem: disposal of the copious sludge produced by the scrubbing. A large power plant may produce enough sludge annually to cover 2.6 square km (1 square mile) of land 0.3 meter (1 foot) deep. Currently, most power plants place the sludge in holding ponds and landfills.

The 1990 Clean Air Act (see Chapter 19) states that the nation's 111 dirtiest coal-burning power plants must cut sulfur dioxide emissions by 1995. This would result in a total annual decrease of 5 million metric tons nationwide. (In comparison, the U.S. total sulfur oxide emissions in 1989 exceeded 20 million metric tons per year.) These

**Figure 10–11**
Dead fir trees enveloped in acid rain clouds on Mt. Mitchell, North Carolina. Forest decline was first documented in Germany and Eastern Europe. More recently, it has been observed in the eastern United States, particularly at higher elevations. The exact reason for forest decline is unknown, but acid precipitation has been implicated. (John Shaw)

power plants may continue to use high-sulfur coal and get a two-year extension of the 1995 deadline if they commit to buying scrubbers. In the second phase of the Clean Air Act, more than 200 additional power plants must make $SO_2$ cuts by the year 2000, resulting in a total annual decrease of 10 million tons nationwide. A nationwide cap on $SO_2$ emissions from coal-burning power plants will be imposed after the year 2000. Utilities must also cut nitrogen oxide emissions by 2 million tons per year, beginning in 1995.

While $CO_2$ emissions remain a significant problem, new methods for burning coal (called **clean coal technologies**) are being developed that will not contaminate the atmosphere with sulfur oxides and will significantly reduce nitrogen oxide production. Clean coal technologies include coal gasification (considered shortly, in the discussion of synfuels) and fluidized-bed combustion.

**Fluidized-bed combustion** mixes crushed coal with particles of limestone in a strong air current during combustion. This coal-burning process has greater efficiency and several additional advantages. Because fluidized-bed combustion takes place at a lower temperature than regular coal burning, fewer nitrogen oxides are produced. (Higher temperatures cause atmospheric nitrogen and oxygen to combine, forming nitrogen oxides.) Also, because the sulfur in coal reacts with the calcium in limestone to form calcium sulfate, which then precipitates out, sulfur is removed from the coal *during* the burning process, so scrubbers are not needed to remove it *after* combustion.

Fluidized-bed combustion is being tested at several large power plants in the United States, and the conversion of other plants to this method will probably begin before the year 2000. The 1990 Clean Air Act provides incentives for utility companies to convert to clean coal technologies.

## OIL AND NATURAL GAS

From the 1600s through the 1800s, wood was the predominant fuel in the United States. By 1900, coal had taken its place as the most important energy source. Beginning in the 1940s, oil and natural gas became increasingly important, largely because they are easier to transport and burn more cleanly than coal. Today, oil and natural gas supply approximately two-thirds of the energy used in the United States. Globally in 1989, 39 percent of the world's energy was provided by oil and 21 percent by natural gas. In comparison, other major energy sources used worldwide include coal (28 percent), hydroelectric power (7 percent), and nuclear power (6 percent) (Figure 10–12).

**Petroleum,** or **crude oil,** is a liquid composed of a number of hydrocarbon compounds. During petroleum refining, the compounds are separated into different products, such as gasoline, heating oil, diesel oil, and asphalt (Figure 10–13). Oil also contains **petrochemicals,** compounds that are used in the production of such diverse products as fertilizers, plastics, paints, pesticides, medicines, and synthetic fibers.

In contrast to petroleum, natural gas contains only a few different hydrocarbons: methane and smaller amounts of propane and butane. The propane and butane are removed from the methane, stored in pressurized tanks as a liquid called **liquefied petroleum gas,** and used primarily as fuel for heating and cooking in rural areas. Methane is used to heat residential and commercial buildings, to generate electricity in power plants, and for a variety of purposes in the organic chemistry industry.

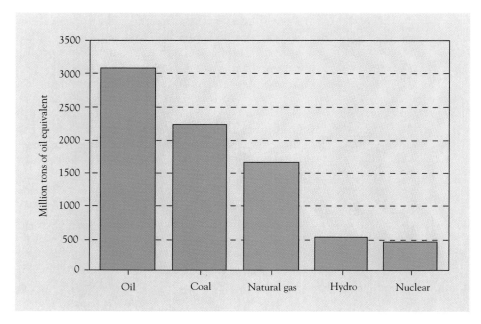

**Figure 10–12**
The world's energy sources in 1989.

**Figure 10–13**
Crude oil is separated into a variety of products during refining. Different components of crude oil have different boiling points; they are separated from one another in a distillation column.

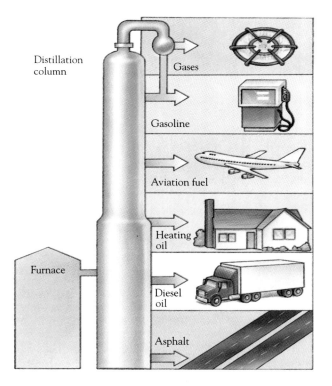

Methane is usually distributed by being pumped through pressurized pipelines or shipped (as a solid) in refrigerated tankers. (At *very* low temperatures, natural gas becomes a solid.)

## Geological Exploration for Oil and Natural Gas

Exploration is continually under way in search of new oil and natural gas deposits, which are usually found together under one or more layers of rock. Usually oil and natural gas deposits are discovered indirectly by the detection of **structural traps,** geological structures that tend to trap any oil or natural gas that is present (recall that oil and natural gas tend to migrate upward until they reach an impermeable rock layer). Two examples of structural traps are anticlines and salt domes (Figure 10–14).

An **anticline** is an upward folding of **strata** (rock layers). Sometimes the strata that arch upward include both porous and impermeable rock. If impermeable layers overlie porous layers, it is possible that any oil or natural gas present will work its way up through the porous rock to accumulate under the impermeable layer.

A number of important oil and natural gas deposits (for example, oil deposits known to exist in the Gulf of Mexico) have been found in association with **salt domes,** underground columns of salt. Salt domes develop when extensive salt deposits form at the Earth's surface as a result of the evaporation of

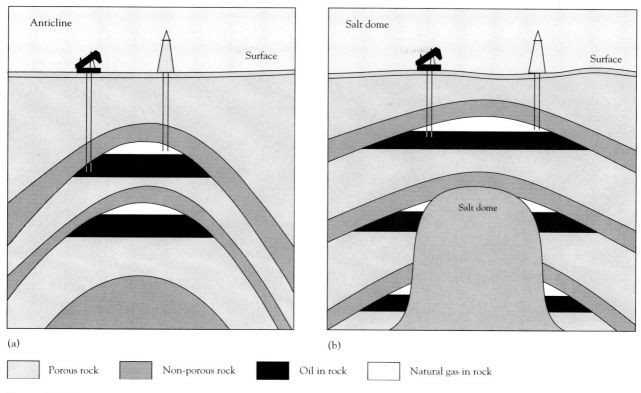

(a)                                                                (b)

☐ Porous rock        ▨ Non-porous rock        ■ Oil in rock        ☐ Natural gas in rock

**Figure 10–14**
Structural traps are underground formations that may contain natural gas or oil. (a) Anticlines form when rock strata buckle, or fold upward. Oil and natural gas seep through porous rock and collect under nonporous, or impermeable, layers. (b) Salt domes are low-density salt formations that tend to rise, whereas high-density rock layers surrounding them sink. Oil and natural gas collect alongside the salt dome under nonporous rock strata.

water. All surface water contains dissolved salts. The salts dissolved in ocean waters are so concentrated that they can be tasted, but even fresh water contains some dissolved material.

If a body of water lacks a passage to the ocean, as an inland lake often does, the salt concentration in the water gradually increases.[3] If such a lake were to dry up, a massive salt deposit called an **evaporite deposit** would remain. Evaporite deposits may eventually be covered by layers of sediment, which convert to sedimentary rock after millions of years. Because salt is less dense than rock, the rock layers settle, and the salt deposit tends to rise in a

column—a salt dome. The ascending salt dome, together with the rock layers that buckle over it, provides a trap for oil or natural gas.

Geologists use a variety of techniques to identify structural traps that might contain oil or natural gas. One method is to drill test holes in the Earth's surface and obtain rock samples. Another method is to produce an explosion at the surface and measure the echoes of sound waves that bounce off rock layers under the surface. These data can be interpreted to determine whether or not structural traps are present. It should be emphasized, however, that many structural traps do not contain oil or natural gas.

Searching for oil and natural gas is very expensive. It costs millions of dollars just to do the basic geological analyses to find structural traps. And once oil or natural gas has been located, drilling and operating the wells cost additional millions.

---

[3] The Great Salt Lake in North America is an example of a salty inland body of water that formed in this way. Although three rivers empty into the Great Salt Lake, water escapes from the lake only by evaporation, accounting for its high salinity—four times as high as that of ocean water.

## Declining Reserves of Oil and Natural Gas

Although oil and natural gas deposits exist on every continent, their distribution is uneven, and a disproportionate share of total oil deposits are clustered relatively close to each other (Figure 10–15). Enormous oil fields containing more than half of the world's total estimated reserves are situated in the politically unstable Middle East. In addition, major oil fields are known to exist in the North Sea, the Gulf of Mexico, and the Arctic (in Alaska and Russia). Because North America has more deposits of natural gas than Europe and other developed areas, use of natural gas is much higher in North America (see Focus On: New Roles for Natural Gas).

It is unlikely that major new oil fields will be discovered in the continental United States, which has been explored for oil more extensively than any other country on Earth. In the last two decades, the success rate of searches for oil has declined, as has the amount of exploration.

There is reason to believe that large oil deposits exist on the **continental shelves,** the relatively flat underwater areas that surround continents. Despite a number of problems, such as storms at sea and the potential for major oil spills, many countries engage in offshore drilling for this oil. Environmentalists generally oppose opening the outer continental shelves for oil and natural gas exploration because of the threat a major oil spill would pose to marine and coastal environments. Coastal industries, including fishing and tourism, also oppose oil and natural gas exploration in these areas.

**How Long Will Oil and Natural Gas Supplies Last?**
A major problem associated with oil and natural gas is their limited supplies. It is difficult to say with certainty how long it will be before the world runs out of oil and natural gas, because there are so many unknowns. We do not know how many additional oil and natural gas reserves will be discovered, nor do we know if or when technological

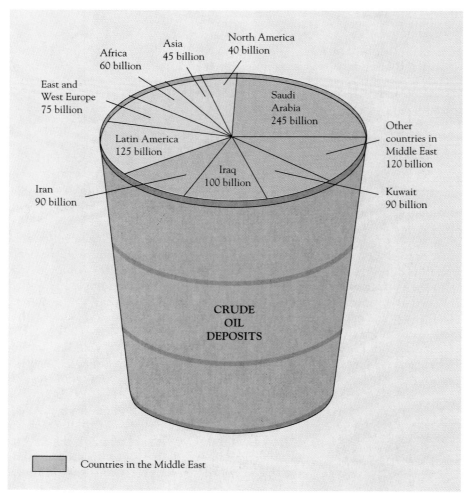

**Figure 10–15**
The world's oil deposits are not evenly distributed. Some regions such as the Middle East contain huge oil deposits in a relatively small area, while other areas have very few. Numbers represent billions of barrels of oil that are proven oil reserves.

## New Roles for Natural Gas

In this age when we must search for alternative fuels, natural gas is gaining new popularity. The general public, researchers, and political leaders have begun to pay greater attention to this relatively clean, efficient source of energy.

Natural gas contains almost no sulfur, a contributor to acid rain. In addition, natural gas produces far less $CO_2$ and fewer hydrocarbons than do gasoline and coal. (Carbon dioxide contributes to global warming, and hydrocarbons produce photochemical smog.) Finally, natural gas does not produce harmful particulate matter the way coal and oil fuels do.

Use of natural gas is increasing in three main areas—generation of electricity, transportation, and commercial cooling. Combined-cycle systems, steam-injected gas turbines, and cogeneration systems (see Chapter 12) are several processes that use natural gas to provide electricity cleanly and efficiently. Natural gas can also be used to help control emissions and costs at already-existing coal-fueled electric power plants. Currently, researchers are studying ways to use natural gas for transportation fuel. Large long-distance trucks seem to travel well on compressed natural gas, for example. Finally, natural gas can efficiently fuel residential and commercial cooling systems. One example is the use of a desiccant-based (air-drying) cooling system, which is ideal for supermarkets, where humidity control is more important than temperature control.

Given the emphasis on reducing dependence on oil and coal, new technologies for the practical use of natural gas should continue to develop in the coming decades.

---

breakthroughs will allow us to extract more fuel from each deposit. The answer to how long these fuels will last also depends on whether worldwide consumption of oil and natural gas increases, remains the same, or decreases. Economic factors influence oil and natural gas consumption; as reserves are exhausted, prices will increase, which can drive down consumption and stimulate greater energy efficiency and the search for additional deposits and alternative energy sources.

Estimates on when we will run out of oil and natural gas vary from several decades to 100 years, but they are only guesses. The only thing we can say with certainty is that, at projected rates of consumption, oil and natural gas reserves will be depleted before coal.

### Global Oil Demand and Supply

One difficult aspect of oil consumption is that the world's major oil producers are not its major oil consumers (Figure 10–16a). In 1989 almost half of the world's oil was consumed by North America and Western Europe, yet these same countries produced only 23 percent of the world's oil. In contrast, the countries in the Persian Gulf region consumed 4.5 percent and produced 26 percent of the world's oil. In the United States, a severe economic burden has resulted from the large amount of oil that is imported; more than half of our huge trade deficit in 1990 was from imported oil.

The imbalance between oil consumers and oil producers will probably worsen in the years to come, because the Persian Gulf region has much higher proven reserves than other countries (Figure 10–16b). For example, at current rates of production, North America's oil reserves would run out in ten years and Western Europe's known reserves would be depleted in 13 years. But the Persian Gulf nations, which have 65 percent of the known world oil reserves, can produce oil at current rates for more than a century.

### Environmental Problems Associated with Oil and Natural Gas

Two sets of environmental problems are associated with the use of oil and natural gas: the problems that result from burning the fuels (combustion) and the problems involved in obtaining them in order to burn them (production and transport). We have already mentioned the $CO_2$ emissions that are a direct result of the combustion of fossil fuels. As with coal, the burning of oil and natural gas produces $CO_2$ that accumulates in the atmosphere, preventing the Earth's heat from radiating into

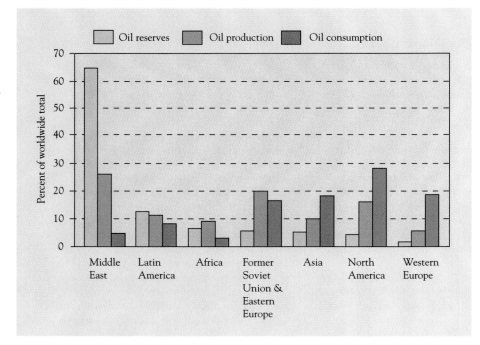

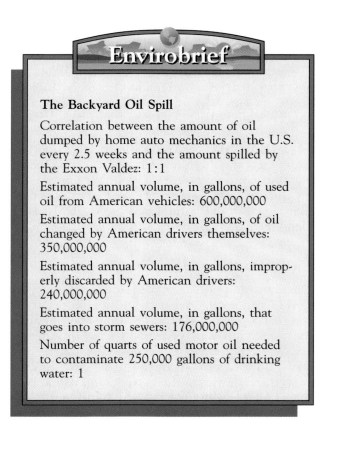

**Figure 10–16**
Oil reserves, production, and consumption in 1989. The world's major oil producers are not the major oil consumers. Proven oil reserves are highest in the politically unstable Middle East.

space. The Earth's climate is calculated to be warming more rapidly now than it did during any of the warming periods following the ice ages. The environmental impact of rapid global climate change could be catastrophic.

Another negative environmental impact of burning oil is acid precipitation. Although oil doesn't produce appreciable amounts of sulfur oxides, it does produce nitrogen oxides, mainly through gasoline combustion in automobiles, which contributes approximately half the nitrogen oxides released into the atmosphere. (Coal combustion is responsible for the other half.) Nitrogen oxides contribute to acid precipitation. The burning of natural gas, on the other hand, doesn't pollute the atmosphere as much as the burning of oil; natural gas is the cleanest of the fossil fuels.

One of the concerns in oil and natural gas production is the environmental damage that may occur during their transport, which is often over long distances by pipelines or by ocean tankers. A serious spill along the route creates an environmental crisis, particularly in aquatic ecosystems, where the oil slick can travel.

One of the most serious oil spills ever in North America took place in Prince William Sound, off the southern coast of Alaska near the town of Valdez, on March 24, 1989. Valdez is at the south end of the oil pipeline that runs through Alaska's interior from Prudhoe Bay in the north. When the supertanker *Exxon Valdez* ran aground, it dumped approximately 11 million gallons of oil into the pristine waters of Prince William Sound. An inten-

sive cleanup campaign was mounted to remove oil from the water, rescue wildlife, protect fish hatcheries, clean the shoreline, and assist local communities that were economically hurt by the accident (Figure 10–17). The long-term environmental

### Envirobrief

**The Backyard Oil Spill**

Correlation between the amount of oil dumped by home auto mechanics in the U.S. every 2.5 weeks and the amount spilled by the Exxon Valdez: 1:1

Estimated annual volume, in gallons, of used oil from American vehicles: 600,000,000

Estimated annual volume, in gallons, of oil changed by American drivers themselves: 350,000,000

Estimated annual volume, in gallons, improperly discarded by American drivers: 240,000,000

Estimated annual volume, in gallons, that goes into storm sewers: 176,000,000

Number of quarts of used motor oil needed to contaminate 250,000 gallons of drinking water: 1

(a)

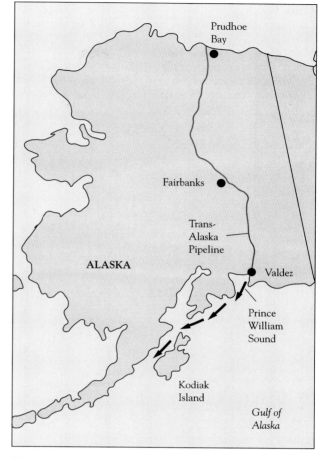

(b)

**Figure 10–17**

(a) An aerial view of the massive oil slick. (b) The extent of the spill (black arrows). Water currents caused it to spread rapidly for hundreds of miles. Countless animals, such as sea otters and ocean birds, died. (*a,* Natalie Fobes/All Stock, Inc.)

consequences of this devastating accident will take years to assess (see Focus On: Alaska's Oil Spill).

The most massive oil spill in history occurred in 1991, during the Persian Gulf War, when about 250 million gallons of crude oil—more than 20 times the amount of the *Exxon Valdez* spill—were deliberately dumped into the Persian Gulf. Many oil wells were also set on fire, and lakes of oil spilled into the desert around the burning oil wells (see Chapter 19). Cleanup efforts along the coastline and in the desert were initially hampered by the war, and environmentalists fear that it may take a century or more for the area to completely recover.

## To Drill or Not To Drill: A Case Study of the Arctic National Wildlife Refuge

In order to understand the complexities of energy issues, let's look at a recent controversy that has pitted environmentalists against oil developers, politicians against politicians, and Americans

against Americans: the proposed opening of the Arctic National Wildlife Refuge to oil exploration. On one side are those who seek to protect rare and fragile natural environments; on the other side are those whose higher priority is the development of the last domestic oil supplies.

**Background of the Arctic National Wildlife Refuge** In 1960 Congress declared a section of northeastern Alaska protected because of its distinctive wildlife. In 1980 Congress expanded this wilderness area to form the Arctic National Wildlife Refuge— 7.3 million hectares (18 million acres) of untouched northern forests, tundra wetlands, and glaciers (Figure 10–18a). The Department of the Interior was given permission to conduct a study of the potential for oil discoveries in the area, but exploration and development could proceed only with congressional approval.

The refuge is home to an extremely diverse fish and wildlife community, including polar bears, arc-

**Figure 10–18**
The Arctic National Wildlife Refuge. (a) Located in the north-eastern part of Alaska, the Arctic National Wildlife Refuge is situated close to the Trans-Alaska pipeline, which begins at Prudhoe Bay and extends south to Valdez. (b) Caribou migrate from Canada, where they winter, to the Arctic National Wildlife Refuge, where calves are born. (David C. Fritts © 1993 Animals Animals, Inc.)

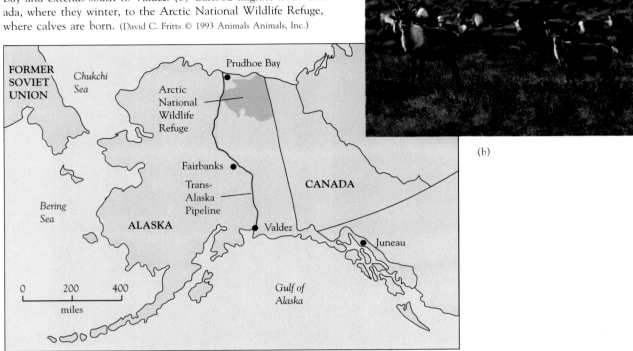

(b)

(a)

tic foxes, peregrine falcons, musk-oxen, Dall sheep, wolverines, and snow geese. It also contains the calving area for a large migrating herd of caribou (Figure 10–18b). The Porcupine caribou herd, named after the Porcupine River in Canada where the herd winters, contains more than 150,000 head. Dominant plants in this coastal plain of tundra include mosses, lichens, sedges, grasses, dwarf shrubs, and small herbs. Under a thin upper layer of soil is the permafrost layer, which contains permanently frozen water.

Although it is biologically rich, the tundra is an extremely fragile ecosystem, in part because of its harsh climate. The organisms living here are adapted to their environment, but they live "on the edge." Any additional stress has the potential to harm or even kill them. Thus, arctic organisms are particularly vulnerable to human activities.

**Support for Oil Exploration in the Refuge** Supporters cite economic considerations as the main reason for searching for oil in the refuge. They point out that the United States is spending a large proportion of its energy budget to purchase foreign oil. Development of domestic oil would help to improve the balance of trade and make us less dependent on foreign countries for our oil.

The oil companies are eager to develop this particular site because it is near Prudhoe Bay, where large oil deposits are already being tapped. Prudhoe Bay has a sprawling industrial complex to support oil production, including roads, pipelines, gravel pads, and storage tanks (Figure 10–19). The Prudhoe Bay oil deposits have peaked in production and will decline in productivity over the next few years. As a result, the oil industry is looking for sites that can make use of the infrastructure already in place.

The study conducted by the Department of the Interior on the possibility of oil in the wildlife refuge was made public in 1987. It concluded that there is a 19 percent chance of finding oil there, which the oil industry considers enough to justify exploration.

Supporters of oil exploration argue that it will have little lasting impact on the environment or on

### Alaska's Oil Spill

On March 24, 1989, the supertanker *Exxon Valdez* slammed into Bligh Reef and spewed 11 million gallons (260,000 barrels) of crude oil into Prince William Sound off the coast of Alaska, creating one of the most massive oil spills in history. As it spread, the black, tarry gunk eventually covered thousands of square kilometers of water and contaminated hundreds of kilometers of shoreline. According to the U.S. Fish and Wildlife Service and the Alaska Department of Environmental Conservation, more than 30,000 birds (sea ducks, loons, cormorants, bald eagles, and other species) and between 3,500 and 5,500 sea otters died as a result of the spill. The area's killer whale population declined, and salmon migration was disrupted. Throughout the area, there was no fishing season that year.

Within hours of the spill, scientists began to arrive on the scene to advise both Exxon and the government on the best way to try to contain and clean up the spill. But it took much longer for any real action to be taken.

Eventually, nearly 12,000 workers took part in the cleanup; their activities included mechanized steam cleaning and rinsing, which actually killed shoreline creatures such as barnacles, clams, mussels, eelgrass, and rockweed. About 2.6 million gallons of oil were recovered and returned to Exxon. Exxon's cost in the cleanup totaled about $1.28 billion. The corporation faced hundreds of lawsuits, including some from fishermen whose livelihoods were lost to the spill; some of these suits will take years to settle. There was also the hidden cost of public outrage in the form of boycotts and shredded Exxon credit cards.

In September 1989, Exxon declared the cleanup "complete." But it left behind contaminated shoreline, reduced reproductivity among eagles, and a reduced commercial salmon catch, among other problems. In October 1991, Exxon agreed to pay the state of Alaska a total of about $1 billion in ten annual installments. This money is going into a trust fund to help restore the clean coastline of Prince William Sound.

wildlife. They say that there has been little environmental disruption or contamination of Prudhoe Bay by the oil industry; further, they point out that the number of caribou in that area has actually increased.

**Opposition to Oil Exploration in the Refuge**  Conservationists think that oil exploration poses permanent threats to the delicate balance of nature in the Alaskan wilderness, in exchange for a very temporary oil supply. They think that the money that would be spent searching for oil would be better used for research into alternative, renewable energy sources and energy conservation—a more permanent solution to the energy problem. They further argue that "Drain America first" policies will only increase our future dependence on foreign oil supplies.

Opponents of oil exploration also refute supporters' claims that Prudhoe Bay has been developed with little environmental damage. Studies such as one conducted by the U.S. Fish and Wild-

**Figure 10–19**
Prudhoe Bay in northern Alaska has a massive industrial complex built to support the production of oil. (Bruce F. Molnia/Terraphotographics)

life Service document considerable habitat damage and declining numbers of wolves and bears in the Prudhoe Bay area;[4] it appears to biologists that the increase in the caribou herd has been the direct result of fewer predators. The oil industry and conservationists *do* agree on one point: it is not financially practical to restore developed areas in the Arctic to their natural states. Thus, development in the Arctic causes permanent changes in the natural environment.

## SYNFUELS AND OTHER POTENTIAL FOSSIL FUEL RESOURCES

**Synthetic fuels,** or **synfuels,** are used in place of oil or natural gas. They are synthesized from coal and other sources and may be liquid or gaseous. Synfuels include tar sands, oil shales, gas hydrates, alcohol fuels, coal gas, and liquefied coal. All synfuels emit $CO_2$ when burned, and many have other negative environmental effects, such as land damage from strip mining. Although synfuels are more expensive to produce than fossil fuels, they will probably become more important as fossil fuel reserves decline.

**Tar sands** are underground sand deposits permeated with tar or oil so thick and heavy that it doesn't move. The oil in tar sands deep in the ground cannot be pumped out unless it is heated underground with steam to make it more fluid. If tar sands are close to the Earth's surface, however, they can be strip mined. Once oil is obtained from tar sands, it must be refined (as crude oil is). Major tar sands are found in Alberta, Canada, and in Venezuela. World tar sand reserves are estimated to contain half again as much fuel as world oil reserves.

"Oily rocks" were discovered by western American pioneers whose rock hearths caught fire and burned. In order to yield their oil, these sedimentary rocks, called **oil shales,** must be crushed, heated, and refined after they are mined. Because the mining and refinement of oil shales require the expenditure of a great deal of energy, it is not cost-efficient to process oil shales that do not yield a significant amount of oil. Large oil shale deposits are located in the United States, Russia, China, and Canada. Wyoming, Utah, and Colorado have the largest deposits in the United States (Figure 10–20). Like tar sands, oil shale reserves may contain half again as much fuel as world oil reserves.

**Figure 10–20**
This huge oil shale deposit in eastern Utah is the site of the first U.S. oil shale processing plant. (Nolan Preece/Biological Photo Service)

**Gas hydrates** are reserves of ice-encrusted natural gas deep underground in porous rock. Deposits have been identified in the Arctic tundra, deep under the permafrost. It is possible that gas hydrates could be found in the deep ocean sediments of the continental slope, as well. The oil industry is not particularly interested in extracting gas from gas hydrates at present, because of the expense involved. In a pilot program in Siberia, Russian scientists have successfully removed natural gas from hydrate deposits by pumping methanol (an alcohol) into the hydrate region. Natural gas that is associated with ice readily dissolves in methanol, which can then be pumped out of the ground.

Most industries have the ability to switch fuels if one type becomes temporarily unavailable or too expensive. For example, most power utilities routinely switch the type of fossil fuel they use. The automobile, however, is completely dependent on gasoline, which is refined from crude oil. The auto industry has been searching for a suitable liquid fuel that can be substituted for gasoline. Methanol ($CH_3OH$) and ethanol ($CH_3CH_2OH$) are **alcohol fuels** that may eventually replace gasoline. These synfuels can be produced from **biomass**—living plant or animal material (see Chapter 12). Methanol can also be produced from either natural gas (the least expensive source) or coal. One of the advantages of alcohol fuels is that they burn more cleanly than gasoline. Although alcohol fuels produce $CO_2$ and therefore contribute to global warming, they produce substantially fewer nitrogen oxides than gasoline does. Technological improvements in their production could bring their costs down, but currently they are not cost-competitive with fossil fuels.

---

[4]The top predators are usually more susceptible to environmental disruption than are the organisms occupying lower positions in a food web.

Coal has also been used to produce a nonalcohol liquid fuel (Figure 10–21). This process, called **coal liquefaction,** was first developed before World War II, but its expense prevented it from replacing gasoline production. Research in the 1980s resulted in a series of technological improvements that have lowered the cost of coal liquefaction, but it is still not cost-competitive. It may become commercially attractive when the cost of gasoline rises or when new innovations reduce the cost of producing liquid coal even further.

Another synfuel is a gaseous product of coal. Coal gas has been produced since the 19th century. As a matter of fact, it was the major fuel used for lighting and heating in American homes until oil and natural gas replaced it in this century. Production of combustible gases (carbon monoxide and hydrogen) from coal is called **coal gasification.** A promising coal gasification technique was developed at Stanford University, and since 1984, a pilot power plant that utilizes coal gas produced by the new technology has operated in southern California.

One advantage of coal gas over solid coal is that it burns cleanly. Because sulfur is removed during coal gasification, no scrubbers are needed when coal gas is burned. Like other synfuels, coal gas is more expensive to produce than fossil fuels.

## Environmental Impacts of Synfuels

Although synfuels are promising energy sources, they have many of the same undesirable effects as fossil fuels. Their combustion releases enormous quantities of $CO_2$ into the atmosphere, thereby contributing to global warming. Some synfuels,

### What Comes from a Car

The regulated phase-out of lead from gasoline began in 1975 because of its toxic and bioaccumulative effects on the environment. Scientists have since discovered that even tiny amounts of lead, ingested regularly by children, can cause serious impairment of their intellectual capacities. To compensate for lost octane, petroleum companies replaced lead in gasoline with a family of substances known as aromatic hydrocarbons. These include benzene, toluene, and xylene. All are toxic. Benzene is water-soluble and spreads through the environment easily. Chronic exposure to it can cause leukemia. Toluene can damage kidneys. In a volatile, partially combusted form, these aromatics are in the air of every city street, emitted from the tailpipes of almost every car, truck, and bus on the road.

such as coal gas, require large amounts of water during production and would therefore be of limited usefulness in arid areas, where water shortages are already commonplace. Also, mining the fossil fuels that are needed to produce synfuels damages the land. Enormously large areas of land would have to be strip mined in order to recover the fuel in tar sands and oil shales.

**Figure 10–21**

An advanced coal liquefaction research and development facility in Alabama. Coal liquefaction produces a liquid fuel from coal. This plant can process 6 tons of coal per day. The price of producing liquid coal is still not competitive with the price of gasoline. However, someday it may replace gasoline as the automotive fuel of choice. (Department of Energy)

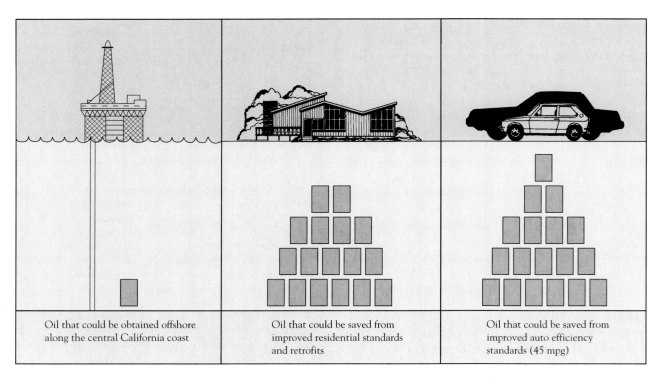

| Oil that could be obtained offshore along the central California coast | Oil that could be saved from improved residential standards and retrofits | Oil that could be saved from improved auto efficiency standards (45 mpg) |

**Figure 10–22**

Energy can be saved by improved energy conservation and efficiency. Shown is a comparison of the amount that could be saved versus the amount of offshore oil that could be obtained from wells drilled along the central California coast. Each block represents 0.6 billion barrels of oil.

## AN ENERGY STRATEGY FOR THE UNITED STATES

The United States needs a comprehensive energy strategy for several reasons: (1) the supply of fossil fuels is limited; (2) fossil fuels pollute the environment; (3) our heavy dependence on foreign oil makes us economically vulnerable. Because of the complex nature of energy issues, any policy that is adopted by our political leaders has to utilize many approaches. Although there is no way to completely eliminate our vulnerability to disruptions in foreign oil supplies and to oil price increases, we can lessen the effects of such events through a comprehensive energy strategy. Such a strategy must provide us with a secure supply of energy, encourage us to use less energy, and protect the environment. The following elements should be included in a comprehensive national energy policy.

**1. Increase Energy Efficiency and Conservation** Between 1975 and 1990, energy efficiency in the United States improved by about 26 percent. There is, however, still room for great improvement on all fronts, from individuals conserving heating oil by weatherproofing their homes, to groups of commut-

ers conserving gasoline by carpooling, to corporations developing more energy-efficient products (Figure 10–22). The automobile industry could be required to increase the average new-car gas mileage, for example.

One way to encourage energy conservation is to eliminate government subsidies that keep energy prices artificially low. When prices reflect the true costs of energy, including the environmental costs incurred by its production, transport, and use, energy will be used more efficiently (see Focus On: Energy Subsidies and the Real Price of Fuel). Gasoline prices in the United States do not reflect the true cost of gasoline and are unrealistically low. During the late 1980s, for example, Europeans paid three to four times more for gasoline than Americans did (Figure 10–23). It has been demonstrated many times that the price of gasoline affects the level of gasoline consumption: lower prices encourage greater consumption. Over the next few years, a more realistic price for gasoline should be introduced to encourage people to buy fuel-efficient automobiles, carpool, and use public transportation.

Other gasoline-conserving measures could be adopted, such as reinstating the 55-mile-per-hour speed limit in areas where the limit is now 60 or 65

## Energy Subsidies and the Real Price of Fuel

The price you pay for gasoline at the fuel pump is not its actual price. Its real cost is subject to your country's energy pricing policy, which attempts to stabilize energy prices and reduce our dependence on sources of foreign energy supplies. Most governments practice some type of energy pricing policy, often in the form of energy subsidies or energy taxes.

An energy subsidy is equal to the difference between the world market price and the domestic market price for a particular fuel. For example, if gasoline trades for $1 per gallon on the world market, but costs only $.75 within a particular country, then that country's government is subsidizing energy at a cost of $.25 per gallon. Energy subsidies are often aimed at reducing the price of fuel for consumers, with the intent of stimulating economic growth. Low costs for consumers encourage a high use of energy, which in turn accelerates the depletion of a nonrenewable energy resource. When a government removes its energy subsidies, consumers have

to pay more for fuel, which encourages them to use energy more efficiently so they can save money. The resulting decrease in energy use reduces a country's dependence on costly fuel imports and lessens the harmful environmental impacts of fuel production and consumption.

Whereas subsidies reduce the cost of fuel for consumers, energy taxes *increase* the cost. Energy taxes serve to raise a government's revenues. In addition, they encourage consumers to conserve energy, thereby helping to reduce a country's dependence on foreign energy supplies. Energy taxes reflect some of the hidden costs of energy consumption, such as pollution control and cleanup.

Energy pricing policies apply not only to gasoline (that is, oil) but also to electricity, natural gas, and coal. So the next time you fill up at the gas station, turn on a light, cook dinner, or adjust the thermostat, keep in mind that the price you pay for energy is determined by more than just the current day's price for a barrel of foreign oil.

---

mph. Automobile fuel efficiency is cut by approximately 30 percent if a car is driven at 65 mph rather than 55 mph. In addition, federal financial support for transportation could be shifted from highway construction to public transportation. We say more about energy conservation in Chapter 12 (see also You Can Make a Difference: Getting Around Town).

**2. Secure Future Energy Supplies** A comprehensive national energy strategy will probably include the environmentally sound and responsible development of domestically produced fossil fuels, especially natural gas.

There are two types of opposition to this element of a national energy strategy; one is economic and the other is environmental. Some think it is better to deplete foreign oil reserves while prices are reasonable and save domestic supplies for the future. Most economists argue against this view, however, because of the United States trade deficit; we do not currently finance our oil imports by exporting goods and services of equal value. Many environmentalists oppose the development and increased use of domestic fossil fuels, largely due to

the environmental problems already discussed.

Everyone, environmentalists included, recognizes the need for dependable energy supplies. Securing a future energy supply is a *temporary* solution, however, because fossil fuels are nonrenewable resources that will eventually be depleted, regardless of how efficient our use or how much we conserve. Having a secure energy supply for the short term will, however, allow us time to develop alternative energy sources for the long term.

**3. Improve Energy Technology** Research and development must be expanded for all possible alternatives to fossil fuels, especially renewable energy sources such as solar and wind energy (see Chapter 12). Our long-term energy policy goal should be to shift to energy sources that do not threaten the environment.

Who should pay for the research costs of improving energy conservation and developing alternative forms of energy? The answer is that we all should share in these costs because we will all share in the benefits. The proceeds of a gasoline tax are being considered as a means of financing programs

## Getting Around Town

Can you imagine getting around town without a car? How would you get to class, the grocery store, the laundromat? Hopping in your car for every errand seems like the natural thing to do. But think about this: from the production of gasoline to the disposal of old automobiles, the car has a significant negative impact on the environment. Acid rain and global warming are just some of the problems caused by the combustion of gasoline. Vehicle exhaust, photochemical smog, and chronic low-level exposure to toxins are all health threats to car owners and to those who live in areas with a high density of cars (see Chapter 19). Dumping of engine oil, fumes from the burning of tires and batteries, and automobile junkyards threaten both our health and our environment.

There is something you can do about it. Granted, you may not be able to give up your car entirely, but you can cut down on its use wherever possible. For example, try the following:

1. Carpool to class, to work, to the grocery store, to social events. One car on the road is better than three or four.
2. Buy a good bicycle; it is less expensive than a car to buy and maintain, and it is great for local transportation. It is also good exercise.
3. Ride the bus or the train whenever possible. Think about how jammed the road would be if all 50 people on a bus were driving individual cars!
4. Walk to class or work if you live within a mile or so. You will need to allow yourself a little extra time, but once you get into the habit, it's easy. Walking is good exercise, too.
5. Modify your driving habits to save gas. Minimize braking, and don't let your engine idle for more than 1 minute. Keep your car well tuned, replace air and oil filters often, and keep your tires inflated at the recommended pressure. Remove any unnecessary weight from your car. All of these measures help boost gasoline mileage.

to achieve the goals in steps 1 through 3. Some policy makers have suggested a tax of as much as 50 cents per gallon. This may sound excessive until U.S. gasoline prices are compared with those in Japan and Europe—which are much higher. More expensive gasoline doesn't seem to have hampered Japan's or Europe's economic competitiveness.

**4. Accomplish the First Three Objectives Without Further Damaging the Environment** The environmental costs of using a particular energy source must be weighed against its benefits when it is considered as a practical component of an energy policy. If domestic supplies of fossil fuels are developed with as much attention to the environment as possible, they will not only help reduce our dependence on foreign oil, but also give us time to develop alternative forms of energy. One suggestion has been to add a 5-cent tax on each barrel of domestically produced oil to establish a reclamation fund for undoing some of the environmental damage caused by mining and production of fossil fuels.

| | Gasoline consumption (gallons per person per year) | Gas price (dollars per gallon, 1988) |
|---|---|---|
| United States | 484 | $0.95 |
| Sweden | 221 | 2.81 |
| West Germany | 206 | 2.18 |
| Norway | 191 | 3.09 |
| France | 180 | 3.04 |
| Britain | 176 | 2.52 |
| Austria | 176 | 2.67 |
| Denmark | 172 | 3.67 |
| Netherlands | 155 | 3.00 |
| Italy | 148 | 3.90 |
| Japan | 133 | 3.47 |

**Figure 10–23**
When gasoline prices and per capita gasoline consumption are compared, it is obvious that higher gasoline consumption is a direct result of lower gasoline prices.

# SUMMARY

**1.** Oil, gas, and coal are fossil fuels that formed several hundred million years ago from plant and animal remains. They are nonrenewable resources. None of them is a perfect energy source; their combustion produces several pollutants—in particular, large amounts of $CO_2$, a greenhouse gas that prevents heat from escaping from the Earth.

**2.** Coal is formed when partially decomposed plant material is exposed to heat and pressure for aeons. There are several grades of coal. Lignite is soft coal that is low in sulfur and produces less heat than other grades; bituminous coal is usually high in sulfur and produces a lot of heat; anthracite is low in sulfur and produces more heat than the other coal types.

**3.** Coal is present in greater quantities than oil or natural gas, but its use has a greater potential to harm the environment. Underground coal mining is dangerous and unhealthful. Strip mining produces large coal spoils that are difficult to restore.

**4.** Coal produces more $CO_2$ emissions per unit of heat than do other fossil fuels, and there is currently no way to reduce or eliminate $CO_2$ emissions at the site where coal is burned. Burning soft coals that contain sulfur contributes to acid precipitation. There are several ways to control sulfur emissions, including fluidized-bed combustion and the installation of scrubbers in smokestacks.

**5.** Oil and natural gas deposits occur in association with structural traps such as anticlines and salt domes. Although world oil and natural gas reserves are large, they will probably be depleted during the 21st century.

**6.** The environmental problems associated with the use of oil and natural gas include damage to the environment where the wells are located, accidental oil spills during transport and storage, and $CO_2$ emissions when the oil is burned.

**7.** Synfuels (tar sands, oil shales, gas hydrates, alcohol fuels, coal gas, and liquid coal) are liquid or gaseous fuels that substitute for oil or natural gas. Synfuels are currently more expensive to produce than oil or natural gas, but they will probably be utilized more in the future, as fossil fuel reserves decline. Synfuels have many of the environmental drawbacks associated with traditional fossil fuels.

**8.** The goal of a national energy strategy should be to ensure adequate energy supplies without harming the environment. Such a policy can be accomplished by increasing energy efficiency, securing future energy supplies, improving energy technology, and avoiding harm to the environment.

# DISCUSSION QUESTIONS

**1.** It has been suggested that the Industrial Revolution was concentrated in the Northern Hemisphere largely because coal is located there. What is the relationship between coal and the Industrial Revolution?

**2.** Several politicians have commented that the United States is "the OPEC of coal." What does this mean?

**3.** Based on what you have learned about coal, oil, and natural gas, which fossil fuel do you think the United States should exploit in the short term (during the next 20 years)? Explain your rationale.

**4.** In your estimation, which fossil fuel has the greatest potential for the long term (the next century)? Why?

**5.** Which of the negative environmental impacts associated with fossil fuels is most serious? Why?

**6.** Which major consumer of oil is most vulnerable to disruption in the event of another energy crisis: electric power generation, motor vehicles, heating and air conditioning, or industry? Why?

**7.** Why are synfuels promising, but not a panacea for our energy needs?

# SUGGESTED READINGS

Corcoran, E. Cleaning up coal. *Scientific American*, May 1992. A review of new technologies that could clean up coal's dirty environmental image.

Dreyfus, D. A., and A. B. Ashby. Fueling our global future. *Environment* 32:4, 1990. The outlook for the energy situation in both the United States and other countries.

Earle, S. A. Persian Gulf pollution: Assessing the damage one year later. *National Geographic* 181:2, February 1992. How researchers started evaluating the environmental destruction in Kuwait and the Persian Gulf. *National Geographic* had a related article, "After the Storm," in its August 1991 issue.

Flavin, C. Conquering U.S. oil dependence. *World*

*Watch* 4:1, 1991. Ways in which we can break our dependence on oil.

Fulkerson, W., R. R. Judkins, and M. K. Sanghvi. Energy from fossil fuels. *Scientific American,* September 1990. Discusses today's technological challenge of extracting as much energy as possible from fossil fuels while harming the environment as little as possible.

Holloway, M. Soiled shores. *Scientific American,* October 1991. An examination of the technologies used to clean up after the *Exxon Valdez* spill.

Hubbard, H. M. The real cost of energy. *Scientific American,* April 1991. Analyzes how the U.S. government has subsidized energy and what should be done to bring market prices in line with true energy costs.

Lee, D. B. Oil in the wilderness: An arctic dilemma. *National Geographic* 174:6, 1988. Discusses the riches of the Arctic National Wildlife Refuge.

Yergin, D. The age of hydrocarbon man. *World Monitor* (The Christian Sciences Monthly Monitor) 4:2, 1991. An examination of past energy crises, which were based on the lack of a secure supply of energy, and a looming future energy crisis—the impact of oil on the environment.

# Chapter 11

*The nuclear power plant at Three Mile Island, Pennsylvania. (Breck P. Kent,
© 1993 Earth Scenes)*

# Nuclear Energy

uclear energy involves changes in the nuclei of atoms that result in the release of large amounts of energy, which is then used to generate electricity. In the United States, 111 nuclear power plants supply 20 percent of the nation's electricity, yet no plants have been ordered since 1976, and only a few are still under construction. Although nuclear power could supply us with long-term electricity, its use is complicated by difficult environmental, economic, social, and public trust problems. Particularly serious are the issues of nuclear waste disposal, expense, and safety. If they can be resolved (and it is not clear that they can be), nuclear energy has the potential to replace fossil fuels as the main energy source for electricity in the 21st century.

## HOW DO WE GET ENERGY FROM ATOMS?

In this chapter we examine the facts and controversies of nuclear power. In order to arrive at conclusions that are intelligent and informed, we must first understand some of the basic science behind nuclear power and how nuclear technology is used to produce energy.

As a way to obtain energy, nuclear power is fundamentally different from the combustion that produces energy from fossil fuels. Combustion is a chemical reaction. In ordinary chemical reactions, atoms of one element do not change into atoms of another element, nor does any of their mass (matter) change into energy. The energy released in combustion and other chemical reactions comes from changes in the chemical bonds that hold the atoms together. Chemical bonds are associations between electrons, so ordinary chemical reactions involve the rearrangement of electrons (see Appendix I).

In contrast, nuclear energy involves changes within the *nuclei* of atoms; small amounts of matter from the nucleus are converted into very large amounts of energy. There are two different reactions that release nuclear energy: fission and fusion. In **fission,** larger atoms of certain elements are split into two smaller atoms, whereas in **fusion,** two smaller atoms are combined to make one larger atom. In each case, the mass of the end product(s) is less than the mass of the starting material(s) because a small quantity of the starting material is converted to energy.

Nuclear reactions produce 100,000 times more energy per atom than chemical reactions such as combustion do. In nuclear bombs this energy is released all at once, producing a tremendous surge of heat and power that destroys everything in its vicinity. On the other hand, in the utilization of nuclear energy to generate electricity, the nuclear re-

**Figure 11–1**

Atomic structure. Matter is composed of atoms. (a) Atoms contain a nucleus made of positively charged particles (protons) and particles with no charge (neutrons). Circling the nucleus is a cloud of small, negatively charged particles called electrons. (b) Particles with like charges tend to repel one another, which would cause the nucleus to fall apart (1). Atomic forces keep the positively charged protons together (2). Neutrons tend to stabilize the nucleus. Some atoms have too few or too many neutrons to maintain nuclear stability (3). These atoms are radioactive.

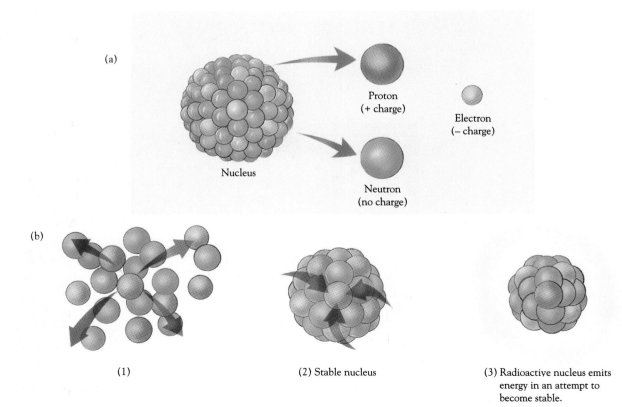

(a)

Nucleus

Proton
(+ charge)

Electron
(– charge)

Neutron
(no charge)

(b)

(1)

(2) Stable nucleus

(3) Radioactive nucleus emits energy in an attempt to become stable.

action is controlled to produce smaller amounts of energy in the form of heat, which can then be converted to electricity.

## Atoms and Radioactivity

All atoms are composed of positively charged protons, negatively charged electrons, and electrically neutral neutrons (Figure 11–1). Protons and neutrons, which have approximately the same mass, are clustered in the center of the atom, making up its nucleus. Electrons, which possess little mass in comparison to protons and neutrons, orbit around the nucleus in distinct regions. Atoms that are electrically neutral possess identical numbers of positively charged protons and negatively charged electrons.

The **atomic mass** of an element is equal to the sum of protons and neutrons in the nucleus. Each element contains a characteristic number of protons per atom, called its **atomic number.** In contrast, the number of neutrons in each atom of a given element may vary, resulting in atoms of one element with different atomic masses. Forms of a single element that differ in atomic mass are known as **isotopes.** For example, normal hydrogen, the lightest element, contains one proton and *no* neutrons in the nucleus of each atom. The two isotopes of hydrogen are **deuterium,** which contains one proton and *one* neutron per nucleus, and **tritium,** which contains one proton and *two* neutrons per

nucleus. Many isotopes are stable, and some are unstable; the unstable ones are said to be **radioactive** because they spontaneously emit **radiation,** a form of energy. The only radioactive isotope of hydrogen is tritium.

As a radioactive element emits radiation, its nucleus changes into the nucleus of a different element, one that is more stable; this process is known as **radioactive decay.** For example, the radioactive nucleus of one isotope of uranium, U-235,[1] decays over time into lead (Pb-207). Each radioactive isotope has its own characteristic rate of decay. The period of time required for one-half of a radioactive substance to change into a different material is known as its **radioactive half-life.** A radioactive material gives off negligible radiation after ten half-lives. There is enormous variation in the half-lives of different radioactive isotopes (Table 11–1). For example, the half-life of iodine (I-132) is only 2.4 hours, whereas the half-life of uranium (U-234) is 250,000 years.

### Mini-Glossary of Nuclear Energy Terms

**nuclear energy:** The energy released by nuclear fission or fusion.

**fission:** The splitting of an atomic nucleus into two smaller fragments, accompanied by the release of a large amount of energy.

**fusion:** Fusing two lightweight atomic nuclei into a single nucleus of heavier mass, accompanied by the release of a large amount of energy.

**isotope:** One of two or more forms of an element that have the same atomic number but different atomic masses. Some isotopes are radioactive.

**radioactive decay:** The emission of energetic particles or rays from unstable atomic nuclei.

**radioactivity:** Radioactive decay.

**half-life:** The amount of time it takes for half of the atoms in a radioactive substance to disintegrate, or decay.

| Table 11–1 Some Common Radioactive Isotopes Associated with the Fission of Uranium | |
|---|---|
| **Radioisotope** | **Half-Life (years)** |
| Cerium 144 | 0.8 |
| Cesium 137 | 30 |
| Iodine 131 | 0.02 (8.1 days) |
| Krypton 85 | 10.4 |
| Neptunium 237 | 2,130,000 |
| Plutonium 239 | 24,400 |
| Plutonium 240 | 6,600 |
| Radon 226 | 1,600 |
| Ruthenium 106 | 1.0 |
| Strontium 90 | 28 |
| Tritium | 13 |
| Xenon 133 | 0.04 (15.3 days) |

## CONVENTIONAL NUCLEAR FISSION

Uranium ore, the mineral fuel used in conventional nuclear power plants, is a nonrenewable resource present in limited amounts in the Earth's crust.

---

[1] Uranium 235 (U-235 or $^{235}$U) is an isotope of uranium with an atomic mass of 235. The atomic number of uranium—the number of protons in each uranium nucleus—is 92. Since the atomic mass is equal to the sum of the protons and neutrons in each nucleus, we can calculate the number of neutrons in each U-235 nucleus to be 235 (atomic mass) − 92 (number of protons) = 143 (number of neutrons).

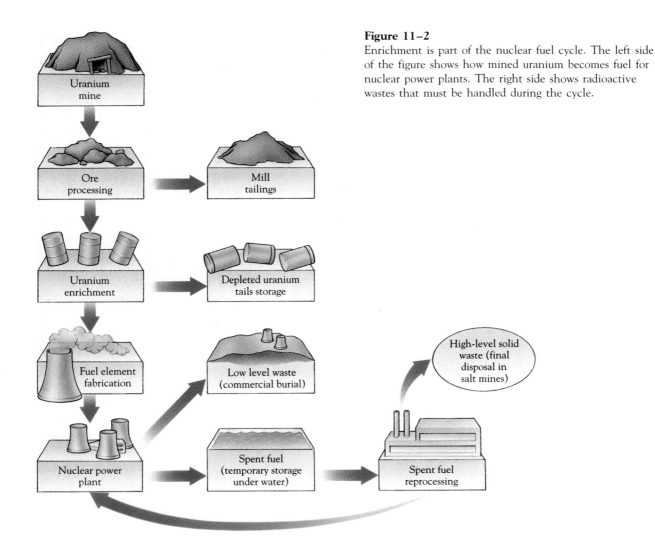

**Figure 11–2**
Enrichment is part of the nuclear fuel cycle. The left side of the figure shows how mined uranium becomes fuel for nuclear power plants. The right side shows radioactive wastes that must be handled during the cycle.

Uranium deposits are usually located in sedimentary rocks, but how they got there is not well understood. It is possible that they accumulated along with the sediments that eventually became sedimentary rock, much as coal deposits accumulated. Another possibility is that groundwater containing dissolved uranium seeped through the sediments, which gradually became infiltrated with uranium.

Substantial deposits of uranium are found in North America (43.8 percent of the world's reserves),[2] Africa (30.2 percent), and Australia (15.6 percent). Uranium ore contains three isotopes, U-238 (99.28 percent), U-235 (0.71 percent), and U-234 (less than 0.01 percent). Because U-235, the isotope that is utilized in conventional fission reactions, is such a minor part (0.71 percent) of ura-

nium ore, uranium ore must be refined after mining to increase the concentration of U-235 to about 3 percent; this refining process is known as **enrichment** (Figure 11–2).

The uranium fuel used in a nuclear reactor is processed into small pellets of uranium dioxide (Figure 11–3), each of which contains the energy equivalent of a ton of coal. The pellets are placed in **fuel rods,** 12-foot-long closed pipes. The fuel rods are then grouped into square **fuel assemblies,** generally of 200 rods each (Figure 11–4). A typical nuclear reactor contains about 250 fuel assemblies.

In nuclear fission U-235 is bombarded with neutrons (Figure 11–5). When the nucleus of an atom of U-235 is struck by and absorbs a neutron, it becomes unstable and splits into two smaller atoms, each approximately half the size of the original uranium atom. In the fission process, two or three neutrons are also ejected from the uranium atom. They collide with other U-235 atoms, creating a chain

[2] Information on the world's reserves of uranium does not include those found in Communist countries or countries recently under Communist rule.

**Figure 11–3**
Uranium dioxide pellets, held in a gloved hand, contain about 3 percent uranium 235, which is the fission fuel in a nuclear reactor. Each pellet contains the energy equivalent of one ton of coal. (Courtesy of Westinghouse Electric Corp., Commercial Nuclear Fuel Division)

reaction as those atoms are split and more neutrons are released to collide with additional U-235 atoms.

The fission of U-235 releases an enormous amount of heat, which is used in a nuclear power plant to transform water into steam, which is, in turn, used to generate electricity. Production of electricity is possible because the fission reaction is *controlled* (recall that an uncontrolled fission reaction results in a nuclear explosion). Fission reactions in the reactor of a nuclear power plant can be started or stopped, increased or decreased, thus allowing the desired amount of heat energy to be produced.

## How Electricity Is Produced from Nuclear Power

A typical nuclear power plant has four main parts: (1) the **reactor core,** where fission occurs; (2) the **steam generator,** where the heat produced by nuclear fission is used to produce steam from liquid water; (3) the **turbine,** which uses the steam to generate electricity; and (4) the **condenser,** which cools the steam, converting it back to a liquid (Figure 11–6).

Fission takes place in the reactor core, which contains the fuel assemblies. Above each fuel assembly is a **control rod** made of a special metal alloy that is capable of absorbing neutrons. The plant operator signals the control rod to move either up out of or down into the fuel assembly. If the control rod is out of the fuel assembly, free neutrons collide with the fuel rods and fission of uranium takes place. If the control rod is completely lowered into the fuel assembly, it absorbs the free neutrons and fission of uranium no longer occurs. By exactly

**Figure 11–4**
The uranium pellets are loaded into long fuel rods, which are grouped into square fuel assemblies (shown). (Courtesy of Westinghouse Electric Corp.)

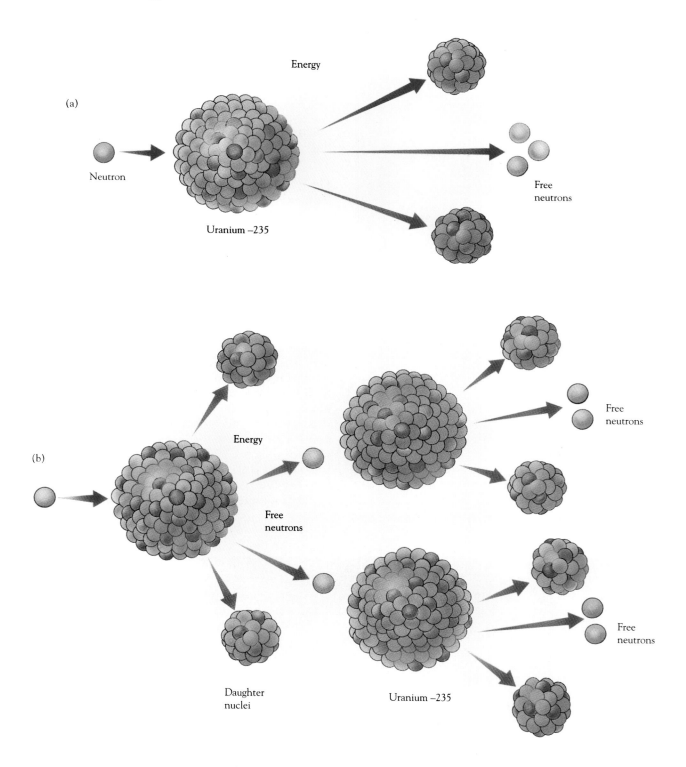

**Figure 11–5**
Nuclear fission. (a) Neutron bombardment of a U-235 nucleus causes it to split into two smaller atomic fragments and several free neutrons. (b) The free neutrons bombard nearby U-235 nuclei, causing them to split and release still more free neutrons (a process called a chain reaction).

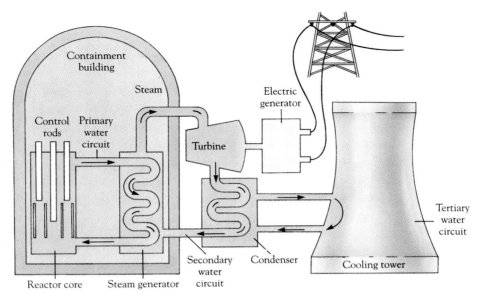

**Figure 11–6**
A nuclear power plant. Fission of U-235 that occurs in the reactor core produces heat, which is used to produce steam in the steam generator. The steam drives a turbine to generate electricity. The steam then leaves the turbine and is pumped through a condenser before returning to the steam generator. Excess heat is controlled by pumping hot water from the condenser to a massive cooling tower. After it is cooled, the water is pumped back to the condenser.

controlling the placement of the control rod, the plant operator can produce the exact amount of fission required.

A typical nuclear power plant has three water circuits. The **primary water circuit** heats water, using the energy produced by the fission reaction. This circuit is a closed system that circulates water under high pressure through the reactor core, where it is heated to about 293°C (560°F). Because it is under such high pressure, this superheated water cannot expand to become steam and so remains in a liquid state.

From the reactor core, the very hot water circulates to the steam generator, where it boils water held in a **secondary water circuit,** converting the water to steam. The pressurized steam goes to and turns the turbine, which in turn spins a generator to produce electricity. After it has turned the turbine, the depleted steam in the secondary water circuit goes to a condenser, where it is converted to a liquid again.

A **tertiary water circuit** provides cool water to the condenser, which cools the spent steam in the secondary water circuit. As the water in the tertiary water circuit is heated, it moves from the condenser to a **cooling tower,** where it is cooled before circulating back to the condenser.

**Safety Features of Nuclear Power Plants**   The reactor core where fission occurs is surrounded by a huge steel pot-like structure called a **reactor vessel**— a safety feature designed to prevent the accidental release of radiation into the environment. The reactor vessel and steam generator are placed in a **containment building,** an additional line of defense against accidental radiation leaks. Containment buildings have steel-reinforced concrete walls 0.9 to 1.5 m (3 to 5 ft) thick and are built to withstand severe earthquakes, the high winds of hurricanes and tornadoes, and even planes crashing into them.

## BREEDER NUCLEAR FISSION

Uranium ore is mostly U-238, which is not fissionable and is therefore a waste product of conventional nuclear fission. In **breeder nuclear fission,** however, U-238 is converted to plutonium, Pu-239, a human-made isotope that is fissionable. Breeder reactors can use either U-235 or Pu-239 as fuel.[3] Some of the neutrons that are emitted in breeder nuclear fission are used to produce addi-

---

[3]Breeder reactors can also use thorium 232 as fuel.

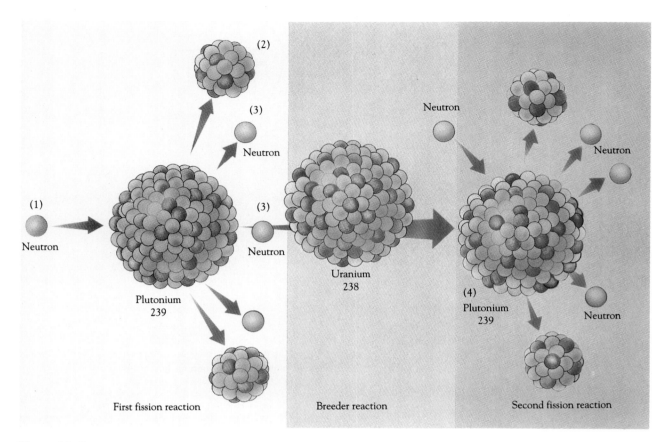

**Figure 11-7**
Breeder nuclear fission. A neutron (1) from a previous fission reaction bombards plutonium
239, causing it to split into smaller radioactive fission fragments (2). Additional neutrons (3)
are ejected in the process, some of which may collide with uranium 238 to form plutonium
239 (4). Neutron bombardment of this plutonium 239 molecule causes it to split, and the process continues.

tional plutonium from U-238 (Figure 11-7). A
breeder reactor thus makes more fissionable fuel
than it uses.

Because breeder fission can utilize U-238, it has
the potential to generate much larger quantities of
energy from uranium ore than traditional nuclear
fission can. For example, if the U-238 stored in
nuclear waste sites across the United States could
be taken out and used in breeder reactors, it would
supply the entire country with electricity for the
next 100 years! When one adds the uranium reserves in the ground to these nuclear waste stockpiles, breeder fission has the potential to supply the
entire country with electrical energy for several
centuries.

Although the first breeder reactor experiments
were performed in the United States, leadership in
developing this technology has been assumed by
Europe and Russia. In the whole world, only several

breeder fission plants are operational, and the development of additional breeder reactors will be a
slow process, as many technical and safety problems
have yet to be resolved. For example, for reasons
too complex to consider here, breeder fission reactors use liquid sodium (a highly reactive metal that
could easily corrode pipes and cause leaks) as a
coolant, rather than water. Should a leak cause the
loss of some of the liquid sodium coolant, the temperature within the reactor might get high enough
to cause an uncontrolled nuclear fission reaction—
that is, a small nuclear explosion. The force of this
explosion would almost certainly rip open the containment building, releasing radioactive materials
into the atmosphere.

Public and governmental distrust of breeder
reactors is greater than misgivings about conventional fission, because plutonium is used not only in
breeder nuclear fission, but also in nuclear weap-

### Economic Issues in Nuclear Energy

Nuclear energy is supposed to be economical, but one of the reasons electrical utilities won't commit to building new nuclear power plants is their high costs. These costs must be paid for by the utility customers in the form of higher prices for their electricity. Nuclear power plants are large and take years to plan and build. (In comparison, electric plants powered by coal or oil can be much smaller, and therefore easier and less expensive to build.) The initial cost estimates for building a nuclear power plant are high, and the actual costs are usually much higher than the forecasts. Cost overruns, which are borne by the utility and its customers, occur partly because building often falls far behind schedule. Consider the Seabrook nuclear reactor in Seabrook, New Hampshire, which has probably been one of the most controversial plants ever licensed.* Seabrook obtained its operat-

*One of the main controversies surrounding Seabrook was concern over the evacuation plan that would be used should an accident occur at the plant. Seabrook is located in a small coastal town without large, multi-lane highways. The evacuation area includes several tourist spots that are crammed with hundreds of thousands of visitors during summer months. In reviewing the evacuation plan for Seabrook, the NRC chose not to count the tourists.

ing license from the Nuclear Regulatory Commission in 1990, after numerous delays had put its construction 11 years behind schedule. The plant cost $6.45 billion, *12 times* the original estimate.

Another economic issue is the regulatory process itself, which is cumbersome and makes it difficult and expensive for new plants to obtain funding. If you then add the cost of decommissioning a nuclear power plant approximately 40 years after it is built (which is difficult to project), the result is an economic nightmare.

In an effort to promote nuclear energy, nuclear and utility executives in late 1990 developed a plan addressing the safety and economic issues associated with nuclear power. They envisioned building a series of "new generation" nuclear reactors, designed to be ten times safer than current reactors. They also gave serious consideration to the financial risks involved in building a nuclear power plant. According to their plan, costs could be held in line by standardizing nuclear power plants rather than custom-building each one. The plan also calls for building schedules to be improved and the regulatory process to be streamlined.

ons. Getting the public to support construction of nuclear breeder reactors is currently very difficult in Europe and virtually impossible in the United States.

## IS NUCLEAR ENERGY A CLEANER ALTERNATIVE THAN COAL?

One of the reasons proponents of nuclear energy argue for the widespread adoption of nuclear power is that nuclear energy is less polluting than fossil fuels, particularly coal. They point out that the combustion of coal to generate electricity is responsible for more than one-third of the air pollution in this country. As discussed in Chapter 10, coal is an extremely dirty fuel, especially since we have used up most of our reserves of cleaner-burning coal.

Today most coal-burning power plants burn soft coal that contains sulfur, which interacts with moisture in the atmosphere to form acid precipitation. In addition, the combustion of coal releases carbon dioxide, a greenhouse gas that traps solar heat in our atmosphere and possibly causes the Earth to warm.

In comparison, nuclear energy emits very few pollutants into the atmosphere. It does, however, generate nuclear waste that is highly radioactive and therefore very dangerous (to be discussed shortly). The extreme health and environmental hazards created by this waste require that special measures be taken in its storage and disposal. Nuclear power plants also produce other nuclear wastes, such as radioactive coolant fluids and gases in the reactor.

Opponents of nuclear energy contend that the fact that coal is a dirty fuel is not so much an argument in favor of nuclear energy as it is an argument in favor of a cleaner alternative to coal. They point out that pollution control devices can significantly lessen the air pollution produced by coal-burning power plants. As provisions of the 1990 Clean Air Act go into effect, more and more coal-fired power plants will install pollution control equipment.

Moreover, the replacement of coal-burning power plants with nuclear power does not significantly lessen the threat of global warming, because only 15 percent of the greenhouse gases come from power plants in the first place. Most greenhouse gases are produced by automobile emissions and industrial processes, which are unaffected by nuclear power. Also, the uranium-mining through uranium-enrichment steps in the nuclear fuel cycle require the combustion of fossil fuels, meaning that nuclear energy indirectly contributes to the greenhouse effect.

## IS ELECTRICITY PRODUCED BY NUCLEAR ENERGY CHEAP?

The cost of producing electricity from a particular energy source varies over time, but in some areas of the country electricity can be produced more cheaply by nuclear energy than by coal-fired plants. As the 1990 Clean Air Act clamps down on air pollution from coal-fired power plants, those utilities will be faced with the purchase of expensive scrubbers to reduce air pollution. This additional expenditure may make nuclear power appear more economical by comparison.

Proponents of nuclear energy usually point to France, which gets 70 to 80 percent of its electricity from nuclear energy. In France, electricity generated by nuclear plants is 30 percent less expensive than that generated from coal. Nuclear energy proponents believe there is no reason why the United States could not achieve the same economic efficiency.

Opponents of nuclear power dismiss the example of France because the French government heavily subsidizes the nuclear industry. Among other things, the French government funds research and development, waste disposal, and insurance coverage for its nuclear industry. The government-owned utility, Electricité de France, currently has a debt of $37 billion, yet government-approved rate increases are hard to obtain.

The true costs of nuclear energy are not always obvious in utility bills, whether in France or in the United States. The generation of electricity using nuclear energy is *very* expensive when *all* costs are taken into account, including the cost of building and maintaining the nuclear power plant, the cost of storing and disposing of spent fuel and other radioactive wastes, and the cost of dismantling the nuclear power plant after it is too old to safely and economically produce energy (to be discussed shortly; also see Focus On: Economic Issues in Nuclear Energy on page 213).

## CAN NUCLEAR ENERGY DECREASE OUR RELIANCE ON FOREIGN OIL?

International crises, such as the oil embargo of the early 1970s and the Persian Gulf crisis in the early 1990s, occasionally threaten the supply of oil to the United States. Supporters of nuclear energy point out that our dependence on foreign oil would be lessened if all oil-burning power plants were converted to nuclear plants. This claim is not as convincing as it seems, however, because oil is responsible for generating only 6 percent of the electricity in the United States; we rely on oil primarily for automobiles and for home heating. Thus, the replacement of electricity generated by oil with electricity generated by nuclear energy would do little to lessen our dependence on foreign oil, because we would still need oil for heating buildings and driving automotive vehicles.

## PROBLEMS ASSOCIATED WITH NUCLEAR POWER

Many people have found the promise of nuclear energy to be false, largely because of concerns over the safety of nuclear power plants and the disposal of radioactive wastes. There are also concerns about terrorism and sabotage because the technology of nuclear power is closely linked to that of nuclear weapons.

### Safety in Nuclear Power Plants

Approximately 390 conventional nuclear power plants are in operation worldwide, and accidents have occurred in some of them. Although conventional nuclear power plants cannot explode like atomic bombs, situations can be created in which dangerous levels of radiation might be released into the environment and result in human casualties. At high temperatures the metal encasing the uranium fuel melts, releasing radiation; this is called a **meltdown.** Also, the water that is used in a nuclear reactor to transfer heat can boil away during an accident, contaminating the atmosphere with radioactivity.

The probability of a major accident occurring is considered low by the nuclear industry, but the consequences of such an accident are drastic and life-threatening, both immediately and long after the accident has occurred. We now consider two relatively recent accidents, one in the United States (Three Mile Island) and the other (first introduced in Chapter 1) at Chernobyl in the former Soviet Union.

**Three Mile Island** The most serious nuclear reactor accident in the United States occurred in 1979 at the Three Mile Island power plant in Pennsylvania. A partial meltdown of the reactor core took place. Had there been a complete meltdown of the fuel assembly, dangerous radioactivity would have been emitted into the surrounding countryside. Fortunately, the containment building kept virtually all the radioactivity released by the core material from escaping. Although a small amount of radiation entered the environment, there were evidently no substantial environmental damages and no immediate human casualties. A study conducted within a 10-mile radius around the plant ten years after the accident concluded that cancer rates were in the "normal" range and that there was no association between cancer rates and radiation emissions from the accident.

The seriousness of the situation at Three Mile Island elevated public apprehension about nuclear power. It took six years for Three Mile Island to be repaired and reopened,[4] and in the aftermath of the accident, public wariness prompted construction delays and cancellations of a number of new nuclear power plants across the United States. On the positive side, the accident at Three Mile Island prompted new safety regulations—including evacuation plans for the areas surrounding nuclear power plants—and reduced the complacency that had been commonplace in the nuclear industry.

**Chernobyl** The worst accident ever to occur at a nuclear power plant took place at the Chernobyl plant in the former Soviet Union on April 26, 1986, when one or possibly two explosions ripped a nuclear reactor apart and expelled large quantities of radioactive material into the atmosphere (Figure

---

[4]The reactor that was involved in the accident at Three Mile Island was destroyed during the partial meltdown. However, a second reactor that was undamaged during the accident is currently in operation.

---

## Envirobrief

### The Legacy of Three Mile Island

America's most notorious nuclear accident, a near-meltdown at the Three Mile Island nuclear generating station near Harrisburg, Pennsylvania, occurred on the morning of March 28, 1979. As with most nuclear accidents, it was caused by a combination of human error and mechanical failure, in this case broken pumps and blocked valves. Had the reactor heated up out of control and "melted down," as did the one at Chernobyl in the former Soviet Union, hundreds of thousands of people in the northeastern United States could have been killed directly or over time by radioactive elements. The reactor was so dangerously contaminated that it could not even be entered by humans for two years.

While it will never be free of radioactivity, the reactor will eventually be entombed in concrete, but not before a "cleanup" job that is estimated to cost over $3 billion and consume the following:

Number of workers required: 2,000
Number of cloth coveralls: 20,000
Number of paper coveralls: 1,000,000
Number of plastic coveralls: 1,000,000
Number of raincoats: 100,000
Number of plastic booties: 100,000
Number of pairs of rubber boots: 100,000
Number of pairs of rubber gloves: 1,000,000
Number of surgical caps: 100,000
Number of square feet of plastic sheeting: 1,000,000

Ironically, the materials used to "clean" the crippled reactor will themselves become contaminated in the process, and require costly disposal.

---

11–8). The effects of this accident were not confined to the area immediately surrounding the power plant: in the atmosphere radiation quickly spread across large portions of the Northern Hemisphere. The Chernobyl accident affected and will continue to affect many nations, especially the former Soviet Union and the countries of Northern Europe.

(a)

(b)

**Figure 11–8**
Chernobyl. (a) The site of the explosion is indicated by the arrow. (b) Since the accident, the damaged area has been completely entombed in concrete. (The Bettmann Archive)

The first task the Soviets faced after the accident was to contain the fire that had broken out after the explosion and prevent it from spreading to other reactors at the power plant. Local fire fighters, many of whom later died from exposure to the high levels of radiation, battled courageously to contain the fire. In addition, 116,000 people who lived within a 30-kilometer (18.5-mile) radius around the plant had to be quickly evacuated and resettled. Ultimately, more than 300,000 people were forced to resettle as a result of the accident.

Once the danger from the explosion and fire had passed, the radioactivity at the power station had to be cleaned up and contained so that it would not spread. Dressed in protective clothing similar to that worn by the people in Figure 11–9, workers were transported to the site in radiation-proof vehicles, and initially the radioactivity was so high that they could stay in the area for only a few minutes at a time. There are few photographs of the cleanup because camera film was quickly ruined by the radiation. After the initial cleanup, the damaged reac-

**Figure 11–9**
Cleaning up highly radioactive contamination requires special clothing and protective breathing masks. (David Doody/Tom Stack & Associates)

tor building was encased in 300,000 tons of concrete. Then the surrounding countryside had to be decontaminated: highly radioactive soil was removed, and buildings and roads were scrubbed down.

Although as much cleanup as can be done in the immediate vicinity of Chernobyl is largely accomplished, the people in Byelorussia (now Belarus) and the Ukraine, the areas immediately adjacent to Chernobyl, face many long-term problems. Twenty percent of the farmland and 15 percent of the forests of Byelorussia, for example, are so contaminated that they cannot be used for more than a century. Loss of agricultural production is one of the largest costs of the Chernobyl accident for the local economy.

Inhabitants in many areas of Byelorussia and the Ukraine still cannot drink the water or consume locally produced milk, meat, fruits, or vegetables. Mothers cannot nurse their babies because their milk is contaminated by radioactivity. Hundreds of children are in hospitals, dying from leukemia (cancer of white cells in the blood and bone marrow). The health of approximately 350,000 Ukrainians is being constantly monitored.

In the investigation that ensued after the accident, it became apparent that there had been two fundamental causes. First, the design of the nuclear reactor was flawed—the reactor was not housed in a containment building and was extremely unstable at low power.[5] Second, many of the Chernobyl

plant operators lacked scientific or technical understanding of the plant they were operating. As a result of the disaster at Chernobyl, the Soviet Union developed a retraining program for operators at all the nuclear power plants in the country. In addition, safety features were added to existing reactors.

One of the disquieting consequences of Chernobyl was the lack of predictability of the course taken by spreading radiation (Figure 11–10). Chernobyl's radiation cloud dumped radioactive fallout over some areas of Europe and Asia, leaving other areas relatively untouched. This unevenness makes it difficult to plan emergency responses for future nuclear accidents.

The health effects of the accident at Chernobyl will be monitored for many years. Although only 31 people initially died from exposure to radiation,[6] it is estimated that as many as 24,000 people received dangerous doses of radiation (see Focus On: The Effects of Radiation on Living Organisms). An increase in mental retardation in newborns has been noted not only in the former Soviet Union but also in parts of Europe; this was expected, because a similar pattern emerged in Japan after the bombings of Hiroshima and Nagasaki at the end of World War II. An increase in cancer deaths directly attributable to Chernobyl is also expected, with 5,000 to 75,000 deaths projected. Psychological injuries to those living under the cloud of Chernobyl are also being assessed.

The world has learned much from this nuclear disaster. Most countries are taking nuclear power more seriously, hoping to prevent more accidents. Safety features that are commonplace in North American and European reactors are being incorporated into new nuclear power plants around the world. In addition, nuclear engineers learned a great deal from the cleanup and entombment of Chernobyl; this knowledge will be useful in the decommissioning of power plants. For example, remote-controlled bulldozers and robots were relatively ineffective in the cleanup process because their electronics failed, probably as a result of exposure to high levels of radiation. Doctors learned more about effective treatment of people who have been exposed to massive doses of radiation. In the years to come, health researchers will learn more about the relationship between cancer and radiation as they follow the health of the thousands who were exposed to radiation from Chernobyl. Also,

[5] This type of reactor, called an RBMK reactor, is not used commercially in either North America or Europe because nuclear engineers consider it too unsafe. The former Soviet Union still has a number of RBMK reactors in operation.

[6] Although the official death count is 31, a citizens' group called Chernobyl Union says it has proof that between 5,000 and 7,000 workers who participated in the cleanup have died.

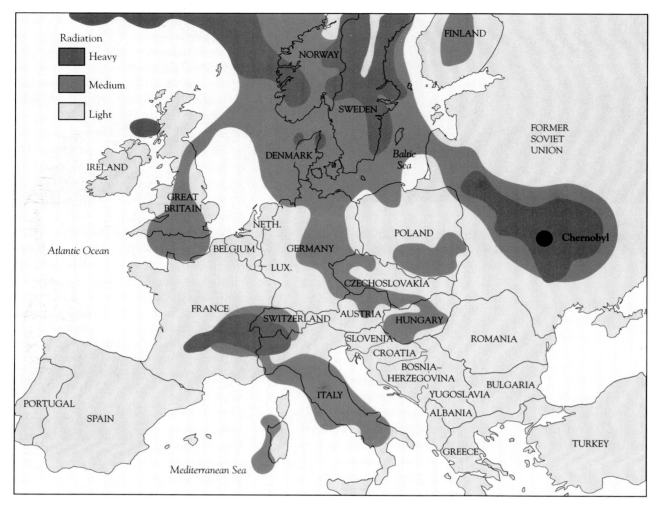

**Figure 11–10**
Fallout from the accident at Chernobyl was quite unpredictable. This map of Europe and part
of the former Soviet Union shows the locations that were hardest hit by radioactive fallout.

the area within a 10-kilometer radius around Chernobyl has been designated an ecological reserve
that will be left to recover on its own. The only
people allowed inside will be scientists who study
the effects of high levels of radiation on plants and
animals and monitor the environment's recovery.

**Proponents of Nuclear Power Say That Newly
Designed Nuclear Power Plants Are Safer**  New
reactors are being designed that are considered by
nuclear experts to be much safer than the current
generation of reactors operating in this country.
Some of the new reactors will not have a meltdown
even in a worst-case scenario. Because the fuel cannot melt, radiation cannot be released into the
environment during the plant's operation. Some of
the designs utilize a gas (helium) rather than water

to turn the turbines and cool the system. Helium,
being much less corrosive than steam, is another
safety feature, because it is less likely to cause leaks
by corroding pipes. According to the nuclear power
industry, the new generation of nuclear power
plants will also be smaller, simpler in design, and
less expensive to build. Nevertheless, they will still
have many of the unresolved problems of traditional designs, including what to do with radioactive wastes.

## Radioactive Wastes

Radioactive wastes are classified as either low-level
or high-level. **Low-level radioactive wastes** are
radioactive solids, liquids, or gases that give off
small amounts of ionizing radiation. Produced by

nuclear power plants, university research, hospitals (nuclear medicine), and industries, low-level radioactive wastes have traditionally been stored in steel drums in six special government-operated landfills. Three of these have leaked and been forced to close. In 1990 the Nuclear Regulatory Commission (NRC) proposed as a cost-saving measure that low-level radioactive wastes be handled as household trash and dumped in local sanitary landfills. Several environmental groups filed a lawsuit against the NRC to block this policy; the negative public reaction caused the NRC to place an indefinite moratorium on its proposal.

**High-level radioactive wastes** are radioactive solids, liquids, or gases that initially give off large amounts of ionizing radiation (spent fuel rods, for example). Produced by nuclear power plants and nuclear weapons facilities, high-level radioactive wastes are the most dangerous hazardous wastes produced by humans. *No disposal methods for high-level radioactive wastes are guaranteed safe and permanent.*

Radioactive wastes produced during nuclear fission include the reactor metals (fuel rods and assemblies), coolant fluids, and air or other gases found in the reactor.[7] Fuel rods, for example, can be used for only about three years, after which they become the most highly radioactive waste on Earth (Table 11–2). Their dangerous level of radioactivity requires that they be handled in special ways. In addition to nuclear wastes produced during fission, considerable **tailings**—piles of loose rock—are produced when uranium is mined and processed. A third type of waste, the nuclear power plant itself once it has outlived its usefulness, will be considered separately.

Some radioactive wastes are created when neutrons are absorbed by the uranium fuel rods, thereby forming radioactive isotopes. As the isotopes decay, they produce considerable heat, are extremely toxic, and remain radioactive for extended periods of time. Examples of radioactive isotopes found in spent fuel rods include plutonium 239 (half-life 24,400 years), neptunium 237 (half-life 2,130,000 years), and plutonium 240 (half-life 6,600 years). Secure storage of these materials must be guaranteed for hundreds of thousands of years, until they can decay sufficiently to be safe.

Clearly, the safe disposal of radioactive wastes is one of the main difficulties that must be overcome if nuclear power is to realize its potential.

[7]Nuclear power plants release several radioactive gases into the atmosphere, which cause a very slight increase in the background radiation to which the general populace is exposed.

**Table 11–2**
**Nuclear Power Plants and Waste in Selected Countries**

| Country (in Alphabetical Order) | Number of Operable Commercial Reactors* | Spent Fuel Inventories (tons of radioactive metal) |
| --- | --- | --- |
| Argentina | 2 | X† |
| Canada | 18 | 11,000 |
| China (Taiwan) | 6 | 900 |
| Czechoslovakia | 8 | X |
| France | 55 | 12,700 |
| Japan | 38 | 5,600 |
| Korea | 8 | 700 |
| South Africa | 2 | 100 |
| Former Soviet Union | 56 | X |
| Sweden | 12 | 1,900 |
| United Kingdom | 40 | 30,900 |
| United States‡ | 108 | 17,606 |

*As of December 31, 1988.
†X = Data not available.
‡The U.S. has 111 commercial reactors as of July 1992.

Many people question whether we can safely guarantee the storage of wastes that must be isolated from living organisms for millennia. High-level radioactive wastes, which have very long half-lives, must be stored in an isolated area where there is no possibility they can contaminate the biosphere. The storage site must also have geological stability (imagine the consequences of storing wastes near an earthquake zone!) and little or no water flowing nearby (which might transport the waste away from its original site).

What are the best sites for the long-term storage of high-level radioactive wastes? Some experts think we could store the wastes in stable rock formations deep in the Earth. However, many experts think that deep salt deposits are the best solution for long-term storage. Salt is an effective barrier to radiation; further, the presence of 200-million-year-old undisturbed salt deposits indicates their stability in the Earth's crust. Another suggestion for the long-term storage of radioactive wastes is mausoleums, which would be aboveground sites built in remote locations such as deserts. If we built mauso-

(text continues p. 223)

## The Effects of Radiation on Living Organisms

Living organisms are continually exposed to low levels of background radiation from several natural sources, including cosmic rays from outer space and radioactive elements in the Earth's crust (see figure; also see the discussion of radon in Chapter 19). This radiation, often referred to as **ionizing radiation,** contains enough energy to eject electrons from atoms, which results in the formation of charged atoms called ions (in this case, the ions are positively charged). In living organisms, ionizing radiation contains enough energy to damage the structure of biologically important molecules. Fortunately, organisms have evolved repair mechanisms that appear

to counteract the damaging effects of low levels of radiation. However, the repair mechanisms cannot handle the cell and tissue damage that results from exposure to higher levels of radiation, such as those accidentally released from nuclear power plants or radioactive waste disposal sites.

One of the most dangerous effects of ionizing radiation is the damage it does to DNA, the genetic material of living organisms. Because DNA provides a blueprint for all characteristics of an organism *and* directs the activities of cells, damage to DNA is almost always harmful to the organism. Changes in DNA are known as **mutations.**

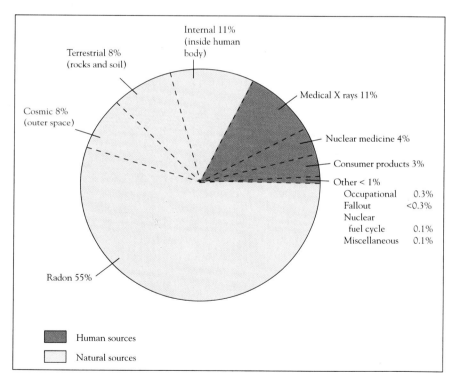

Average sources of ionizing radiation in the United States. Most of the radiation (82%) to which people are exposed comes from natural (and unavoidable) sources. The nuclear fuel cycle is responsible for an average of only 0.1 percent of the radiation in the environment.

When mutations occur in reproductive cells (that is, eggs or sperm), the changes can be passed on to the next generation, where they might result in birth defects, mental retardation, or genetic disease. Pronounced effects of high doses of radiation on subsequent generations have been documented in experimental animals (mice, for example), but no such evidence exists for humans.* Extensive studies of the offspring of the survivors of the atomic bombs at Hiroshima and Nagasaki, for example, suggest that there has been some genetic damage, but the results are not statistically significant.

If mutations occur in nonreproductive cells of the body, they may alter the functioning of those cells. This can lead to health problems during an individual's lifetime. Exposure to very high levels of radiation may cause such severe physiological damage that death occurs. Radiation exposure that is extensive but not great enough to cause death may create numerous medical problems, including burns on the skin, an increased chance of developing cataracts, temporary male sterility, and a number of types of cancer such as leukemia and cancers of the bone, thyroid, skin, lung, and breast (see figure). For example, a higher incidence of leukemia was correlated with greater exposures to radiation in the survivors of the World War II atomic bombings of Hiroshima and Nagasaki.

A question that remains to be answered definitively is whether low-level radiation such as that around nuclear power plants causes a higher incidence of cancer in people who live and work nearby. Some recent studies, such as one on the incidence of leukemia in adults living near a nuclear power plant in Massachusetts, show a direct correlation.

*However, in 1990 a study concluded there was a higher-than-normal incidence of leukemia in children whose fathers work in a nuclear waste reprocessing plant in the United Kingdom. The study is under review.

Other U.S. and French studies of cancer rates near nuclear power plants do not confirm a cancer risk from exposure to low levels of radiation.

Exposure of the human body to a high dose of radiation can have a number of immediate effects, some of which may lead to death.

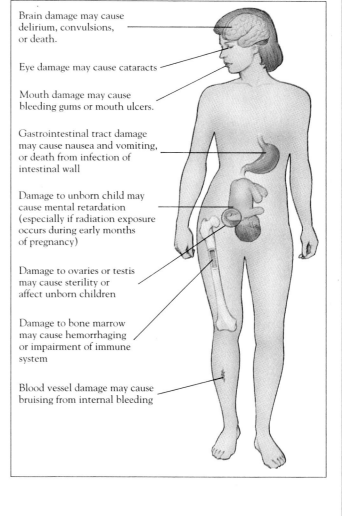

Brain damage may cause delirium, convulsions, or death.

Eye damage may cause cataracts

Mouth damage may cause bleeding gums or mouth ulcers.

Gastrointestinal tract damage may cause nausea and vomiting, or death from infection of intestinal wall

Damage to unborn child may cause mental retardation (especially if radiation exposure occurs during early months of pregnancy)

Damage to ovaries or testis may cause sterility or affect unborn children

Damage to bone marrow may cause hemorrhaging or impairment of immune system

Blood vessel damage may cause bruising from internal bleeding

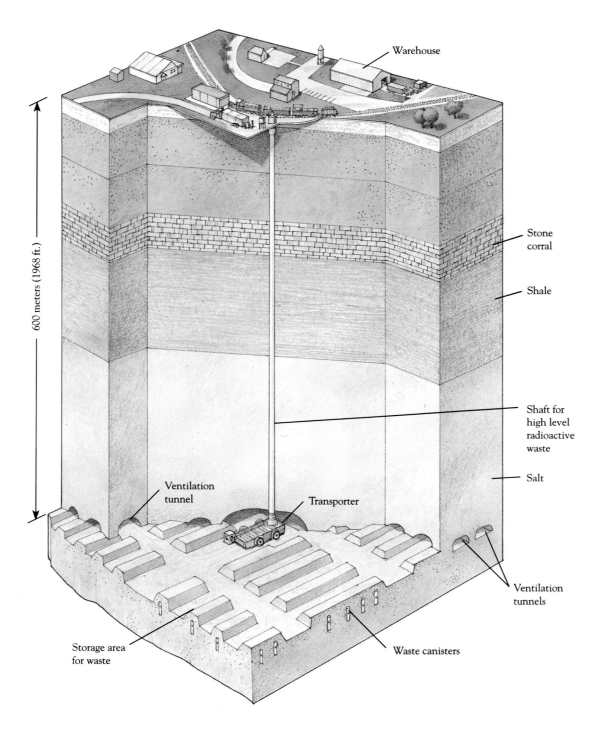

Warehouse

Stone corral

Shale

Shaft for high level radioactive waste

Salt

Ventilation tunnel

Transporter

Ventilation tunnels

Waste canisters

Storage area for waste

600 meters (1968 ft.)

**Figure 11–11**
In the United States, permanent storage sites for high-level radioactive wastes will probably be deep underground in rock or salt formations.

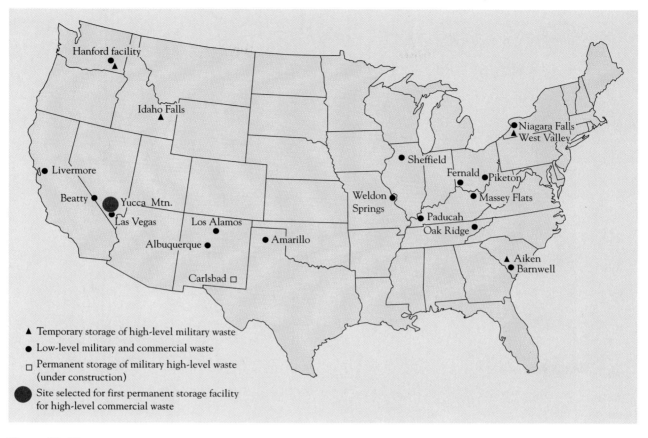

**Figure 11–12**
Some of the major nuclear waste storage sites.

leums, however, we would not be able to simply store the wastes and forget about them. Mausoleums would have to have cooling systems (to remove the excess heat produced during radioactive decay) and adequate security to guarantee their safety. Other long-term possibilities that have been considered include storage in ice sheets, burial beneath the ocean floor, and storage in deep ocean trenches. Most experts today support underground geologic disposal in either rock formations or salt deposits (Figure 11–11). The selection of these sites is further complicated by people's reluctance to have radioactive wastes stored near their homes (see Focus On: Human Nature and Nuclear Energy).

The enormity of this problem is demonstrated by the fact that there are *no* permanent storage facilities for radioactive wastes in *any* country. In the United States, there are about 100 sites at which our radioactive wastes have been "temporarily" stored for decades (Figure 11–12). In 1982 the passage of the Nuclear Waste Policy Act put the burden of developing permanent waste sites on the federal government and required the first site to be operational by 2010.[8] The federal government has selected Yucca Mountain in Nevada as a storage site for high-level nuclear wastes from commercially operated power plants. High-level nuclear wastes from military weapons are be stored at Carlsbad, New Mexico. Neither site is perfect. The Yucca Mountain site is controversial because it is near a young volcano and active fault lines; earthquakes might possibly disturb the site and raise the water table, which could result in radioactive contamination of air and groundwater. There is concern that oozing brine at the Carlsbad site, a network of rooms carved out of rock salt, could corrode the steel waste drums and contaminate groundwater with radioactivity.

Selecting and building the sites doesn't solve the entire nuclear waste problem, as there are also concerns about transporting the high-level nuclear

[8] So far, the deadline for completion of an operational high-level radioactive waste repository has been postponed from 1985 to 1989 to 1998 to 2003 to, most recently, 2010.

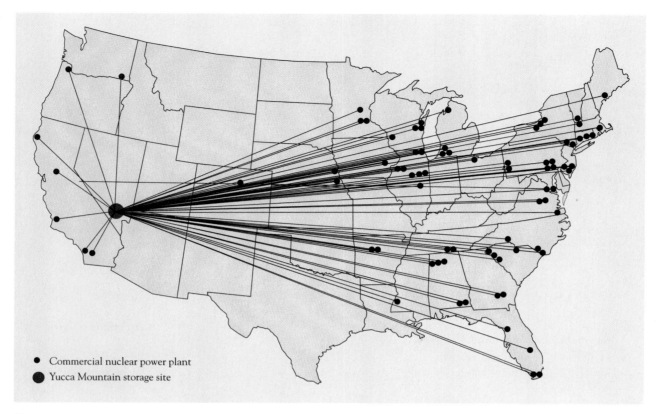

**Figure 11–13**
High-level nuclear waste from commercial nuclear power plants will be transported to Yucca
Mountain for long-term storage. Residents of some areas along major routes to Nevada are
concerned about the danger of spills if an accident occurs during transport.

wastes to these sites (Figure 11–13). Wastes would
be transported by truck or train across the country,
and residents of areas through which waste disposal
vehicles would pass are naturally worried about
what would happen if an accident caused the
wastes to be spilled.

**Radioactive Wastes with Relatively Short Half-
Lives** Some radioactive wastes are produced directly from the fission reaction. U-235, the reactor
fuel, may split in several different ways, forming a
number of smaller atoms, many of which are radioactive. Most of these, including strontium 90 (half-
life 28 years), cesium 137 (half-life 30 years), and
krypton 85 (half-life 10.4 years), have *relatively*
short-term radioactivity. In 300 to 600 years they
will have decayed to the point where they are safe.
(Recall that a radioactive material produces negligible radiation after ten half-lives.)

The safe storage of fission products with relatively short half-lives is of concern because fission
produces larger amounts of these materials than of

the materials with extremely long half-lives. Also,
health concerns exist because many of the shorter-
lived fission products mimic essential nutrients and
tend to concentrate in the body, where they continue to decay, with harmful effects. For example,
one of the common fission products, strontium 90,
is chemically similar to calcium. If strontium 90
were to be accidentally released into the environment from radioactive waste that had not been
stored properly, it could be incorporated into
human and animal bones and teeth in place of calcium. In like manner, cesium 137 replaces potassium in the body and accumulates in muscle tissue,
and iodine 131 concentrates in the thyroid gland.

**Decommissioning Nuclear Power Plants**

Nuclear power plants can operate for only 25 or 30
years before certain critical sections, such as the
reactor vessel, become brittle or corroded. At the
end of their operational usefulness, however, nuclear power plants cannot simply be abandoned or

## Focus On

### Human Nature and Nuclear Energy

The NIMBY response casts a pall over the promise of nuclear energy. NIMBY stands for "**n**ot **i**n **m**y **b**ackyard." As soon as people hear that a nuclear power plant or a nuclear waste disposal site may be situated nearby, the NIMBY response rears its head. Part of the reason NIMBYism is so prevalent in nuclear power issues is that, despite the assurances given by experts that a site will be safe, no one can guarantee *complete* safety, with no possibility of an accident.

A sister response to NIMBY is the NIMTOO response, which stands for "**n**ot **i**n **m**y **t**erm **o**f **o**ffice." Politicians are sensitive to their constituents' concerns and are not likely to support the construction of a nuclear power plant or nuclear waste disposal site in their district.

Given human nature, the NIMBY and NIMTOO responses are not surprising. But emotional reactions, however reasonable they might be, do little to constructively solve complex problems. Consider the disposal of nuclear waste. There is universal agreement that we need to safely isolate nuclear waste until it can decay enough to create little danger. But NIMBY and NIMTOO, with their associated demonstrations, lawsuits, and administrative hearings, prevent us from effectively dealing with nuclear waste disposal. Every potential disposal site is near someone's home, in some politician's state.

We all agree that our generation has the responsibility to dispose of nuclear waste generated by the nuclear power plants we have already built. Only we want to put it in someone else's state, in someone else's backyard.

---

demolished, because many parts have become contaminated with radioactivity. In addition, highly radioactive spent fuel, which is usually placed in water-filled storage ponds in the plant throughout its operation, must be safely disposed of.

Three options exist when a nuclear power plant is closed: storage, entombment, and decommissioning. If an old plant is put into storage, it is simply guarded by the utility company for 50 to 100 years, during which time some of the radioactive materials decay. This decrease in radioactivity makes it safer to dismantle the plant later, although accidental leaks during the storage period are still a threat.

**Entombment,** in which the entire power plant is permanently encased in concrete, is not considered a viable option by most experts because the tomb would have to remain intact for thousands of years. It is likely that accidental leaks would occur during that time. Also, we cannot guarantee that future generations would inspect and maintain the "tomb."

The third option for the retirement of a nuclear power plant is to **decommission,** or dismantle, the plant immediately after it closes (Figure 11–14). The workers who dismantle the plant must wear protective clothing and masks. Some portions of the plant are too "hot" (radioactive) to be safely dismantled by workers, although advances in robotics may make it feasible to tear down these sections. (Such advances would have to overcome the problem experienced at Chernobyl, in which high levels of radiation caused the electronics of remote-controlled robots to malfunction, rendering them useless.) As the plant is torn down, small sections of it are transported to a permanent storage site.

Several small nuclear power plants have been decommissioned. Shippingport, the nation's first commercial nuclear power plant, was dismantled in 1989 and transported by barge more than 8,000 miles from its working site in Pennsylvania to Hanford Military Reservation, a military dump site in Washington State. The decommissioning of a large nuclear power plant will not be possible, however, until advances in robotics provide the technology to safely dismantle it (Shippingport's reactor vessel was small enough to be kept intact) and until there are permanent storage sites for all the radioactive pieces.

Decommissioning nuclear power plants is the responsible thing to do once a plant is no longer operable. There are risks, including dangers to workers during the decommissioning process and accidental discharges of radiation into the environ-

(a)

(b)

**Figure 11-14**
The Elk River nuclear power plant in Elk Rapids, Minnesota, is one of the earliest that has been successfully decommissioned. It was a very small plant without a full-sized nuclear reactor, so its decommissioning was not as involved as decommissioning larger plants has been. (a) The plant before dismantling had begun. (b) Dismantling the reactor building, which was done by the Atomic Energy Commission, a predecessor to the Department of Energy. Much of the structure of nuclear power plants is radioactive and must be safely stored for many years.

ment either during dismantling or during transport of radioactive debris to a permanent site.

Worldwide, many nuclear power plants are nearing retirement age. In 1990 approximately 35 plants were 25 years old or older; by 1995 there will be 66, and by 2000 there will be 150 retirement-age plants. As we enter the 21st century, we may find that we are paying more in our utility bills to close old plants than to have new plants constructed.

## The Link Between Nuclear Energy and Nuclear Weapons

Fission is involved in both the production of electricity by nuclear energy and the destructive power of nuclear weapons. Uranium 235 and plutonium 239 are the two fuels commonly used in atomic fission weapons. As you know, plutonium is produced in breeder reactors. It is also possible to reprocess spent fuel from conventional fission reactors to make weapons-grade plutonium.

Many countries are using or contemplating using nuclear power to generate electricity. The possession of nuclear power plants gives these countries relatively easy access to the fuel needed for atomic weapons. Many world leaders are concerned about the proliferation of nuclear warheads and the consequences of terrorist groups and nations building atomic weapons (see Focus On: The Effects of Nuclear War). Also, the transport of nuclear wastes to storage sites increases the chance that terrorist groups will steal the wastes and use them to make nuclear weapons. These concerns have caused many people to shun nuclear energy, particularly breeder fission, and to seek alternatives that are not so intimately connected with nuclear weapons.

## FUSION: NUCLEAR ENERGY FOR THE FUTURE

The atomic reaction that powers the stars, including our sun, is fusion. In fusion, two lighter atomic nuclei are brought together under conditions of high heat and pressure in such a way that they fuse, producing a larger nucleus. The energy produced by fusion is considerable; it makes the energy produced by the burning of fossil fuels seem trifling by comparison: 30 ml (1 oz) of fusion fuel has the energy equivalent of 266,000 l (70,000 gal) of gasoline.

Isotopes of hydrogen are the fuel of fusion (Figure 11–15). In the fusion reaction, the nuclei of deuterium and tritium combine to form helium, releasing huge amounts of energy in the process. Deuterium, also called heavy hydrogen, is present in water and is relatively easy to separate from normal hydrogen. Tritium, which is weakly radioactive, is not found in nature; this human-made hydrogen isotope can be formed during the fusion reaction by bombarding another element, lithium, with neutrons. The isotope of lithium that is required, Li-6, is found in seawater and in certain types of surface rocks.

Supporters of nuclear energy view fusion as the best possible form of energy, not only because its fuel, hydrogen, is available in virtually limitless supply, but also because fusion will apparently produce little in the way of radioactive pollution. Unfortunately, many technological difficulties have been encountered in efforts to stage a controlled fusion reaction. It takes a phenomenally high temperature (100,000,000°C) to make atoms fuse, and once the reaction starts, no one knows whether it can be regulated. Fusion research is going on in several countries, particularly the United States and the former Soviet Union.

**Figure 11–15**

Two possible fusion reactions. (a) When deuterium fuses with tritium, helium-4 is produced, along with a neutron and considerable energy. (b) When two deuterium nuclei fuse, helium-3 is produced, along with a neutron and considerable energy.

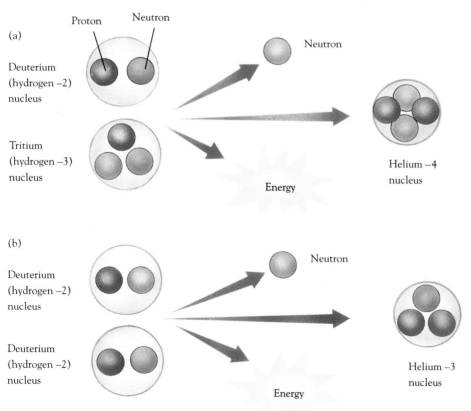

## The Effects of Nuclear War

The proliferation of nuclear weapons, along with current international tensions, has led to global concern about the effects of nuclear war. Aside from the obvious consequence—immediate human death and suffering—nuclear war would have environmental consequences that are even worse.

In trying to imagine the effects of a nuclear war, consider the massive volcanic eruption that occurred in Indonesia in 1815. It spewed approximately 25 cubic miles of dust and debris into the atmosphere. The effects of this eruption were global, for the dust clouds prevented sunlight from warming the Earth in its usual manner. The Northern Hemisphere experienced a drastic climate change that produced crop failures across New England and Europe. It was a "year without a summer."

The effects of nuclear war might be similar to, but more drastic than, the 1815 volcanic explosion. The copious dust, debris, and sooty smoke that would be ejected into the atmosphere would come not only from nuclear explosions, but also from urban wildfires caused by the thermal radiation of those explosions (see figure). The dust, debris, and smoke would screen sunlight to the extent that the light available at noon might be equivalent only to the light of a full moon. The darkened Earth would grow even colder than it did after the 1815 volcanic explosion, causing widespread crop failures and consequential famine. Numerous plant and animal species would die in great numbers, and some that were already endangered would probably become extinct. **Nuclear winter** is the name

given to these hypothetical global climatic effects of nuclear war.

More recently, some scientists have reassessed the effects of nuclear war on global climate and think that they would amount to a **nuclear autumn,** a less drastic situation than a nuclear winter.

The reason scientists do not agree on how pronounced the environmental effects of nuclear war would be is that there are many unknowns. For example, more knowledge is needed about (1) the optical properties of smoke particles, (2) the height to which smoke particles from large fires caused by atomic explosions will rise in the atmosphere, and (3) the effectiveness of rainfall in removing smoke from the air.

Other environmental repercussions of nuclear war include further damage to the ozone layer in the upper atmosphere, which could have undesirable effects on plants and animals as well as humans (see Chapter 20). Radioactivity would be widespread in the environment as a result of radiation fallout. Interestingly, insects would probably be the real winners of a nuclear war because they are quite resistant to the damaging effects of radiation, whereas their predators (birds, amphibians, and so on) would be eliminated or reduced in numbers.

The world's leaders are increasingly in agreement that, given the destruction and devastation nuclear weapons can unleash on the entire world, including nuclear winter or nuclear autumn and its effects, no country can "win" a nuclear war.

In a nuclear war, nuclear explosions would fill the atmosphere with massive amounts of debris and sooty smoke, resulting in nuclear winter as the sun's energy is blocked from the Earth.

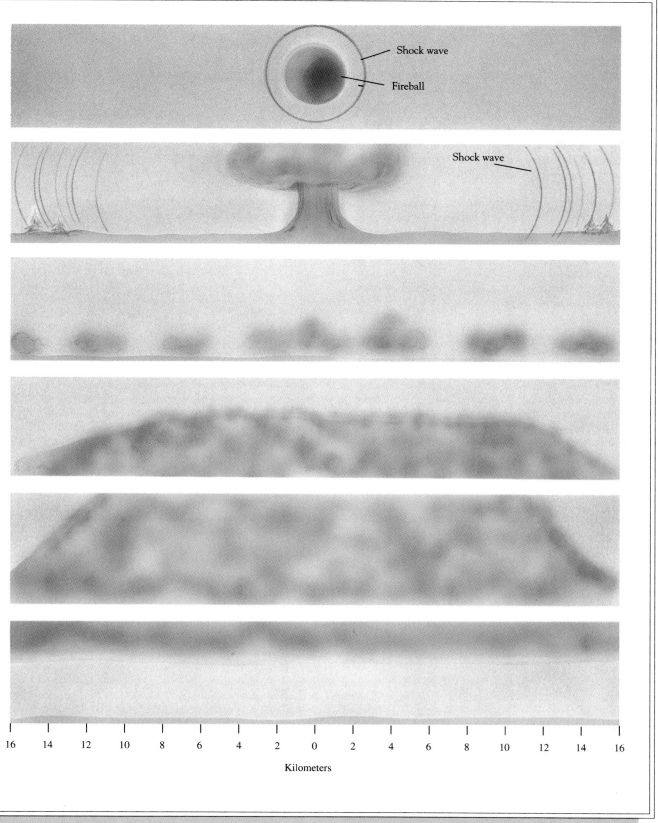

Shock wave

Fireball

Shock wave

Kilometers

# SUMMARY

1. Nuclear power involves the nucleus (center) of an atom. In fission atomic nuclei are split apart, whereas in fusion atomic nuclei are fused. Both fission and fusion result in a significant release of energy in comparison to the chemical combustion of fossil fuels.

2. A typical nuclear power plant contains a reactor core, where fission occurs; a steam generator; a turbine; and a condenser. Safety features include a steel reactor vessel and a containment building. In a nuclear reaction, U-235 is bombarded with neutrons, which split the nucleus into two smaller atoms plus additional neutrons. These neutrons, in turn, collide with additional U-235 atoms. The fission of U-235 releases heat that converts water to steam, which is used to generate electricity.

3. In breeder nuclear reactors, both U-235 and Pu-239 are split to release energy. The large quantity of U-238 in uranium fuel is converted into Pu-239; thus, breeder reactors make more fuel than they use.

4. Supporters say nuclear energy is better than alternatives because it is less polluting and more economical (which is not true when all costs are considered), and its fuel, uranium, is plentiful. They also suggest that increased use of nuclear power will decrease our dependence on foreign oil.

5. Problems associated with nuclear power include questions about safety in nuclear power plants, disposal of radioactive wastes, decommissioning nuclear power plants, and the link to nuclear weapons. Radioactive fallout from the 1986 accident at the Chernobyl power plant resulted in widespread environmental contamination as well as serious local contamination. Chernobyl stimulated a worldwide reassessment of nuclear power.

6. Low-level ionizing radiation is a natural part of the environment. Data linking human cancer to exposure to lower levels of radiation, such as those in areas near nuclear power plants, are inconclusive. Exposure to high-level ionizing radiation (such as the radiation that escapes from a nuclear power plant in an accident like the one at Chernobyl) may result in death, physiological damage, or genetic defects.

7. In all countries that utilize nuclear power, permanent waste disposal sites are urgently needed to house spent fuel, which is highly radioactive, as well as radioactive parts of dismantled power plants.

8. The proliferation of nuclear weapons has led to concerns over the environmental consequences of nuclear war. If there were a worldwide nuclear war, dust and debris ejected into the atmosphere could cause global climate changes known as nuclear winter (the worst-case scenario) or nuclear autumn (a slightly less catastrophic scenario).

9. Commercial fusion as a source of energy has yet to become a reality, but fusion may have the potential to produce unlimited energy with very little radioactive waste in the future.

# DISCUSSION QUESTIONS

1. Compare the environmental effects of fossil fuel combustion and conventional fission.

2. Breeder reactors produce more fuel than they consume. Does this mean that if we use breeder reactors, we will have a perpetual supply of plutonium for breeder fission? Why or why not?

3. Why is the permanent storage of high-level radioactive wastes such a problem?

4. Can accidents at nuclear power plants be prevented in the future? Why or why not?

5. Why is decommissioning nuclear power plants such a major task?

6. Are you in favor of the United States developing additional nuclear power plants to provide us with electricity in the 21st century? Why or why not?

# SUGGESTED READINGS

Bojcun, M. The legacy of Chernobyl. *New Scientist*, vol. 130, 20 April 1991. How people in the Ukraine are living with the continuing effects of Chernobyl. Another article in this issue discusses the international effects of Chernobyl.

Charles, D. Nuclear safety: Some like it hot. *New Scientist*, vol. 124, 1989. New designs in fission reactors promise to make nuclear power safer and more reliable.

Flavin, C. How many Chernobyls? *World Watch* 1:1,

1988. The safety problems of nuclear power in the aftermath of Chernobyl are addressed.

Hafele, W. Energy from nuclear power. *Scientific American*, September 1990. Examines the current situation of nuclear power and addresses its future.

Lenssen, N. Facing up to nuclear waste. *World Watch*, March/April 1992. The fate of nuclear power cannot be determined until the problem of nuclear waste is resolved.

Miller, P. Our electric future. *National Geographic* 180:2, August 1991. Discusses whether a comeback is possible for nuclear power.

Pollock, C. "Decommissioning Nuclear Power Plants," in *State of the World: 1986*, ed. L. R. Brown. W. W. Norton & Company, New York, 1986. All aspects of decommissioning nuclear power plants, from decontamination to waste disposal, are considered.

Shulman, S. When a nuclear reactor dies, $98 million is a cheap funeral. *Smithsonian*, October 1989. A readable account of the decommissioning of Shippingport nuclear power plant.

Slovic, P., M. Layman, and J. H. Flynn. Lessons from Yucca Mountain. *Environment* 33:3, April 1991. Examines perceived risk, public trust, and nuclear waste.

Warner, S. The environmental effects of nuclear war: Consensus and uncertainties. *Environment* 30:5, 1988. A well-researched, balanced account of nuclear winter.

# Chapter 12

*A sailing vessel with solar-powered electronics. (Courtesy of Sunelco)*

# Renewable Energy and Conservation

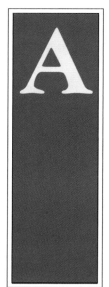

lternative energy sources are becoming increasingly important, both because they are renewable and because their use generally has less environmental impact than the use of fossil fuels or nuclear power. The most promising alternative energy technologies utilize the sun's energy. Direct solar energy can be used to heat water, heat buildings, and generate electricity (using solar thermal methods and photovoltaic solar cells). Biomass (such as wood, agricultural wastes, and fast-growing plants), wind, and hydropower (the energy of flowing water) are all examples of indirect solar energy. Currently, the most important nonsolar renewable energy source is geothermal energy—the heat of the Earth. In addition, tides, ocean currents, and salinity gradients have the potential to provide energy.

## ALTERNATIVES TO FOSSIL FUELS AND NUCLEAR POWER

In the Philippines, 100-hectare (247-acre) wood plantations are being managed for present and future wood-fired power plants. Denmark has built large clusters of windmills to generate electricity, and Brazil has significantly reduced its reliance on imported oil by converting sugarcane crops into alcohol fuels. Photovoltaic solar cells are being manufactured in India, Algeria, and Yugoslavia. Hungary and Mexico use increasingly large amounts of geothermal energy. In the United States, a power plant in California generates electricity solely from the combustion of old tires. As these examples show, most countries are trying new approaches in the endless quest for energy.

Alternatives to fossil fuels and nuclear power are receiving a great deal of attention these days, and for good reason. Although alternative energy sources are not pollution-free, they have fewer environmental problems than fossil fuels or nuclear power. We have seen that there are a number of concerns about using fossil fuels for energy. Reserves of oil, gas, and even coal are limited and will eventually be depleted, and burning these fossil fuels for energy has negative environmental consequences such as global warming, air pollution, acid rain, and oil spills. Some people have suggested that nuclear energy can be used when we run out of fossil fuels, but as we saw in Chapter 11, a number of extremely serious problems are associated with the use of nuclear fission. In addition, uranium, the fuel for nuclear fission, is a nonrenewable resource.

Given the rapidly expanding world population, future energy needs will probably demand the exploitation of most energy sources. The recognized need for a long-term solution has prompted worldwide interest in **renewable energy sources,** those sources that are replenished by natural processes so that they can be used indefinitely. Among the most attractive renewable energy sources is solar energy, in part because its use has little negative impact on the environment. **Solar energy** can be used directly to heat water and buildings and generate electricity (Figure 12–1). In addition, the energy of the sun may be harnessed indirectly as wind, biomass (wood, agricultural wastes, and fast-growing plants), and hydropower (the energy of flowing water). Other renewable energy sources include geothermal energy and ocean tides. Renewable energy is currently more expensive than energy produced by fossil fuels and nuclear power (Table 12–1), but as technological advances are made, costs will probably decrease.

Renewable forms of energy are not the complete and final solution to our energy problem, however. All energy resources must be used with conservation in mind, and we must continue to find ways to enhance energy efficiency and decrease waste. *Energy conservation is our single most important long-term energy solution.*

**Figure 12–1**
As our traditional forms of energy are used up, we will have to rely increasingly on energy alternatives such as solar energy.

| Table 12–1 Capacities and Generating Costs of Electric Power Plants* | | |
|---|---|---|
| **Renewable Energy Sources** | **Production Capacity (megawatts)** | **Generating Costs (cents per kilowatt-hour)†** |
| Hydropower | 87,901 | 2–9 |
| Biomass | 5,150 | 6–9 |
| Geothermal | 2,530 | 4–9 |
| Wind | 1,525 | 7–9 |
| Solar thermal | 200 | 12–25 |
| Photovoltaics | 12 | 25–35 |

*As of January 1, 1989. Source: Investor Responsibility Research Center.

†For comparison, natural gas-fired plants produce electricity at 4.5 to 6.5 cents/kWh, coal-fired plants at 5 to 10 cents/kWh, and nuclear power plants at 5 to 25 cents/kWh.

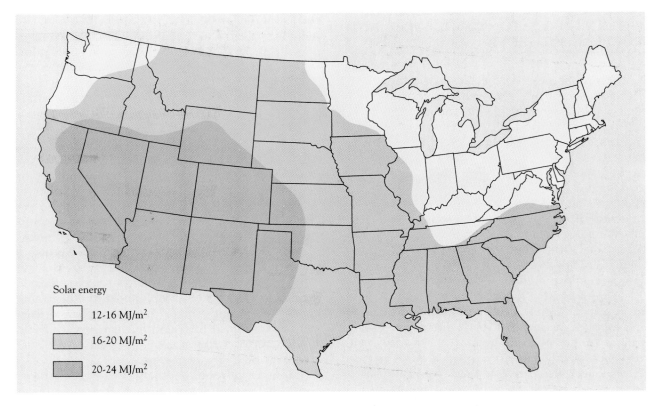

**Figure 12–2**
Solar energy distribution over the United States. This map shows the average daily total of
solar energy that would be received on a solar collector that tilts to compensate for latitude.
The units are in MJ/m². The Southwest is the best area in the United States for year-round
solar energy collection.

## DIRECT SOLAR ENERGY

As you know from earlier chapters of this book, the
sun produces a tremendous amount of energy, most
of which dissipates into outer space and a very
small portion of which is radiated to the Earth (see
Chapter 5, Figure 5–8). Solar energy is different
from fossil and nuclear fuels, not only because it is
perpetually available (we will run out of solar energy only when the sun dies), but also because it is
dispersed over the Earth's entire surface rather than
concentrated in highly localized areas as are coal,
oil, and uranium deposits. In order to make solar
energy useful to us, we must concentrate it.

As you recall from Chapter 5, solar radiation
varies in intensity depending on the latitude, season of the year, time of day, and degree of cloud
cover.

Areas at lower latitudes (closer to the equator)
receive more solar radiation annually than

do latitudes closer to the North and South
poles.

More solar radiation is received by the Earth
during summer than during winter, because
the sun is directly overhead in the summer
and lower on the horizon in winter (see
Figure 5–9).

Solar radiation is most intense when the sun is
high in the sky (noon) than when it is low
in the sky (dawn or dusk).

Clouds absorb some of the sun's energy, thereby
reducing its intensity.

Because of its lack of cloud cover and lower latitude, the southwestern United States receives the
greatest annual solar radiation, whereas the Northeast receives the least (Figure 12–2).

Although the technology exists to directly utilize solar energy, it has not been adopted widely,
largely because the initial costs associated with
converting to solar power are high. However, the
long-term energy savings of solar power offset the

## Focus On

### Cooking with Sunlight

Maybe you tried to fry an egg on a hot summer sidewalk when you were a child. Or maybe you used a magnifying glass to burn a hole in a leaf. Probably neither technique worked very well, but you had the right idea: solar energy can be harnessed for cooking.

Recent designs for solar ovens literally capture solar energy in a box. The new solar cooker, which has a glass top, transmits solar light into the box, but does not transmit out the infrared wavelengths (heat) that would normally escape. Pots containing the food to be cooked are placed inside the box on a black metal plate. The solar oven can reach a temperature of 177°C (350°F) and can be used to boil, bake, simmer, and sauté foods. In average sunlight, a person can cook a full meal in 2 to 4 hours. The solar oven works like a Crock Pot in that the foods retain moisture and more vitamins and minerals than they would if cooked by conventional methods.

The solar cooker can be built from inexpensive, easily available materials. This is a crucial factor for the people who stand to gain the most from the use of the new technology. Communities in areas such as Central America, where fragile ecosystems are continually being destroyed by deforestation, are already learning how to use the solar box cooker, which reduces the number of trees that are destroyed for fuel. People can build the structure in a few hours and use it for the majority of their cooking.

Given that more than 23 percent of the world's people use wood fuel for cooking, the solar cooker may very well prove to be one of the most important developments in the use of solar energy.

high start-up costs. Trapping the sun's energy using current technology is also inefficient, meaning that relatively little of the sun's energy that hits the solar panels is actually utilized. With new technological developments, the efficiency of solar energy collection could increase, making it a more cost-effective alternative source of energy.

Despite these limitations, the United States now obtains more than 5 percent of its total energy from direct and indirect solar sources. Solar energy is projected to become increasingly important in the future (see Focus On: Cooking with Sunlight). As you might expect from examining Figure 12–2, the greatest potential for directly utilizing solar energy in the United States is in the Southwest.

### Heating Buildings and Water

You've probably noticed that, in winter or summer, the air inside a car that is sitting in the sun with its windows rolled up becomes much hotter than the surrounding air. Similarly, the air inside a greenhouse remains warmer than the outside air during cold months. (Greenhouses usually require additional heating in cold climates, but far less than might be expected.) This kind of warming, known as the **greenhouse effect,** occurs partly because the material—such as glass—that envelops the air inside the enclosure is transparent to visible light but impenetrable to infrared radiation (heat). Thus, visible light from the sun penetrates the glass and warms the surfaces of objects inside, which in turn give off **infrared radiation**—invisible waves of heat energy. Because infrared radiation cannot penetrate glass, heat does not escape, and the area surrounded by glass grows continuously warmer.[1]

In **passive solar heating,** the greenhouse effect is used to heat buildings without the need for pumps or fans to distribute the collected heat. Certain design features can be put to use in a passive solar heating system to warm buildings in winter and help them remain cool in summer (Figure 12–3). For example, large south-facing windows (or, in the Southern Hemisphere, north-facing windows) receive more total sunlight during the day than windows facing other directions. The sunlight entering through the windows provides heat that is then stored in floors, walls, or containers of water. This stored heat can be transmitted throughout the building naturally by convection (the circulation that occurs because warm air rises and cool air

[1] Chapter 20 discusses the greenhouse effect in relation to global climate.

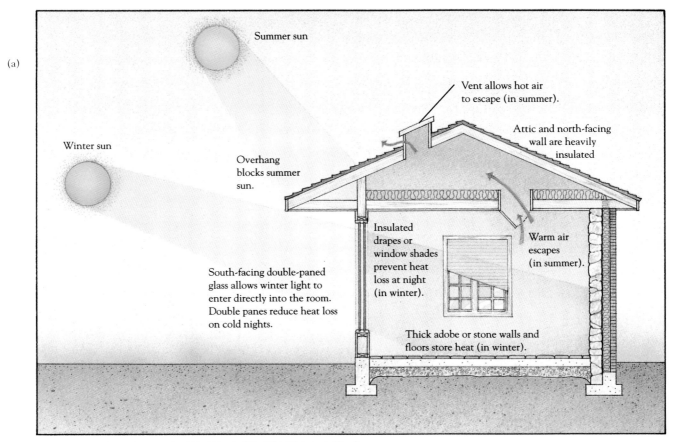

(a)

Summer sun

Winter sun

Vent allows hot air to escape (in summer).

Attic and north-facing wall are heavily insulated

Overhang blocks summer sun.

Warm air escapes (in summer).

Insulated drapes or window shades prevent heat loss at night (in winter).

South-facing double-paned glass allows winter light to enter directly into the room. Double panes reduce heat loss on cold nights.

Thick adobe or stone walls and floors store heat (in winter).

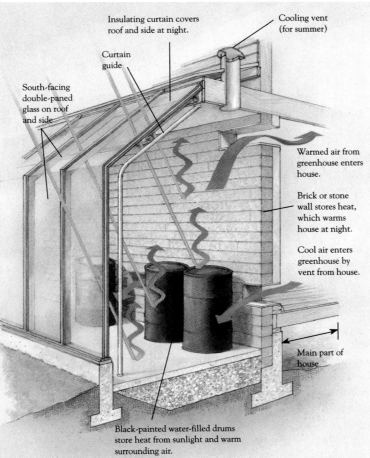

Insulating curtain covers roof and side at night.

Cooling vent (for summer)

Curtain guide

South-facing double-paned glass on roof and side

Warmed air from greenhouse enters house.

Brick or stone wall stores heat, which warms house at night.

Cool air enters greenhouse by vent from house.

Main part of house

Black-painted water-filled drums store heat from sunlight and warm surrounding air.

(b)

**Figure 12–3**

Passive solar heating designs. (a) Several passive designs are incorporated into this home. The south-facing window on the right admits light, which warms the room. The roof overhang allows sunlight to enter the room in winter when the sun is lower in the sky, but blocks the sun's rays in summer. Insulation is a must, particularly for the attic and north-facing walls. Insulating curtains or blinds are drawn over windows at night to help reduce heat loss. (b) A solar greenhouse can be added to existing homes. Light entering the greenhouse is generally used to warm water-filled containers. The heat from these containers provides warmth to the rest of the house at night. The glass in the greenhouse walls and roof is double-paned for additional insulation at night.

sinks). Buildings with passive solar heating systems must be well insulated so that accumulated heat does not escape. Passive solar heating is most effective in sunny climates; in cloudy regions it must be augmented by some other form of energy.

In **active solar heating,** a series of collection devices mounted on a roof or in a field is used to gather solar energy (Figure 12–4a). The most common collection device is a flat solar panel or plate of black metal (which absorbs the sun's energy), enclosed in an insulated box (Figure 12–4b). The absorbed heat is transferred to liquid or air inside the panel, which is then pumped to a building or a storage tank (Figure 12–5). Active solar heating is especially effective for water (Figure 12–6). Because approximately 8 percent of the energy consumed in the United States goes toward heating water, active solar heating has the potential to supply a significant amount of the nation's energy demand.

The use of solar energy for space heating will undoubtedly become more important as other energy supplies dwindle. Furthermore, when diminishing supplies of fossil fuels force their prices higher, solar heating, now usually costlier than more conventional forms of space heating, will become more competitive. Solar air conditioning is not yet commercially feasible.

### Storing the Summer Sun for Winter

The University of Massachuestts at Amherst has a heating system in its sports arena that stores and re-uses summer heat in winter. Seven acres of solar collectors gather the heat and transfer it to liquid antifreeze. The antifreeze is piped into a 100-foot-deep bed of clay, which retains the heat. In winter, the clay transfers heat to the antifreeze, which transfers it to forced-air blowers that warm the arena. The system returns 85% of the stored energy.

## Solar Thermal Electric Generation

Electricity can be produced by **solar thermal electric generation,** in which trough-shaped mirrors (guided by computers), tracking the sun for optimum efficiency, center sunlight on oil-filled pipes and heat the oil to 390°C (735°F) (Figure 12–7). The hot oil is circulated to a water storage system

**Figure 12–4**
Active solar heating. (a) Solar collection devices are mounted on the roof of a building. The solar collectors shown here are used to heat water for an ice cream store's operations. (Courtesy of the National Renewable Energy Laboratory, formerly the Solar Energy Research Institute) (b) Each solar panel is a box with a black metal base and glass covering. Sunlight enters the glass and warms the pipes and the liquid or air that is flowing through them.

(a)

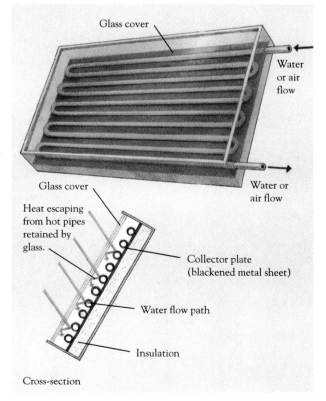

(b)

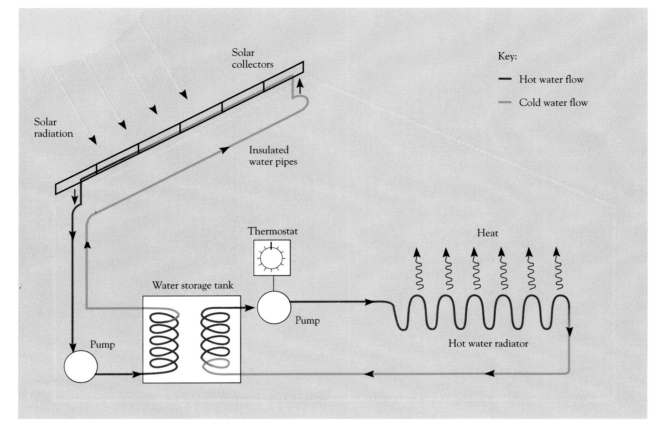

**Figure 12–5**

A simplified diagram of active solar heating. In this example, water is heated in solar collectors on the roof and then pumped to a water storage tank where it heats water for a hot water radiator. The cooled water is then pumped back to the roof to be warmed again.

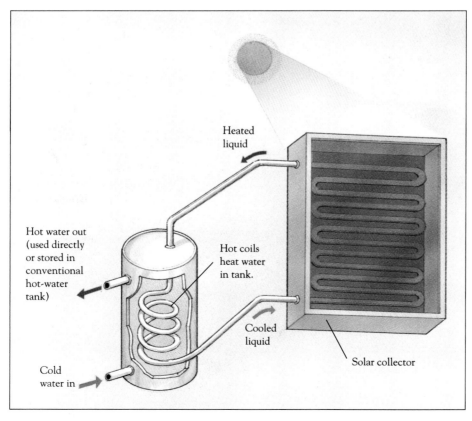

**Figure 12–6**

Active solar water heating. Although active solar heating (to heat buildings) is only needed during cooler months, active solar water heating can be used year-round. The liquid that is heated in solar collectors is used to heat water. Passive solar water heating is also possible.

(a)

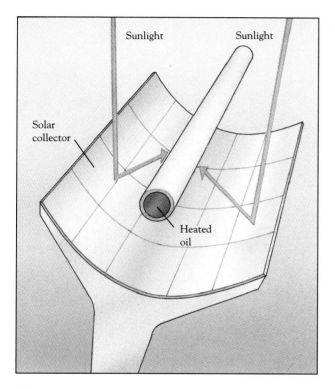

Sunlight          Sunlight

Solar
collector

Heated
oil

(b)

**Figure 12–7**
Solar thermal electric generation. (a) A solar thermal plant in a desert in southern California. (b) How electricity is generated using solar thermal technology. Sunlight is concentrated onto oil-filled pipes. The heated oil is circulated to a water tank where it provides the heat to generate steam, which is then used to produce electricity. (a, Courtesy of Luz International, Ltd.)

**Figure 12–8**
A solar power tower ("Solar One") in the Mojave desert near Barstow, California. The concentrated sunlight, from over 1800 mirrors that focus the sun's energy on a boiler at the top of the tower, is used to heat water to steam, which then generates electricity. "Solar Two," a newly designed tower that stores solar energy as molten salt at the top of the tower, is under construction at the same location. When electricity is needed, the molten salt is used to heat water to steam. (Courtesy of Southern California Edison Co.)

and used to change water to steam, which turns a turbine to generate electricity. Alternatively, the heat can be used in industrial processes or for desalinization (removal of salt from water) and water purification.

Solar thermal energy systems are efficient at trapping the sun's energy; currently, up to 22 per-cent of the energy that hits the collector is converted to electricity. With improved engineering, manufacturing, and construction methods, solar thermal energy is becoming cost-competitive with fossil fuels. One problem with this form of energy is that a backup system must be available to generate electricity when solar power is not operating—at night and during cloudy days. LUZ International Corporation, the Los Angeles–based company that has a virtual monopoly on solar thermal systems in the United States, uses natural gas to augment solar thermal heat. LUZ has installed several solar thermal systems in California and is planning larger ones in Nevada and Brazil.

**Solar Power Towers: An Alternative Solar Thermal Technique**   The **solar power tower,** also called **central receiver energy generation,** is a tall building or tower surrounded by numerous mirrors (Figure 12–8). The mirrors move to follow the sun, focusing solar radiation on a central receiver at the top of the tower; there a liquid is heated to produce steam, which is used to generate electricity. Some of the heat may be stored (as hot liquid) to be used

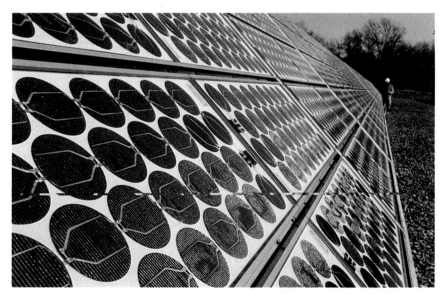

**Figure 12–9**
Arrays of photovoltaic cells are used to generate electricity at Beverly High School in Beverly, Massachusetts. About 10 percent of the electricity needed by the school is supplied by its photovoltaic system. (Department of Energy)

for electricity generation during the night, when solar energy is unavailable.

Solar power towers are being tested in pilot plants in New Mexico and California, and Europeans are planning the Phoebus project, a massive central receiver to be built in Jordan. Whether we will be able to use solar power towers to generate increased amounts of electricity in the future is uncertain. The construction and maintenance of this type of facility are prohibitively expensive and require large tracts of land. Also, using current technology, it is very costly to store electrical energy for use at night.

### Photovoltaic Solar Cells

It is possible to convert sunlight directly into electricity by using **photovoltaic (PV) solar cells** (Figure 12–9). Photovoltaic cells are wafers or thin films of crystalline silicon that are treated with certain metals in such a way that they generate electricity (that is, a flow of electrons) when solar energy is absorbed. Photovoltaic solar cells are arranged on large panels that are set up to absorb sunlight.

Our current photovoltaic solar cell technology, which is also used to power satellites, watches, and calculators, has several limitations that prevent the cells' widespread use to generate electricity. Photovoltaic solar cells are not very efficient at converting solar energy to electricity;[2] they are extremely

expensive to produce; and the number of solar panels needed for large-scale use requires a great deal of land. On the positive side, PVs generate electricity with no pollution and minimal maintenance. They can be used on any scale, from small, portable modules to multi-megawatt power plants. Also, the cost of producing electricity from photovoltaics declined more rapidly from 1970 to 1990 than had been predicted. Additional technological progress may eventually make photovoltaics economically competitive with conventional energy sources. For example, the production of a new type of solar cell from a thin film of a semiconductor material promises to decrease costs. New ways to grow silicon crystals are also being investigated.

One of the main benefits of photovoltaic devices for utility companies is that they can be purchased in small modular units that become operational in a short amount of time. A utility company can purchase photovoltaic elements to increase its generating capacity in small increments, rather than committing a billion dollars (or more) and a decade (or more) of construction for a massive conventional power plant. Used in this supplementary way, the photovoltaic units can provide the additional energy needed, for example, to power air conditioners and irrigation pumps on hot, sunny days.

Although there has been steady improvement in the design and materials of photovoltaic solar cells, until their efficiency is improved it is unlikely that they can be used to generate electricity on a large scale, because we simply don't have enough space. At current efficiencies, for example, several *thousand* acres of solar panels would be required to

---

[2] The average PV cell today has an efficiency of 10 to 15 percent. Some prototype photovoltaics have been developed that have efficiencies around 30 percent.

absorb enough solar energy to produce the electricity generated by a single conventional power plant.[3] The use of photovoltaic solar cells for smaller generating requirements is promising, however. In remote areas that are not served by any electrical power plant (such as rural areas of developing countries), it is more economical to utilize photovoltaic solar cells for electricity than to extend power lines. Photovoltaics is the energy choice to pump water, refrigerate vaccines, grind grain, charge batteries, and supply rural homes with lighting. Thousands of people in developing countries of Asia, Latin America, and Africa have installed photovoltaic solar cells on the roofs of their homes. A PV panel the size of two pizza boxes can supply a rural household with enough electricity for five lights, a radio, and a television.

[3] However, some analysts consider the space requirements for a PV plant to be equivalent to the total space requirements for a coal-fired plant, once mining and waste disposal sites are taken into account.

Several pilot programs in California are successfully generating electricity using photovoltaics. Although efficiency is improving and costs are coming down, the total electricity produced by photovoltaics worldwide is about one-tenth of that produced by a single large nuclear power plant. Energy experts project that photovoltaic solar cells will not become a significant source of energy until well into the 21st century.

**Solar Hydrogen** Solar electricity generated by photovoltaics can be used to split water into the gases oxygen and hydrogen.[4] Hydrogen is a clean fuel (it produces water and heat when it is burned) and produces no sulfur oxides, no carbon monoxide, no hydrocarbon particulates, and no $CO_2$ emissions. It does produce some nitrogen oxides, but the amounts of these pollutants are fairly easy to control. Hydrogen has the potential to provide energy for transportation (in the form of hydrogen-powered electric automobiles) as well as for heating buildings and producing electricity.

It may seem wasteful to use electricity generated from solar energy to make hydrogen, which can then be used to generate more electricity. However, electricity that is generated by existing photovoltaic cells cannot be stored long-term; it must be used immediately. Hydrogen offers a convenient way to store solar energy as chemical energy. It can be transported by pipeline, possibly less expensively than electricity can be transported by wire (Figure 12–10).

Production of hydrogen from solar electricity currently has an efficiency of 8 percent, which means that only 8 percent of the solar energy absorbed by the photovoltaic cells is actually converted into the chemical energy of hydrogen fuel. Scientists are working to improve the efficiency, which will decrease costs and make solar hydrogen fuel more attractive. A pilot plant that makes solar hydrogen fuel is being tested in Saudi Arabia under the joint sponsorship of Germany and Saudi Arabia.

## Solar Ponds

Because water absorbs solar energy, it is possible to build a pond of water specifically to collect solar energy. **Solar ponds** are generally dug 1 to several meters deep and are frequently lined with black

[4] Hydrogen can also be produced using conventional energy sources such as fossil fuels, but then, of course, the environmental and energy security problems of fossil fuels are not avoided. For that reason, we limit this discussion to hydrogen fuel production using solar electricity.

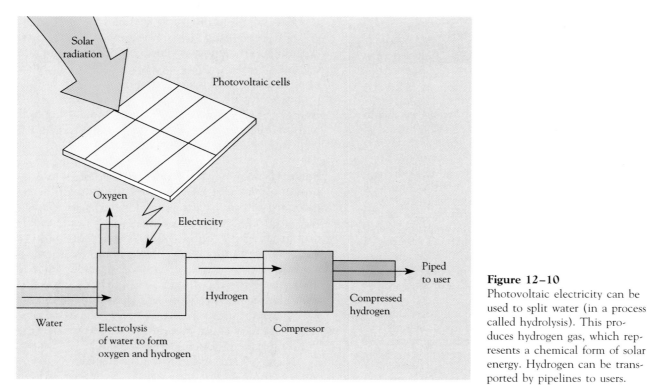

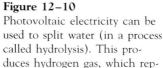

**Figure 12–10**
Photovoltaic electricity can be used to split water (in a process called hydrolysis). This produces hydrogen gas, which represents a chemical form of solar energy. Hydrogen can be transported by pipelines to users.

plastic. Because a large percentage of solar radiation penetrates to the bottom of the pond, the temperature near the bottom may be as high as 100°C; the water near the surface remains at air temperature (Figure 12–11).

Under normal conditions, warm water rises to the top of a body of water because it is less dense than cool water. However, the water at the bottom of a solar pond is made denser than the surface water by the addition of salt (brackish water—water in which salt is dissolved—is denser than fresh water). Therefore, minimal mixing occurs between the warm, dense bottom layer and the cooler, less dense surface layer.

**Figure 12–11**
Solar ponds can be used to generate electricity or provide heat for buildings.

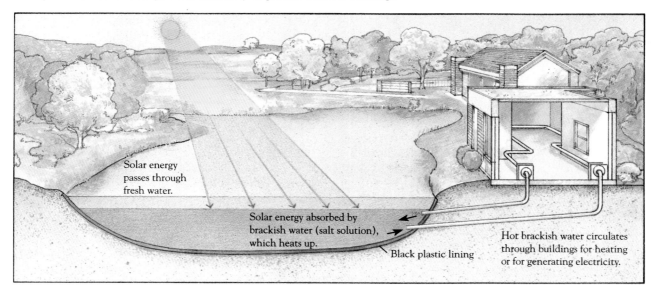

One of the problems associated with solar ponds is the large amount of land needed. Also, there is a potential for environmental contamination: if the brackish water leaked into the surroundings, it would harm plants and other wildlife.

The use of solar ponds is still in the experimental stage. One experiment is taking place in Israel; a small power plant there generates electricity from a solar pond that contains salt water from the Dead Sea.

### Orbiting Solar Power Stations: A Future Technology?

It has been suggested by scientific visionaries that satellites containing large collections of photovoltaic solar cells could orbit the Earth, collecting solar energy far above the clouds. This energy could be concentrated by solar cells and then beamed to Earth as microwaves. On Earth large antennas would receive the microwaves and convert their energy into electrical power.

Obviously, orbiting power stations are highly speculative and not yet technologically possible. The idea is intriguing, though, because this method could be used to supply solar energy to areas under cloud cover and to all areas at night. However, scientists do not know how the microwave radiation would affect plants, animals, or humans in its vicinity. The microwaves might interfere with radio transmissions. Worse, the microwaves might affect atmospheric properties, which in turn could alter climate.

## INDIRECT SOLAR ENERGY

There are a number of ways to utilize the sun's energy indirectly. Combustion of **biomass**—wood and other organic matter—is an example of indirect solar energy, because the energy contained in biomass is produced by green plants that use solar energy for photosynthesis. Windmills can harness the energy of **wind**—surface air currents that are caused by the solar warming of air. The damming of rivers and streams to generate electricity is a type of **hydropower**—the energy of flowing water—which exists because of the hydrologic cycle that is driven by solar energy (see Chapter 5).

### Biomass Energy

Biomass consists of such materials as wood, fast-growing plant and algal crops, crop wastes, and animal wastes (Figure 12–12). Biomass contains

**Figure 12–12**
Biomass is organic material from animals or plants that is used as a source of energy. (a) Women in Rwanda gather firewood. Firewood is the major energy source for most of the developing world. (b) Energy cane (left), a high-biomass form of sugarcane (right), is harvested in Puerto Rico for its energy. (c) An Ethiopian woman makes cow dung patties, which will be dried and burned for fuel. Animal dung is an important energy source in Ethiopia and many parts of the world. (a, Robert E. Ford/Terraphotographics; b, photo by Gene Elle Calvin, courtesy of Melvin Calvin, Lawrence Berkeley Laboratory; c, Mike Andrews © 1993 Earth Scenes)

(b)

(a)

(c)

chemical energy, the source of which can be traced back to radiant energy from the sun, which was used by photosynthetic organisms to form the organic molecules of biomass. Biomass is a renewable form of energy as long as it is managed properly.

Biomass fuel, which can be a solid, liquid, or gas, is burned to release its energy. Solid biomass such as wood is burned directly to obtain energy. Biomass—particularly firewood, charcoal (wood that has been turned into coal by partial burning), animal dung, and peat (partly decayed plant matter found in bogs and swamps)—supplies a substantial portion of worldwide energy. At least half of the world's population relies upon biomass as their main source of energy. In developing countries, for example, wood is the primary fuel for cooking.

Biomass can also be converted to liquid fuels (see Chapter 10), especially **methanol** (methyl alcohol) and **ethanol** (ethyl alcohol), which can then be used in internal combustion engines. However, a major disadvantage of alcohol fuels, whether they are produced from biomass, natural gas, or coal, is that 30 to 40 percent of the energy in the starting material is lost in the conversion to alcohol (Figure 12–13).

It is also possible to convert biomass, particularly animal wastes, into **biogas.** Biogas, usually composed of a mixture of gases, can be stored and transported easily like natural gas. It is a clean fuel whose combustion produces fewer pollutants than either coal or biomass. In China, several million family-sized **biogas digesters** use microbial decomposition of household and agricultural wastes to produce biogas that is used for heating and cooking (Figure 12–14). When biogas conversion is complete, the solid remains are removed from the digester and used as fertilizer.

**Advantages of Biomass Use**  Biomass is attractive as a source of energy because it reduces dependence on fossil fuels and because it can make use of wastes, thereby reducing our waste disposal problem (see Chapter 23). For example, the Mesquite Lake Resource Recovery Project in southern California burns cow manure in special furnaces to generate electricity for thousands of homes. The manure is too salty and contaminated with weed seeds to be used as fertilizer, so its use as an energy source helps solve the problem of its disposal.

Biomass is usually burned to produce energy, so the pollution problems caused by fossil fuel combustion, particularly carbon dioxide emissions, are not completely absent in biomass combustion. However, the low levels of sulfur and ash produced

**Figure 12–13**
Alcohol production from biomass is energy-inefficient. Not only is a substantial input of energy required to grow and harvest the crop, but much of the original energy in biomass is lost during its conversion into alcohol.

Produce and spread pesticides and fertilizers

Plow, plant, cultivate, and harvest the crop

Cook the grain, ferment the mash, and extract the alcohol

Gasoline added

30–40% energy lost in converting biomass to alcohol

Gasohol

by biomass combustion compare favorably with the high levels produced when bituminous coal is burned. It is possible to offset the $CO_2$ that is released into the atmosphere from biomass combustion by increasing tree planting. As trees photosynthesize, they absorb atmospheric $CO_2$ and lock it up in organic molecules that make up the body of the tree, thereby providing a carbon "sink" (see Chapter 5 discussion of carbon cycle). Thus, if biomass is regenerated to replace the biomass used, there is no net contribution of $CO_2$ to the atmosphere and to global warming.

**Disadvantages of Biomass Use**   Some problems are associated with use of biomass, especially from plants. For one thing, biomass production requires land and water. The use of agricultural land for energy crops competes with the growing of food crops, so shifting the balance toward energy production might decrease food production, leading to higher food prices. For this reason, some scientists

are interested in the commercial development of certain desert shrubs, which produce oils that could be used for fuel. The shrubs do not require prime agricultural land, although care would have to be taken to ensure that the desert soils were not degraded or eroded by overuse.

As mentioned earlier, at least half of the world's population relies on biomass as its main source of energy. Unfortunately, in many areas people burn wood faster than they replant it. Intensive use of wood for energy has resulted in severe damage to the environment, including soil erosion (see Chapter 14), deforestation and desertification (see Chapter 17), air pollution (see Chapters 19 and 20), and degradation of water supplies (see Chapter 21).

Crop residues, another category of biomass that includes cornstalks, wheat stalks, and wood wastes at paper mills and sawmills, are increasingly being used for energy. At first glance, it may seem that crop residues, which normally remain in the soil after harvest, would be a good source of energy if they were collected and burned. After all, they're just waste material that will eventually decompose. As it turns out, however, crop residues left in the ground prevent erosion by helping to hold the soil

**Figure 12–14**

This concrete biogas digester in Chu Zhang Village, near Beijing, has proven more reliable than earlier clay models that cracked and were prone to gas and water leakage. The 135 digesters in this village of 600 people offer the benefit of improved sanitation as well as fuel for cooking. Human and animal wastes, in addition to crop residues, charge the digester with material to stimulate bacterial action, which eventually converts the material to clean methane gas. Note the tubing that leads directly to individual homes in the village. (Marc Sherman)

### Fuel from Corn

Agricultural fuels like ethanol, which is distilled from mashed starchy vegetables such as corn, will be used increasingly as a replacement for gasoline. The Clean Air Act requires greater use of fuels that reduce pollutants from vehicle exhaust. Ethanol burns cleaner than gasoline. But corn isn't the perfect source of ethanol. Growing corn and converting it to ethanol consumes more energy than it yields. Sweet sorghum may be a better choice. It can yield 40% more ethanol per acre and there is no net energy loss in growing it and converting it to fuel. Corn is a more efficient fuel source in solid form. Either shelled or on the cob, it is being used increasingly to heat farm buildings and homes thanks to a new generation of efficient corn-burning stoves. One bushel of cobs or 13 pounds of shelled corn has the heating potential of one gallon of propane.

in place. Also, their decomposition serves to enrich the soil by making the minerals that were originally locked up in the plant residues available for new plant growth. If all crop residues were removed from the ground, the soil would eventually be depleted of minerals and its future productivity would decline. Forest residues, which remain in the soil after trees are harvested, fill similar ecological roles.

The use of crop and forest residues for biomass would have to be carefully managed: some of the residues could be removed for energy, and the rest would have to be left in the soil to maintain soil fertility.

## Wind Energy

Wind, which results from the warming of the atmosphere by the sun, is an indirect form of solar energy in which the radiant energy of the sun is transformed into mechanical energy—the movement of air molecules. Wind is sporadic over much of the Earth's surface, varying in direction and magnitude; and, like direct solar energy, wind power is a highly dispersed form of energy. Harnessing wind energy to generate electricity has great potential, however, and it is likely that someday wind will have a greater role in supplying our energy needs. It is currently the most cost-competitive of all forms of solar energy.

For many centuries people have used windmills to pump water, irrigate fields, and grind grain. The Dutch designs for large, slow windmills, which were developed by the 16th century, remained unchanged until wind was first used to generate elec-

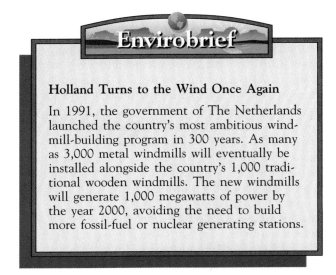

### Envirobrief

**Holland Turns to the Wind Once Again**

In 1991, the government of The Netherlands launched the country's most ambitious windmill-building program in 300 years. As many as 3,000 metal windmills will eventually be installed alongside the country's 1,000 traditional wooden windmills. The new windmills will generate 1,000 megawatts of power by the year 2000, avoiding the need to build more fossil-fuel or nuclear generating stations.

tricity in the 19th century. Improvements in wind machine design in the 20th century have resulted in much greater efficiency, and further development of this new, highly sophisticated technology promises even more increases in efficiency.

Harnessing wind energy is most profitable in areas that receive fairly continual winds, such as islands, coastal areas, mountain passes, and grasslands. The world's largest cluster of wind turbines is located on the northern side of the island of Oahu in Hawaii. In California mountain passes, the number of **wind farms,** arrays of wind turbines (Figure 12–15), increased significantly during the 1980s. California currently produces 85 percent of the world's wind-generated electricity. Other wind

**Figure 12–15**
A wind farm operated by U.S. Windpower in Livermore, California. (Courtesy of U.S. Windpower, Inc.)

farms operate in Denmark, the Netherlands, and India.

In the continental United States, the best locations for large-scale electricity generation from wind energy are off the coasts of New England and the Pacific Northwest and on the western Great Plains. Some experts envision in the future large collections of wind turbines scattered across the Great Plains, supplying much of America's electrical needs.

The use of wind power does not cause major environmental problems (although one concern, currently under study, is reported bird kills). Because it produces no waste, it is a clean source of energy. A major problem with increasing our use of wind power is aesthetics: wind machines detract from the beauty of the landscape. Fortunately, most of the locations that are most appropriate for large-scale wind power are not densely populated. Combining wind farms with cattle grazing, as is done in Altamont, California, for example, is a very productive and profitable use of land.

## Hydropower

The sun's energy drives the hydrologic cycle: evaporation from land and water and transpiration from plants, precipitation, and drainage and runoff. As water flows from higher elevations back to sea level, we can harness its energy. Unlike the sun's energy, which is highly dispersed, hydropower is a *concentrated* energy. The potential energy of water held back by a dam is converted to kinetic energy as the water falls over a spillway (Figure 12–16), where it turns turbines to generate electricity.

Currently, hydropower produces approximately one-fourth of the world's electricity, making it the form of solar energy in greatest use. Developed countries have already built dams at most of their potential sites, but this is not the case in many developing nations. There—particularly in undeveloped, unexploited parts of Africa and South America—hydropower is still a great potential source of electricity. However, even if all potential sites worldwide were used, the energy generated would be less than 15 percent of the total energy needed.

One of the problems associated with hydropower is that building a dam changes the natural flow of a river. A dam causes water to back up, flooding large areas of land and forming a reservoir, which destroys plant and animal habitats. Below the dam, the once-powerful river is reduced to a relative trickle. The natural beauty of the country-side is affected, and certain forms of wilderness recreation are made impossible or less enjoyable.

In arid regions, the creation of a reservoir behind a dam results in greater evaporation of water, because the reservoir has a larger surface area in contact with the air than the stream or river did. As a result, serious water loss and increased salinity of the remaining water may occur (see Chapter 13).

Dams destroy farmlands and displace people. When a dam breaks, people and property downstream may be endangered. In addition, waterborne diseases such as schistosomiasis may spread throughout the local population. **Schistosomiasis** is a tropical disease, caused by a parasitic worm, that can damage the liver, urinary tract, nervous system, and lungs. It is estimated that half the population of Egypt suffers from this disease, largely as a direct result of the Aswan Dam, built on the Nile River in 1902 to control flooding but used since 1960 to provide electrical power.

The ecological, environmental, and personal impacts of a dam may not be acceptable to the people living in a particular area. Laws have been passed to prevent or restrict the building of dams in certain locations. In the United States, for example, the Wild and Scenic Rivers Act prevents the hydroelectric development of 37 rivers.

Dams cost a great deal to build but are relatively inexpensive to operate. A dam has a limited life span, usually 50 to 200 years, because over time the reservoir fills in with silt until it cannot hold enough water to generate electricity. This trapped silt, which is rich in nutrients, is prevented from enriching agricultural lands downstream. For example, the gradual depletion of agricultural productivity downstream from the Aswan Dam in Egypt is well documented.

**Ocean Temperature Differences** In the future it may be possible to generate power using **ocean temperature gradients,** the differences in temperature at various ocean depths. There may be as much as a 24°C difference between warm surface water and very cold, deeper ocean water. Ocean temperature gradients, which are greatest in the tropics, are the result of solar energy warming the surface of the ocean.

Some people visualize giant power plants that would float on the ocean and make use of this temperature differential (Figure 12–17). Warm surface water would be pumped into the power plant, where it would heat a liquid, such as ammonia, to the boiling point. (Because liquid ammonia has a very low boiling point, −33.3°C, the heat from

(a)

**Figure 12–16**
Hydroelectric power may be obtained from water held back by a dam. (a) The McNary Dam supplies some of the electricity used in the Pacific Northwest. (b) How a hydroelectric power plant works. A controlled flow of water released down the spillway turns a turbine, which generates electricity. (a, Department of Energy)

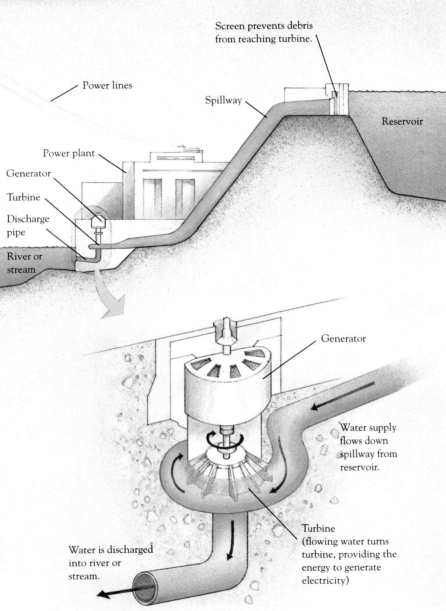

(b)

## The Power of Waves

Three-fourths of the Earth is covered by ocean water, whose movement represents a huge amount of energy. So it stands to reason that ocean waves could somehow be harnessed to create electrical power. That's what British engineers are working on at sites off the coasts of Ireland and Scotland. They are testing the effectiveness of a simple wave power device in producing electric current for two small communities. So far, the experiment looks promising.

Scientists think that the stormy west coast of Europe is ideal for ocean power because the surface swell there is enormous, and deeper waters could yield five times the power that coastal waters could. The wave power station uses a simple technology. Essentially, the concrete, hollow power-plant box is sunk into a gully off the coast to catch waves. As each new wave enters the chamber (about every 10 seconds), the rising water in the chamber pushes air into a vent that contains a turbine, causing the turbine

to spin. In turn, the spinning turbine drives a generator. When the wave recedes, it draws the air back into the chamber and the moving air continues to drive the turbine.

More testing and improvements in design need to be made, and engineers cautiously warn the public about past failures of wave power stations. In addition, whereas a great deal of power can be obtained from the ocean, the power plants themselves need to be able to harness that energy at a rate that is efficient and productive. Currently, harnessing wave power is very inefficient. For example, to deliver 1,000 megawatts of wave-produced electricity (enough for 250,000 households) to consumers would require a string of hollow power-plant boxes spread out along about 48 km (30 mi) of coastline. But as the general public becomes more interested in using renewable resources, funding may increase for the research needed to make wave power a practical resource.

---

warm ocean water would cause it to boil.) The ammonia steam would drive a turbine and thus generate electricity. The ammonia would then be cooled and condensed back to liquid form by colder

water drawn up from a depth of, say, 1,000 meters.[5] The generation of electricity from ocean temperature differences is known as **ocean thermal energy conversion (OTEC).**

To see if such an ideas was workable, a tiny OTEC plant was designed and operated near Hawaii in 1979. Although this plant required enormous amounts of energy to operate (for example, to pump water), it did generate more energy than it used. Thus, we know that OTEC is technologically possible. However, the potential impact of bringing massive quantities of cold water to the surface in a tropical area needs to be considered carefully. Such water properties as dissolved gases, turbidity (cloudiness), nutrient levels, and salinity gradients (differences in salt concentrations) are bound to be altered along with the temperature, and these changes would probably have a profound effect on marine organisms. Even so, some experts believe

**Figure 12–17**
Energy can be tapped from temperature gradients at different oceanic depths using OTEC, Ocean Thermal Energy Conversion. This mini-OTEC project generated the world's first OTEC power in 1979 off the island of Hawaii. (Dillingham Construction)

[5] It may seem wasteful to cool the heated ammonia after it has been used to drive the turbine—after all, once it's cooled, it has to be heated again to be useful. Condensation of the ammonia steam is necessary, however, to convert ammonia to a liquid form so it can be pumped. Otherwise, it couldn't be used again. The same process of heating and then cooling occurs in traditional steam power plants, with water.

that OTEC will have fewer environmental consequences than other sources of energy, such as nuclear power.

**Ocean Waves**   Ocean waves are produced by winds, which are caused by the sun, so wave energy is considered an indirect form of solar energy. Like other types of flowing water, wave power has the potential to turn a turbine, thereby generating electricity. Norway, Japan, and several other countries are investigating the production of electricity from ocean waves (see Focus On: The Power of Waves on page 249). Norway's pilot wave power station at Tostestallen is testing several different ways of tapping wave energy. Although harnessing wave power is technologically feasible, more research will have to be done to make generation of electricity from wave motion practical.

## OTHER RENEWABLE ENERGY SOURCES

Geothermal energy and tidal energy are renewable energy sources that are not direct or indirect results of solar energy. **Geothermal energy,** which is actually a form of nuclear energy, is the heat produced in the Earth by the natural decay of radioactive elements. In locations where radioactive elements are close the surface, this heat can be tapped. **Tidal energy,** which is caused by the change in water level between high and low tides, has also been exploited to generate electricity. In addition, ocean currents and salinity gradients are potential renewable energy sources that may become practical after further research.

### Geothermal Energy

Geothermal energy, the heat of the Earth's interior, is carried to the surface through volcanoes and groundwater. Heated groundwater flows upward as hot water or steam; thus, natural hot springs are frequently found in areas where radioactive elements (whose decay produces the heat) are close to the Earth's surface. Because it takes many years for natural processes to replace groundwater, geothermal energy is sometimes considered nonrenewable or very slowly renewable.

Geothermal energy from hot springs has been exploited for thousands of years for bathing, cooking, and space heating. A number of small geothermal power plants are in use today, including one in Italy that has been operational since 1904. The plants use the steam from hot springs to turn turbines and generate electricity (Figure 12–18).

Iceland and New Zealand are able to make optimum use of geothermal energy. Iceland, situated on the mid-Atlantic ridge, a boundary between two

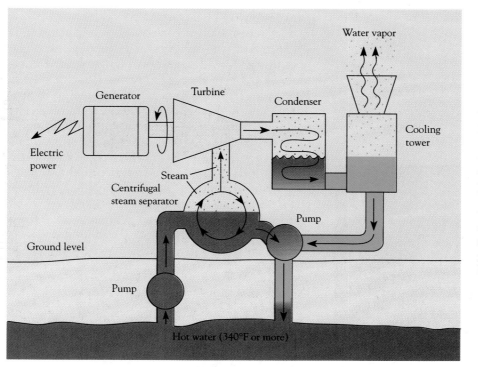

**Figure 12–18**
Direct flash geothermal power plant. There are several different designs for geothermal power plants. In the direct flash design, steam is separated from hot water that was pumped from beneath the ground. The steam is used to turn a turbine and generate electricity. After its use, the steam is condensed and pumped back into the ground. By reinjecting spent water into the ground, geothermal energy remains a renewable energy source because the cooler, reinjected water can be reheated by the Earth and used again.

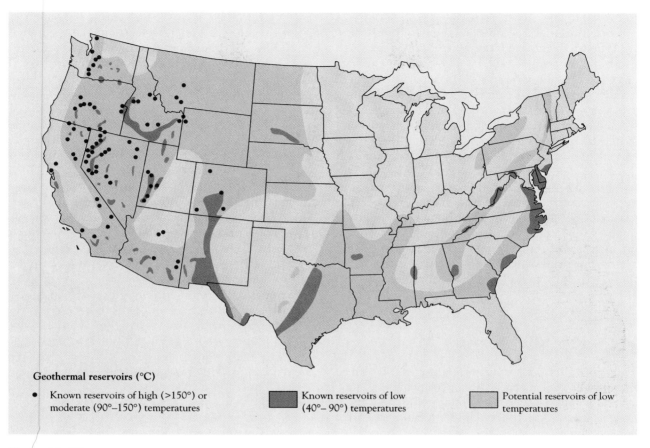

**Figure 12–19**

Areas in the United States with the greatest geothermal potential. Dozens of small power plants in the United States produce electricity using geothermal energy.

continental plates, is an island of intense volcanic activity and numerous hot springs.[6] Because of its location, Iceland uses geothermal energy to meet a substantial portion of its energy requirements. Two-thirds of Icelandic homes are heated with geothermal energy. In addition, most of the fruits and vegetables required by the people of Iceland are grown in geothermally heated greenhouses. New Zealand, which is on a plate boundary in the South Pacific, also uses geothermal energy to generate much of its electricity.

Countries that are increasing their use of geothermal energy include Italy, Japan, Mexico, the Philippines, and the United States (Figure 12–19). Currently, the world's total geothermal energy pro-

duction is fairly small. A large nuclear power plant would produce more energy than all sources of geothermal energy combined.

One of the environmental concerns associated with geothermal energy is that the surrounding land may subside, or sink, as the water from hot springs and their connecting underground reservoirs is removed. Conventional geothermal energy also produces several pollutants, including hydrogen sulfide and carbon dioxide. Also, geothermal energy becomes nonrenewable if the hot groundwater is tapped faster than it can be replaced by natural processes.

**Geothermal Energy from Dry Rocks**    Conventional use of geothermal energy relies on hot springs—that is, on groundwater bringing geothermal energy to the surface from subsurface hot rocks. Scientists at the Los Alamos National Laboratory

[6] Volcanoes are commonly located at the boundary between two geologic plates. Like hot springs, volcanic activity is a product of geothermal energy and is associated with radioactive elements close to the Earth's surface.

in New Mexico are investigating the feasibility of utilizing subsurface geothermal energy in dry areas. In 1986 they reported the production of enough geothermal energy from dry hot rock to produce electricity for a town of 2,000 people. They drilled a shaft to the subsurface hot rock, used hydraulic pressure to fracture the rock, and then pumped water into the fractured area under high pressure. When the pressurized water returned to the surface by way of a second well, it turned to steam, which drove an electric turbine.

The dry hot rock system has an additional benefit over conventional geothermal energy because it produces less pollution. Although this method requires the development of sophisticated technology if it is to become a practical reality, scientists are optimistic that it could be used in many locations that cannot employ conventional geothermal energy.

## Tidal Energy

Tides, the alternate rising and falling of the surface waters of oceans and seas that occur twice each day, are the result of the gravitational pull of the moon and the sun. Normally, the difference in water level between high and low tides is about 0.5 meter (1 or 2 feet). However, certain coastal regions with narrow bays have extremely large differences in water level between high and low tides. The Bay of Fundy in Nova Scotia, for example, has the largest tides in the world, with up to 16 meters (53 feet) difference between high and low tides.

By building a dam across a bay, it is possible to harness the energy of large tides to generate electricity. In one type of system, the dam's floodgates are opened as high tide raises the water on the bay side. Then the floodgates are closed. As the tide falls, water flowing back out to the ocean over the dam's spillway is used to turn a turbine and generate electricity.

Currently, three power plants that make use of tidal power are in operation, in the former Soviet Union, France, and Canada. The dam at La Rance power station on the Rance River in France utilizes the water-movement energy of *both* rising and falling tides (Figure 12–20). As the tide rises, the dam's gates are opened. Water passing upstream through the gates turns 24 separate turbines to generate electricity. At high tide, the water levels on both sides of the dam are the same, so no electricity can be generated, and the gates are closed. As the tide recedes, the gates are opened again, and water moving back toward the ocean turns the turbines and produces electricity. At low tide, the water lev-

els are again the same on both sides of the dam, the gates are closed, and no electricity is generated.

Tidal energy cannot become a significant resource worldwide, because few geographical locations have large enough differences in water level between high and low tides to make power generation feasible. The most promising locations for tidal power in North America include the Bay of Fundy in Nova Scotia, Passamaquoddy Bay in Maine, Puget Sound in Washington, and Cook Inlet in Alaska.

Other problems associated with tidal energy include the high cost of building a tidal power station and the potential environmental problems. The greatest tidal energy is found in estuaries, areas where river currents meet ocean tides. Because the mixing of fresh and salt waters creates a nutrient-rich environment, estuaries are the most productive aquatic environments in the world. Fish and countless invertebrates migrate there to spawn. Building a dam across the mouth of an estuary would prevent these animals from reaching their breeding habitats. Estuaries are also popular sites for recreation, which would be severely curtailed by a tidal dam.

## Ocean Currents and Salinity Gradients

Oceans cover 71 percent of the Earth's surface and are in constant motion. Great currents, which are driven by the Earth's rotation and by wind, circle around the edges of the oceans. A tremendous amount of energy occurs in these currents because of their size; however, the currents flow slowly, so the amount of energy in any given small portion of the ocean is unimpressive. In other words, the ocean currents have low energy density. This makes it difficult to harness their energy.

**Salinity gradients,** or differences in salt concentration at different depths in the ocean, have only recently been recognized as a potential energy source. The technology to harness this energy is not currently available, but some scientists are interested in developing it, both to provide energy and to desalinize brackish water (see Chapter 13).

## ENERGY SOLUTIONS: CONSERVATION AND EFFICIENCY

Human energy needs will continue to increase, if only because the human population is growing. We must therefore place a high priority not only on developing alternative energy sources but also on

(a)

**Figure 12–20**
Tidal energy. (a) The world's largest plant for electrical generation from tidal power is at La Rance Power Station in France, which has been operational since 1968. (b) The La Rance Power Station generates electricity from rising and falling tides.
(a, Visuals Unlimited/Science VU–API)

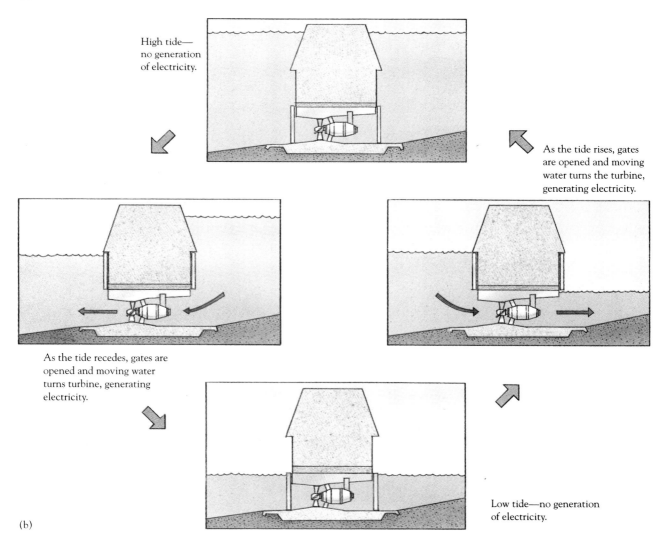

High tide—
no generation
of electricity.

As the tide rises, gates are opened and moving water turns the turbine, generating electricity.

As the tide recedes, gates are opened and moving water turns turbine, generating electricity.

Low tide—no generation of electricity.

(b)

energy conservation (moderating or eliminating wasteful or unnecessary energy-consuming activities) and energy efficiency (using technology to accomplish a particular task with less energy). As an example of the difference between energy conservation and energy efficiency, consider gasoline consumption by automobiles. Energy *conservation* measures to reduce gasoline consumption would include carpooling and reducing driving speed, whereas energy *efficiency* measures would include making more fuel-efficient automobiles. Conservation and efficiency accomplish the same goal—saving energy.

Energy conservation and efficiency could be considered the most promising energy "sources" available to humans, because they not only save energy for future use but also buy us time to explore new energy alternatives. Energy conservation and efficiency also cost less than development of new sources or supplies of energy. A system that supports energy conservation and efficiency makes good economic sense, as well. The adoption of energy-efficient technologies creates new business opportunities, including the development, manufacture, and marketing of those technologies.

In addition to economic benefits and energy resource savings, there are important environmental benefits from greater energy efficiency and conservation. For example, using more energy-efficient appliances could cut our carbon dioxide emissions by millions of tons each year, thereby slowing global climate change. Energy conservation and efficiency also reduce air pollution, acid precipitation, and other environmental damage related to energy production and consumption.

## Energy Consumption Trends and Economics

Energy consumption in developed countries did not increase between 1975 and 1990, as had originally been projected; it actually decreased during that period, partly because of improved energy efficiency. Technological improvements in the paper-making industry, for example, make it possible for us to use less energy to manufacture paper today than we used just a few years ago. Similarly, new aircraft are much more fuel-efficient than older models. The energy savings from such improvements in efficiency translate into greater profits for the companies employing them. It has been estimated that the United States is approximately 25 percent more efficient in its energy use today than it was in the early 1970s. This means that the United States uses 25 percent less energy to generate each dollar of its gross national product.

A country's or region's total energy consumption divided by its gross national product gives its energy intensity (Figure 12–21). Despite gains in

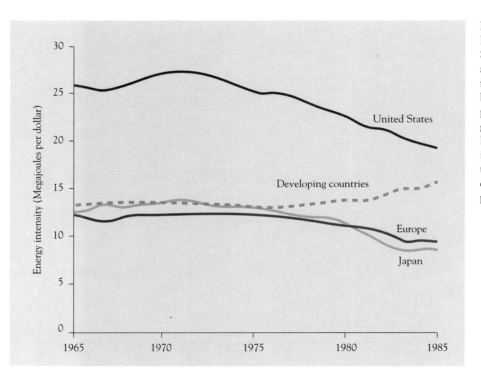

**Figure 12–21**
Energy intensity of the United States, Western Europe, Japan, and developing countries. The four areas indicated here have been equalized for comparison purposes by conversion of all gross national products to dollars. The increase in energy intensity in developing countries is a result of economic development and expanding populations.

efficiency, the energy intensity of the United States is still considerably higher than that of Japan or Western Europe.

**The Challenge in Developing Countries** Per-capita consumption of energy in developing nations is substantially less than it is in industrialized countries, although the fastest increase in energy consumption today is occurring in the developing nations. As developing nations boost their economic development, their energy demands will continue to increase. This is partly because the "new" industrial and agricultural processes being adopted in developing countries often represent older technologies that are less energy-efficient. Also, the burgeoning populations in developing countries will raise energy demands.

Developing countries are faced with the need for economic development and the need to control environmental degradation. At first glance, these two goals appear to be mutually exclusive. However, both goals can be realized by the use of technology now being developed in industrialized nations to achieve greater energy efficiency. For example, it would cost Brazil $44 billion to build power plants to meet its projected electricity needs for the near future; this cost could be avoided by investing $10 billion in more efficient refrigerators, lighting, and electric motors. The energy efficiency approach in Brazil would not only cause fewer environmental problems but also foster and expand the growth of manufacturing industries devoted to energy-efficient products.

## Energy-Efficient Technologies

The development of more efficient appliances, automobiles, and buildings has been a major factor in the recent reduction of energy consumption in developed countries. Compact fluorescent light bulbs, introduced in the mid-1980s, require 25 percent of the energy used by regular incandescent bulbs and last nine times longer; the energy-efficient bulbs do cost more, but they more than pay for themselves in energy savings. New condensing furnaces require approximately 30 percent less fuel than conventional gas furnaces. "Superinsulated" homes in Sweden and the United States use 68 to 90 percent less heat than do homes insulated with standard methods (Figure 12–22).

The National Appliance Energy Conservation Act sets national appliance efficiency standards for refrigerators, freezers, washing machines, clothes dryers, dishwashers, air conditioners, and water heaters. The energy cost that consumers will save

as a direct result of energy savings required by this law is estimated at $28 billion by the year 2000.

Automobile efficiency has improved dramatically as a result of the use of lighter materials and designs that reduce air drag. A Japanese prototype automobile with special design features and materials, for example, achieved 98 miles per gallon. Using current technology, automobiles with fuel efficiencies of 60 to 65 miles per gallon could be routinely manufactured before the year 2000. (Some strategies for reducing gasoline consumption by automobiles and other vehicles are discussed in Chapter 10. See also Meeting the Challenge: New Cars, New Fuel on page 257.)

**Cogeneration** One nontraditional energy technology with a bright future is **cogeneration,** which is a way of recycling waste heat. Cogeneration is currently being used on a small scale, but its use is increasing. Modular cogeneration systems enable hospitals, hotels, restaurants, and other businesses to harness steam that would otherwise be wasted. In a typical cogeneration system, electricity is produced in a traditional manner—that is, some type of fuel provides heat to form steam from water. Normally, the steam used to turn the electricity-generating turbine would be cooled before being pumped back to the boiler to be reheated. In cogeneration, after the steam is used to turn the turbine, it supplies energy to heat the building, cook food, or operate machinery before it is cooled and pumped back to the boiler as water (Figure 12–23).

Cogeneration can also be accomplished on a large scale. Prince Georges County, Maryland, plans to build a natural gas-fired power plant that will sell electricity to the local utility. The waste steam produced in the generation of electricity will be used to operate a soft-drink carbonation plant adjacent to the power plant. A similar cogeneration plant in North Carolina supplies electricity to utilities in two states and steam to an adjoining textile company.

**Energy Savings in Commercial Buildings** Energy costs often account for 30 percent of a company's operating budget. Unlike cars, which are traded in every few years, buildings are usually used for 50 or 100 years, so a company housed in an older building normally does not have the benefits of new energy-saving technologies. It makes good economic sense for these businesses to invest in energy improvements, which often pay for themselves in a few years.

To get businesses to install new energy-efficient technologies, many energy-service companies (com-

*(Text continues on p. 258.)*

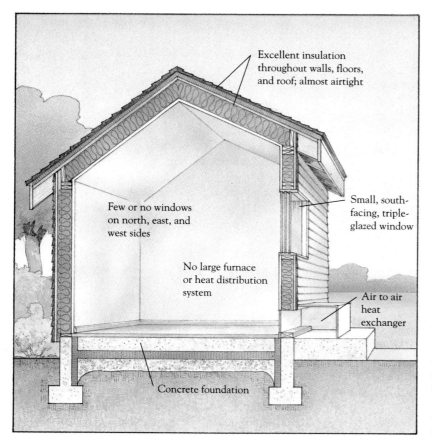

Few or no windows on north, east, and west sides

No large furnace or heat distribution system

Excellent insulation throughout walls, floors, and roof; almost airtight

Small, south-facing, triple-glazed window

Air to air heat exchanger

Concrete foundation

**Figure 12–22**
Some of the characteristics of a superinsulated home, which is so well insulated and airtight that it does not require a furnace in winter! Heat from the inhabitants, light bulbs, the stove, and other appliances provides almost all the necessary heat.

**Figure 12–23**
In a cogeneration system, electricity is produced in the usual manner. In this example, fuel combustion occurs in a boiler (1). The heat produced is used to make steam, which turns a turbine (2) that generates electricity in a generator (3). The electricity that is produced (4) is used in-house or sold to a local utility. Cogeneration involves using the waste heat (leftover steam) from electricity generation to do useful work, such as cooking, space heating, or operating machinery (5). Any residual steam that is not used is piped into a condenser (6) and recycled back to the boiler.

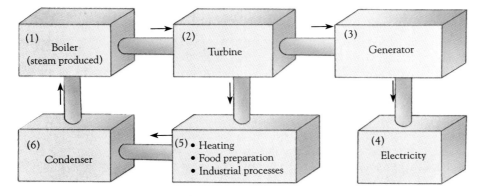

# New Cars, New Fuel

Nearly two-thirds of all oil consumed in the United States is used for transportation. Thus, it makes sense that research into alternative fuels for motor vehicles is a must for reducing this country's dependency on foreign oil and nonrenewable resources. Another advantage is that alternative fuels usually produce smaller amounts of pollutants than gasoline does. Several alternative fuels—and the cars to go with them—are currently being developed.

Methanol and ethanol, which are alcohol fuels that are much cleaner than gasoline (they don't release as much of the smog-causing hydrocarbons and nitrogen oxides), can be made from renewable resources such as wood. They already exist, but scientists are working on ways to reduce their costs. American and foreign car manufacturers are experimenting with engines designed specifically for these fuels, and many manufacturers will market methanol-powered cars in California within the next few years.

An electric engine is also much cleaner (and quieter) than a gasoline engine. However, if its electricity comes from a coal-fired power plant, an electric car actually produces *more* emissions than a gasoline-powered car; electric cars are cleaner than gasoline-powered cars only when their electricity comes from natural gas, solar energy, or hydrogen. Also, the vehicle cost, performance, and refueling convenience of electric cars still lag far behind those of gasoline-powered cars. But General Motors is building a two-seater (which should be available in the mid-1990s) that can be run for 120 miles before needing an overnight recharge. Chrysler has been testing an electric mini-van that can travel from 110 to 120 miles between recharges and can reach speeds of about 70 miles per hour. In addition, a company based in California is marketing affordable electric cars that have a 50-mile traveling range. These vehicles can be recharged by household electrical sockets or solar cells.

Liquid hydrogen is a clean but expensive fuel, so scientists are looking for ways to use photovoltaic cells to produce hydrogen less expensively. BMW has already built a car that uses liquid hydrogen, although it is not expected to be a viable alternative until sometime in the 21st century. Mercedes has a car that is powered by hydrogen generated by solar electricity (it will reach speeds of up to 105 mph), but it has a traveling range of only around 75 miles. A conventional engine that is fueled by burning hydrogen emits no carbon dioxide or sulfur dioxide.

Scientists at New Mexico State University are studying saline groundwater as a possible source of fuel. They hypothesize that algae could be grown in the water and that their lipids could be harvested and converted to gasoline at a competitive cost.

Much of the current interest in alternative fuels is the result of California's tough tail-pipe emissions standards. A California law passed in 1990, for example, mandates that at least 2 percent of the cars sold in that state by 1998 must have zero tailpipe emissions; by 2003, 10 percent of new cars must meet that criterion. Because more automobiles are bought in the state of California than in any other state, auto manufacturers cannot ignore the California mandate, and almost all of the domestic and foreign manufacturers are experimenting with a variety of batteries for electric cars. A number of states, including Texas, New York, Connecticut, and Massachusetts, have followed California's lead by toughening their emission standards.

Alternative fuels for transportation are clearly the wave of the future, and it is important for the government and the general public to support these research efforts. After all, we're the ones who will benefit from them.

General Motors' electric vehicle prototype, *Impact*, produces zero emissions. The company plans to mass-produce and market these cars in the late 1990s. *Impact*'s battery can be fully recharged in about three hours on a 220-volt circuit, it has a practical range of 80 miles per charge, and can go from zero to 60 mph in eight seconds. (Courtesy of General Motors Corp.)

panies that specialize in improving energy efficiency) offer their assistance in such a way that the business makes little or no financial outlay. Here's how it works. An energy-services company makes a detailed assessment of how a business can improve its energy efficiency. In developing its proposal, the energy-services company guarantees a certain amount of energy savings. It also provides the funding to accomplish the improvements, which may be as simple as fine-tuning existing heating, ventilation, and air conditioning (HVAC) systems or as major as replacing all existing windows and lights. The reduction in utility costs is used to pay the energy-services company, but once the bill is paid, the business benefits from substantial energy savings.

**Energy Savings in Homes**   When buying a new home, a smart consumer should demand energy efficiency. Although a more energy-efficient house might cost a little more, depending on the technologies employed, the improvements usually pay for themselves in two or three years. Any time spent in the home after the payback period means substantial energy savings. Energy efficiency will almost certainly be an important part of the design of homes of the future.

Some energy-saving improvements, such as thicker wall insulation, are easier to install while the home is being built. Other improvements, such as installing thicker attic insulation, installing storm windows and doors, caulking cracks around windows and doors, replacing inefficient furnaces, and adding heat pumps, can be made in older homes to improve energy efficiency and, as a result, reduce the cost of heating the homes.

Many of the same improvements also provide energy savings when a home is air conditioned. Additional cooling efficiency is achieved by insulating the ductwork for the air conditioner (especially in the attic), buying an energy-efficient air conditioner, and shading the south and west sides of a house with trees.

Other home improvements that result in substantial energy savings include replacing incandescent bulbs with energy-efficient compact fluorescent light bulbs and wrapping insulation around water heaters and water pipes.

How does a homeowner learn which improvements will result in the most substantial energy savings? In addition to reading the many articles on energy efficiency that appear in newspapers and magazines, a good way to learn about your home is to have an energy audit done. Most local utility companies can send an energy expert to your home to perform an audit for little or no charge.

## Electric Power Companies and Energy Efficiency

Recent changes in the regulations governing electric utilities have made it possible for utilities to make more money by generating *less* electricity. Such programs provide incentives to save energy and thereby reduce power plant emissions that contribute to environmental problems.

Traditionally, to meet future power needs, electric utilities planned to build new power plants or purchase additional power from alternative sources. Now they can avoid these massive expenses by helping their consumers save energy. Some utilities support energy conservation and efficiency by offering cash awards to consumers who install energy-efficient technologies. The reward might be a $100 rebate to a homeowner who purchases an energy-efficient refrigerator, or thousands of dollars to a large industry that overhauls its entire production to increase energy efficiency.

Some utilities also give customers highly efficient fluorescent light bulbs, air conditioners, or other appliances. They then charge higher rates, but the greater efficiency results in savings for both the utility company and the consumer. The utility company makes more money from selling less electricity because rates are higher, and the consumer saves because less energy is used by the efficient light bulbs or appliances.

## What About Energy Conservation?

Simple measures such as lowering your thermostat during the winter and raising it during the summer, turning off lights when you leave a room, and driving more slowly result in small energy (and cost!) savings. You also contribute to energy conservation by making use of carpools or public transportation. The cumulative effect of many people taking similar measures is substantial.

Energy conservation and efficiency are sound ideas for all of us. Energy saved today will be available for our grandchildren. The energy we save now will help to slow down climate change and environmental degradation so that our consumption will not become an overwhelming burden on future generations. The energy we save now will give us additional time to develop and improve alternative energy sources (see You Can Make a Difference: Using Your Government).

## You Can Make a Difference

### Using Your Government

One of the most important ways you can help decrease energy consumption is to become an involved citizen. Let your elected officials know that you support legislative measures that result in energy conservation. Governments generally respond to energy shortages by making it easier to find and exploit additional sources of energy, rather than by conserving energy. Don't let this happen.

1. Write to your senators and representatives. Ask them to pass laws requiring all energy-consuming technologies to be rated for energy efficiency. The ratings should be visibly displayed, whether the equipment or appliance is for domestic, commercial, or industrial purposes.
2. Encourage elected officials to support tax credits for homeowners and businesses that make energy-conserving improvements in their residences or workplaces.
3. Tell officials that you want them to support an increase in available funding for research into renewable energy technologies such as solar, wind, and ocean power.
4. Encourage state legislators to pressure public utilities to spend more of their research time and budget on renewable energy technology.

Go ahead and make yourself heard. Be a squeaky wheel. After all, you have the power of your vote when election day rolls around. You have nothing to lose, but an elected official who turns a deaf ear to energy concerns could easily lose at the polls.

## SUMMARY

**1.** Alternative energy sources are increasingly being examined with our future energy needs in mind, because they are renewable and generally cause less environmental impact than fossil fuels or nuclear energy. However, using alternative energy sources is generally feasible only in restricted locations.

**2.** Many forms of renewable energy are dispersed and therefore tend to be inefficiently utilized. Their inefficiency is partly a result of the need to collect or concentrate the energy so that it is useful.

**3.** Solar energy can be used directly or indirectly. Direct solar energy is the sun's radiant energy, whereas indirect solar energy is energy produced by phenomena that have already incorporated the sun's radiant energy.

**4.** Direct solar energy can be used either actively or passively to heat buildings and water. Solar thermal electric generation and photovoltaic cells are two methods that convert solar energy into electricity.

**5.** Forms of indirect solar energy include wind, biomass (wood, agricultural wastes, and fast-growing plants), and hydropower (the energy of flowing water). Biomass can be burned directly or converted to gas or liquid fuels. Biomass is already being used for energy on a large scale, particularly in developing nations.

**6.** Wind and hydropower have potential as indirect sources of solar energy in certain areas. Ocean thermal energy conversion (OTEC), which is under investigation, is a special type of hydropower that makes use of the temperature gradient of the ocean.

**7.** Geothermal energy can be obtained from hot rocks near the Earth's surface. The technology for extracting geothermal energy from hot rocks that are associated with groundwater is well established. Obtaining energy from dry hot rocks is currently under investigation.

**8.** Other forms of renewable energy that show some potential are tides, ocean currents, and salinity gradients.

**9.** The most promising energy "sources" are energy conservation and energy efficiency. Using technology to increase energy efficiency results in energy savings that conserve our fuel supplies.

# DISCUSSION QUESTIONS

**1.** Solar energy in all its forms could be considered an indirect form of nuclear energy. Why?

**2.** Explain the following statement: Unlike fossil fuels, solar energy is not resource-limited, but is technology-limited.

**3.** Give an example of how one or more of the alternative energy sources discussed in this chapter could have a negative effect on each of the following aspects of ecosystems: soil preservation, natural water flow, production of foods used by natural plant and animal populations, and maintenance of the diversity of living organisms found in an area.

**4.** Biomass is considered an example of indirect solar energy because it is the result of photosynthesis. Given that plants are the organisms that photosynthesize, why are animal wastes also considered biomass?

**5.** Why is it easier to obtain energy from a small river with a steep grade than from the vast ocean currents?

**6.** One advantage of the various forms of renewable energy, such as tidal and wind energy, is that they cause no net increase in carbon dioxide. Is this true for biomass? Why or why not?

**7.** Explain how energy conservation and efficiency are major "sources" of energy.

**8.** Evaluate which forms of energy other than fossil fuels and nuclear power have the greatest potential where you live.

**9.** List energy conservation measures that you could adopt for each of the following aspects of your life: washing laundry, lighting, bathing, buying a car, cooking, driving a car.

# SUGGESTED READINGS

Bevington, R., and A. H. Rosenfeld. Energy for Buildings and Homes. *Scientific American*, September 1990. Presents the new technologies and strategies that can substantially reduce energy bills.

de Groot, P. Plant Power: Fuel for the Future. *New Scientist*, vol. 124, December 1989. Examines the use of biomass as an energy source.

Dostrovksy, I. Chemical fuels from the sun. *Scientific American*, December 1991. Discusses the need to harvest solar energy on a large scale and convert it into chemical fuels that can be easily stored and transported.

Fickett, A. P., C. W. Gellings, and A. B. Lovins. Efficient Use of Electricity. *Scientific American*, September 1990. Examines the advanced technologies that improve the efficiency of machines and equipment using electricity.

Flavin, C., and N. Lenssen. "Designing a Sustainable Energy System," in *State of the World 1991*, Worldwatch Institute, W.W. Norton, New York, 1991. Examines the projected transition from fossil fuels to solar energy, including environmental benefits, economic considerations, and policy needs.

Hirst, E. Boosting U.S. energy efficiency. *Environment* 33:2, March 1991. Improvement of energy efficiency would save money, increase economic productivity, increase international competitiveness, reduce oil imports, and decrease environmental pollution.

Reddy, A. K. N., and J. Goldemberg. Energy for the developing world. *Scientific American*, September 1990. The challenge of providing affordable energy to developing countries can be met without degrading the environment.

Weinberg, C. J., and R. H. Williams. Energy from the sun. *Scientific American*, September 1990. Considers the advantages and disadvantages of various forms of solar energy.

*World Resources 1992–93: A Guide to the Global Environment*. Oxford University Press, New York, 1992. Focuses on trends in environmental issues. Of particular relevance to this chapter is the section on energy, which includes a discussion of energy efficiency in both developed and developing countries.

A black bear cub. Bear populations are declining precipitously worldwide, in part because of the demand for their gallbladders for Chinese folk medicine.

Our Precious Resources

Part 5

# Russell Train

## Preserving Biological Diversity in the Developing World

*Russell Train is Chairman of the World Wildlife Fund in Washington, D.C.; he was a founding trustee of WWF-U.S. in 1961, and was its President from 1978–1985. Mr. Train received his undergraduate degree from Princeton University, and his law degree from Columbia University in New York City. In 1972, he headed the U.S. delegation to the United Nations Conference on the Human Environment in Stockholm, the predecessor of the 1992 Earth Summit Conference in Rio de Janeiro. Also in 1972, he was the President's representative to the International Whaling Convention, which approved the moratorium (still in effect) on commercial whaling. He has been Undersecretary of Interior, chairman of the Council on Environmental Quality, and from 1973–1977, was Administrator of the Environmental Protection Agency. Currently, he also chairs the National Commission on the Environment.*

**How does the World Wildlife Fund target an endangered or threatened species? Currently, what are the most common causes of endangerment?**

Destruction of habitat is the primary cause of endangerment. Therefore, the emphasis of the World Wildlife Fund's activities is far more on habitats and ecosystems today than

on any one particular endangered species. We can only guess at how many species there are on Earth; there may be tens of millions. Most are still unidentified, a high percentage are probably threatened or endangered, and most are probably insects. For example, we may know only 10% of the species that inhabit a square mile of the tropical rain forests, the largest repositories of biological diversity. But if you concentrate on protecting the ecosystem, you will be protecting all of the species involved, whether or not you are even aware of their existence.

**How did this shift in focus occur?**

It evolved over the years. In the early 1970s we launched a major campaign in Nepal to save the Bengal tiger. The development of that program hinged on the identification of important tiger habitats, and establishing them as protected reserves. It was apparent that the tiger could not be protected in isolation from its supporting ecosystem. Preserving the grasslands and forests, as well as the wide diversity of animal species, including the tiger's prey species, became as important as protecting the tiger itself.

**Do you still maintain your programs directed at specific species, which have attracted so much popular interest?**

We are still concerned with certain specific species, which I call "flagship species," and we have programs devoted to them. Animals such as the tiger, the elephant, the whale, the black rhinoceros, and the mountain gorilla are big, dramatic creatures that can become symbols of world conservation and concern. They can attract funding and public support. But by preserving the habitats of flagship species, we are also preserving that same habitat for countless, less glamorous species, like insects.

**If destruction of habitat is the leading cause of endangerment, what are the chief causes of habitat destruction?**

I would put human population growth at the top of the list of threats to wildlife and associated habitats all over the world. Habitat is generally destroyed by people, and it is usually the pressure of population growth that pushes people into new areas. Human settlement, slash and burn cultivation, the timber industry, and other forms of economic development can all wipe out habitat, and all are often the result of expanding human populations.

**To assist developing countries in preserving biodiversity, while at the same time addressing the problems of overpopulation, hunger, and poverty, the WWF has established several innovative programs. Explain the Wildlands and Human Needs program.**

One of our highest priorities is building local conservation organizations around the world. Sometimes this means starting from scratch, working with local people to create an organization. More often, it means identifying an organization existing that shows a certain amount of capability, and then building that organization by training its staff. We bring some of them here to Washington, where we instruct them in fundraising, communications, program management, grant writing, public relations, media relations, etc.

We also provide financial help, sometimes with matching funds, in order to encourage these organizations to raise money locally. The "debt-for-nature" swap, which we pioneered, is another mechanism for generating funds for a variety of conservation needs in developing countries. We've done debt-for-nature swaps in Madagascar, the Philippines, Zambia, Ecuador, Costa Rica, and a number of other countries, generating over $30 million for conservation activities. These elements are all part of the Wildlands and Human Needs program.

**The program also directly assists the people who live and work in and around threatened habitats.**

Yes, we have come to the conclusion that if conservation programs are going to work, there has to be a marriage, if you will, of conservation objectives with benefits for local people. Plans imposed from the outside are never successful in the long run.

It's very important that conservation be seen by local people as important to their livelihood, and that the wildlife and protected areas are not perceived as antagonistic entities. We recognize that, in many cases, the local population must utilize some portion of the wildlife for their own benefit.

**What are some of the developing countries in which your efforts have been particularly effective?**

In the West African and Central American forests, the development of honey production as a form of livelihood has become an important reason for maintaining the forests. In the Caribbean, we had a program to determine why the conch, a commercially valuable species of shellfish, had disappeared in many areas. We investigated their breeding habits, and helped the local people to understand the problems of over-fishing. They succeeded in restoring conch populations and reviving the conch industry. In Zimbabwe we're working with communal groups who have taken on the job of wildlife management themselves. They harvest a certain portion of the wildlife for their own benefit, but recognize that this has to be done on a sustainable yield basis to protect the wildlife stock. Once you get something like that going, then you've really got a basis for long-term conservation.

**Are there some countries that need more protective legislation, and others that need stronger enforcement of established laws?**

By and large, most countries, both developed and developing, have reasonably good legislation on the books. We have helped a lot of countries write their environmental laws. The problem tends to be more in the implementation of the legislation.

**What impedes most countries from implementing their laws and regulations?**

In most of the developing world there is a lack of funds, infrastructure, trained personnel, education, and research. Conservation and national park areas are given low budget priority almost everywhere around the world, especially in developing countries where there is so much economic pressure. Economic development is extremely important, but it is a two-edged sword. On the one hand, economic development is one of the chief reasons for the destructive exploitation of natural areas. On the other hand, it is the means of generating the resources for a country to develop and turn its attention to matters of conservation—thus, the long-term, critical importance of *sustainable* development.

# Chapter 13

Little water porters in Guatemala. (E. R. Degginger)

# Water: A Fragile Resource

resh water is one of our most important natural resources because it's needed for survival and there is no substitute for it. Indeed, unlike energy, which comes in a variety of forms, water offers no alternatives. Most of our planet is covered by water, but only a tiny fraction of it is available as fresh water, and the amount of accessible fresh water varies from country to country and region to region. For example, although water is abundant overall in the United States, local and regional shortages exist. Local fresh water supplies are increased by collecting water in reservoirs, diverting water from areas with plentiful water supplies, and removing salt from ocean water. Water conservation can extend the supply of fresh water available for agriculture, industry, and domestic use, and is practiced seriously in areas with severe water shortages.

# THE IMPORTANCE OF WATER

The view of planet Earth from outer space reveals that it is different from other planets in the Solar System. Earth is a predominantly blue planet because of the water that covers three-fourths of its surface. Water has a tremendous effect on our planet: it helped shape the continents, it moderates our climate, and it allows living organisms to evolve and survive.

Life on planet Earth would be impossible without water. All life forms, from simple bacteria to complex multicellular plants and animals, contain water. Humans are composed of approximately 70 percent water (by weight), and we depend on water for our survival as well as for our convenience: we drink it, cook with it, wash with it, travel on it, and use an enormous amount of it for agriculture, manufacturing, mining, energy production, and waste disposal.

Although Earth has plenty of water, it is distributed unevenly, and serious water supply problems exist. In regions where fresh water is in short supply, such as deserts, obtaining it becomes critically important; and because the use of water for one purpose decreases the amount available for other purposes, serious conflicts often arise over how water should be used. Even regions with readily available fresh water are not without problems, however, and maintaining the quality and quantity of water is a top priority.

Worldwide, we are using increasingly more water, in part because our population is increasing and in part because, on the average, each person is using more water. The World Resources Institute estimates that water use has increased 4 to 8 percent each year since 1950. The *rate* of increase is now slowing because water use has stabilized in developed nations, although it is still increasing in developing countries.

To meet the growing need for water, we try to augment our supply by building dams to create reservoirs (artificial lakes in which water is stored for later use) and by diverting river water. In many areas, the quantity of water is not as critical as its quality, and steps must be taken to ensure a supply of clean water. All of these efforts to obtain and maintain a steady supply of clean water involve considerable expense.

This chapter examines some of the ecological processes and human activities that affect the availability of water. Water quality, including water pollution, is such a significant issue that it is covered separately, in Chapter 21.

# PROPERTIES OF WATER

Water ($H_2O$) is a molecule, consisting of 2 atoms of hydrogen and 1 atom of oxygen (Figure 13–1a), that can exist in any of three forms: solid (ice), liquid, and vapor (water vapor or steam). Water is a polar molecule; that is, one end of the molecule has a positive electrical charge, and the other end has a negative charge. The negative (oxygen) end of one water molecule is attracted to the positive (hydrogen) end of another water molecule, forming a **hydrogen bond** (Figure 13–1b) between the two molecules. Hydrogen bonds are the basis for a number of water's physical properties including its high melting/freezing point (0°C, 32°F) and high boiling point (100°C, 212°F). Because most of the Earth has a temperature between 0°C and 100°C, most water exists as the liquid on which living organisms depend.

**Figure 13–1**

Many of the unusual properties of water are a result of its chemistry. (a) Water, which consists of two hydrogen atoms and one oxygen atom, is a polar molecule with positively and negatively charged ends. (b) The polarity of water molecules causes them to form hydrogen bonds between the positive ends of one molecule and the negative ends of others.

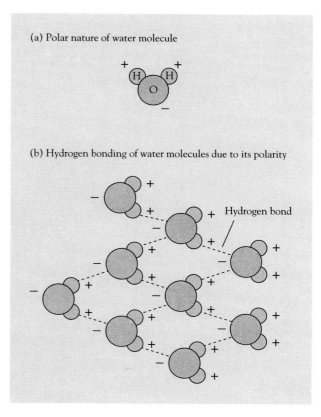

(a) Polar nature of water molecule

(b) Hydrogen bonding of water molecules due to its polarity

Hydrogen bond

Water absorbs a great deal of solar heat without its temperature rising substantially. It is this high heat capacity that allows the oceans to have a moderating influence on climate, particularly of coastal areas. Another consequence of water's high heat capacity is that oceans do not experience the wide temperature fluctuations that are common on land.

Water must absorb a lot of heat before it **vaporizes,** or changes from liquid to vapor. When it does evaporate, it carries the heat (called its heat of vaporization) with it. Thus, evaporating water has a cooling effect. That is why your body is cooled when perspiration evaporates from your skin.

Water is sometimes called the "universal solvent," and although this is an exaggeration, many materials do dissolve in water. In nature, water is never completely pure, because it contains dissolved gases from the atmosphere and dissolved mineral salts from the Earth. Seawater, for example, contains a variety of dissolved salts including sodium chloride, magnesium chloride, magnesium sulfate, calcium sulfate, and potassium chloride. Water's dissolving ability has a major drawback, however: many of the substances that dissolve in water also pollute it.

Water partially obeys the general physical rule that heat expands and cold contracts. As water cools, it contracts and becomes denser until it reaches 4°C (39°F), the temperature at which it is densest. When the temperature of water falls *below* 4°C, however, it becomes less dense. This is why ice (at 0°C) floats on denser, slightly warmer liquid water. Because water freezes from the top down rather than from the bottom up, aquatic organisms can survive beneath a frozen surface.

## The Hydrologic Cycle

Water continuously circulates through the physical environment, from the oceans to the atmosphere to the land and back to the oceans, in a complex cycle known as the hydrologic cycle (see Chapter 5). The result is a balance among water in the oceans, water on the land, and water in the atmosphere. The cycle thus continually renews the supply of purified water on land, which is essential to terrestrial organisms.

## OUR WATER SUPPLY AND ITS RENEWAL

Approximately 97 percent of the Earth's water is in the oceans and contains a high amount of dissolved salts (Figure 13–2). Seawater is too salty for human consumption and for most other uses. (For example, if you watered your garden with seawater, your plants would die.) Most fresh water is unavailable for easy human consumption because it is frozen as polar or glacial ice or is in the atmosphere or soil. Lakes, creeks, streams, rivers, and groundwater account for only a small portion (0.52 percent) of the Earth's fresh water.

**Surface water** is fresh water found on the Earth's surface in streams and rivers, lakes, ponds,

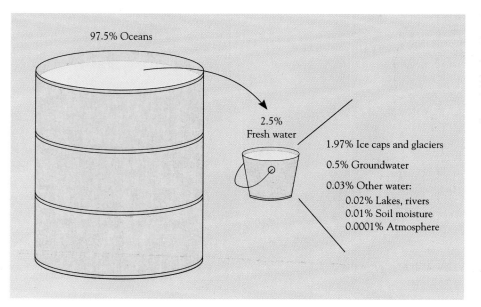

97.5% Oceans

2.5% Fresh water

1.97% Ice caps and glaciers

0.5% Groundwater

0.03% Other water:
0.02% Lakes, rivers
0.01% Soil moisture
0.0001% Atmosphere

**Figure 13–2**
Although three-fourths of the Earth's surface is covered with water, only a fraction of one percent is available for humans. Most water is salty, frozen, or inaccessible in the soil and atmosphere.

reservoirs, and **wetlands**—areas of land that are covered with water for at least part of the year. Surface waters are replenished by the runoff of precipitation from the land and are therefore considered a renewable resource. A **drainage basin,** or **watershed,** is the area of land that is drained by a river.

The Earth contains underground formations that collect and store water in the ground. This water originates as rain or melting snow that seeps into the soil and finds its way down through cracks and spaces in rock until it is stopped by an impenetrable layer; there it accumulates as **groundwater.** The porous layers of underground rock in which the groundwater is stored are called **aquifers** (Figure 13–3), and they can be either unconfined or confined. Those whose contents are replaced by surface water directly above them are **unconfined aquifers.** The upper limit of an unconfined aquifer, below which the ground is saturated with water, is the **water table.** A **confined aquifer** (also called an **artesian aquifer**) is a groundwater storage area between impermeable layers of rock. The water in a confined aquifer is trapped and often under pressure, and its recharge area (the land from which water percolates to replace groundwater) may be hundreds of miles away. Most groundwater is considered a nonrenewable resource, because it has taken millions of years to accumulate, and usually only a small portion of it is replaced each year by percolation of precipitation. The recharge of confined aquifers is particularly slow.

## Water Supply in the United States

Compared to many countries, the United States has a plentiful supply of fresh water. According to the U.S. Geological Survey, approximately 15.9 trillion liters (4.2 trillion gallons) of water fall as precipitation per day in the continental United States. Approximately two-thirds of this water returns to the atmosphere through evaporation and transpiration, leaving roughly 5.3 trillion liters (1.4 trillion gallons) per day in the soil, surface waters, and groundwater (Figure 13–4). Most of this remaining water, approximately 4.9 trillion liters

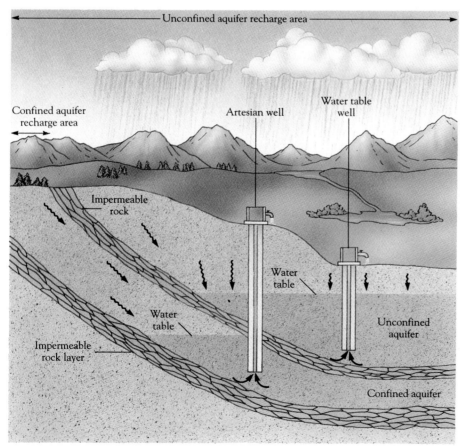

**Figure 13–3**
Excess water in the soil seeps downward through soil and porous rock layers until it reaches impermeable rock or clay. Groundwater that is recharged by surface water directly above is known as an unconfined aquifer. In a second type of aquifer, known as a confined aquifer, groundwater is stored between two impermeable layers and is often under pressure. Artesian wells, which produce water from confined aquifers, often do not require pumping because of this pressure.

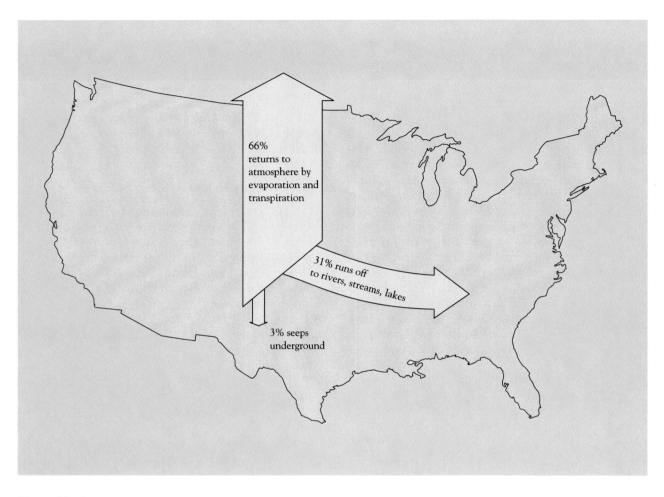

66%
returns to
atmosphere by
evaporation and
transpiration

31% runs off
to rivers, streams, lakes

3% seeps
underground

**Figure 13–4**
Water budget for the United States, excluding Alaska and Hawaii. Approximately 66 percent
of the 15.9 trillion liters (4.2 trillion gallons) that fall on the continental United States as
precipitation recycles into the atmosphere almost immediately by evaporation from wet surfaces
and transpiration by plants. Most of the rest (about 31 percent) flows in rivers and streams to
the oceans or Canada and Mexico. Only a small percentage (about 3 percent) of the precipitation seeps into underground aquifers. Most water withdrawn for human use is cycled back to
rivers and streams.

(1.3 trillion gallons) per day, makes its way into the
oceans without ever being used. The relatively
small amount of water that we borrow for our own
purposes eventually returns to rivers and streams—
for example, as treated or untreated sewage or industrial wastes.

Almost one-fourth of the water used in the
United States is groundwater, the largest supplies of
which are in the Midwest (Figure 13–5) and the
Southeast—areas that also have large supplies of
surface water.

Despite the overall abundance of fresh water in
the United States, many areas have severe water

shortages because of geographical variations (Figure
13–6). For example, arid and semiarid areas of the
western United States normally receive little rainfall. Annual fluctuations in precipitation also
occur; even areas of the country that usually receive adequate precipitation sometimes experience
droughts.

In the United States there are eleven major
rivers or river systems that either flow into U.S.
coastal waters or cross U.S. boundaries (Table 13–1
and Figure 13–7). The largest river system by far is
the Mississippi River, which receives water from
seven other large rivers and empties into the Gulf

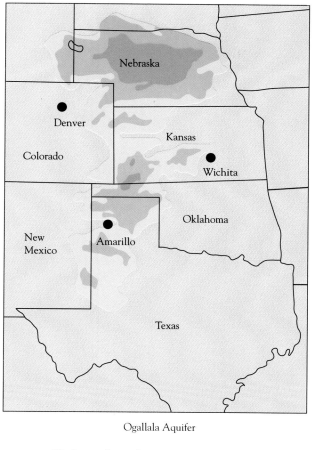

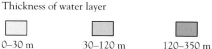

Thickness of water layer

| | | |
|---|---|---|
| 0–30 m | 30–120 m | 120–350 m |

**Figure 13–5**

A massive deposit of groundwater, known as the Ogallala Aquifer, lies under eight midwestern states. Extensive portions of the Ogallala Aquifer are located in Texas, Kansas, and Nebraska.

of Mexico. An estimated 1.5 trillion liters (400 billion gallons) of water enter the Gulf of Mexico from the Mississippi River *each day.*

## Global Water Supply

Data on global water availability and use indicate that, overall, the amount of fresh water on the planet is adequate to meet human needs, even taking population growth into account. These data do not, however, consider the distribution of water resources in relation to human populations. Citizens of Bahrain, the tiny island nation in the Persian Gulf, for example, have *no* fresh water supply

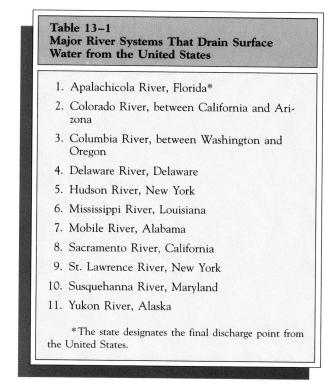

**Table 13–1**
**Major River Systems That Drain Surface Water from the United States**

1. Apalachicola River, Florida*
2. Colorado River, between California and Arizona
3. Columbia River, between Washington and Oregon
4. Delaware River, Delaware
5. Hudson River, New York
6. Mississippi River, Louisiana
7. Mobile River, Alabama
8. Sacramento River, California
9. St. Lawrence River, New York
10. Susquehanna River, Maryland
11. Yukon River, Alaska

*The state designates the final discharge point from the United States.

and must rely completely on desalinization (removing salt) of salty ocean water for their fresh water.

Large variations in per-capita use of water exist from country to country and from continent to continent, depending on the size of the human population and the available water supply. South America and Asia are the two continents with the greatest total water supply (Table 13–2): together they receive more than one-half of the world's renewable fresh water (by precipitation). Although South America has more available water per person than Asia does, it does not have the potential to support as many people as its water supply would suggest. That is because most of the precipitation received by South America falls in the Amazon River basin, which has poor soil and is therefore unsuitable for agriculture. In contrast, most of the precipitation in Asia falls on land that *is* suitable for agriculture; therefore, the water supply can support more people.

Humans need an adequate supply of water year round. Global water supply is complicated by the fact that **stable runoff,** the portion of runoff (from precipitation) that is available throughout the year, can be low despite the fact that total runoff is quite high. India, for example, has a wet season—June to September—during which 90 percent of its annual

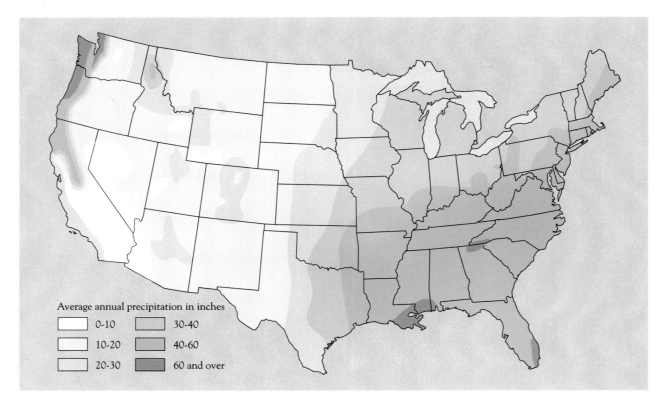

**Figure 13–6**
The average annual precipitation varies greatly across the United States.

**Figure 13–7**
Eleven major river systems drain surface water from the United States. Most of these rivers empty into the Atlantic or Pacific Oceans, although some cross into Canada and Mexico.

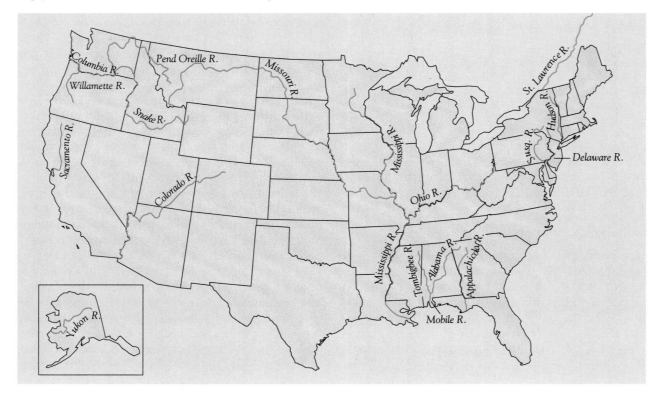

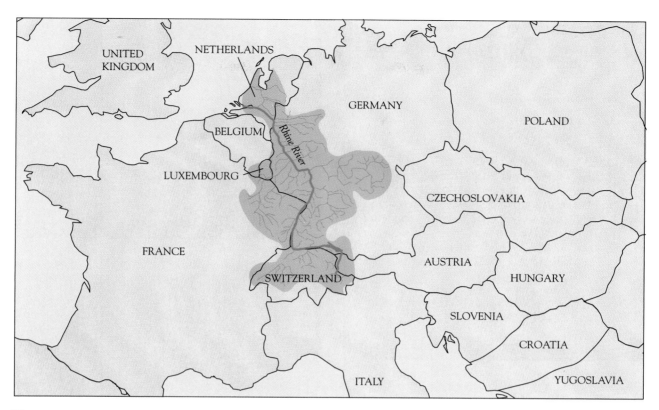

**Figure 13–8**
The Rhine River drains four European countries—Switzerland, France, Germany, and the Netherlands. Water management of such a river requires international cooperation.

| Table 13–2 Fresh Water Resources in 1990 | | |
|---|---|---|
| **Region** | **Total Renewable (Surface) Water (cubic kilometers)** | **1990 Per-Capita Water Use (cubic meters)** |
| Africa | 4,184 | 6.46 |
| Asia | 10,485 | 3.37 |
| Australia–Oceania | 2,011 | 75.96 |
| Europe | 2,321 | 4.66 |
| North and Central America | 6,945 | 16.26 |
| South America | 10,377 | 34.96 |
| Former U.S.S.R. | 4,384 | 15.22 |
| World Total | 40,707 | 7.69 |

Source: World Resources Institute.

precipitation occurs. Most of the water that falls during India's wet season quickly drains away into rivers and is unavailable during the rest of the year; thus, India's stable runoff is low.

Variation in *annual* water supply is an important factor in certain areas of the world. The African Sahel region has wet years and dry years, for example, and the lack of water during the dry years limits human endeavors during the wet years.

Global water supply is also complicated by the fact that surface water is often an international resource. The management of rivers that cross international boundaries requires international cooperation. The river basin for the Rhine River in Europe, for example, is in Switzerland, Germany, France, and the Netherlands (Figure 13–8). Traditionally, Switzerland, Germany, and France used water from the Rhein for industrial purposes and then discharged polluted water back into the river. The Dutch then had to clean up the water so they could drink it. Today, these countries recognize that international cooperation is essential if the supply

(a)

(b)

**Figure 13–9**
Agricultural use of water. (a) Center-pivot irrigation produces massive green circles. (b) The circular shape of center-pivot irrigation is caused by a long irrigation pipe that slowly rotates around the circle, spraying the crop.
(Gene Alexander/USDA/Soil Conservation Service)

and quality of the Rhine River are to be conserved and protected.

## HOW WE USE WATER

Water consumption varies among countries. In some countries, acute water shortages limit water use per person to several gallons per day. In developed countries, per-capita water use may be as high as several *hundred* gallons per day, an amount that encompasses agricultural and industrial uses as well as direct individual consumption.

Worldwide, more water—approximately 70 percent of total withdrawals—is used for agricultural purposes than for any other reason (Table 13–3). Irrigation of arid and semiarid lands has become increasingly important worldwide in efforts to produce enough food for burgeoning populations (Figure 13–9), and since 1955 the amount of land being irrigated worldwide has tripled. The World Resources Institute estimates that three-fourths of the agricultural land that is currently irrigated, some 200 million hectares (494 million acres), is in developing countries. Asia has more agricultural land under irrigation than do other regions, with China, India, and Pakistan accounting for most of it. It is projected that water use for irrigation will continue to increase in the 21st century, but at a slower rate than in the last half of this century.

## WATER RESOURCE PROBLEMS

Water resource problems fall into three categories: too much, too little, and quality/contamination (Chapter 21 addresses the third category). Floods and droughts are part of natural climate variations and cannot be prevented. Human activities sometimes exacerbate them, however, and humans often

**Table 13–3**
**Water Usage (cubic kilometers) in 1980s and Year 2000 (Projected)**

| Region | Irrigation | | Industry | | Domestic/Municipal | |
|---|---|---|---|---|---|---|
| | 1980s | 2000 | 1980s | 2000 | 1980s | 2000 |
| Africa | 120 | 160 | 6.5 | 30–35 | 10 | 30 |
| Asia | 1,300 | 1,500 | 118 | 320–340 | 88 | 200 |
| Australia–Oceania | 16 | 20 | 1.4 | 3.0–3.5 | 4.1 | 5.5 |
| Europe | 110 | 125 | 193 | 200–300 | 48 | 56 |
| North and Central America | 330 | 390 | 294 | 360–370 | 66 | 90 |
| South America | 70 | 90 | 30 | 100–110 | 24 | 40 |
| Former U.S.S.R. | 260 | 300 | 117 | 140–150 | 23 | 35 |
| World total | 2,206 | 2,585 | 759.9 | 1,153–1,308.5 | 263.1 | 456.5 |

Source: World Resources Institute.

court disaster when they make environmentally unsound decisions, such as building in an area that is prone to flooding.

## Too Much Water

Not all floods are bad. Many ancient civilizations (ancient Egypt, for example) developed near rivers that periodically spilled over, inundating the surrounding land with water. When the water receded, a thin layer of sediment that was rich in organic matter remained and enriched the soil. These civilizations flourished partly because of their agricultural productivity, which in turn was the result of floods replenishing the soil's nutrients.

Modern floods can cause widespread destruction of property and sometimes loss of life. Today's floods are more disastrous in terms of property loss than those of the past, but not because they involve more water. Human activities, such as the removal of water-absorbing plant cover from the soil and the construction of buildings on **flood plains** (areas bordering a river that are subject to flooding), increase the likelihood of both floods and flood damage.

Forests, particularly on hillsides and mountains, provide nearby lowlands with some protection from floods by trapping and absorbing precipitation. When woodlands are cut down, particularly when they are clearcut (stripped of all trees), the area cannot hold water nearly as well. Heavy rain-

fall then results in rapid runoff from the exposed, barren hillsides. This not only causes soil erosion (see Chapter 14), but puts lowland areas at extreme risk of flooding.

When a natural area—that is, an area undisturbed by humans—is inundated with heavy precipitation, the plant-protected soil absorbs much of the excess water. What the soil cannot absorb runs off into the river, which may spill over its banks in the flood plain. However, because rivers meander (that is, they aren't straight), the flow is slowed, and the swollen waters rarely cause significant damage to the surrounding area. (For a discussion of a river whose natural course was altered, see Focus On: Untwisting and Twisting the Kissimmee River.)

When an area is developed for human use, much of the water-absorbing plant cover is removed. Buildings and paved roads do not absorb water, so runoff (usually in the form of storm sewer runoff) is significantly greater (Figure 13–10). People who build homes or businesses on the flood plain of a river will most likely experience flooding at some point.

It is easier and more economical to ban or restrict development in a flood plain than to build on it and then try to prevent flooding with such structures as retaining walls and levees. Increasingly, local governments, both in the United States and in the rest of the world, now zone flood plains to curtail development.

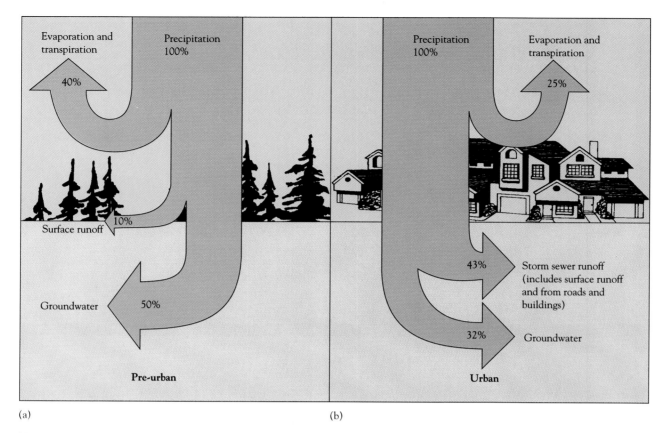

(a)                                                    (b)

**Figure 13–10**
The development of an area changes the natural flow of water. The fate of precipitation in Ontario, Canada (a) before and (b) after urbanization. After Ontario was developed, runoff increased substantially, from 10 percent to 43 percent.

## Too Little Water

**Arid** lands, or deserts, are fragile ecosystems in which plant growth is limited by lack of precipitation. **Semiarid** lands receive more precipitation than deserts but are subject to frequent and prolonged droughts. Forty percent of the world's population lives in arid or semiarid lands, primarily in Asia and Africa. These people spend substantial amounts of time and effort obtaining water. Each day they may have to walk many miles to a stream or river and carry back the heavy water.

Overpopulation in arid and semiarid regions intensifies the problem of water shortage. The immediate need for food prompts people to remove natural plant cover in order to grow crops on marginal land (land subject to frequent drought and subsequent crop losses), and their livestock overgraze the small amount of plant cover in natural pastures. As a result of the lack of plants, when the rains do come, runoff is greater, for the soil cannot

absorb the water. Because the soil is not replenished by the precipitation that does fall, crop productivity is poor and the people are forced to cultivate food crops on additional marginal land.

**Overdrawing Surface Waters** Removing too much fresh water from a river or lake can have disastrous consequences in local ecosystems. Humans can remove perhaps 30 percent of a river's flow without greatly affecting the natural ecosystem. In some places, however, considerably more than that amount is withdrawn for human use. In the American Southwest, for example, it is not unusual for 70 percent or more of surface water to be removed.

When surface water is overdrawn, wetlands dry up. Natural wetlands play many roles (see Chapter 17), not the least of which is serving as a breeding ground for many species of birds. Estuaries, where rivers empty into seawater, become saltier when surface waters are overdrawn, and this change in

## Untwisting and Twisting the Kissimmee River

For thousands of years, Florida's Kissimmee River wound its way from its headwaters near Orlando to its mouth at Lake Okeechobee, in the south-central part of the state. The trip from start to finish covered about 157 km (98 mi).

During heavy rains, the Kissimmee flooded its banks, creating 18,211 hectares (45,000 acres) of wetlands that supported shrubs, forests, waterfowl, wading birds, bald eagles, mammals, reptiles, and fish. This natural flooding caused problems for developers and ranchers who wanted to use the land for their own purposes, so between 1961 and 1971 the Army Corps of Engineers diverted the river by bulldozing a straight course from Orlando to Lake Okeechobee. The canal, 91 m (300 ft) in width and more than 9 m (30 ft) in depth, cut the river's length from 157 km (98 mi) to 83 km (52 mi). The river, now under human control, no longer flooded the valuable land.

But upon the opening of the canal, tens of thousands of gallons of polluted water flowed into Lake Okeechobee, which is the main source of fresh water for nearly half the population of Florida. Since the marsh grasses were gone, there was nothing to absorb the waste (mostly from cattle grazing along the river), and the lake quickly became contaminated. Mercury and phosphorus rose to dangerous levels in the water, and fish, ducks, and other wildlife populations were decimated.

By 1983 it was clear that something had to be done. A small demonstration project proved that the river's course could be restored, and by 1986 Congress approved $2.3 million to undertake partial restoration of the river. Flooding would be dealt with by purchasing approximately 65,000 acres of the flood plain. Several models for the restoration have been proposed, and whereas the simplest method would be to build weirs (low dams) to guide the river back to its old course, the most effective method probably would be to have the Army Corps of Engineers undo its original project and fill in the ditch it spent 10 years digging.

---

salinity greatly affects the productivity that is associated with estuaries.

Mono Lake, a salty lake in eastern California, is one of the most striking examples of the effects of humans' removing too much surface water (Figure 13–11). This lake is replenished by rivers and streams that are largely formed from snowmelt in the Sierra Nevada range. Evaporation provides the only natural outflow from the lake.

Much of the surface water that would naturally feed Mono Lake is now diverted to Los Angeles. As a result, Mono Lake's water level has subsided about 12.2 m (40 ft), and its salinity has increased dramatically (Table 13–4).[1] The organisms living in and around Mono Lake—including rabbits, shrews, muskrats, mink, mule deer, porcupines, and

[1]Over time, Mono Lake is becoming saltier as rivers deposit dissolved salts (recall that fresh water contains some salt) and as water, but not salt, is removed by evaporation.

**Table 13–4**
**Lake Elevation and Salinity Measurements for Mono Lake**

| Lake Elevation meters (feet) | Salinity (grams/liter total dissolved solids) |
|---|---|
| 1,927 (6,417) | 51.3 |
| 1,925 (6,410) | 56.3 |
| 1,923 (6,403) | 60.2 |
| 1,915 (6,376) | 89.3 |
| 1,914 (6,372) | 99.4 |

Source: Mono Basin Ecosystem Committee, *The Mono Basin Ecosystem: Effects of Changing Lake Level,* National Academy Press, Washington, D.C., 1987.

**Figure 13-11**
Mono Lake in California. The water draining into Mono Lake has been increasingly diverted for public consumption, causing major changes in the lake's water level and salinity. A recent study concluded that further changes would be disastrous to the lake's wildlife, including its migratory bird population. (W. Kleck/Terraphotographics)

shoreline vegetation—have been adversely affected by these changes. If Mono Lake becomes much more saline as a result of further drops in water level, it will be unable to support the brine shrimp and aquatic brine flies that are, in turn, consumed by a large migratory bird population that includes Arctic loons, grebes, cormorants, egrets, sandpipers, gulls, and bitterns. In all, there are 293 different species of birds in the Mono Lake ecosystem. If Mono Lake becomes more saline, it will be a dead lake, unable to support wildlife. The state of California is currently reviewing the Mono Lake situation and should rule on its own water rights in 1993.

The Aral Sea in Kazakhstan, part of the former Soviet Union, is suffering from the same problem as Mono Lake. Like Mono Lake, it has no outflow other than evaporation. In recent years, much of the inflow has been diverted for irrigation of fertile farmland around the lake. Since 1960, the Aral Sea has declined in area by 40 percent (Figure 13–12). Much of its biological diversity has disappeared, and, like Mono Lake, it runs the risk of becoming a lifeless brine lake. Millions of people living near the Aral Sea have developed health problems rang-

**Figure 13-12**
The Aral Sea in Kazakhstan has shrunk from water diversion for irrigation. As the water receded, these fishing boats became stranded far from the water's edge. (© 1989 David Turnley/Black Star)

**Figure 13–13**
Aquifer depletion can cause the ground to sink. Sinkholes, which often appear in a matter of minutes, are caused when the top of an underground cavern collapses from aquifer depletion. This sinkhole, which is 122 meters (400 feet) wide and 38 meters (125 feet) deep, occurred in Winter Park, Florida. (M. Timothy O'Keefe/Tom Stack & Associates)

ing from allergies to throat cancer, presumed to be caused by winds that whip the salt on the receding shoreline into the air, causing blinding salt storms. Moreover, the salt is carried by the wind hundreds of miles from the Aral Sea, and where it is deposited it causes soil pollution, which reduces the productivity of the land.

**Aquifer Depletion**  The removal by humans of more groundwater than can be recharged by precipitation or melting snow—called **aquifer depletion**—has several serious consequences. Prolonged aquifer depletion drains an aquifer dry, effectively eliminating it as a water resource. The depth to which one must go to find water in the Ogallala aquifer in Texas, for example, has increased steadily since the 1940s, a result of aquifer depletion. In addition, aquifer depletion from porous rock causes **subsidence,** or sinking, of the land on top (Figure 13–13). For example, the San Joaquin Valley in California has sunk almost 10 m (32.8 ft) in the past 50 years due to aquifer depletion. **Salt water intrusion,** the movement of seawater into a freshwater aquifer, can occur along coastal areas when groundwater is depleted faster than it can be replenished (Figure 13–14).

**Irrigation**  Semiarid lands usually provide marginal crops, but farmers can increase the agricultural productivity of semiarid lands with irrigation. Almost any crop can be grown in the desert if enough water is supplied to the soil.

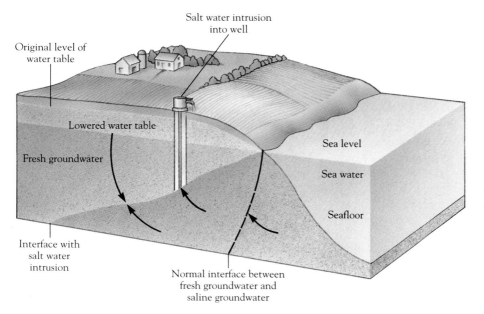

**Figure 13–14**
In coastal regions, aquifer depletion can cause salt water intrusion, which makes the aquifer water salty and unfit to drink. Normally the interface between fresh groundwater and salty groundwater is far from the site of the well. However, the removal of large amounts of fresh groundwater has caused the interface to migrate inland so that the well now draws up salty groundwater.

**Figure 13–15**
Irrigation water contains dissolved mineral salts. As the water evaporates from the soil's surface, the salts are left behind and gradually accumulate. Eventually, the salty soil is unfit for agriculture. (USDA/Soil Conservation Service)

Although irrigation improves the agricultural productivity of arid and semiarid lands, it can also cause salt to accumulate in the soil, a process called **salinization** (Figure 13–15). In a natural scenario, as a result of precipitation runoff, rivers carry salt away. Irrigation water, however, normally soaks into the soil and does not run off the land into rivers, so when it evaporates, the salt remains behind and accumulates in the soil. Salty soil results in a decline in productivity and in extreme cases renders the soil completely unfit for crop production. Chapter 21 discusses the problem of soil salinization in greater detail.

## Water Problems in the United States

Americans consume less than one-fourteenth of the fresh water that is available in the continental United States. However, this general picture of our water supply overlooks the regional and seasonal differences in distribution of water, amount of groundwater, climate, and consumption rates that make acquisition of water a challenge for many regions. Droughts, higher-than-average precipitation rates, and other natural conditions cause problems in water availability throughout the country, and human activities sometimes exacerbate the difficulties.

**Surface Water**  The increased use of U.S. surface water for agriculture, industry, and personal consumption during the past 35 years has caused many water supply and quality problems. Some U.S. regions that have grown in population during this period (for example, California, Arizona, and Florida) have placed correspondingly greater burdens on their water supplies. If water consumption in these and other areas continues to increase, the availability of surface waters could become a regional problem in many places that have never before experienced water shortages. Areas that derive their water from the Arkansas River, for example, may soon experience critical water shortages.

Nowhere in the country are water problems as severe as they are in the West and Southwest. Much of this large region is arid or semiarid and receives less than 50 cm (about 20 in) of precipitation annually. The West and Southwest consume an average of 44 percent of their renewable water, as compared to an average consumption of 4 percent of renewable water elsewhere in the United States (Figure 13–16). Historically, water in the West was used primarily for irrigation. However, with the rapid expansion of population in that region during the past 25 years, municipal, commercial, and industrial uses now compete heavily with irrigation for available water.

Until recently, the development of new sources of water met expanding water needs in the West and Southwest. Water was diverted from distant sources and transported via aqueducts—large conduits—to areas that needed it. As long ago as 1913, for example, Los Angeles started bringing in water from an area of California 400 km (250 miles) north, along the east side of the Sierra Nevada. Dams were built and water-holding basins created to ensure a year-round supply. These solutions are no longer viable, because the closest, most practical water sources have already been used and because the public, which has come to expect inexpensive water, opposes paying for costly solutions such as dams and aqueducts.

**The Colorado River Basin**  One of the most serious water supply problems in the United States is in the Colorado River basin. The river's headwaters are in Colorado, Utah, and Wyoming, and major tributaries—often collectively called the upper Colorado—extend throughout these states. The lower Colorado River runs through part of Arizona and then along the border between Arizona and both Nevada and California. It then crosses into Mexico and empties into the Gulf of California.

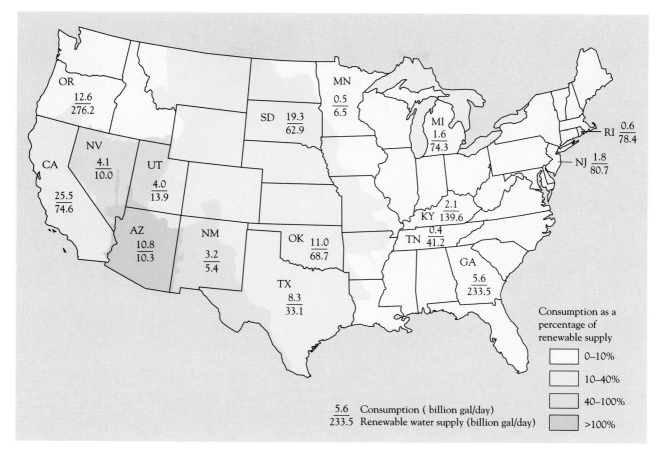

**Figure 13–16**
A comparison of the amount of water actually used (consumption) with the total renewable
supply gives an indication of areas with severe water problems. Note that consumption actually
exceeds renewable supply in the Southwest (the entire Colorado River basin).

An international agreement with Mexico,
along with federal and state laws, severely restricts
the use of the Colorado's waters. Traditionally, the
upper Colorado region appropriated little of the
water to which it was entitled, because it had few
people and little development. This made more
water available to the faster-developing lower Colorado region, but it also gave that area a false impression of the size of its water supply. Water is
diverted from the lower Colorado for the cities of
Tucson and Phoenix in Arizona as well as San
Diego and Los Angeles in California (Figure 13–
17a). Recent development in the upper Colorado
region is now threatening the lower Colorado region's water supply. Further, so much water is taken
from the lower Colorado by people in the states
through which it flows that the remainder is insufficient to meet Mexico's needs as set forth by international treaty (Figure 13–17b). To compound the

problem, as more and more water is used, the water
quality deteriorates, because the lower Colorado
becomes increasingly saline as it flows toward
Mexico.

**Groundwater** Roughly half the population of the
United States uses groundwater for drinking. Many
large cities, including Tucson, Miami, San Antonio, and Memphis, depend entirely or almost entirely on groundwater for their drinking water.
Groundwater is also used for industry and agriculture. Approximately 40 percent of the water used
for irrigation in the United States comes from
groundwater reservoirs.

Between 1945 and 1980, groundwater consumption in the United States increased fourfold,
from 79 billion to 334 billion liters (21 billion to
89 billion gallons) per day. Groundwater levels
have dropped in many areas of heavy use across the

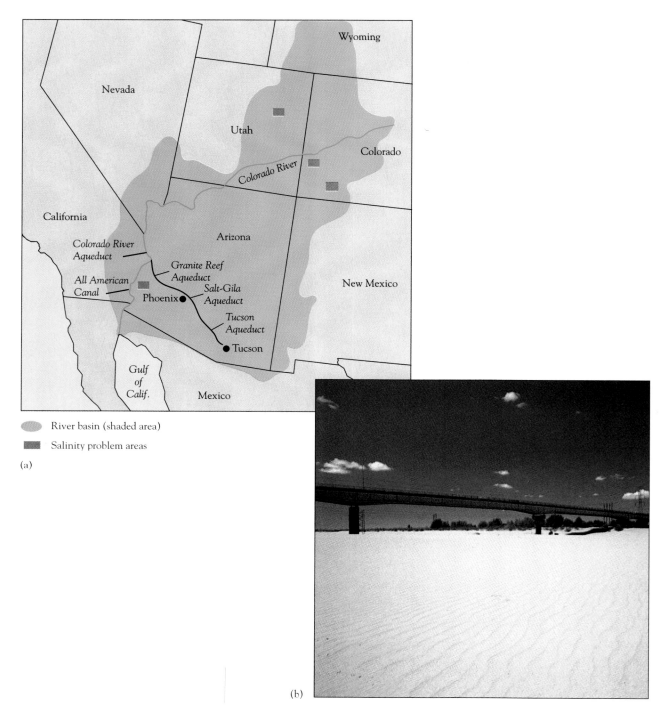

(a)

River basin (shaded area)

Salinity problem areas

(b)

**Figure 13–17**

Water resource problems in the Colorado River. The Lower Colorado River is extensively diverted (by way of aqueducts) to provide water for irrigation and for cities such as San Diego, Los Angeles, Phoenix, and Tucson. In addition, development of the Upper Colorado is decreasing water supplies to the Lower Colorado. Add Mexico's right by treaty to its fair share and you have an intense conflict over who gets what portion of this limited resource. (a) The location of aqueducts that draw water away from the Colorado River. Salinity is also a significant problem throughout the Colorado River basin. (b) As a result of diversion for irrigation and other uses in the United States, the Colorado River often dries up before reaching the Gulf of California in Mexico. (© 1990 Dan Lamont/Matrix)

United States and are predicted to drop even farther if high consumption continues. Aquifer depletion is particularly critical in three regions: southern Arizona, California, and the Great Plains (a band of states extending from Montana and North Dakota south to Texas) because so much groundwater has been withdrawn for irrigation (Figure 13–18).

In certain coastal areas of Louisiana and Texas, the removal of too much groundwater has resulted in the intrusion of salt water from the Gulf of Mexico. Salt water intrusion from the Pacific Ocean has occurred along parts of the California coast, along coastal areas of Puget Sound in northwestern Washington, and in certain areas of Hawaii. Florida and many coastal regions in the Northeast and Mid-Atlantic states also have salt water intrusion.

The quality of an area's groundwater varies from fresh to saline depending on the soil and rock characteristics of the area as well as the age of the water and its rate of recharge. Generally, groundwater is more saline in areas that have high evaporation rates and where saline water infiltrates (penetrates) through rocks such as carbonate and sandstone, as it does in parts of North Dakota and Minnesota. Certain areas in the United States have poor groundwater quality due to naturally high localized concentrations of salts of such chemicals as fluoride and arsenic. Some areas of Nevada and Utah have unusually high levels of toxic minerals in their groundwater as a result of natural conditions, rather than pollution caused by humans. (Chapter 21 discusses groundwater contamination caused by pollution.)

## Global Water Problems

Many developing countries have insufficient water to meet the most basic drinking and household needs of their people (Figure 13–19). The World Health Organization (WHO) estimates that 1 billion people lack access to safe drinking water and almost 2 billion are without access to a satisfactory means of domestic wastewater and fecal waste disposal. They risk disease because the water they consume is contaminated by sewage or industrial

**Figure 13–18**
Aquifer depletion is a widespread problem in the United States, particularly in the Great Plains, California, and southern Arizona.

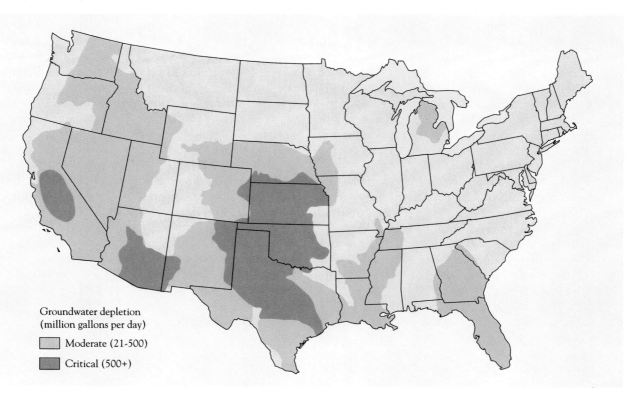

Groundwater depletion
(million gallons per day)

▢ Moderate (21-500)

▣ Critical (500+)

**Figure 13–19**
Girls carrying water pots in Burkina Faso, a country in west Africa. Obtaining water that is safe to drink is a time-consuming, labor-intensive job in many developing countries. Installation of a public water pipe is a relatively inexpensive improvement that can have a far-reaching impact on the community. (Robert E. Ford/Terraphotographics)

wastes. Many of these people have to travel great distances to secure the water they need; this practice consumes large amounts of time, particularly for women and children, and tends to perpetuate poverty. WHO also estimates that 80 percent of human illness results from insufficient water supplies and poor water quality caused by lack of sanitation. Although many developing countries have installed or are installing public water systems, population increases tend to overwhelm efforts to improve the water supply.

The United States is involved in efforts to improve water quality and supply in countries with critical water problems. The Agency for International Development (AID) manages projects in areas vulnerable to prolonged drought, such as the Sahel region in Africa. AID has assisted in well digging and other measures that alleviate the effects of drought in the Sahel. In addition to contributions by individual governments such as the United States, both the United Nations and the World Bank sponsor water management projects in developing countries.

As the world's population continues to increase, global water problems will become more se-

rious. Population growth is already outstripping water supplies in countries such as India, where approximately 8,000 villages have no local water. The water supply to some Indian cities—Madras, for example—has been so severely depleted that water is rationed from a public tap. Water supplies are also precarious in much of China, owing to population pressures. One-third of the wells in Beijing, for example, have gone dry, and the water table continues to drop. Mexico is facing the most serious water shortages of any country in the Western Hemisphere. The main aquifer supplying Mexico City, for instance, is dropping by as much as 3.5 m (11.2 ft) per year.

Shortages in global water supplies may also affect humans by limiting the amount of food they can grow. Recall that the main use of fresh water is for irrigation. As fresh water supplies are depleted by a growing human population, less water will be available for crops. The availability problem is compounded by salinization of agricultural soil as a result of irrigation practices. Local or even widespread famines from water shortages are a very real danger.

The decade of the 1990s may well see countries facing one another in armed conflict over water rights. One particularly troublesome spot is between Israel and Jordan, neighboring arid countries who have never been friendly. Both countries obtain fresh water from the Jordan River basin. Israel and Jordan both anticipate large population increases during the 1990s, which could make their water situation critical.[2] Neither country approves of the other increasing its allotment of water from the Jordan River, because an increase for one country would mean a smaller supply for the other.

The nations of Ethiopia and Sudan plan to divert some of the Nile River's flow to increase their water supplies. Because almost all of Egypt's water supply comes from the Nile, the actions of Ethiopia and Sudan could imperil Egypt's fresh water supply at a time when its population is increasing. Egypt may be able to increase its water supply through conservation. Reducing the water wasted during irrigation, for instance, would help save water; but installing technologically advanced water-saving irrigation systems is prohibitively expensive for a developing country. Efforts are under way at the United Nations to persuade the Nile River countries to develop a water-use agreement.

[2]Israel anticipates an influx of large numbers of emigrants from the former Soviet Union, whereas Jordan's high fertility rate will result in a large population increase.

# WATER MANAGEMENT

People have always thought of water as different from other resources. Resources such as coal or gold may be owned privately and sold as free-market goods, but we view water as public property.

Historically, both in the United States and in other countries, water rights were bound with land ownership. As more and more users compete for the same water, however, allocation decisions are increasingly being made by state or provincial governments. This is true in both developed and developing countries, although the details of state-level trusteeships vary. Some countries have separated land and water ownership so that water rights are sold separately.

Because rivers usually flow through more than one government jurisdiction, jurisdictions or states must create agreements with each other about the management of a river's resources. Such interstate cooperation permits comprehensive rather than piecemeal management of a river. In addition, these arrangements allow the water to be divided fairly between the jurisdictions, which then apportion their respective shares to individual users according to an established set of priorities.

Groundwater management is more complicated, in part because the extent of local groundwater supplies is not known. Some states manage groundwater, particularly where demand exceeds supply. Groundwater management includes issuing permits to drill wells, limiting the number of wells in a given area, and restricting the amount of water that may be pumped from each well.

The price of water varies, depending on how it is used. Historically, domestic use is most expensive and agricultural use is least expensive. In any case, the consumer rarely pays directly for the entire cost of water (including its transportation, storage, and treatment). State and federal governments heavily subsidize water costs, so we pay for some of the cost of water indirectly, through taxes.

The main goal of water management is to provide a sustainable supply of high-quality water. Increasingly, state and local governments are considering the *price* of water as a mechanism to help ensure an adequate *supply* of water. Raising the price of water to users so that it reflects the actual cost generally promotes more efficient use of water.

France is a model of water management because it has an effective, comprehensive regional water plan. France is divided into eight separate regions that are managed by government agencies with public and individual user representation. Water taxes finance additional water projects such as the construction of water treatment plants. Industrial users pay a pollution tax, but they can be exempted from this tax if they clean up the water after they use it and before they discharge it. This policy has substantially reduced water pollution in France. (For a discussion of federal water management policies in the United States, see Focus On: Landmarks in Federal Water Management Policy.)

## Dams and Reservoirs

Dams ensure a year-round supply of water in areas that have seasonal precipitation or snowmelt. Dams confine water in reservoirs, from which the flow is regulated (Figure 13–20). Dams have other

**Figure 13–20**
The Grand Coulee Dam and its reservoir, the Franklin D. Roosevelt Lake. Dams help to regulate water supply, storing water that is produced in times when precipitation is plentiful to be used during dry periods. The many beneficial uses of dams include electricity generation and flood control, but they also destroy the natural river habitat and are expensive to build. (U.S. Department of Energy)

## Landmarks in Federal Water Management Policy

Ever since the early 1960s, the U.S. government has attempted to shape a comprehensive water management policy. Following are some of the events that have signified changes in direction and attitude.

The Water Resources Research Act of 1964 was passed. This act established a Federal Office of Water Resources and Technology, as well as a water-resources research institute in each state to deal with local water management problems.

The Water Resources Planning Act of 1965 was passed. This act established the U.S. Water Resources Council and encouraged a cohesive, comprehensive federal attitude toward water resources management. The act also served as a base from which states could begin their own water resources planning.

The National Environmental Policy Act of 1970 and the Federal Water Pollution Control Act of 1972 (later amended as the Clean Water Act) shifted the focus of federal funding from dam and canal construction to environmental protection and restoration.

The National Water Commission released a 1973 report, *Water Policies for the*

*Future*, which focused on the need for protection of water resources rather than just water supply.

In the late 1970s President Carter released his water policy initiatives, which emphasized more efficient planning, management, and conservation of water resources; building of economical water projects; better cooperation between state and federal governments; and more attention to water quality.

In 1981 President Reagan's administration sought to shift most water resources planning and management from the federal government to state governments.

In 1987 the U.S. Bureau of Reclamation announced that water-resources programs would shift from irrigation development to management.

The United States is vast and encompasses widely divergent terrains and climates. While states and cities must deal with local water problems in the ways that are most appropriate for them, they must also be aware of the regional, and sometimes global, ramifications of their actions.

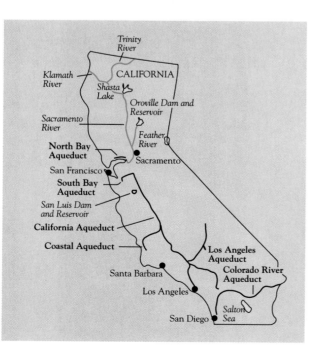

### Figure 13–21

Southern California, largely desert, relies on water diversion for the water needs of its millions of inhabitants. (a) The California Water Plan includes an extensive network of aqueducts to transfer large quantities of water to southern California. (b) An aqueduct in California. (Courtesy of the California Department of Water Resources)

(a)

(b)

benefits, including the generation of electricity (see Chapter 12). They provide flood control for areas downstream, because a reservoir can hold a large amount of excess water during periods of heavy precipitation and then release it gradually. Some of the reservoirs formed by dams also have recreational benefits: people swim, boat, and fish in them.

Many people, however, feel that the drawbacks of dams, including the cost of building them, far outweigh any benefits they provide (see Focus On: Damming the Big Muddy). Chapter 12 examines other problems associated with dams. (For a discussion of the impact of dams on natural fish communities, see Meeting the Challenge: The Columbia River—A Case Study in Water Management.)

## Water Diversion Projects

One way to increase the natural supply of water to a particular area is to divert water from areas where it is in plentiful supply. This is done by pumping water through a system of aqueducts. Much of southern California receives its water supply via aqueducts from northern California (Figure 13–21). Water from the Colorado River is also diverted into southern California.

Large-scale water diversion projects are controversial and expensive. The Central Arizona Project, recently completed at a cost of almost $4 billion, pumps water from the Colorado River to Phoenix and Tucson. As we saw earlier, damage is done to a river or other body of water when a major portion of its water is diverted. Pollutants, which would have been diluted in the normal river flow, reach higher concentrations when much of the flow has been removed. Wildlife, including fish, may decline in number and diversity. Although no one denies that people must have water, opponents of water diversion projects contend that serious water conservation efforts would eliminate the need for additional large-scale water diversion.

## Harvesting Icebergs

For well over a decade, visionaries in arid areas such as Saudi Arabia, California, and Australia have toyed with the idea of towing icebergs from the Antarctic or Arctic so they could be melted down to supply fresh water (Figure 13–22). Skeptics voice concerns about the ownership of icebergs, particularly in Antarctica. The effects on marine wildlife of introducing an iceberg into warm tropical waters is unknown.

The Department of Natural Resources in Alaska has begun issuing permits that allow icebergs to be collected. One entrepreneur plans to lift small icebergs from the sea, chop them into smaller pieces, and ship them to Japan, where they will be used as "gourmet" ice cubes. (Glacial ice makes better ice cubes than freezer ice because it is extremely pure and takes longer to melt.)

Australians are also serious about harvesting icebergs. They plan to wrap Antarctic icebergs in strong, lightweight fabrics to hold the water from melting during transport. Glacial ice will likely be cost-competitive with other fresh water sources in Australia, even taking into account the expense of harnessing, wrapping, and towing the icebergs.

## Desalinization

Seawater and saline groundwater can be made fit to drink through the removal of salt, called **desalinization** (or **desalination**), by several methods. One of the most common is **distillation**—heating the salt water until the water evaporates, leaving behind a crust of salt; the water vapor is then condensed to produce fresh water. Another method, called **reverse osmosis,** involves forcing salt water through a membrane that is permeable to water but not to salt.

Desalinization is expensive because it requires a large input of energy. Recent advances in reverse osmosis technology have increased its efficiency so that it requires much less energy than distillation.

(*text continues on p. 288.*)

**Figure 13–22**
Icebergs represent a source of very pure fresh water that certain arid nations wish to utilize. (Carolina Biological Supply Company)

# The Columbia River— A Case Study in Water Management

The Columbia River is the fourth largest river in North America. Its river basin, which covers an area the size of France, includes seven states and two Canadian provinces (see map). Such a large, complex river system has multiple uses. There are more than 100 dams within the Columbia River system, 19 of which are major generators of hydroelectric power. The Columbia River system supplies municipal and industrial water to several major urban areas, including Boise, Portland, Seattle, and Spokane. More than 1.2 million hectares (3 million acres) of agricultural land are irrigated with the Columbia's waters. Commercial ships navigate 805 kilometers (500 miles) of the river. Recreational uses of the river include boating, windsurfing, and swimming. The Columbia River system also offers sport and commercial fishing for salmon, steelhead trout, and other fish.

As is often the case in natural resource management, a particular use of the Columbia River system may have a negative impact on other uses. For example, the dam impoundments along the Columbia River that generate electricity and control floods have adversely affected fish populations, including salmon. Salmon are migratory fish. They spawn in the upper reaches of freshwater rivers and streams. The young offspring, called smolts, migrate to the ocean, where they spend most of their adult lives. Salmon complete their life cycle by returning to their place of birth to reproduce and die. The salmon population in the Columbia River system is only a fraction of what it was before the basin was developed. While numerous factors have contributed to their decline, the many dams that impede their migrations are the most significant.

Recognizing the need for more balanced management of the Columbia River system, the U.S. Congress passed the Northwest Power Act in 1980. This act provided the framework for the development of long-term plans to meet energy needs and rebuild fish populations. It is managed by federal and

The Columbia River basin covers seven states and two Canadian provinces.

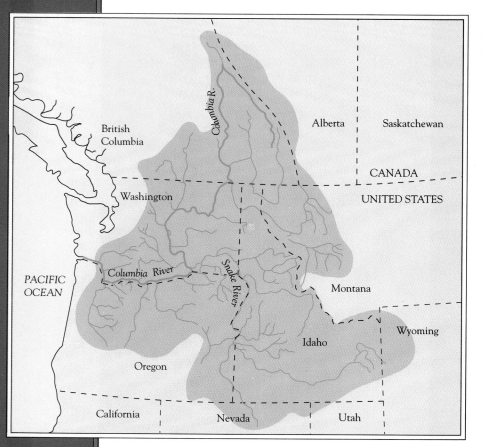

state agencies, North American Indian tribes with a vested interest in fish populations, and utility companies that generate hydroelectric power.

A number of projects to rebuild the fish populations were implemented during the 1980s and early 1990s, and much more action is planned. Many of the dams already had fish ladders, which enable some of the adult salmon to bypass the dams and continue their upstream migration (see photo). The downstream migration of the young smolts is even more perilous, however: of every five fish that start the trip, only about one makes it to the Pacific. Prior to extensive dam building, smolts made the downstream trip to the Pacific in a few days, carried by the heavy spring flush of water. Now the same trip may take several months, with many of the fish becoming stranded in the tranquil waters of the reservoirs. A substantial number of smolts are consumed by large predators in the reservoirs or destroyed by the turbine blades as they fall over the dams. The salmon population has shrunk so much that many of the old spawning grounds have been abandoned.

To increase the number of salmon, several new hatcheries are to be built *upstream* of the dams, in tributaries of the Columbia. Young fish produced at these hatcheries will be released to imprint the "smell" of the streams, enabling them to return there as adults to reproduce. It is hoped that this effort will reestablish natural spawning grounds. In addition, to protect some of the remaining natural habitats, a number of streams in the Columbia River system have been designated off limits for dam development. Special underwater screens and passages are being installed at dams to steer the smolts away from the turbines. Trucks transport some of the young fish around dams, while others swim safely over the dam because the electrical generators are periodically turned off to allow passage. One of the more interesting approaches is the establishment of a "water budget" for the young fish. Extra water, simulating spring snowmelt, is released from the dams to help wash the smolts downstream.

The Columbia River is a vital natural resource for the Pacific Northwest. Current efforts to manage it, which take into account such diverse interests as power production and fish populations, are commendable for several reasons. First, they demonstrate that groups with opposing interests can come together and develop balanced, sustainable solutions to complex issues. Second, the policies being formulated and implemented will provide a working model for the future management of other resources.

Fish ladder at the Bonneville Dam in Oregon. Fish ladders help migratory fish to bypass dams in their migration upstream. (Alan Pitcairn, from Grant Heilman)

## Damming the Big Muddy

The "Big Muddy" flows from western Montana to St. Louis, Missouri, where it joins the Mississippi River and flows on to the Gulf of Mexico. The Big Muddy, also known as the Missouri River, is 3,704 km (2,315 mi) long—the longest river in the United States. It contains North America's three largest reservoirs and the world's largest compacted-earth dam. In fact, the river has a series of dams, built by the Army Corps of Engineers. They have provided a number of benefits and caused a number of problems for people living along the river.

Since 1987 the Corps has opened up the northern dams in order to protect downstream navigation, including the shipping of $92 million of grain each year. In addition, people who live downstream count on the river for irrigation, electrical power, and individual water consumption. But the area along the northern Missouri River has suffered from several years of drought, and the shrinking northern reservoirs now threaten the region's multimillion-dollar fishing and tourism industry; in short, the livelihood of the northerners is on the line.

North Dakota, South Dakota, and Montana sued the Army Corps of Engineers for discharging water in an "arbitrary and capricious way." They won the decision, which was to have kept the Corps from releasing water from one of the large reservoirs in North Dakota, but the U.S. Court of Appeals reversed the ruling, and water flowed through the dams again.

The battle over water rights is becoming more heated and entangled as those who live upstream and those who live downstream fight to protect their interests. Although each side says it is willing to share, neither side seems willing to give up much. South Dakota, for example, wants the Corps to shorten the navigation season during years of drought so that upstream states will have enough. But residents downstream counter that South Dakota wants to hold the water in the reservoirs in order to maintain its recreation industry. Downstream states claim that because their water supply depends upon the dams, their needs should be favored over those of the fishing and tourism industry up north.

The Army Corps of Engineers has tried measures such as shortening the navigation season by about five weeks and providing extended boat ramps for those who use the reservoirs for recreation. But these measures are unlikely to appease either the upstreamers or the downstreamers.

---

Other expenses involved in desalinization projects include the cost of transporting the desalinized water from the site of production to where it will be used.

The disposal of salt produced by desalinization is also a concern. Simply dumping it back into the ocean, particularly near coastal areas that are highly productive, could cause a localized increase in salinity that would harm marine wildlife.

Water desalinization is a huge industry in the Middle East and North Africa, particularly in Saudi Arabia, because it is cost-competitive with alternative methods of obtaining fresh water in that arid region. The United States is also a major producer of desalted water—for example, at the Yuma Desalting Plant in Arizona. At present, however, desalinization is not a viable solution for the water supply problems of many developing nations, because it is too expensive.

## WATER CONSERVATION

The right to an unlimited supply of water at a reasonable cost has always been assumed automatically by most Americans. However, population and economic growth have placed an increased demand on

## Saving Water in The Magic Kingdom

In 1991, as part of its ongoing effort to conserve water, The Disney Corporation equipped nearly 200 washrooms in the Magic Kingdom of Disneyland in Anaheim, California, with infrared faucet sensors and toilet flushometers. The measure will save over 27 million gallons per year. In addition, all waterways in the theme park re-use their own water; gardens and landscapes are irrigated with low-volume sprinklers and drip irrigators; and grassed areas are gradually being replaced with "xeriscapes"—areas landscaped with rockery and plants that require little water.

**Figure 13–23**

Microirrigation, which delivers a precise amount of water directly to the plants' roots, eliminates much of the waste associated with traditional methods of irrigation. (USDA/Soil Conservation Service)

our water supply. Today there is more competition among water users whose priorities differ than ever before. Water conservation measures are necessary to guarantee sufficient water supplies. Most water users use more water than they really need, whether it is for agricultural, industrial, or direct personal consumption. With incentives, these users will lower their rates of water usage. Many studies have shown that higher prices for water provide the motivation to conserve water. For example, farmers are more likely to invest in water-saving irrigation technologies if the money saved from decreased water consumption covers the expense of the initial installation.

## Reducing Agricultural Water Waste

Irrigation generally makes inefficient use of water. Traditional irrigation methods, which have been practiced for more than 5,000 years, involve flooding the land or diverting water to fields through open channels. Water flow must be increased in order to guarantee that the far end of the field or higher elevations of the field receive water. Less than 50 percent of the water applied to the soil by such methods is absorbed by plants; the rest usually evaporates into the atmosphere.

One of the most important innovations in agricultural water conservation is **microirrigation,** also called **drip** or **trickle irrigation,** in which pipes with tiny holes bored in them convey water directly to individual plants (Figure 13–23). This reduces the water needed to irrigate crops by a substantial amount, usually 40 to 60 percent. Microirrigation also reduces the amount of salt left in the soil by irrigation water.

Another important water-saving measure in irrigation is the use of lasers to level fields. As a laser beam sweeps across a field, a field grader receives the beam and scrapes the soil, leveling it. Because farmers must use extra water to ensure that plants on higher elevations of a nonlevel field receive enough, laser leveling of the field reduces the water required for irrigation.

The use of sound water management principles in agriculture reduces water consumption. Traditionally, farmers have been allotted specific amounts of water at specific times, with a "use it or lose it" philosophy. This approach encourages waste. If, instead, water needs are carefully monitored (often through computer controls), water can be applied in small, regulated quantities, thereby reducing overall consumption.

Although advances in irrigation technology are improving the efficiency of water use, many challenges remain. For one thing, sophisticated irrigation techniques are prohibitively expensive. Few farmers in developed countries, let alone subsis-

## Conserving Water

Personal consumption of water has been increasing steadily in the United States for the past 30 years (see graph). Part of this increase stems from the greater use of appliances that need water, including dishwashers, garbage disposals, and washing machines. The growth of suburbs, with their expansive landscaping that requires watering, has also been responsible for increased water use.

The cumulative effect of many people practicing personal water conservation measures has a significant impact on overall water consumption. You can practice these yourself.

Many manufacturers make low-flow toilets. The Kohler model shown flushes with only 1.5 gallons of water. (Courtesy of Kohler Co.)

1. Install water-saving shower heads and faucets to cut down significantly on water flow.
2. Install a low-flush toilet or use a water displacement device in the tank of a conventional toilet (see photo). (Low-flush toilets use about 30 percent less water than conventional toilets.)
3. If you're in the market for a washing machine, front-loading washing machines require less water than top-loading models.
4. Modify your personal habits to conserve water. For example, allowing the faucet to run while shaving consumes an average of 20 gallons of water; you'll use only 1 gallon if you simply fill the basin with water or run the water only to rinse your razor. You may save as much as 10 gallons of water a day by wetting your toothbrush and then turning off the tap while you brush your teeth, as opposed to running the water during the entire process.

tence farmers in developing nations, can afford to install them.

## Reducing Water Waste in Industry

Electric power generators and many industries require water in order to function (recall that power plants heat water to form steam, which turns the turbines). In the United States, five major industries consume almost 90 percent of industrial water (not including water used for cooling purposes): chemical products, paper and pulp, petroleum and coal, primary metals, and food processing.

Stricter pollution control laws in many countries provide some incentive for industries to conserve water. Industries usually reduce their water use, and therefore their water treatment costs, by recycling water. The National Steel Corporation plant in Granite City, Illinois, for example, recycles approximately two-thirds of the 62 million gallons of water it uses daily; the used water is cleaned up before being discharged into a lake that spills into the Mississippi River.

It is likely that water scarcity, in addition to more stringent pollution control requirements, will encourage further industrial recycling. The poten-

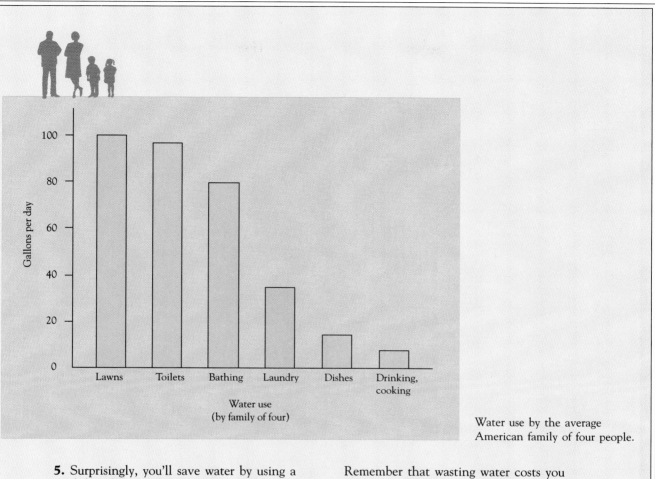

Water use by the average American family of four people.

**5.** Surprisingly, you'll save water by using a dishwasher instead of washing dishes by hand with the tap running—but only if you run the dishwasher with a full load of dishes.

Remember that wasting water costs you money. Conserving water at home reduces not only your water bill but also your heating bill: if you are using less hot water, you are also using less energy to heat that water.

tial for industries to conserve water by recycling is enormous.

### Reducing Municipal Water Waste

Like industries, regions and cities can reduce their water consumption by recycling or reusing water before it is discharged. For example, individual homes and buildings can be modified to collect and store "gray water"—water that has already been used in sinks and showers (Figure 13–24). The "gray water" can then be reused to flush toilets, wash the car, or sprinkle the lawn.

Israel probably has the world's most highly developed system of treating and reusing municipal wastewater. Israel does this out of necessity, because all of its possible fresh water sources have already been tapped. The reclaimed water is used for irrigation, which allows higher-quality fresh water to be channeled to cities. Used water contains pollutants, but most of these are nutrients from treated sewage and are therefore beneficial to crops.

Automated systems to purify and recycle wastewater have been developed and are cost-competitive with fresh water. In Tokyo, for example, the

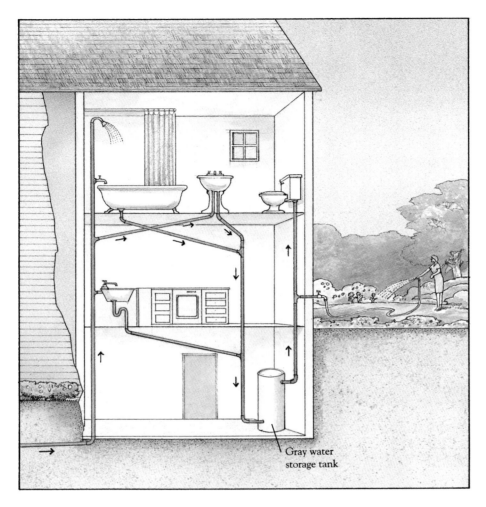

Gray water
storage tank

**Figure 13–24**
Individual homes and buildings can be modified to collect and store "gray water," water that has already been used in sinks and showers. This "gray water" can be used when clean water is not required, for example, in flushing toilets, washing the car, and sprinkling the lawn.

wastewater in the Mitsubishi office building is purified and recycled.

In addition to recycling and reuse, cities can decrease water consumption through other conservation measures, including consumer education, the use of water-saving household fixtures, and the development of economic incentives to save (see You Can Make a Difference: Conserving Water on pages 290–291). These measures have been used successfully to pull cities through dry spells; they are effective because individuals are willing to conserve for the common good during water crisis periods.

Increasingly, however, cities are examining ways to encourage individual water conservation methods all the time. The installation of water meters in residences in Boulder, Colorado, reduced water consumption by one-third. Before the installation, homeowners were charged a flat fee, regardless of their water use. In addition to installing water meters, a city might encourage water conservation by offering a rebate to any homeowner who installs a conserving device such as a water-saving toilet. The city of Tucson, Arizona, for example, has broken its citizens' trend toward increased water consumption and now needs 25 percent less water per person than it did in the 1970s. In arid towns such as Tucson, simply replacing grass lawns with native desert plants substantially reduces water use by making lawn irrigation unnecessary.

## INNOVATIVE APPROACHES TO WATER RESOURCE MANAGEMENT

Water management presents complex problems. We have seen that water resources are scarce in the western United States, where rapidly expanding urban populations have increased the need for water. At the same time, western farmers need irrigation water. Although everyone recognizes that outdated irrigation methods waste large quantities of water, farmers often cannot afford up-to-date irrigation technologies.

Sometimes an innovative approach to water management can provide solutions for several different parties. In January 1989, the Metropolitan Water District of Southern California (MWDSC) and the Imperial Irrigation District (IID) reached a mutually beneficial agreement on water needs. (MWDSC supplies water to growing coastal communities in southern California, whereas IID provides irrigation water to the highly productive Imperial Valley.) MWDSC agreed to pay for an updated irrigation system for IID. In exchange, the water that will be saved as a result of more efficient irrigation will be given to MWDSC for municipal use; it will probably be enough to meet the needs of 1 million people.

# SUMMARY

**1.** Water has a number of special properties, including a high heat capacity and a high dissolving ability. Many of the properties of water are the result of its polarity, which causes hydrogen bonding between water molecules.

**2.** Although a large portion of the Earth is covered by water, only a small percentage of it is available as fresh water. Fresh water occurs as groundwater, which is stored in aquifers, and as surface water.

**3.** Flooding occurs as a result of too much fresh water entering a particular area. Flood damage is exacerbated by the deforestation of hillsides and mountains and by the development of flood plains.

**4.** Although there is enough fresh water to support all the living organisms on Earth, it is not evenly distributed. Some areas are barely able to support human life because of the shortage of water.

**5.** Water shortages can be dealt with in two ways. The amount of available water can be increased by damming rivers, tapping additional groundwater, diverting additional water from other areas, or desalinization. Alternatively, water conservation measures can make the present supply adequate.

**6.** Irrigation is the single largest use of water. Although irrigation increases agricultural productivity, it can lead to water pollution, soil salinization, and depletion of water supplies. New agricultural techniques, including microirrigation and field leveling, have significantly cut agricultural water consumption.

**7.** After agriculture, the two largest users of water are industry and domestic/municipal water use. Recycling and reuse can reduce both industrial and municipal water consumption.

**8.** Water management of both surface water and groundwater should have the long-term goal of developing a sustainable resource rather than the short-term goal of providing water in limitless supply.

# DISCUSSION QUESTIONS

**1.** Discuss the significance of the dissolving ability of water, for (a) ocean salinity and (b) water pollution.

**2.** Are our water supply problems simply the result of too many people? Give reasons why you support or negate this idea.

**3.** Discuss the problems that result from withdrawing too much groundwater from a particular area.

**4.** Imagine you are a water manager for a southwestern metropolitan district with a severe water shortage. What strategies would you use to develop a sustainable water supply?

**5.** How is water used in agriculture? Discuss two ways agricultural water can be conserved.

**6.** Which industries consume the most water? Discuss one way in which water can be conserved in industry.

**7.** Water used by cities and industries is not considered consumed, but irrigation water *is* consumed. Explain the distinction and its significance.

**8.** Create a detailed water conservation plan for your own personal daily use. How do you think you could save the most water?

# SUGGESTED READINGS

Babbitt, B. Age-old Challenge: Water and the West. *National Geographic*, June 1991. A comprehensive examination of water problems in the western United States.

Brown, L. The Aral Sea: Going, Going. *World Watch* 4:1, January/February 1991. The removal of water that once emptied into the Aral Sea is becoming increasingly costly from environmental and human viewpoints.

Duplaix, N. South Florida Water: Paying the Price. *National Geographic*, July 1990. How recent development in southern Florida has adversely affected water supply and quality.

Glantz, M. H. Drought in Africa. *Scientific American*, June 1987. Discusses the recurring droughts of the sub-Saharan region and considers ways in which they might be managed.

Hunt, C. E., and V. Huser. *Down By the River*, Island Press, Washington, D.C., 1988. Examines the impact of federal water policies and management on biological diversity in a number of important U.S. river systems.

Mono Basic Ecosystem Study Committee, Board on Environmental Studies and Toxicology, Commission on Physical Sciences, Mathematics and Resources; and National Research Council. *The Mono Basic Ecosystem: Effects of Changing Lake Level*. National Academy Press, Washington, D.C., 1987. A case study of the ecological effects of water diversion from Mono Lake.

Muckleston, K. W. Striking a Balance in the Pacific Northwest. *Environment* 32:1, 1990. Examines the salmon-versus-hydropower issue of the Columbia River system.

Postel, S. Trouble on Tap. *World Watch* 2:5, 1989. Discusses water issues in the Middle East, Egypt, India, China, and the former Soviet Union.

Viessman, W., Jr. A Framework for Reshaping Water Management. *Environment* 32:4, May 1990. Considers how water management policies in the United States should be changed to meet future water needs.

World Resources Institute. *World Resources 1992–93: A Guide to the Global Environment*. Oxford University Press, New York, 1992. Includes tables of freshwater data for all countries.

# Chapter 14

*Amish farmers protect the soil by utilizing many soil conservation practices. (Larry Lefever, from Grant Heilman)*

# Soils and Their Preservation

Soil, which is composed of mineral particles, organic material, water, and air, is a valuable natural resource upon which humans depend for food. Different soil types form as a result of complex interactions among rock, climate, living organisms, surface topography, and time. Many human activities create or accentuate soil problems including erosion and mineral depletion. Farming in arid or semiarid areas, for example, can increase wind erosion. Although all soils that are farmed experience mineral depletion, this problem is particularly acute when tropical rain forests are removed for agriculture. The goals of soil conservation are to minimize soil erosion and maintain soil fertility so that this resource can be used in a sustainable fashion.

**Table 14–1
Essential Elements for Plant Growth**

| Element | Source | Significant Function in Plants |
|---|---|---|
| Carbon | Air (as $CO_2$) | Part of important biological molecules |
| Hydrogen | Water | Part of important biological molecules |
| Oxygen | Water, air (as $O_2$) | Part of important biological molecules |
| Nitrogen | Soil | Part of important biological molecules |
| Phosphorus | Soil | In genetic molecules and energy molecules |
| Calcium | Soil | Part of cell walls |
| Magnesium | Soil | In chlorophyll molecules |
| Sulfur | Soil | In proteins and vitamins |
| Potassium | Soil | Involved in ionic balance of cells |
| Chlorine | Soil | Involved in ionic balance of cells |
| Iron | Soil | Involved in photosynthesis and respiration |
| Manganese | Soil | Involved in respiration and nitrogen metabolism |
| Copper | Soil | Involved in photosynthesis |
| Zinc | Soil | Involved in respiration and nitrogen metabolism |
| Molybdenum | Soil | Involved in nitrogen metabolism |
| Boron | Soil | Exact role unclear |

## WHAT IS SOIL?

Soil is the ground underfoot, a thin layer of the Earth's crust that has been modified by the natural actions of agents such as weather and organisms. It is easy to take soil for granted. We walk on and over it throughout our lives, but rarely stop to think about how important it is to our survival.

Vast numbers and kinds of organisms inhabit soil and depend on it for shelter and food. Plants anchor themselves in soil, and from it they receive essential minerals and water. Thirteen of the 16 different elements essential for plant growth are obtained directly from the soil (Table 14–1). Terrestrial plants could not survive without soil, and because we depend on plants for our food, humans could not exist without soil, either.

### How Soils Are Formed

Soil is formed from rock (called parent rock) that is slowly broken down into smaller and smaller particles by chemical and physical **weathering processes** in nature. It takes a very long time, sometimes thousands of years, for rock to disintegrate into finer and finer mineral particles. Time is also required for organic material to accumulate in the soil. Soil formation is a continuous process that involves interactions between the Earth's solid crust and the biosphere (Figure 14–1). The weathering of parent rock beneath soil that has already formed continues to add new soil.

Living organisms and climate both play essential roles in weathering, sometimes working together. For example, soil organisms produce acids that etch tiny cracks in the rock. In temperate climates, water from precipitation seeps into these cracks, which enlarge when the water freezes. Over many seasons, alternate freezing and thawing cause small pieces of the rocks to break off.

**Topography,** a region's surface features—such as the presence or absence of mountains and valleys—is also involved in soil formation. Steep slopes often have very little or no soil on them because soil and rock are continually transported down the slopes by gravity; runoff from precipitation tends to amplify erosion on steep slopes. Moderate slopes, on the other hand, may encourage the formation of deep soils.

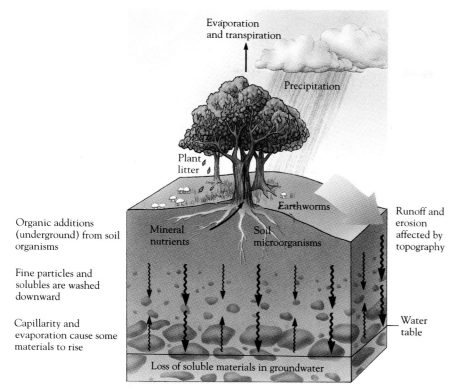

Evaporation
and transpiration

Precipitation

Plant
litter

Earthworms

Organic additions
(underground) from soil
organisms

Mineral
nutrients

Soil
microorganisms

Runoff and
erosion
affected by
topography

Fine particles and
solubles are washed
downward

Capillarity and
evaporation cause some
materials to rise

Water
table

Loss of soluble materials in groundwater

**Figure 14–1**
Weather, climate, topography, and
living organisms interact with the
Earth's crust to form soil, the complex material that supports life on
land.

## SOIL STRUCTURE

Soil is composed of four distinct elements—
mineral particles, organic matter, water, and air—
and occurs in layers, each of which has a certain
composition and special properties. The plants,
animals, and microorganisms that inhabit soil interact with it, and minerals are continually cycled
from the soil to living organisms and back to the
soil.

### Components of Soil

The mineral portion, which comes from weathered
rock, forms the basic soil material. It provides anchorage and essential minerals for plants, as well as
pore space for water and air. Because the mineral
compositions of rocks vary at different locations,
the soils that develop from them vary in mineral
composition and chemical properties.[1]

The age of a soil also affects its mineral composition. In general, older soils are more weathered
and lower in certain essential minerals. Large por-

tions of Australia, South America, and India have
old, infertile soils. In contrast, in geologically recent time, glaciers passed across much of the
Northern Hemisphere, pulverizing bedrock and
forming fertile soils.[2] Essential minerals are readily
available in these geologically young soils and in
young soils formed in areas of volcanic activity.

The litter, droppings, and remains of plants,
animals, and microorganisms in various stages of
decomposition constitute the organic portion of
soils. Microorganisms, particularly bacteria and
fungi, gradually decompose this material, and the
black or dark brown organic material that remains
after much decomposition has occurred is called
**humus** (Figure 14–2). Certain components of
humus may persist in the soil for hundreds of years.
Although humus is somewhat resistant to further
decay, a succession of microorganisms gradually
reduces it to carbon dioxide, water, and minerals.
Detritus-feeding animals such as earthworms, termites, and ants also help break down humus.

As organic material is decomposed, essential
minerals are released into the soil, where they may

[1] Rocks that are rich in aluminum form acidic soils, for
example, whereas rocks that contain silicates of magnesium and
iron form soils that may be deficient in calcium, nitrogen, and
phosphorus.

[2] The Pleistocene Epoch, which began approximately 2 million years ago, was marked by four periods of glaciation. At their
greatest extent, these ice sheets covered nearly 4 million square
miles of North America, extending south as far as the Ohio and
Missouri rivers.

**Figure 14–2**
Humus is partially decomposed organic material (primarily from plant and animal remains). Soil that is rich in humus has a loose, somewhat spongy structure with several properties (such as increased water holding capacity) that are beneficial for plants and other organisms living in it. (USDA/Soil Conservation Service)

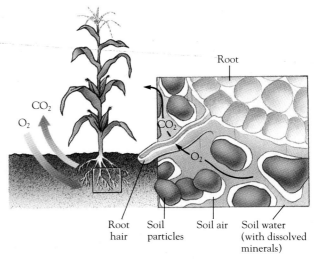

**Figure 14–3**
Pore space in soil is occupied by varying amounts of soil air (including oxygen and carbon dioxide) and soil water. Both oxygen and water are absorbed and carbon dioxide is released by plant roots.

be absorbed by plant roots. Organic matter also increases the soil's water-holding capacity by acting like a sponge.

The pore space between soil particles occupies roughly 35 to 60 percent of a soil's volume and is filled with varying proportions of water (called **soil water**) and air (called **soil air**) (Figure 14–3); both are necessary to sustain all the organisms living in the soil.

Soil water contains low concentrations of dissolved mineral salts that enter the roots of plants as they absorb the water. Soil water not absorbed by plants moves down through the soil, carrying dissolved minerals with it. The removal of dissolved materials from the soil by water percolating downward is called **leaching.** The deposit of leached material in the lower layers of soil is known as **illuviation.** Iron and aluminum compounds, humus, and clay are some illuvial materials that can gather in the subsurface portion of the soil. Some substances completely leach out of the soil because they are so soluble that they migrate all the way down to the groundwater. It is also possible for water to move *upward* in the soil, transporting dissolved materials with it, as when the water table rises.

Soil air contains the same gases as atmospheric air, although they are usually present in different proportions. Generally, as a result of respiration by soil organisms, there is more carbon dioxide and

less oxygen in soil air than in atmospheric air. Among the important gases in soil air are (1) oxygen, required by soil organisms for respiration; (2) nitrogen, used by nitrogen-fixing soil organisms (see Chapter 5); and (3) carbon dioxide, involved in soil weathering.[3]

## Soil Horizons

A deep vertical slice, or section, through many soils reveals that they are organized into horizontal layers called **soil horizons.** A **soil profile** is a section from surface to parent rock, showing the horizons (Figure 14–4).

The uppermost layer of soil, the **O-horizon,** is rich in organic material. Plant litter, including dead leaves and stems, accumulates in the O-horizon and gradually decays. In desert soils the O-horizon is often completely absent, but in certain organic-rich soils it may be the dominant layer.

Just beneath the O-horizon is the topsoil, or **A-horizon,** which is dark and rich in accumulated humus. The A-horizon has a granular texture and is somewhat nutrient-poor due to the gradual loss of many nutrients to deeper layers by leaching.

The **B-horizon,** the light-colored layer beneath the A-horizon, is often a zone of illuviation in

[3]$CO_2$ dissolves in water to form carbonic acid, $H_2CO_3$, a weak acid that helps to weather rock.

**Figure 14-4**
Many soils have multiple horizons or layers that are revealed in a soil profile. Each horizon has its own chemical and physical properties. This particular soil, located on a farm in Virginia, has no O-horizon because it is used for agriculture; the surface litter that would normally comprise the O-horizon was plowed into the A-horizon. The shovel gives an idea of the relative depths of each horizon. (Photo by Ray Weil, courtesy of Martin Rabenhorst)

which minerals that leached out of the topsoil and litter accumulate. It is typically rich in iron and aluminum compounds and clay.

Beneath the B-horizon is the **C-horizon,** which contains weathered pieces of rock and borders the solid parent rock. The C-horizon is below the extent of most roots and is often saturated with groundwater.

## Soil Organisms

Although soil organisms are usually hidden underground, their numbers are huge. Millions of soil organisms may inhabit just 1 teaspoon of fertile agricultural soil! In the soil ecosystem, bacteria, fungi, algae, worms, protozoa, insects, plant roots, and larger animals such as moles, snakes, and groundhogs all interact with each other and with the soil (Figure 14-5).

Earthworms, probably one of the most familiar soil inhabitants, eat soil and obtain energy and raw materials by digesting humus. **Castings,** bits of soil that have passed through the gut of an earthworm, are deposited on the soil surface. In this way, minerals from deeper layers in the soil are brought to upper layers. Earthworm tunnels serve to aerate the soil, and the worms' waste products and corpses add organic material to deeper layers of the soil.

Ants live in the soil in enormous numbers, constructing tunnels and chambers that help to aerate the soil. Members of soil-dwelling ant colonies forage on the surface for bits of food, which they carry back to their nests. Not all of this food is eaten, however, and its eventual decomposition helps increase the organic matter in the soil. Many ants are also indispensable in plant reproduction because they bury plant seeds in the soil. Seeds buried by ants typically have special structures, called oil bodies, that are very nutritious. Ants bring the seeds underground, eat the oil bodies, and dispose of the rest of the seeds in underground refuse piles, along with their droppings and members of the colony who have died. Thus, the ants not only bury the seeds away from animals that might eat them, but also place the seeds in well-fertilized soil that is ideal for seed germination and seedling growth.

Plants are greatly affected by the properties of soil, although most plants can tolerate a wide range of soil types. Soil, in turn, is affected by the types of plants that grow on it (Figure 14-6). As a result of the complex interactions among plants, climate, and soil, it is hard to specify cause and effect in their relationships. For example, are the plants growing in a certain locality because of the soil that is found there, or is the soil's type determined by the plants?

**Figure 14–5**
The diversity of life in fertile soil is remarkable. It includes plants, algae, fungi, earthworms, flatworms, roundworms, insects, spiders and mites, bacteria, and burrowing animals such as moles and groundhogs.

One very important symbiotic relationship in the soil occurs between fungi and the roots of vascular plants. These associations, called **mycorrhizae,** enable plants to absorb adequate amounts of essential minerals from the soil. The threadlike body of the fungal partner, called a **mycelium,** ex- tends into the soil well beyond the plant root. Min- erals absorbed from the soil by the fungus are trans- ferred to the plant, while food produced by photosynthesis in the plant is delivered to the fun- gus. Mycorrhizae have been demonstrated to en- hance the growth of plants (Figure 14–7).

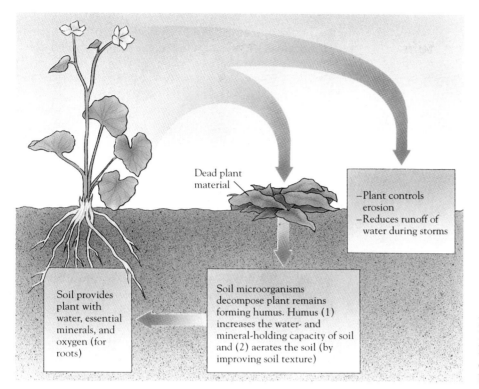

Sand - large
Silt - med
clay - small

Dead plant material

−Plant controls erosion
−Reduces runoff of water during storms

Soil provides plant with water, essential minerals, and oxygen (for roots)

Soil microorganisms decompose plant remains forming humus. Humus (1) increases the water- and mineral-holding capacity of soil and (2) aerates the soil (by improving soil texture)

**Figure 14–6**
Plants and the soils they grow in have a mutual relationship, with each contributing to the well-being of the other.

### Mini-Glossary of Soil Terms

**humus:** Partly decomposed organic material in the soil; brown or black in color.

**leaching:** The movement of water and dissolved material downward through the soil.

**illuviation:** The deposit of a material into a lower layer of the soil from a higher layer. Illuviation is the result of leaching.

**sand:** Large (0.05- to 2-mm-diameter) inorganic particles in the soil.

**silt:** Medium-sized (0.002- to 0.05-mm-diameter) inorganic particles in the soil.

**clay:** Small (<0.002-mm-diameter) inorganic particles in the soil.

**Figure 14–7**
Mycorrhizae, which are associations between a fungus and the roots of vascular plants, enhance the growth of plants. The two soybeans on the right have formed mycorrhizae and the soybean on the left has not. All three plants were grown under identical conditions. Mycorrhizae apparently absorb nutrients from the soil very efficiently and transfer these essential minerals to the plant. (Visuals Unlimited)

## Nutrient Recycling

In a balanced ecosystem, the relationship between soil and the organisms that live in and on it ensures soil fertility. As we saw in Chapter 5, essential minerals such as nitrogen and phosphorus are cycled from the soil to organisms and back again to the soil (Figure 14–8). Microorganisms such as bacteria and fungi decompose plant and animal detritus and wastes, releasing nutrients into the soil to be used again. Although leaching causes some minerals to be lost from the soil ecosystem to groundwater, the weathering of the parent rock replaces much or all of them.

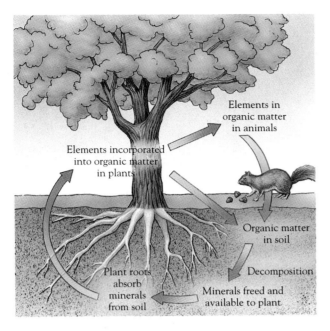

**Figure 14–8**
In a balanced ecosystem, nutrients cycle from the soil to living organisms and then back again to the soil.

## PHYSICAL PROPERTIES OF SOIL

Texture and acidity are two parameters that help to characterize soils.

### Soil Texture

The texture, or structural character, of a soil is determined by the amounts (percentages by weight) of different-sized inorganic particles—sand, silt, and clay. Large particles (0.05 to 2 mm in diame-

ter) are called **sand,** medium-sized particles (0.002 to 0.05 mm in diameter) are called **silt,** and small particles (<0.002 mm in diameter) are called **clay.** Sand particles are large enough to be seen easily with the eye, silt particles (about the size of flour particles) are barely visible with the eye, and clay particles are too small to be seen with an ordinary light microscope. These size assignments for sand, silt, and clay are arbitrary; they give soil scientists a way to classify soil texture. Obviously, soil particles form a continuum of sizes from very large to very small. Soil always contains a mixture of different-sizes particles, but the proportions vary from soil to soil. **Loam,** which makes ideal agricultural soil, has approximately equal portions of sand, silt, and clay.

A soil's texture affects many of that soil's properties, which, in turn, influence plant growth (Table 14–2). Generally, the larger particles provide structural support, aeration, and permeability to the soil, whereas the smaller particles bind together into aggregates, or clumps, and hold nutrients and water.

Clay is particularly important in determining many soil characteristics because it has the greatest surface area for chemical reactions. (If the surface areas of about 450 g [1 lb] of clay particles were laid out side by side, they would occupy 2.5 hectares [1 acre].) Each grain of clay has negative charges on its outer surface that attract and bind positively charged mineral ions (Figure 14–9).[4] Many of these mineral ions, such as potassium ($K^+$) and magnesium ($Mg^{2+}$), are essential for plant growth and are "held" in the soil for plant use by their interactions with clay particles.

---

[4]Soil minerals are often present in charged forms called **ions.** Mineral ions may be positively charged ($K^+$, for example) or negatively charged ($NO_3^-$, for example).

| | Soil Textural Type | | |
|---|---|---|---|
| **Table 14–2** **Soil Properties Affected by Soil Texture** | | | |
| **Soil Property** | *Sandy soil* | *Loam* | *Clay soil* |
| Aeration | Excellent | Good | Poor |
| Drainage | Excellent | Good | Poor |
| Nutrient-holding capacity | Low | Medium | High |
| Water-holding capacity | Low | Medium | High |
| Workability (tillage) | Easy | Moderate | Difficult |

can be absorbed by the plant, whereas insoluble forms cannot. At a low pH, for example, the aluminum and manganese in soil water are more soluble, sometimes becoming toxic to plants because they are absorbed by the roots in greater concentrations than are good for the plant. Other mineral salts that are essential for plant growth, such as calcium phosphate, become less soluble at higher pH's.

Soil pH greatly affects the availability of nutrients. An acidic soil has less ability to bind positively charged ions to it. As a consequence, certain mineral ions that are essential for plant growth, such as potassium ($K^+$), are leached more readily from acidic soil. The optimum soil pH for most plant growth is 6.5 to 7.5, because most nutrients needed by plants are available to the plants in that pH range.

Soil pH is affected by the types of plants growing on it. Soil litter composed of the needles of conifers, for example, contains acids that leach into the soil, lowering its pH.

## MAJOR SOIL TYPES

Variations in climate, vegetation, parent rock, topography, and soil age throughout the world result in many soil types that differ in color, depth, acidity, and other properties. Here we focus on five soil types that are very common: spodosols, alfisols, mollisols, aridisols, and oxisols (Figure 14–10).

Regions with colder temperate climates, ample precipitation, and good drainage typically have soils called **spodosols,** with very distinct layers. A spodosol usually forms on the floor of a coniferous forest and has a thin layer of acidic litter composed primarily of needles; an ash-gray acidic, leached A-horizon; and a dark brown illuvial B-horizon. Spodosols do not make good farmland because they are too acidic and are nutrient-poor due to leaching.

Temperate deciduous forests grow over **alfisols,** soils with a brown to gray-brown A-horizon. Precipitation is great enough to leach much of the mineral matter out of the O- and A-horizons. When the deciduous forest is intact, soil fertility is maintained by a continual supply of plant litter such as leaves and twigs. When the soil is cleared for farmland, however, fertilizers (which contain nutrients such as nitrogen, potassium, and phosphorus) must be used to maintain fertility.

**Mollisols,** found primarily in temperate, semiarid grasslands, are very fertile soils. They possess a thick, dark brown to black A-horizon that is rich in humus. Much of the mineral matter remains in the upper layers, because precipitation is not great enough to leach it into lower layers. Most of the world's grain crops are grown on mollisols.

**Aridisols** are found in arid regions of North America, South America, and Africa. The lack of precipitation in these deserts precludes much leaching, and the lack of lush vegetation precludes the accumulation of organic matter. As a result, aridisols do not have distinct layers. Some aridisols provide rangeland for grazing animals, and crops can be grown on aridosols if water is supplied by irrigation.

**Oxisols,** which are low in nutrients, exist in tropical and subtropical areas with ample precipitation. Little organic material accumulates in the A-horizon because humus is rapidly decomposed. The B-horizon, which is quite thick, is highly leached, acidic, and nutrient-poor. Oddly enough, tropical rain forests, with their lush vegetation, grow on oxisols. Most of the minerals in tropical rain forests are locked up in the vegetation rather than in the soil. As soon as plant and animal remains touch the forest floor, they promptly decay, and the minerals are quickly reabsorbed by plant roots. Even wood, which may take years to be recycled in temperate soils, is decomposed rapidly (in a matter of months) in tropical rain forests by subterranean termites.

### Overview of Soil Types in the United States

Figure 14–11 illustrates the major soil types in the United States. Spodosols occur across the northern Great Lakes region and the extreme Northeast, areas dominated by coniferous forests. The eastern temperate region south of the spodosol area has deciduous forests and mixed forests of deciduous and coniferous trees, which grow on alfisols. Mollisols, which support grassland prairies, are found across the central part of the United States. In a large area to the west of the mollisol region, the soils are predominantly aridisols, supporting sagebrush communities to the north and other types of desert plants to the south.

## SOIL PROBLEMS

Human activities often create or exacerbate soil problems including erosion and mineral depletion of the soil, both of which occur worldwide.

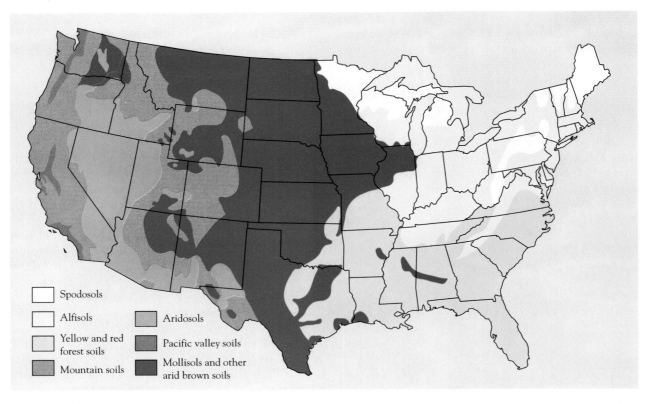

**Figure 14–11**
Generalized soil map of the United States. Spodosols, alfisols, mollisols, and aridisols are discussed in the text.

## Soil Erosion

Water, wind, ice, and other agents promote **soil erosion,** the wearing away or removal of soil from the land. Water and wind are particularly effective in removing soil. Rainfall loosens soil particles, which can then be transported away by moving water (Figure 14–12). Wind loosens soil and blows it away, particularly if the soil is barren and dry.

Because erosion reduces the amount of soil in an area, it limits the growth of plants. Erosion also causes a loss of soil fertility, because essential minerals and organic matter that are part of the soil are also removed. As a result of these losses, the productivity of eroded agricultural soils drops, and more fertilizer must be used to replace the nutrients lost to erosion.

Soil erosion is a national and international problem that rarely makes the headlines. To get a feeling for how serious the problem is, consider that approximately 2.7 billion metric tons (3.0 billion tons) of topsoil are lost *each year* from U.S. farmlands as a result of soil erosion. The U.S. Department of Agriculture estimates that approximately one-fifth of U.S. cropland is vulnerable to soil erosion damage.

Humans often accelerate soil erosion with poor soil management practices. Here we consider soil erosion caused by agriculture in some detail, but it is important to realize that agriculture is not the only culprit. Removal of natural plant communities during the construction of roads and buildings also accelerates erosion. Unsound logging practices, such as clearcutting large forested areas, cause severe erosion (Figure 14–13).

Soil erosion has an impact on other resources as well. Sediment that gets into streams, rivers, and lakes affects water quality and fish habitats. If the sediment contains pesticide and fertilizer residues, they further pollute the water. Also, when forests within the watershed for a hydroelectric power facility are removed, accelerated soil erosion can cause the reservoir behind the dam to fill in with sediment much faster than usual. This process results in a reduction of electricity production at that facility.

Sufficient plant cover limits the amount of soil erosion: leaves and stems cushion the impact of

(a)

**Figure 14–12**
Soil erosion caused by water. (a) Runoff from an open field carries soil sediments, fertilizers, and pesticides with it. (b) Gullies form as a result of extensive erosion damage from water. Gullies grow rapidly because they are prone to greater rates of erosion as they provide channels for the runoff of water when it rains. (a, Visuals Unlimited/T. McCabe; b, USDA/Soil Conservation Service)

(b)

rainfall, and roots help to hold the soil in place. Although soil erosion is a natural process, abundant plant cover makes it negligible in many natural ecosystems.

**Wind Erosion in Grasslands**  Semiarid lands, such as the Great Plains of North America, have low annual precipitation and are subject to periodic droughts. Prairie and steppe grasses, the plants that grow best in semiarid lands, are adapted to survive droughts.[5] Although the aboveground portions of the plant may die, the root systems can survive several years of drought. When the rains return, the root systems send up new leaves. Soil erosion is minimal because the dormant but living root systems hold the soil in place and resist the assault by water and wind.

The soils of semiarid lands are of very high quality, due largely to the accumulation of a thick, rich humus over many centuries. These lands are excellent for grazing and for growing crops on a small scale. Problems arise, however, when large areas of land are cleared for crops or when the land is overgrazed by animals (see Chapter 17). The

removal of the natural plant cover opens the way for climatic conditions to "attack" the soil, and it gradually deteriorates from the onslaught of hot summer sun, occasional violent rainstorms, and wind. If a prolonged drought occurs under such conditions, disaster can strike.

**Figure 14–13**
Erosion of hillsides in Ecuador occurred after the forest was removed. Clear cutting forests on steep slopes causes soil erosion in the drainage basin of the slopes. (Doug Wechsler)

---

[5] Prairies are tallgrass communities, whereas steppes, or plains, are shortgrass communities. Differences in annual precipitation account for the differences in species composition. Prairies receive more precipitation than steppes.

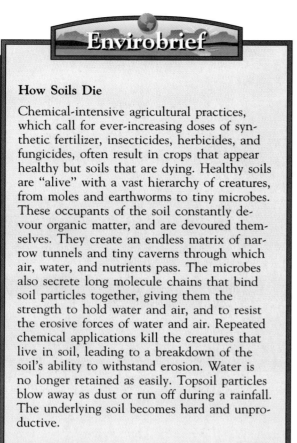

### How Soils Die

Chemical-intensive agricultural practices, which call for ever-increasing doses of synthetic fertilizer, insecticides, herbicides, and fungicides, often result in crops that appear healthy but soils that are dying. Healthy soils are "alive" with a vast hierarchy of creatures, from moles and earthworms to tiny microbes. These occupants of the soil constantly devour organic matter, and are devoured themselves. They create an endless matrix of narrow tunnels and tiny caverns through which air, water, and nutrients pass. The microbes also secrete long molecule chains that bind soil particles together, giving them the strength to hold water and air, and to resist the erosive forces of water and air. Repeated chemical applications kill the creatures that live in soil, leading to a breakdown of the soil's ability to withstand erosion. Water is no longer retained as easily. Topsoil particles blow away as dust or run off during a rainfall. The underlying soil becomes hard and unproductive.

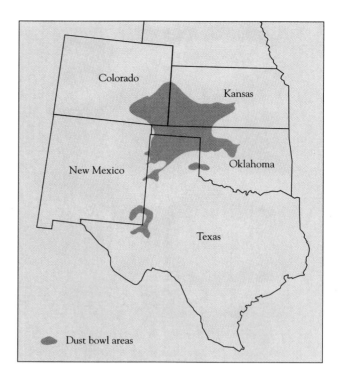

**Figure 14–14**
More than thirty million hectares (74 million acres) of land in the Great Plains were damaged during the Dust Bowl years. Shaded parts of Colorado, Kansas, Oklahoma, and Texas suffered the most extensive damage.

**The American Dust Bowl**  The effects of wind on soil erosion were vividly experienced throughout several western states during the 1930s (Figure 14–14). Throughout the late 19th and early 20th centuries, much of the native grasses had been removed to plant wheat. Then, between 1930 and 1937, the semiarid lands stretching from Oklahoma and Texas into Canada received 65 percent less annual precipitation than was normal. The rugged prairie and steppe grasses that had been replaced by crops could have survived these conditions, but not the wheat. The prolonged drought caused crop failures, which left fields barren and particularly vulnerable to wind erosion.

Winds from the west swept across the barren, exposed soil, causing dust storms of incredible magnitude (Figure 14–15). Topsoil from Colorado and Oklahoma was blown eastward for hundreds of miles. Women hanging out clean laundry in Georgia went outside later to find it dust-covered. Bakers in New York City had to keep freshly baked bread away from open windows so it wouldn't get dirty. The dust even discolored the Atlantic Ocean several hundred miles off the coast.

The Dust Bowl occurred during the Great Depression, and ranchers and farmers quickly went bankrupt. Many abandoned their dust-choked land and dead livestock and migrated west to the promise of California (Figure 14–16); their plight is movingly portrayed in the novel *The Grapes of Wrath* by John Steinbeck.

Although the Midwest is no longer a dust bowl, soil erosion is still a major problem there and elsewhere. We discuss the extent of the problem later in this chapter.

### Mineral Depletion of the Soil

In a natural ecosystem, essential minerals cycle from the soil to living organisms, and back again to the soil when those organisms die and decay. An agricultural system disrupts this pattern when the crops are harvested. Much of the plant material, containing minerals, is removed from the cycle, so it fails to decay and release its nutrients back to the

(a)

(b)

**Figure 14–15**
The Dust Bowl years. (a) A family running for shelter during a dust storm. (b) An abandoned Oklahoma farm in 1937. Total devastation was often the aftermath of dust storms. (USDA)

**Figure 14–16**
Many farming families from Oklahoma migrated to California after dust storms destroyed their farms. This photograph of one such family (taken in a resettlement camp in San Jose, California, in the 1930s) starkly conveys the hopelessness that comes from having lost everything. (USDA)

soil. Thus, over time, soil that is farmed inevitably loses its fertility (Figure 14–17).

**Mineral Depletion in Tropical Rain Forest Soils** In tropical rain forests, the climate, the typical soil type, and the removal by humans of the natural forest community result in a particularly severe type of mineral depletion. Recall that oxisols, soils found in tropical rain forests, are nutrient-poor because the nutrients are stored in the vegetation. Any minerals that are released as dead organisms decay in the soil and are promptly reabsorbed by plant roots and their mutualistic fungi. If this did not occur, the heavy rainfall would quickly leach the nutrients away. Nutrient reabsorption by vegetation is so effective that oxisols can support luxuriant rain forest growth despite the relative infertility of the soil, *as long as the forest remains intact.*

When the rain forest is cleared, whether to sell the wood or to make way for crops or rangeland, its efficient nutrient recycling is disrupted. Removal of the vegetation that so effectively stores the forest's nutrients allows minerals to leach out of the system.

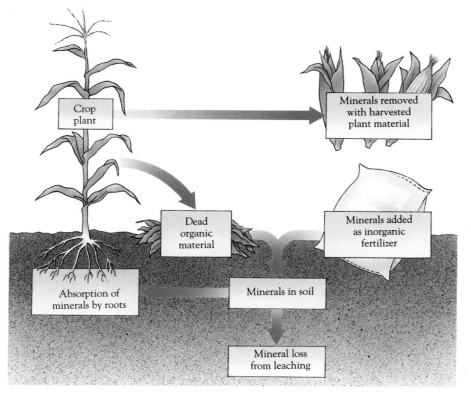

**Figure 14-17**
As plant and animal detritus decomposes in natural ecosystems, nutrients are cycled back to the soil for reuse. In agriculture, much of the plant material is harvested. Because the mineral nutrients in the harvested portions are unavailable to the soil, the nutrient cycle is broken. For this reason, fertilizer must be added to the soil periodically.

Crops can be grown on these soils for only a few years before the small mineral reserves in the soil are depleted. When cultivation is abandoned, a secondary forest develops, but it is never as luxuriant or biologically diverse as the primary forest, because most of the original nutrients have left the system. If the secondary forest is later cleared for cultivation, the soil becomes even more impoverished. Eventually, only a very few species of plants are capable of growing on the compacted, exposed soil. (Chapter 17 discusses aspects of deforestation other than soil degradation.)

**Laterization of Tropical Soils** When a forest is eliminated in tropical regions, **laterization,** a soil process that produces a rock-hard soil, may occur. (The term "laterization" comes from the Latin word for brick, *later*.) Laterized soil is so hard that in tropical areas it has been cut into bricks, allowed to dry, and used to construct temples and shrines. Although the removal of the forest causes most minerals to be washed away, iron and aluminum compounds, which don't leach readily, can be present in high concentrations, giving laterite soil a red or yellow color. As the remaining humus decays, the soil hardens in the sun.

Large areas of laterized soils, which are often called "red deserts," are common in parts of India and Southeast Asia. Some scientists initially expressed concern that South America's recent tropical deforestation would lead to widespread laterite

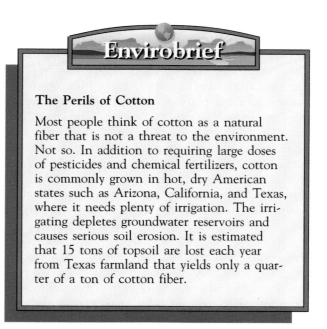

**Envirobrief**

**The Perils of Cotton**

Most people think of cotton as a natural fiber that is not a threat to the environment. Not so. In addition to requiring large doses of pesticides and chemical fertilizers, cotton is commonly grown in hot, dry American states such as Arizona, California, and Texas, where it needs plenty of irrigation. The irrigating depletes groundwater reservoirs and causes serious soil erosion. It is estimated that 15 tons of topsoil are lost each year from Texas farmland that yields only a quarter of a ton of cotton fiber.

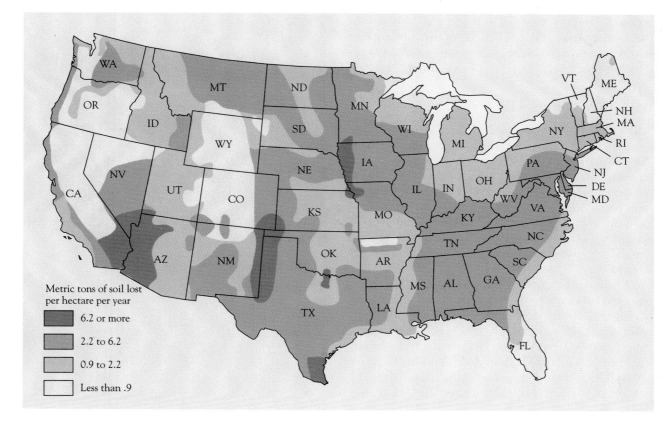

**Figure 14–18**
Average annual erosion of cropland soils in the United States.

formation. However, there is little current evidence that laterization is occurring on a large scale there.

## Soil Problems in the United States

In spite of 50 years of government-supported soil conservation programs, erosion is still a serious threat to cultivated soils in many regions throughout the United States (Figure 14–18). One contributor to the problem is the fact that federal agricultural policies are inconsistent with one another and with the goal of soil conservation. In many federal policies, for example, increased food production is a greater priority than is the protection of soil and other resources. Some farm programs (price support programs, for example) offer incentives for farmers to produce crops at the expense of fragile lands and soils. The problems caused or exacerbated by government policies, however, can be mitigated by reforming the policies. The Food Security Act of 1985 (see Chapter 17) is a good example because it contains provisions that eliminate inconsistencies.

The plains and deserts are particularly vulnerable to wind erosion. When this land is irrigated, crops can be grown without danger of failure, but without irrigation the frequent and prolonged droughts increase the likelihood of crop failures, which result in bare, easily eroded soil. Because of persistent water shortages—particularly in the Southwest—many farmers are abandoning farming altogether. It may take centuries for the abandoned barren land to return to its natural state; until then, it will be susceptible to erosion, especially by wind.

Erosion of soil by water is particularly severe along the Mississippi and Missouri rivers, as well as the central valley of California and the hilly Palouse River region of Washington State. The Soil Conservation Service estimates that about 25 percent of agricultural land in the United States is losing topsoil faster than it can be regenerated by natural soil-forming processes. This loss is often so gradual that even farmers fail to notice it. For example, a big rainstorm may wash away 1 mm (0.04 inch) of soil, which seems insignificant until the cumulative effects of many storms are taken into account. Twenty years of soil erosion amounts to the loss of about 2.5 cm (1 inch) of soil, an amount that would take 500 years to replace by natural soil-forming processes.

## Worldwide Soil Problems

Soil erosion and mineral depletion are significant problems worldwide. More than 1 billion people depend upon agricultural lands that are not productive enough to adequately support them. A combination of factors has created this situation, including unsound farming methods, extensive soil erosion, and expanding deserts (see Chapter 17). Along with these factors, the needs of a rapidly expanding population exacerbate soil problems worldwide.

Local and regional soil problems have been reported for many years. The first global assessment of soil conditions, released in 1992, was the summary of a three-year study of global soil degradation sponsored by the United Nations Environment Program. It reported that 1.96 billion hectares (4.84 billion acres) of soil—an area equal to 17 percent of the Earth's total vegetated surface area—have been degraded since World War II. Eleven percent of the Earth's vegetated surface—an area the size of China and India—has been degraded so badly that it will be very costly (or in some cases impossible) to reclaim it. The main causes of soil degradation are farming, overgrazing, and deforestation. (Overgrazing and deforestation are discussed in Chapter 17.)

Asia and Africa have the largest land areas with extensive soil damage, and in both places the problem is compounded by rapid population growth. The Sahelians in Africa, for example, must use their land to grow crops and animals for food or they will starve, but the soil is so overexploited that it is able to support fewer and fewer people. The day is approaching when the Sahel will be utterly unproductive desert (Figure 14–19). To reclaim the land would require restricting its use for many years so it could recover; but if these measures were taken, the Sahelians would have no means of obtaining food.

Attempts to develop highly productive, sustainable "temperate" agriculture in tropical areas have often failed, particularly in humid areas. When a rain forest is cut down and burned to prepare the land for crops or pastures, all the nutrients tied up in the vegetation are released at one time instead of being released slowly and reabsorbed quickly by the plants. The nutrients are then rap-

**Figure 14–19**
Cattle in Burkina Faso have eaten all the ground cover; the trees that remain will probably be stripped of branches to feed the hungry cattle. Over-exploitation of the Sahel, a semiarid region south of the Sahara Desert in Africa, is increasing the amount of unproductive desert area. (Robert E. Ford/Terraphotographics)

idly leached, and the soil quickly becomes unproductive.

## SOIL CONSERVATION AND REGENERATION

Conservation tillage, crop rotation, contour plowing, strip cropping, and terracing all help to minimize erosion and mineral depletion of the soil. Land that has been badly damaged by soil erosion can be successfully restored, but it is a costly, time-consuming process.

## Conservation Tillage

Conventional methods of tillage, or working the land, include spring plowing, in which the soil is cut and turned in preparation for planting seeds; and harrowing, in which the plowed soil is leveled, seeds are covered, and weeds are removed. Conventional tillage prepares the land for crops, but in removing all plant cover it greatly increases the likelihood of soil erosion (Figure 14–20). Fields that are conventionally tilled also contain less organic material and generally hold less water than does undisturbed soil.

Since the early 1980s, many farmers have adopted a new approach called **conservation tillage,** in which residues from previous crops are left in the soil, partially covering it and helping to hold it in place. The several types of conservation tillage include reduced tillage and no-tillage. **Reduced tillage,** in which the subsurface soil is tilled without disturbing the topsoil, greatly reduces the amount of soil erosion. **No-tillage** leaves even the subsurface soil undisturbed, because special machines punch tiny holes in the soil for seeds (Figure 14–21).

Conservation tillage increases the organic material in the soil, which in turn improves water-holding capacity. Decomposing organic matter re-

**Figure 14–20**
Conventional tillage leaves almost no plant residues on the soil surface. If heavy spring rains come before the young plants become established, significant soil loss can occur. (Grant Heilman)

**Figure 14–21**
Conservation tillage reduces soil erosion because plant residues from the previous season's crops are left in the soil. (a) Loading seeds into a no-till planter. A no-till machine plants seeds and applies fertilizer by punching small holes in the ground. (b) New plants (soybean) grow surrounded by decaying residues from the previous year's crop (rye). (USDA/Soil Conservation Service)

(a)                                                (b)

leases nutrients more gradually than when conventional tillage methods are employed. Although conservation tillage is an effective way of reducing soil erosion, it requires greater use of herbicides to control weeds. Research is needed to develop alternative methods of weed control for use with conservation tillage.

## Crop Rotation

Farmers who practice effective soil conservation measures often use a combination of conservation tillage and **crop rotation,** the planting of a series of different crops in the same field over a period of years. When the same crop is grown continuously, pests tend to accumulate to destructive levels, so crop rotation lessens damage by insects and disease. Also, many scientific studies have shown that continuously growing the same crop over a period of years depletes the soil of certain essential nutrients faster and makes soils more prone to erosion. Crop rotation is therefore effective in maintaining soil fertility and in reducing erosion.

A typical crop rotation would be corn–soybeans–oats–alfalfa. Soybeans and alfalfa, both members of the legume family, actually increase soil fertility through their association with bacteria that fix atmospheric nitrogen into the soil. Thus, soybeans and alfalfa help produce higher yields of the grain crops with which they alternate in crop rotation.

## Contour Plowing, Strip Cropping, and Terracing

Hilly terrain must be cultivated with care because it is more prone than flat land to soil erosion. Contour plowing, strip cropping, and terracing help control erosion of farmland with variable topography.

In **contour plowing,** fields are plowed and planted in curves that conform to the natural contours of the land, rather than in straight rows. Furrows run *around,* rather than up and down, hills (Figure 14–22).

**Strip cropping,** a special type of contour plowing, produces alternating strips of different crops along natural contours. For example, alternating a row crop such as corn with a closely sown crop such as wheat reduces soil erosion. Even more effective control of soil erosion is achieved when strip cropping is done in conjunction with conservation tillage.

**Figure 14–22**
Contour plowing and strip cropping are evident in this well-managed farm. Quite often crop rotations in such strips include a legume, which reduces the need for nitrogen fertilizers. (USDA)

In mountainous terrain, **terracing** produces level areas and thereby reduces soil erosion (Figure 14–23). Nutrients and soil are retained on the horizontal platforms instead of being washed away. Soils are preserved in a somewhat similar manner at low lying areas that are diked to make rice paddies. The water forms a shallow pool, retaining sediments and nutrients.

**Figure 14–23**
Terracing hilly or mountainous areas reduces the amount of soil erosion. However, some slopes are so steep they are totally unsuitable for agriculture. These areas should be left covered by natural vegetation to prevent extensive erosion. (David Cavagnaro)

## Preserving Soil Fertility

The two main types of fertilizer are organic and inorganic. Organic fertilizers include such natural materials as animal manure, crop residues, bone meal, and compost. (For two very different discussions of compost, see You Can Make a Difference: Practicing Environmental Principles, and Meeting the Challenge: Municipal Solid Waste Composting.) Organic fertilizers are complex, and their exact compositions vary. The nutrients in organic fertilizers become available to plants only as the organic material decomposes. For that reason, organic fertilizers are slow-acting and long-lasting.

Inorganic fertilizers are manufactured from chemical compounds, and thus their exact compositions are known. Because they are soluble, they are immediately available to plants. However, inorganic fertilizers are available in the soil for only a short period of time because they quickly leach away.

It is environmentally sound to avoid or limit the use of manufactured fertilizers, for several reasons. First, because of their high solubility, inorganic fertilizers are very mobile and often leach into groundwater or surface runoff, polluting the water (see Chapter 21). Second, manufactured fertilizers do not improve the water-holding capacity of the soil as organic fertilizers do. Another advantage of organic fertilizers is that, in ways that are not yet completely understood, they change the types of organisms that live in the soil, sometimes suppressing the microorganisms that cause certain plant diseases.

**Figure 14–24**

Shelterbelts on a Minnesota farm. Trees reduce the ability of the wind to pick up soil from farmland. (Grant Heilman, from Grant Heilman Photography)

## Soil Reclamation

It is possible to reclaim land that is badly damaged from erosion. The United States has largely reversed the effects of the 1930s Dust Bowl, for example, and China has reclaimed badly eroded land in Inner Mongolia (northern China). Soil reclamation involves two steps: (1) stabilizing the land to prevent further erosion and (2) restoring the soil to its former fertility. In order to stabilize the land, the bare ground is seeded with plants; they eventually grow to cover the soil, holding it in place. For example, after the Dust Bowl, land in Oklahoma and Texas was seeded with drought-resistant grasses. One of the best ways to reduce the effects of wind on soil erosion is by planting **shelterbelts,** rows of trees that lessen the impact of wind (Figure 14–24).

The plants that have been established to stabilize the land start to improve the quality of the soil almost immediately, as dead portions are converted to humus. The humus holds mineral nutrients in place and releases them a little at a time; it also improves the water-holding capacity of the soil.

Restoration of soil fertility to its original level is a slow process, however. During the soil's recovery, use of the land must be restricted: it cannot be farmed or grazed. Disaster is likely if the land is put back to use before the soil has completely recovered. But restriction of land use for a period of several to many years is sometimes very difficult to accomplish. How can a government tell landowners that they may not use their own land? How can land use be restricted when people's livelihoods and maybe even their lives depend upon it?

## Soil Conservation Policies in the United States

The disastrous effects of the Dust Bowl years on U.S. soils focused attention on the fact that soil is a valuable natural resource. Upon passage of the Soil Conservation Act in 1935, the Soil Conservation Service was formed; its mission is to assess soil damage and develop policies to improve and sustain our soil resource.

Historically, farmers have been more likely to practice soil conservation during hard financial times and periods of agricultural surpluses, both of which translate into lower prices for agricultural products. When prices are high, with a good market for agricultural products, farmers have more incentive to put every parcel of land into production, including marginal, highly erodible lands. During

*(Text continues on p. 319.)*

# Municipal Solid Waste Composting

The sanitary landfill, a modern replacement for the city dump, has been the recipient of most of the nation's solid waste for the past several decades, but it is not a long-term solution to waste disposal. For one thing, landfills often leak, and when this happens all sorts of nasty chemicals leach from them into surrounding soil and waterways. Also, most landfills are filling up, and it is unlikely that enough replacement sites can be found. NIMBYism—the not in my backyard syndrome—frequently takes hold when government officials attempt to find new places for sanitary landfills.

Much of the bulky waste in sanitary landfills—paper, yard refuse, food wastes, and such—is organic and, given the opportunity, could decompose into compost. However, in sanitary landfills little of this material breaks down. Rapid and complete decomposition requires the presence of oxygen, and in a sanitary landfill, garbage is buried under a layer of soil so that little oxygen is available.

Of the several options for decreasing the quantity of trash and garbage in sanitary landfills, which include both recycling and incineration for energy, one—**municipal solid waste composting**—has only recently received serious attention in this country. Municipal solid waste composting is the large-scale composting of the entire organic portion of a community's garbage. Since approximately 75 percent by weight of household garbage is organic, municipal solid waste composting, should it prove technically and economically feasible, would substantially reduce demand for sanitary landfills.

Numerous city and county governments are currently composting leaves and yard wastes in an effort to reduce the amount of solid waste sent to landfills. Although this endeavor alone is undeniably beneficial, municipal solid waste composting encompasses much more than yard wastes.

Consider Recomp, Inc., a solid waste center in St. Cloud, Minnesota. Recomp is paid by St. Cloud to pick up garbage, 100 tons each day, which it takes to its facility. There, bulky items such as old appliances are removed and taken to the sanitary landfill. Some paper is removed and sent to a local utility, where it is incinerated to produce energy, and large magnets remove steel cans and other metals that contain iron. Everything else—food wastes, glass, plastic bottles, and yard wastes—is placed in giant drums for composting. (The inorganic materials such as plastic and glass do not decompose; they are screened out at a later time.)

At another facility in Lee's Summit, Missouri, municipal yard wastes, such as branches and shrub clippings, are ground for composting (a). The resulting compost can then be used for fertilizer and mulch. (b) The compost is picked up for use in the city's park beautification program. (a, © Eric R. Berndt/Unicorn Stock Photos; b, ©Aneal Vohra/Unicorn Stock Photos)

(a)

(b)

Initial composting at Recomp occurs quickly—in 3 to 4 days—because conditions such as moisture and the carbon-nitrogen ratio are continually monitored and adjusted (by adding water or fertilizer, for example) for maximum decomposition. The decay process is carried out by billions of bacteria and fungi, which convert the organic matter into carbon dioxide, water, and humus. So many decomposers eat, reproduce, and die in the compost heap that the drum heats up, killing off potentially dangerous organisms such as disease-causing bacteria.

When the material emerges from the drums, it is screened to remove items such as plastic and glass that didn't decompose. Then it is placed outside for several months to cure, during which time additional decomposition occurs. Finally it is sold as compost.

The potential market for compost is largely unexploited. Compost can be used by professional nurseries, landscapers, greenhouses, and golf courses. The largest potential market, however, is the American farmer. Tons of compost could be used to reclaim the 167 million hectares (413 million acres) of badly eroded farmland in the United States; 59 metric tons (65 tons) of compost would be needed to apply 1 inch of compost to a single acre. Compost could also improve the fertility of badly eroded rangeland, forest land, and strip mines. There is no shortage of markets for compost. Thus, it appears that, should certain technical problems be resolved, composting could become economically feasible.

Technical problems include concerns over the presence of pesticide residues and heavy metals in the compost. Pesticides sprayed on urban and suburban landscapes would naturally find their way into compost material on leaves, grass clippings, and other yard wastes. However, several scientific studies indicate that most pesticides are either decomposed by bacteria and fungi during composting or broken down by the high temperatures in the compost heap.

More troubling is the concern over heavy metals, such as lead and cadmium, that can enter compost from sewage sludge—which may contain industrial wastes—or consumer products such as batteries. (Sewage sludge is often added to compost because it is a rich source of nitrogen for the decomposing microorganisms.) One way to reduce heavy metal contamination in municipal compost is by sorting out heavy metal sources before everything is dumped into the composting drum. Yet Recomp, Inc., has found that it costs too much to mechanically sort garbage before it is composted. This problem could be solved if consumers could be taught to separate products high in heavy metals before they put out their garbage.

Municipal solid waste composting facility near St. Cloud, Minnesota.
(Courtesy of Recomp of Minnesota)

## Practicing Environmental Principles

You can maintain and improve the soil in your own lawn and garden by using compost and mulch. Gardeners often dispose of grass clippings, leaves, and other plant refuse by either bagging it for garbage collection or burning it. But these materials don't have to be treated like wastes; they can be a valuable resource for making **compost,** a natural soil and humus mixture that improves not only soil fertility but also soil structure. Grass clippings, leaves, weeds, sawdust, coffee grounds, ashes from the fireplace or grill, shredded newspapers, potato peels, and eggshells are just some of the materials that can be **composted,** or transformed by microbial action to compost.

To make a compost heap, spread a 6- to 12-inch layer of grass clippings, leaves, or other plant material in a shady area, sprinkle it with an organic garden fertilizer or a thin layer of animal manure, and cover it with several inches of soil. Add layers as you collect more organic debris. Water the mixture thoroughly, and turn it over with a pitchfork each month to aerate it. (Although it is possible to make compost by just heaping it on the open ground in layers, it is more efficient to construct an enclosure. An enclosed com-

Mulches discourage the growth of weeds and help keep the soil damp. Organic mulches have the added benefit of gradually decaying, thereby increasing soil fertility. (USDA/Soil Conservation Service)

post heap is also less likely to attract animals.) When the compost is uniformly dark in color, is crumbly, and has a pleasant, "woodsy" odor, it is ready to use. The time it takes for decomposition will vary from one to six months depending on the climate, the materials you are using, and how often you turn it and water it.

Whereas compost is mixed into soil to improve the soil's fertility, **mulch** is placed on the surface of soil, around the bases of plants. Mulch helps control weeds and increases the amount of water in the upper levels of the soil by reducing evaporation. It lowers the soil temperature in the summer and extends the growing season slightly by providing protection against cold in the fall. Mulch also decreases erosion by lessening the amount of precipitation runoff.

Although mulches can consist of inorganic material such as plastic sheets or gravel, natural mulches of compost, grass clippings, straw, chopped corncobs, or shredded bark have the added benefit of increasing the organic content of the soil. Grass clippings are a very effective mulch when placed around the bases of garden plants because they mat together, making it difficult for weeds to become established. You must replace grass mulches often, however, because they decay rapidly. Some gardeners prefer mulches of more expensive materials such as shredded bark, because they take longer to decompose and are more attractive.

A compost heap. Composts form faster when the heap is located in a shady spot, kept damp, and aerated (turned over) frequently. (USDA/Soil Conservation Service)

**Figure 14–25**
This field of drought-resistant grass was planted on land that was previously cultivated. Farmlands with highly erodible soils are being removed from cultivation and allowed to return to their natural state under the provisions of the Conservation Reserve Program. (USDA/Soil Conservation Service)

times when the farm economy has been strong, federal soil conservation programs have actually contributed to production on marginal lands by relying on voluntary rather than mandatory compliance. The federal government has traditionally used incentives, rather than penalties for noncompliance, to encourage soil conservation practices.

In a different approach, the Conservation Reserve Program (CRP), which is part of the Food Security Act of 1985, pays farmers to stop producing crops on highly erodible farmland. It requires planting grasses or trees on such land and then "retiring" it from further use for ten years (Figure 14–25). This land may not be grazed, nor may the grass be harvested for hay during that period. The CRP required that, as of 1990, all land designated as highly erodible be placed either in the CRP or in a locally approved soil conservation program. The provisions of the CRP must be fully implemented by 1995 or farmers will not be eligible for any federal program benefits, such as diversion payments, Farmers Home Administration loans, federal crop insurance, and conservation reserve payments. When the CRP is fully implemented, soil erosion from the most vulnerable lands should be reduced considerably.

The U.S. government is also trying to reduce soil erosion in drought-prone areas of the West and Southwest by retiring marginal farmlands and allowing them to revert to their natural states. The U.S. Department of Agriculture (USDA) shares with farmers the cost of planting ground covers to stabilize this land.

# SUMMARY

**1.** The formation of soil, the complex material in which plants root, involves interactions among parent rock, climate, living organisms, time, and topography. Soil, which is composed of inorganic minerals, organic materials, soil air, and soil water, is often organized into layers called horizons. The texture of a soil depends on the relative amounts of sand, silt, and clay in it. The soil properties of texture and acidity affect a soil's water-holding capacity and nutrient availability, which in turn determine how well plants grow.

**2.** Living organisms are important not only in forming soil, but also in recycling nutrients. In a balanced ecosystem, the minerals removed from the soil by plants are returned when plants or animals that eat the plants die and are decomposed by microorganisms.

**3.** Five important soil types are spodosols (northern, highly layered soils), alfisols (gray-brown forest soils), mollisols (rich prairie soils), aridisols (desert soils), and oxisols (tropical rain forest soils).

**4.** Soil erosion, the removal of soil from the land by the actions of water, wind, ice, and other agents, is a natural process that is often accelerated by human activities such as farming on arid land and deforestation. The Dust Bowl that occurred in the western United States during the 1930s is an example of accelerated wind erosion caused by human exploitation of marginal land for agriculture.

**5.** Mineral depletion occurs in all soils that are farmed. It is a particularly serious problem when tropical rain forests are removed, because the minerals in the soil are quickly leached out. Excessive leaching of tropical soils sometimes results in the formation of rock-hard laterized soils.

**6.** Conservation tillage, crop rotation, contour plowing, strip cropping, and terracing can be used to help control erosion and mineral depletion. Soil that has been badly damaged by erosion can be reclaimed by stabilizing the land (providing plant cover and shelterbelts) and by restoring soil fertility (restricting land use until the soil has time to recover).

# DISCUSSION QUESTIONS

**1.** Charles Darwin once wrote that the land is plowed by earthworms. Explain.

**2.** Describe three ways in which nutrients can be lost from the soil.

**3.** It could be said that, unlike other communities, tropical rain forests live *on* the soil rather than *in* it. What does this statement imply about tropical soils?

**4.** The American Dust Bowl is sometimes portrayed as a "natural" disaster brought on by drought and high winds. Present a case for the point of view that this disaster was not caused by nature as much as by humans.

**5.** Why doesn't laterization of tropical rain forest soil occur when the forest is undisturbed?

**6.** How is human overpopulation related to worldwide soil problems?

**7.** Conservation tillage has many benefits, including reduction of soil erosion. However, certain pests that cause plant disease can reside in the plant residues left on the ground with conservation tillage. Knowing that disease-causing organisms are often quite specific for the plants they attack, recommend a way to control such disease organisms. Base your answer on the soil conservation methods you have learned in this chapter.

# SUGGESTED READINGS

Arden-Clarke, C., and D. Hodges. Soil erosion: The answer lies in organic farming. *New Scientist,* vol. 113, 1987. Considers the soil erosion problem in Great Britain and how it can be curbed.

Lal, R. Managing the Soils of Sub-Saharan Africa. *Science,* vol. 236, 1987. The growing population and arid climate of sub-Saharan Africa are putting extreme pressure on their soil resources.

Marinelli, J. Composting: From Backyards to Big-Time. *Garbage,* July/August 1990. Explains municipal solid waste composting—why it's needed, how it's done, and problems that need to be resolved.

McDermott, J. Some heartland farmers just say no to chemicals. *Smithsonian,* April 1990. How natural farming practices help maintain a healthy soil.

Postel, S. "Halting Land Degradation," in *State of the World 1989: A Worldwatch Institute Report on Progress Toward a Sustainable Society,* ed. L. R. Brown et al. W. W. Norton & Company, New York, 1989. An in-depth examination of the causes of land degradation in many countries, along with ways to restore soil productivity.

Reganold, J. P., R. I. Papendick, and J. F. Parr. Sustainable agriculture. *Scientific American,* June 1990. Discusses alternative methods of agriculture that reduce a farmer's dependence on chemicals *and* improve and maintain the soil.

World Resources Institute, United Nations Environment Program, and United Nations Development Program. *World Resources 1992–93.* Oxford University Press, New York, 1992. Chapter 8 reports on a major global study of the extent of soil degradation.

# Chapter 15

*Reverse vending machines are automated recycling machines in which the consumer deposits old beverage cans and receives cash. (Courtesy of USX Corporation)*

# Minerals: A Nonrenewable Resource

The distribution and abundance of minerals vary widely. Chile and the United States are rich in copper, for example, whereas South Africa and Russia are rich in manganese. Because minerals are generally located in underground deposits, obtaining and processing them causes environmental damage, including land disturbance and air, soil, and water pollution. In the future, as mineral reserves are depleted, it may make increasing economic sense to exploit minerals in areas that are currently inaccessible (such as Antarctica), in newly discovered mineral deposits (primarily in developing countries), and in deep deposits. Recycling is an increasingly important "source" of many minerals because it requires less energy and produces less waste and pollution than the production and consumption of virgin ore.

## REVERSE VENDING MACHINES FOR STEEL AND ALUMINUM

During the period from 1989 to 1991, a new kind of vending machine appeared in many cities of Ohio, Georgia, Florida, Alabama, and North and South Carolina. Known as a **reverse vending machine,** it accepts used steel and aluminum cans and dispenses a penny for each can. Participants are also eligible for bonus groceries and other prizes. Once the cans are collected, they are transported to steel companies, where they are melted down and used to make new steel and aluminum.

Why would steel companies take the trouble to install reverse vending machines in convenience stores? Why should consumers take the trouble to return cans for a penny each? Reverse vending machines are a sign of the future. They teach people to recycle and serve as a temporary recycling system until curbside recycling is adopted. Scrap metal is increasingly in demand by manufacturers because it is less expensive to recycle used materials than to obtain and process virgin ores from the Earth. Recycling has the added benefits of reducing the amount of solid waste in landfills and of producing less pollution and using far less energy than the processing of virgin ores.

## USES OF MINERALS

**Minerals,** elements or compounds of elements that occur naturally in the Earth's crust, are such a part of our daily lives that we often take them for granted (Figure 15–1, Table 15–1). Indeed, it is probably impossible for you to spend a day without encountering or using minerals. Steel, an essential building material, is a blend of iron and other metals. Beverage cans, aircraft, automobiles, and buildings all contain aluminum. Copper, which readily conducts electricity, is used for electrical and communications wiring. The concrete used in buildings and roads is made from sand, gravel, and cement, which contains crushed limestone. Sulfur, a com-

(a)

(b)

(c)

**Figure 15–1**
Minerals are an important part of our lives. (a) Concrete highways are made of sand, gravel, and crushed limestone. (b) Table salt is a nonmetallic mineral. (c) Many everyday items important in our lives are composed of metals. (a and b, Dennis Drenner; c, Visuals Unlimited/Rich Treptow)

**Table 15–1**
**Some Important Minerals and Their Uses**

| Mineral | Type | Some Uses |
|---|---|---|
| Aluminum (Al) | Metal element | Structural materials (airplanes, automobiles), packaging (beverage cans, toothpaste tubes), fireworks |
| Borax ($Na_2B_4O_7$) | Nonmetal | Diverse manufacturing uses—glass, enamel, artificial gems, soaps, antiseptics |
| Chromium (Cr) | Metal element | Chrome plate, pigments, steel alloys (tools, jet engines, bearings) |
| Cobalt (Co) | Metal element | Pigments, alloys (jet engines, tool bits), medicine, varnishes |
| Copper (Cu) | Metal element | Alloy ingredient in gold jewelry, silverware, brass, and bronze; electrical wiring, pipes, cooking utensils |
| Gold (Au) | Metal element | Jewelry, money, dentistry, alloys |
| Gravel | Nonmetal | Concrete (buildings, roads) |
| Gypsum ($CaSO_4 \cdot 2H_2O$) | Nonmetal | Plaster of Paris, soil treatments |
| Iron (Fe) | Metal element | Basic ingredient of steel (buildings, machinery) |
| Lead (Pb) | Metal element | Pipes, solder, battery electrodes, pigments |
| Magnesium (Mg) | Metal element | Alloys (aircraft), firecrackers, bombs, flashbulbs |
| Manganese (Mn) | Metal element | Steel, alloys (steamship propellers, gears), batteries, chemicals |
| Mercury (Hg) | Liquid metal element | Thermometers, barometers, dental inlays, electric switches, streetlamps, medicine |
| Molybdenum (Mo) | Metal element | High-temperature applications, lamp filaments, boiler plates, rifle barrels |
| Nickel (Ni) | Metal element | Money, alloys, metal plating |
| Phosphorus (P) | Nonmetal element | Medicine, fertilizers, detergents |
| Platinum (Pt) | Metal element | Jewelry, delicate instruments, electrical equipment, cancer chemotherapy, industrial catalyst |
| Potassium (K)* | Metal element | Salts used in fertilizers, soaps, glass, photography, medicine, explosives, matches, gunpowder |
| Common salt (NaCl) | Nonmetal | Food additive, raw material for synthetics |
| Sand (largely $SiO_2$) | Nonmetal | Glass, concrete (buildings, roads) |
| Silicon (Si) | Metal element | Electronics, solar batteries, ceramics, silicones |
| Silver (Ag) | Metal element | Jewelry, silverware, photography, alloys |
| Sulfur (S) | Nonmetal element | Insecticides, rubber tires, paint, matches, papermaking, photography, rayon, medicine, explosives |
| Tin (Sn) | Metal element | Cans and containers, alloys, solder, utensils |
| Titanium (Ti) | Metal element | Paints; manufacture of aircraft, satellites, and chemical equipment |
| Tungsten (W) | Metal element | High-temperature applications, light bulb filaments, dentistry |
| Zinc (Zn) | Metal element | Brass, metal coatings, electrodes in batteries, medicine (zinc salts) |

*Potassium, which is very reactive chemically, is never found free in nature; it is always combined with other elements.

ponent of sulfuric acid, is an indispensable industrial mineral with many applications in the chemical industry. It is used to make plastics and fertilizers and to refine oil. Other important minerals include platinum, mercury, manganese, and titanium.

Human need and desire for minerals have influenced the course of history. Phoenicians and Romans explored Britain in a search for tin. One of the first metals to be used by humans, tin came into its own during the Bronze Age (3500 to 1000 B.C.). The desire for gold and silver was directly responsible for the Spanish conquest of the New World. A gold rush in 1849 led to the settlement of California; more recently, the lure of gold in Amazonian and Indonesian rain forests has contributed to their destruction.

The Earth's minerals are elements or (usually) compounds of elements and have precise chemical compositions. For example, **sulfides** are mineral compounds in which certain elements are combined chemically with sulfur, and **oxides** are mineral compounds in which elements are combined chemically with oxygen.

**Rocks** are aggregates, or mixtures, of minerals and have varied chemical compositions. An **ore** is rock that contains a large enough concentration of a particular mineral that the mineral can be profitably mined and extracted. **High-grade ores** contain relatively large amounts of particular minerals, whereas **low-grade ores** contain lesser amounts.

Minerals can be metallic or nonmetallic. **Metals** are minerals such as iron, aluminum, and copper, which are malleable, lustrous, and good conductors of heat and electricity. **Nonmetallic minerals** lack these characteristics; they include sand, stone, salt, and phosphates.

# MINERAL DISTRIBUTION AND ABUNDANCE

Certain minerals, such as aluminum and iron, are relatively abundant in the Earth's crust. Others, including copper, chromium, and molybdenum, are relatively scarce. Abundance does not necessarily mean that the mineral is easily accessible or profitable to extract, however. It is possible, for instance, that you have gold and other expensive minerals in your own backyard. However, unless the concentrations are large enough to make them profitable to mine, they will remain there.

Like other natural resources, mineral deposits in the Earth's crust are distributed unevenly around the Earth. Some countries have extremely rich mineral deposits, whereas others have few or none. Although iron is widely distributed in the Earth's crust, for example, Africa has less than the other continents. Many copper deposits are concentrated in North and South America, particularly in Chile, whereas Asia has a relatively small amount of copper. The distribution of nickel is surprising in that a substantial portion of the world's known supply is found in the tiny island nation of Cuba. Much of the world's tin is in Malaysia and Indonesia, and most of the chromium reserves are in South Africa.

## Mini-Glossary of Mineral Terms

**minerals:** Elements and compounds that occur naturally in the Earth's crust. Minerals are either metals or nonmetals.

**rock:** A mixture of minerals that has varied chemical concentrations.

**ore:** Rock that contains a large enough concentration of a particular mineral for the mineral to be profitably mined and extracted. High-grade ores contain relatively large amounts of the desired minerals, and low-grade ores, relatively small amounts.

**metals:** Minerals that are malleable, lustrous, and good conductors of heat and electricity. Examples are gold, copper, and iron.

**nonmetals:** Minerals that are nonmalleable, nonlustrous, and poor conductors of heat and electricity. Examples are sand, salt, and phosphates.

## Formation of Mineral Deposits

Concentrations of minerals within the Earth's crust are apparently caused by a number of mechanisms, including magmatic concentration, hydrothermal processes, sedimentation, and evaporation.

As molten rock cools and solidifies deep in the Earth's crust, it often separates into layers, with the heavier iron- and magnesium-containing rock settling on the bottom and the lighter silicates (rocks containing silicon) rising to the top. Higher concentrations of different minerals are often found in the various rock layers. This layering, which is thought to be responsible for some deposits of iron, copper, nickel, chromium, and other metals, is called **magmatic concentration.**

**Hydrothermal processes** involve groundwater that has been heated in the Earth. This water seeps through cracks and fissures and dissolves certain minerals in the rocks, which are then carried along in the hot water solution. The dissolving ability of the water is greater if chlorine or fluorine is present, because these elements react with many metals (such as copper) to form salts (copper chloride, for example) that are soluble in water. When the hot solution encounters sulfur, a common element in the Earth's crust, a chemical reaction between the metal salts and the sulfur produces metal sulfides. Because metal sulfides are not soluble in water, they form deposits by settling out of the solution. Hydrothermal processes are responsible for deposits of minerals such as gold, silver, copper, lead, and zinc.

The chemical and physical weathering processes that break rock into finer and finer particles are important not only in soil formation (as we saw in Chapter 14) but in the production of mineral deposits. Weathered particles can be transported by water and deposited as sediment on riverbanks, deltas, and the sea floor in a process called **sedimentation.** During their transport, certain minerals in the weathered particles dissolve in the water. They later settle out of solution—for example, when the warm water of a river meets the cold water of the ocean—because less material dissolves in cold water than in warm water. Important deposits of iron, manganese, phosphorus, sulfur, copper, and other minerals have been formed by sedimentation.

Significant amounts of dissolved material can accumulate in inland lakes and in seas that have no outlet or only a small outlet to the ocean. If these bodies of water dry up by **evaporation,** a large amount of salt is left behind; over time, it may become incorporated into rock layers. (There is geological evidence that the Mediterranean Sea once dried up completely, for example. When water spilled back into the sea, its vast salt deposits were covered with sediment.) Significant worldwide deposits of common salt ($NaCl$), borax ($Na_2B_4O_7$), potassium salts, and gypsum ($CaSO_4 \cdot 2H_2O$) have been formed by evaporation.

### How Large Is Our Mineral Supply?

**Mineral reserves** are mineral deposits that have been identified and are currently profitable to extract. In contrast, **mineral resources,** deposits of low-grade ores, are potential sources of minerals that are currently unprofitable to extract but may be profitable to extract in the future. Resources also include estimates of as-yet-unidentified deposits of minerals that may be confirmed in the future. The

combination of a mineral's reserves and resources is called its **total resources** or its **world reserve base.** The World Resources Institute estimates the 1990 reserves of copper, for example, at 321 million metric tons, and its world reserve base is 549 million metric tons. Based on the 1990 rate of consumption, copper *reserves* will last until the year 2028.

Estimates of mineral reserves and resources fluctuate with economic, technological, and political changes. If the price of a mineral on the world market falls, for example, certain borderline mineral reserves may slip into the resource category; increasing prices may restore them to the reserve category. When new technological methods decrease the cost of extracting mineral ores, mineral deposits that have been ranked in the resource category are reclassified as reserves. If the political situation in a country becomes so unstable that reserves cannot be mined, the reserves are reclassified as resources; another change in the political situation at a later time may cause the minerals to be placed in the reserve list again.

## HOW MINERALS ARE FOUND AND EXTRACTED

The process of making mineral deposits available for human consumption occurs in several steps. First, a particular mineral deposit is located. Second, the mineral is extracted from the Earth by mining. Third, the mineral is processed, or refined, by concentrating it and removing impurities. Processing often involves **smelting**—melting the ore at high temperatures to help separate impurities from molten metal. During the fourth and final step, the purified mineral is used to make a product.

### Discovering Mineral Deposits

Geologists employ a variety of instruments and measurements to help locate valuable mineral deposits (Figure 15–2). Aerial or satellite photography sometimes discloses geological formations that are associated with certain types of mineral deposits. Aircraft and satellite instruments that measure the Earth's magnetic field and gravity can reveal certain types of deposits. Geological knowledge of the Earth's crust and how minerals are formed is also used to estimate locations of possible mineral deposits. Once these sites are identified, geologists drill or tunnel for mineral samples and analyze their

**Figure 15–2**
Vibrator trucks produce vibrations that bounce off underlying rock formations. Such geological surveys of inaccessible areas often provide information about mineral deposits.
(George Steinmetz)

**Mining with Microbes**

"Ecological engineering" with microorganisms may be an economical and environmentally preferable way to mine metals in the future. Mining faces a problem of diminishing returns. As prime veins have been exhausted over the last century, ore grades have declined steadily. To obtain the necessary volumes, mining companies have increased their scale of operation. But this has also increased the burden they impose on the environment, through leaching waste dumps, massive open pit excavations, smelter emissions, and the need for vast quantities of cheap energy. The processing of one ton of copper requires five tons of coal and emits two tons of sulfur gases. And over half the energy used in conventional copper mining goes into crushing and moving rock that averages 99.4% waste. Microorganisms are far more efficient. When mixed with sulfuric acid, a microscopic creature called *Thiobacillus ferrooxidans* promotes a chemical reaction that leaches copper into an acidic solution. The process is basically a matter of spraying a heap of ore-containing rock with the solution, collecting it after it drains through, and extracting the metal from the liquid. *Thiobacillus ferrooxidans* has helped the copper industry lower its per pound production cost from 90 cents to less than 30 cents. Phelps Dodge, the largest copper producer in the U.S., expects to derive half its copper through bioleaching by the mid-1990s.

composition. Seismographs, which are used to detect earthquakes, also provide valuable clues about mineral deposits.

Deposits on the ocean floor cannot be estimated until detailed three-dimensional maps of the sea floor are produced, usually with the aid of depth-measuring devices. Sophisticated computer analysis is necessary to evaluate the complex data recorded by such devices.

## Extracting Minerals

The depth of a particular deposit determines whether it will be extracted by **surface mining,** in which minerals are extracted near the Earth's surface, or **subsurface mining,** in which minerals that are too deep to be removed by surface mining are extracted. Surface mining is more common because it is less expensive than subsurface mining. However, because even surface mineral deposits occur in rock layers beneath the Earth's surface, the overlying layers of soil and rock (called **overburden**) must first be removed, along with the vegetation

growing in the soil. Then giant power shovels scoop the minerals out of the Earth.

There are two kinds of surface mining. Iron, copper, stone, and gravel are usually extracted by **open-pit surface mining,** in which a hole is dug in the Earth's surface (Figure 15–3). Large holes that are formed by open-pit surface mining are called quarries. In **strip mining,** a trench is dug to extract the minerals (see Figure 10–5). Then a new trench is dug parallel to the old one; the overburden from the new trench is put into the old trench, creating a hill of loose rock known as a **spoil bank.**

(a)

(b)

**Figure 15–3**

Open-pit surface mining creates giant holes in the Earth. (a) This copper open-pit mine is near Tucson, Arizona. (b) The "Big Muskie" power scoop removing overburden from an underlying coal deposit in Ohio. (a, Bruce F. Molnia/Terraphotographics; b, Connie Toops)

Subsurface mining, which is done underground and is more complex, may be done with a shaft mine or a slope mine; both types are shown in Figure 10–5. A **shaft mine** is a direct vertical shaft to the vein of ore. The ore is broken apart underground and then hoisted through the shaft to the surface in buckets. A **slope mine** has a slanting passage that makes it possible to haul the broken ore out of the mine in cars rather than hoisting it up in buckets.

Subsurface mining disturbs the land less than surface mining, but it is more expensive and more hazardous for miners. There is always a risk of death or injury from explosions or collapsing walls, and prolonged breathing of dust in subsurface mines can result in lung disease.

## ENVIRONMENTAL IMPLICATIONS

There is no question that the extraction, processing, and disposal of minerals harm the environment. Mining disturbs and damages the land, and processing and disposal of minerals pollute the air, soil, and water. As noted in the discussion of coal in Chapter 10, pollution can be controlled and damaged lands can be partially restored, but these remedies cost money. In most cases, the environ-

### Envirobrief

**Blue Jeans and the Environment**

Fashion is having an adverse effect on the environment in New Mexico. To achieve the already-washed look and soft feel of "stone-washed" jeans, clothing manufacturers tumble denim fabric with pumice, a lightweight and abrasive volcanic rock found only in a few locations in the United States. One of those locations is the Sante Fe National Forest in New Mexico, where as of 1991, with permission from the U.S. Forest Service, a 33-acre site was being strip mined for the mineral. Strip mining is an aesthetically unpleasant form of mining that involves cutting away successive layers of topsoil and substratums in a large open trench—a process that clearly disrupts habitats in protected wilderness areas such as national forests. Because of high demand for pumice in the jeans industry, a cubic yard of the rock has risen in price from around $7 to over $60. This prompted the mining company to apply for a patent that would give it access to an additional 1,700 acres, a patent potentially worth $300 million to the company.

mental "costs" of minerals are not made a part of their actual price to consumers (see Chapter 7).

Most developed countries have regulatory mechanisms in place to minimize environmental damage from mineral consumption, and many developing nations are in the process of putting them in place. These regulatory programs include policies to prevent or reduce pollution, restore mining sites, and exclude certain recreational and wilderness sites from mineral development.

## Mining and the Environment

Mining, particularly surface mining, disturbs huge areas of land. In the United States, current and abandoned metal and coal mines occupy an estimated 9 million hectares (22 million acres). Because any vegetation that has grown there is destroyed by mining, this land is particularly prone to erosion, with wind erosion causing air pollution and water erosion polluting nearby waterways and damaging aquatic habitats.

Acids and other toxic substances in the spoil banks of mines are washed into soil and water by runoff (see Figure 10–10) and also contribute to air pollution. When such toxic compounds make their way into nearby lakes and streams, they adversely affect the numbers and kinds of aquatic life; groundwater can also be contaminated. This type of pollution is called **acid mine drainage.**

## Environmental Impacts of Refining Minerals

Approximately 80 percent or more of mined ore consists of impurities that become wastes after processing. These wastes, also called **tailings,** are usually left in giant piles on the ground or in ponds near the processing plants (Figure 15–4a). Toxic

**Figure 15–4**

Processing crude ores can adversely affect the environment. (a) Copper ore tailings from a mine in Anaconda, Montana. Toxic materials from mine tailings that are left in mountainous heaps can pollute the air, soil, and water. (b) A blast furnace is used to smelt iron. Iron ore, limestone rock, and coke (modified coal used as an industrial fuel) are added at the top of the furnace, while heated air or oxygen is added at the bottom. The iron ore reacts with coke to form molten iron and carbon dioxide. The limestone reacts with impurities in the ore to form a molten mixture called slag. Both molten iron and slag collect at the bottom; slag floats on molten iron because it is less dense than iron. Note the vent near the top of the iron smelter for exhaust gases. If air pollution control devices are not installed, many dangerous gases are emitted during smelting. (a, Visuals Unlimited/R. Ashley)

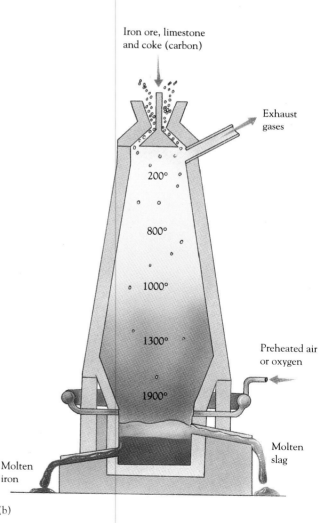

(a)

(b)

substances and dust from tailings left exposed in this way can contaminate the air, soil, and water.

Unless expensive pollution control equipment has been added to smelting plants, they emit large amounts of air pollutants during mineral processing (Figure 15–4b). (See Focus On: Copper Basin, Tennessee, for a specific example of environmental degradation caused by smelting.) Lead, arsenic, and cadmium are some of the toxic pollutants that may be discharged into the atmosphere. Cadmium, for example, is found in zinc ores, and emissions from zinc smelters are a major source of environmental cadmium contamination. In humans, cadmium is linked to high blood pressure; diseases of the liver, kidneys, and heart; and certain types of cancer. (The health effects of lead, another common emission from smelters, are discussed in Chapter 21.) In addition to airborne pollutants, smelters emit hazardous liquid and solid wastes that can cause soil and water pollution.

## Restoration of Mining Lands

When a mine is no longer profitable to operate, the land can be reclaimed, or restored to a seminatural condition (Figure 15–5). Approximately two-thirds of the Copper Basin in Tennessee has been partially reclaimed, for example. The goals of reclamation include preventing further degradation and erosion of the land, eliminating or neutralizing local sources of toxic wastes, and making the land productive for purposes other than mining. Restoration can also make such areas visually attractive.

A great deal of research is available on techniques of restoring lands that have been degraded by mining, called **derelict lands.** Restoration involves filling in and grading the land to its natural contours, then planting vegetation to hold the soil in place. The establishment of plant cover is not as simple as throwing a few seeds on the ground. Often the topsoil is completely gone or contains toxic levels of metals, so special types of plants that can tolerate such a challenging environment must be used. According to experts, the main limitation on the restoration of derelict lands is not lack of knowledge but lack of funding.

Reclamation of areas that were surface mined for coal is required by the Surface Mining Control and Reclamation Act of 1977. This law orders coal companies to restore areas that have been surface mined, beginning in 1977. Surface-mined land that was damaged prior to 1977 is gradually being restored as well, using money from a tax on currently mined coal.

No law is in place to require restoration of derelict lands produced by mines other than coal mines, however. A mining law that was passed in 1872 and that currently makes no provision for reclamation may be revised by Congress in the next several years.

**Figure 15–5**
Part of a phosphate mine in Florida has been reclaimed and is currently used as a pasture (background). Restoration of derelict lands makes them usable once again, or at the very least, stabilizes them so that further degradation does not occur. (William Felger, from Grant Heilman)

**Focus On**

## Copper Basin, Tennessee

Travel to the southeast corner of Tennessee, near its borders with Georgia and North Carolina, and you'll progress from lush forests to a panorama of red, barren hills baking in the sun. No trees, no green plants, no animals—just 145 square kilometers (56 square miles) of hills with deep ruts gouged into them. The ruined land has a stark, otherworldly appearance. What is this place, and how did it come to be?

During the middle of the 19th century, copper ore was discovered in southeastern Tennessee. Copper mining companies extracted the ore from the Earth and dug vast pits to serve as open-air smelters. They cut down the surrounding trees and burned them in the smelters to create the high temperatures needed for the separation of copper metal from other contaminants in the ore. The ore contained great quantities of sulfur, which reacted with oxygen in the air to form sulfur dioxide. As sulfur dioxide from the open-air smelters billowed into the atmosphere, it reacted with water, forming sulfuric acid that fell to the Earth as acid rain.

Between deforestation and acid rain, ecological ruin of the area was assured in a few short years. Any plants attempting a comeback after removal of the forests were quickly killed by the acid rain, which acidified the soil. Because plants no longer covered the soil and held it in place, soil erosion cut massive gullies in the gently rolling hills. Of course, the forest animals disappeared with the plants, which had provided their shelter and food. The damage didn't stop here: soil eroding from the Copper Basin, along with acid rain, ended up in the Ocoee River, killing its entire aquatic community.

In the 1930s a number of government agencies, including the Tennessee Valley Authority and the U.S. Soil Conservation Service, tried to replant a portion of the area. They planted millions of loblolly pine and black locust trees as well as shorter ground-cover plants that tolerate acid conditions, but most of the plants died. The success of such efforts was marginal until the 1970s, when land reclamation specialists began using new techniques such as application of seed and time-released fertilizer by helicopter. These plants had a greater sur-

vival rate, and as they became established, their roots held the soil in place. Leaves dropping to the ground contributed organic material to the soil. The plants provided shade and food for animals such as birds and field mice, which slowly began to return.

Today approximately two-thirds of the Copper Basin has one sort of vegetation or another. If funding continues, it may be possible to have the entire area under plant cover in 10 to 20 years. Of course, the return of the complex forest ecosystem that originally covered the land before the 1850s will take a century or two—if it ever occurs.

Botanists and land reclamation specialists have learned a lot from the Copper Basin, and they will put this knowledge to use in future reclamation projects around the world.

Ducktown, Tennessee. Air pollution from a copper smelter in Tennessee killed the vegetation, and then water erosion carved gullies into the hillsides. (Visuals Unlimited/Pat Armstrong)

## MINERAL RESOURCES: AN INTERNATIONAL PERSPECTIVE

The economies of industrialized countries require the extraction and processing of large amounts of minerals to make products. Most of these developed countries rely on the mineral reserves in developing countries, having long since exhausted their own supplies. But as developing countries become more industrialized, their own mineral requirements increase correspondingly, adding further pressure to an already dwindling supply.

### U.S. and World Use

At one time, most of the developed nations had rich resource bases, including abundant mineral deposits, that enabled them to industrialize. In the process of industrialization, they have largely depleted their domestic reserves of minerals so that they must increasingly turn to developing countries. This is particularly true for Europe, Japan, and, to a lesser extent, the United States.

As with the consumption of other natural resources, there is a large difference in consumption of minerals between developed and developing countries. Four countries—the United States, Japan, the former Soviet Union,[1] and Germany—consume approximately three-fourths of many of the world's metals (Figure 15–6). In other words, about 14 percent of the world's population consumes more than 70 percent of some important minerals. It is too simplistic, however, to divide the world into two groups, the mineral consumers (developed countries) and the mineral producers (developing countries). For one thing, four of the world's top five mineral producers *are* developed countries: the United States, Canada, Australia, and the former Soviet Union.[2] Furthermore, many developing countries lack *any* significant mineral deposits.

Because industrialization increases the demand for minerals, countries that at one time met their mineral needs with domestic supplies become increasingly reliant on foreign supplies as development occurs. South Korea is one such nation. Dur-

---

[1] Although the former Soviet Union now comprises a number of independent nations, information about the distribution of mineral resources in the individual nations is largely unavailable as we go to press.

[2] South Africa, a moderately developed, middle-income country, is the other mineral producer in the top five.

---

### Envirobrief

#### User-Unfriendly Minerals

To the great disappointment of artists everywhere, "heavy metals" such as lead, cadmium, chromium, lithium, and manganese are rapidly being banned or phased out as ingredients in art supplies. Historically, these metals have been used to lend stability and brilliance to paints; one traditional bright yellow hue is actually known as Cadmium Yellow. Such metals, however, are toxic and tend to accumulate in the environment. Some researchers have hypothesized that a number of famous painters—including Vincent van Gogh, Auguste Renoir, Peter Paul Rubens, and Paul Klee—might well have suffered damage to their circulatory and respiratory systems from exposure to the toxic metals in their paints. Van Gogh achieved renowned qualities of brightness in various portrayals of the sun and stars, and in his immortal painting *Sunflowers*, with a lead-containing paint called Naples Yellow.

---

ing the 1950s it exported iron, copper, and other minerals. South Korea experienced dramatic economic growth from the 1960s to the present and, as a result, must now import iron and copper to meet its needs.

### Distribution Versus Consumption

Chromium, a metallic element, provides a useful example of worldwide versus national distribution and consumption. Chromium is used to make vivid red, orange, yellow, and green pigments for paints, chrome plate, and (combined with other metals) certain types of hard steel. There is no substitute for chromium in many of its important applications, including jet engine parts. Therefore, industrialized nations that lack significant chromium deposits, such as the United States, must import essentially all of their chromium. South Africa is one of only a few countries with significant deposits of chromium. Zimbabwe and Turkey also export chromium. Albania and the former Soviet Union produce significant amounts of chromium but do not export it. Although worldwide reserves of chromium are adequate for the immediate future, the

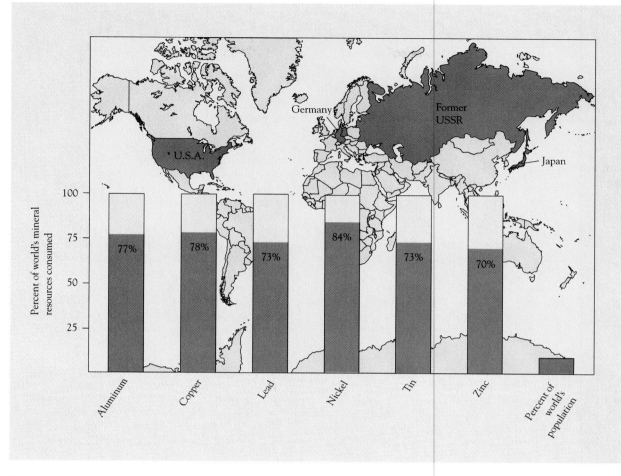

**Figure 15–6**
The United States, Japan, the former Soviet Union, and Germany, which together make up
13.7 percent of the world's population, consume a disproportionate share of many of the
world's metals.

United States and several other industrialized countries are utterly dependent on a few, often politically volatile, countries for their chromium supplies.

Many industrialized nations have stockpiled strategically important minerals to reduce their dependence on politically unstable suppliers. The United States and others have stockpiles of such metals as titanium, tin, manganese, chromium, platinum, and cobalt, mainly because these metals are critically important to industry and defense. The stockpiles are supposed to be large enough to provide strategic metals for a period of three years.

## Will We Run Out of Important Minerals?

A mineral's **depletion time** is an estimate of the time it will take for 80 percent of the known reserves of that mineral to be expended. Depletion

times are often meaningless because it is extremely difficult to forecast future mineral supplies. In the 1970s, projections of escalating demand and impending shortages of many important minerals were commonplace. None of these shortages actually materialized, in part because metals in many products were replaced by plastics, synthetic polymers, ceramics, and other materials, and in part because a worldwide economic slump resulted in lower consumption of minerals. Today on the world market there is even a glut of some minerals, which has caused their value to spiral downward. However, there is always the possibility that changes in the worldwide economic situation will contribute to mineral shortages.

One of the most significant economic factors in mineral production is the cost of energy. Mining and refining of minerals require a great deal of energy, particularly if the mineral is being refined

### Antarctica—Should We Mine It or Leave It Alone?

To date, no substantial mineral deposits have been found in Antarctica, although smaller amounts of valuable minerals such as gold and platinum have been discovered. Geologists think it likely that major deposits of valuable metals and oil are present and that they will be discovered in the future. Many nations have been involved in negotiations on the future of Antarctica and its possible mineral wealth.

The issue of ownership of mines in Antarctica would have to be resolved if mining were to become a reality there. Some of the world's poorer nations insist that all nations should share Antarctica's mineral wealth equally. In addition, conflicting territorial claims to most of Antarctica by Chile, Argentina, Australia, New Zealand, Great Britain, France, and Norway further complicate the ownership issue (see map).

The Antarctic Treaty, an international agreement that has been in effect since 1961, limits activity in Antarctica to scientific studies (and thus excludes military operations). During the 1980s, nearly a decade of delicate negotiations among 33 nations resulted in the writing of a pact that would have permitted strictly controlled exploitation of Antarctica's minerals. Each mining operation was to be allowed only if *all* signatory countries unanimously approved it. The pact, called the Convention on the Regulation of Antarctic Mineral Resource Activities, or CRAMRA, also required unanimous agreement in order to be ratified.

In 1989, several countries refused to support the pact because of concerns that *any* mineral exploitation would damage Antarctica's environment. These countries, led by Australia and France, contend that a new agreement should be developed that bans all mining in Antarctica forever, in effect turning Antarctica into an ecological park.

Why is there concern about Antarctica's environment? Polar regions such as Antarctica are extremely vulnerable to human activities. Even scientific investigations, with their trash, pollution, and noise, have negatively affected the wildlife in Antarctica. No one doubts that large-scale mining operations would wreak havoc on such a fragile environment.

The issue that must be resolved, then, is whether we want to sacrifice the pristine environment of Antarctica for the economic gain associated with mineral exploitation. The reason this matter needs to be resolved now—*before* major mineral deposits are discovered—is so that uncontrolled "gold rush" mining operations, with their associated environmental destruction, can be prevented.

---

from low-grade ore. For example, gold is currently being extracted from low-grade ores in Nevada. For every 0.9 metric ton (2,000 lb) of rock that is dug and crushed, as little as 0.7 g (0.025 oz) of gold is refined. Huge amounts of energy are required to dig and crush countless tons of rock. Higher energy prices result in higher production costs, which may cause the market price of minerals to increase. In turn, the higher prices decrease mineral consumption, extending supplies.

Economic factors aside, prediction of future mineral needs is also difficult because it is impossible to know when or if there will be new discoveries of mineral reserves or replacements for minerals (such as plastics). It is also impossible to know when or if new technological developments will make it economically feasible to extract minerals from low-grade ores.

With these reservations in mind, it is safe to say that, assuming present rates of consumption, minerals now in short supply will probably last between 20 and 100 years. During your lifetime, then, several important minerals will become increasingly scarce, and it is likely that current world reserves of silver, copper, mercury, tungsten, and tin will be exhausted.

Another reasonable projection is that the prices of even relatively plentiful minerals, such as iron and aluminum, will increase during your lifetime. The depletion of large, rich, and easily acces-

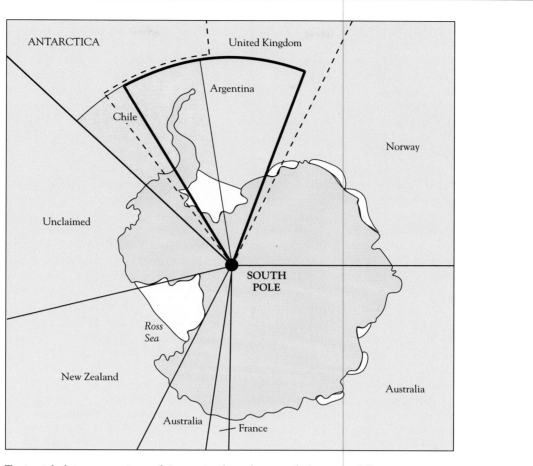

Territorial claims to portions of Antarctica have been made by seven different countries. Some of these claims overlap.

sible deposits of these metals will require mining and refining of low-grade ores, which will be more expensive.

## INCREASING OUR MINERAL SUPPLIES

Some economists consider minerals to be an inexhaustible resource. Even though minerals are non-renewable in the sense that there is only a limited supply of each kind in the Earth's crust, this economic view has some validity. As a resource becomes scarce, efforts intensify to discover new supplies, to conserve existing supplies of that resource, and to develop new substitutes for it. Although many reserves have been discovered and exploited, others that are as yet unknown may be found. In addition, the development of advanced mining technologies may make it possible to exploit known resources that are too expensive to develop using existing techniques.

### Locating and Mining New Deposits

Many known mineral reserves have not yet been exploited. For example, although Indonesia is known to have many rich mineral deposits, its thick forests and malaria-carrying mosquitos have made accessibility to these deposits very difficult.

Both northern and southern polar regions have had little mineral development. This is due in part to a lack of technology for mining in frigid environments. For example, normal offshore drilling rigs cannot be used in Antarctic waters, because the shifting ice formed during the harsh winter would tear the rigs apart. As new technologies become available, increasing pressure will be exerted to mine in northern Canada, Siberia, and Antarctica (see Focus On: Antarctica—Should We Mine It or Leave It Alone? on page 334).

Plans are afoot to exploit some of the rich mineral deposits in Siberia, although new technologies will have to be developed to make this feasible. Some of the ore deposits in Siberia have unusual combinations of minerals (for example, potassium combined with aluminum) that cannot be separated using existing technology.

Is there a possibility that currently unknown mineral deposits will be discovered at some future time? The U.S. Geological Survey thinks that undiscovered mineral deposits may exist, particularly in developing countries where detailed geological surveys have not been performed. It is likely that a detailed survey of the western portion of South America, along the Andes Mountains, will reveal mineral deposits. Geologists also presume that minerals will be found in the Amazon Basin, although in many ways the rain forest makes these deposits as inaccessible as those in Antarctica. Examination of certain areas deep in the rain forest to ascertain the likelihood of deposits being present creates logistical problems and poses a grave environmental threat.

Geologists also consider it likely that deep deposits, those buried 1,000 or more meters in the Earth's crust, will someday be discovered and exploited. The special technology required to mine deep deposits is not yet available.

## Minerals from the Oceans

The mineral reserves of the oceans may also provide us with future supplies. The sea floor may be mined, particularly where minerals have accumulated in the loose ocean sediments. Alternatively, minerals could be extracted from seawater.

**Ocean Floor**  Although large deposits of minerals lie on the ocean floor, the expense of obtaining them is prohibitive, given the current technology. These deposits may never be economical to acquire, and the environmental impacts of mining them are another obstacle to their exploration and development (see Focus On: Safeguarding the Seabed).

Significant deposits of iron, copper, and zinc and lesser amounts of gold and silver were discovered in the Red Sea in the 1960s. The governments of Saudi Arabia and Sudan, both of which border the Red Sea, are investigating the possibility of a joint venture to mine these deposits by dredging the floor of the Red Sea and vacuuming the loose sediments up a tube to a ship. The minerals in this slurry would then be refined on land.

**Manganese nodules**—small rocks the size of potatoes that contain manganese and other minerals, such as iron, copper, and nickel—are widespread on the ocean floor, particularly in the Pacific (Figure 15–7). According to the Marine Policy Center at the Woods Hole Oceanographic Institute, the estimates of these reserves are quite large. However, dredging manganese nodules from the ocean floor would adversely affect sea life. Further, it is not clear which country, if any, has the legal right to these minerals, which are in international waters. Monitoring and policing their removal would almost certainly require international cooperation.

**Seawater**  Seawater, which covers approximately three-fourths of our planet, contains many different dissolved minerals. The total amount of minerals available in seawater is staggering, but their concentrations are very low. Currently, sodium chloride (common table salt), bromine, and magnesium

**Figure 15–7**
Manganese nodules on the ocean floor. These nodules have enticed miners, but the commercial feasibility of obtaining them is out of reach. (Visuals Unlimited/Science VU)

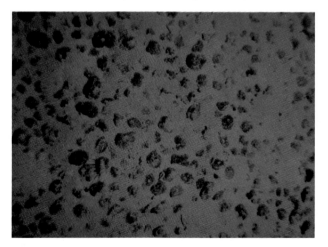

Focus On

### Safeguarding the Seabed

The oceans cover about two-thirds of our planet and constitute a vast resource that is tapped by humans for food and scientific research. Its potential to supply us with minerals has been dreamed of for many years, but exploiting the floor of the oceans is controversial. Many people think it is inevitable that minerals will be mined from the floor of the deep sea. Others think the seabed should be declared off limits because of the potential ecological havoc mining could wreak on the diverse life forms inhabiting the oceans. In addition to ecological problems, mining the loose sediments of the deep sea also poses ethical and legal questions. An international conference held in Berlin in 1991, the Dahlem Workshop, attempted to address some of these issues. The participants, mostly scientists, discussed such topics as how the mineral resources of the ocean floor would be harvested, to whom they belong, and how ocean mining should be regulated to protect the ocean's life forms.

These problems have been grappled with since the 1960s, when a number of developed countries, including the United States, expressed an interest in removing manganese nodules from the ocean floor. Their interest triggered the formation of an international treaty called the **United Nations Convention on the Law of the Sea** (UNCLOS), which becomes effective when 60 countries have ratified it. This treaty views the minerals and organisms in the open seas as belonging to all of humankind. It requires the establishment of an international group to oversee seabed mining. This group would have the power to sell mining rights to a particular country or investor. The money thus obtained would be used to help developing countries. Another requirement of the UNCLOS is that any initial investor must bring in a developing country as a joint partner. So far, 45 countries have ratified the treaty, although the United States has not.

It seems likely that ocean mining will not become technologically feasible or profitable until sometime in the 21st century. Nevertheless, the fact that experts are trying to address the complex issues associated with undersea mining well in advance shows foresight. Their discussions may eventually result in international agreements on who can mine the ocean floor; when, where, and how they can extract minerals; and how the ocean environment will be protected for the sake of its many life forms. Had such talks been instigated in advance of mining on land, perhaps the environmental destruction that inevitably accompanies surface mining would have been minimized.

---

can be profitably extracted from seawater. It may be possible in the future to profitably extract other minerals from seawater and concentrate them, but current mineral prices and technology make this impossible now.

### Advanced Mining Technology

We have already mentioned that special technologies will be needed to mine minerals in inaccessible areas such as polar regions and deep deposits. Capitalizing on large, low-grade mineral deposits throughout the world will also require the development of special techniques. As minerals grow scarcer, economic and political pressure to exploit low-grade ores will increase. Obtaining high-grade metals from low-grade ores is an expensive proposition, in part because a great deal of energy must be expended to obtain enough ore. Future technology may make such exploitation more energy-efficient, thereby reducing costs.

Even if advanced technology makes obtaining minerals from low-grade ores feasible, other factors may limit exploitation of this potential source. In arid regions, the vast amounts of water required during the extraction and processing of minerals may be the limiting factor. Also, the environmental costs may be too high, because obtaining minerals from low-grade ores causes greater land disruption and produces far more pollutants than does the development of high-grade ores.

# EXPANDING OUR SUPPLIES THROUGH SUBSTITUTION AND CONSERVATION

Because much of our civilization's technology depends upon minerals, and because certain minerals may be unavailable or quite limited in the future, our society should extend existing mineral supplies as far as possible through substitution and conservation.

## Finding Mineral Substitutes

The substitution of more abundant materials for scarce minerals is an important goal of manufacturing. The search for substitutes is driven in part by economics; one effective way to cut production costs is to substitute an inexpensive or abundant material for an expensive or scarce one. In recent years, plastics, ceramic composites, and high-strength glass fibers have been substituted for scarcer materials in many industries.

Because of the use of substitutes, the amount of steel in American automobiles has been decreasing for a number of years. The typical 1988 model contained almost 227 kg (500 lb) less plain carbon steel than the 1978 model. High-strength alloys and synthetics such as plastic have been substituted for steel, in part because their lighter weight results in better gasoline mileage.

Earlier in this century, tin was a critical metal for can-making and packaging industries; since then, other materials have been substituted for tin, including plastic, glass, and aluminum. The amounts of lead and steel used in telecommunications cables have decreased dramatically during the past 35 years, while the amount of plastics has had a corresponding increase. In addition, glass fibers have replaced copper wiring in telephone cables.

Although substitution can extend our mineral supplies, it is not a cure-all for dwindling resources. Certain minerals have no known substitutes. Platinum, for example, catalyzes many chemical reactions that are important in industry. So far, no other substance has been found that possesses the catalyzing abilities of platinum.

## Mineral Conservation

Our mineral supplies can be extended by conservation. In **recycling,** used items such as beverage cans and scrap iron are collected, remelted, and reprocessed into new products. The **reuse** of items such as beverage bottles (which can be collected, washed, and refilled) is another way to extend mineral resources. In addition to the introduction of specific conservation techniques such as recycling

(a)

(b)

**Figure 15–8**

Recycling of scrap metal. (a) The metal in these old, discarded automobiles will be recycled. (b) Iron and steel scrap, obtained from ground-up automobiles, will be fashioned into new products. (Courtesy of the Institute of Scrap Recycling Industries, Inc.)

and reuse, public awareness and attitudes about resource conservation can be modified to encourage low waste.

**Recycling**   A large percentage of the products made from minerals—including cans, bottles, chemical products, electronic devices, and batteries—are typically discarded after use. The minerals in some of these products—batteries and electronic devices, for instance—are difficult to recycle. Minerals in other products, such as paints containing lead, zinc, or chromium, are lost through normal use.

However, there is no question that we have the technology to recycle many other mineral products. Recycling of certain minerals is already a common practice. Significant amounts of gold, lead, nickel, steel, copper, silver, zinc, and aluminum are now being recycled (Figure 15–8).

Recycling has several advantages in addition to extending mineral resources. It saves unspoiled land from the disruption of mining, reduces the amount of solid waste that must be disposed (see Chapter 23), and reduces energy consumption and pollution. For example, recycling an aluminum beverage can saves the energy equivalent of about 180 ml (6 oz) of gasoline. Recycling aluminum also reduces the emission of aluminum fluoride, a toxic air pollutant produced during aluminum processing.

More than half the aluminum cans in the United States are currently being recycled. The aluminum industry, local governments, and private groups have established more than 5,000 recycling centers across the country. People who turn in aluminum cans receive a small refund (usually a penny or a nickel) for each can. It takes approximately six weeks for a can that has been returned to be melted, reformed, filled, and put back on a supermarket shelf. Clearly, however, more recycling is possible. It may be that today's sanitary landfills will become tomorrow's mines, as valuable minerals and other materials are extracted from them.

**Reuse**   When the same product is used over and over again, as when beverage containers are collected, washed, and refilled, both mineral consumption and pollution are reduced. The benefits of reuse are even greater than those of recycling. For example, to recycle a glass bottle requires crushing it, melting the glass, and forming a new bottle. Reuse of a glass bottle simply requires washing it, which obviously expends less energy than recycling. Reuse is a national policy in Denmark,

where nonreusable beverage containers are prohibited.

A number of countries and states have adopted beverage container deposit laws, which require consumers to pay a deposit, usually a nickel, for each beverage bottle or can they purchase. The nickel is refunded when the container is returned to the retailer or to special redemption centers. In addition to encouraging recycling and reuse, thereby reducing mineral resource consumption, beverage container laws save tax money by reducing litter and solid waste. Countries that have adopted beverage container deposit laws include the Netherlands, Norway, and Sweden. Parts of Canada and the United States also have deposit laws.

**Changing Our Mineral Requirements**   We can reduce our mineral consumption by becoming a low-waste society. Americans have developed a "throwaway" mentality in which damaged or unneeded articles are discarded (Figure 15–9). This attitude has been encouraged by industries looking for short-term economic profits, even though the long-term economic and environmental costs of such an attitude are high. Products that are durable and repairable enable us to consume fewer resources. Laws such as those requiring a deposit on beverage containers also reduce consumption by encouraging recycling and reuse.

**Figure 15–9**

The throw-away mentality of our industrial society is evident in this heap of discarded items. Much of this material can be recycled. (Courtesy of the Institute of Scrap Recycling Industries, Inc.)

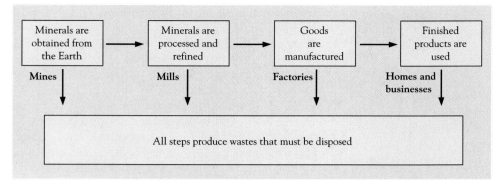

(a)

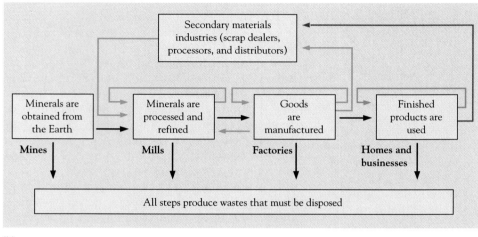

(b)

Key:

━━━  Sustainable manufacturing

━━━  Consumer reuse

━━━  Consumer recycling

**Figure 15–10**
The flow of minerals in an industrial society. (a) The traditional flow of minerals is a one-way direction from the Earth to solid waste production. Massive amounts of solid waste are produced at all steps from mining the mineral to discarding the used-up product. (b) The flow of minerals in a low-waste society is more complex, with recycling and reuse practiced at intermediate steps.

The throwaway mentality has also been evident in manufacturing industries. Traditionally, industries consumed raw materials and produced not only goods but a large amount of waste that was simply discarded (Figure 15–10a). Increasingly, however, manufacturers are finding that the waste products from one manufacturing process can be used as raw materials in another industry. By selling these "wastes," industries gain additional profits and lessen the amounts of materials that must be thrown away. For example, ARCO's oil refinery in

Los Angeles now sells its used alumina catalysts to Allied Chemical. Until this market was found, used alumina catalysts were classified as a hazardous waste that required special disposal procedures. Such minimization of waste by industry is known as **sustainable manufacturing** (Figure 15–10b).

**Dematerialization**    As products evolve, they tend to become lighter in weight and often smaller. Washing machines manufactured in the 1960s were much heavier than comparable machines manufac-

**Figure 15–11**
Dematerialization. The first general-purpose electronic calculator/computer was built at the University of Pennsylvania in 1946. The power source for this calculator required another room. Today a computer that is a fraction the size of this one could perform the calculations done by this massive computer. (UPI/Bettmann Archive)

tured in the 1990s, for example. The same is true of other household appliances, automobiles, and electronic items (Figure 15–11). This decrease in the weight of products over time is called **dematerialization.**

Although dematerialization gives the appearance of reducing consumption of minerals and other materials, it may have the opposite effect. Often products that are smaller and lighter are of lower quality. Because repairing broken lightweight items is difficult and may cost more than the original products, consumers are encouraged by retailers and manufacturers to replace rather than repair the items. Thus, although the weight of materials being used to make each item has decreased, the *number* of such items being used in a given period of time may have actually increased.

## SUMMARY

**1.** Minerals, which are essential to our industrial society, are naturally occurring elements and compounds. They may be metallic or nonmetallic. The Earth's crust has an uneven distribution of minerals, which are often found in concentrated, highly localized deposits due to magmatic concentration, hydrothermal processes, sedimentation, or evaporation.

**2.** Mineral reserves are deposits that have been identified and are profitable to extract, whereas mineral resources are deposits of low-grade ores that may or may not be profitable to extract in the future. Depletion time is an estimate of the time remaining before 80 percent of known reserves of a particular element will be depleted. It is extremely difficult to predict depletion times accurately, because mineral consumption changes over time. Economic factors influence all aspects of mineral consumption.

**3.** Steps in converting a mineral deposit into a usable product include locating the deposit, mining, and processing or refining. If the deposit is near the Earth's surface, it is extracted by open-pit surface mining or strip mining. Subsurface mines, either shaft mines or slope mines, are used to obtain minerals located deep in the Earth's crust.

**4.** Any form of mineral extraction and processing has negative effects on the environment. Surface mining disturbs the land more than subsurface mining, but subsurface mining is more expensive and dangerous. Mineral processing can cause air, soil, and water pollution. Derelict lands, which are extensively damaged due to mining, can be restored to prevent further degradation and to make the land productive for other purposes; land reclamation is, however, extremely expensive.

**5.** Developed nations consume a disproportionate share of the world's minerals, but as developing countries industralize, their need for minerals increases. The richest concentrations of minerals in highly industrialized countries have largely been exploited. As a result, these nations have increasingly turned to developing countries for the minerals they require. Sometimes developed nations must rely on politically volatile nations for strategically important minerals.

6. Mineral supplies can be extended in several ways. It is possible that new deposits will be identified. Advanced mining technology may make it possible to profitably extract minerals from inaccessible regions or from low-grade ores.

7. Substitution and conservation extend mineral supplies. Manufacturing industries continually try to substitute more common, less expensive minerals for those that are scarce and expensive. Mineral conservation includes recycling, in which discarded products are collected and reprocessed into new products, and reuse, in which a product is collected and used over again. In addition to conserving minerals, recycling and reuse cause less pollution and save energy when compared to the extraction and processing of virgin ores.

8. If our mineral supplies are to last and if our standard of living is to remain high, consumers must decrease their consumption. To accomplish this, manufacturers must make high-quality, durable products that can be repaired.

# DISCUSSION QUESTIONS

1. Distinguish among the following ways in which mineral deposits may form: magmatic concentration, hydrothermal processes, sedimentation, and evaporation.

2. Explain why it is difficult to obtain an accurate appraisal of our total mineral resources.

3. Discuss several harmful environmental effects of mining and processing minerals.

4. Explain why obtaining minerals from low-grade ores is more environmentally damaging than extracting them from high-grade ores.

5. Some people in industry argue that the planned obsolescence of products, which means they must be replaced often, creates jobs. Others think that the production of smaller quantities of durable, repairable products would create jobs and stimulate the economy. Explain each viewpoint.

6. What is sustainable manufacturing, and whom does it benefit?

7. Outline the benefits of beverage container deposit laws. Why do you think such laws are not in place throughout the United States? Who might oppose such laws?

# SUGGESTED READINGS

Barnhardt, W. The Death of Ducktown. *Discover,* October 1987. Discusses the environmental history and future of the Copper Basin area of Tennessee.

Frosch, R. A., and N. E. Gallopoulos. Strategies for Manufacturing. *Scientific American,* September 1989. Creative solutions that will enable us to keep our industrialized standard of living and at the same time stop poisoning the environment.

Gillis, A. M. Bringing back the land. *BioScience* 41:2, February 1991. Scientists at Utah State University are pioneering rapid reclamation of arid land that has been surface mined.

Kimball, L. A. *Southern Exposure: Deciding Antarctica's Future.* World Resources Institute and the Tinker Foundation, Washington, D.C., 1990. The author builds a case for the importance of Antarctica and the need for a strong international agreement on its management.

World Resources Institute in collaboration with the U.N. Environment and Development Programs. *World Resources 1992–93.* Oxford University Press, New York, 1992. Table 21.5 lists the production, consumption, and reserves of selected metals from 1975 to 1990, and Table 21.6 shows the 1990 world reserves of major metals.

Young, J. E. "Reducing Waste, Saving Materials," in *State of the World: 1991, a Worldwatch Institute Report.* W.W. Norton & Company, New York, 1991. Examines all aspects of the throwaway society, from our increasing need for minerals to our disposal of mineral wastes in landfills.

Young, J. E. Aluminum's real tab. *World Watch* 5:2, March/April 1992. Although cheap and invaluable to society, aluminum has hidden environmental costs that make it very expensive.

Young, J. E. "Mining the Earth," in *State of the World: 1992, a Worldwatch Institute Report on Progress Toward a Sustainable Society.* W.W. Norton & Company, New York, 1992. Discusses the importance of minerals in the global economy and their environmental costs.

A mated pair of bald eagles. (Stan Osolinski/Dembinsky Photo Associates)

# Wildlife: Our Plant and Animal Resources

lthough biological diversity (the number and variety of Earth's life forms) is one of our most important natural resources, it is not fully appreciated and is not utilized to its full potential. The term "biological diversity" encompasses three different aspects: genetic diversity within a species, species diversity (number of different species), and ecosystem diversity (variety within and between ecosystems). Biological diversity contributes toward a sustainable environment—one that provides the life support system which enables humans as well as other species to survive. The Earth's species and their genetic materials also provide us with industrial, agricultural, and medicinal products. The reduction in biological diversity caused by extinction of the Earth's living organisms results in lost opportunities and lost solutions to future problems.

## THE BALD EAGLE: MAKING A COMEBACK

The American bald eagle—the symbol of the United States and an emblem of strength—was a common sight throughout colonial North America. More recently, however, the bald eagle fell on hard times. Its numbers dropped precipitously—to fewer than 5,000 nationwide in 1979—until it was in danger of extinction. Several factors contributed to its decline. As European settlers pushed across North America, they cleared many thousands of square kilometers of forest near lakes and rivers, thus destroying the bald eagle's habitat. Eagles were hunted for sport and because it was thought they preyed on livestock and commercially important fish. In fact, bounties were offered for dead bald eagles as recently as 1952. In addition, eagles' numbers dwindled because they could not reproduce at high enough levels to ensure their population growth or their survival. This reproductive failure was the direct result of the eagles ingesting prey contaminated with the pesticide dichloro-diphenyl-trichloroethane (DDT). DDT caused the eagles' eggs to be so thin-shelled that they cracked open before the embryos could mature and hatch. Mercury, lead, and selenium were other environmental pollutants that harmed bald eagles. More recently, the 1989 Exxon oil spill in Prince William Sound caused the demise of many Alaskan bald eagles.

Conservation efforts have helped the bald eagle make a remarkable comeback. In the mid-1970s, the first eagles to be bred in captivity were released in the wild. In addition to raising birds in captive breeding programs, biologists also remove eagle eggs from their nests in the wild, raise the baby eagles in wildlife refuges, and return them to the wild. (Removal of eggs actually helps increase the number of eagles, because nesting eagles commonly lay more eggs to replace those that were removed.) As a result of continuing efforts, the number of nesting pairs in the continental United States doubled from 1,000 to 2,000 between 1975 and 1990.

Today many states are reintroducing bald eagles to the wild. Save The Eagle Project (STEP), a private, nonprofit conservation organization, works with the National Fish and Wildlife Foundation to restore bald eagle populations in the United States. Federal and state governments have supported such efforts, and private and corporate donors have also been generous in their support.

Although bald eagles are still low in number in every state but five (Washington, Oregon, Minnesota, Wisconsin, and Michigan), it is clear that they have a fighting chance for survival. Studies are being conducted to determine the precise habitat requirements for these birds so that, once their numbers have been restored, they can be sustained.

The world is a richer place because of the bald eagle. Today it symbolizes more than a country, for the bald eagle demonstrates that our biological heritage can be preserved if enough people care to do something about it. In this chapter we examine the importance of all forms of plant and animal wildlife and consider extinction, which has become an increasing threat to so many species. Finally, we explore what can be done to preserve our wildlife resources and save the many endangered species from disappearing from the Earth forever.

## HOW MANY ANIMAL AND PLANT SPECIES ARE THERE?

We do not know exactly how many species exist, but most biologists estimate that there are at least 5 to 10 million different species.[1] Some biologists think there may be as many as 30 to 100 million, or even more, species. So far, approximately 250,000 flowering plant species, 800,000 lower plant species, and 1.5 million animal species have been identified.

The variation among living organisms is referred to as **biological diversity** or **biodiversity** (Figure 16–1), but the concept includes much more than simply the *number* of different species (called **species diversity**). It also takes into account **genetic diversity,** the genetic variety *within* a species—that is, the different populations that make up a particular species (Figure 16–2). Biological diversity also encompasses **ecosystem diversity,** the variety of interactions among living things in natural communities. For example, a forest community, with its trees, shrubs, vines, herbs, insects, worms, vertebrate animals, fungi, bacteria, and other microorganisms, has greater ecosystem diversity than does a cornfield (Figure 16–3). Ecosystem diversity also means the variety of ecosystems found on Earth: the forests, prairies, deserts, coral reefs, lakes,

(Text continues on page 346.)

[1] A species is a group of more or less distinct organisms that are capable of interbreeding with one another in the wild but do not interbreed with other sorts of organisms.

(a)

(b)

(c)

(d)

(f)

**Figure 16–1**

There is an incredible array of biological diversity on Earth, as shown in this sampling of life in tropical rain forests. (a) *Herrania* blooms in eastern Ecuador. (b) A dart-poison frog (*Dendrobates pumilio*) in Costa Rica. (c) A katydid (*Panalanthus*) in Ecuador. (d) Mushrooms on a fallen tree in Panama. (e) A pair of Geoffrey's marmosets in Brazil. (f) A rufous-backed kingfisher in Borneo. (a, c, e, and f, Doug Wechsler; b, James L. Castner; d, Norbert Wu)

**Figure 16–2**
The variation in corn kernels and ears is evidence of the genetic diversity in corn. The diversity in corn and many other crop plants is preserved at the Seed Savers Exchange in Decorah, Iowa. (David Cavagnaro)

**Figure 16–3**
Tropical rain forests such as this one in Micronesia exhibit a high degree of ecosystem diversity. A single acre of lush tropical rain forest may contain more different species than are found in some countries. (G. J. James/Biological Photo Service)

coastal estuaries, and other ecosystems of our planet.

## WHY WE NEED WILDLIFE

Humans depend on the contributions of thousands of different plant and animal species for their survival. In primitive societies, these contributions are direct: plants and animals are the sources of food, clothing, and shelter. In more advanced societies most people do not hunt for their morning breakfasts or cut down trees for their shelter and firewood; nevertheless, we still depend on living organisms.

Although all societies make use of many different kinds of plants, animals, and microorganisms, most living things have never been evaluated for their potential usefulness to humans. There are approximately 250,000 different plant species, but 225,000 of them have *never* been evaluated with respect to their industrial, medicinal, or agricultural potential. The same is true of most of the millions of animal species. Most people don't think of insects as an important biological resource, for example, but insects are instrumental in several important environmental and agricultural processes including pollination of crops, weed control, and insect pest control. Microorganisms such as bacteria and fungi provide us with foods, medicines, and important environmental services such as soil enrichment. As long as biological diversity remains high, it represents a rich, untapped resource for future uses and benefits: many as-yet-unknown species may someday provide us with products, for example. A reduction in biological diversity decreases this "treasure" permanently.

### Ecosystem Stability and Species Diversity

You may recall from Chapters 3, 4, and 5 that the living world functions much like a complex machine. Each ecosystem is composed of many separate parts, the functions of which are organized and integrated to maintain the ecosystem's overall performance. The activities of all living things are interrelated; we are bound together and dependent upon one another, often in subtle ways (Figure 16–4).

Plants, animals, and microorganisms are instrumental in a number of environmental processes without which humans could not exist. Forests are not just a potential source of lumber; they provide

**Figure 16–4**

The alligator plays an integral, but often subtle, role in its natural ecosystem. It helps maintain populations of smaller fish by eating the gar, a fish that preys on them. Alligators dig underwater holes that other aquatic organisms use during periods of drought when the water level is low. The nest mounds they build are enlarged each year and eventually form small islands that are colonized by trees and other plants. In turn, the trees on these islands support heron and egret populations. The alligator habitat is maintained in part by underwater "gator trails," which help to clear out aquatic vegetation that might eventually form a marsh.

(Connie Toops)

watersheds from which we obtain fresh water, and they reduce the number and severity of local floods. Many species of flowering plants depend upon insects to transfer pollen for reproduction. Animals and microorganisms help to keep the populations of various species in check so that the numbers of one species do not increase enough to damage the stability of the entire ecosystem. Soil dwellers, from earthworms to bacteria, develop and maintain soil fertility for plants. Bacteria and fungi perform the crucial task of decomposition, which allows nutrients to recycle in the ecosystem.

You might think that the loss of some species from an ecosystem would not endanger the rest of the living organisms, but this is far from true. Imagine trying to assemble an automobile if some of the parts were missing. You might be able to piece it all together so that it resembled a car, but it probably wouldn't run. Similarly, the removal of organisms from a community makes an ecosystem run less smoothly. If enough organisms are removed, the entire ecosystem can collapse.

**Envirobrief**

**The Megadiverse Countries**

A relative few of the world's nations contain a surprising percentage of its biological diversity. Most of these "megadiverse" countries are located in the Southern Hemisphere and have neither the funds nor the political will to adequately protect the organisms within their borders from unsustainable commercial exploitation. Four of them (Brazil, Zaire, Madagascar, and Indonesia) account for two-thirds of all primate species; four (Mexico, Brazil, Indonesia, and Australia) are home to more than one-third of all reptiles; seven (Brazil, Colombia, Mexico, Zaire, China, Indonesia, and Australia) have more than half of all flowering plants; and three (Brazil, Zaire, and Indonesia) contain more than half of the world's tropical rain forests. Following is a sampling of the numbers of animal species in some of these countries:

Brazil—428 species of mammals; 1,622 species of birds; 467 species of reptiles

Zaire—409 species of mammals

Indonesia—515 species of mammals; 1,519 species of birds; more than 600 species of reptiles

Mexico—449 species of mammals; 717 species of reptiles

Australia—686 species of reptiles

Colombia—359 species of mammals; 1,721 species of birds; 383 species of reptiles

China—394 species of mammals; 1,195 species of birds

## Genetic Reserves

The maintenance of a broad genetic base for economically important plants and animals is critical. During the **green revolution** of the 20th century, plant scientists developed genetically uniform, high-yielding varieties of important food crops such as wheat. It quickly became apparent, however, that genetic uniformity resulted in increased susceptibility to pests and disease. By crossing the "super strains" with more genetically diverse relatives, disease and pest resistance can be reintroduced into such plants. For example, a corn blight fungus that ruined the corn crop in the United States in 1970 was brought under control by crossing the cultivated, highly uniform U.S. corn varieties with genetically diverse ancestral varieties from Mexico. When some of the genes from Mexican corn were incorporated into the U.S. varieties, the latter became resistant to the corn blight fungus.

## Scientific Importance of Genetic Diversity

Genetic engineering, the incorporation of genes from one organism into an entirely different species (see Chapter 18), makes it possible to use the genetic resources of living organisms. The gene for human insulin, for example, has been placed in bacteria, which subsequently become tiny chemical factories, manufacturing insulin that can be used by diabetics (Figure 16–5). Genetic engineering has the potential to provide us with new vaccines, safer pesticides, and more productive farm animals and food crops.

Although we have the skills to transfer genes from one organism to another, we do not have the ability to *make* genes. Genetic engineering depends upon a broad base of genetic diversity from which it can obtain genes. It has taken hundreds of millions of years for evolution to produce the genetic diversity found in organisms living on our planet today, a diversity that may hold solutions not only to problems we have today but to problems we have not even begun to conceive. It would be very unwise to allow such an important part of our heritage to disappear.

## Medicinal, Agricultural, and Industrial Importance of Wildlife

The genetic resources of living organisms, particularly plants, are vitally important to the pharmaceutical industry, which incorporates into its medicines many hundreds of chemicals derived from plants. From extracts of cherry and horehound for cough medicines to certain ingredients of periwinkle and mayapple for cancer therapy, derivatives of plants play important roles in the treatment of illness and disease (Figure 16–6).

The agricultural importance of plants and animals is indisputable, since humans must eat to survive. However, the number of different kinds of foods we eat is limited when compared with the total number of edible species on Earth. There are probably many plant and animal species that are nutritionally superior to our common foods. For example, quinoa (*Chenopodium quinoa*), a plant native to the Andes Mountains in South America, looks and tastes somewhat like rice but has a much higher concentration of protein and is more nutritionally balanced. Winged beans (*Psophocarpus tetragonolobus*) are a tropical legume from Southeast Asia and Papua New Guinea (Figure 16–7). Because the seeds of the winged bean contain large

**Figure 16–5**
Genetic engineering, which results in new combinations of genes, can produce novel traits in organisms. Human insulin for diabetics is now produced by genetically altered bacteria. (Visuals Unlimited/© SIU)

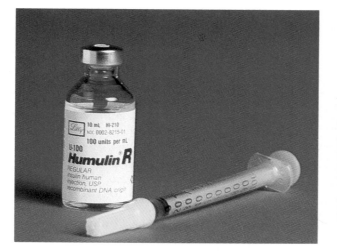

**Figure 16–6**
The rosy periwinkle, *Catharanthus roseus*, produces chemicals that are effective against certain cancers. (Doug Wechsler)

**Figure 16–7**
A large number of edible plants have not been utilized to any great extent, including the winged bean, a tropical legume that is nutritionally superior to many other foods. (Sigrid Salmela © 1993 Animals Animals/Earth Scenes)

**Figure 16–8**
The armadillo is used in research on leprosy because it is one of only two species known to be susceptible to that disease (the other species is *Homo sapiens*, or humans). (E. R. Degginger)

quantities of protein and oil, they may be the tropical equivalent of soybeans. Almost all parts of the plant are edible, from the young, green seedpods to the starchy storage roots. The European fallow deer (*Cervinae dama*) could become a more nutritious replacement for beef in our diets, because its meat is extremely low in cholesterol.

Modern industrial technology depends upon a broad range of genetic material from living organisms, particularly plants, that are used in many products. Plants supply us with oils and lubricants, perfumes and fragrances, dyes, paper, lumber, waxes, rubber and other elastic latexes, resins, poisons, cork, and fibers. Animals provide wool, silk, fur, leather, lubricants, waxes, and transportation, and they are important in scientific research (Figure 16–8).

Insects secrete a large assortment of chemicals that represent a wealth of potential products. Certain beetles produce steroids with birth-control potential, and fireflies produce an antiviral compound that may be useful in treating viral infections. Centipedes secrete a fungicide over the eggs

of their young that could help control the fungi that attack crops. Because biologists estimate that perhaps 90 percent of all insects have not yet been identified, insects represent a very important potential biological resource.

## Aesthetic and Ethical Value of Wildlife

Wildlife not only contributes to human survival and physical comfort; it also provides recreation, inspiration, and spiritual solace. Our natural world is a thing of beauty largely because of the diversity of living forms found in it. Artists have attempted to capture this beauty in drawings, paintings, sculpture, and photography, and it has inspired poets, writers, architects, and musicians to create works reflecting and celebrating the natural world.

The strongest ethical consideration involving the value of wildlife is how humans perceive themselves in relation to other living things. Traditionally, humankind has viewed itself as the "master" of the rest of the world, subduing and exploiting other forms of life for its benefit. An alternative view is that we humans are *stewards* of the life forms on Earth and that we should watch over and protect their existence. The conviction that all creatures have the right to exist and that humans should not cause the extinction of other living things is known as **deep ecology**. The basic tenets of deep ecology are not new; the belief in the sacredness of life held by Eastern religions such as Buddhism and Taoism is similar to that of deep ecology.

## ENDANGERED AND EXTINCT SPECIES

**Extinction,** the death of a species, occurs when the last individual member of a species dies. Extinction is an irreversible loss—once a species is extinct it can never reappear. Biological extinction is the eventual fate of all species, much as death is the eventual fate of all living organisms. Biologists estimate that for every 2,000 species that have ever lived, 1,999 of them are extinct today.

During the span of time in which living organisms have occupied Earth, there has been a continuous, low-level extinction of species, known as **background extinction.** At certain periods in the Earth's history, maybe five or six times, there has been a second kind of extinction, **mass extinction,** in which numerous species disappeared during a relatively short period of geological time (Figure 16–9). The course of a mass extinction episode may have taken millions of years, but that is a short time compared with the age of the Earth (4.6 billion years).

The causes of background extinction and past mass extinctions are not well understood, but it appears that biological and environmental factors were involved. A major climate change could have triggered the mass extinction of species. Marine organisms are particularly vulnerable to temperature changes; if the Earth's temperature changed by just a few degrees, for example, it is likely that many marine species would have become extinct.

It is also possible that mass extinctions of the past were triggered by catastrophes, such as the collision of the Earth and a large meteorite. The impact of a meteorite could have forced massive quantities of dust into the atmosphere, blocking the sun's rays and cooling the planet.

### Extinctions Today

Although extinction is a natural biological process, it can be greatly accelerated by human activities. The burgeoning human population has forced us to spread into almost all areas of the Earth, and when-

**Figure 16–9**

Several mass extinctions have occurred during the course of the history of the Earth. One of the largest occurred at the end of the Cretaceous period, approximately 65 million years ago. It was during this period of mass extinction that the archosaurs, one of five main groups of reptiles, largely became extinct. The only archosaur descendants surviving today are crocodiles and birds.

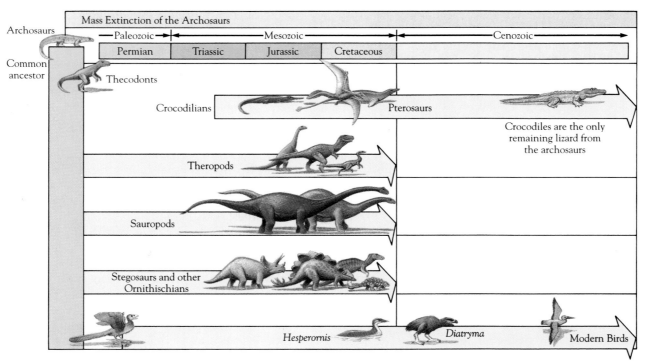

ever humans invade an area, the habitats of many plants and animals are disrupted or destroyed, which can lead to their extinction. For example, the dusky seaside sparrow, a small bird that was found only in the marshes of St. Johns River in Florida, became extinct in 1987, largely due to human destruction of its habitat.

Currently, the Earth's biological diversity is disappearing at an alarming rate (Table 16–1). Conservation biologists estimate that at least one species becomes extinct each day and that it is likely that a substantial portion of the Earth's biological diversity will be eliminated within the next few decades. As many as one-fourth of the higher plant

**Table 16–1**

WILDLIFE AT RISK

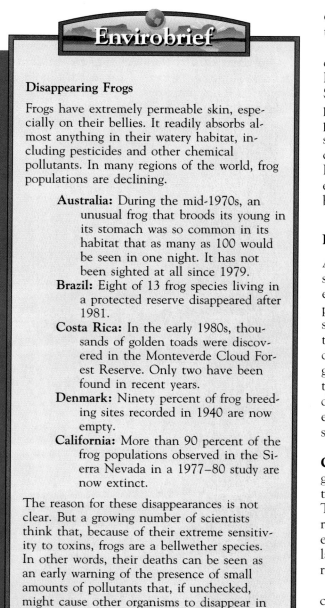

**Disappearing Frogs**

Frogs have extremely permeable skin, especially on their bellies. It readily absorbs almost anything in their watery habitat, including pesticides and other chemical pollutants. In many regions of the world, frog populations are declining.

**Australia:** During the mid-1970s, an unusual frog that broods its young in its stomach was so common in its habitat that as many as 100 would be seen in one night. It has not been sighted at all since 1979.

**Brazil:** Eight of 13 frog species living in a protected reserve disappeared after 1981.

**Costa Rica:** In the early 1980s, thousands of golden toads were discovered in the Monteverde Cloud Forest Reserve. Only two have been found in recent years.

**Denmark:** Ninety percent of frog breeding sites recorded in 1940 are now empty.

**California:** More than 90 percent of the frog populations observed in the Sierra Nevada in a 1977–80 study are now extinct.

The reason for these disappearances is not clear. But a growing number of scientists think that, because of their extreme sensitivity to toxins, frogs are a bellwether species. In other words, their deaths can be seen as an early warning of the presence of small amounts of pollutants that, if unchecked, might cause other organisms to disappear in the future.

families[2] may be extinct by the end of the 21st century, and countless animal species that depend upon those plants for food and habitat will probably become extinct. Some biologists fear that we are

[2] A family is a level of organization in the scientific classification of living organisms. A family consists of related genera, each of which consists of related species. When a family becomes extinct, all the species of all the genera in that family no longer exist.

entering the greatest period of mass extinction in the Earth's history.

The current mass extinction differs from previous periods of mass extinction in several respects. First, it is directly attributable to human activities. Second, it is occurring in a tremendously compressed period of time (just a few decades as opposed to millions of years). Perhaps even more sobering, larger numbers of plant species are becoming extinct than in previous mass extinctions. Because plants are the base of the food chain, the extinction of animals that depend on plants cannot be far behind.

## Endangered and Threatened Species

A species is **endangered** when its numbers are so severely reduced that it is in danger of becoming extinct. When extinction is less imminent but the population of a particular species is quite low, the species is said to be **threatened.** Endangered and threatened species represent a decline in biological diversity, because as their numbers decrease, their genetic variability is severely diminished. Long-term survival and evolution depend upon genetic diversity, so its loss adds to the risk of extinction for endangered and threatened species as compared to species that have greater genetic variability.

**Characteristics of Endangered Species**    Endangered species share certain characteristics that seem to have made them more vulnerable to extinction. These include (1) an extremely small (localized) range or (2) a large (extended) range that has been extensively modified by humans; (3) living on islands; and (4) low reproductive success, usually the result of a small population size.

The area of the Earth in which a particular species is found is its **range.** Many endangered species have a very limited natural range, which makes them particularly prone to extinction if their habitat is altered. The Tiburon mariposa lily, for example, is found nowhere in nature except on a single hilltop near San Francisco. Development of that area would almost certainly cause the extinction of this species.

Species that require extremely large territories in order to survive may be threatened with extinction when all or part of their territory is modified. For example, the California condor, a scavenger bird that lives off of carrion and requires a large, undisturbed territory (hundreds of square kilometers) in order to find adequate food, is on the brink of extinction. During the five-year period from 1987 to 1992, it was no longer found in the wild

**Figure 16-10**
California condors, which are scavengers, require large territories in order to obtain enough food. They are critically endangered largely because development has reduced the size of their wilderness habitat. (M. A. Chappell © 1993 Animals Animals)

**Figure 16-11**
The giant panda, which is the symbol of the World Wildlife Fund, is threatened with extinction. (Sharon Cummings/ Dembinsky Photo Associates)

(Figure 16-10). In 1992 two zoo-bred California condors were released in the Los Padres National Forest north of Los Angeles. These young condors had been raised by a hand-operated condor puppet so that they would not come into direct contact with humans. Initially they had lived on a netted platform to protect them from falling while they learned to fly; the net was then dropped.

Many island species that are endemic to certain islands (that is, they are not found anywhere else in the world) are endangered. These organisms often have small populations that cannot be replaced by immigration should their numbers be destroyed. Because they evolved in isolation from competitors, predators, and disease organisms, they have few defenses when such organisms are introduced (usually by humans) to their habitat.

In order for a species to survive, its members must be present within their range in large enough numbers for males and females to mate. The minimum population density and size that ensure reproductive success vary from one type of organism to another. However, for all organisms, if the population density and size fall below a critical minimum level, the population declines, becoming susceptible to extinction.

Endangered species often share other characteristics. Some have low reproductive rates (the female blue whale, for example, produces a single calf every other year). Some endangered species breed only in very specialized areas (the green sea turtle, for example, lays its eggs on just a few beaches). Highly specialized feeding habits can also endanger a species. The giant panda (Figure 16-11) eats only bamboo, a plant all of whose members periodically flower and die together; when this occurs, panda populations face starvation. (Like many other endangered species, giant pandas are also endangered because of habitat destruction.)

## HUMAN CAUSES OF ENDANGERED SPECIES AND EXTINCTION

Most species facing extinction today are endangered because of the destruction of habitats by human activities (Figure 16-12; Table 16-2). We demolish habitats when we build roads, parking lots, and buildings; clear forests to grow crops or graze domestic animals; and log forests for timber. We drain marshes to build on aquatic habitats, thus converting them to terrestrial ones, and we flood terrestrial habitats when we build dams. Because most organisms are utterly dependent on a particular type of environment, habitat destruction reduces their biological range and ability to survive.

**Figure 16–12**
Baby loggerhead sea turtles crawl toward the ocean after hatching. Loggerhead sea turtles are endangered because development of coastal shorelines in the Southeastern United States has destroyed their nesting sites. Female loggerheads come ashore on sandy beaches in South Carolina to lay their eggs. (Connie Toops)

## Envirobrief

**Homeless Birds**

Sixty to eighty percent of all birds that live in North America during the summer spend the winter in Central and South America. Their journey south is often difficult, requiring some species to burn off stored fat during long, uninterrupted flights. If these small birds cannot find food and shelter because of the disappearance of their rain-forest habitat, they perish. At least 12 migrant species are in serious decline because of habitat destruction: the rose-breasted grosbeak, the ovenbird, the white-throated sparrow, the veery, the gray catbird, the northern waterthrush, the wood thrush, the gray-cheeked thrush, Swainson's thrush, the Nashville warbler, the brown thrasher, and the rufous-sided towhee.

Even habitats that are left "totally" undisturbed and natural are modified by human-produced acid rain, ozone depletion, and climate change (see Chapter 20). Acid rain is thought to have contributed to the decline of large stands of forest trees and the biological death of many freshwater lakes. Because ozone in the upper atmosphere shields the ground from a large proportion of the sun's harmful ultraviolet radiation, ozone depletion in the upper atmosphere represents a very real threat to all terrestrial life. Global climate change, which is caused in part by carbon dioxide released when fossil fuels are burned, is another threat. Such habitat modifications particularly reduce the biological diversity of species with extremely narrow and rigid environmental requirements. Wildlife is also affected by other types of pollutants, including industrial and agricultural chemicals, organic pollutants from sewage, acid wastes seeping from mines, and thermal pollution from the heated wastewater of industrial plants (Figure 16–13). (See Chapters 19 through 23 for additional discussion of the adverse effects of pollutants on the environment.)

### The Problem of Exotic Species

The introduction of a foreign, or exotic, species into an area where it is not native often upsets the balance among the organisms living in that area. The foreign species may compete with native species for food or habitat or may prey on them. Generally, an introduced competitor or predator has a greater negative effect on local organisms than do native competitors or predators. Although exotic species may be introduced into new areas by natural means, humans are usually responsible for such introductions, either knowingly or unknowingly. The blue water hyacinth, for example, was deliberately

| Table 16–2 Some Human Causes of Endangered Species and Extinction | |
|---|---|
| **Cause** | **Some Examples Mentioned in Chapter** |
| Habitat destruction | Dusky seaside sparrow (extinct), California condor |
| Commercial hunting | Snow leopard, Imperial Amazon macaw |
| Sport hunting | American bison, passenger pigeon (extinct) |
| Predator and pest control | Carolina parakeet (extinct), prairie dogs |
| Pollution (e.g., pesticides) | Bald eagles |
| Introduction of exotic species | Catalina mahogany |

**Figure 16–13**
A common loon gives her chicks a ride. Pollution, including mercury contamination, has been strongly implicated in the decline of the common loon. (Jean F. Stoick/Dembinsky Photo Associates)

brought from South America to the United States because it has lovely flowers. Today it has become a nuisance in Florida waterways, clogging them so that boats cannot easily move and crowding out native species. Another exotic species, the zebra mussel, is a small mollusk that was accidentally introduced into the United States from Europe during the 1980s when it hitched a ride on a ship. Since that time, it has been growing out of control in U.S. waterways.

In 1977 a carnivorous snail was introduced in Moorea, an island in French Polynesia, as a way to control another snail species that had been introduced by humans and had become a pest (Figure 16–14a, b). The newly introduced species started consuming native snail species in large numbers. As a result, all of the seven native species originally present in Moorea are no longer found in the wild; five exist only as small captive populations and two have become extinct. Likewise, herbivorous mammals such as goats and sheep were introduced to Santa Catalina Island, California, where they overgrazed the Catalina mahogany (*Cercocarpus traskiae*) until only seven trees remained (Figure 16–14c).

(a)

(c)

(b)

**Figure 16–14**
The introduction of exotic species often threatens native species. (a) *Euglandina rosea* hunting for prey, which may include (b) *Partula aurantia*. After this carnivorous snail was introduced on Moorea, it started consuming native *Partula* (snail) species. (c) A closeup of a branch of *Cercocarpus traskiae*, the Catalina mahogany. This species was overgrazed by herbivorous mammals introduced to Santa Catalina Island, California. Today the Catalina mahogany is the rarest tree in the United States and has a total population of 6 or 7 small trees. (a and b, Dr. James J. Murray, Univ. of Virginia; c, Courtesy of Robert Thorne, Claremont Graduate School, Botanical Gardens)

Islands are particularly susceptible to the introduction of exotic species. For example, Abingdon Island, one of the Galapagos Islands off the coast of South America, was home to an endemic giant tortoise. In 1957 several fishermen introduced goats to Abingdon Island, and within five years the Abingdon tortoise was extinct. The goats, with no natural predators on the island, had greatly increased in number and had eaten the tortoises' food. In Hawaii, the introduction of mouplan sheep has imperiled the mamane tree—because the sheep eat it— and the honeycreeper, an endemic bird that relies on the tree for food.

## Hunting

Sometimes species become endangered or extinct as a result of deliberate efforts to eradicate or control their numbers. Many of these species prey on game animals; some prey on livestock. Populations of large predators such as the wolf, mountain lion, and grizzly bear have been decimated by ranchers, hunters, and government agents. Predators of game animals and livestock are not the only animals vulnerable to human control efforts. Some animals are killed because their life styles cause problems for humans. The Carolina parakeet, a beautiful green, red, and yellow bird endemic to the southeastern United States, was extinct by 1920, exterminated by farmers because it ate fruit. Prairie dogs and pocket gophers, two more examples of animals killed by humans because of their life styles, have been poisoned and trapped because their burrows weaken the ground on which unwary cattle graze, and if the cattle step into the burrows, they may be crippled. As a result of sharply decreased numbers of prairie dogs and pocket gophers, the black-footed ferret, the natural predator of these animals, has not been found in the wild in the United States since 1986. (A successful captive breeding program enabled scientists to release 50 black-footed ferrets to the Wyoming prairie in September 1991.)

In addition to predator and pest control, hunting is done for three other reasons: (1) **commercial hunters** kill animals for profit—for example, by selling their fur; (2) **sport hunters** kill animals for recreation; and (3) **subsistence hunters** kill animals for food. Subsistence hunting has caused the extinction of certain species in the past but is not a major cause of extinction today, mainly because so few human groups still rely on hunting for their food supply. Sport hunting, also a major factor in the extinction (in the case of the passenger pigeon) or near extinction (in the case of the American

**Figure 16–15**
A game warden in Kenya carries one of many elephant tusks confiscated from poachers. Tusks were later destroyed by the Kenyan government. (Steve Turner/Oxford Scientific Films © 1993 Animals Animals)

bison)[3] of animals in the past, is now strictly controlled in most countries.

Commercial hunting, however, continues to endanger a number of larger animals including the tiger, cheetah, and snow leopard, whose beautiful furs are quite valuable. Rhinoceroses are slaughtered for their horns (used for dagger handles in the Middle East and for purported medicinal purposes and as an aphrodisiac in Asia) and bears for their gallbladders (used in Asian medicine purportedly to treat ailments from indigestion to hemorrhoids). Although these animals are legally protected, the demand for their products on the black market has caused them to be hunted illegally (Figure 16–15). The American black bear's gallbladder, for example, can fetch thousands of dollars on the black market.

In contrast to commercial hunting, in which the target organism is killed, **commercial harvest** is the removal of the *live* organism from the wild.

[3] Decimation of the bison in the 19th century also disrupted the food supply of the Plains Indians.

### Vanishing Tropical Forests

Tropical forests are found in South and Central America, central Africa, and Southeast Asia. Although only 7 percent of the Earth's surface is covered by tropical forests, at least 50 percent of the Earth's species of plants, animals, and other organisms inhabit them. Habitat destruction is occurring throughout the Earth, but tropical forests are being destroyed faster than any other ecosystem. Using remote sensing surveys, scientists have determined that approximately 1 percent of tropical forests are being cleared or severely degraded *each year*. The forests are making way for human settlements, banana plantations, oil and mineral explorations, and other human activities.

Many species in tropical forests are endemic, which means they are found nowhere else in the world. The clearing of tropical forests therefore leads to their extinction. These species are important in their own right, but the mass extinction that is currently taking place in tropical forests has indirect ramifications as far away as North America. Birds that migrate from North America to Central America and the Caribbean have been declining in numbers. Not all migratory birds are declining at the same rate: those birds that winter in tropical forests are declining at a much greater rate than birds that winter in tropical grasslands or tropical shrubs. Thus, tropical deforestation is affecting organisms of the temperate region.

Tropical forests provide important ecological services that help to maintain their ecosystem. For example, much of the rainfall in tropical forests is generated by the forest itself. If half of the existing rain forest in the Amazon region of South America were to be destroyed, precipitation in the remaining forest would decrease. As the land became drier, organisms adapted to moister conditions would be replaced by organisms able to tolerate the drier conditions. Many of the original species, being endemic and unable to tolerate the drier conditions, would become extinct.

Perhaps the most unsettling outcome of tropical deforestation is its effect on the evolutionary process. In the Earth's past, mass extinctions were followed during the next several million years by the formation of many new species to replace those that died out. For example, after the dinosaurs became extinct, ancestral mammals evolved into the variety of running, swimming, flying, and burrowing mammals that exist today. The evolution of a large number of related species from an ancestral organism is called **adaptive radiation.** In the past, warm ecosystems such as tropical forests have supplied the base of ancestral organisms from which adaptive radiations could occur. By destroying tropical forests, we may be reducing or eliminating nature's ability to restore its species through adaptive radiation.

Commercially harvested organisms end up in zoos, aquaria, research laboratories, and pet stores. Several million birds are commercially harvested each year for the pet trade, but unfortunately many of them die in transit, and many more die from improper treatment after they are in their owners' homes. At least nine bird species are now threatened or endangered because of commercial harvest. Although it is illegal to capture endangered animals from the wild, there is a thriving black market, mainly because collectors in the United States, Europe, and Japan are willing to pay extremely large amounts to obtain rare tropical birds. Imperial Amazon macaws, for example, fetch up to $30,000 each.

Animals are not the only organisms threatened by commercial harvest. Many unique and rare plants have been collected from the wild to the point that they are classed as endangered. These include carnivorous plants, certain cacti and orchids.

## Where Is Declining Biological Diversity the Greatest Problem?

As many as 40 percent of the world's species are concentrated in tropical forests, areas that are increasingly threatened by habitat destruction (see Focus On: Vanishing Tropical Forests). This means that a few countries, primarily developing nations, hold most of the biological diversity that is so ecologically and economically important to the entire world. The situation is complicated by the fact that these countries are least able to afford the protective measures needed to maintain biological diversity. International cooperation will clearly be needed to preserve our biological heritage.

## WILDLIFE CONSERVATION

Two types of efforts are being made to save wildlife: in situ and ex situ. **In situ conservation,** which includes the establishment of parks and reserves, concentrates on preserving biological diversity *in the wild*. A high priority of in situ conservation is the identification and protection of sites with a great deal of biological diversity. With increasing demands on land (see Chapter 17), however, in situ conservation cannot guarantee the preservation of all types of biological diversity. **Ex situ conservation** involves conserving biological diversity *in human-controlled settings*. The breeding of captive species in zoos and the seed storage of genetically diverse plant crops are examples of ex situ conservation.

### Protecting Wildlife Habitats

Many nations are beginning to appreciate the need to protect their biological heritage and have set aside areas for wildlife habitats. There are currently more than 3,000 national parks, sanctuaries, refuges, forests, and other protected areas throughout the world. Some of these have been set aside to protect specific endangered species. The first such refuge was established in 1903 at Pelican Island, Florida, to protect the brown pelican. Today the National Wildlife Refuge System of the United States has land set aside in more than 400 refuges; the bulk of the protected land is in Alaska.

Many protected areas have multiple uses. National parks may serve recreational needs, for example, and national forests may be open for logging, grazing, and farming operations. The mineral rights to many refuges are privately owned, and some refuges have had oil, gas, and other mineral development. For example, the D'Arbonne Wildlife Refuge in Louisiana, a sanctuary for 145 species of birds, has soil and water pollution and vegetation damage from natural gas wells. Hunting is allowed in more than half of the wildlife refuges in the United States, and military exercises are conducted in several of them. The Air Force, for example,

conducts low-flying jet exercises and live-fire exercises over portions of the Prieta Wildlife Refuge, an Arizona refuge for bighorn sheep.

Certain parts of the world are critically short of protected areas. Protected areas are urgently needed in tropical rain forests, the tropical grasslands and savannahs of Brazil and Australia, and dry forests that are widely scattered around the world. The wildlife of tropical deserts is underprotected in northern Africa and Argentina, and the wildlife of many islands and lake systems needs protection.

### Restoring Damaged or Destroyed Habitats

Scientists can reclaim disturbed lands and convert them into areas with high biological diversity. The most famous example of ecological restoration has been carried out since 1934 by the University of Wisconsin–Madison Arboretum (Figure 16–16). During that time, several different communities that are native to Wisconsin were carefully developed on damaged agricultural land. These communities include a tallgrass prairie, a dry prairie, and several types of pine and maple forests.

Restoration of disturbed lands not only creates wildlife habitats but has additional benefits such as the regeneration of soil that has been damaged by agriculture or mining (see Chapter 15). The disadvantages of restoration include the expense and the amount of time it requires to restore an area. Even so, restoration is an important aspect of wildlife conservation.

### Zoos, Aquaria, Botanical Gardens, and Seed Banks

Zoos, aquaria, and botanical gardens often make attempts to save species that are on the brink of extinction. Eggs may be collected from the wild, or the remaining few animals may be captured and bred in zoos and other research environments. Special techniques, such as artificial insemination, embryo transfer, and foster parenting, are used to increase the number of offspring (Figure 16–17).

There have been a few spectacular successes in captive breeding programs, in which large enough numbers of a species have been produced to reestablish small populations in the wild (see Focus On: Reintroducing Endangered Species to the Wild). Whooping cranes, which had declined to the critically low population of 15 in 1941, now number over 100. Conservation biologists are hoping to have the whooping crane removed from the

(a)

(b)

**Figure 16–16**
The University of Wisconsin Arboretum at Madison has pioneered restoration ecology. (a) The restoration of the prairie was at an early stage in November, 1935. (b) The prairie as it looks today. This picture was taken at approximately the same location as the 1935 photograph. (a, Courtesy of University of Wisconsin-Madison Arboretum; b, Courtesy of Virginia Kline, University of Wisconsin-Madison Arboretum)

endangered species list and classified as only threatened by the year 2000.

Attempting to save a species on the brink of extinction is prohibitively expensive. Moreover, zoos, aquaria, and botanical gardens do not have the space to mount efforts to save all endangered species. This means that conservation biologists must prioritize which species to attempt to save. Clearly, it is more cost-effective to maintain natural habitats so that species will never become endangered in the first place.

**Seed Banks**  Seed collections called **seed banks** exist around the world. They offer the advantage of storing a large amount of plant genetic material in a very small space. Seeds stored in seed banks are safe from habitat destruction. There have even been some instances of seeds from seed banks being used to reintroduce to the wild a plant species that was eliminated by habitat destruction.

There are also some disadvantages to seed banks, however. First, many types of plants, such as potatoes and orchids, cannot be stored as seeds. Second, seeds don't remain viable (alive) indefinitely, so periodically they must be germinated and new seeds collected. Accidents such as fires or power failures can result in the permanent loss of the genetic diversity represented by the seeds. Perhaps the most important disadvantage of seed banks is that plants stored in this manner remain

stagnant in an evolutionary sense; they do not evolve in response to changes in their natural environments. As a result, they may be less fit for survival when they are reintroduced into the wild.

**Figure 16–17**
A gaur calf and its surrogate mother, a Holstein cow. The gaur, a wild ox native to India, is an endangered species. This young calf was transferred as an embryo to the uterus of the Holstein, where it completed development. The young animal is cared for by its foster mother. (Photo by King's Island Wild Animal Habitat, courtesy of Dr. Betsy L. Dresser, Cincinnati Zoo and Botanical Garden)

## Reintroducing Endangered Species to the Wild

The ultimate goal of captive breeding programs is to produce offspring in captivity to release into nature so that wild populations can be restored. But what guarantees that a reintroduced population will survive?

Whether such species reintroductions actually succeed has been scientifically studied only in recent years. For example, the Hawaiian goose called the nene (pronounced "nay-nay") was reintroduced to the islands of Hawaii and Maui in the 1970s, but although more than 1,600 birds were released, a self-sustaining nene population failed to develop on either island. Apparently, some of the same factors that originally caused the nene's extinction in the wild, including habitat destruction, hunters, and non-native predators, were responsible for the failure of the reintroduced birds.

Now, before attempting reintroduction to the wild, conservation biologists make a feasibility study. This includes determining (1) what factors originally caused the species to become extinct in the wild, (2) whether these factors still exist, and (3) whether any suitable habitat still remains.

If the animal to be reintroduced is a social animal, a small herd is usually released together. This is accomplished by first placing the herd in a large, semi-wild enclosure that is somewhat protected from predators but requires the herd to obtain its own food. When the herd's behavior begins to resemble the behavior of wild herds, it is released.

Once animals are released into the wild, they must continue to be monitored. If any animals die, their cause of death is determined in order to search for ways to prevent unnecessary deaths in future reintroductions. The nene was not monitored after its reintroduction, so biologists do not know for sure why only 4 individuals out of more than 1,600 survived. Without this information, a second reintroduction of nenes would be risky.

## Conservation Organizations

Conservation organizations are an essential part of the effort to maintain biological diversity (see Appendix II). They help to educate policy makers and the public about the importance of biological diversity. In certain instances they serve as catalysts by galvanizing public support for important wildlife preservation efforts. They also provide financial support for conservation projects, from basic research to the purchase of land that is a critical habitat for a particular organism or group of organisms.

The International Union for Conservation of Nature and Natural Resources (IUCN) assists countries with wildlife conservation projects. It and other conservation organizations are currently assessing how effective established wildlife areas are in *maintaining* biological diversity (Figure 16–18). In addition, IUCN and the World Wildlife Fund have identified major conservation priorities by determining which biomes and ecosystems are not represented by protected areas. IUCN maintains a data bank on the status of the world's species; its material is published in *Red Data Books* about plants, animals, and habitats.

## Policies and Laws

In 1973 the **Endangered Species Act** was passed in the United States, authorizing the U.S. Fish and Wildlife Service to protect from extinction endangered and threatened species in the United States and abroad. Since the passage of the act, approximately 600 species in the United States and more than 500 species worldwide have been listed as endangered or threatened. The Endangered Species Act makes it illegal to sell or buy any product made from an endangered or threatened species.

The Endangered Species Act, which was updated in 1982, 1985, and 1988 and is scheduled to be reauthorized by Congress in 1992, is considered one of the strongest pieces of environmental legislation in the United States, in part because species

**Figure 16–18**
When a protected area is set aside, it is important to know what the minimum size of that area must be which will not be affected by encroaching species from surrounding areas. Shown are 1 hectare and 10 hectare plots that are part of the Minimum Critical Size of Ecosystems Project, which is being conducted in Brazil by the World Wildlife Fund and Brazil's National Institute for Amazon Research. It is a long-term study of the effects of fragmentation on Amazonian rain forest. (© 1990 R. O. Bierregaard)

are designated as endangered or threatened entirely on biological grounds—economic considerations cannot influence the designation. The Endangered Species Act has also been one of the most controversial pieces of environmental legislation because it has interfered with several federally funded development projects.

Some critics—notably business interests—view the Endangered Species Act as an impediment to economic progress. The construction of Tennessee's Tellico Dam was halted in 1977, for example, because it would have altered the habitat of an endangered fish called a snail darter. More recently, to protect the habitat of the spotted owl, the timber industry has been blocked from logging

old-growth forest in certain parts of the Pacific Northwest (see Chapter 7). Those who defend the act point out that, of 34,000 past cases of endangered species versus "development," only 21 cases could not be resolved by some sort of compromise. The route of a new Illinois highway that would have destroyed a small population of the endangered prairie bush clover, for example, was changed to accommodate the plant's habitat. Also, when the black-footed ferret was reintroduced on the Wyoming prairie, it was classified as an "experimental, nonessential species" so that its reintroduction would not block ranching and mining in the area. Thus, the ferret release program obtained the support of local landowners, support that was deemed crucial to its survival in the wild.

Defenders of the Endangered Species Act agree that it is not perfect. Few endangered species have recovered enough to be removed from protection. The law is geared more to saving a few popular or unique endangered species rather than the much larger number of less glamorous species that perform valuable ecosystem services; yet it is the less glamorous organisms such as fungi and insects that dominate ecosystems and contribute most to their functioning. Conservationists would like to see the Endangered Species Act strengthened in such a way as to preserve whole ecosystems and maintain complete biological diversity rather than attempting to save individual endangered species.

**International Policies and Laws** The **World Conservation Strategy,** a plan designed to conserve biological diversity worldwide, was formulated in 1980 by the IUCN, the World Wildlife Fund, and the United Nations Environment Program. In addition to conserving biological diversity, the World Conservation Strategy seeks to preserve the vital ecosystem processes upon which humans depend for survival and to develop sustainable uses of living organisms and ecosystems. Many countries are in varying stages of developing a **national conservation strategy,** a detailed plan of wildlife conservation for a specific country.

The exploitation of endangered species can be somewhat controlled through legislation. At the international level, 87 countries now participate in the Convention on Trade in Endangered Species of Wild Flora and Fauna, which bans hunting, capturing, and selling of endangered and threatened species. Unfortunately, enforcement of this treaty varies from country to country, and even where enforcement exists, the penalties are not very severe (Figure 16–19). As a result, illegal trade in rare, commercially valuable species continues.

**Figure 16–19**
A Kenyan park ranger stalks poachers. Increasingly, African governments are taking action against poachers. (Steve Turner/Oxford Scientific Films © 1993 Animals Animals)

## WILDLIFE MANAGEMENT

Efforts to handle wildlife populations and their habitats in order to ensure their sustained welfare are part of the science of **wildlife management.** Wildlife managers must know when and how to protect species that are endangered or of economic importance, and they must be able to set priorities, often in the face of conflicting goals. For example, a wildlife manager must sometimes decide whether it is better to manage an area to maintain maximum biological diversity or to protect a single "important" species within that area. Wildlife managers regulate an area by population control and habitat manipulation.

### Population Control

The natural predators of many game animals have largely been eliminated in the United States. As a result of the disappearance of wolves, coyotes, and mountain lions, the populations of animals such as squirrels, ducks, and deer sometimes exceed the carrying capacity of their environment (see Chapter 8). When this occurs, the habitat deteriorates and many animals starve to death.

Sport hunting can help control overpopulation of game animals, provided restrictions are observed to prevent overhunting. Laws in the United States determine the times of year and lengths of hunting seasons for various species as well as the number, sex, and size of each species that may be killed.

### Habitat Management

Wildlife managers affect a particular species by manipulating the plant cover, food, and water supplies of its habitat. Because different animals predominate in different stages of ecological succession (see Chapter 4), controlling the stage of ecological succession of an area's vegetation encourages the presence of certain animals and discourages others. For example, quail and ring-necked pheasant are found in weedy, open areas that are characteristic of early-successional stages. Moose, deer, and elk predominate in partially open forest, such as an abandoned field or meadow adjacent to a forest; the field provides food, and the forest provides protective cover. Other animals, such as grizzly bears, California condors, and big-horn sheep, require undisturbed climax vegetation. Wildlife managers control the stage of succession with techniques such as planting certain types of vegetation, burning the undergrowth with controlled fires, and building artificial ponds.

### Management of Migratory Animals

International agreements must be established to protect migratory animals. Ducks, geese, and shorebirds, for example, spend their summers in Canada and their winters in the United States and Central America. During the course of their annual migrations, which usually follow established routes called **flyways,** they must have areas in which to rest and feed. Wetlands, the habitat of these animals, also must be protected in both their winter and summer homes.

### Management of Aquatic Organisms

Fish with commercial or sport value must be managed to ensure that they are not overexploited to the point of extinction. Freshwater fish such as trout and salmon are managed in several ways. Fishing laws regulate the time of year, size of fish,

### When Birds Meet Oil

U.S. federal wildlife officials estimate that more than 500,000 migratory birds die each year in five southwestern states after landing on open oil pits, storage tanks, and ponds used to store industrial wastes. The birds mistake the glossy surfaces of the pits and tanks for water. Once they land, they cannot extricate themselves and die by either drowning, exposure, starvation, dehydration, or poisoning. An order from the Fish and Wildlife Service in 1988 resulted in New Mexico covering most of its 56,049 known pits.

and maximum allowable catch. Natural habitats are maintained to maximize population size; this includes pollution control. Ponds, lakes, and streams may be restocked with young hatchlings from hatcheries.

Traditionally, the ocean's resources have been considered common property, available to the first people to exploit them. As a result, many marine fish have been severely reduced in numbers by commercial fishing. To protect this dwindling resource, the national sovereignty of coastal waters now extends 200 miles from the shoreline, meaning that these waters are owned, or under the control of, whatever nation borders the coastline. This ruling provides governments with the opportunity to regulate and control fishing, which should help the populations of endangered and threatened fish to increase. Habitats for marine organisms can be constructed by dumping such benign wastes as old tires and building materials offshore. Such refuse—sometimes called artificial reefs—provides cover for fish and points of attachment for algae, corals, and other sedentary organisms.

**Whales** During the 19th and 20th centuries, many whale species were harvested to the point of **commercial extinction,** meaning that so few remain that it is unprofitable to hunt them. Although commercially extinct species still have living representatives, their numbers are so reduced that they are endangered.

In 1946 the International Whaling Commission set an annual limit on killed whales for each whale species in an attempt to secure sustainable whale populations. Unfortunately, these limits were set too high, resulting in further population declines during the next 20 years. Conservationists began to call for a worldwide ban on commercial whaling; such a ban went into effect in 1986. Whaling still occurs, however, because of a loophole in the ban that allows countries to harvest whales for research purposes and then use the meat for domestic consumption. Although a few countries still use this loophole and allow the harvest of whales, they may soon bow to international pressure and discontinue whaling. Unfortunately, it may be too late for some whales, because their numbers are so low that they may become extinct in the next few years.

## WHAT CAN WE DO ABOUT DECLINING BIOLOGICAL DIVERSITY?

Although it appears likely that our children and grandchildren will inherit a biologically impoverished world, we should view this problem as a challenge. People who are dedicated to preserving our biological heritage can reverse the trend toward extinction. It is important to realize that you don't have to be a wildlife biologist to make a contribution. Some of the most important contributions come from outside the biological arena. Following is a partial list of actions that can be taken to help maintain the wild plants and animals that are our heritage.

### Increase Public Awareness

The consciousness of both the public and legislators must be increased so that they understand the importance of biological diversity. A political commitment to protect wildlife is necessary because no immediate or short-term economic benefit is obtained from conserving species. This commitment will have to take place at all political levels, from local to international. Law-making will not ensure the protection of wildlife without strong public support. Thus, increasing public awareness of the benefits of biological diversity is critical.

Providing publicity on wildlife conservation issues costs money. Funds raised by organizations such as the Sierra Club, the Nature Conservancy, and the World Wildlife Fund support such endeavors, but clearly more money is needed. As an individual, you can help preserve biological diversity by joining and actively supporting wildlife conservation organizations.

## Support Research in Wildlife Conservation

Before an endangered species can be "saved," its numbers, range, ecology, biological nature, and vulnerability to changes in its environment must be determined; basic research provides this information. We cannot preserve a given species effectively until we know how large a protected habitat must be established and what characteristics are essential in its design.

There are acute shortages of trained specialists in tropical forestry, conservation genetics, taxonomy, resource management, and similar disciplines. Many young people who are interested in these careers have selected others because of the dearth of funding for such research. The funding covers training and salaries of skilled personnel, research equipment and supplies, and miscellaneous expenses such as transportation costs.

As an individual, you can inform local and national politicians of your desire to have conservation research funded with tax dollars. When more funds are available, colleges, universities, and other research institutions will be able to justify adding faculty and research positions. As a result, more young people with interest in conservation research will be able to undertake the necessary education.

## Organize an International System of Parks

A worldwide system of protected parks and reserves that includes every major ecosystem must be established. Conservationists estimate that a minimum of 10 percent of the world's land should be set aside for this purpose. The protected land would provide humans with benefits in addition to the preservation of biological diversity. It would safeguard the watersheds that supply us with water, and it would serve as a renewable source of important biological products in areas with multiple uses. It would also provide people with unspoiled lands for aesthetic and recreational enjoyment. In addition to the establishment of new parks and reserves, particularly in developing nations, parks and reserves in developed nations must be expanded. As an individual, you can help establish parks by writing to national lawmakers.

## Control Pollution

The establishment of wildlife parks and refuges will not be enough to prevent biological impoverishment if we continue to pollute the Earth, because it is impossible to protect parks and refuges from threats such as acid rain, ozone depletion, and climate change. Strong steps must be taken to curb the toxins we dump into the air, soil, and water—not only for human health and well-being but also for the well-being of the organisms that are so important to ecosystem stability. (Specific recommendations on how you as an individual can help reduce pollution are discussed in Chapters 19 through 23.)

## Provide Economic Incentives to Tropical Nations

There are few economic incentives to encourage the preservation of biological diversity (see Meeting the Challenge: Wildlife Ranching as a Way to Preserve Biological Diversity in Africa). This issue is particularly critical because developing nations in the tropics, the repositories of most of the Earth's genetic diversity, do not have money to spend on conservation. Their governments are consumed with human problems such as overpopulation, disease, and crushing foreign debts.

One way to help such countries appreciate the importance of the genetic resources they possess is to allow them to charge fees for the use of that genetic material. Much of the money thus earned could be used to help alleviate human problems. And some of the money generated by genetic resources could be used to provide protection for wildlife, thus preserving biological diversity for continued, sustained exploitation. Genetic and biological diversity has always been considered common property, like air, belonging to all nations. Traditionally, developed nations have had free use of genetic diversity to develop products that, in some cases, have had great economic value. In a sense, developing countries are exploited when developed nations use their biological resources at no cost and then profit by selling the products made from those resources back to developing countries.

The idea of a country selling its genetic resources, much as it sells its mineral resources, is controversial. For the most part, developing nations support genetic commerce because they stand to benefit financially from it. The users of genetic resources, primarily developed nations, oppose their sale for several reasons. First, they view genetic materials as a "shared" resource that cannot be sold because it belongs to everyone. Also, it is difficult to assign a market value to such materials. Despite the objections to selling genetic resources, it is indisputable that such a step, if conducted

*(Text continues on page 366.)*

# Wildlife Ranching as a Way to Preserve Biological Diversity in Africa

The scenario of clearing the wildlife from a section of land to make room for cattle and crops is increasingly enacted in Africa as the expanding human population strives to find a better way of life. This course began during Africa's colonial days, when European settlers claimed the best lands for themselves, leaving the more arid, infertile lands for the native Africans. Cattle and European crops do well in moister areas of Africa. But native Africans, who emulated the European colonists, discovered that cattle and crops such as onions, tomatoes, and peppers don't succeed in dry lands. Cattle crop the grass so short that it cannot survive dry spells, and the soil is left exposed and prone to erosion. The cattle themselves don't fare well because they are susceptible to diseases. And nontraditional European crops require irrigation, an expensive proposition for native farmers.

Where does wildlife fit into this picture? It doesn't. Farmers do not like wildlife because wildlife kills or spreads disease among cattle and tramples crops. Game animals are dangerous neighbors for farmers.

Zimbabwe has grappled with the cattle–people–wildlife issue and come up with an unorthodox solution—wildlife ranching. Beginning in 1975 with the passage of Zimbabwe's Wildlife Act, private landowners may own wild animals. More and more farmers are converting from cattle and crops to wildlife ranching. African game earns more money than cattle in several ways. For one thing, tourists and photographers are willing to pay to observe and photograph wildlife. Hunters pay to stalk and kill wildlife. Game yields beautiful hides and leather as well as lean, low-cholesterol meat.

Wildlife ranching, besides being financially attractive, is less harmful to the environment than traditional agriculture. Game animals, unlike cattle, eat a variety of plants and don't permanently damage the vegetation, so soil erosion is less of a problem. In addition, some wild animals require less water and are more resistant to disease than cattle.

As game farming takes hold in Zimbabwe, African attitudes about wildlife are changing. People are more tolerant of wildlife since discovering its economic value: game animals earn up to three times more than cattle in dry areas.

Environmentalists who are also animal preservationists object to any use of exotic animals and therefore oppose game ranching. However, wildlife ranching represents a compromise between people (who must use the land to earn a living) and wildlife (whose populations are declining). Africans engaged in wildlife ranching are earning more than they thought was possible, and wildlife herds are increasing in size. As long as wildlife is economically profitable, it will survive.

Many species of hoofed animals, such as these impala, are increasingly being raised by wildlife ranchers. Some ranchers are trying crocodile farming as well. (Stan Osolinski/Dembinsky Photo Associates)

**Figure 16–20**
The ruffled lemur is an endangered species endemic to Madagascar. The U.S. government has arranged a debt-for-nature swap with Madagascar. It has agreed to purchase $1 million of the country's national debt in exchange for government support of local conservation management, including the protection of endangered species. (Stouffer Enterprises, Inc. © 1993 Animals Animals)

properly, could help to protect the wildlife that is most threatened.

A second way of providing economic incentives to developing nations is for developed countries to forgive or reduce debts owed by such nations. In exchange, the developing countries would agree to protect their biological diversity. Such for-giveness of debts provides a tangible reward for preserving a nation's wildlife resources (Figure 16–20).

Once again, you can help in the formulation of such policies. Let your lawmakers know where you stand. Join and support conservation groups. Campaign to preserve our biological heritage for future generations.

## Summary

1. Biological diversity, the number and variety of living organisms, encompasses genetic diversity (the variety within a species), species diversity (the number of different species), and ecosystem diversity (variety within and between ecosystems). A reduction in biological diversity is occurring worldwide, but the problem is most critical in tropical areas.

2. Living organisms are an important natural resource. They provide us with essential ecosystem services (bacteria and fungi, for example, perform the important task of decomposition) and serve as sources of medicinal, agricultural, and industrial products. Genetic reserves are used for domesticated plant and animal breeding, both through traditional breeding methods and through genetic engineering. Living things give us recreation, inspiration, and spiritual comfort.

3. When the last individual member of a species dies, it is said to be extinct. Extinction represents a permanent loss in biological diversity; once an organism is extinct, it can never exist again.

4. An organism whose numbers are severely reduced so that it is in danger of extinction is said to be endangered. When extinction is less imminent but numbers are quite low, a species is said to be threatened.

Endangered and threatened species have limited natural ranges and low population densities. They may also have low reproductive rates or very specialized eating or reproducing requirements.

5. Human activities that contribute to a reduction in biological diversity include habitat destruction and disturbance, pollution, introduction of foreign species, pest and predator control, hunting, and commercial harvest. Of these, habitat destruction is the most significant cause of declining biological diversity.

6. Efforts to preserve biological diversity in the wild are known as in situ conservation. Ex situ conservation, which includes captive breeding and the establishment of seed banks, occurs in human-controlled settings.

7. Wildlife management includes the regulation of hunting and fishing and the management of food, water, and habitat. Different wildlife management programs often have different priorities—for example, maintaining the population of a specific species versus managing a community to ensure maximum biological diversity.

8. There are several ways to reverse the trend of declining biological diversity. Both the public and lawmakers must become more aware of the importance of our biological heritage. Funding must be found for addi-

tional research in both basic and applied fields relating to wildlife conservation. A worldwide system of protected parks and reserves must be established, hopefully encompassing a minimum of 10 percent of the land area. Pollution, which is damaging to wildlife and humans alike, must be brought under control. Economic incentives must be created to help the developing nations that are the repositories of much of the world's biological diversity realize the value of their living resources.

## Discussion Questions

**1.** If you had the assets and authority to take any measure to protect and preserve biological diversity, but could take only one, what would it be?

**2.** Does being pro-nature mean that you're also antidevelopment?

**3.** Is wildlife a renewable or nonrenewable resource? Why could it be seen both ways?

**4.** Would controlling the growth of the human population have any effect on biological diversity? Why or why not?

**5.** This chapter discussed two types of economic incentives that would help tropical countries realize the value of their biological resources. These were (1) charging a fee for the use of genetic resources and (2) having their debts to other countries forgiven. Can you suggest any other economic incentives that might help tropical countries preserve biological diversity and at the same time provide them with much-needed foreign capital?

**6.** According to this chapter, one of the ways in which you can help the world preserve its biological diversity is by talking about the problem, even to your friends. How will talking about the problem of biological impoverishment contribute to its reversal?

## Suggested Readings

Ackerman, D. Last refuge of the monk seal. *National Geographic* 181:1, January 1992. The plight of Hawaiian monk seals, which are on the brink of extinction.

Cheater, M. Death of the ivory market. *World Watch* 4:3, May/June 1991. The worldwide ban on ivory and other elephant products has stopped illegal poaching of the African elephant but has also eliminated an important source of revenue for developing African nations.

Cohn, J. P. Reproductive biotechnology. *BioScience* 41:9, October 1991. Techniques developed to help infertile couples have children are being adapted for endangered species.

Cole, M. A farm on the wild side. *New Scientist,* September 1990. Wildlife ranching is improving the lots of people and wildlife in Zimbabwe.

McNeely, J. A., K. R. Miller, W. V. Reid, R. A. Mittermeier, and T. B. Werner. Strategies for conserving biodiversity. *Environment* 32:3, April 1990. A good review article of the importance of biological diversity and what can be done to conserve it.

Potten, C. J. America's illegal wildlife trade: A shameful harvest. *National Geographic* 180:3, September 1991. Illegal hunting and the black market for animal products are decimating America's wildlife.

Shell, E. Seeds in the bank could stave off disaster on the farm. *Smithsonian,* January 1990. How seed banks are used to improve important crop plants.

Solbrig, O. T. The origin and function of biodiversity. *Environment* 33:5, June 1991. How evolution produces biological diversity and why the latter is so important for ecosystem stability.

Stolzenburg, W. The fragment connection. *Nature Conservancy,* July/August 1991. The pros and cons of using wildlife corridors to connect small wilderness areas that are becoming increasingly fragmented by human activities.

Terborgh, J. Why American songbirds are vanishing. *Scientific American,* May 1992. Migratory songbirds continue to decline in numbers, in part because of deforestation in tropical areas where they spend the winter.

# Chapter 17

*Banner and Ritter Mountains above Thousand Island Lake, Sierra Nevada, California. (© David Muench 1993)*

# Land Resources and Conservation

**L**and has multiple uses and benefits. Natural areas—wilderness, forests, grasslands, and wetlands—provide many essential environmental services. These include controlling soil erosion, protecting water quality, regulating water flow in rivers and streams, and providing wildlife habitat. Human uses of land include harvesting timber, grazing livestock, growing crops, and building cities. Unfortunately, human activities sometimes degrade land to the point where it cannot provide essential environmental services. Two extremely serious land problems are removal of forest, or deforestation, and expansion of deserts, or desertification. In this chapter we examine the importance of land resources, as well as the various ways in which land is put to use.

# IMPORTANCE OF NATURAL AREAS

Most of the Earth's land area has a low density of humans. These sparsely populated areas, known as **nonurban** or **rural lands,** include wilderness, forests, grasslands, and wetlands. The many environmental services that are performed by rural lands enable the majority of humans to live in concentrated urban environments.

As you will see throughout this chapter, undisturbed land benefits us in many ways of which we are often not even aware. Consider the higher elevations of mountains, above the tree line, which have a distinctive ecosystem known as **alpine tundra** (Figure 17–1). The alpine tundra, which has strong winds, cold temperatures, and snow, is a harsh environment inhabited by few kinds of plants. Sparse populations of Tibetan herdsmen in the Himalayas are among the few groups of people living in alpine tundra. The western United States has approximately 1.2 million hectares (3 million acres) of mostly uninhabited alpine tundra. Although this ecosystem is worth preserving simply because it is unique and interesting, it also has an ecological value that is not readily apparent to most people. For example, melting snows in the highest portions of Utah's alpine tundra furnish 60 percent of the water in the state's streams, even during the hot days of summer.

Unfortunately, the amount of U.S. alpine tundra that can be classified as disturbed by human activities is increasing. Sheep grazing, recreation, and mining all take their toll, particularly because the soil found in such regions is extremely thin and fragile.

Maintaining parcels of undisturbed land adjacent to agricultural and urban areas provides vital environmental services such as pest control, flood and erosion control, and groundwater recharge (see Chapters 13 and 14). Undisturbed land also breaks down pollutants and recycles wastes.

Natural environments provide homes for Earth's plant and animal species. One of the best ways to maintain wildlife and to protect endangered and threatened species (see Chapter 16) is by preserving or restoring the natural areas to which wildlife is adapted.

Ecologists who conduct research on the complexities of ecosystems frequently use natural areas as outdoor laboratories. Geologists, zoologists studying behavior, botanists studying plant diversity, and soil scientists are just a few of the other scientists who use natural sites for scientific en-

**Figure 17–1**
Alpine tundra, which is mostly wilderness area in the United States, performs a valuable service by regulating the flow of water from snowmelt in the mountains to lowland areas. (Lynn and Sharon Gerig/Tom Stack and Associates)

quiry. Natural areas provide perfect settings for educational experiences not only in science, but also in history, because they demonstrate the way the land was when our ancestors settled here.

Certain unspoiled natural areas are also important for their recreational value, providing places for hiking, swimming, boating, rafting, sport hunting, and fishing (Figure 17–2).

Wild areas are also important to the human spirit. Forest-covered mountains, rolling prairies, barren deserts, and other undeveloped areas are not only aesthetically pleasing, but help us to recover from the stresses of urban and suburban living. We can escape the tensions of the civilized world by retreating, even temporarily, to the solitude of natural areas.

**Figure 17–2**
Horse Ridge in the Wind River Range, Wyoming. Natural areas provide us with various forms of recreation and escape from the daily pressures of civilization. (© David Muench 1993)

## CURRENT LAND USE IN THE UNITED STATES

Thirty-five percent of the land in the United States, which encompasses all types of ecosystems from tundra to desert, is owned by the federal government (Figure 17–3). This includes land that contains important resources such as minerals and fossil fuels, land that possesses historical or cultural significance, and land that provides critical biological habitat. Most of the federally owned land is in Alaska and the western states and is managed by several agencies in the U.S. Department of the Interior and the Department of Agriculture (Table 17–1).

Of the remaining land, about 55 percent is privately owned by citizens, corporations, and non-

**Table 17–1**
**Administration of Federal Lands**

| Agency | Land Held | Area in Millions of Hectares (Acres) |
|---|---|---|
| Bureau of Land Management (Dept. of Interior) | National resource lands | 130 (321) |
| U.S. Forest Service (Dept. of Agriculture) | National forests | 76 (188) |
| U.S. Fish and Wildlife Service (Dept. of Interior) | National wildlife refuges | 34 (84) |
| National Park Service (Dept. of Interior) | National Park System | 29 (72) |
| Other—includes Department of Defense, Corps of Engineers (Dept. of the Army), and Bureau of Reclamation (Dept. of Interior) | | 29 (72) |

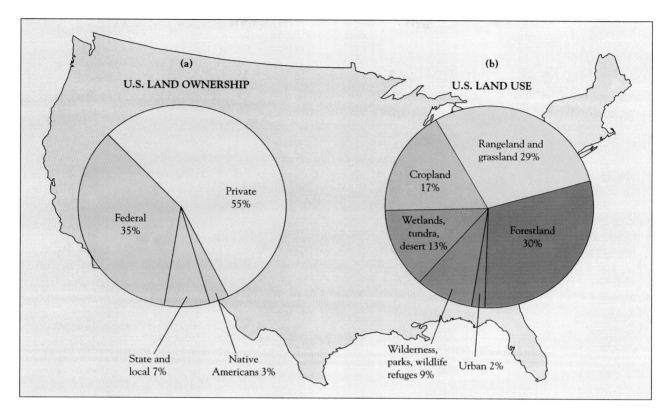

**Figure 17–3**
Land in the United States. (a) Land ownership. (b) How land is used.

profit organizations, and about 3 percent by native American tribes. The rest (approximately 7 percent) is owned by state and local governments.

## WILDERNESS

Regions of the Earth that have not been greatly disturbed by human activities and that humans visit but do not inhabit are known as **wilderness.** The Wilderness Act of 1964 authorized the U.S. government to set aside public wilderness areas, ranging from tiny islands to national parks that are several million hectares in size, as part of the National Wilderness Preservation System. Although mountains are the most common terrain to be safeguarded by this system, representative examples of a number of other ecosystems have been set aside, including tundra, desert, and wetlands.

The Wild and Scenic Rivers Act was passed in 1968 to protect rivers with outstanding beauty, recreational value, or important wildlife. As of mid-1992, 147 rivers (0.33 percent of the nation's total river systems) were protected by this act, with others being considered for inclusion. Rivers that have been given this designation have little or no development along their banks; most have no dams. Camping, swimming, boating, sport hunting, and fishing are permitted, but development of the shoreline is prohibited. Mining claims are permitted, however.

Millions of people visit U.S. wilderness areas each year, and some areas are overwhelmed by this use: soil and water pollution, litter and trash, and human congestion predominate in place of quiet, unspoiled land. Government agencies now restrict the number of people allowed into each wilderness area at one time, but it is likely that some of the most popular wilderness areas will require more intensive management. This would include the development of trails, outhouses, cabins, and campsites, amenities that are not encountered in true wilderness.

## Do We Have Enough Wilderness?

Large tracts of wilderness, most of it in Alaska, have been added to the National Wilderness Preservation System since the passage of the Wilderness Act in 1964. The designation of wilderness areas is supported by people who view wilderness as a nonrenewable resource. They think it is particularly important to preserve additional land in the lower 48 states, where currently less than 2 percent of the total land area is specified as wilderness. Increasing the amount of land in the National Wilderness Preservation System is usually opposed by groups who operate businesses on public lands, including timber, mining, ranching, and energy companies.

## PARKS AND WILDLIFE REFUGES

The National Park System was originally composed of large, scenic areas in the West such as Grand Canyon and Yosemite Valley (Figure 17–4) (see

Chapter 7). Today, however, the National Park System has more cultural and historical sites—battlefields and historically important buildings and towns, for instance—than scenic wilderness. More than 300 different sites are part of the National Park System, and many of them have been purchased with money provided by the Land and Water Conservation Fund Act of 1965. Urban parks, such as the Golden Gate National Recreation Area near San Francisco, have also been established.

One of the primary roles of the Park Service is to teach people about the natural environment and management of natural resources by providing nature walks and guided tours of its parks. The popularity and success of national parks in the United States (Table 17–2) have encouraged many other nations to establish national parks (see Focus On: A National Park in West Africa). As in the United States, these parks usually have multiple roles, from providing wildlife habitat to facilitating human recreation.

The National Wildlife Refuge System (see Chapter 16) contains more than 400 different parcels of land that represent all major ecosystems found in the United States, from tundra to temperate rain forest to desert. Newer acquisitions are

**Figure 17–4**
Arches National Park in Utah has breathtaking scenery.
(© David Muench 1993)

| Table 17–2 The Ten Most Popular National Parks | |
|---|---|
| **National Park** | **Number of Recreation Visitors in 1990** |
| Great Smoky Mountains (North Carolina and Tennessee) | 8,151,800 |
| Grand Canyon (Arizona) | 3,776,700 |
| Yosemite (California) | 3,124,900 |
| Yellowstone (Wyoming, Montana, and Idaho) | 2,823,600 |
| Olympic (Washington) | 2,794,900 |
| Rocky Mountain (Colorado) | 2,647,300 |
| Acadia (Maine) | 2,393,591 |
| Zion (Utah) | 2,102,400 |
| Glacier (Montana) | 1,986,700 |
| Mammoth Cave (Kentucky) | 1,924,500 |
| Total visitors to all 50 parks | 57,610,278 |

## A National Park in West Africa

**Focus On**

Deforestation is occurring at an unprecedented rate throughout the tropics. In most of West Africa the forests have disappeared. Cameroon, a West African country, is fortunate in that its government is committed to forest preservation. Recently, three national parks were established in the southern rain forest area of the country. One of these, Korup National Park, has been extensively surveyed and has the richest biological diversity in Africa. It contains more than 400 tree species, 50 mammal species (including almost 25 percent of the primate species found in Africa), and more than 250 bird species.

Plans for Korup National Park during the next few decades include many aspects of conservation and sustainable utilization of forest resources. Villages now located in the park will be resettled outside its boundaries. Local people will be educated about environmental issues and the importance of conservation. Some of these people will become park staff members. A scientific research program is being established to help develop sustainable management practices for the park. Tourism, which depends upon preservation of the area's unique biological diversity, will be developed to help provide income for the local economy. Many international organizations are cooperating with the Cameroonian

people in the establishment of Korup National Park. They include the British Overseas Development Administration, USAID, Parcs Canada, and the World Wide Fund for Nature. Korup National Park is a model of land conservation that other developing countries can emulate.

Korup National Park is a national treasure belonging to the Cameroonian people and to the world. (Visuals Unlimited/Jane Thomas)

## Threats to Parks

National parks are even more overcrowded than wilderness areas. All of the problems plaguing urban areas are found in popular national parks during peak seasonal use, including crime, vandalism, litter, traffic jams, and pollution of the soil, water, and air. Many people think that more funding is needed to maintain and repair existing parks and that, in addition, new parks should be created to meet current and projected demands.

Some national parks have imbalances in wildlife populations. For example, in the 1920s wolves were exterminated in Yellowstone National Park because of their threat to herbivorous mammals

sometimes made to preserve the habitats of threatened or endangered species.

such as elk. As a result, the elk population at Yellowstone has increased to the point where today it is endangering many plant species upon which it feeds. A similar elk problem exists at the Rocky Mountain National Park, where wolves have been absent since the turn of the century. Because national parks are supposed to preserve natural ecosystems, many conservationists think that wolves should be reintroduced at Yellowstone and Rocky Mountain national parks.

Additional examples of wildlife imbalances on national park lands involve declining populations of many other types of mammals, including bears, white-tailed jackrabbits, and red foxes (see Chapter 1). For example, the number of grizzly bears in national parks in the western United States has greatly diminished. Grizzly bears require large areas of wilderness undisturbed by humans. Clearly, human influences in national parks are a factor in their decline. More important, the parks may be too small to support grizzlies. Fortunately, grizzly bears have survived in sustainable numbers in Alaska and Canada.

National Parks are also affected by human activities beyond their borders. Pollution doesn't respect park boundaries. Also, parks are increasingly becoming islands of natural habitat surrounded by human development. Development on the borders of national parks limits the area in which wildlife may range, forcing wildlife into isolated populations. Ecologists have found that, when environmental stresses occur, several small "island" populations are more likely to become threatened than a single large population occupying a sizable range.

## FORESTS

Approximately one-third of the earth's total land area is covered by forests. Timber harvested from forests is used for fuel, construction materials, and paper products. Forests also influence local climate conditions. If you walk into a forest on a hot summer day, you'll notice that the air is cooler and moister than it was outside the forest. This is the result of a biological cooling process called **transpiration** in which water from the soil is absorbed by roots, transported through the plant, and then evaporated from their leaves and stems. Transpiration also provides moisture for clouds, eventually resulting in precipitation (Figure 17–5).

Forests play an essential role in global biogeochemical cycles such as those for carbon and nitrogen (see Chapter 5). For example, photosynthesis by trees removes large quantities of carbon dioxide from the atmosphere and fixes it into carbon compounds. At the same time, oxygen is released into the atmosphere. Tree roots hold vast tracts of soil in place, reducing erosion. Forests are effective watersheds because they absorb, hold, and slowly release water; this provides a more regulated flow of water, even during dry periods, and helps to control floods (see Chapter 13). In addition, forests provide essential wildlife habitat.

### Forest Management and Harvesting Trees

When forests are managed, their species composition and other characteristics are altered. Specific varieties of trees are planted, not only to conserve the forest in areas where trees have been removed

75%
Water recycled by transpiration and evaporation

25%
Water lost in runoff

**Figure 17–5**
Forests play an important role in the hydrologic cycle by returning most of the water that falls as precipitation to the atmosphere by transpiration. In contrast, most precipitation is lost from deforested areas as runoff.

by fire or commercial harvest, but to prevent soil erosion and preserve watersheds. Forest management also includes thinning out or removing trees that are not as commercially desirable.

Forest management often results in low-diversity forests. In the southeastern United States, for example, forests of young pine that are grown for timber and paper production are all the same age and are planted in rows a fixed distance apart (Figure 17–6). These forests are essentially **monocultures**—areas covered by one crop, like a field of corn. One of the disadvantages of monocultures is that they are more prone to damage by insect pests and disease-causing microorganisms. Consequently,

**Figure 17–6**
Tree plantations such as this pine plantation are monocultures, with trees of uniform size and age planted in rows.
(Visuals Unlimited/Kirtley-Perkins)

pests and diseases must be controlled in managed forests, usually by applying pesticides (see Chapter 22).

Trees are harvested in several ways—by selective cutting, shelterwood cutting, seed tree cutting, clearcutting, and whole-tree harvesting (Figure 17–7). **Selective cutting,** in which mature trees are cut individually or in small clusters while the rest of the forest remains intact, allows the forest to regenerate naturally. The trees left by selective cutting produce seeds that germinate to fill the void. Selective cutting has fewer negative effects on the forest environment than other methods of tree harvest, but it is not as profitable because timber is not removed in great enough quantities.

The removal of all mature trees in an area over a period of time is known as **shelterwood cutting.** In the first year of harvest, undesirable tree species and dead or diseased trees are removed. The forest is then left alone for perhaps a decade, during which the remaining trees continue to grow and new seedlings become established. During the second harvest, many mature trees are removed. The forest is then allowed to regenerate on its own for perhaps another decade. A third harvest removes the remaining mature trees, but by this time a healthy stand of younger trees is replacing the mature ones. Little soil erosion occurs with this method of tree removal, even though more trees are removed than in selective cutting.

In **seed tree cutting,** almost all trees are harvested from an area; a scattering of desirable trees is left behind to provide seeds for the regeneration of the forest. **Clearcutting** is the removal of all trees from an area. After the trees have been removed by clearcutting, the area is either allowed to reseed and regenerate itself naturally or is planted with specific varieties of forest trees. Timber companies prefer clearcutting because it is the most cost-effective way to harvest trees; also, very little road building has to be done to harvest a large number of trees. However, clearcutting is ecologically unsound. It destroys wildlife habitat that takes many years to restore itself. On sloping land, clearcutting increases soil erosion. Obviously, the recreational benefits of forests are lost when clearcutting occurs.

A special type of clearcutting is **whole-tree harvest,** in which machines harvest the entire tree, including roots and small branches, and cut it into small chips, which are processed for paper products or fuel. Whole-tree harvest has the short-term economic benefit of making maximum use of all parts of every tree in the forest. However, over time, the nutrients in soils where whole-tree harvest has occurred become depleted because no part of the tree remains to decompose and recycle essential nutrients back to the soil.

## U.S. Forests

The U.S. Forest Service, an agency in the Department of Agriculture, manages approximately 10 percent of the land in the United States. Forest Service lands have multiple uses, including timber harvest, livestock forage, water resources, recreation, and providing habitat for fish and wildlife. The Bureau of Land Management (BLM), an agency in the Department of the Interior, oversees some public forest lands, two-thirds of which lie in Alaska. Outside Alaska there are approximately 2 million hectares (5 million acres) of public forest, which require intensive management to produce

a. Selective cutting

Original forest of varied species and ages

Mature trees selectively cut

Forest regenerates naturally

b. Shelterwood cutting

Undesirable and dead trees removed from original forest

After a decade or so of growth, many mature trees are removed

After another decade, more mature trees are removed, forest continues to grow and regenerate

c. Seed tree cutting

Original forest

Most trees removed; remaining trees provide seeds for regeneration

Forest regenerates slowly

d. Clearcutting

Original forest

All trees removed

Forest regenerates naturally or is replanted

e. Whole tree harvest

Original forest

Trees and roots removed completely leaving nothing to decompose and recycle nutrients to soil

Soil depleted of nutrients and can support replanted trees only if fertilized

high-quality timber. The most commercially valuable public forest land is 1 million hectares (2.4 million acres) in western Oregon; it produces more than 90 percent of the timber that is harvested from public lands managed by the BLM.

Recreation in the national forests ranges from camping at specially built campsites to backpacking in the wilderness. Visitors to national forests also swim, boat, picnic, and observe nature. Effective maintenance and management of these areas for public use and enjoyment include trash removal, trail maintenance, and repair of damage caused by vandalism, but the funds allotted for these purposes do not keep pace with the increasing public usage of national forests.

Slightly more than half of the forest land in the United States is privately owned, and three-fourths of this private land is in the eastern part of the country. Projected conversion of forest to agricultural, urban, and suburban land over the next 40 years is expected to have the greatest impact in the South, although there will be considerable losses in other regions as well.

## Tropical Forests and Deforestation

There are two types of tropical forests: tropical rain forests and tropical dry forests. In places where the climate is very moist throughout the year—on the order of 200 to 450 cm (79 to 177 inches) of precipitation annually—**tropical rain forests** prevail. Tropical rain forests are found in Central and South America, Africa, and Southeast Asia, but almost half of them are in just three countries: Brazil in South America, Zaire in Africa, and Indonesia in Southeast Asia (Figure 17–8).

In other tropical areas where annual precipitation is less but is still enough to support trees—including regions subjected to a wet season and a prolonged dry season—**tropical dry forests** occur. During the dry season, tropical trees shed their leaves and remain dormant, much as temperate trees do during the winter. India, Kenya, Zimbabwe, Egypt, and Brazil are a few of the countries that have tropical dry forests.

The importance of tropical forests, particularly tropical rain forests, as the repositories of most of the world's biological diversity was discussed in Chapter 16. Tropical forests also provide such important environmental services as forming and holding the soil, cleansing the water and air, and providing shelter and food for countless animals and humans. When a forest is intact, it regulates surface water and thereby controls floods and droughts. In addition, tropical forests have a profound effect on the global carbon cycle (see Chapter 5), because much of the world's photosynthesis occurs there.

## What Happens When Tropical Forests Disappear?

The destruction of all tree cover in an area is **deforestation.** When tropical forests are harvested or destroyed, they no longer make valuable contributions to the environment or to the people who depend upon them. Tropical forest destruction particularly threatens native people whose cultural and physical survival depends upon the forests (Figure 17–9). When these people come into conflict with developers and colonizers of forests, they have little political strength or legal recourse and are usually forced off their land. Even beyond their right to exist, when indigenous tribes disappear, fields as diverse as anthropology and botany suffer the loss.

Deforestation results in decreased soil fertility and increased soil erosion. Because poor rural people, both natives and colonizers, depend upon the soil for their livelihood, tropical deforestation contributes to the downward spiral of poverty in which many of these people find themselves. Uncontrolled soil erosion, particularly on steep deforested slopes, can affect the production of hydroelectric power if silt builds up behind dams. Increased sedi-

**Figure 17–7**
There are five major systems of tree harvesting. (a) In selective cutting, the older, mature trees are selectively harvested from time to time and the forest regenerates itself naturally. (b) In shelterwood cutting, less desirable and dead trees are harvested. As younger trees mature, they produce seedlings, which continue to grow as the now-mature trees are harvested. (c) Seed tree cutting involves the removal of all but a few trees, which are allowed to remain, providing seeds for natural regeneration. (d) In clearcutting, all trees are removed from a particular site. Clearcut areas may be reseeded or allowed to regenerate naturally. (e) In whole-tree harvest, the entire tree (including its roots and branches) is placed into a chipper. Whole-tree harvest produces small pieces of wood chips, but results in nutrient-poor soil.

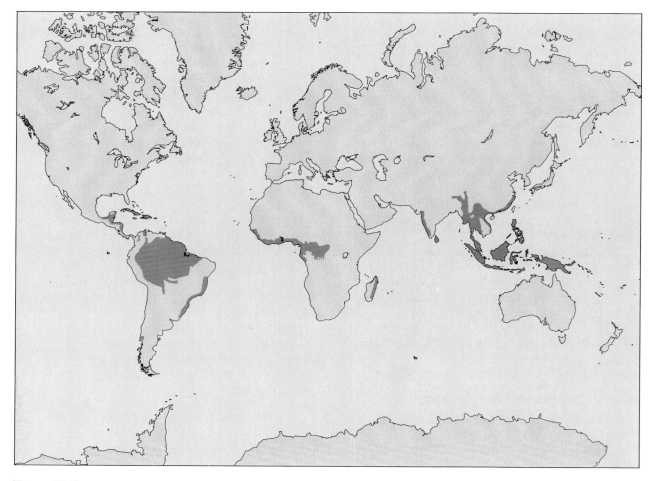

**Figure 17–8**
The distribution of tropical rain forests (green areas). Rain forests are located in South and
Central America, Africa, and Asia.

mentation of waterways caused by soil erosion can also harm fisheries. In drier areas, deforestation can lead to the formation of deserts (to be discussed shortly).

When forest is removed, the total amount of surface water that flows into rivers and streams actually increases. However, because this water flow is no longer regulated by forest, the affected region experiences alternating periods of flood and drought.

Deforestation causes the extinction of plant and animal species. Many tropical species, in particular, have very limited ranges within a forest, so they are especially vulnerable to habitat modification and destruction. Migratory species based in temperate areas, including birds and butterflies, also suffer from tropical deforestation.

Deforestation induces regional and global climate changes. Trees release substantial amounts of

moisture into the air; about 97 percent of the water that a plant's roots absorb from the soil is evaporated directly into the atmosphere. This moisture falls back to the Earth in the hydrologic cycle (see Chapter 5). When forest is removed, rainfall declines and droughts become common in that region. Tropical deforestation contributes to an increase in global temperature (see Chapter 20) by causing a release of stored carbon into the atmosphere as carbon dioxide, which in turn enables the air to retain solar heat.

## Where and Why Are Tropical Forests Disappearing?

In the past thousand years, forests in temperate areas were largely cleared for housing and agriculture. Today, however, deforestation in the tropics is occurring much more rapidly and over a much

**Figure 17–9**
Native Indians from the Amazon River Basin live in harmony with the rain forest. (Visuals Unlimited/G. Prance)

larger area. Most of the remaining undisturbed tropical forest, which lies in the Amazon and Congo river basins of South America and Africa, is being cleared and burned at a rate that is unprecedented in human history. Tropical forests are also being destroyed at an extremely rapid rate in southern Asia, Indonesia, Central America, and the Philippines. Ten countries account for 76 percent of tropical deforestation: Brazil, Indonesia, Zaire, Burma, Colombia, India, Malaysia, Mexico, Nigeria, and Thailand.

Although exact figures on rates of forest destruction are unavailable, the Food and Agriculture Organization (FAO) of the United Nations released in late 1991 its second global assessment of worldwide tropical deforestation. A total of 87 tropical countries were evaluated in this study. The FAO estimated an annual rate of change in forest cover of −0.9 percent per year from 1981 to 1990, and some areas, such as West Africa, experienced a loss of forest estimated at greater than 2 percent per year. If this rate of deforestation—which represents a worldwide annual loss of 16.9 million hectares (41.8 million acres) of forest—continues, tropical forests will be all but gone within the next several decades.[1]

The three main causes of tropical deforestation are subsistence agriculture, commercial logging,

[1] A preliminary assessment by the FAO shows that deforestation from 1980 to 1990 was not as acute in temperate and boreal regions as in the tropics.

and cattle ranching. Subsistence agriculture, in which enough food is produced by a family to feed itself (see Chapter 18), is by far the most important cause, accounting for 60 percent of tropical deforestation. Other reasons for the destruction of tropical forests include mining and the development of hydroelectric dams.

**Subsistence Agriculture** In many developing countries where tropical rain forests occur, the majority of people do not own the land on which they live and work. Land ownership (and therefore profit) is in the hands of a few; for example, in Brazil 5 percent of the farmers own 70 percent of the land. Most subsistence farmers were displaced from traditional farmlands because of the inequitable distribution of land ownership. They have no place to go except into the forest, which they clear to grow food. Tropical deforestation by subsistence farmers is also affected by population pressures (recall that these countries have some of the highest fertility rates in the world) and pervasive poverty. Land reform would make the land owned by a few available to everyone, thereby easing the pressure on tropical forests by subsistence farmers. This scenario is unlikely, however, because wealthy landowners have more economic and political clout than landless peasants.

Subsistence farmers often follow loggers' access roads until they find a suitable spot. They first cut down the forest and allow it to dry, then they burn the area and plant crops immediately after burning; this is known as **slash-and-burn agriculture** (discussed further in Chapter 18). The yield from the first crop is often quite high because the nutrients that were in the burned trees are now available in the soil. However, soil productivity declines at a rapid rate and subsequent crops are poor. In a very short time, the people farming the land must move to a new part of the forest and repeat the process. Cattle ranchers often claim the abandoned land for grazing, because land that is not rich enough to support crops can still support livestock.

Slash-and-burn agriculture done on a small scale with periods of 20 to 100 years between cycles is actually sustainable. But when *several hundred million people* try to obtain a living in this way, the land is not allowed to lie uncultivated long enough to recover, and disaster results.

**Commercial Logging** Twenty-one percent of tropical deforestation is the result of commercial logging; vast tracts of tropical rain forests are being harvested for export abroad. Most tropical coun-

**Figure 17-10**
Cattle graze on Panamanian land that was formerly rain forest. Tropical forests are often burned to provide grazing land for cattle that are exported to countries such as the United States. (David Cavagnaro)

**Figure 17-11**
Masai women carrying firewood in Kenya. Half of the world's people use wood fires to cook food. (E. R. Degginger)

tries allow commercial logging to proceed at a much faster rate than is sustainable. For example, in Sabah and Sarawak (both part of Malaysia), logging is currently removing the forest at almost twice the sustainable rate. If this continues, Malaysia will soon experience shortages of timber and will have to start importing logs and other forest products, losing the potential revenues from its own newly vanished forests. In the final analysis, tropical deforestation does not lead to economic development; rather, it destroys a valuable resource.

**Cattle Ranching and Agriculture for Export** Approximately 12 percent of tropical forest destruction is carried out to provide open rangeland for cattle (Figure 17-10). After the forests are cleared, cattle can graze on the land for six to ten years, after which time shrubby plants, known as **scrub savannah,** take over the range. Much of the beef raised on these ranches, which are often owned by foreign companies, is exported to fast-food restaurant chains.

A considerable portion of the land cleared for plantation-style agriculture produces crops such as citrus fruits and bananas for export.[2] Because cash crops are usually grown on land owned by a few wealthy landowners, who rely on foreign compa-

nies for shipping and processing, this type of agriculture does little to alleviate the poverty of the local people.

**Why Are Dry Tropical Forests Disappearing?** Dry tropical forests are also being destroyed at an alarming rate, primarily for fuel. Wood—perhaps half of the wood consumed worldwide—is used as heating and cooking fuel by much of the developing world (Figure 17-11).

Often the wood cut for fuel is converted to charcoal,[3] which is then used to power steel, brick, and cement factories. Charcoal production is extremely wasteful: 3.6 metric tons (4 tons) of wood produce enough charcoal to fuel an average-sized iron smelter for only 5 minutes.

## RANGELANDS

Rangelands are grasslands, in both temperate and tropical climates, that serve as important areas of food production for humans by providing fodder for domestic animals such as cattle, sheep, and goats (Figure 17-12). The predominant vegetation of rangelands includes grasses, forbs (small herbaceous plants other than grasses), and shrubs. Many of the

[2] Plantation-style agriculture is sustainable on forest soil as long as expensive fertilizers and other treatments are applied.

[3] Partially burning wood in a large kiln from which air is excluded converts the wood into charcoal.

**Figure 17–12**
Rangeland in Montana. When the carrying capacity of rangeland is not exceeded, it is a renewable resource. (Grant Heilman, Grant Heilman Photography)

world's temperate rangelands have been stripped of their natural vegetation and plowed for cultivation of food crops.

## U.S. and World Rangelands

Rangelands make up approximately 30 percent of the total land area in the United States and occur mostly in the western states and Alaska. Of this, approximately one-third is publicly owned and

two-thirds are privately owned. Excluding Alaska, there are at least 89 million hectares (220 million acres) of public rangelands. Approximately 69 million hectares (170 million acres) are managed primarily by the Bureau of Land Management, which is guided by the Taylor Grazing Act of 1934, the Federal Land Policy and Management Act of 1976, and the Public Rangelands Improvement Act of 1978. The U.S. Forest Service manages an additional 20 million hectares (50 million acres).

The federal government issues permits to private livestock operators that allow them to use public rangelands for grazing in exchange for a small fee. Public rangelands are also mined for minerals and energy resources, used for recreation, and preserved for wildlife habitat and soil and water resources.

Worldwide, animals graze on approximately 40 percent of rangelands. The remaining rangelands have been either converted to cropland or degraded, mostly by overgrazing, to the extent that they are no longer useful.

Intensive livestock practices in which animals are kept in small enclosures for some or all of their lives, rather than allowed to roam freely while grazing, are increasingly being adopted in many countries. Although these practices reduce the requirement for open spaces, rangelands are still important in the production of domesticated animals.

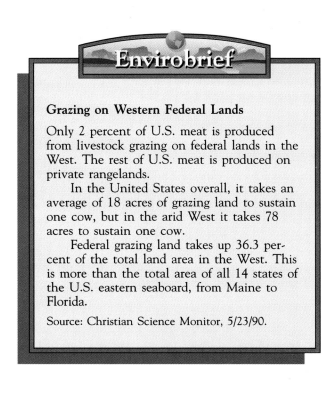

**Grazing on Western Federal Lands**

Only 2 percent of U.S. meat is produced from livestock grazing on federal lands in the West. The rest of U.S. meat is produced on private rangelands.

In the United States overall, it takes an average of 18 acres of grazing land to sustain one cow, but in the arid West it takes 78 acres to sustain one cow.

Federal grazing land takes up 36.3 percent of the total land area in the West. This is more than the total area of all 14 states of the U.S. eastern seaboard, from Maine to Florida.

Source: Christian Science Monitor, 5/23/90.

## Rangeland Deterioration and Desertification

Grasses, the predominant vegetation of rangelands, have a **fibrous root system,** in which many roots form a diffuse network in the soil to anchor the plant. Plants with fibrous roots hold the soil in place quite well, thereby reducing soil erosion. If only the upper portion of the grass is eaten by animals, the roots can continue to develop, allowing the plant to recover and grow to its original size.

The **carrying capacity** of a rangeland is the maximum number of animals the rangeland plants can sustain. When the carrying capacity of a rangeland is exceeded, grasses and other plants are **overgrazed;** that is, so much of the plant is consumed by the grazing animals that it cannot recover, and it dies. Overgrazing results in barren, exposed soil that is susceptible to erosion.

Most of the world's rangelands lie in semiarid areas that have natural extended periods of drought. Under normal conditions, native grasses can survive a severe drought: the aboveground portion of the plant dies back, but underground the

**Figure 17–13**
Overgrazing can lead to nonproductive desert, as occurred in eastern New Mexico and western Arizona. (Visuals Unlimited/J. Alcock)

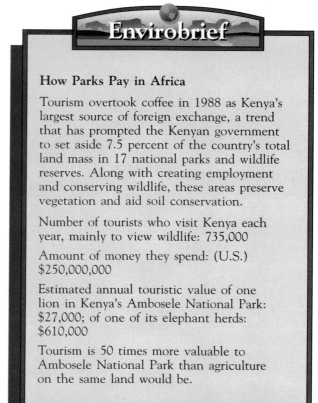

**Envirobrief**

**How Parks Pay in Africa**

Tourism overtook coffee in 1988 as Kenya's largest source of foreign exchange, a trend that has prompted the Kenyan government to set aside 7.5 percent of the country's total land mass in 17 national parks and wildlife reserves. Along with creating employment and conserving wildlife, these areas preserve vegetation and aid soil conservation.

Number of tourists who visit Kenya each year, mainly to view wildlife: 735,000

Amount of money they spend: (U.S.) $250,000,000

Estimated annual touristic value of one lion in Kenya's Ambosele National Park: $27,000; of one of its elephant herds: $610,000

Tourism is 50 times more valuable to Ambosele National Park than agriculture on the same land would be.

extensive root system remains alive and holds the soil in place. When the rains return, the roots send forth new aboveground growth.

When overgrazing occurs in combination with an extended period of drought, however, once-fertile rangeland can be converted to desert. The lack of plant cover due to overgrazing allows winds to erode the soil. Even when the rains return, the land is so degraded that it cannot recover. Water erosion removes the little bit of remaining topsoil, and the sand that is left behind forms dunes. This process, which converts rangeland (or tropical dry forest) to desert, is called **desertification** (Figure 17–13). It ruins economically valuable land, forces out wildlife, and threatens endangered species.

Desertification is related to overpopulation. In the 1970s, a devastating drought occurred in the African Sahel, an area south of the Sahara from Senegal to Sudan. Then, from 1980 to 1986, a disastrous drought struck the arid lands of East Africa, particularly those in Ethiopia, Sudan, and Mozambique. In both cases, the people living in the affected regions suffered greatly, with many children and adults starving after their crops failed. The arid lands of Africa have always had periodic

droughts, but what made these particular droughts so devastating was the tremendous number of people (and livestock) attempting to live on the ecologically fragile land. They overwhelmed the land and degraded it; most of the trees were chopped down for firewood, and rangeland was severely overgrazed by livestock. Had many fewer people been living in such a marginal area, they might have been spared the horrors of starvation.

**Ranchers Versus Conservationists** Conservationists have become increasingly vocal about the ecological damage caused by overgrazing of public rangelands and want to restrict the number of animals allowed to graze. They want public rangelands to be managed primarily for wildlife habitat, recreation, and scenic value rather than for privately owned livestock. Ranchers counter by pointing to the tradition and history of the American cowboy as a reason for them to continue using public rangelands.

One way to decrease the livestock on public rangelands would be to increase the fee that private operators must pay for each 454 kg (1,000 lb) of livestock (equal to one cow or five sheep). In 1992 the monthly fee was $1.92 per cow, a price that is considerably less than the $8 to $12 fee for each animal grazing on privately owned rangelands. Taxpayers subsidize the rancher who grazes his livestock on public rangelands by paying an estimated $6 for each animal; this money is used to maintain the rangeland and fix the damage done by the livestock. Every year or so, a proposal is made in Congress to increase the grazing fee, but ranchers are a powerful political lobby and always succeed in stopping the increase.

## Rangeland Management

Overall, the condition of public rangelands in the United States has slowly improved since the low point of the Dust Bowl in the 1930s (see Chapter 14). Much of this improvement can be attributed to fewer livestock being permitted to graze the rangelands after the passage of the Taylor Grazing Act in 1934. Better livestock management practices, such as controlling the *distribution* of animals on the range, as well as conservation measures have contributed to rangeland repair. But restoration has been slow and costly, and more is needed: in 1984, 42 percent of public rangelands were designated in fair condition and 18 percent in poor condition. These figures were supported by a 1988 study by the General Accounting Office, the investigative arm of Congress, that concluded that more than 50 percent of public rangeland was in either fair or poor condition.

Rangeland management includes seeding in places where plant cover is sparse or absent, constructing fences to prevent overgrazing (Figure 17–14), controlling weeds, and developing wildlife habitat. Special incentives are sometimes offered to livestock operators to use public rangelands in a way that results in their overall improvement.

Thousands of wild horses and burros roam on western public rangelands. Because they symbolize America's pioneer history, they are protected. At the same time, they must be managed so that they don't contribute to the deterioration of the rangeland. To prevent their populations from exceeding the carrying capacity of their rangeland home, the Bureau of Land Management started the Adopt-a-Horse program, in which between 5,000 and 9,000 wild horses and burros are annually removed from the range and given away.

## FRESHWATER WETLANDS

**Wetlands** are lands that are transitional between aquatic and terrestrial ecosystems. They are usually covered by shallow water and have characteristic soils and water-tolerant vegetation. Freshwater wetlands may be marshes, in which grass-like plants

**Figure 17–14**
Fences play an important role in preventing overgrazing. The land to the left of the fence was ungrazed, whereas the grazed land on the right has sparser vegetation. (USDA/Soil Conservation Service)

(a)

(b)

**Figure 17–15**
Freshwater wetlands. (a) Marshes are wetlands composed largely of grasses, while swamps (b) are forested wetlands. (a and b, © David Muench 1993)

dominate, or swamps, in which woody plants (trees or shrubs) dominate (Figure 17–15). Wetlands also include hardwood bottomland forests (lowlands along streams and rivers that are periodically flooded) in the Southeast, prairie potholes (small, shallow ponds that formed when glacial ice melted at the end of the last ice age) in the Midwest, and peat moss bogs (peat-accumulating wetlands where mosses dominate) in the northern states.

At one time wetlands, which occupy 6 percent of the world's land surface, were thought of as wastelands—areas that needed to be filled in or drained so that farms, housing developments, or industries could be built on them. Wetlands are also a breeding place for mosquitoes and therefore were viewed as a menace to public health. Today, however, the crucial environmental services that wetlands provide are widely recognized, and wetlands are somewhat protected by law.

Wetland plants, which are highly productive, provide enough food to support a wide variety of organisms. Wetlands are valued as wildlife habitat for migratory waterfowl and many other bird species, beaver, otters, muskrats, and game fish. For example, more than 50 different fish species spawn or feed in the swamps of the lower Mississippi River.

Wetlands help control flooding by acting as holding areas for excess water when rivers flood their banks. The floodwater stored in wetlands then drains slowly back into the rivers, providing a steady flow of water throughout the year. Wetlands

also serve as groundwater recharging areas. One of their most important roles is to help cleanse and purify water runoff, even water that is polluted. They do this by acting as a **sink,** a reservoir capable of trapping and holding pollutants in the flooded soil. Other pollutants, such as nitrogen from fertilizer runoff, are absorbed by wetland plants.

Freshwater wetlands produce many commercially important products including wild rice, blackberries, cranberries, blueberries, and peat moss. They are also sites of fishing, hunting, boating, photography, and nature study.

Wetlands are increasingly threatened by agriculture, pollution, engineering (such as dams), and urbanization. In the United States, wetland areas have been steadily shrinking by an estimated 81,000 to 162,000 hectares (200,000 to 400,000 acres) per year. In the contiguous 48 states, of the more than 81 million hectares (200 million acres) of wetlands that originally existed, only 38 million hectares (95 million acres) remain. Most of the loss since the 1950s has been the result of farmers' converting wetlands to cropland. Urban and suburban development, dredging, and mining account for most of the remainder of the loss.

The loss of wetlands is legislatively controlled by a section of the 1972 Clean Water Act (currently up for renewal); this legislation does a reasonably good job of protecting coastal wetlands, but a poor job of protecting inland wetlands, which is what most wetlands are. The Emergency Wetlands Resources Act of 1986 authorizes the U.S. Fish and

Wildlife Service to designate and acquire critically important wetlands; the Service is making an inventory and map of wetlands in the United States. The inventory is scheduled to be completed by 1998.

Currently, the United States is attempting to prevent any *new net loss* of wetlands. This means that development of wetlands will be allowed only if a corresponding amount of previously converted wetlands is restored. The policy is complicated by two factors: (1) confusion and dissent about the definition of wetlands (which was not spelled out in the Clean Water Act) and (2) the question of who owns wetlands. In 1989 a team of government scientists developed a comprehensive, scientifically correct definition of wetlands. It provoked an outcry from farmers and real estate developers, who perceived it as a threat to their property values. Largely in response to their criticisms, the Bush administration narrowed the definition of wetlands in 1991, removing marginal wetlands that are not as wet as swamps or marshes. This narrower definition excludes approximately one-third of the wetlands in the United States from protection.

The federal government owns less than 5 percent of wetlands in the United States; the remaining 95 percent is privately owned. This means that private citizens control whether wetlands are protected and preserved or developed and destroyed. Because of the traditional rights of private land-ownership in the United States, landowners resent the federal government's telling them what they may or may not do with their lands. It is therefore important that private landowners become informed of the environmental importance of wetlands and the critical need to maintain them. Although some private owners do recognize the value of wetlands and voluntarily protect them, others don't. The federal government is examining proposals such as tax incentives and the outright purchase of wetlands to encourage their conservation.

In 1990 Congress passed the Food, Agriculture, Conservation, and Trade Act (a new version of the Food Security Act of 1985). One of the provisions of this act is the establishment of the Wetlands Reserve Program, which seeks to restore, in a five-year period, 405,000 hectares (1 million acres) of privately owned freshwater wetlands that have previously been drained and converted to cropland. However, the Wetlands Reserve Program is funded annually by Congress and is therefore subject to budget cuts.

## COASTLINES AND ESTUARIES

**Estuaries** are coastal bodies of water that connect to oceans; they include tidal marshes and tidal rivers. In estuaries, fresh water from the land mixes with salt water from the oceans, resulting in high productivity. Many ocean fish and shellfish spend all or portions of their lives in estuaries, supported by the many producers, which range from microscopic algae to seaweeds and marsh grasses (recall the Chesapeake Bay salt marsh described in Chapter 3). Coastal estuaries, which provide food and protective habitat, could be considered the ocean's nurseries because so many different marine organisms spend the first parts of their lives there.

Historically, coastal wetlands (also called saltwater wetlands) have been regarded as wasteland, good only for breeding large populations of mosquitoes. Coastal wetlands throughout the world have been drained, filled in, or dredged out to turn them into "productive" structures such as industrial parks or marinas. In the United States, people have belatedly recognized the importance of coastal wetlands and have passed legislation to slow their destruction. (See Meeting the Challenge: The Montezuma Wetlands Project for a description of a wetlands restoration project.)

The Coastal Barrier Resources Act of 1982 abolished most federal assistance programs, including federal flood insurance, for new development ventures on undeveloped coastal barriers (Figure 17–16). This law has helped to eliminate some of the contradictions in governmental policies regarding coastal wetlands.

### Mangrove Swamps

Most of the shoreline in the tropics consists of densely vegetated wetlands called **mangrove swamps.** Mangroves—certain trees and shrubs that require salty water—grow best in the intertidal zone, where they are alternately submerged to their trunks at high tide and exposed to their roots at low tide.

Mangroves help to build soil along the shoreline by holding sediments in place. In some places, as the soil accumulates, other plants invade the area and the mangroves continue their slow expansion into the ocean. Mangrove roots provide habitat for oysters, fiddler crabs, and other marine organisms, and mangrove branches provide nesting sites for many shorebirds. Mangrove swamps are

**Figure 17–16**
Overdevelopment of coastal areas often leads to their degradation. The past availability of federally-funded insurance against damage has led to improper development of these environmentally-sensitive areas. (Mark S. Wexler)

**Figure 17–17**
Homes in Napa, California, built on land that was formerly a vineyard. Loss of farmland to urban and suburban development is a problem in certain areas of the country. (David Cavagnaro)

often destroyed to provide firewood, space for coastal development, and agricultural land.

## AGRICULTURAL LANDS

The United States has more than 121 million hectares (300 million acres) of **prime farmland**—land that has the soil type, growing conditions, and available water to produce food, forage, fiber, and oilseed crops. Certain areas of the country have large amounts of prime farmland; for example, 90 percent of the Corn Belt—a corn-growing region in the Midwest encompassing parts of six states—is considered prime farmland.

Keep in mind that not all prime farmland is used to grow crops. Approximately one-third contains roads, pastures, rangeland, forests, feedlots, and farm buildings.

## SUBURBAN SPRAWL AND URBANIZATION

In Chapter 9 we discussed **urbanization,** the concentration of humans in cities, relative to human population growth. Urbanization strains natural resources and particularly affects land use. There was considerable concern a few years ago that much of our prime agricultural land was falling victim to urbanization and the accompanying suburban sprawl by being converted to parking lots, housing developments, and shopping malls. However, most agricultural land is not in danger of urban intrusion. This does not mean that rural land conversion is of no consequence. Unquestionably, rural areas adjacent to urban areas are being developed. In certain areas of the country, loss of rural land, including prime agricultural land, is a significant problem (Figure 17–17). For example, more than three-fourths of the agricultural land in the Northeast is near urban areas and is therefore in danger of being developed.

The most important thing to remember regarding land resources and urbanization is that the support of cities and suburbs requires much more than simply the land they occupy. The needs of people in urban areas for food, drinking water, energy, and minerals are met by large areas of agricultural land, rangeland, forests, wilderness, and wetlands.

## The Environmental Effects of the Conservation Reserve Program

The federal government has long grappled with the challenge of improving the condition of degraded farmlands that are privately owned. Although farmers may understand the environmental benefits of abandoning agriculture on fragile land, they are most likely to actually do so when given financial incentives.

In 1985 Congress passed the Food Security Act, which contained several programs and provisions for reducing crop production on marginal, highly erodible farmland. One of the significant parts of the Food Security Act is the Conservation Reserve Program (CRP), whose purpose is to reduce soil erosion on the country's most marginal farmland. Unlike similar programs offered in the past, which retired land for one or two years, the CRP is a long-term program; farmers with lands that qualify agree to establish grass or trees on the land and then leave it alone for ten years. As an incentive to participate in the CRP, farmers are provided with additional income from the federal government. As of July 1991, more than 14 million hectares (35 million acres) were enrolled in the CRP; the ultimate goal is to enroll 16 to 18 million hectares (40 to 45 million acres).

The Conservation Reserve Program has benefited the environment. Annual loss of soil on CRP lands that have been planted with grasses or trees has been reduced from an average of 7.7 metric tons of soil per hectare (20.9 tons per acre) to 0.6 metric tons per hectare (1.6 tons per acre). Because the vegetation is not disturbed once it is established, it provides habitat for wildlife; ground-nesting birds, small mammals, and birds of prey have increased in number and kind on CRP lands. The reduction in soil erosion has also improved water quality and enhanced fish habitats in surrounding rivers and streams.

Has the Conservation Reserve Program had any detrimental results? Some scientists are concerned that the retired land may provide a breeding ground for insect pests, plant disease organisms, and nematodes—worms that infect the roots of plants. The pests could then invade nearby agricultural land, causing problems. To avoid such a scenario, scientists are carefully monitoring the CRP lands; if a pest outbreak occurs, measures will be taken to control it and prevent subsequent outbreaks.

## LAND USE

Many environmental concerns converge in the issue of land use. Pollution, population issues, preservation of our biological resources, mineral and energy needs, and production of food are all tied to land use.

### Economic Pressures for Land Use

The way privately owned land is taxed affects its use. For example, sometimes forest or agricultural land that is located near urban and suburban areas is taxed as potential urban land. Because of the higher taxes on this land, its owners fall under greater pressure to sell it, which ultimately hastens its development. However, if such land is taxed as forest or farmland, the lower taxes are an incentive for owners to hold onto the land and maintain it in its undeveloped condition. Thus, land use is largely controlled by economic factors.

### Public Planning of Land Use

Examine the use of land where you live. You may be surrounded by high rises and factories or by tree-lined streets interspersed with open parkland. Regardless of your surroundings, it is likely that they got that way by accident. Most areas have a land-use plan that includes zoning, but rarely do land-use plans take into account all aspects of land as a resource both *before* and after development. The philosophy of most land-use plans is that development is good because it increases the tax base (even though the revenue from these taxes is usually consumed providing services to the developed area).

Land-use decisions are complex because they have multiple effects. For example, if a tract of land is to be developed for housing, then roads, sewage lines, and schools must be built nearby to accommodate the influx of people. This usually results in the opening of restaurants and shopping areas, which take up more land.

Public planning of land use must take into account all repercussions of the proposed land use, not just its immediate effects. It is helpful to begin with an inventory of the land, including its soil type, topography, types of plants and animals, endangered or threatened organisms, and historical or archaeological sites.

At this stage, the public planning commission attempts to understand the value of the land *as it currently exists* as well as its potential value after any proposed change. In addition to providing people with open space for recreation and mental health, undeveloped land provides environmental services that must be recognized. All of these benefits should be compared with the possible economic benefits of development. In the long-term, the best use of land may not be the use that provides immediate economic gain.

If the land will ultimately be developed, the development plan should be comprehensive. It should indicate which areas will remain open space, which will remain agricultural, and which will be zoned for high-, medium-, and low-density housing.

**Table 17–3**
**Selected Observations of Worldwide Land Degradation**

| Country/Source | Observation |
|---|---|
| Mali<br>Patricia A. Jacobberger, geologist, Smithsonian Institution, 1986 | On the Landsat maps, there is now—and there wasn't in 1976—a bright ring of soil around villages. Those areas are now 90% devoid of vegetation, the topsoil is gone, and the surface is disrupted and cracked. |
| Mauritania<br>Sidy Gaye, *Ambio*, 1987 | There were only 43 sand-storms in the whole country between 1960 and 1970. The number increased tenfold in the following decade, and in . . . 1983 alone a record 240 sandstorms darkened the nation's skies. |
| Tunisia<br>UNEP, 1987 | Rangelands have been overgrazed with three heads of cattle where only one could thrive . . . Two-thirds of the land area of Tunisia is being eaten away by desertification. |
| China<br>*Beijing Review*, interview with Zhu Zhenda, Chinese Academy of Sciences, 1988 | Unless urgent measures are taken, desertification will erode an additional 75,300 square kilometers . . . by the year 2000, more than twice the area of Taiwan. |
| Indonesia<br>Ronald Greenberg and M. L. Higgins, USAID Jakarta, 1987 | Thirty-six watersheds . . . have critical erosion problems. . . . In Kalimantan, the silt load in streams has increased 33 fold in some logging areas. |
| Thailand<br>D. Phantumvanit and K. S. Sathirathai, Thailand Development Research Board, 1988 | The pace of deforestation has been accelerating since the early 1900s, but it has moved into a higher gear since the 1960s. . . . [Between 1961 and 1986,] Thailand lost about 45 percent of its forests. |
| Brazil<br>Mac Margolis, interview with geologist Helio Penha, *Washington Post*, February 1988 | Every year, rains slash deeper into the bared soil, dumping tons of silt in waterways, causing rivers to overflow into the city's streets. Now 'people flee the drought in the Northeast only to die in floods in Rio.' |

Source: Worldwatch Institute, based on various sources.

# CONSERVATION OF OUR LAND RESOURCES

Our ancestors looked upon natural areas as a resource to exploit. They appreciated prairies as valuable agricultural land and forests as immediate sources of lumber and eventual farmland. This outlook was practical as long as there was more land than people needed. But as the population increased and the amount of available land decreased, it was necessary to consider land as a limited resource. Increasingly, the emphasis has shifted from exploitation to preservation of the remaining natural areas.

Although all types of ecosystems must be conserved, several are in particular need of protection. Deforestation in the tropics has become an international problem, along with desertification and erosion (Table 17–3). In the United States, the amounts of natural wetlands and agricultural lands are of greatest concern. (See Focus On: The Environmental Effects of the Conservation Reserve Program, page 387, for an example of how federal legislation can reverse the effects of land degradation.)

Government agencies, private conservation groups, and private citizens have begun to set aside natural areas for permanent preservation. Unfortunately, different federal and state policies on land use have often contradicted one another. For example, some programs are geared toward preserving wetlands while other projects encourage drainage and development. Agricultural price supports boost the profits from food produced on converted wetlands, and farmers are encouraged to drain wetlands by federally supported, low-interest loans and technical assistance.

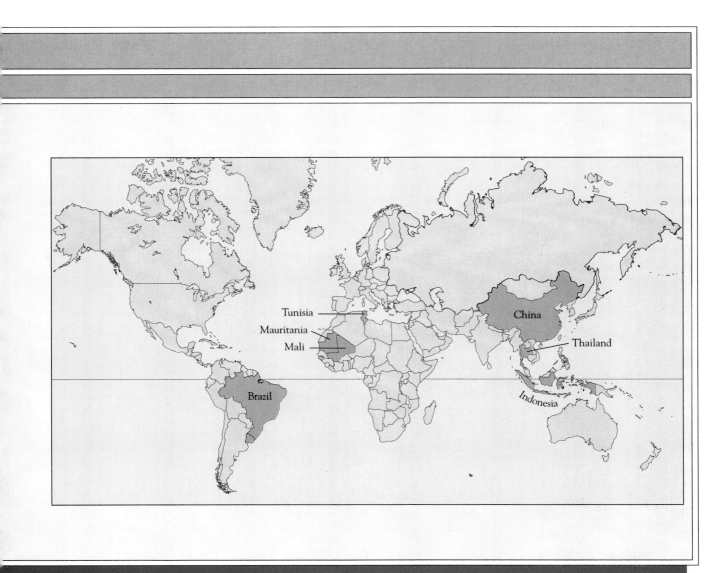

# The Montezuma Wetlands Project

Two environmental concerns in the San Francisco Bay area are the loss of wetlands and the need for safe disposal of dredging sediments. The San Francisco Bay has lost more than 90 percent of its wetlands to housing, airports, industrial parks, farms, and sanitary landfills. These wetlands had held shoreline erosion in check by regulating the flow of water into the bay. In addition, wetlands had purified the water entering the bay, provided critical habitat for wildlife (including a large stopover area for migratory birds), and performed many other valuable environmental services.

Shipping channels, marinas, and ports in the bay gradually fill in with sediment and must be dredged periodically to prevent ships from running aground. But what can be done with the tons and tons of sediment that are removed from the bay floor? In the past, this material was dumped either in other parts of the bay or in the Pacific Ocean. Neither location was satisfactory because of the harmful effects on estuarine and oceanic wildlife. Increasingly, environmental organizations and regulatory agencies demanded an alternative disposal site that would not have so many harmful environmental effects.

In 1991 an environmental consulting firm and a real estate developer jointly announced the Montezuma Wetlands Project, a proposal to restore wetlands in San Francisco Bay using the dredging sediments from the bay. The project, now under way, involves the restoration of approximately 809 hectares (2,000 acres) near the mouth of the Sacramento River. It is the largest private wetlands restoration endeavor ever undertaken in the western United States. The cost of the restoration is funded entirely by dredging disposal fees.

The original wetlands around the bay were drained and diked in the 1880s; since then the land has subsided several feet below high-tide levels. In the restoration project, approximately 13.8 to 18.3 million cubic meters (15 to 20 million cubic yards) of sediment are being transported and dumped at the site. The sediments will be engineered to resemble the channels and land contours of natural tidal and seasonal wetlands. Then native plants, birds, fish, and small mammals, many of which are endangered or threatened species, will be reintroduced to the site.

The Montezuma Wetlands Project, which is strongly supported by many environmental organizations and local, state, and federal agencies, will take at least a decade to complete. Once the site is restored, it will become a wildlife refuge under the permanent protection of either the federal government or an appropriate environmental organization.

A section of the vast Montezuma Wetlands Project in Solano County, California. (Lawrence S. Burr)

# SUMMARY

**1.** Natural areas provide us with many environmental services, including watershed management, soil erosion protection, climate regulation, and wildlife habitat. People benefit directly from natural areas by using them for scientific study, recreation, and renewal of the human spirit.

**2.** Areas of the Earth that have not been greatly disturbed by human activities are called wilderness. Wilderness includes many different ecosystems, such as tundra, forest, grassland, desert, and wetland. Public lands designated as part of the National Wilderness Preservation System are protected from development. National parks have multiple roles, including recreation, ecosystem preservation, and wildlife habitat.

**3.** Forests provide many environmental services as well as commercially important timber. Forests that are intensively managed for commercial harvest have little species diversity. Trees may be harvested by selective cutting, shelterwood cutting, seed tree cutting, clearcutting, and whole-tree harvest. Selective cutting and shelterwood cutting have little negative impact on the forest but are not as economically profitable as the other methods in the short term.

**4.** The greatest problem facing world forests today is deforestation, which is the permanent removal of forest. In the tropics, forests are destroyed to (1) provide colonizers with temporary agricultural land; (2) obtain timber, particularly for developed nations; (3) provide open rangeland for cattle; and (4) supply people with fuel wood. Tropical deforestation is, to a great extent, irreversible in that deforested land cannot regenerate forest with the species diversity it originally had.

**5.** In rangelands, grasses, forbs, and shrubs predominate. Rangelands are often grazed by cattle, sheep, goats, and other domesticated mammals. Provided the number of animals grazing on a particular area of rangeland is kept below the area's carrying capacity, the rangeland remains a renewable resource. However, overgrazing can result in barren, exposed soil that is susceptible to erosion. Overuse of rangeland or dry forest results in desertification, the development of desert conditions.

**6.** Wetlands are transitional areas between aquatic and terrestrial ecosystems and may be either freshwater or coastal (saltwater). Wetlands may be occupied by trees (swamps) or grasses (marshes); they provide habitat for wildlife, purify natural bodies of water, and recharge groundwater. Despite their environmental contributions, wetlands are often drained or dredged for other purposes. In the United States, wetlands are primarily converted to agricultural land.

**7.** Agricultural lands are forests or grasslands that have been plowed for cultivation. In certain areas they are threatened by expanding urban and suburban areas. Loss of prime farmland is particularly serious in the northeastern part of the United States.

**8.** The enlightened view of the interaction between humans and land is that humans are stewards of the land, managing it to achieve sustainable use. It is possible to become involved in local land-use planning by becoming familiar with land-use policies and laws.

# DISCUSSION QUESTIONS

**1.** Do you think more land should be added to the wilderness system? Why or why not?

**2.** Give at least five environmental benefits that nonurban lands provide. Why is it difficult to assign economic values to many of these benefits?

**3.** Suppose a valley contains a small city surrounded by agricultural land. The valley is encircled by mountain wilderness. Explain why the preservation of the mountain ecosystem would help support both urban and agricultural land in the valley.

**4.** Why do commercial logging, cattle ranching, and other forms of agribusiness in Central and South America deprive local people of land and income?

**5.** Explain the relationship between eating a hamburger at a fast-food restaurant and tropical deforestation.

**6.** If the world's tropical forests were all destroyed, how would it affect your life?

**7.** Explain how tax and zoning laws can increase the conversion of prime farmland to urban and suburban development.

**8.** Should private landowners have the right to do whatever they wish to their land? Present arguments for both sides of this issue.

**9.** Why can Korup National Park in Cameroon be considered a model park for tropical Africa?

# SUGGESTED READINGS

Brough, H. B. A new lay of the land. *World Watch* 4:1, January/February 1991. How land reform in the developing world would stem environmental degradation, including deforestation and desertification.

Conniff, R. RAP: On the fast track in Ecuador's tropical forests. *Smithsonian*, June 1991. Biologists evaluate the biological diversity of tropical forests using "rapid assessment"—often only one step ahead of commercial loggers.

Fearnside, P. M. A prescription for slowing deforestation in Amazonia. *Environment* 31:4, 1989. How the Brazilian government could save the remaining forests of Amazonia.

Gillis, A. M. Should cows chew cheatgrass on commonlands? *BioScience* 41:10, November 1991. A new range war pits cattle growers against conservationists over degraded rangelands.

*IUCN Bulletin*, vol. 20, nos. 4–6, April–June 1989. Contains a special report on wetlands.

McNeely, J. A. The future of national parks. *Environment* 32:1, January/February 1990. The problems in national parks around the world are addressed in this solutions-oriented article.

Postel, S. Halting land degradation. *State of the World: 1989*, W.W. Norton & Company, New York, 1989. A detailed evaluation of worldwide land degradation and ways to control it.

Repetto, R. Deforestation in the tropics. *Scientific American*, April 1990. Examines government policies that encourage the destruction of tropical forests.

Stegner, W. It all began with conservation. *Smithsonian*, April 1990. The history of the conservation movement in the United States.

World Resources Institute and the United Nations Environment and Development Programs. *World Resources 1992–93*. Oxford University Press, New York, 1992. Chapter 8 contains the preliminary results of the second global assessment of deforestation by the Food and Agriculture Organization of the United Nations.

A woman harvesting tomatoes in Mali. Women are the farmers in most of Africa.
(Larry C. Price)

# Food Resources: A Challenge for Agriculture

World food problems are many and varied. One of the greatest challenges facing humanity today is to produce and distribute enough food; millions of people, particularly in developing nations, lack adequate nutrition. Most farmers in developing countries usually produce barely enough food to feed themselves and their families, with little left over as a reserve. Developed countries, for their part, have energy-intensive agricultural methods that produce high yields of food but cause serious environmental problems such as soil erosion and pollution. Overshadowing and exacerbating all food problems is the increase in human population. The production of adequate food for the world's people will be an impossible goal until population growth is brought under control.

# HUMAN NUTRITIONAL REQUIREMENTS

Haunting pictures in the news of starving African children became commonplace during the 1980s and early 1990s, and most people are upset by the human suffering and misery that these pictures portray. Positive news about hunger relief did occasionally appear, such as when medical treatment and balanced nutrition resulted in a reversal of nutritional deficiency diseases in certain populations. But it is clear that the specter of famine, which has been with humanity for thousands of years, has not disappeared. Despite the advances made in many fields related to food production, numerous people—adults and children—still do not get enough food. In this chapter we examine some of the progress that has been made in food production, as well as the challenges that still confront us.

Unlike plants, which manufacture their own food, humans and other animals must obtain their nutrients by consuming other organisms. The foods humans eat are composed of several major types of organic molecules that are necessary to maintain health—carbohydrates, proteins, and lipids—which are digested by enzymes in the gastrointestinal tract.

**Carbohydrates,** organic molecules such as sugars and starches, are important primarily because they are metabolized readily by the body in **cell respiration,** a process in which the energy of organic molecules is transferred to a molecule called adenosine triphosphate (ATP). The body uses the energy in ATP to produce heat, repair damaged tissues, grow, fight off infections, and reproduce. That is, carbohydrates supply the body with the energy required to sustain life.

**Proteins** are large, complex molecules composed of repeating subunits called **amino acids.** Proteins perform several critical roles in the body. First, when the plant and animal proteins in food are digested, the body absorbs the amino acids, which are then reassembled in different orders to form human proteins. A substantial part of the human body, from hair and nails to muscles, is made up of protein. In addition, proteins are metabolized in cell respiration to release energy.

There are approximately 20 different amino acids. The human body manufactures 12 of these for itself, using starting materials such as carbohydrates. However, human cells lack the ability to synthesize the other eight amino acids, called **es-**sential amino acids. These must be obtained from food.[1]

**Lipids,** a diverse group of organic molecules that includes fats and oils, are metabolized by cell respiration to provide the body with a high level of energy. Pound for pound, lipids deliver more energy when metabolized than either carbohydrates or proteins. In addition, lipids have a number of important roles in the body; some are hormones (chemical messengers that regulate many of the body's functions), and others are essential components of cell membranes.

In addition to carbohydrates, proteins, and lipids, we humans require minerals, vitamins, and water in our diets. **Minerals** are inorganic elements, such as iron and calcium, and are essential for the normal functioning of the human body. **Vitamins** are complex organic molecules that are required in very small quantities by living cells. Vitamins help to regulate metabolism and the normal functioning of the human body. Whereas plants synthesize most vitamins, humans and other animals must obtain vitamins from food.

# WORLD FOOD PROBLEMS

The average adult human must consume enough food to get approximately 2,600 kilocalories per day.[2] If a person consumes less than this over an extended period of time, his or her health and stamina decline, even to the point of death. People who receive fewer calories than they need are said to be **undernourished.**

The total number of calories consumed is not the only measure of good nutrition, however. People can receive enough calories in their diets but still be **malnourished** because they are not receiving enough of specific, essential nutrients such as proteins or vitamins. For example, a person whose primary food is rice can obtain enough calories, but a diet of rice lacks sufficient amounts of proteins, lipids, minerals, and vitamins to maintain normal body functions. People suffering from malnutrition are more susceptible to disease and have less strength to function productively than those who are well fed.

---

[1] The eight essential amino acids are isoleucine, leucine, lysine, methionine, phenylalanine, threonine, tryptophan, and valine.

[2] The average man requires 3,000 kilocalories per day, whereas the average woman requires 2,200 kilocalories per day.

The two most common diseases of malnutrition are marasmus and kwashiorkor. **Marasmus** (from the Greek work *marasmos,* meaning "a wasting away") is progressive emaciation caused by a diet low in both total calories and protein. Marasmus is most common among children in their first year of life—particularly children of very poor families in developing nations. Symptoms include a pronounced slowing of growth and extreme atrophy (wasting) of muscles. It is possible to reverse the effects of marasmus with an adequate diet.

**Kwashiorkor** (a native word in Ghana, meaning "displaced child") is malnutrition resulting from protein deficiency. It is common among children in all poor areas of the world. The main symptoms include edema (fluid retention); dry, brittle hair; apathy; and stunted growth. One of the most typical features of kwashiorkor is a pronounced swelling of the abdomen. Kwashiorkor can be treated by gradually restoring a balanced diet.

Crop failures caused by drought, war, flood, or some other catastrophic event may result in **famine,** a severe food shortage. The worst African famine in history, which was caused in part by widespread drought, occurred from 1983 to 1985. Hardest hit were Ethiopia and Sudan, in which 1.5 million people died of starvation. The people living in this region lacked sufficient money to purchase food and did not have stored food reserves to protect them against several bad years of crop failures. International efforts to provide relief were complicated by the nations' governments and by the fact that much of this part of Africa lacks road and transportation systems to facilitate the distribution of the food. Famines get a great deal of media attention because of the huge and obvious amount of human suffering they cause. However, many more people die worldwide from undernutrition and malnutrition than from the starvation associated with famine.

Eating food in excess of that required is called **overnutrition.** Overnutrition is most common among people in developed nations. Generally, a person suffering from overnutrition has a diet high in saturated (animal) fats, sugar, and salt. Overnutrition results in obesity, high blood pressure, and an increased likelihood of such disorders as diabetes and heart disease.

## Producing Enough Food

Producing enough food to feed the world's people is the largest challenge in agriculture today (Figure 18–1), and the challenge grows more difficult each

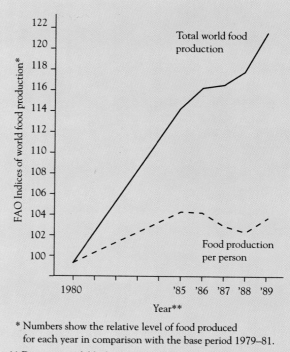

* Numbers show the relative level of food produced for each year in comparison with the base period 1979–81.

** Data not available for years 1981–1984.

**Figure 18–1**

A comparison of world food production and food production per person, 1980 to 1989. Food production has increased greatly over the past decade, but because the human population has also expanded, the amount of food produced per person has not increased by much.

year because the human population is continually expanding. Currently, 1.2 billion metric tons (1.3 billion tons) of grains such as wheat, corn, rice, and barley are required to feed the world's population for one year. Each year, an *additional* 250 million metric tons (280 million tons) of grain must be produced to account for that year's increase in population. Although food production can be increased in the short term, one fact remains: *the long-term solution to the food supply problem is control of the human population.*

## Poverty and Food: Making Food Affordable for the Poor

The main cause of undernutrition and malnutrition is poverty. The world's poorest people—those living in developing countries in Asia, Africa, and Latin America—do not own land on which to grow food and do not have the money to purchase food. At least 1 billion people are so poor that they cannot afford to eat enough food or enough of the right kinds of food.

Poverty and hunger are not restricted to developing nations. A task force of doctors in 1985 reported that 1 out of every 11 Americans was malnourished or undernourished and that at least half of these hungry Americans were children.

## Cultural and Economic Effects on Human Nutrition

Even if the major food-producing countries could produce enough grain to feed the rest of the world, there would still be an enormous economic problem. It costs money to produce, transport, and distribute food. African, Latin American, and Asian countries, which have the greatest need for imported food, are least able to pay for it. On the other hand, the food-producing nations cannot afford to simply give food away and absorb the costs indefinitely. Thus, getting food to the people who need it is mainly an economic problem. In addition, government inefficiency and bureaucratic red tape can add to food problems, sometimes making it difficult to distribute the food to the hungriest people and to ensure that it is not eaten by those who do not need it.

Cultural acceptance of food is another important food problem. If, for example, a particular group of people have been eating rice for hundreds of generations, it is unlikely that they will eat corn even if they are *very* hungry. This may sound strange, but think how reluctant you might be to eat foods such as insect grubs or dog meat, which are exotic to you but are common fare in Africa and Asia. It is human nature to be suspicious of foods to which we are unaccustomed.

Thus, world food problems are many, as are their solutions. We need to increase the production of food, improve overall economic development, improve food distribution, and overcome cultural barriers to acceptance of foods. Developed nations can help developing countries become agriculturally self-sufficient by providing economic assistance and technical aid. But the ultimate solution to world hunger is tied to achieving a stable population in each nation at a level that it can support.

Having a general grasp of world food problems, we now examine the history of agriculture from its beginnings to the present and consider some of the complex challenges facing agriculture today.

# HISTORY OF AGRICULTURE

The earliest human societies were family and tribal units that relied on hunting and gathering to obtain food. This type of society was continually on the move, for they left an area as soon as it was depleted of food resources. The nomadic existence of hunter-gatherer groups prevented them from establishing large, materialistic societies.

The development of agriculture—the raising of plants and animals for food—approximately 10,000 years ago had significant effects on human societies. Agriculture provided a more reliable food supply, so a society's population could grow without depleting food in the area. In addition, cultivation of plants required that people give up their nomadic existence. This led to the establishment of permanent dwellings and eventually to the growth of villages and cities.

Although agriculture was more labor-intensive than hunting and gathering, not every member of the tribe had to be directly involved in obtaining food. In an agricultural society, the farmers produced enough to feed both themselves and some additional people, who could pursue other endeavors and barter with the farmers to obtain their food. This new freedom led to the development of the arts and sciences, trades, and other aspects of human culture. The success of agriculture in supporting people is evidenced by the fact that few societies reverted to hunting and gathering once they developed agricultrue.

## Where Did Agriculture Begin?

Archaeological evidence supports the idea that plants were first domesticated in semiarid regions of the world—grasslands with an adequate water supply but periodic droughts. It seems likely that agriculture evolved independently in three main places.

The **Near East center,** along the eastern end of the Mediterranean Sea (present-day Iran, Iraq, Syria, and Turkey), is sometimes called the *Fertile Crescent* or the *cradle of agriculture.* There is evidence that wheat was domesticated here from wild grasses as long as 10,000 years ago. The maize (corn) culture developed in the **Central/South America center,** although the exact location is difficult to pinpoint. There is evidence of maize, squash, chili pepper, bean, and gourd cultivation in the central highlands of Mexico as early as 7,000 years ago. People living in the highlands of Peru, however, may have had agriculture even earlier. Rice was domesticated in the **Far East center,** but the date is unknown. Rice grows in wetter environments than wheat and maize. Because organic material in these areas tends to be decayed rather than preserved, the survival of good archaeological evidence is unlikely. There are signs that plants other than rice—possibly peas or some type of beans—were domesticated in the Far East center as much as 9,000 years ago.

Agriculture may have developed even earlier than the dates associated with wheat, maize, and rice cultivation. Wheat, maize, and rice are all propagated by seeds and require relatively sophisticated agricultural knowledge to cultivate. Growing other plants, such as bananas, yams, and potatoes, by vegetative propagation is easier and requires less agricultural knowledge. We have no way of knowing whether these plants were cultivated earlier than wheat, maize, or rice, because their propagation would have occurred in moist areas where archaeological evidence could not be preserved.

Regardless of exactly when agriculture arose and which plants were first domesticated, it is indisputable that agriculture led to profound changes in human society.

## Other Early Advances in Agriculture

Agricultural societies in the Near East probably began to irrigate their fields as many as 7,000 years ago. Irrigation in arid and semiarid regions made it possible to produce larger crop yields on a given amount of land. This meant that fewer acres had to be cultivated to feed a population and, therefore, even fewer people had to be involved in producing food. As a result, more people were free to pursue the arts, religion, and trades.

Another important advance in agriculture came with the domestication of animals. When Stone Age humans began to pen wild animals, they were ensuring a dependable food supply for themselves. It is thought that goats were the first animals to be domesticated.

Some domesticated animals were used to prepare fields for planting—another development that freed more people to follow paths other than agriculture. This wasn't the case in North and South America, however, because in those regions there were no native animals that were both suitable for domestication *and* capable of plowing.

## What Is a Domesticated Plant?

A plant that is **cultivated** is protected from natural competition with other plants and from plant-eating animals. The farmer selects desirable traits in a cultivated plant species and encourages their transmission by choosing the seeds of the plants having those traits to save for planting. The plants without the desired traits are not allowed to reproduce, so over time, such artificial selection brings about changes in the genetic makeup of a strain of plants. They may become so different from their indigenous ancestors that it is doubtful they could survive and compete successfully in the wild. Such plants are said to be **domesticated.**

The most extreme example of plant domestication is corn. The ears of maize found at the oldest archaeological sites were scarcely 1 inch long (Figure 18–2). Selective breeding over thousands of years is responsible for the development of large,

**Figure 18–2**

A size comparison between a prehistoric corn cob (right) and a kernel of modern corn (left). (Visuals Unlimited/ Science VU)

compact ears covered by husks that prevent the seeds from being easily dispersed. The original wild grass from which maize apparently evolved still exists in the highlands of Mexico but barely resembles modern-day corn. Its seeds, for example, are borne not compactly on a covered ear, but loosely on an uncovered stalk. And unlike modern corn, which is an **annual** plant (it grows, reproduces, and dies in one growing season), the wild grass ancestor is a **perennial** (it grows year after year).

## AGRICULTURE TODAY

If you were to travel around the world, you would find many different kinds of agriculture and types of food. Despite this diversity, there has been an overall trend—especially during the 20th century—toward greater uniformity in the plants and animals we eat; humans have come to rely on fewer and fewer types of plants and animals for the bulk of food production. There has also been an overall trend toward uniformity in agricultural practices, as farmers in developing nations have adopted techniques used in developed countries.

### Plants and Animals That Stand Between People and Starvation

Although we do not know precisely how many different plant species exist on the earth, biologists estimate that there may be more than 1 million.[3] Of these plants, probably only 3,000 have ever been used for human consumption, although many thousands more are probably edible. Only 150 plants have been cultivated extensively, and of these, 30 provide the bulk of food for humans (Table 18–1). More wheat, rice, corn, and potatoes are produced than all other food crops combined.

Our dependence on so few species of plants for the bulk of our food puts us in an extremely vulnerable position. Should disease or some other factor wipe out one of the important food crops, humans would be threatened by severe famine.

Animals provide us with foods that are particularly rich in protein, such as fish, shellfish, meat, eggs, milk, and cheese. Cows, sheep, pigs, chickens, turkeys, geese, ducks, goats, and water buffalo are the most important types of livestock. Although

[3] This number includes both flowering plant species and lower plants.

| Table 18–1 The 30 Most Important Food Crops | |
|---|---|
| **Food, in Descending Order of Production** | **Type of Crop** |
| 1. Wheat | Cereal grain |
| 2. Rice | Cereal grain |
| 3. Corn | Cereal grain |
| 4. Potato | Ground crop* |
| 5. Barley | Cereal grain |
| 6. Sweet potato | Ground crop |
| 7. Cassava (Manioc) | Ground crop |
| 8. Grape | Fruit, wine |
| 9. Soybean | Legume |
| 10. Oat | Cereal grain |
| 11. Sorghum | Cereal grain |
| 12. Sugarcane | Sugar plant |
| 13. Millet | Cereal grain |
| 14. Banana | Fruit |
| 15. Tomato | Fruit |
| 16. Sugar beet | Sugar plant |
| 17. Rye | Cereal grain |
| 18. Orange | Fruit |
| 19. Coconut | Seed |
| 20. Cottonseed | Oil plant |
| 21. Apple | Fruit |
| 22. Yam | Ground crop |
| 23. Peanut | Legume |
| 24. Watermelon | Fruit |
| 25. Cabbage | Leaf crop |
| 26. Onion | Ground crop |
| 27. Bean | Legume |
| 28. Pea | Legume |
| 29. Sunflower seed | Oil plant |
| 30. Mango | Fruit |

*Ground crops include root, tuber, and bulb crops.

nutritious, livestock is an expensive source of food because animals are inefficient converters of plant food. For example, of every 100 calories of plant material a cow consumes, it burns off approximately 90 in its normal metabolic functioning. That means that only 10 calories out of 100 (10 percent) are stored in the cow to be consumed by humans. Meat consumption is high in affluent societies, so large portions of the crops grown and the fish harvested in developed countries are used to produce livestock animals for human consumption. For example, almost half of the cereal grains grown in developed countries are used to feed livestock (see You Can Make a Difference: Vegetarian Diets).

**Figure 18–3**
Combines harvest wheat. High-input agriculture is very productive, but requires large inputs of fossil fuels. As a result, modern farming practices are very expensive. (Inga Spence/Tom Stack and Associates)

## The Principal Types of Agriculture

Agriculture can be roughly divided into two types: high-input agriculture and subsistence agriculture. Most farmers in developed countries and some in developing countries practice **high-input agriculture,** also called **industrialized agriculture.** It relies on large inputs of energy, in the form of fossil fuels, to produce and run machinery, irrigate crops, and produce chemicals such as fertilizers and pesticides (Figure 18–3). High-input agriculture produces high yields of food per unit of farmland area, but not without costs. A number of problems, including soil degradation and increases in pesticide resistance in agricultural pests, are caused by high-input agriculture (Figure 18–4).

Most farmers in developing countries practice **subsistence agriculture,** the production of enough food to feed oneself and one's family, with little left over to sell or reserve for hard times. Subsistence agriculture, too, requires a large input of energy, but from humans and animals rather than from fossil fuels (Figure 18–5).

Some types of subsistence agriculture require large tracts of land. For example, **shifting agricul-**

**Figure 18–4**
Colorado potato beetles eating potato leaves. As a result of being exposed to heavy applications of pesticides over the years, Colorado potato beetles are resistant to most registered insecticides. (Grant Heilman, from Grant Heilman Photography)

**Figure 18–5**
A Chinese farmer uses oxen to plow his field. Humans and animals provide the energy requirements for subsistence agriculture. (Mike Barlow/Dembinsky Photo Associates)

### Vegetarian Diets

People adopt vegetarian diets for many reasons. Balanced vegetarian diets provide good nutrition without high levels of saturated fats or cholesterol, both of which cause health problems such as heart disease and obesity. Some people become vegetarians because they are philosophically opposed to killing animals, even for food. Others convert because of a sense of responsibility for land use and its wide repercussions. Fewer plants are required to support vegetarians than to support meat eaters. The amount of usable energy in the food chain is decreased by approximately 90 percent by adding an additional level (i.e., the animals we eat) to the chain (see figure). Simply stated, if everyone were to become a vegetarian, much more food would be available for human consumption.

Some people are reluctant to switch to a vegetarian diet because they fear they will not get enough protein (plant foods have a lower percentage of protein than do animal foods). However, the problem is usually not lack of protein (meat eaters in developed countries usually consume much more protein than they need), but obtaining the proper balance of essential amino acids. Some vegetarians consume animal *products*, such as eggs and dairy products, because these foods contain all the essential amino acids. Other vegetarians eat no animal products and rely exclusively on plant products to obtain their protein.

Although the human body can store excess lipids (as fat) and excess carbohydrates (as glycogen in the liver), it has no way to store excess amino acids. Therefore, *all* the amino acids essential for protein manufacture must be eaten together; it does no good to eat some of the essential amino acids for lunch and others for dinner.

A nutritious vegetarian diet includes a combination of foods that contains all the essential amino acids. A meal of rice and beans or corn and beans, for example, provides the proper complement of essential amino acids. Cookbooks and other references contain menus and recipes that give the vegetarian diet adequate amounts of high-quality protein. The following list provides an overview of combinations of foods that offer a proper balance of essential amino acids. At least one food from each column should be consumed at every meal.

---

**ture,** also called slash-and-burn agriculture, involves clearing small patches of tropical forest to plant crops (see Chapter 17). Because tropical soils lose their productivity very quickly when they are cultivated, farmers using shifting agriculture must move from one area of forest to another every three years or so. **Nomadic herding,** in which livestock is supported by land that is too arid for successful crop growth, is another type of land-intensive subsistence agriculture. Nomadic herders must continually move their livestock to find adequate food for them.

### The Effect of Domestication on Genetic Diversity

Wild plant and animal populations usually have great genetic diversity (see Chapter 16), which contributes to species' long-term survival by providing the variation that enables each population to adapt to changing environmental conditions. When plants and animals are domesticated, much of this genetic diversity is lost, because the farmer selects for propagation only those plants and animals with the most desirable characteristics (from the farmer's point of view). At the same time, other traits that are not of obvious value to humans are selected against. Hence, many of the high-yielding crops produced by modern agriculture are genetically uniform. For example, most of the corn grown in the United States is of only a few different varieties (Table 18–2). Likewise, dairy cattle and poultry in the United States have low genetic diversity.

The loss of genetic diversity that accompanies modern agriculture usually doesn't prove disastrous to crop plants, because they do not have to survive in the wild under natural conditions. Under cultivation, they are watered, fertilized, and protected as

**Column I**

**Grains**
Barley, corn, oats, rice, rye, wheat

**Nuts and Seeds**
Almonds, beech-nuts, Brazil nuts, cashews, filberts, pecans, pumpkin seeds, sunflower seeds, walnuts

**Column II**

**Legumes**
Fleshy beans, peas*

**Dairy Products**
Cheese, cottage cheese, eggs, milk, yogurt

*There is a wide variety of beans and peas, including black beans, fava beans, kidney beans, lima beans, pinto beans, mung beans, navy beans, and soybeans (which are usually eaten as bean curd, or tofu).

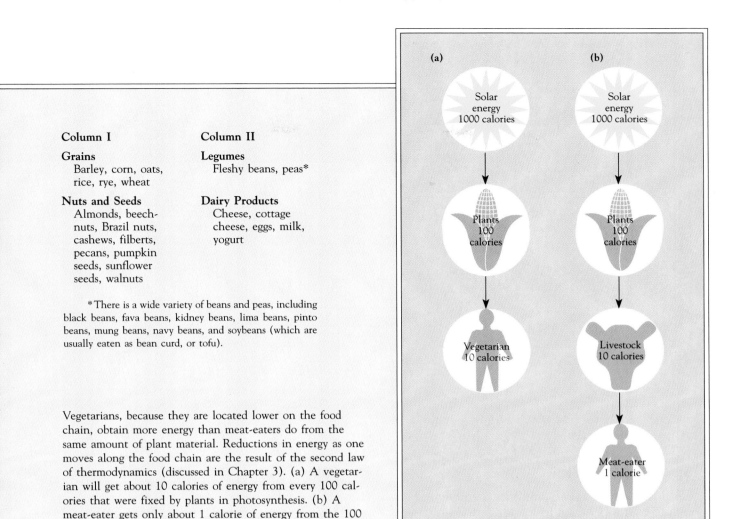

Vegetarians, because they are located lower on the food chain, obtain more energy than meat-eaters do from the same amount of plant material. Reductions in energy as one moves along the food chain are the result of the second law of thermodynamics (discussed in Chapter 3). (a) A vegetarian will get about 10 calories of energy from every 100 calories that were fixed by plants in photosynthesis. (b) A meat-eater gets only about 1 calorie of energy from the 100 calories fixed by the plants that the cattle consumed.

much as possible from pests, including weeds, insects, and disease organisms. Domesticated animals are also protected from the challenges of nature.

However, the lower genetic diversity of domesticated plants and animals does increase the likelihood that they will succumb to new strains of disease organisms, which include bacteria, fungi, and viruses. Disease organisms evolve quite rapidly. When a disease breaks out in a domesticated plant or animal population, the *whole* uniform population is susceptible; thus, the loss is greater than it would be in a natural, varied population, in which at least some individuals would contain genes to resist the pathogen.

**The Global Decline in Domesticated Plant and Animal Varieties** Although domestication leads to less genetic diversity than is found in wild relatives, the many farmers worldwide who have been select-ing for specific traits have developed multiple varieties of each domesticated plant and animal. A given variety, the legacy of the hundreds of farmers who developed it over thousands of years, is adapted to the climate where it was bred and contains a unique combination of traits conferred by its unique combination of genes.

A trend is under way to replace the many local varieties of a particular crop or farm animal with just a few kinds worldwide. When farmers abandon their traditional varieties in favor of more modern ones, the former frequently become extinct. This represents a great loss in genetic diversity, because each variety's characteristic combination of genes gives it distinctive nutritional value, size, color, flavor, resistance to disease, and adaptability to different climates and soil types. The gene combinations of local varieties are potentially valuable to plant breeders because they can be transferred to other

| Table 18–2 Agricultural Diversity in the United States | | |
|---|---|---|
| Crop | Main Varieties in Production | Percentage of Total Crop Produced from These Varieties |
| Corn | 6 | 71 |
| Wheat | 10 | 55 |
| Soybeans | 6 | 56 |
| Rice | 4 | 65 |
| Potatoes | 4 | 72 |
| Peanuts | 9 | 95 |
| Peas | 2 | 96 |

Source: Reichart, W., "Agriculture's Diminishing Diversity," *Environment* 24:9, 1982, pp. 6–11, 39–44.

varieties, either by traditional breeding methods or by genetic engineering.

## Increasing Crop Yields

Until the 1940s, agricultural yields among various countries, both developed and developing, were generally equal. However, advances made by research scientists have caused a dramatic increase in food production in developed countries. Greater knowledge of plant nutrition has resulted in fertilization that promotes optimum yields (Figure 18–6). The use of pesticides to control insects, weeds,

**Figure 18–6**
Worldwide fertilizer use. The increasing use of fertilizers has resulted in higher crop yields.

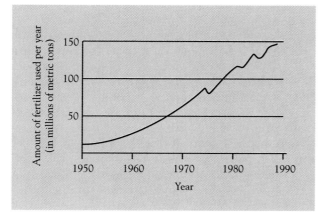

and disease-causing organisms has also improved crop yields. Animal production has been increased by the use of antibiotics to control disease. Selective breeding programs have resulted in agricultural plants and animals with more desirable features. For example, breeders developed wheat plants with larger, heavier grain heads (for higher yield). Because of the weight of the heads, other traits were gradually incorporated into wheat, such as shorter and thicker stalks, which prevent the plants from falling over during storms (Figure 18–7).

**The Green Revolution**  The introduction during the 1960s of new high-yielding varieties of wheat and rice to Asian and Latin American countries gave these nations the chance to provide their people with adequate supplies of food. But the high-yielding varieties required intensive cultivation methods, including the use of fertilizers, pesticides, and mechanized machinery, in order to realize their potential. The production of more food per acre of cropland by using the new, high-yielding varieties and modern cultivation methods has been called the **green revolution.** Some of the success stories of the green revolution have been remarkable. For example, Indonesia used to import more rice than any other country in the world. Today Indonesia produces not only enough rice to feed its people, but also some for export.

**Problems with the Green Revolution**  The two most important problems associated with higher crop production are damage to the environment, which will be considered shortly, and the high energy costs that are built into this form of agriculture. Inorganic fertilizer requires a great deal of energy to manufacture, so its production costs are tied closely to the price of energy. Significant amounts of fossil fuels are required to provide power for farm equipment such as tractors and combines. The installation and operation of irrigation systems, including construction of dams and canals, also requires a substantial energy input.

Other problems are associated with the green revolution. Although rice and wheat are not the only crops to have been improved—high-yielding varieties of crops such as potatoes, barley, and corn have also been developed—many important food crops remain to be improved by selective breeding and scientific research. People in Africa, for example, eat sorghum, millet, cassava, and sweet potatoes, none of which has been greatly improved by green revolution technology. Also, subsistence farmers, who represent a substantial segment of the

**Figure 18–7**
A high-yielding, short-stemmed variety of wheat is growing in rows at the University of Minnesota. The shorter, sturdier stalks enable this variety to remain standing straight during storms. (University of Minnesota Agricultural Experiment Station)

agricultural community in most developing nations, have not benefited from the green revolution. They need improved crops that respond to labor-intensive agriculture (*human and animal labor*) and do not require large outlays of energy and capital.

## Food Processing: Food Additives

Most of the food we eat is not harvested and used directly. Instead, after harvest it is processed, then offered for sale in grocery stores. There are two aspects to food processing. The first encompasses procedures such as drying, freezing, canning, pasteurizing, curing, irradiating (see Chapter 22), and refrigerating food to retard spoilage. The second involves adding **food additives**—chemicals that enhance the taste, color, or texture of the food; improve its nutrition; reduce spoilage and prolong shelf life; or maintain the food's consistency.

Sugar and salt are the two most common food additives. Although they are added to food primarily to make it taste better, in large amounts sugar and salt can be used to help preserve food—for example, fruit jelly and salted meat. Salt and sugar have been used as preservatives for centuries.

Essential amino acids and vitamins are sometimes added to foods to make them more nutritious than they would be naturally or to replace nutrients that are lost during processing. Both natural and synthetic **coloring agents** are used to make food visually appealing. Sodium propionate and potassium sorbate are two examples of **preservatives,** which are chemicals added to food to retard the

growth of bacteria and fungi that cause food spoilage. Food can also spoil when lipids (fats and oils) undergo oxidation. Food additives that prevent oxidation are called **antioxidants** and include butylated hydroxyanisole (BHA) and butylated hydroxytoluene (BHT).

**Protection of the Consumer** The Food and Drug Administration (FDA) is charged with the responsibility of monitoring food additives. In 1958 regulations were passed that require any new food additive to undergo extensive toxicity testing by the manufacturer. The results of such tests are then evaluated by the FDA to determine whether the additive is safe.

Additives that were in use prior to 1958, however, do not have to undergo such testing. In 1959 the FDA designated these chemicals "generally recognized as safe" and made up a list of them—usually called the GRAS (pronounced *grass*) list. All substances on the GRAS list have subsequently been reviewed. Some have been banned, including cyclamates (used for sweetening) and brominated vegetable oil. BHA, BHT, and several other substances on the GRAS list are undergoing further tests.

**Are Food Additives Bad?** The Center for Science in the Public Interest has compiled a list of food additives that may be harmful and should be avoided or eaten sparingly (Table 18–3). These include the preservatives BHA and BHT, a number of coloring agents (especially the red dyes), and ni-

**Table 18-3**
**Food Additives That May Be Harmful**

| Food Additive | Food | Possible Effect |
|---|---|---|
| BHA, BHT | Oils, potato chips, chewing gum | May be carcinogenic; allergic reactions in some |
| Citrus red dye #2 | Skin of some oranges | May be carcinogenic |
| Nitrates, nitrites | Bacon, corned beef, hot dogs, ham, smoked fish, luncheon meats | Formation of carcinogenic N-nitroso compounds |
| Red dyes #3, 8, 9, 19, 37 | Cherries (maraschino, fruit cocktail, candy) | May be carcinogenic |
| Saccharin | Diet foods | Known carcinogen in animal tests |
| Sulfur dioxide | Dried fruit, canned and frozen vegetables, some beverages, bread, salad dressings, and more | Severe allergic reactions |

**Focus On**

## Nitrates and Associated Compounds

**Nitrates** (compounds containing $NO_3^-$) and **nitrites** (compounds containing $NO_2^-$) occur in water and food as well as in the environment. Nitrates come both from natural sources (for example, they are produced by some soil bacteria) and from chemical fertilizers. Nitrates are present naturally in many vegetables, particularly the leafy green ones, such as spinach and lettuce, and the root crops, such as radishes and beets. Nitrates and nitrites are used as food additives and for curing meats. In addition, nitrates sometimes contaminate drinking water, with fertilizer being the most common source. When nitrates get into the body, they are converted to nitrites, which reduce the blood's ability to transport oxygen. This condition is one of the causes of cyanosis (the "blue baby" syndrome), a serious disorder in very young children.

Some **N-nitroso compounds** (related nitrogen-containing compounds) are carcinogenic. N-nitrosodimethylamine, for example, is a potent carcinogen found in cured meats such as bacon, in smoked and salted fish, and in tobacco. Cheese and beer (brewed with water containing high levels of nitrates) are also sources of N-nitroso compounds.

Ingested nitrites are capable of reacting with other chemicals in food and in tobacco to form N-nitroso compounds. We do not know how this occurs. A great deal remains to be learned about the N-nitroso compounds found naturally in foods, as well. Interestingly, most fresh vegetables contain substances that inhibit the production of N-nitroso compounds. All of these uncertainties about N-nitroso compounds mean that we cannot be sure whether these carcinogenic compounds are formed primarily in the body or are already present in our food and water when we ingest them.

What does all this mean to you? At this point, no one has an accurate picture of the risks involved. The presence of nitrates in fresh vegetables should probably not concern you since these plants also contain anti–N-nitroso compounds. You also needn't worry about drinking water from municipal systems, because the level of nitrate in drinking water is monitored. If you drink well water, however, you should probably have its nitrate level checked periodically. In addition, it might be prudent to avoid excessively high amounts of cured meats such as bacon, ham, hot dogs, and luncheon meats.

trates and nitrites (see Focus On: Nitrates and Associated Compounds).

There are two opposing viewpoints about the safety of our food. Some people are *very* concerned about the large number and amounts of food additives in processed food. The main fear is that some additives may be carcinogenic. Supporters of this view worry that, although the risk of developing cancer from exposure to each individual chemical may be quite small, the effects of hundreds of different food additives added together may be significant. Also, there is concern that these chemicals may interact synergistically.

Others think that the health hazards from food additives are greatly exaggerated. They think it is important to put concerns about food additives and even pesticide residues in proper perspective. Everyone acknowledges that the chemicals in our foods do not pose anywhere near the threat of cancer that smoking does, for example.

Moreover, much larger quantities of *natural* carcinogens are present in food. Nature is not benevolent; plants have evolved natural chemical defenses to discourage insects from consuming them. Some of these compounds, which can be present in large amounts, are toxins; some are carcinogenic. Thus, certain people contend that food

additives pose a much smaller threat than do the natural, unavoidable chemicals already present in food.

## THE ENVIRONMENTAL IMPACT OF AGRICULTURE TODAY

The practices of high-input farming have resulted in a number of environmental problems (Figure 18–8). According to the Environmental Protection Agency, agricultural chemicals, including fertilizers and pesticides, are the single largest cause of water pollution in the United States. Some of these chemicals have been detected in groundwater deep underground as well as in surface waters; fish and other aquatic organisms are killed by pesticide run-off into lakes, rivers, and estuaries.

To complicate matters, many insects, weeds, and disease-causing organisms have developed resistance to pesticides, forcing farmers to apply progressively larger quantities (see Chapter 22). Residues of pesticides and antibiotics contaminate our food supply and reduce the number and diversity of beneficial microorganisms in the soil.

Soil erosion and loss of soil fertility persist as important problems; improper irrigation has re-

**Figure 18–8**
An overview of environmental problems associated with high-input agriculture. (Inga Spence/
Tom Stack and Associates)

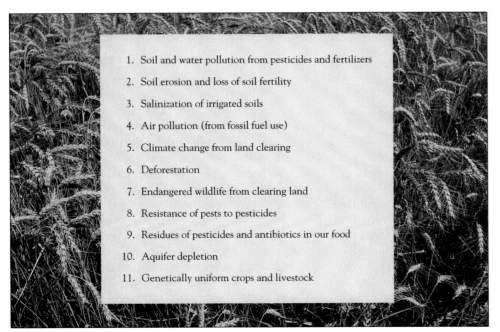

1. Soil and water pollution from pesticides and fertilizers

2. Soil erosion and loss of soil fertility

3. Salinization of irrigated soils

4. Air pollution (from fossil fuel use)

5. Climate change from land clearing

6. Deforestation

7. Endangered wildlife from clearing land

8. Resistance of pests to pesticides

9. Residues of pesticides and antibiotics in our food

10. Aquifer depletion

11. Genetically uniform crops and livestock

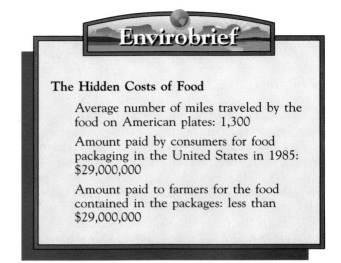

## Envirobrief

**Moving Food**

We take for granted the presence of fresh fruits and vegetables in northern regions of the country during the winter. Some come from warm-climate states, many others from foreign countries. Food in general moves great distances before making its way to a supermarket shelf. The U.S. Department of Agriculture has estimated that the average processed food travels 1,300 miles before being eaten. Trucks move 99% of all livestock, 88% of fresh fruits and vegetables, and 80% of fresh and frozen meats, dairy products, bakery goods, and beverages. The environmental cost is considerable. More than 4.25 million trucks are used primarily to transport food each year in the United States. They travel almost 50 million miles (a distance equal to more than 250 trips to the sun), burning almost $6 billion worth of fuel; they emit well over 4 million tons of pollutants into the air; and they cause hundreds of millions of dollars of damage to federal and state highways because of their extreme weight (one fully loaded 80,000-pound truck causes more damage to roads than 9,600 automobiles). Yet our food processing and distribution system compounds the problem. Large food processors often make deals with large growers in other parts of the country. So it is not uncommon, for example, to see tomatoes shipped all the way from California to a soup factory in New Jersey. Even if suitable tomatoes were grown outside the factory gate, corporate contracts preclude them from being purchased. This is one way in which economic arrangements clearly work against the environment.

## Envirobrief

**The Hidden Costs of Food**

Average number of miles traveled by the food on American plates: 1,300

Amount paid by consumers for food packaging in the United States in 1985: $29,000,000

Amount paid to farmers for the food contained in the packages: less than $29,000,000

## Using More Land for Cultivation

In the temperate areas of the world, almost all the fertile land with an adequate supply of water is used for agriculture. Although in some regions the loss of prime farmland to urbanization is of concern (see Chapter 17), in temperate areas very little prime farmland not already under cultivation remains. Tropical areas, on the other hand, have little prime agricultural land to start with.

One reason the United States had agricultural surpluses during the 1980s is that farmers brought large amounts of previously unused land into production. Unfortunately, much of this land was marginal as agricultural land, because it was prone to soil erosion—subjected to either intermittent floods or frequent droughts (and therefore wind erosion when the ground cover was removed). Harvesting crops from highly erodible land is ecologically unsound and cannot be done indefinitely.

Other countries have paid a high price for cultivating land highly prone to erosion. The former Soviet Union, for example, began to cultivate a large area of marginal land during the 1950s. Although the initial production of cereal crops was high, by the 1980s much of this land had to be abandoned. The annual per-capita food production in Haiti, a very poor Caribbean nation, is half what it was in 1950 as a result of both population growth and loss of production on eroded soils.

From the 1940s to the 1980s, the amount of agricultural land was greatly increased through irrigation of dry land. During the last few years, however, there has been a worldwide decline in the rate of expansion of irrigation, and in the United States the amount of irrigated land has actually decreased. This change is due to the depletion of aquifers, the

sulted in declining soil productivity as salts have accumulated in the soil. Aquifers have been depleted by irrigation withdrawals.

Clearing grasslands and forests and draining wetlands to grow crops have resulted in habitat losses that reduce biological diversity. Many plant and animal species have become endangered or threatened as a result of habitat loss caused by agriculture. A reduction of genetic diversity in agriculturally important crops and livestock has also occurred as a result of selective breeding.

## U.S. Agricultural Policies

Following World War II, American agricultural productivity grew at a phenomenal rate for about 35 years. We not only produced enough food to feed ourselves, but had plenty to spare for export. But the 1980s introduced severe economic hardship to many American farmers. The increasing agricultural productivity of other countries, including some developing countries, reduced their dependence on U.S. food exports; many of these countries began to have food surpluses. Economic factors that contributed to the U.S. farm problem included a worldwide recession and the high international value of the dollar. All this led to a sudden decrease in crop prices, which left many farmers unable to pay their debts.

Starting in 1983, the federal government increased federal income and price support programs to help farmers survive. These supports have made crop prices higher than the world market dictates. More recently, the demand for U.S. food exports has also grown. Although the situation has improved for farmers, many are still experiencing financial difficulties.

Several federal farm policies, including commodity programs, income and price supports, pesticide regulations, and tax policy, discourage the introduction of innovative techniques such as alternative agriculture. For example, farmers who incorporate crop rotation may actually be penalized for doing so, because the commodity program pays farmers who grow certain crops based on the acres of a given crop planted every year for five years. The farmer who substitutes another crop on those acres during that period is financially penalized even though crop rotation improves soil fertility, reduces the need for pesticides, and increases overall crop yields.

In summary, federal programs have an important effect on farmers' agricultural practices. Generally speaking, the traditional goal of federal agricultural policies has been higher production at the expense of protection of farm resources. However, some recent farm legislation at the federal level has placed a higher priority on environmental and conservation concerns (see Chapters 14 and 17), and the trend may continue.

---

abandonment of salty soil, and the diversion of irrigation water to residential and industrial use.

## SOLUTIONS TO AGRICULTURAL PROBLEMS

Food production poses an environmental quandary: We must increase food production in order to eliminate world hunger, but growing more food damages the environment, which lessens our chances of increasing food production in the future. Fortunately, the dilemma is not as hopeless as it seems. Farming practices and techniques exist that can ensure sustainable agriculture. Farmers who have been practicing high-input agriculture can adopt these alternative agricultural methods, which cost less and are less damaging to the environment. Advances are even being made in sustainable shifting agriculture. In addition, new technologies, such as genetic engineering and the widespread adoption

of "new" crop plants, promise greater productivity and variety in nutritious foods.

### Alternative Agriculture, a Substitute for High-Input Agriculture

More and more farmers are trying forms of agriculture that cause fewer environmental problems than high-input agriculture (see Focus On: U.S. Agricultural Policies for a discussion of how federal farm programs affect farming). **Alternative agriculture,** also called **sustainable** or **low-input agriculture,** relies on beneficial biological processes and environmentally friendly chemicals (those that disintegrate quickly and do not persist as residues in the environment) rather than conventional agricultural techniques.

In alternative agriculture, enhancement of natural predator-prey relationships can substitute for heavy pesticide use (see Chapter 22). For example, apple growers in Maryland monitor and encourage the presence of black ladybird beetles in their orchards because these insects feed voraciously on European red mites, a major pest of apples. Animal manure added to soil decreases the need for high levels of chemical fertilizers. Also, employment of biological nitrogen fixation to convert atmospheric nitrogen into a form that can be utilized by plants lessens the need for nitrogen fertilizers. The breeding of disease-resistant crop plants is an important part of alternative agriculture, as is the maintenance of animal health rather than the continual use of antibiotics to prevent disease.

An important goal of alternative agriculture is to sustain the quality of agricultural soil. For example, crop rotation, minimum tillage, and contour plowing help control erosion and maintain soil fertility (see Chapter 14). Water and energy conservation are also practiced.

Alternative agriculture is not a single program but rather a series of programs that are adapted for specific land, climate, and farming requirements. For example, some alternative farmers—those who practice **organic agriculture**—use no pesticide chemicals; others use a system of **integrated pest management (IPM).** In IPM a limited use of pesticides is incorporated with such practices as crop rotation, continual monitoring for potential pest problems, use of disease-resistant varieties, and biological pest controls.

## Making Shifting Agriculture Sustainable

Traditional slash-and-burn agriculture is sustainable as long as there are few farmers and large areas of forest. Since relatively small patches of forest are cleared for raising crops, the trees quickly return when the land is abandoned and the farmer has moved on to clear another plot of forest. If the abandoned land lies fallow (idle) for a period of 20 to 100 years, the forest recovers to the point where subsistence farmers can again clear the forest for planting. Burning the trees releases another flush of nutrients into the soil so that crops can again be grown there.

Some researchers have been seeking ways to make land retain its productivity for a longer period than is usual in shifting agriculture. Consider Papua New Guinea, a small island nation in which approximately 80 percent of the people are subsistence farmers. Research scientists in this country have developed methods to deal with some of the most troublesome problems associated with shifting agriculture: soil erosion, declining fertility, and attacks by insects and diseases. Their research, which is part of the Shifting Agriculture Improvement

Program in Papua New Guinea, has helped forest plots remain productive for longer periods of time. Heavy mulching with organic material, such as weed and grass clippings, has lessened soil infertility and erosion. The composted mulch is then piled into rows that follow the contours of the land, further reducing erosion. Several crops are planted together, reducing insect damage. One of the crops is always a legume (such as beans), which helps restore nitrogen fertility to the soil. An extension program demonstrates these methods to farmers, and a book, *Subsistence Agriculture Improvement Manual*, assists in educating rural farmers.

## Genetic Engineering

The ability to take a specific gene from one cell and place it into another cell, where it is expressed, is called **genetic engineering** or **biotechnology** (although some people define biotechnology more broadly, to include any use of living organisms to produce products). Genetic engineering has great potential to revolutionize not only agriculture but other fields such as medicine.

The goals of genetic engineering in agriculture are not new. Using traditional breeding methods, farmers and scientists have developed desirable characteristics in crop plants and agricultural animals for centuries. It takes time to develop such genetically improved organisms, however. For example, using traditional breeding methods, it might take 15 years or more to incorporate genes for disease resistance into a particular crop plant. Genetic engineering has the potential to accomplish the same goal in a fraction of that time (Figure 18–9). Moreover, it differs from traditional breeding methods in that desirable genes from *any* organism can be used, not just those from the species of the plant or animal that is being improved. For example, if a gene for disease resistance found in petunias would be beneficial in tomatoes, the genetic engineer can splice the petunia gene into the tomato plant. This could never be done by traditional breeding methods, because petunias and tomatoes belong to separate groups of plants and don't interbreed.

Genetic engineering could produce food plants that would be more nutritious because they would contain *all* the essential amino acids (currently, no single food crop has this trait). Crop plants that were resistant to insect pests and disease or that could tolerate drought, heat, cold, herbicides, or salty soil could be developed (Figure 18–10).

Genetic engineering could be used to develop more productive farm animals, as well. For exam-

**Figure 18–10**
The genetically engineered tomato on the right possesses resistance to insect damage. The unaltered, "normal" tomato on the left, which was also exposed to ravenous caterpillars, shows severe damage. (Courtesy of Monsanto Company)

**Figure 18–9**
The development of disease-resistant varieties of peaches is accomplished in a fraction of the time it would take using traditional breeding methods. This research scientist developed these plants using cell and tissue culture methods, which are sophisticated technologies used to a great extent by genetic engineers. (Agricultural Research Service, USDA)

ple, the common intestinal bacterium *Escherichia coli* has been genetically engineered to produce a cow hormone that increases milk production when injected into cows (Figure 18–11). Perhaps the greatest potential contribution of genetic engineering in the animal arena, however, is in the production of vaccines against disease organisms that harm agricultural animals.

Not everyone greets the potential of genetic engineering with enthusiasm. Opponents regard these new techniques as dangerous because of their ability to alter organisms (See Focus On: The Safety of Genetic Engineering). Proponents of genetic engineering point out that traditional breeding methods have accomplished some of the same changes for centuries. Both sides agree, however, that genetic engineering can do things that traditional breeding methods cannot. It is not possible to use traditional breeding methods to remove a gene from a cow and transfer it to a bacterium, but

this type of gene transferral is common in genetic engineering. Concerns about genetically altered organisms wreaking havoc in the natural environment have led to efforts to strictly control the types of experiments and testing that are allowed (Figure 18–12).

Although genetic engineering has the potential to revolutionize agriculture, the changes will not occur overnight. Several hundred private biotechnology firms, as well as many scientists in colleges, universities, and government research labs, are involved in agricultural genetic engineering. However, a great deal of research must be done before most of the envisioned benefits from genetic engineering are realized.

## Eating New Foods

A trip to the fresh food section of most supermarkets in North America today reveals a number of "new" fruits and vegetables. Many of these crops are gaining acceptance, despite our natural reluctance to try foods that are not part of our cultural background. Malanga, carambola, chayote, and winged beans are just some of the "exotic" foods that are beginning to grace our tables.

Some of these plants are more nutritious than our more common foods. **Quinoa** (pronounced *keen-wa*) is an ancient grain crop of the Inca civilization in the Andes Mountains of South America

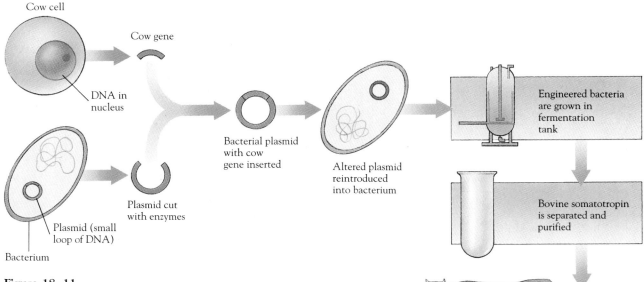

**Figure 18–11**

The cow hormone known as bovine somatotropin has been produced in large amounts by genetically engineered bacteria. The steps involved in inserting a cow gene into bacteria include isolating the gene (DNA) that codes for the specific cow hormone and inserting it into a bacterial plasmid (a small loop of bacterial DNA). The altered plasmid is then placed in a bacterial cell, from which are grown large numbers of bacteria. Each of these cells is able to make cow hormone because it contains the cow gene that codes for this hormone. Cows injected with bovine somatotropin are able to produce larger quantities of milk. The milk obtained from such cows is virtually indistinguishable from milk taken from untreated cows.

**Figure 18–12**

Before any genetically altered organism is released into the environment, extensive tests are performed to determine if it will cause any adverse effects. (Courtesy of Monsanto Company)

**Figure 18–13**

Seeds imported from South America were used to plant this field of quinoa in Colorado. (Grant Heilman, from Grant Heilman Photography)

## Focus On

### The Safety of Genetic Engineering

While acknowledging the potential uses of genetic engineering as important and beneficial, some people are concerned about potential misuses and unpredictable side effects. An organism might be produced that would have undesirable ecological or other effects—not by design, but by accident. New strains of bacteria or other organisms, with which the world of life has no previous experience, might be similarly difficult to control.

Most of what we humans do is quite temporary, however severe the consequences of our actions may be in the short term. Real "permanence" in a strain of organisms involves their genetic information, for that is actively self-propagating in a way that our day-to-day environmental manipulations cannot match. To date, only the extinction of plants and animals has had permanent conse-

quences for us, and of course that also involves genetic information—specifically its loss. Now, however, it is possible to *create* new combinations of genetic information and incorporate them into self-reproducing life forms that, once created, might well persist for geological epochs. Thus, the most serious problem associated with genetic manipulation may be the permanence of its results.

Recent history has failed to bear out these genetic worries. Experiments over more than a decade have demonstrated that genetic engineering experiments can be carried out in complete safety. It must be acknowledged, however, that initial concerns about the safety of genetic engineering contributed to the development of safe experimental design.

(Figure 18–13). Quinoa is receiving a great deal of attention today because it is more nutritious than grains such as corn, wheat, and rice. Not only does quinoa have more protein than any other grain, it also contains a balanced amino acid composition, including the essential amino acids lysine and methionine.

Although the nutritional characteristics of such exotic foods are fairly well known, methods for profitably growing them in large quantities must be developed. Their adaptability to different climates, soil types, and water availability must be determined before farmers will be willing to risk growing them, and marketing research will have to be conducted to guarantee a wide market for these crops, should they be cultivated on a large scale.

## FISHERIES OF THE WORLD

The world's oceans contain a valuable food resource. Just under 90 percent of the world's total marine catch is fish, with clams, oysters, squid, octopus, and other mollusks representing an additional 6 percent of the total catch. Crustaceans,

including lobsters, shrimp, and crabs, make up about 3 percent, and marine algae constitute the remaining 1 percent (Figure 18–14).

Fish and other seafood are highly nutritious because they contain high-quality protein (protein with a good balance of essential amino acids) that is easily digestible. Worldwide, approximately 5 percent of the total protein in the human diet is obtained from fish and other seafood; the rest is obtained from milk, eggs, meat, and plants. However, in certain countries, particularly in developing nations that border the oceans, seafood makes a much larger contribution to the total protein in the human diet.

Most of the world's marine catch is obtained by fleets of fishing vessels (Figure 18–15). In addition, numerous fish are captured in shallow coastal waters of developing nations. The world's total marine catch has increased substantially during the past 40 years. The rate of increase has slowed in recent years, however, making it obvious that the oceans can yield only a limited number of fish. Furthermore, each type of fish has a maximum sustainable harvest level; if a particular species is overharvested, its numbers drop and harvest is no longer economically feasible.

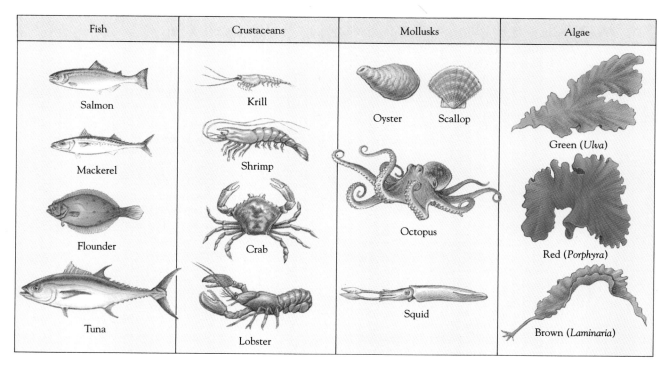

| Fish | Crustaceans | Mollusks | Algae |
|---|---|---|---|

**Figure 18–14**
The major types of seafood that are commercially harvested include fish, crustaceans, mollusks, and algae.

**Figure 18–15**
A full fish net is pulled on board a fishing vessel in the Bering Sea off the coast of Alaska. Fleets of fishing vessels are responsible for most of the world's fish harvest. Some of these ships are quite large and actually process the seafood on board. (Jack D. Swenson/Tom Stack and Associates)

## Problems and Challenges for the Fishing Industry

No nation lays legal claim to the oceans. Consequently, resources in the ocean are more susceptible to overuse and degradation than are resources on the land, which individual nations own and for which they feel responsible (see Focus On: The Tragedy of the Commons in Chapter 2). The most serious problem for marine fisheries is that many marine species have been overharvested to the point that their numbers are severely depleted. Fishermen tend to concentrate on a few fish species with high commercial value, such as menhaden, salmon, tuna, and flounder, while other fish species are underutilized. In response to overharvesting, many nations have adopted a policy of **ocean enclosure,** which puts the living organisms within 200 miles of land under the jurisdiction of the country bordering the ocean (Figure 18–16). This allows nations to regulate the amounts of fish and other seafood harvested from their waters, thereby preventing overexploitation.

One of the great paradoxes of human civilization is that the same oceans that are used to provide food to a hungry world are also used as dumping grounds. Pollution increasingly threatens the fisheries of the world. Everything from accidental oil spills to the deliberate dumping of litter pollutes the water. Heavy metals such as lead, mercury, and cadmium are finding their way into the aquatic food chain, where they are highly toxic to both fish and the humans who eat fish.

Between 60 and 80 percent of all commercially important ocean fish spend at least part of their lives in coastal areas, which include tidal marshes, mangrove swamps, and estuaries. These areas are also in high demand for recreational and residential development (see Chapter 17).

## Aquaculture: Fish Farming

**Aquaculture,** the rearing of aquatic organisms, is more closely related to agriculture on land than it is to the fishing industry just described (Figure 18–17). Aquaculture is carried out both in fresh water and in marine water near the shore; the cul-tivation of marine organisms is sometimes called **mariculture.**

Although aquaculture is an ancient practice that probably originated in China several thousand years ago, its enormous potential to provide food has only recently been appreciated. Aquaculture can contribute variety to the diets of people in de-veloped countries. Inhabitants of developing na-tions can benefit even more from aquaculture: it may provide them with much-needed protein and even serve as a source of foreign exchange when they export such delicacies as aquaculture-grown shrimp.

Aquaculture differs from fishing in several re-spects. For one thing, although the developed na-tions harvest more fish from the oceans, the devel-oping nations produce much more seafood by aquaculture. One reason for this is that developing nations have an abundant supply of cheap labor, which is a requirement of aquaculture because it,

**Figure 18–16**
Many coastal nations have adopted a policy of open enclosure, in which they declare the 200 miles of ocean bordering their land (designated dark blue) as belonging to them. This allows these nations to regulate the amount of fish and other seafood that can be harvested from these waters.

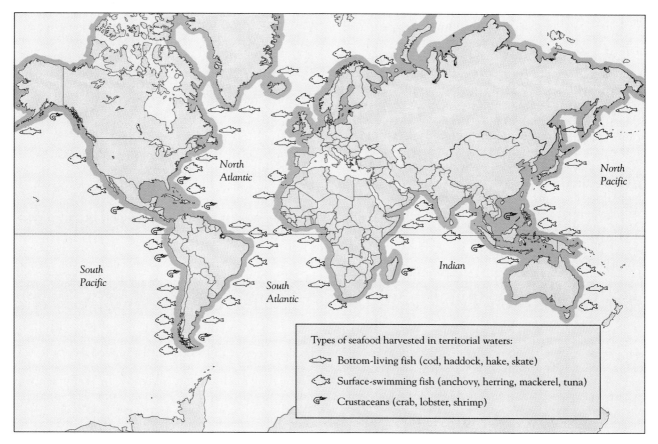

Types of seafood harvested in territorial waters:

- Bottom-living fish (cod, haddock, hake, skate)
- Surface-swimming fish (anchovy, herring, mackerel, tuna)
- Crustaceans (crab, lobster, shrimp)

**Figure 18–17**

*Tilapia* grown at a large aquaculture facility in Hawaii. Aquaculture is becoming increasingly important in providing the world's seafood. (Greg Vaughn/Tom Stack and Associates)

which have a tendency to spread rapidly in the crowded conditions that are characteristic of aquaculture.

One of the most important limits on aquaculture's potential is the receptivity of animals to the domestication process itself. Land animals such as cows, pigs, and sheep were domesticated over a period of thousands of years. During this time, there were undoubtedly failed attempts to domesticate many other animals, which for one reason or another could not be domesticated. The same is true of aquaculture: it is not simply a matter of observing a need for more tuna, for instance, and therefore opening an aquaculture facility that produces tuna. The organisms that are to be produced profitably by aquaculture must have certain traits that make their domestication possible. For example, aquatic organisms that are social by nature and do not exhibit territoriality or aggressive behavior are possible candidates for domestication.

like land-based agriculture, is labor-intensive. Also, the limit on the size of a catch in fishing is the population of fish found in nature, whereas the limit on aquacultural production of fish and other seafood is largely the size of the area in which they are grown. Aquaculture is practiced close to land, either in estuaries or in open ocean near the shore; therefore, other uses of shorelines compete with aquaculture for available space.

Although interest in aquaculture is increasing worldwide, several factors, besides the need for area along coastlines, are slowing its expansion. Setting up and running an aquaculture facility is expensive. Also, scientific research is necessary in order to make aquaculture of certain organisms profitable. For example, an organism's requirements for breeding must be established, and ways to control excessive breeding must be available (so that the population doesn't overbreed and produce many stunted individuals rather than fewer large ones). The population must be continually monitored for diseases,

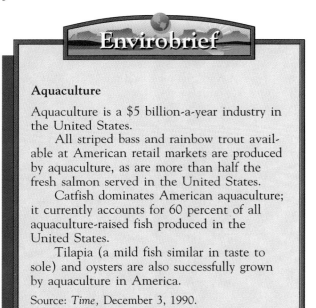

**Envirobrief**

**Aquaculture**

Aquaculture is a $5 billion-a-year industry in the United States.

All striped bass and rainbow trout available at American retail markets are produced by aquaculture, as are more than half the fresh salmon served in the United States.

Catfish dominates American aquaculture; it currently accounts for 60 percent of all aquaculture-raised fish produced in the United States.

Tilapia (a mild fish similar in taste to sole) and oysters are also successfully grown by aquaculture in America.

Source: *Time*, December 3, 1990.

# SUMMARY

**1.** Humans require a balanced diet that includes carbohydrates, proteins, and lipids in addition to vitamins, minerals, and water. People who consume fewer calories than they need are said to be undernourished, whereas people who consume enough calories but whose diets are lacking in some specific nutrient are said to be malnourished.

**2.** The greatest challenge in agriculture today is producing enough food to feed the world's population. Complicating factors include poverty, the problems of distributing food where it is needed, and cultural acceptance of nutritious but unfamiliar foods. The long-term solution to the problem of producing adequate food is control of human numbers.

**3.** Agriculture developed independently at several centers beginning about 10,000 years ago. Early advances included domestication of plants and animals, irrigation, and use of animals for plowing. Although thousands of plant species are edible, today only about 30 plants provide the bulk of the world's food.

**4.** Developed countries and some developing countries rely on high-input agriculture, which uses large amounts of fossil fuels to power machinery and produce fertilizers and pesticides. In high-input agriculture, yields of food per area unit of farmland are high, but so are environmental costs, including soil erosion and pollution.

**5.** Many of the world's farmers rely on subsistence agriculture, in which human and animal energy is used instead of fossil fuels. Subsistence farming produces less food per unit of farmland than high-input agriculture.

**6.** The challenges confronting agriculture are being met in a variety of ways. New methods are being developed to make both high-input and subsistence agriculture sustainable. Genetic engineering is the high-technology answer to some of agriculture's problems. Our food base is also increasing through the introduction and general acceptance of "new" foods, and aquaculture is supplementing traditional fishing in the supply of high-quality protein from seafood.

## DISCUSSION QUESTIONS

**1.** What age group in human populations is usually most adversely affected by undernutrition, malnutrition, and famine? Why?

**2.** If modern-day domesticated corn were to be abandoned by agriculture, it is unlikely it would survive in the wild. Why? Give at least one trait of corn that is desirable from an agricultural viewpoint, but disastrous for survival in nature.

**3.** Why does decreased genetic diversity in farm plants and animals increase the likelihood of economic disaster from disease?

**4.** Describe the environmental problems associated with farming each of these areas: tropical rain forests; hillsides; arid regions.

**5.** Some scientists are genetically engineering herbicide resistance into crop plants so that when they apply herbicides, only the weeds die and not the crops. How might this specific example of genetic engineering have very severe negative impacts on the environment?

**6.** What is the most fundamental solution to world food problems? Explain your answer.

**7.** Explain why aquaculture is more like agriculture than it is like traditional fishing.

**8.** It has been suggested that toxicity testing of food additives should be performed by a research group independent of the food industry. Do you think this is a good idea? Why or why not?

## SUGGESTED READINGS

Barbier, E. B. Sustaining agriculture on marginal land. *Environment* 31:9, November 1989. Examines social, economic, and environmental factors that relate to subsistence agriculture.

Bardach, J. Aquaculture: Moving from craft to industry. *Environment* 30:2, March 1988. Examines the potential of aquaculture as a business.

Brown, L. Feeding six billion. *World Watch* 2:5, September/October 1989. Examines whether we can increase food production to keep up with population growth.

Brown, L. "Reexamining the World Food Prospect," in *State of the World 1989.* W.W. Norton & Company, New York, 1989. Production trends, cropland base, water for irrigation, and land productivity potential are considered in detail.

Brown, L. R., and J. E. Young. "Feeding the World in the Nineties," in *State of the World 1990.* W.W. Norton & Company, New York, 1990. Examines the effects of land degradation, fertilizers, and biotechnology on food production.

Crosson, P. R., and N. J. Rosenberg. Strategies for agriculture. *Scientific American,* September 1989. Social and economic changes are as important as scientific and technological advancements in increasing food production without further harming the environment.

Ost, L. The world begins to savor exotic flavors from the past. *National Research Council News Report* 39:8, August/September 1989. Some of the ancient crops of the Incas are being introduced to the world.

Reganold, J. P., R. I. Papendick, and J. F. Parr. Sustainable agriculture. *Scientific American,* June 1990. A review of alternative agriculture, which is more conservation-minded than traditional high-input agriculture.

Rhoades, R. E. The world's food supply at risk. *National Geographic* 179:4, April 1991. Documents the loss of genetic diversity in the world's food crops.

# Richard J. Mahoney

## Delivering Biotechnology to Developing Nations

*Richard J. Mahoney is chairman and chief executive officer of Monsanto Company. In his capacity as CEO of America's 53rd largest company, he has concentrated on guiding the firm away from its origins as a traditional chemical company into a technology-based global firm focusing on specialty chemicals, pharmaceuticals, industrial controls and the agricultural life sciences. Trained as a chemist with a B.S. from the University of Massachusetts, Mr. Mahoney joined Monsanto as a product development specialist in 1962. He has worked in virtually every part of the company, including five years in the agricultural division, where he gained hands-on experience as sales director before being promoted to managing director of the division. He was named CEO in 1983 and became chairman in 1986.*

**Mr. Mahoney, are we going to be able to feed a global population predicted to double in the next 40 years?**

We are not even feeding five billion people properly now, much less the 10 billion that are projected. Perhaps those projections will prove wrong. Perhaps educational programs in population control will have a significant effect. But, we certainly must not count on those things happening. We have to prepare for the worst case, the greatest challenge. Developing nations are intent on economic development; as individual incomes rise, the demand for a better diet will rise. Thus, a doubled population will almost surely need more than twice the food.

**How are we going to do it?**

I think the single most important event that will put us on the road to agricultural sufficiency is the conversion to a global market economy. The collapse of communism not only eased the nuclear threat, but it ended a tragic experiment in political management. The style of economic management tried in the Soviet Union was emulated in many nations across the world, and all it succeeded in doing was to destroy the mechanisms by which nations and people support themselves and thrive. For a nation like Cuba, blessed with perfect climate and soil and an energetic people, to slide into a state of near collapse can only be called a failure of political management. The collapse of communist economics demonstrates the reliability of the marketplace where far better decisions are made than in government-controlled systems. Market mechanisms that reach deep down into food-short nations will be the essential first step.

Second in importance is the advance of technology. The United States gets three times more food out of an acre of land than it did at the beginning of the century. The green revolution (see Chapter 18) turned India into a food exporter and did wonders for other countries. But, American agriculture is highly mechanized, high energy and high input. The green revolution hinged on dramatically improved strains of key crops, but ones that depended heavily on fertilizers and chemicals. Today's technologies are not for everybody. The price, the demands on skill, the access to supplies, is beyond the reach of those who need the help the most. We have pushed today's technologies about as far as they will go.

**If we need a step change in agricultural technology, is "genetic engineering" the answer?**

*I think the single most important event that will put us on the road to agricultural sufficiency is the conversion to a global market economy.*

Let me first say that the worst thing about genetic engineering is the phrase. The word "biotechnology" is not much better. They both suggest some other-worldly, rather scary kind of tinkering with nature. It is not. It is nothing more than an advance in the third leg of the technologies that have been available to us in agriculture for many years—the plant breeding leg that complemented mechanization and chemicals. Without repeating what is in these pages, let me say that I think the essence of this modernization of agricultural science is its power to make crops grow successfully, and to grow *where they are needed,* with no special skill, costly inputs or equipment needed. Thus, the technology jumps over the barriers to the effective delivery of an adequate food supply. It is a relatively easy task to supply a bag of seed to a subsistence farmer who needs it. The seeds will be custom-tailored to resist whatever makes crops fail from region to region—insects, viruses, cold, drought, salty or acidic soil. Agricultural sufficiency begins at the bottom of the economic ladder, and DNA technology can reach to that level. It is not a panacea, but I believe it is the most powerful agricultural tool yet developed.

**Can you cite specific examples of how biotechnology will have the reach you described?**

As we talk now, there are many programs under way of importance to developing nations, but let me cite one from our own experience. Cassava and sweet potatoes are dietary staples for many developing nations. They provide a major portion of the caloric energy that African villagers need to live and work. But, both are very vulnerable to viruses that cause the plants to wilt and die. One strain of cassava virus typically wipes out 80 percent of the crop—nothing short of a calamity for a village dependent on the crop for existence. We know how to make crops resistant to viruses—we can in effect immunize them. In 1990, with the support of the Agency for International Development, we invited a biologist from Kenya to help estab-

lish a program at our Life Sciences Research Center. We housed her, equipped her, and put her together with our research team to develop strains of cassava and sweet potatoes that will be able to resist viruses. We expect that after three years of work, she will go back to Kenya with a perfected technology in hand that her country and international institutions such as the World Bank can put directly into the hands of farmers in developing countries.

**But, how can they, or we, be sure this technology is safe?**

No scientist will ever say "never." But, if there is a risk, it has not shown up yet after two decades of extremely intense research and field testing. All we have succeeded in doing is to cross the sexual barrier in plant breeding. Instead of tomato-to-tomato breeding, the donor of the beneficial gene can now be any organism with a characteristic that a tomato could use. That includes the simplest of organisms, bacteria, which possess an immense number of beneficial traits. Nature has its bad genes, to be sure, but they are easy to spot, and we exhaustively test any improved crop strains we develop—and then go through a rigorous regulatory approval process. No, the risk is not in the technology. The risk is in failing to develop and transfer the technology to areas where it is needed. The politics and the patent systems to enable that to happen are not yet fully developed. The technology is in hand. We now need the mechanisms to deliver it.

**What needs to be done?**

Let me answer you this way. We could always use more Ph.D.s in the agricultural and life sciences. But, that is not the greatest need. If you look at the road map from our research center in St. Louis to a subsistence farmer's field in Africa, there are many points of entry, many places where today's students can make a difference. In the end, it is not only our scientists who are going to help feed the world; it is the people who are dotted along that road, making the global agricultural system work. They will be in the World Bank or the U.N., making the global market work better. They will be teaching farmers how to get more productivity out of existing land rather than plowing up new land. Or, in the Peace Corps. Much of that work does not take a Ph.D. It just takes commitment. And, commitment to the world's largest and most important industry is my idea of a useful and important career.

**Earth Day, 1990, Washington, D.C.** (Joanna Pinneo)

Environmental Concerns

Part 6

# Frank Press

## Transnational Cooperation in Environmental Science

*Frank Press has been President of the National Academy of Sciences in Washington, D.C., since 1981. He received his undergraduate degree in physics from the City College of New York, and his doctorate in geophysics from Columbia University in New York City. Dr. Press is recognized internationally for his pioneering contributions in geophysics, oceanography, lunar and planetary sciences, and natural resource exploration. He helped to conceive and organize the International Geophysical Year, the first coordinated worldwide attempt to measure and map geophysical phenomena. A leader in major national and international projects, he has served as a science policy advisor during the Kennedy, Nixon, and Ford administrations and as Science Advisor to President Carter. He was a member of the U.S. delegation to the nuclear test ban negotiations in Geneva from 1959–61, and Moscow in 1963, and was largely responsible for the 1979 scientific cooperation agreements between China and the United States.*

**What are the ecological priorities of the National Academy of Sciences, and how are they addressed?**

The National Academy of Sciences is primarily an advisory organization to Congress and the Executive Branch, and to the American public at large. When public policy recommendations are needed, we identify the issues, examine all of the research data available, and then respond with our recommendations. One of our current priorities is that Congress has asked us to look into the Endangered Species Act. We also issue reports on specific species, such as the Pacific Northwest salmon, or an endangered bird species in Hawaii whose sole habitat is a large private ranch on the big island of Hawaii. We sent a team in to examine the problem and made recommendations for the survival of that species.

**What kind of a role does the Academy play internationally?**

We have global concerns about the problems of preserving biodiversity around the world. As a scientific resource and an important public service organization, we have a responsibility to document the state of affairs, to express concern when needed, and to spur governments to action, if we think that's necessary. We work with other countries, like China, Russia, Japan, and England, in a number of different ways, such as preparing surveys and joint policy statements on a host of subjects regarding world ecology. For example, one recent policy statement issued by our Academy and the Royal Society of London dealt with the effects of human population growth on biodiversity.

**What may be the future prospects for greater international scientific cooperation, especially in view of the Soviet Union's demise?**

By and large, the world acknowledges that basic science, especially ecological and environmental science, is a transnational endeavor. Now that we are also recognizing the growing ascendancy of global scientific, environmental, and ecological issues, I would predict that the trend toward relatively free exchange of scientists and ideas across the world will increase greatly in the years ahead.

**Will the United States need to make any legislative or policy changes in order to take full advantage of this exchange?**

In general, the United States does encourage international scientific dialogue. We have budgets that support it, although they could be increased. The United States has educated literally hundreds of thousands of scientists from other countries who have taken that knowledge home, along with a friendly disposition toward us. If there are any actions needed on the part of the U.S. government, they are to increase the allocation of resources supporting science, recognizing that international collaboration increases scientific productivity and benefits human society in general.

**In your work as a geophysicist, what are some of the most significant advances you've seen in our understanding of planetary functions and forces? How do they relate to ecological concerns and the mechanisms of pollution?**

The biggest innovation in the geosciences has been the plate tectonics revolution, which for the first time explained geological phenomena on a global basis. Previously, geologists mapped their own regions with little idea of how their findings connected with those in other continents. But now that we understand the dynamics of continental drift, of sea floor spreading, and of plate tectonics, all the data can be tied together.

We have developed new tools for understanding the nature of climate. For instance, from small air bubbles trapped in ice cores we can sample the atmospheric composition going back hundreds of thousands of years. And deep sea cores are able to tell us what temperatures were like many millions of years ago. Unraveling the history of global environmental change caused by natural processes enables us to better understand what's occurring now, when human influences are becoming as important as natural effects.

**What do your studies of the geological record reveal?**

We've discovered that it is possible for climatic changes to take place over much shorter periods of time than expected. It was previously thought that such changes could only occur over thousands of years; now it seems as though climates can be altered in only 100 years. To me, as a scientist, that means that the introduction of human activities can trigger profound changes that are completely unanticipated. For example, the ozone hole over the Antarctic is something that nobody ever predicted. It never came

out of any of our scientific models. What nature tells us is that there have been rapid natural changes in the past, and that human interferences may cause major climatic excursions.

**As a scientific advisor to presidents and policy makers, you have promoted joint research ventures amongst industry, universities, and governments. What is the best way, both for those potentially involved, and for the general public, to encourage such efforts?**

The first step is to promote a public awareness that for America to improve its competitive position, it has to build on its strengths. One of those strengths is its fine university system interacting with industry, particularly in areas of emerging technology, which could prove to be very important economically.

There are many examples: IBM, AT&T, and MIT cooperate in a very important technological area dealing with high-temperature superconductivity. Ten years from now, this could have very profound commercial applications. The companies contribute their own money, and MIT receives government funding that it uses to support this research. In this cooperative arrangement, the industry people, who know the practical applications, work side by side with the university people, who do the basic science. It makes for a very strong team.

**What earth science disciplines are most in need of further study and of more committed scientists?**

There's an emerging new profession to be found in the environmental sciences, a new aspect of the earth sciences. People entering this field could be scientists or engineers who have a background in geology, chemistry, physics, and biology. They would understand both the nature of industrial processes and the nature of environmental change (and how humankind interferes with the environment), in order to mitigate such changes and interferences. So, an increasing number of universities are offering this type of interdisciplinary degree in environmental science. I think there will be demand for people with this training in industry and in government. Anyone working in this important area would be making a substantial contribution to public service.

# Chapter 19

*The "ocean of air" is actually an extremely thin layer compared to the size of the Earth. (NASA)*

# Air Pollution

The air we breathe is often dirty and contaminated with pollutants, particularly in urban areas. Although thousands of different chemicals pollute the air as a result of human activities, there are only six main types of air pollutants: particulates, nitrogen oxides, sulfur oxides, carbon oxides, hydrocarbons, and ozone. These pollutants and others, such as lead, cause a great many health and environmental problems. Air pollution also extends indoors, and the air we breathe at home, at the workplace, and in our automobiles may be more polluted than the air outdoors. Because air pollution is dangerous to human health, harms plants and animals, damages and corrodes materials from fabrics to stonework, and reduces visibility, most developed nations and many developing nations have established air quality standards for numerous pollutants.

# THE ATMOSPHERE AS A RESOURCE

The atmosphere is an invisible layer of gases that envelops the Earth. Oxygen (21 percent) and nitrogen (78 percent) are the predominant gases in the atmosphere, accounting for about 99 percent of dry air; other gases, including argon, carbon dioxide, neon, and helium, make up the remaining 1 percent. In addition, water vapor and dust particles are present in the air. The atmosphere becomes less dense as it extends outward into space; most of its mass is found near the Earth's surface.

The atmosphere performs several ecologically important functions. It protects the Earth's surface from most of the sun's ultraviolet radiation and x rays and from lethal amounts of cosmic rays from space. Without this shielding by the atmosphere, life as we know it would cease to exist. While the atmosphere protects the Earth from high-energy radiations, it allows visible light and some infrared radiation to penetrate, and they warm the Earth's surface and the lower atmosphere. This interaction between the atmosphere and solar energy is responsible for the Earth's climate and weather (see Chapter 5).

The two atmospheric gases most important to living things are carbon dioxide and oxygen (see Chapter 3). During photosynthesis, plants, algae, and certain bacteria use carbon dioxide to manufacture sugars and other organic molecules. During cell respiration, living things use oxygen to break down food molecules and supply themselves with chemical energy.

Living organisms depend on the atmosphere, but they also maintain and, in certain instances, modify its composition. For example, atmospheric oxygen is thought to have increased to its present level as a result of millions of years of photosynthesis. The level is maintained by a balance between oxygen-producing photosynthesis and oxygen-using respiration.

We think of air as an unlimited resource, but perhaps we should reconsider. Ulf Merbold, a German space shuttle astronaut, felt very differently about the atmosphere after viewing it in space. "For the first time in my life, I saw the horizon as a curved line. It was accentuated by the thin seam of dark blue light—our atmosphere. Obviously, this was not the 'ocean' of air I had been told it was so many times in my life. I was terrified by its fragile appearance."

## Layers of the Atmosphere

The atmosphere is composed of a series of five concentric layers—the troposphere, stratosphere, mesosphere, thermosphere, and exosphere—which vary in altitude and temperature with latitude and season (Figure 19–1). The **troposphere** is the layer of atmosphere closest to the Earth's surface. It extends up to a height of approximately 10 kilometers (6.2 miles), where the atmosphere reaches a temperature minimum well below 0°C. Weather, including turbulent wind, storms, and most clouds, occurs in the troposphere.

In the next layer of atmosphere, the **stratosphere,** there is a steady wind (but no turbulence) and the temperature is more or less uniform (−45°C to −75°C); commercial jets fly here. The stratosphere extends from 10 km to 45 km (6.2 to 28 miles) above the Earth's surface and contains a layer of ozone—the much publicized ozone shield— that is critical to life because it absorbs much of the sun's damaging ultraviolet radiation.

The **mesosphere,** the layer of atmosphere directly above the stratosphere, extends from 45 km to 80 km (28 to 49.6 miles) above the Earth's surface. Temperatures drop steadily in the mesosphere to the lowest in the atmosphere—as low as −138°C.

The **thermosphere,** which extends from 80 km to 500 km (49.6 to 310 miles), is characterized by steadily rising temperatures. Gases in the thermosphere absorb x rays and short-wave ultraviolet radiation; the latter heats this layer to as much as 1,000°C. The aurora, a colorful display of lights in dark polar skies, is produced when charged particles from the sun hit oxygen or nitrogen molecules in the thermosphere. The thermosphere is important in long-distance communication because it reflects outgoing radio waves back toward the Earth.

The outermost layer of the atmosphere is the **exosphere,** which continues to thin until it converges with outer space.

# TYPES, SOURCES, AND EFFECTS OF AIR POLLUTION

**Air pollution** consists of gases, liquids, or solids present in the atmosphere in high enough levels to harm humans, other animals, plants, or materials. Although air pollutants can come from natural

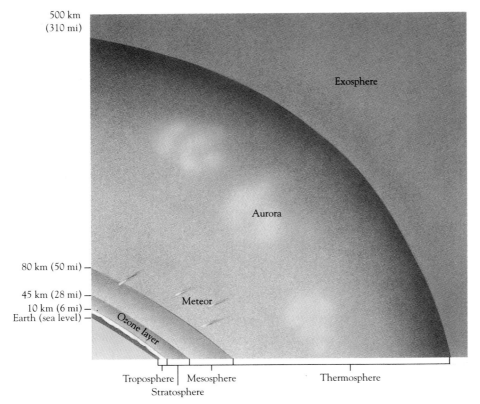

500 km
(310 mi)

Exosphere

Aurora

80 km (50 mi) —

45 km (28 mi) —
10 km (6 mi) —
Earth (sea level) —

Ozone layer

Meteor

Troposphere | Mesosphere                Thermosphere

Stratosphere

**Figure 19–1**
The atmosphere is composed of several layers, formed by temperature changes at different altitudes. Our weather patterns occur in the troposphere. The stratosphere contains a thin ozone layer that absorbs ultraviolet radiation. The mesosphere is extremely cold, while the thermosphere, the largest layer, is very warm because it absorbs high energy radiation such as x rays. The outermost layer, the exosphere, borders interplanetary space.

sources—as, for example, when lightning causes a forest fire or when a volcano erupts—human activities make a major contribution to global air pollution. (See Focus On: Air Pollution in Kuwait for some of the environmental effects of the recent war in the Persian Gulf.) From the standpoint of human health, probably more significant than the overall human contribution of air pollution is the fact that much of the air pollution released by humans is concentrated in densely populated urban areas.

Although many different air pollutants exist, we will focus our attention on the six most important types: particulates, nitrogen oxides, sulfur oxides, carbon oxides, hydrocarbons, and ozone (Table 19–1). Air pollutants are often divided into two categories, primary and secondary (Figure 19–2). **Primary air pollutants** are harmful chemicals that enter directly into the atmosphere. The major ones are carbon oxides, nitrogen oxides, sulfur dioxide, particulates, and hydrocarbons. **Secondary air pollutants** are harmful chemicals that form from other substances which have been released into the atmosphere. Ozone and sulfur trioxide are secondary air pollutants because both are formed by chemical reactions that take place in the atmosphere.

The two main human sources of primary air pollutants are motor vehicles and industries (Figure 19–3). Automobiles and trucks release significant quantities of nitrogen oxides, carbon oxides, particulates, and hydrocarbons as a result of the combustion of gasoline. Electric power plants and other industries emit most of the particulate matter and sulfur oxides released in the United States; they also emit sizable amounts of nitrogen oxides, hydrocarbons, and carbon oxides. The combustion of fossil fuels, especially coal, is responsible for most of these emissions. The top three indus-

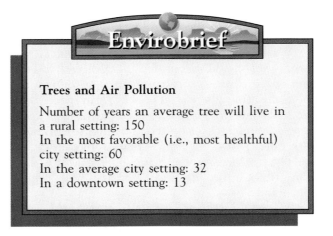

**Envirobrief**

**Trees and Air Pollution**

Number of years an average tree will live in a rural setting: 150
In the most favorable (i.e., most healthful) city setting: 60
In the average city setting: 32
In a downtown setting: 13

**Table 19–1**
**Major Air Pollutants**

| Pollutant | Composition | Class | Primary or Secondary | Characteristics |
|---|---|---|---|---|
| Dust | Variable | Particulates | Primary | Solid particles |
| Lead | Pb | Particulates | Primary | Solid particles |
| Sulfuric acid | $H_2SO_4$ | Particulates | Secondary | Liquid droplets |
| Nitrogen dioxide | $NO_2$ | Nitrogen oxide | Secondary (mainly) | Reddish brown gas |
| Sulfur dioxide | $SO_2$ | Sulfur oxide | Primary | Colorless gas with strong odor |
| Carbon monoxide | CO | Carbon oxide | Primary | Colorless, odorless gas |
| Methane | H—C—H or $CH_4$ (with H above and below C) | Hydrocarbon | Primary | Colorless, odorless gas |
| Benzene | $C_6H_6$ (ring structure) | Hydrocarbon | Primary | Liquid with sweet smell |
| Ozone | O—O or $O_3$ (with O below) | Photochemical oxidant | Secondary | Pale blue gas with sweet smell |

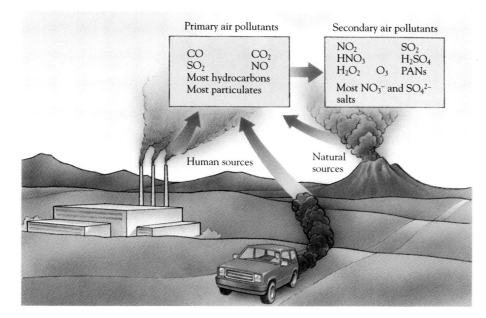

**Primary air pollutants**

| | |
|---|---|
| CO | $CO_2$ |
| $SO_2$ | NO |
| Most hydrocarbons | |
| Most particulates | |

**Secondary air pollutants**

| | |
|---|---|
| $NO_2$ | $SO_2$ |
| $HNO_3$ | $H_2SO_4$ |
| $H_2O_2$ | $O_3$ PANs |
| Most $NO_3^-$ and $SO_4^{2-}$ salts | |

Human sources

Natural sources

**Figure 19–2**
Primary air pollutants are emitted, unchanged, from a source directly into the atmosphere, while secondary air pollutants form from chemical reactions involving primary air pollutants.

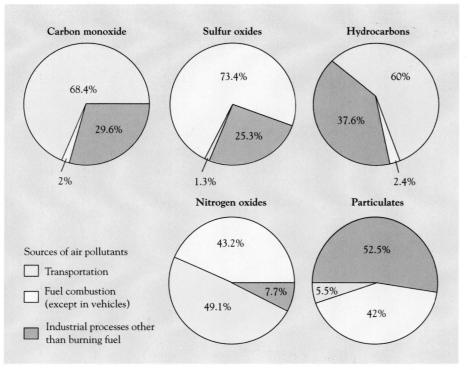

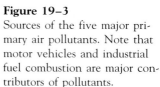

**Carbon monoxide**

68.4%

29.6%

2%

**Sulfur oxides**

73.4%

25.3%

1.3%

**Hydrocarbons**

60%

37.6%

2.4%

**Nitrogen oxides**

43.2%

7.7%

49.1%

**Particulates**

52.5%

5.5%

42%

Sources of air pollutants

☐ Transportation

☐ Fuel combustion
(except in vehicles)

▨ Industrial processes other
than burning fuel

**Figure 19–3**
Sources of the five major primary air pollutants. Note that motor vehicles and industrial fuel combustion are major contributors of pollutants.

trial sources of *toxic* air pollutants (that is, chemicals released into the air that are fatal to humans at specified concentrations) are the chemical industry, the metals industry, and the paper industry.

Air pollution damages living organisms, reduces visibility, and attacks and corrodes materials.

**Figure 19–4**
Plants exposed to air pollution exhibit a variety of symptoms, including a lowered productivity. Compare the size and overall vigor of the radish grown in clean air (left) with the radish damaged by air pollution (right).
(Science VU/Visuals Unlimited)

The respiratory tracts of animals, including humans, are particularly harmed by air pollutants, which also exacerbate existing medical conditions such as chronic lung disease, pneumonia, and cardiovascular problems. The overall productivity of crop plants is reduced by most forms of air pollution (Figure 19–4), and, when combined with other environmental stresses (such as low winter temperatures or prolonged droughts), air pollution causes plants to decline and die. Air pollution is involved in acid deposition, global temperature changes, and stratospheric ozone depletion (all discussed in Chapter 20).

## Major Air Pollutants

**Particulate matter** consists of thousands of different solid and liquid particles that are suspended in the atmosphere. **Solid particulate matter** is generally referred to as dust, whereas liquid suspensions are commonly called mists. Particulates include soil, soot, lead, asbestos, and sulfuric acid droplets. All particulate matter eventually settles out of the atmosphere, but it is possible for small particles, some of which are especially harmful to humans, to remain suspended in the atmosphere for weeks or even years.

Particulates reduce visibility by scattering and absorbing sunlight. Urban areas receive less sunlight than rural areas, partly as a result of greater

**Focus On**

## Air Pollution in Kuwait

During February 1991, it was impossible to read a news magazine or watch a news show on television without seeing pictures of thick, black smoke pouring out of hundreds of Kuwait's oil fields, ignited by Iraqi soldiers before their withdrawal from Kuwait. The 650 fires burned for months before being completely extinguished on November 6, 1991, and the media filled the news with dire predictions of the long-term effects of the fires. Doctors feared serious health effects for people in the immediate region, and some

An oil well burns in the Kuwaiti desert in the aftermath of the Persian Gulf War. Many oil wells were set on fire, and lakes of oil spilled into the desert around the burning oil wells.

(Visuals Unlimited/James H. Ripley)

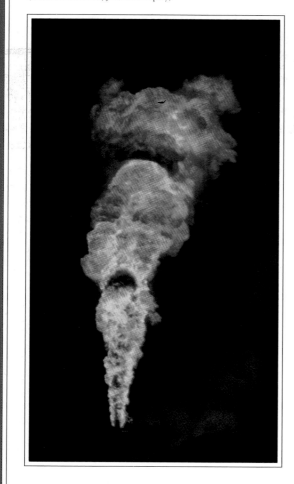

scientists predicted global and regional problems such as severe acid rain, absence of a monsoon season in Asia, and global climate change.

Although the total effects of the fires in Kuwait will take years to study and evaluate, an understanding of their impact has begun to emerge. Some people who live in the immediate vicinity of the fires and who inhaled the smoke suffered health problems, including respiratory illnesses, and some people have even died from ailments associated with the fires. Certain groups living in Kuwait—children, old people, and people with respiratory and heart diseases—were more at risk from breathing the smoke than the rest of the population.

Research groups in the United States, Canada, Germany and Great Britain have developed models of the atmospheric effects of the fires. All models agree that the fires will *not* have a large, long-term impact on global climate. Fortunately, the smoke plumes appear to have stayed in the troposphere, where they were subjected to wind and precipitation. (Had much of the smoke risen into the stratosphere, where weather does not occur, the smoke particles could have remained for many years, exerting a cooling influence on global climate.) Further, the fires produced too little carbon dioxide over too short a period to contribute much to global warming.

In Kuwait and nearby areas, the climate models predicted that the smoke would prevent some of the sun's warming rays from penetrating to the surface and would thereby cause a drop in local temperatures. Indeed, temperatures in Kuwait were unseasonably low—sometimes as much as 20 degrees below normal—during the spring of 1991, when the sky was still black with smoke.

Acid deposition, including acid rain, caused by sulfur oxides in the smoke fell in areas downwind from the fires. Acid deposition was severe in Kuwait and may have fallen as far away as China. Russian scientists, for example, noted an exceptionally high level of acid rain across southern Russia in the spring of 1991; however, it is difficult to prove a direct cause-and-effect relationship between the oil fires in Kuwait and acid deposition in Russia or China.

quantities of particulate matter in the air. Particulate matter corrodes metals, erodes buildings and sculpture when the air is humid, and soils clothing and draperies. Smaller particles are inhaled into the respiratory system and can cause health problems; lead (see Chapter 21) and asbestos particles are especially harmful.

**Nitrogen oxides** are gases produced by the chemical interactions between nitrogen and oxygen. They consist mainly of nitric oxide (NO), nitrogen dioxide ($NO_2$), and nitrous oxide ($N_2O$). Nitrogen oxides inhibit plant growth and, when breathed, aggravate a number of health problems. They are involved in the production of photochemical smog (to be discussed shortly), acid deposition, and global warming. Nitrogen oxides cause metals to corrode and textiles to fade and deteriorate.

**Sulfur oxides** are gases produced by the chemical interactions between sulfur and oxygen. Sulfur dioxide ($SO_2$), a colorless, nonflammable gas with a strong, irritating odor, is a major sulfur oxide emitted as a primary air pollutant. Another major sulfur oxide is sulfur trioxide ($SO_3$), a secondary air pollutant that forms when sulfur dioxide reacts with oxygen in the air. Sulfur trioxide, in turn, reacts with water to form another secondary air pollutant, sulfuric acid.

Sulfur oxides are very important in acid deposition, and they corrode metals and damage stone and other materials. Sulfuric acid and sulfate salts that are produced in the atmosphere from sulfur oxides damage plants and irritate the respiratory tracts of animals, including humans.

**Carbon oxides** are the gases carbon monoxide (CO) and carbon dioxide ($CO_2$). Carbon monoxide, a colorless, odorless, and tasteless gas produced in the largest quantities of any atmospheric pollutant except carbon dioxide, is poisonous and reduces the blood's ability to transport oxygen. Carbon dioxide, also colorless, odorless, and tasteless, traps heat in the atmosphere and is therefore involved in global climate change.

**Hydrocarbons** are a diverse group of organic compounds that contain only hydrogen and carbon. Some hydrocarbons are straight or branched chains, and some are cyclic (form rings); the simplest is methane ($CH_4$). The smaller hydrocarbons are gaseous at room temperature; methane, for example, is a colorless, odorless gas that is the principal component of natural gas.[1] Medium-sized hy-

drocarbons are liquids at room temperature, although many are volatile, or evaporate readily. The largest hydrocarbons are actually solids at room temperature—the waxy substance paraffin, for example.

Because there are so many different hydrocarbons, they have a variety of effects on human and animal health; some appear to cause no adverse effects, some injure the respiratory tract, and others cause cancer. All except methane are important in the production of photochemical smog. Methane is involved in global climate change.

**Ozone** ($O_3$) is a form of oxygen considered a pollutant in one part of the atmosphere but an essential component in another. In the stratosphere, oxygen reacts with ultraviolet radiation coming from the sun to form ozone. As already discussed, stratospheric ozone prevents much of the solar ultraviolet radiation from penetrating to the Earth's surface. Unfortunately, stratospheric ozone is being attacked by human-made pollutants.

Unlike stratospheric ozone, tropospheric ozone is a human-made air pollutant. (It doesn't replenish the ozone that has been depleted from the stratosphere because it is converted back to oxygen in a few days.) Tropospheric ozone is a secondary air pollutant that forms when sunlight catalyzes a reaction between nitrogen oxides and volatile hydrocarbons. The most harmful component of photochemical smog, it reduces air visibility and causes health problems. The Environmental Protection Agency (EPA) estimates that about 67 million people in the United States are routinely exposed to higher ozone levels than the maximum established by the Clean Air Act. Ozone also stresses plants and is a possible contributor to forest decline, which will be considered in more detail in Chapter 20. In addition, tropospheric ozone is involved in global climate change.

## Other Air Pollutants

Most of the hundreds of other air pollutants—which include lead, hydrochloric acid, formaldehyde, radioactive substances, and fluorides—are present in very low concentrations, although it is possible to have high local concentrations of specific pollutants. Some of these air pollutants are extremely toxic and may pose long-term health risks to people who live and work around factories or other facilities that produce them. Complicating the situation is the fact that little is known about the health effects of many of these compounds. As a result of this lack of scientific detail, legal air quality standards have not been established for

---

[1] The odor of natural gas comes from sulfur compounds that are deliberately added so that humans can detect the presence of the gas.

many pollutants, and their emissions are not regulated.

## URBAN AIR POLLUTION

Air pollution that is localized in urban areas, where it reduces visibility, is often called **smog.** The word "smog" was coined at the beginning of the 20th century for the smoky fog that was so prevalent in London because of coal combustion. Today there are several different types of smog. Traditional London-type smog (that is, smoke pollution) is sometimes called **industrial smog.** The principal pollutants in industrial smog are sulfur oxides and particulate matter. The worst episodes of industrial smog typically occur during winter months, when household fuel combustion is high.

Another important type of smog is **photochemical smog,** a brownish orange haze formed by chemical reactions involving sunlight (Figure 19–5). First noted in Los Angeles in the 1940s, photochemical smog is worst during the summer months. Both nitrogen oxides and hydrocarbons are involved in its formation. One of the photochemical reactions occurs between nitrogen oxides (largely from automobile exhaust), volatile hydrocarbons, and oxygen in the atmosphere to produce ozone, a reaction that requires solar energy (Figure 19–6). The ozone formed in this way then reacts with other air pollutants, including hydrocarbons, to form more than 100 different secondary air pollutants (**peroxyacyl nitrates** [PANs], for example) which can injure plant tissues, irritate eyes, and aggravate respiratory illnesses in humans.

**Figure 19–5**
Photochemical smog in New York City. A thermal inversion has trapped the polluted air near the ground.
(William E. Ferguson)

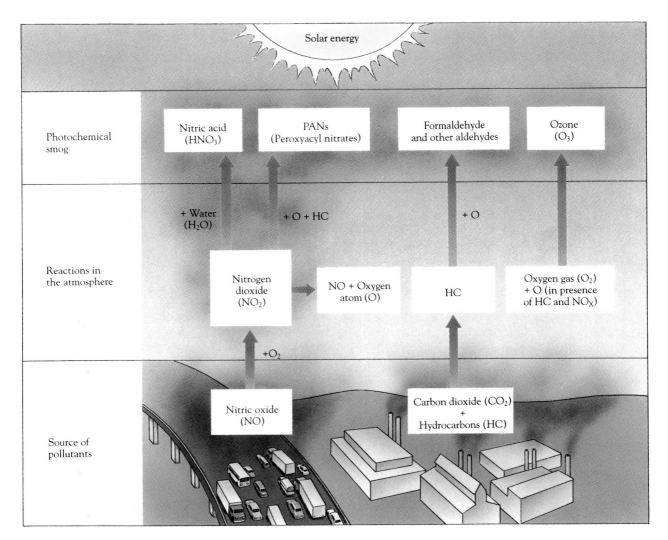

**Figure 19–6**
Photochemical smog consists of a complex mixture of pollutants that includes ozone, peroxyacyl nitrates (PANs), nitric acid, and organic compounds such as formaldehyde.

Although the main source of photochemical smog is the automobile, you may be surprised to know that bakeries and dry cleaners are also significant contributors of the air pollutants that cause photochemical smog. When bread is baked, yeast byproducts are released that are converted to ozone by sunlight. The volatile fumes from dry cleaners also contribute to photochemical smog.

## How Climate and Topography Affect Air Pollution

Variation in temperature during the day usually results in air circulation patterns that help to dilute and blow away air pollutants. As the sun increases surface temperatures, the air near the ground is warmed. This heated air expands and rises to higher levels in the atmosphere, creating a low-pressure area near the ground; the surrounding air then moves into the low-pressure area. Thus, under normal conditions, air circulation patterns prevent toxic pollutants from increasing to dangerous levels near the ground.

However, during periods of **thermal inversion,** in which the air near the ground is colder than the air at higher levels (Figure 19–7), polluting gases and particulate matter remain trapped in high concentrations close to the ground, where people live and breathe. Thermal inversions usually persist for only a few hours before they are broken up by atmospheric turbulence. Sometimes, however, atmospheric stagnation caused by a stalled high-pressure air mass allows a thermal inversion to persist for several days.

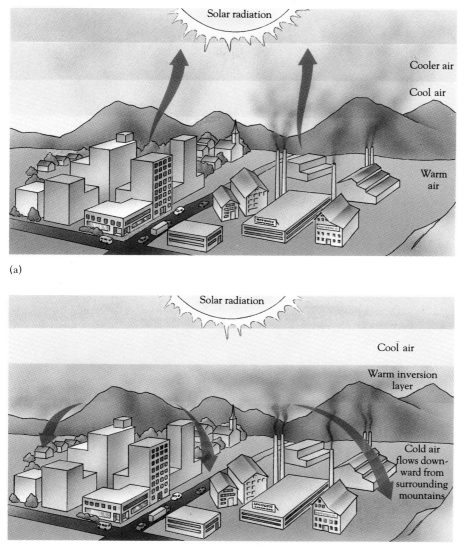

**Figure 19–7**
How a thermal inversion occurs. (a) Normally, warm, polluted air rises, diluting pollutants in the process. (b) When a thermal inversion occurs, a layer of warm air covers cooler air near the ground and the polluted air is trapped. Pollutants can increase to dangerous levels during thermal inversions.

Certain types of topography (surface features) increase the likelihood of thermal inversions. Cities located in valleys, near the coast, or on the leeward side of mountains (the side toward which the wind blows) are prime candidates for thermal inversions. Los Angeles, California, has some of the worst smog in the world. Mountains surround the Los Angeles basin on three sides, and the ocean is on its fourth side. This location, combined with a sunny climate, large population, and high automobile density, is conducive to the formation of thermal inversions that trap photochemical smog near the ground.

**Dust Domes**

Heat released by human activities such as fuel combustion is also highly concentrated in areas of high population density, known as **urban heat islands.** As a result, the air in these urban areas is warmer than the air in the surrounding suburban and rural countrysides. Urban heat islands affect local air currents and contribute to the buildup of pollutants, especially particulates, with which they form **dust domes** over cities.

## CONTROLLING AIR POLLUTANTS

Many of the measures we've already discussed for energy efficiency and conservation (see Chapter 12) also help to reduce air pollution. Smaller, more fuel-efficient automobiles produce fewer emissions, for example. Appropriate technologies exist to control all the forms of air pollution discussed in this chapter except carbon dioxide.

Smokestacks that have been fitted with electrostatic precipitators, fabric filters, wet scrubbers, or other technologies remove particulate matter (Figure 19–8). In addition, particulates are controlled by careful land-excavating activities, such as sprinkling water on dry soil that is being moved during road construction.

Several methods exist for removing sulfur oxides from flue gases, but it is often less expensive to simply switch to a low-sulfur fuel such as natural gas or even to a non–fossil fuel energy source such as solar energy. Sulfur can also be removed from fuels before they are burned, as in coal gasification (see Chapter 10).

Reduction of combustion temperatures in automobiles lessens the formation of nitrogen oxides. Use of mass transit helps reduce automobile use, thereby decreasing nitrogen oxide emissions. Nitrogen oxides produced during high-temperature combustion processes in industry can be removed from smokestack exhausts.

Modification of furnaces and engines to provide more complete combustion helps control the

**Figure 19–8**
Electrostatic precipitators are very effective at removing particulates. A comparison of emissions from a Delaware Valley steel industry with the electrostatic precipitator turned off (a) and on (b). (a and b, Visuals Unlimited/John D. Cunningham)

(a)

(b)

production of both carbon monoxide and hydrocarbons. Catalytic afterburners, used immediately following combustion, oxidize any unburned gases; the use of catalytic converters to treat auto exhaust, for example, can reduce carbon monoxide and volatile hydrocarbon emissions by about 85 percent over the life of the car. Careful handling of hydrocarbons such as solvents and petroleum reduces air pollution from spills and evaporation.

## Taxing Polluters

Sweden levies taxes on carbon dioxide and sulfur dioxide emissions to encourage power plants and factories to reduce emissions. The money generated by the pollution taxes goes into the government coffer. Because the amounts of these two pollutants vary with the type of fossil fuel burned (see Chapter 10), the tax encourages power plants and manufacturing firms to burn cleaner fuels. A tax on nitrogen oxides produced by power plants is also being phased in.

## AIR POLLUTION IN THE UNITED STATES

The Clean Air Act was first passed in 1970 and has been reauthorized (updated) twice since then, in 1977 and 1990. Overall, air quality in the United States has slowly improved since 1970 (Figure 19–9). The most dramatic improvement has been in the amount of lead in the atmosphere, which showed a 96 percent decrease between 1970 and 1989, primarily because of the switch from leaded to unleaded gasoline. Levels of sulfur oxides, ozone, carbon monoxide, and particulates have also been reduced. For example, between 1980 and 1989, sulfur dioxide emissions declined by 10 percent. Although air quality has been gradually improving, the atmosphere in many urban areas still contains higher levels of pollutants than are recommended based on health standards, and photochemical smog continues to be a major problem in metropolitan areas (Table 19–2).

**Figure 19–9**

A comparison of 1970 and 1989 emissions in the United States: particulate matter, sulfur oxides, carbon monoxide, nitrogen oxides, and volatile organic compounds. All showed decreases except for nitrogen oxides, which increased by 8 percent.

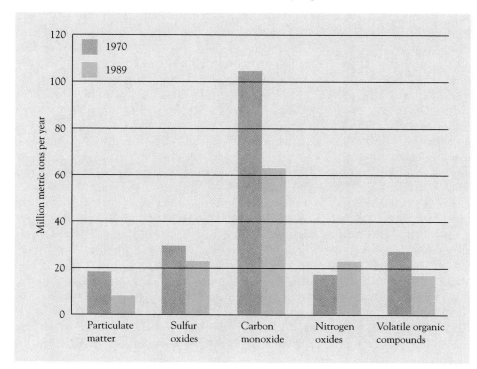

**Table 19–2**
**The Ten U.S. Urban Areas with the Worst Air Quality in 1990 (Ozone Classified Nonattainment Areas)**

**Extreme (0.28 ppm ozone)**
Los Angeles and south coast air basin, California

**Very Severe (0.19 to 0.28 ppm ozone)**
Chicago; Gary and Lake County, Indiana
Houston, Galveston, and Brazoria, Texas
Milwaukee and Racine, Wisconsin
New York City and Long Island; Northern New
    Jersey; Connecticut
Southeast Desert, California

**Severe (0.18 to 0.19 ppm ozone)**
Baltimore, Maryland
Philadelphia; Wilmington, Delaware; Trenton,
    New Jersey; Maryland
San Diego, California
Ventura County (between Santa Barbara and Los
    Angeles), California

The Clean Air Acts of 1970, 1977, and 1990 have required progressively stricter controls of motor vehicle emissions. For example, the provisions of the 1990 Clean Air Act include the development of superclean cars, which emit lower amounts of nitrogen oxides and hydrocarbons, and the use of cleaner-burning gasolines in the country's most polluted cities; these changes will be phased in gradually by the year 2000. Although more recent automobile models do not produce as many pollutants as older models, air quality has not improved in some areas of the United States because of the large increase in the number of cars being driven. It is doubtful that the 1990 Clean Air Act will greatly alter the number of cars on the road. However, some consumers may be discouraged from purchasing new automobiles because of the cost increase that is expected to result from installing pollution control devices.

The 1990 Clean Air Act focuses on industrial airborne toxic chemicals in addition to motor vehicle emissions. Between 1970 and 1990, the airborne emissions of only seven toxic chemicals were regulated. In comparison, the 1990 Clean Air Act authorizes a 90 percent reduction in the atmospheric emissions of 189 toxic and cancer-causing chemicals by 2003. To comply with this requirement, both small businesses such as dry cleaners and large manufacturers such as chemical compa-

nies will have to install pollution-control equipment. Sulfur dioxide and nitrogen oxide emissions from coal-fired power plants will be substantially reduced by the year 2000 (see Chapter 10); these provisions of the Clean Air Act are the first U.S. legislation to target acid rain (see Chapter 20). The provisions for stratospheric ozone depletion require the United States to phase out the production of ozone-destroying chemicals by the year 2000, which will keep the country in compliance with an international agreement to be discussed in Chapter 20.

## AIR POLLUTION IN DEVELOPING COUNTRIES

As developing nations become more industrialized, they also produce more air pollution. The leaders of most developing countries believe they must become industrialized rapidly in order to compete economically with developed countries. Environmental quality is usually a low priority in the race to develop. Thus, while air quality is slowly improving in the United States and other developed countries, it is deteriorating in developing nations.

In particular, the growing number of automobiles in developing countries is contributing to air pollution in those countries' urban areas. Many vehicles in these countries are old and have no pollution control devices. During the 1980s the most rapid proliferation of motor vehicles worldwide occurred in South America and Asia. Lead pollution from heavily leaded gasoline is an especially serious problem in developing nations.

Environmental quality has declined so much in certain cities that some developing nations have been forced to deal with air pollution. Mexico City, the world's largest city (see Chapter 9), has the most polluted air of any major metropolitan area in the world. This is due in part to its great population growth in the past 40 years and in part to the city's location. Mexico City is in an elevated valley that is ringed on three sides by mountain peaks; winds coming in from the open northern end are trapped in the valley. The city has almost 3 million motor vehicles and thousands of businesses, which the Mexican government says spew 3.94 million metric tons (4.35 million tons) of pollutants into the air each year. In 1990 Mexico embarked on an ambitious plan to improve Mexico City's air quality by gradually replacing old buses, taxis, delivery trucks, and cars with "cleaner" vehicles and by switching to unleaded gasoline. In addition, businesses that violate anti-pollution laws are being closed.

## AIR POLLUTION AND HUMAN HEALTH

Generally speaking, exposure to low levels of pollutants such as ozone, sulfur oxides, nitrogen oxides, and particulates irritates the eyes and causes inflammation of the respiratory tract (Table 19–3). Evidence exists that many air pollutants also suppress the immune system, increasing susceptibility to infection. In addition, evidence continues to accumulate indicating that exposure to air pollution during respiratory illnesses may result in the development later in life of chronic respiratory diseases, such as emphysema and chronic bronchitis. Some other health problems that can result from long-term exposure to toxic air pollutants are cancer, chronic obstructive pulmonary disease, asthma, respiratory infections, and cardiovascular disease.

### Health Effects of Specific Air Pollutants

Both sulfur dioxide and particulates irritate the respiratory tract and, because they cause the airways to constrict, actually impair the lungs' ability to exchange gases. People suffering from asthma and emphysema are very sensitive to sulfur dioxide and particulate pollution. Nitrogen dioxide also causes airway constriction and, in people suffering from asthma, oversensitivity to pollen and dust.

Carbon monoxide combines with the blood's hemoglobin, reducing its ability to transport oxygen. At medium concentrations, carbon monoxide causes headaches and fatigue. As the concentration increases, reflexes slow down and drowsiness occurs; at a certain high level, carbon monoxide causes death. People at greatest risk from carbon monoxide include pregnant women, infants, and those with heart or respiratory diseases.

Ozone and the volatile compounds in smog cause a variety of health problems, including burning eyes, coughing, and chest discomfort. Ozone also brings on asthma attacks and suppresses the immune system.

## INDOOR AIR POLLUTION

The air in enclosed places such as automobiles, homes, schools, and offices may have significantly higher levels of air pollutants than the air outdoors. In congested traffic, for example, levels of harmful pollutants such as carbon monoxide, benzene, and airborne lead may be several times higher *inside* an automobile than in the air immediately outside. The concentrations of certain air pollutants indoors may be five times greater than those outdoors. Indoor pollution is of particular concern to

**Table 19–3**
**Effects of Several Major Air Pollutants**

| Pollutant | Source | Effects |
|---|---|---|
| Particulate matter | Industries, motor vehicles | Aggravates respiratory illnesses; long-term exposure may cause increased incidence of chronic conditions such as bronchitis |
| Sulfur oxides | Electric power plants and other industries | Irritate respiratory tract; same effects as particulates |
| Nitrogen oxides | Motor vehicles, industries | Respiratory irritants; aggravate respiratory conditions such as asthma and chronic bronchitis |
| Carbon monoxide | Motor vehicles, industries | Reduces blood's ability to transport $O_2$; headache and fatigue at lower levels; mental impairment or death at high levels |
| Ozone | Formed in atmosphere (secondary air pollutant) | Irritates eyes; irritates respiratory tract; produces chest discomfort; aggravates respiratory conditions such as asthma and chronic bronchitis |

### Smoking

Smoking, which causes serious diseases such as lung cancer, emphysema, and heart disease, has killed more people than died in all the wars of the 20th century up to the early 1990s. In the United States it is estimated that, of the 140,000 deaths from lung cancer each year, about 120,000 of them are caused by cigarette smoking.

Passive smoking—nonsmokers' chronic breathing of smoke from cigarette smokers—also increases the risk of cancer. In addition, passive smokers suffer more respiratory infections, allergies, and other chronic respiratory diseases than other nonsmokers. Passive smoking is particularly harmful to infants and young children, pregnant women, the elderly, and people with chronic lung disease.

There is good news and bad news about smoking. The good news is that fewer people in developed nations are smoking. Seventeen million fewer people in the United States smoke now than smoked a few decades ago, when the U.S. surgeon general began warning of the dangers of tobacco. A poll taken in 1986 found that fewer than 27 percent of American adults said they were currently smoking, compared with more than 40 percent two decades before. Smoking has also declined in most European countries and in Japan.

The bad news is that more and more people are taking up the habit in Brazil, Pakistan, and other developing nations. In some countries, the smoking habit costs as much as 20 percent of a worker's annual income. To-

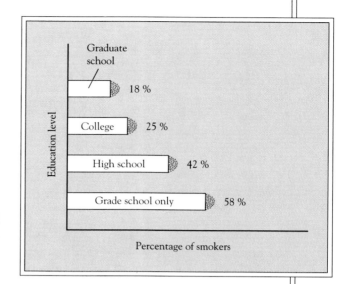

An inverse relationship exists between smoking and education level.

bacco companies in the United States promote smoking abroad because a substantial portion of our tobacco crop is exported.

Although fewer Americans are smoking, certain groups in our society still have high numbers of tobacco addicts (see figure), including minorities and blue-collar workers. A need exists to continue educating these groups, as well as all young people, about the dangers of smoking.

urban residents because they typically spend 90 percent or more of their time indoors.

Because illnesses caused by indoor air pollution usually resemble common ailments such as colds, influenza, or upset stomachs, they are often not recognized. The most common contaminants of indoor air are radon, cigarette smoke, carbon monoxide, nitrogen dioxide (from gas stoves), formaldehyde (from carpeting, fabrics, and furniture), household pesticides, cleaning solvents, ozone (from photocopiers), and asbestos (Figure 19–10) (see Focus On: Smoking). Viruses, bacteria, fungi (yeasts, molds, and mildews), dust mites (micro-

scopic animals found in household dust), pollen, and other living organisms or their toxic parts found in heating, air conditioning, and ventilation ducts are additional important forms of indoor air pollution.

Health officials are paying increasing attention to the **sick building syndrome,** the presence of air pollution inside office buildings that can cause eye irritations, nausea, headaches, respiratory infections, depression, and fatigue. The EPA estimates that the annual medical costs for treating the health effects of indoor air pollution in the United States exceed $1 billion.

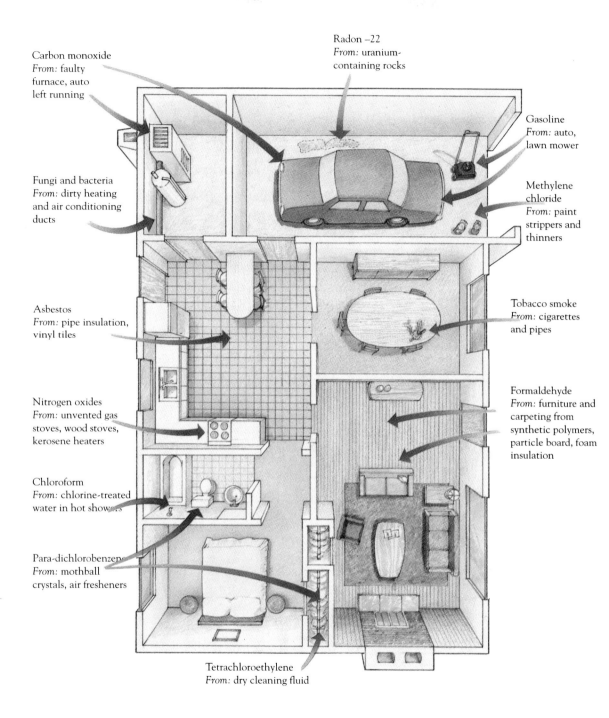

Carbon monoxide
*From:* faulty furnace, auto left running

Fungi and bacteria
*From:* dirty heating and air conditioning ducts

Asbestos
*From:* pipe insulation, vinyl tiles

Nitrogen oxides
*From:* unvented gas stoves, wood stoves, kerosene heaters

Chloroform
*From:* chlorine-treated water in hot showers

Para-dichlorobenzene
*From:* mothball crystals, air fresheners

Radon –22
*From:* uranium-containing rocks

Gasoline
*From:* auto, lawn mower

Methylene chloride
*From:* paint strippers and thinners

Tobacco smoke
*From:* cigarettes and pipes

Formaldehyde
*From:* furniture and carpeting from synthetic polymers, particle board, foam insulation

Tetrachloroethylene
*From:* dry cleaning fluid

**Figure 19–10**
Indoor air pollution comes from a variety of sources.

**Figure 19–11**
How radon infiltrates a house. Cracks in basement walls or floors, openings around pipes, and pores in concrete blocks provide some of the entries for radon.

## Radon

Not all environmental health hazards are the result of human activities. The most serious indoor air pollutant is **radon,** a colorless, tasteless radioactive gas produced naturally during the radioactive decay of uranium in the Earth's crust (see Chapter 11). The threat of dying from radon exposure is greater than that from any other environmental hazard, including pesticides and **carcinogenic,** or cancer-causing, air pollutants such as benzene. Radon is capable of seeping through the ground and entering buildings, where it sometimes accumulates to dangerous levels (Figure 19–11). Although radon is also emitted into the atmosphere, it gets diluted and dispersed and is of little consequence outdoors.

Radon and its decay products emit alpha particles, a form of ionizing radiation that is very damaging to living tissue but cannot penetrate very far into the body. Consequently, radon can harm the body only when it is ingested or inhaled. The radioactive particles lodge in the tiny passages of the lungs and damage lung tissue. There is compelling evidence, based primarily on several studies of uranium miners, that inhaling radon increases the risk of lung cancer. Of the 140,000 deaths from lung cancer in the United States each year, the EPA attributes about 13,000 to radon.[2]

[2] The actual number of deaths from lung cancer that can be attributed to radon is estimated to be between 5,000 and 20,000. A reasonable middle-of-the-range estimate is 13,000.

One of the interesting conclusions to emerge as the health effects of radon have been studied is that smoking greatly increases the risks from radon exposure. That is, the lung cancer rate of people who smoke *and* are exposed to radon is much greater than the sum of the two risks.

Ironically, efforts to make our homes more energy-efficient have increased the radon hazard. Drafty homes waste energy but allow radon to escape outdoors so it doesn't build up inside. According to the EPA, the number of American homes with high enough levels of radon to warrant corrective action may be as great as 8 million (close to 10 percent of all homes). The highest radon levels in this country have been found in homes on a geological formation called the **Reading Prong,** which runs across Pennsylvania into northern New Jersey and New York. The state with the most pervasive radon problem appears to be Iowa, where 71 percent of the homes tested in early 1989 had radon levels high enough to warrant corrective action.

Every home should be tested for radon because levels vary widely from home to home, even in the same neighborhood. Testing is inexpensive, and corrective actions, if required, are reasonably priced. Radon concentrations in homes can be minimized by sealing basement concrete floors and by ventilating crawl spaces and basements.

## Asbestos

Asbestos is a natural mineral that does not burn or conduct heat or electricity (Figure 19–12). These properties make asbestos valuable in construction and industry for fire retardant materials, electrical insulation, and roofing and pipe insulation. It is also used in automobile brake linings.

Asbestos has the ability to separate into long, thin fibers, which are so small and lightweight that they can remain suspended in air almost indefinitely. Asbestos fibers are easily inhaled into the lungs. The mucous membranes of the respiratory system remove some asbestos, but if more asbestos is inhaled than the body can handle, some of it lodges in the lungs.

Definite links exist between exposure to asbestos and several serious diseases, including lung cancer and **mesothelioma,** a rare and almost always fatal cancer of the body's internal linings. The lag time between exposure to asbestos and development of cancer is anywhere from 20 to 40 years. The danger from inhaling high levels of asbestos is well documented, but it is not known what, if any, danger exists from inhaling small amounts of asbestos.

**Figure 19–12**
Asbestos is a mineral that occurs naturally in the Earth's crust. Note its fibrous nature. When asbestos crumbles, microscopic fibers are released into the air. These fibers sometimes accumulate indoors, creating a health hazard when inhaled into the lungs. (Runk/Schoenberger, from Grant Heilman; specimen from North Museum, Franklin and Marshall College)

Although asbestos has been known to be a health hazard since the 1920s, it wasn't until 1979 that the EPA began regulating its use. During the 1980s concern increased about the widespread presence of asbestos in schools, offices, commercial buildings, and homes. Federal legislation (the Asbestos Hazard and Emergency Response Act of 1986) now requires the removal of asbestos from all school buildings. Under a mandate from the EPA, all uses of asbestos must be phased out by 1997.

Recently experts have generally agreed that the presence of asbestos in a building does not necessarily mean that people in the building are at risk, unless the asbestos is exposed and crumbling. A much greater danger probably comes from the fibers that are released to the air when asbestos is removed improperly. Sometimes it is safer to seal the asbestos instead of trying to remove it.

The risk of developing cancer from exposure to asbestos fibers is extremely small for most people. The EPA estimates that between 3,000 and 12,000 Americans die each year from asbestos-related cancer. Construction workers, firemen, and building custodians are more likely than other people to be exposed to asbestos fibers. Smoking also greatly increases the risk of developing disease from asbestos exposure.

## NOISE POLLUTION

**Sound** is caused by vibrations in the air (or some other medium) that reach the ears and stimulate a sensation of hearing. Sound is called **noise** when it becomes loud or disagreeable, particularly when it results in physiological or psychological harm. Of the 28 million Americans with hearing impairments, as many as 10 million could attribute their impairments at least in part to noise. Like other kinds of pollution in the environment, noise pollution can be reduced, although there is a cost associated with its reduction.

Most of the noise produced in the environment is of human origin. Vehicles of transportation, from trains to power boats to snowmobiles, produce a great deal of noise. Power lawn mowers, jets flying overhead, chain saws, jackhammers, and heavy traffic are just a few examples of the outside noise that assails our ears. Indoors, dishwashers, trash compactors, washing machines, televisions, and stereos add to the din.

### Measuring Noise

The **intensity** (loudness) of sound is measured relative to a reference sound that is so low it is almost inaudible to the human ear. Relative loudness is expressed numerically using the **decibel (db)** scale or a modified decibel scale called the **decibel-A (dbA)** scale, which takes into account high-pitched sounds to which the human ear is more sensitive (Table 19–4). Sound that is barely audible, such as rustling leaves or breathing, is rated at 10 dbA. A quiet neighborhood during the day might have a background sound level equivalent to 50 dbA. Noise at 90 dbA (such as a motorcycle at close range) impairs hearing, and noise at 120 dbA (such as a chain saw) causes pain.

### Effects of Noise

Prolonged exposure to noise damages hearing. The part of the ear that perceives sound is the **cochlea,** a spiral tube that resembles a snail's shell (Figure 19–13). Inside the cochlea are approximately

**Table 19–4**
**The Decibel-A Scale (dbA)**

| dbA | Example | Perception/General Effects |
|-----|---------|----------------------------|
| 0 | | Hearing threshold |
| 10 | Rustling leaves, breathing | Very quiet |
| 20 | Whisper | Very quiet |
| 30 | Quiet rural area (night) | Very quiet–quiet |
| 40 | Library | Quiet |
| 50 | Quiet neighborhood (daytime) | Quiet–moderately loud |
| 60 | Average office conversation | Moderately loud |
| 70 | Vacuum cleaner, television | Moderately loud |
| 80 | Washing machine, typical factory | Very loud, intrusive |
| 90 | Motorcycle at 8 meters | Very loud; impaired hearing with prolonged exposure |
| 100 | Dishwasher (very close), jet flyover at 300 meters | Very loud–uncomfortably loud |
| 110 | Rock band, boom box held close to ear | Uncomfortably loud |
| 120 | Chain saw | Uncomfortably loud–painfully loud |
| 130 | Riveter | Painfully loud |
| 140 | Deck of aircraft carrier | Painfully loud |
| 150 | Jet at takeoff | Painfully loud—ruptured eardrum |

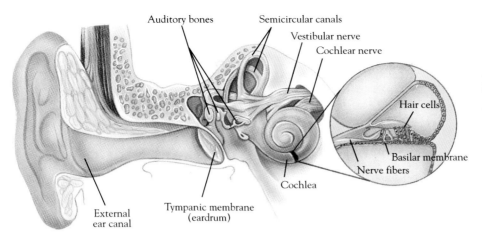

**Figure 19–13**
The structure of the human ear. Note the spiral cochlea. (Inset) Inside the cochlea are thousands of hair cells that change their shape as the basilar membrane to which they're attached vibrates in response to sound. The hair cells initiate nerve impulses along the auditory nerve to the brain.

24,000 **hair cells** that detect differences in pressure caused by sound waves. When the hairlike projections of the hair cells move back and forth in response to sound, the auditory nerve sends a message to the brain. Loud, high-pitched noise injures the hair cells in the cochlea. Because injured hair cells are not replaced by the body, prolonged exposure to loud noise results in permanent hearing impairment.

In addition to hearing loss, noise produces a number of physiological effects in the body. It increases the heart rate, dilates the pupils, and causes muscle contraction. Evidence exists that prolonged exposure to high levels of noise causes a permanent constriction of blood vessels, which can increase the blood pressure, thereby leading to heart disease. Other physiological effects associated with noise pollution include migraine headaches, nausea, dizziness, and gastric ulcers. Noise pollution also causes psychological stress.

## Controlling Noise

Obviously, noise pollution can be reduced by producing less noise. This can be accomplished in a variety of ways, from restricting the use of sirens and horns on busy city streets to engineering motorcycles, vacuum cleaners, jackhammers, and other noisy devices so that they produce less noise. The engineering approach is technologically feasible but is often avoided because consumers associate loud noise with greater power.

Putting up shields between the noise producer and the hearer can also help control noise pollution (Figure 19–14a). Examples of sound shields include the noise barriers erected along heavily traveled highways and the noise-absorbing material in-

**Figure 19–14**
Controlling noise. (a) Noise barriers along highways do not reduce the amount of noise but direct the sound waves away from the residential areas that are behind the barriers. (b) Many occupations, such as airline workers, require the use of sound-reducing headphones or earplugs. (a and b, Dennis Drenner)

(a)

(b)

stalled around dishwashers. Earplugs are an effective way to protect oneself from unwanted noise (Figure 19–14b); they differ from the preceding examples in that they shield the receiver rather than the noise producer.

## ELECTROMAGNETISM AND HUMAN HEALTH

Concern is increasing that electric and magnetic fields associated with such common things as power lines, electric blankets, video displays, and radar emitters may be health hazards (Figure 19–15). We are subjected to electric and magnetic fields outdoors (high-voltage transmission lines) as well as indoors (when electricity is used in household appliances, electrical wiring, and light fixtures).

Recent research has indicated that living cells are altered by electric and magnetic fields. Even very weak electric fields trigger biological responses, such as altered enzyme activity in cells, and modify animals' internal biological clocks. It is not clear exactly how these fields cause biological changes in living cells.

Several studies have demonstrated that children who are exposed to low-level magnetic fields from power lines may have an increased risk of leukemia. In addition, a study of men who died from brain tumors revealed that a disproportionate number of them had careers (such as electricians, telephone workers, and electronics engineers) in which they would have received higher-than-normal exposure to electromagnetic fields. Although the health risk associated with exposure to electric and magnetic fields has not been proven, the evidence is sufficiently suggestive of possible risk to require further study.

**Figure 19–15**
The strong electromagnetic field produced under high voltage transmission lines is a purported health hazard.
(Runk/Schoenberger, from Grant Heilman)

## SUMMARY

1. Air pollution comes from both natural sources and human activities, particularly motor vehicles and industries. The six main types of air pollutants produced by human activities are particulates, nitrogen oxides, sulfur oxides, carbon oxides, ozone, and hydrocarbons.

2. Particulates are solid particles and liquid droplets suspended in the air. Nitrogen oxides, sulfur oxides, carbon oxides, and ozone are gaseous air pollutants that contain oxygen. Hydrocarbons may be solids, liquids, or gases, depending on their molecular size. Hundreds of other chemicals, some of which are toxic, are emitted into the air.

3. Two kinds of smog occur, industrial smog and photochemical smog. Industrial smog is composed primarily of sulfur oxides and particulates. The formation of photochemical smog involves a complex series of chemi-

cal reactions involving nitrogen oxides, hydrocarbons, carbon monoxide, ozone, and sunlight.

**4.** Certain climates and topographies cause thermal inversions, in which the lower layers of air are cooler than higher layers. Thermal inversions that persist in congested urban areas cause air pollutants to build up to dangerous levels.

**5.** An area of local heat production associated with high population density is known as an urban heat island. Heat islands affect air currents and can cause pollutants to accumulate in dust domes over cities.

**6.** Air pollutants can have adverse effects on humans, animals, and plants. They can also damage materials and reduce visibility. All major forms of air pollution except carbon dioxide can be prevented or controlled with current technologies, although control often involves considerable expense.

**7.** The effect of air pollution on human health is a concern both outdoors—particularly in cities—and in enclosed places. In general, air pollutants irritate the eyes, inflame the respiratory tract, and suppress the immune system. People at greatest risk from air pollution include those with heart and respiratory diseases.

**8.** The most serious indoor air pollutant is radon, a radioactive gas that is produced naturally during the radioactive decay of uranium in the Earth's crust. Although loose asbestos fibers in indoor air are an extremely dangerous threat to human health, asbestos is usually not in this form and thus is normally not a major health hazard.

**9.** Noise pollution not only has the potential to cause hearing impairment, but also alters many physiological processes in the body and causes psychological stress. Electric and magnetic fields produced by power lines, electrical wiring, and electrical appliances may be hazardous to human health and should be evaluated further.

## DISCUSSION QUESTIONS

**1.** Which is a more stable atmospheric condition, cool air layered over warm air or warm air layered over cool air? Explain. Which condition is a thermal inversion?

**2.** How have energy conservation efforts contributed to indoor air pollution and the sick building syndrome?

**3.** A recent study suggests that it might be safer for rock collectors to keep their home collections of minerals in display cases that are vented to the outdoors. Can you offer an explanation for this recommendation, based on what you have learned in this chapter?

**4.** One of the most effective ways to reduce the threat of radon-induced lung cancer is to quit smoking. Explain.

**5.** Give one or two occupations that would probably be associated with higher-than-average levels of each of the following:
**a.** Radon-induced lung cancer
**b.** Asbestos-induced lung cancer
**c.** Electromagnetism-induced cancer

**6.** Why are some places considering restricting the loudness of music in public places, including nightclubs and concerts, to 85 dbA?

**7.** Some people have proposed that all smoking in public places be banned. Do you agree or disagree? Why?

## SUGGESTED READINGS

Hester, G. L. Electric and magnetic fields: Managing an uncertain risk. *Environment* 34:1, January/February 1992. Examines the biological effects of extremely low-frequency radiation from electric and magnetic fields.

Kerr, R. A. Indoor radon: The deadliest pollutant. *Science,* vol. 240, 29 April 1988. An overview of the indoor pollution problem from radon.

Monitoring the global environment: An assessment of urban air quality; a report from the United Nations Environment Programme and the World Health Organization. *Environment* 31:8, October 1989. Contains portions of the Global Environment Monitoring System's 1988 report on urban air quality.

Stone, R. No meeting of the minds on asbestos. *Science,* vol. 254, 15 November 1991. Scientists disagree on the health risk to the general public from asbestos fibers.

Stranahan, S. Q. It's enough to make you sick. *National Wildlife,* February–March 1990. An overview of pollution problems. An accompanying article by B. Lawren provides step-by-step instructions on how to make communities cleaner and safer places to live.

Thompson, J. East Europe's dark dawn. *National Geographic* 179:6, June 1991. A compelling look at the overwhelming environmental problems—including high levels of air pollution—that plague Eastern Europe.

Vaughan, C. Streetwise to the dangers of ozone. *New Scientist*, 26 May 1990. How ozone emissions can be reduced at ground level.

# Chapter 20

*Trees are being planted in Guatemala to absorb the extra carbon dioxide emitted into the atmosphere by a fossil-fuel burning power plant in Connecticut. (CARE Photo by Rudolph von Bernuth)*

# Global Changes

Air pollutants cause regional and global changes: acid deposition (commonly called acid rain), global climate change, and depletion of the atmosphere's ozone shield. Acid deposition kills aquatic organisms and corrodes metals and building materials, but its effect on plants is unclear. The production of atmospheric pollutants that trap solar heat in the atmosphere will affect the Earth's climate during the 21st century. Global warming could alter food production, destroy forests, submerge coastal areas, and displace millions of people. The chemical destruction of the stratospheric ozone shield by gaseous air pollutants is permitting large amounts of solar ultraviolet radiation to penetrate to the Earth's surface. Plants and animals, including humans, will be harmed by increased exposure to ultraviolet radiation.

## PLANTING FORESTS TO OFFSET CARBON DIOXIDE EMISSIONS

During the 1990s, farmers in Guatemala are planting trees, 52 million of them, paid for by an American company that is also building a coal-burning power plant in Connecticut. This project is being undertaken as a novel way to mitigate the adverse effects of carbon dioxide, an air pollutant produced in large quantities when coal and other fossil fuels are burned. The trees being planted in Guatemala will, if allowed to grow to maturity, remove the same amount of carbon dioxide from the air that the new power plant will release into the air during its 40-year life span.

Why would the proprietors of a power plant feel an obligation to combat the effects of the plant's carbon dioxide emissions? How do carbon dioxide and certain other pollutants affect the regional and global atmosphere? The climate? In this chapter we examine the problems of air pollution—acid deposition, global temperature changes, and depletion of the stratospheric ozone shield—in terms of regional and worldwide environments.

## ACID DEPOSITION

What do lakes without fish in the Adirondack Mountains, recently damaged Mayan ruins in southern Mexico, and stunted trees in the mountains of North Carolina have in common? The answer is that their damage may be the result of acid precipitation or, more properly, **acid deposition** (recall the discussion of acid precipitation in Chapter 1). Acid deposition is a type of air pollution. It includes acid that falls to the Earth in precipitation (sometimes called **wet deposition**) as well as dry, sulfate-containing particles that settle out of the air (sometimes called **dry deposition**).

Acid deposition is not a new phenomenon. It has been around since the Industrial Revolution began. The term "acid rain" was coined in the 19th century by a chemist who noticed that buildings in areas with heavy industrial activity were being worn away by rain.

Acid precipitation, including rain, sleet, and snow, poses a serious threat to the environment. The Northern Hemisphere has been hurt the most—especially the Scandinavian countries, central Europe, Russia, China, and North America. In the United States alone, the damage from acid deposition has been estimated at $8 billion each year.

### Measuring Acidity

The relative degree of acidity or alkalinity of a substance is expressed using the pH scale, which runs from zero to 14 (see Appendix I). A pH of 7 is neither acidic nor alkaline, whereas a pH less than 7 indicates an acidic solution. The pH scale is logarithmic, so a solution with a pH of 6 is ten times more acidic than a solution with a pH of 7, a solution with a pH of 5 is ten times more acidic than a solution with a pH of 6, and so on. Solutions with a pH greater than 7 are alkaline, or basic.

For purposes of comparison, distilled water has a pH of 7, tomato juice has a pH of 4, vinegar has a pH of 3, and lemon juice has a pH of 2. Normally, rainfall is slightly acidic (with a pH from 5 to 6) because carbon dioxide and other naturally occurring compounds in the air dissolve in rainwater, forming dilute acids. However, the pH of precipitation in the northeastern United States averages 4 and is often 3 or even lower.

### How Acid Deposition Develops

Acid deposition occurs when sulfur oxides and nitrogen oxides are released into the atmosphere (Figure 20–1). Motor vehicles are a major source of nitrogen oxides. Electrical power plants, large smelters (see Figure 15–4b), and industrial boilers are the main sources of sulfur dioxide emissions and produce substantial amounts of nitrogen oxides. Sulfur and nitrogen oxides, released into the air from tall smokestacks, can be carried long distances by atmospheric winds. For example, tall smokestacks allow England to "export" its acid deposition problem to the Scandinavian countries, and the midwestern United States to "export" its acid emissions to New England and Canada.

During their stay in the atmosphere, sulfur oxides and nitrogen oxides are transformed into an assortment of secondary pollutants that make dilute solutions of sulfuric and nitric acids. Precipitation such as rain or snow returns these acids to the ground, causing the pH of surface waters and soil to decrease.

### Effects of Acid Deposition

The link between acid deposition and declining aquatic animal populations is well established (Figure 20–2). A report published by the National Academy of Sciences concluded, for example, that acid deposition is responsible for the decline of fish in lakes of the Adirondack Mountains and in Nova Scotia rivers. Heavy metals such as cadmium and

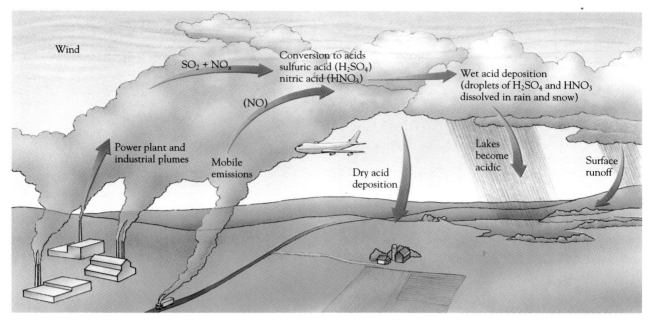

**Figure 20–1**
How acid deposition occurs.

**Figure 20–2**
Different aquatic organisms have varying sensitivities to low pH (higher acidity) caused by acid deposition. Some organisms, like the freshwater mussel, can tolerate very little acid, whereas others, like the water boatman, have a wide pH tolerance.

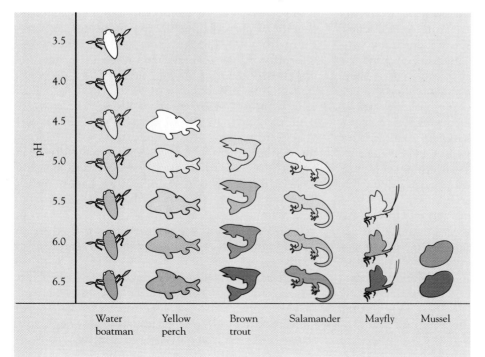

**The Increase in Acid**

Number of strongly acidic lakes in the United States in 1990: 1,130

Total number of lakes in the United States: approximately 28,000

Number of strongly acidic lakes in Canada in 1990: 31,000

Total number of lakes in Canada: approximately 1,140,000

Proportion of European forests showing damage attributable in part to acid rain in 1982: 8 percent

Proportion of European forests showing damage attributable in part to acid rain in 1988: 52 percent

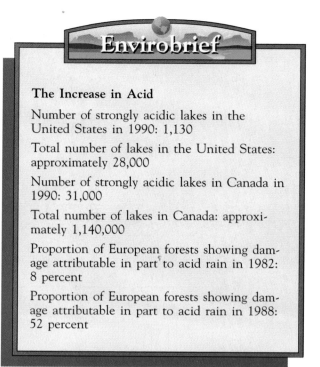

**Figure 20–3**

An ecologist at the Argonne National Laboratory monitors the effects of acid deposition on plants. (U.S. Department of Energy)

mercury become soluble in acidic lakes and streams and enter the food chain; this may explain the adverse effect that acidic water has on fish.

Although fish have received the lion's share of attention in regard to the acid deposition issue, other animals are also adversely affected. A 1989 study of the effects of acid deposition on birds found that birds living in areas with pronounced acid deposition were much more likely to lay eggs with thin, fragile shells. The inability to produce strong eggshells was attributed to reduced calcium in the birds' diets. Calcium is unavailable to them because in acidic soils it becomes soluble and is washed away, with little left for absorption by plant roots. A lesser amount of calcium in plant tissues means a lesser amount of calcium in the insects that eat the plants; moving along the food chain, the birds that eat these insects have less calcium in their diets.

The largest unanswered question regarding acid deposition is its effect on plants, particularly forests (Figure 20–3). More than half of the red spruce trees in the mountains of the northeastern United States have died since the mid-1970s, and other tree species (sugar maples, for example) are also dying. Many trees that are not dead are exhibiting symptoms of **forest decline,** characterized by gradual deterioration and often death of trees. The general symptoms of forest decline are reduced vigor and growth, but some plants exhibit specific symptoms, such as yellowing of needles in conifers.

Forest ecosystems are so complex that it is impossible to trace forest decline to a single cause (Figure 20–4). Although acid deposition correlates well with areas that are experiencing tree damage, it is only one of several possible causes, including insects and other forms of human-made air pollution such as tropospheric ozone ($O_3$) (see Chapter 19). Weather factors such as drought and severe winters (cold can injure susceptible plants) may also be important. To complicate matters further, the actual causes of forest decline may vary from one tree species to another and from one location to another. Thus, forest decline appears to result from the *combination* of multiple stresses—acid deposition, insect attack, drought, and so on. When one or more stresses weaken a tree, then an additional stress may be decisive in causing its death.

One way in which acid deposition harms plants is well established: acid deposition alters the chemistry of soils, which affects the development of plant roots as well as their uptake of nutrients and water from soil. Essential plant minerals such as

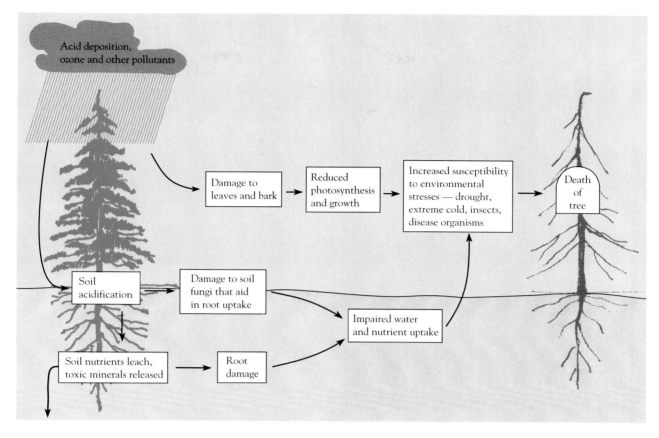

**Figure 20–4**
Air pollutants are one of several stresses that interact together, possibly leading to decline and death of trees.

calcium and magnesium wash readily out of acidic soil, whereas others, such as nitrogen, become available in larger amounts (see Chapter 14). Also, heavy metals such as manganese and aluminum dissolve in acidic soil, becoming available for absorption in toxic amounts by the roots of plants. A study completed in 1989 in Central Europe, which has experienced greater forest damage than North America, concluded that there was a strong relationship between forest damage and soil chemistry altered by acid deposition.

Acid deposition corrodes metals and building materials (Figure 20–5). It eats away at important monuments, such as the Washington Monument in Washington, D.C., and the Statue of Liberty in New York harbor. Historic sites in Venice and Rome and all across Europe are being worn away by acid deposition. Emissions from uncapped oil wells (owned by the Mexican government) in the Gulf of Mexico cause acid deposition that has been tied to

the destruction of ancient Mayan ruins in southern Mexico.

## The Politics of Acid Deposition

One of the factors that makes acid deposition so hard to combat is that it does not occur only in the locations where the gases that cause it are emitted. Acid deposition doesn't recognize borders between states or between countries; it is entirely possible for sulfur and nitrogen oxides that were released in one spot to return to the Earth hundreds of miles from their source.

The United States has wrestled with this issue. Several states in the Midwest and East—Illinois, Indiana, Missouri, Ohio, Pennsylvania, Tennessee, and West Virginia—produce between 50 and 75 percent of the acid deposition that contaminates New England and southeastern Canada. When legislation was formulated to deal with the problem,

**Figure 20–5**
Acid deposition eats away at stonework. Most of the writing on this English tombstone, carved in 1817, has been eaten away by acid deposition. (Bruce F. Molnia/Terraphotographics)

## Controlling Acid Deposition

Although the science and the politics surrounding acid deposition are complex, the basic concept of control is straightforward: reducing emissions of sulfur and nitrogen oxides reduces acid deposition. Simply stated, if sulfur and nitrogen oxides are not released into the atmosphere, they cannot come down as acid deposition. Installation of scrubbers in the smokestacks of traditional coal-fired power plants and utilization of clean-coal technologies to burn coal without excessive emissions effectively diminish acid deposition (see Chapter 10).

In turn, a decrease in acid deposition prevents surface waters and soil from becoming more acidic than they already are. But the effect of the reduction of acid-producing emissions on lakes, streams, and soils that have *already* been damaged by a low pH is unknown.

## GLOBAL TEMPERATURE CHANGES

Imagine a world in which beautiful tropical islands, such as the Maldives in the Indian Ocean, disappear forever under the waves. Closer to home, consider the permanent flooding of Louisiana's bayous and the Florida Keys. Imagine the forests in the southern United States dying from high temperatures and drought. Think of the effects of an annual hurricane season that lasts a month longer than it does now. Consider the consequences of tropical pests and diseases spreading northward into the United States and remaining throughout the winter.

Although none of these events has occurred, there is a chance that some or all of them will take place, possibly even during your lifetime. That is because the Earth may become warmer during the next century than it has been for the past 1 million years (recall the brief discussion of global climate change in Chapter 1). Unlike climate changes in the past, which occurred over thousands of years or longer, this change would take place in a matter of decades.

Some experts think the warming trend has already begun. Eight of the ten warmest years on record (records have been kept since the mid-19th century) occurred during the 1980s and early 1990s: 1980, 1981, 1983, 1987, 1988, 1989, 1990, and 1991. 1990 was the warmest year on record, and 1991 was the second warmest. Although most scientists agree that the world will continue to

however, arguments ensued about who should pay for the installation of expensive air pollution devices to reduce emissions. Should the states emitting the gases be required to pay all the expenses to clean up the air, or should some of the cost be absorbed by the areas that stand to benefit most from a reduction in pollution?

Pollution abatement issues are quite complex within one country, but are magnified even more in international disputes. For example, England, which has large reserves of coal, uses coal to generate electricity. Gases from coal-burning power plants in England move eastward with prevailing winds and return to the Earth as acid deposition in Sweden and Norway. England is not the only country responsible for acid deposition that crosses international borders. Emissions from the United States produce acid deposition in Canada, whereas Japan receives acid deposition from China.

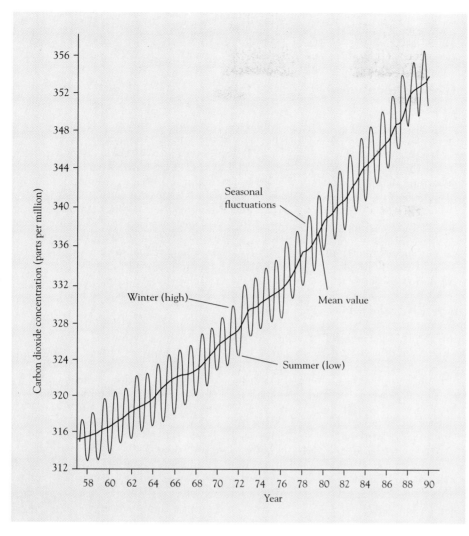

**Figure 20–6**
The concentration of carbon dioxide in the atmosphere has shown a slow but steady increase for many years. These measurements were taken at the Mauna Loa Observatory, far from urban areas where carbon dioxide is emitted by factories, power plants, and motor vehicles. (The seasonal fluctuation each year corresponds to winter—a high level of $CO_2$—and summer—a low level of $CO_2$—and is caused by greater photosynthesis in summer.)

warm, they disagree over how rapidly the warming will proceed, how severe it will be, and where it will be most pronounced. One of the complications that makes global warming difficult to predict is that other air pollutants, from both natural and human sources, actually exert a cooling influence (see Focus On: Cooling the Global Greenhouse). As a result of these uncertainties, many people, including policy makers, are confused about what should be done.

## The Causes of Global Warming

Carbon dioxide and certain other trace gases, including methane ($CH_4$), nitrous oxide ($N_2O$), chlorofluorocarbons (CFCs), and tropospheric ozone ($O_3$), are accumulating in the atmosphere as a result of human activities (Table 20–1 and Figure 20–6). The concentration of atmospheric carbon dioxide has increased from about 280 parts per mil-

**Table 20–1**
**Increase in Atmospheric Greenhouse Gases, Pre–Industrial Revolution to 1989**

| Gas | Concentration in Air | |
| --- | --- | --- |
| | *Preindustrial* | *1989* |
| Carbon dioxide | 274 ppm* | 354 ppm |
| Methane | 0.7 ppm | 1.7 ppm |
| Chlorofluorocarbon-12 | 0 ppb† | 0.5 ppb |
| Chlorofluorocarbon-11 | 0 ppb | 0.3 ppb |
| Nitrous oxide | 280 ppb | 306 ppb |
| Tropospheric ozone | NA§ | 35 ppb |

*ppm = parts per million.
†ppb = parts per billion.
§Data not available.

### Cooling the Global Greenhouse

A tug of war is being waged between two different types of pollution in the atmosphere. At stake is the Earth's climate. On one side are the greenhouse gases, which have the ability to warm the planet by preventing heat from radiating out into space. On the other side is sulfur haze, a type of air pollution that not only causes acid deposition, but cools the planet by reflecting sunlight away from the Earth. Climatologists think that sulfur haze may have significantly counteracted greenhouse warming in the Northern Hemisphere during the last few decades.

Sulfur emissions, which produce sulfur haze, come from the same smokestacks that pour forth carbon dioxide. In addition, volcanic eruptions can inject sulfur-containing particles into the atmosphere. The explosion of Mount Pinatubo in the Philippines in June 1991 was the largest volcanic eruption in the 20th century. The force of this eruption injected massive amounts of sulfur into the stratosphere, where sulfur tends to remain longer than it does when emitted into the troposphere. Because sulfur in the stratosphere reduces the amount of sunlight that reaches the planet, this eruption may cause the Earth to enter a period of global cooling that could last for several years. The overall effect of such cooling is projected to be a reduction in the Earth's temperature by about 0.5°C (1°F). The sulfur will gradually fall to the ground as acid deposition.

Human-produced sulfur emissions should not be viewed as a panacea for the greenhouse effect, despite their cooling effect. For one thing, they are produced in heavily populated industrial areas, primarily in the Northern Hemisphere. Because they do not remain in the atmosphere for months, they do not disperse globally. Thus, sulfur pollution may cause regional, but not global, cooling.

In the grand scheme of things, the greenhouse gases will probably win out over sulfur haze. Greenhouse gases remain in the atmosphere for hundreds of years, whereas human-produced sulfur emissions remain for only days or weeks. And carbon dioxide and other greenhouse gases help warm the planet 24 hours a day, whereas sulfur haze cools the planet only during the daytime. In addition, because sulfur emissions also cause acid deposition, most nations are trying to reduce their sulfur emissions, not maintain or increase them.

---

lion approximately 200 years ago (before the Industrial Revolution began) to 360 parts per million today. And it is still rising, as are the levels of the other trace gases associated with global warming. For example, every time you drive your car, the combustion of gasoline in the car's engine releases carbon dioxide and nitrous oxide and triggers the production of tropospheric ozone. Every day, as tracts of rain forest are burned in the Amazon, carbon dioxide is released. CFCs are released into the atmosphere from old, leaking refrigerators and air conditioners.

Global warming occurs because these gases retain infrared radiation (that is, heat), which normally would dissipate into space from the Earth, in the atmosphere. Thus, the atmosphere warms (Figure 20–7). Some of the heat from the atmosphere is transferred to the oceans and raises their temperature as well. As the atmosphere and oceans warm, the overall temperature of the Earth rises. Because carbon dioxide and other gases trap the sun's radiation in much the same way that glass does in a greenhouse, global warming produced in this manner is known as the **greenhouse effect** (see Chapter 12).

### The Effects of Global Warming

Because the interactions among the atmosphere, the oceans, and the land are too complex and too large to study in a laboratory, climatologists develop models by using computer simulations. A model, however, is only as good as the data and assumptions upon which it is based, and a number of uncertainties are built into our models of global warming. For example, if global warming causes

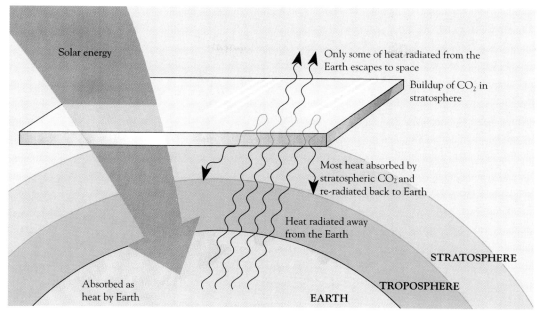

**Figure 20–7**
How carbon dioxide and the greenhouse effect promote global warming.

more low-lying clouds to form, they will block some sunlight and decrease the warming trend. On the other hand, global warming may cause more high, thin cirrus clouds to form, which would actually intensify the greenhouse effect. As new data about these uncertainties become available, they are used to make the models' predictions more precise.

Current models predict that a doubling of carbon dioxide concentration in the atmosphere will cause the average temperature of the Earth to increase by 2 to 5°C before the end of the 21st century, although the warming will not be uniform from region to region. At current rates of fossil fuel combustion and deforestation, scientists expect the doubling of carbon dioxide to occur within the next 50 years. However, the warming trend will be slower than the increasing carbon dioxide might indicate, because the oceans take longer than the atmosphere to absorb heat. Greater warming will probably take place in the second half of the 21st century than in the first half.

**Rising Sea Level**  If the overall temperature of the Earth increases by just a few degrees, there could be a major thawing of glaciers and the polar ice caps (Figure 20–8). In addition, thermal expansion of the oceans will probably occur, because water, like other substances, expands as it warms. Scientists estimate that these two changes may cause the sea level to rise by as little as 0.2 meter or as much as

**Figure 20–8**
A glacier in New Zealand. A great deal of water currently frozen in glaciers and the polar icecaps could melt if global warming were to occur. (G. R. Roberts)

**Figure 20–9**
A home in Aptos, California, was wrecked by a powerful ocean storm. A rise in sea level would inundate many coastal areas.
(William E. Ferguson)

**Figure 20–10**
One possible scenario of how precipitation might be altered by warmer global temperatures. This map is based on precipitation rates that occurred thousands of years ago when the Earth was warmer.

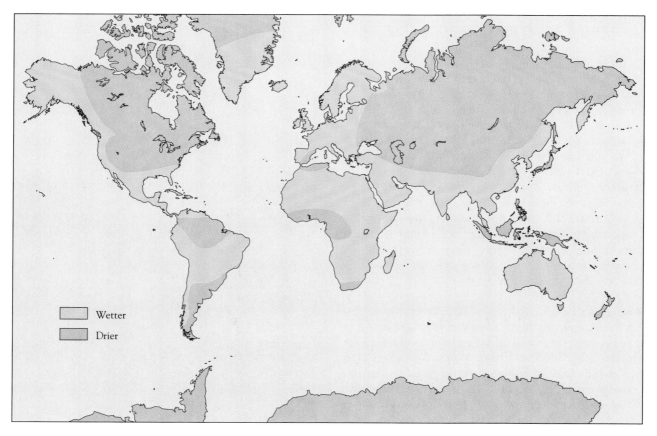

## Focus On

### Two Species Affected by Global Warming

Biologists are only beginning to assess the possible impacts of global warming on the world's living things, but it is clear that in many areas there will be major changes in both species diversity and species distribution. Two organisms—a mountain butterfly and an Asian vine widely introduced in the United States—provide us with a glimpse of what may be in store for many of the world's organisms.

A rare mountain butterfly may become one of the first species to face extinction solely because of global warming. This little creature (*Boloria acrocnema*), a member of a group of butterflies called fritillaries, lives in an extremely restricted area: a few high-altitude, northeast-facing, snow-moistened slopes in the San Juan Mountains of western Colorado. Its habitat is surrounded by warmer, drier areas that the butterfly cannot tolerate. During the hot, dry summers of the late 1980s, one of the two known populations of this fritillary species disappeared, and less than half of the other population survived the decade. Researchers have concluded that the only reason for the butterfly's declining

numbers was the unusually warm, dry weather. In the summer of 1991 the fritillary was listed as an endangered species, but this federal protection will not keep it from becoming extinct if the climate continues to be as warm and dry in the 1990s as it was in the late 1980s.

Kudzu is a fast-growing, purple-flowered vine that has become a weed throughout much of the southern United States. It was imported to the United States from Japan and was planted widely throughout the South during the 1930s in an effort to halt soil erosion. Because it grows rapidly—as much as 1 foot per day under ideal growing conditions—kudzu has choked out many other plants, including trees, and has become a major nuisance from eastern Texas to Florida; its current range extends north through most of Maryland. Botanists who have studied kudzu conclude that low winter temperatures are the only reason the plant hasn't spread farther north. Should global warming cause average winter temperatures to increase by 3°C, the vine may well spread north to Michigan, New York, and Massachusetts.

---

2.2 meters, with resultant flooding of low-lying coastal areas (Figure 20–9), by 2050. Because of climatic changes, coastal areas that are not inundated will be more likely to suffer damage from hurricanes and typhoons. These likely effects are certainly a cause for concern.

The countries that are most vulnerable to a rise in sea level—countries such as Bangladesh, Egypt, Vietnam, and Mozambique—have dense populations in low-lying river deltas. For example, a rising sea level could cause Bangladesh to lose as much as 25 percent of its land. Since 1970, the flooding and high waves accompanying tropical storms have resulted in the deaths of at least 300,000 people in Bangladesh. A rising sea level caused by global warming would put even more people at risk in this densely populated nation.

**Changes in Precipitation Patterns** Global warming is also expected to change precipitation patterns,

causing some areas to have more frequent droughts (Figure 20–10). At the same time, in other areas flooding may become more likely. The frequency and intensity of storms may also increase. All of these factors could affect the availability and quality of fresh water in many locations. It is projected that arid and semiarid regions will have the most troublesome water shortages.

**Effects on Living Organisms** Biologists have started to examine some of the potential consequences of global warming for wildlife. Each species reacts to changes in temperature differently (see Focus On: Two Species Affected by Global Warming). Some species will undoubtedly become extinct, particularly those with narrow temperature requirements, those confined to small reserves and parks, and those living in fragile ecosystems; other species may survive in greatly reduced numbers and ranges. Ecosystems considered to be at greatest risk of loss of species in the short term are polar seas, coral reefs, mountain ecosystems, coastal wetlands, tundra, and boreal and temperate forests.

Some species may be able to migrate to new environments or adapt to the changing conditions in their present habitats. Also, some species may be unaffected by global warming, whereas others may come out of global warming as winners, with greatly expanded numbers and range. Those considered most likely to prosper include weeds, pests, and disease-carrying organisms that are already common in a wide range of environments.

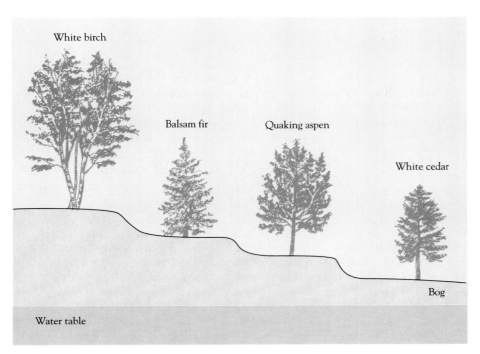

(a)

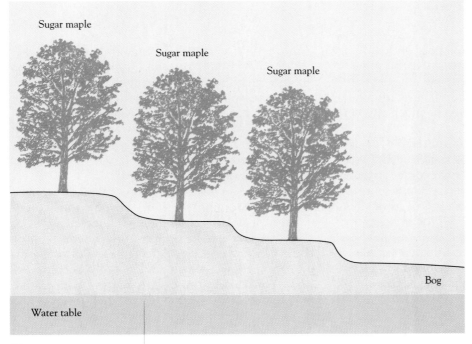

(b)

**Figure 20–11**
The species composition of a forest in northern Minnesota (a) now and (b) after global warming occurs.

Biologists generally agree that global warming will have an especially severe impact on plants because they cannot move about when conditions change (Figure 20–11). Although the plants' seeds are dispersed, sometimes over long distances, seed dispersal has definite limitations in terms of speed of species migration. Moreover, soil characteristics, water availability, competition with other plant species, and human alterations of natural habitats all affect the rate at which plants can move into a new area.

The increase in carbon dioxide in the atmosphere and the resulting increase in temperature will probably not have a direct, adverse effect on human health. Human health will be indirectly affected, however, as such disease carriers as malarial mosquitos and encephalitis-infected flies expand their range into the newly warm areas. People can also expect more frequent and more severe heat waves during summer months.

**Effects on Agriculture** Global warming will increase problems for agriculture, which is already beset with the challenge of providing enough food for a hungry world without doing irreparable damage to the environment. The rise in sea level will cause water to inundate river deltas, which are some of the world's best agricultural lands. Certain agricultural pests and disease-causing organisms

will probably proliferate. Global warming will also increase the frequency and duration of droughts. To get an idea of how devastating droughts can be, consider the one in the American grain belt during the summer of 1988. Wheat yields were reduced by 40 percent, and the United States, which is usually the world's largest exporter of grain, used more grain than it produced.

## International Implications of Global Warming

Global warming will be complicated by a variety of social and political factors. For example, how will we deal with the environmental refugees produced by global warming? Where will they go? Who will assist them to resettle? It will be difficult for all countries to develop a consensus on dealing with global warming, partly because global warming will clearly have greater impacts on some nations than on others. However, *all* nations must cooperate if we are to effectively deal with global warming.

**Developed Nations Versus Developing Nations** Although greenhouse gases are produced primarily by the developed countries (Figure 20–12), the repercussions of climate change will be widely felt by both developed and developing nations. Furthermore, the developing nations will probably

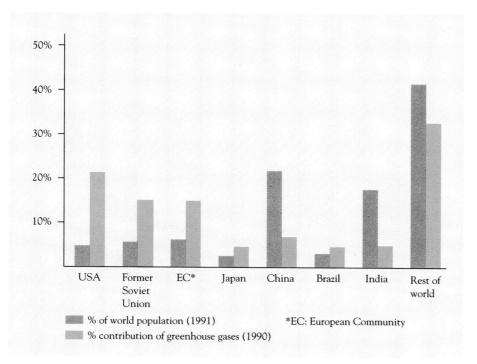

**Figure 20–12**
A comparison of the population of people in selected countries with the amount of carbon dioxide each country or group of countries emits. Note that the developed nations produce a disproportionate share of carbon dioxide emissions.

■ % of world population (1991)
■ % contribution of greenhouse gases (1990)

*EC: European Community

experience the greatest impacts of global warming. Unfortunately, these countries are the very ones that are least able to respond to the challenges of global warming, because they lack the technical expertise and economic resources.

Tensions are sure to mount between nations, especially between the developed and developing countries. The latter see increased use of fossil fuels as their route to industrial development; they will likely resist pressure from developed nations to decrease fossil fuel consumption. And when countries that have already burned their own forests (during their periods of development) tell developing countries not to burn theirs, the natural response is resentment.

## How We Can Deal with Global Warming

Even if we were to immediately stop polluting the atmosphere with greenhouse gases, the Earth would still have some climate change because of the greenhouse gases that have accumulated during the past 100 years. The amount and severity of global warming depend upon how much additional greenhouse gas pollution we add to the atmosphere.

There are three basic ways to manage global warming—prevention, mitigation, and adaptation. *Prevention* of global warming can be accomplished by developing ways to prevent the buildup of greenhouse gases in the atmosphere. Prevention is the ultimate solution to global warming because it is permanent. *Mitigation* of global warming is moderation or postponement, which buys us time to pursue other, more permanent solutions. It also gives us time to understand more fully how global warming operates so we can avoid some of its worst consequences. *Adaptation* is responding to changes brought about by global warming. Developing strategies to adapt to climate change implies an assumption that global warming is unavoidable.

**Prevention of Global Warming**    The development of alternatives to fossil fuel offers a permanent solution to the challenge of global warming caused by carbon dioxide emissions.[1] (Alternatives are also *necessary*, given that fossil fuels are present in limited amounts.) Some alternatives to fossil fuels—solar energy and nuclear energy—were discussed in Chapters 11 and 12.

The invention of technological innovations that trap the carbon dioxide being emitted from

[1] Alternatives to the other greenhouse gases will also have to be developed, but carbon dioxide is the focus here because it is produced in the greatest quantities and has the largest total effect of all the greenhouse gases.

smokestacks would help prevent global warming and yet allow us to continue using fossil fuels for energy (while they last). Incentives may be necessary to inspire such innovation. For example, several nations have imposed taxes on greenhouse gases. The taxes stimulate emitters to improve efficiency and to develop carbon dioxide–free technologies.

A number of developed nations (notably Japan and the European Community) have established limits on carbon dioxide emissions. Germany intends to reduce its carbon dioxide emissions by 25 percent (from its 1987 level) by 2005, for example. The United States has so far resisted adopting specific carbon dioxide limits.

In recent years, several unusual proposals have surfaced for reducing atmospheric carbon dioxide and thereby preventing global warming. One proposal is to liquefy carbon dioxide and then pump it deep into the ocean, where oceanic pressure might solidify it, allowing it to remain there. Some scientists have even suggested fertilizing the ocean around Antarctica with iron. It has been hypothesized that algal growth in this area is limited by a lack of iron, which is needed by algae to manufacture chlorophyll. Adding iron to the water might stimulate massive growth of algae, which would remove dissolved carbon dioxide from the water for use in photosynthesis. The unknown effects of massive algal growth on the complex Antarctic ecosystem have caused scientists to largely reject this proposal.

One of the most unusual suggestions for helping to prevent global warming is to place giant mirrors in orbit around the Earth. The mirrors would supposedly reflect some of the sun's rays away from the earth, thereby reducing the amount of heat reaching our planet.

Most scientists reject these unusual proposals to "fix" the climate because of concerns that they could backfire. Atmospheric scientists caution that tinkering with a system as complex as the global climate is risky business; unexpected chemical reactions might occur that could intensify the very problem they were intended to prevent.

**Mitigation of Global Warming**    One of the most effective ways to mitigate global warming involves forests. As you know, atmospheric carbon dioxide is removed from the air by growing forests, which incorporate the carbon into leaves, stems, and roots through the process of photosynthesis (see Chapter 3). In contrast, deforestation releases carbon dioxide into the atmosphere as trees are burned or decomposed. We can mitigate global warming by planting and maintaining new forests, and some environmentalists have even suggested that developed nations should pay developing nations to maintain their own tropical forests.

Increasing the energy efficiency of automobiles and appliances—and thus reducing the output of carbon dioxide—would also help mitigate global warming.

**Adaptation to Global Warming** Government planners and social scientists are developing a number of strategies to help us adapt to global warming. One of the most pressing issues is rising sea level. People living in coastal areas could be moved inland, away from the dangers of storm surges. This solution would have high societal and economic costs, however. An alternative, also extremely expensive, is the construction of dikes and levees to protect coastal land. Rivers and canals that spill into the ocean would have to be channeled to protect freshwater and agricultural land from salt water intrusion (see Chapter 13). The Dutch, who have been building dikes and canals for several hundred years, are offering their technical expertise to several developing nations that are threatened by a rise in sea level.

We will also need to adapt to shifting agricultural zones. A number of countries with temperate climates are evaluating semitropical crops to determine the best ones to substitute for traditional crops if and when the climate warms. Drought-resistant strains of trees are being developed by large lumber companies now, because the trees

planted today will be harvested in the middle of the next century, when global warming may already be well advanced.

## OZONE DEPLETION IN THE STRATOSPHERE

You may recall from Chapter 19 that ozone ($O_3$) is a form of oxygen that is a human-made pollutant in the troposphere but a naturally produced, essential component in the stratosphere, which encircles our planet some 10 to 45 km (6 to 28 mi) above the Earth's surface. The stratosphere contains a layer of ozone that shields the Earth from much of the ultraviolet radiation coming from the sun (Figure 20–13). Should ozone disappear from the stratosphere, the Earth would become uninhabitable for most forms of life.

The problem of ozone depletion was dramatically demonstrated beginning in 1985, with the discovery of a large "hole," or thin spot, in the ozone layer over Antarctica (recall the introduction to the ozone depletion problem in Chapter 1) (Figure 20–14). There, ozone levels decrease by as much as 67 percent each year. A smaller hole has been detected in the stratospheric ozone layer over the Arctic. Probably the most disquieting news is that worldwide levels of stratospheric ozone have been decreasing for several decades. In 1991 a United Nations panel of experts reported that during the 1980s the ozone shield was significantly depleted in

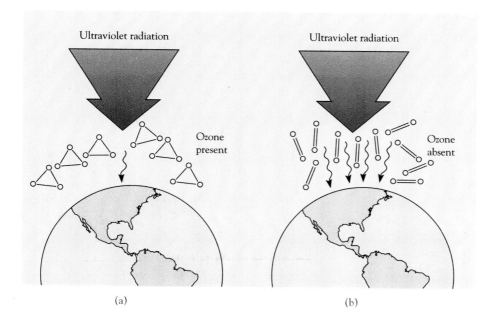

(a)        (b)

**Figure 20–13**

Ultraviolet radiation and the ozone layer. (a) Ozone absorbs ultraviolet radiation, effectively shielding the Earth. (b) When ozone is absent, more high-energy ultraviolet radiation penetrates the atmosphere to the Earth's surface, where its presence harms living things.

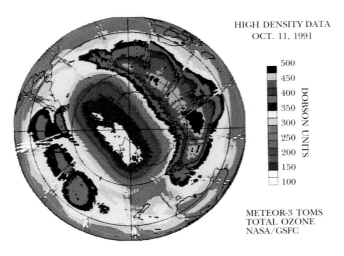

HIGH DENSITY DATA
OCT. 11, 1991

500
450
400
350
300
250
200
150
100

DOBSON UNITS

METEOR-3 TOMS
TOTAL OZONE
NASA/GSFC

**Figure 20–14**

A computer-generated image of the Southern Hemisphere in October 1991 reveals the ozone "hole" over Antarctica. Similar holes were noted in 1989 and 1990. (NASA)

sphere, chlorine is capable of attacking ozone and converting it into oxygen (Figure 20–15). The chlorine is not altered by this process; as a result, a single chlorine atom can break down many thousands of ozone molecules.

The hole in the ozone layer that was discovered over Antarctica occurs annually between September and November, when the **circumpolar vortex,** a mass of cold air, circulates around the southern polar region, in effect isolating it from the warmer air in the rest of the world. The cold causes stratospheric clouds to form; these clouds contain ice crystals to which chlorine and other chemicals adhere, enabling them to attack ozone. When the circumpolar vortex breaks up, the ozone-depleted air spreads northward, diluting ozone levels in the stratosphere over South America, New Zealand, and Australia.

the mid-latitudes even during the summer, when the rate of ozone depletion is lowest. The rate of ozone depletion during the 1980s was roughly three times the rate during the 1970s.

## The Causes of Ozone Depletion

The primary culprits responsible for ozone loss in the stratosphere are a group of commercially important compounds called chlorofluorocarbons, or CFCs. CFCs are used as propellants for aerosol cans, as coolants in air conditioners and refrigerators (e.g., Freon), as foam for insulation and packaging (e.g., Styrofoam), and as medical sterilizers (see You Can Make a Difference for things you can do in your everyday life to help protect the ozone layer). Additional compounds that also attack ozone include (1) halons (found in many fire extinguishers), (2) methyl bromide (used as a fumigant in agriculture), (3) methyl chloroform (used to degrease metals), and (4) carbon tetrachloride (used in many industrial processes, including the manufacture of pesticides and dyes).

## How Ozone Depletion Takes Place

CFCs and similar human-made compounds drift up to the stratosphere, where ultraviolet radiation breaks them down into chlorine, fluorine, and carbon.[2] Under certain conditions found in the strato-

**Figure 20–15**

The destruction of ozone by free chlorine. Chlorine (1) pulls one of the oxygen atoms away from ozone (2), creating an oxygen molecule, $O_2$ (3), and a molecule of chlorine oxide (4). The chlorine oxide then reacts with a second molecule of ozone (5) to form two molecules of oxygen (6) and free chlorine (1), which is then available to attack yet another ozone molecule. Thus, for every two ozone molecules that chlorine destroys, three oxygen molecules are produced. Although ozone is continually being formed by natural processes, its rate of formation does not equal its faster rate of destruction.

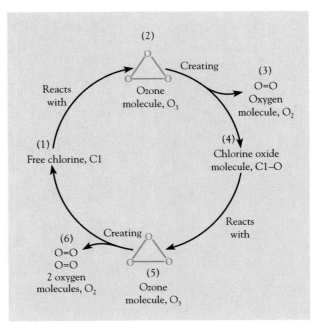

(2)
Creating
Ozone
molecule, $O_3$

Reacts
with

(3)
O=O
Oxygen
molecule, $O_2$

(1)
Free chlorine, Cl

(4)
Chlorine oxide
molecule, Cl–O

Reacts
with

(6)
Creating
O=O
O=O
2 oxygen
molecules, $O_2$

(5)
Ozone
molecule, $O_3$

[2] Humans are not the only cause of ozone depletion. Evidence is accumulating that implicates volcanic eruptions (such as that of Mount Pinatubo in June 1991) in the acceleration of ozone loss.

## How You Can Help Solve the Stratospheric Ozone Depletion Problem

Although CFC production in the United States will be terminated by the end of 1995, great amounts of it and other ozone-destroying chemicals are still around. You can do five things in your daily life that will reduce the use of CFCs and other ozone-destroying gases.

1. If you have an automobile air conditioner, insist on CFC recycling when you have it serviced (see figure).
2. Limit your use of foam plates, cups, and the like.
3. Purchase dry-chemical, rather than halon, fire extinguishers for home use.

4. Read the labels on electronic component cleaners, do-it-yourself car air conditioner refills, boat warning horns, drain cleaners, spray confetti, and film-negative cleaners to avoid purchasing products that contain CFCs.
5. When building a home, avoid CFC foam insulation; use fiberglass or cellulose insulation instead.

In addition, you can support (1) the establishment of CFC recycling or reclamation centers and (2) the mandatory recovery of CFCs from discarded refrigerators and automobile air conditioners in your community.

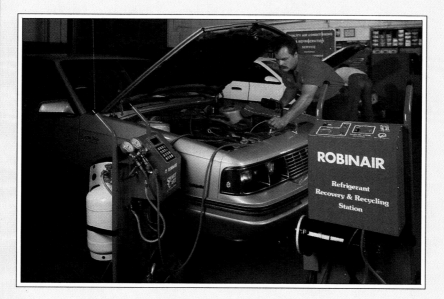

The "vampire" machine sucks air-conditioning coolant out of automobile air conditioners when they are being serviced. (Courtesy of Robinair Division, SPX Corporation)

## CFCs and the Electronics Industry

Despite terrible damage to the Earth's protective ozone layer caused by chlorine-based chemicals such as chlorofluorocarbons (CFCs), many electronics companies fought against the elimination of the chemicals in the 1980s. They insisted there was no effective replacement for these chemicals in the cleaning of solder flux from circuit boards. Interestingly, by 1991, faced with a legally required phaseout of CFCs, the same companies had adopted a number of harmless and inexpensive alternatives.

Consider the savings in CFCs seen at IBM's huge computer disk-drive factory in San Jose, California. In 1987, it was emitting more CFC-113 than any other single source in the United States. But by 1991, its emissions of CFC-113 were down by 95 percent from the 1987 level because IBM, rather than cleaning circuit boards with CFC-113, was dunking them in soapy water, shaking them, and blow-drying them with hot air.

## The Effects of Ozone Depletion

Ozone molecules in the stratosphere absorb incoming solar ultraviolet radiation. With depletion of the ozone layer, higher levels of ultraviolet radiation reach the surface of the Earth. Excessive exposure to ultraviolet radiation is linked to a number of health problems in humans, including eye cataracts, skin cancer, and weakened immunity. There is also substantial scientific documentation of crop damage from high levels of ultraviolet radiation.

Scientists are concerned that the ozone hole over Antarctica could possibly damage the plankton that forms the base of the food chain in the southern oceans. A 1992 study confirmed that increased ultraviolet radiation is penetrating the sur-face waters around Antarctica and that the productivity of Antarctic phytoplankton has declined by at least 6 to 12 percent as a result of increased exposure to ultraviolet radiation. If the phytoplankton continue to decline, the food chain of Antarctica, which includes fish, seals, penguins, whales, and vast populations of birds, will collapse.

## Protecting the Ozone Layer

In 1978 the United States, the world's largest user of CFCs, banned the use of CFC propellants in products such as antiperspirants and hair sprays. Although this ban was a step in the right direction, it did not solve the problem. Most nations did not follow suit, and besides, propellants represent only the tip of the iceberg in terms of CFC use.

In 1987, representatives from a number of countries met in Montreal to sign the Montreal Protocol, an agreement to significantly reduce CFC production by 50 percent by 1998. Then, in 1990, more than 90 countries signed an agreement to phase out all use of CFCs by 2000. Despite these agreements, the environmental news about CFCs has continually worsened since 1990. In 1992, for example, scientists reported that decreases in stratospheric ozone are occurring over the heavily populated mid-latitudes of the Northern Hemisphere in all seasons. As a result, some nations have taken even stricter measures to limit CFC production. CFCs are now scheduled to be completely phased out in the United States by the end of 1995.[3] Industrial companies that manufacture CFCs are quickly developing substitutes.

Unfortunately, CFCs are extremely stable, and those being used today will continue to deplete stratospheric ozone well into the 21st century. Another complication of the ozone problem is that other ozone-destroying compounds, including methyl chloroform and carbon tetrachloride, are currently not covered by the international agreement. It will be very difficult to lower chlorine levels in the stratosphere if these chemicals are not controlled in addition to CFCs.

[3] "Limited exceptions," which allow continued production of CFCs for existing equipment such as refrigerators and air conditioners, will be granted after the complete phaseout. The chemical industry estimates that 15 percent of the 1986 level of CFC production will be needed for existing equipment after the 1995 deadline.

# SUMMARY

**1.** Acid deposition, commonly called acid rain, is a serious regional problem caused by sulfur and nitrogen oxides. Acid deposition kills aquatic organisms and may contribute to forest decline by changing soil chemistry. Acid deposition also attacks materials such as metals and stone.

**2.** Carbon dioxide pollution from the combustion of fossil fuels causes the air to retain heat (infrared radiation), which warms the Earth. The increase in carbon dioxide and other greenhouse gases (methane, nitrous oxide, CFCs, and tropospheric ozone) in the atmosphere is causing concerns about possible major climate changes during the next century.

**3.** Global warming could cause a rise in sea level, changes in precipitation patterns, death of forests, extinction of animals and plants, and problems for agriculture. It could result in the displacement of thousands or even millions of people, thereby increasing international tensions.

**4.** The challenge of global warming can be met by prevention (stop polluting the air with greenhouse gases), mitigation (slow down the rate of global warming), and adaptation (make adjustments to live with global warming).

**5.** The ozone layer in the stratosphere helps to shield the Earth from harmful ultraviolet radiation. Plants and other organisms are damaged by exposure to increased ultraviolet radiation. In humans, excessive exposure to ultraviolet radiation causes cataracts, weakened immunity, and skin cancer.

**6.** The total amount of ozone in the stratosphere is slowly declining, and large ozone holes develop over Antarctica and the Arctic each year. The attack on the ozone layer is caused by chlorofluorocarbons (CFCs) and similar chlorine-containing compounds.

# DISCUSSION QUESTIONS

**1.** The wisest way to "use" fossil fuels might be to leave them in the ground. How would this affect air pollution? Global warming? Energy supplies?

**2.** Discuss some of the possible causes of tree damage—both decline and death.

**3.** Why might developing countries be more reluctant than developed countries to curtail their use of fossil fuels in the interest of solving the global climate problem?

**4.** Discuss and give examples of the three approaches to global warming—prevention, mitigation, and adaptation.

**5.** The following statement was overheard in an elevator: "CFCs cannot be the cause of stratospheric ozone depletion over Antarctica because there are no refrigerators in Antarctica." Criticize the reasoning behind this statement.

**6.** Distinguish between the benefits of the ozone layer in the stratosphere and the harmful effects of ozone at ground level.

**7.** Why will adaptation to global warming be easier for developed nations than for developing nations?

**8.** Distinguish between global warming and stratospheric ozone depletion.

# SUGGESTING READINGS

Bazzaz, F. A., and E. D. Fajer. Plant life in a CO$_2$-rich world. *Scientific American*, January 1992. How plants will be affected by increased CO$_2$ in the atmosphere.

Can we repair the sky? *Consumer Reports*, May 1989. A look at the gases that attack the Earth's ozone shield; includes what the government, industries, and individual consumers can do.

Fellman, B. Sacrificial lakes. *International Wildlife*, July/August 1990. A lake in western Ontario has been deliberately acidified to study the effects of and recovery from acid deposition; this research may help save other lakes from acidification.

Gleick, P. H. Climate change and international politics: Problems facing developing countries. *Ambio* 18:6, 1989. Some of the worst effects of global climate change will fall on developing nations. This issue also contains several other articles on climate change.

Graedel, T. E., and P. J. Crutzen. The changing atmosphere. *Scientific American*, September 1989.

How trace gases in the atmosphere are altered by human activities and what their short- and long-term effects are. This issue also includes an article on global climate change.

Jones, P. D., and T. M. L. Wigley. Global warming trends. *Scientific American*, August 1990. An evaluation of global temperatures and projections for global warming.

Magnuson, J. J., and J. A. Drury. Global change ecology. *The World and I*, April 1991, 304–311. Scientists from many disciplines and many countries are working together to unravel the effects of human activities on the global environment.

Mathews, J. T., ed. *Preserving the Global Environment: The Challenge of Shared Leadership*. W. W. Norton & Company, New York, 1991. Examines environmental problems such as ozone depletion and global climate change from different perspectives—those of U.S. strategies, international cooperation, and economic policies.

Mungall, C., and D. J. McLaren, eds. *Planet Under Stress: The Challenge of Global Change*. Oxford University Press, Toronto, 1990. A comprehensive evaluation of global climate change and other environmental problems. Published for the Royal Society of Canada.

Silver, C. S., and R. S. DeFries. *One Earth, One Future: Our Changing Global Environment*. National Academy Press, Washington, D.C., 1990. Examines global climate change, ozone depletion, and acid deposition.

# Chapter 21

*The wastewater treatment system for Arcata, California, provides a habitat for wildlife and is a popular area for hiking. (Ted Streshinsky/Photo 20–20).*

# Water and Soil Pollution

**W**ater pollution comes in many different types: sediments, sewage, disease-causing agents, inorganic plant nutrients, organic compounds, inorganic chemicals, radioactive substances, and thermal pollution. Although water pollution has both natural and human causes, pollution caused by human activities is generally more widespread. The benefits of effective water pollution control include improved human health, greater recreational value, and—most important—sustainable aquatic ecosystems. Although the United States has made visible progress in cleaning up its water during the past 20 years, much remains to be done. In some areas, water quality has actually deteriorated, whereas in other areas, strong cleanup efforts have allowed us only to hold our ground, not make any gains.

## NONTRADITIONAL WATER TREATMENT

Fifteen thousand people inhabit the town of Arcata on the coast of northern California. Like many small towns, Arcata has an annual festival that attracts visitors to the city. Arcata's gala is very unusual, however. It is called the Flush with Pride Festival, and it celebrates the town's unique water treatment system.

Arcata's treatment of wastewater initially follows the steps used in most municipalities. The solid contaminants are allowed to settle out, and then the dissolved organic wastes are biologically degraded. However, conventional treatment does not remove other pollutants, such as nitrogen and phosphorus, because it is too expensive; such pollutants are usually left in the treated wastewater. Unfortunately, when treated wastewater is discharged into rivers, streams, or oceans, these contaminants sometimes cause problems.

Arcata developed a way to remove such contaminants from treated water for a fraction of the cost of a normal advanced treatment plant and, at the same time, increase the amount of ecologically important wetlands in the town's vicinity. Essentially, since 1986 Arcata has used cattails, bulrushes, and other marsh plants to remove the contaminants. The town hired biologists, who worked with city engineers to create a wastewater refuge that occupies about 60.7 hectares (150 acres). After water spends some time being purified in the marshes, it is pumped to a treatment center, where it is chlorinated (to kill bacteria), dechlorinated (to remove the chlorine), and finally released into nearby Humboldt Bay. Many forms of wildlife have moved into the wetlands, including fish, muskrats, otters, ducks, seabirds, osprey, and falcons.

Arcata is not the only town that uses wetlands to filter municipal wastewater. In 1987 Orlando, Florida, restored a 486-hectare (1,200-acre) wetland that had been drained and used as a cow pasture since the late 1800s. This wetland now removes phosphorus and nitrate contaminants from 49 million liters (13 million gallons) of treated city wastewater each day. The success of the Orlando project has led Florida state officials to propose a similar project to reverse the decline in water quality in the Everglades National Park. Treated wastewater from densely populated southern Florida will be cleansed as it meanders through 14,200 hectares (35,000 acres) of newly restored wetlands.

As we saw in Chapter 13, water is required by all living things for their survival. However, having water of good *quality* is just as important as having *enough* water. This chapter discusses some of the pollutants found in water and how we can improve water quality.

## TYPES OF WATER POLLUTION

Water pollutants are divided into eight categories.

1. **Sediment pollution** is caused by soil particles that enter the water as a result of erosion.
2. **Sewage** is wastewater carried off by drains or sewers (from toilets, washing machines, and showers) and includes water that contains human wastes, soaps, and detergents.
3. **Disease-causing agents,** such as the organisms that cause typhoid and cholera, may be present in water; they come from the wastes of infected individuals.
4. **Inorganic plant nutrients** such as nitrogen and phosphorus come from animal wastes and plant residues as well as fertilizer runoff.
5. **Organic compounds,** most of which are synthetic, are often toxic to aquatic organisms. Because many of them have unusual structures that are difficult for microorganisms to degrade, these compounds persist in the environment for a long time.
6. **Inorganic chemicals** include such contaminants as mercury compounds, road salt, and acid drainage from mines.
7. **Radioactive substances** include the wastes from the mining and refinement of radioactive metals as well as the pollution caused by their use.
8. **Thermal pollution** occurs when heated water, produced during many industrial processes, is released into waterways. Thermal pollution affects a wide variety of wildlife, not just aquatic organisms.

### Sediment Problems in the Aquatic Environment

Sediments are particles suspended in a body of water that eventually settle out and accumulate on the bottom. Sediment pollution comes from agricultural lands, forest soils exposed by logging, overgrazed rangelands, strip mines, and construction. Control of soil erosion (discussed in detail in Chapter 14) reduces sediment pollution in waterways. Sediment pollution causes problems by reducing

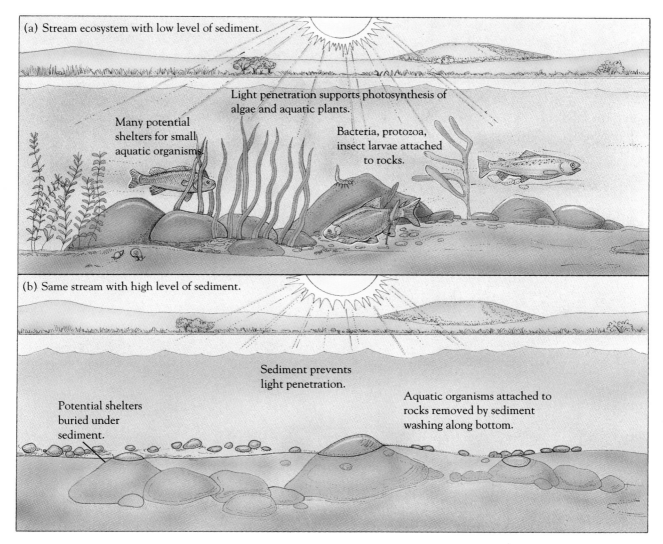

(a) Stream ecosystem with low level of sediment.

Light penetration supports photosynthesis of algae and aquatic plants.

Many potential shelters for small aquatic organisms.

Bacteria, protozoa, insect larvae attached to rocks.

(b) Same stream with high level of sediment.

Sediment prevents light penetration.

Potential shelters buried under sediment.

Aquatic organisms attached to rocks removed by sediment washing along bottom.

**Figure 21–1**
The effects of sediment pollution on a stream ecosystem. (a) A stream without sediment pollution can support diverse aquatic life. (b) How the same stream would appear after prolonged exposure to heavy sediment pollution.

light penetration, covering aquatic organisms, bringing insoluble toxic pollutants into the water, and filling waterways.

When sediment particles are suspended in the water, they make the water turbid (cloudy), which in turn decreases the distance that light can penetrate through the water. Because the base of the food chain in an aquatic ecosystem is photosynthetic algae and plants and because light is a requirement of photosynthesis, turbid water lessens the ability of producers to photosynthesize. Extreme turbidity also reduces the *number* of photosynthesizing organisms, which in turn causes a decrease in the number of aquatic organisms that feed

on the primary producers (Figure 21–1). When sediments build up to the point where they envelop coral reefs and shellfish beds, they can clog the gills and feeding structures of many aquatic animals.

Sediments adversely affect water quality by carrying toxic chemicals, both inorganic and organic, into the water. The sediment particles provide surface area to which some insoluble, toxic compounds adhere, so when sediments get into water, so do the toxic chemicals. Disease-causing agents can also be transported into water via sediments.

When sediments settle out of solution, they fill in waterways. This problem is particularly serious in reservoirs and in channels through which ships must pass. Thus, sediment pollution adversely affects the shipping industry.

## Water Pollution Due to Sewage

The release of sewage into water causes a number of pollution problems. First, because it carries disease-causing agents, water polluted with sewage poses a threat to public health (see the next section, on disease-causing agents). Sewage also creates two serious environmental problems in water, enrichment and oxygen demand. **Enrichment,** the fertilization of a body of water, is caused by the presence of high levels of plant nutrients such as nitrogen and phosphorus. Because these nutrients get into waterways not only from sewage but also from other sources, we will consider enrichment later in the chapter and confine this discussion to oxygen demand.

Sewage and other organic materials are decomposed into carbon dioxide, water, and similar inoffensive materials by the action of microorganisms.

**Figure 21–2**
The effect of raw sewage on the dissolved oxygen, biological oxygen demand, and living organisms in a stream. Note how the stream gradually recovers as the sewage is diluted and degraded. (Organisms not drawn to scale.)

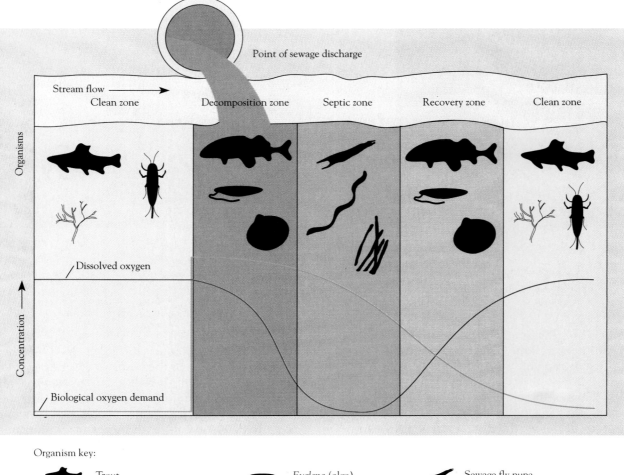

This degradation process, known as **cell respiration,** requires the presence of oxygen. Dissolved oxygen is also used by most organisms living in healthy aquatic ecosystems, including fish. But oxygen has a limited ability to dissolve in water, and when an aquatic ecosystem contains high levels of sewage or other organic material, the decomposing microorganisms use up most of the dissolved oxygen, leaving little for fish or other aquatic animals. At extremely low oxygen levels, fish begin to die off.

Sewage and other organic wastes are measured in terms of their **biological oxygen demand (BOD),** the amount of oxygen needed by microorganisms to decompose the wastes. A large amount of sewage in water creates a high BOD, which robs the water of dissolved oxygen (Figure 21–2). When dissolved oxygen levels are low, anaerobic (without-oxygen) microorganisms also produce compounds that have very unpleasant odors, further deteriorating water quality.

## Disease-Causing Agents

Municipal wastewater usually contains many bacteria, viruses, protozoa, parasitic worms, and other infectious agents that cause human or animal diseases (Table 21–1). Typhoid, cholera, bacterial dysentery, polio, and infectious hepatitis are some of the more common diseases caused by bacteria or viruses that are transmissible through contaminated food and water. However, many human diseases are not transmissible through water (the AIDS virus, for example).

**Monitoring for Sewage** Because sewage-contaminated water is a threat to public health, periodic tests are made for the presence of sewage in our water supplies. Although many different microorganisms thrive in sewage, the common intestinal bacterium *Escherichia coli* is typically used as an indication of the amount of sewage present in water and as an indirect measure of the presence of

**Table 21–1**
**Some Human Diseases Transmitted by Polluted Water**

| Disease | Infectious Agent | Type of Organism | Symptoms |
|---|---|---|---|
| Cholera | *Vibrio cholerae* | Bacterium | Severe diarrhea, vomiting; fluid loss of as much as 20 quarts per day causes cramps and collapse |
| Dysentery | *Shigella dysenteriae* | Bacterium | Infection of the colon causes painful diarrhea with mucus and blood in the stools; abdominal pain |
| Enteritis | *Clostridium perfringens*, other bacteria | Bacterium | Inflammation of the small intestine causes general discomfort, loss of appetite, abdominal cramps, and diarrhea |
| Typhoid | *Salmonella typhi* | Bacterium | Early symptoms include headache, loss of energy, fever; later, a pink rash appears along with (sometimes) hemorrhaging in the intestines |
| Infectious hepatitis | Hepatitis virus A | Virus | Inflammation of liver causes jaundice, fever, headache, nausea, vomiting, severe loss of appetite; aching in the muscles occurs |
| Poliomyelitis | Poliovirus | Virus | Early symptoms include sore throat, fever, diarrhea, and aching in limbs and back; when infection spreads to spinal cord, paralysis and atrophy of muscles |
| Amoebic dysentery | *Entamoeba histolytica* | Amoeba | Infection of the colon causes painful diarrhea with mucus and blood in the stools; abdominal pain |
| Schistosomiasis | *Schistosoma* sp. | Fluke | Tropical disorder of the liver and bladder causes blood in urine, diarrhea, weakness, lack of energy, repeated attacks of abdominal pain |
| Ancylostomiasis | *Ancylostoma* sp. | Hookworm | Severe anemia, sometimes symptoms of bronchitis |

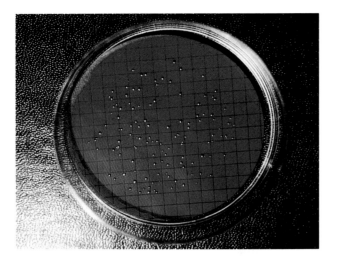

**Figure 21–3**
The fecal coliform test is used to indicate the likely presence of disease-causing agents in water. A water sample is first passed through a filtering apparatus. The filter disk is then placed on a medium that supports coliform bacteria for a period of 24 hours. After incubation, the number of bacterial colonies are counted. Each colony arose from a single coliform bacterium in the original water sample. (Visuals Unlimited/K. Talaro)

disease-causing agents. *E. coli* is perfect for monitoring sewage because it is not present in the environment except from human and animal feces, where it is found in large numbers.

To test for the presence of *E. coli* in water, the **fecal coliform test** is performed (Figure 21–3). A small sample of water is passed through a filter to trap all bacteria. The filter is then transferred to a petri dish that contains nutrients. After an incubation period, the number of greenish colonies present indicates the number of *E. coli*. Safe drinking water should contain no more than 1 coliform bacterium per 100 ml of water (about $\frac{1}{2}$ cup), safe swimming water should have no more than 200 per 100 ml of water, and general recreational water (for boating) should have no more than 2,000 per 100 ml. In contrast, raw sewage may contain several million coliform bacteria per 100 ml of water. Although the coliform bacteria themselves do not cause disease, the fecal coliform test is a reliable way to indicate the likely presence of pathogens, or disease-causing agents, in water.

## Inorganic Plant Nutrients

Fertilizer runoff from agricultural and residential land is a major contributor of inorganic plant nutrients such as nitrogen and phosphorus to water, where they encourage excessive growth of algae and aquatic plants. Although algae and aquatic plants are the base of the food chain in aquatic ecosystems, their excessive growth disrupts the natural balance between producers and consumers and causes other problems, including enrichment, bad odors, and a high biological oxygen demand. The high BOD occurs when the excessive numbers of algae die and are decomposed by bacteria.

High nitrate levels caused by inorganic fertilizers are also dangerous in drinking water, particularly for infants and small children (see Focus On: Is It Safe To Drink the Water?).

## Organic Compounds

Most of the thousands of organic (carbon-containing) compounds found in water are synthetic chemicals that are created for human activities; these include pesticides, solvents, industrial chemicals, and plastics. (Several examples of organic compounds sometimes found in polluted water are given in Table 21–2.) Some organic compounds find their way into surface water and groundwater by seeping from landfills, whereas others, such as pesticides, leach downward through the soil into groundwater or get into surface water by runoff from farms and residences. Many organic compounds are dumped directly into waterways by industries.

Many synthetic organic compounds are toxic (see Chapter 23). Although very few have been tested extensively, some have been shown to cause cancer or birth defects, as well as a variety of other health disorders, in laboratory animals. The long-term health effects of drinking minute amounts of these substances are unknown.

There are several ways to control the presence of organic compounds in our water. Care should be taken by everyone, from individual homeowners to large factories, to prevent organic compounds from ever finding their way into water. Alternative organic compounds, which are less toxic and degrade more readily so that they are not as persistent in the environment, can be developed and used. Also, tertiary water treatment, considered later in this chapter, effectively eliminates many organic compounds in water.

## Inorganic Chemicals

A large number of inorganic chemicals find their way into both surface water and groundwater from sources such as industrial plants, mines, irrigation runoff, oil drilling, and municipal storm runoff. Some of these inorganic pollutants are toxic to aquatic organisms. Their presence may make water unsuitable for drinking or other purposes.

**Table 21–2**
**Some Synthetic Organic Compounds Found in Polluted Water**

| Compound | Some Reported Health Effects |
| --- | --- |
| Aldicarb (pesticide) | Attacks nervous system |
| Benzene (solvent) | Blood disorders, leukemia |
| Carbon tetrachloride (solvent) | Cancer, liver damage; may also attack kidneys and vision |
| Chloroform (solvent) | Cancer |
| Dioxin (TCDD) (chemical contaminant) | Cancer, birth defects |
| Ethylene dibromide (EDB) (fumigant) | Cancer; attacks liver and kidneys |
| Polychlorinated biphenyls (PCBs) (industrial chemical) | Attacks liver and kidneys; may cause cancer |
| Trichloroethylene (TCE) (solvent) | Induces liver cancer in mice |
| Vinyl chloride (plastics industry) | Cancer |

## Envirobrief

### Powerful Poisons

One of the more insidious forms of soil and water pollution comprises toxic elements and compounds that do not break down, but accumulate over time. Mercury and cadmium, two heavy metals that are toxic to humans, behave this way. Dry cell batteries, like those that power portable stereos and flashlights, are by far the greatest source of environmental mercury and cadmium. More than 2.5 billion dry cell batteries are purchased in the United States each year. Although they account for only .005 percent of the U.S. waste stream by weight, they contribute more than 50 percent of the mercury and cadmium found in trash. Approximately 14 million pounds of mercury enter U.S. landfills each year, mostly in batteries. Once released into the environment through incineration or as landfill leachate, mercury and cadmium accumulate in soil, water, plants, and animals.

Technically, batteries are recyclable. But the associated costs and complications have discouraged a recycling market. Battery companies have reduced the amount of mercury in their products since 1990, but they have yet to remove the toxic metals altogether. If they do not do so voluntarily by the mid-1990s, they may be forced to do so by law.

Here we consider three inorganic chemicals that sometimes contaminate water: mercury, road salt, and acid drainage from mines. Mercury is used in a variety of industrial processes. When the industrial plants release their wastewater, some metallic mercury may enter natural bodies of water along with the wastewater. The combustion of coal releases small amounts of mercury into the air; this mercury then moves from the atmosphere to the water via precipitation. Mercury sometimes enters water by precipitation after household trash containing batteries, paints, and plastics has been burned in solid waste incinerators (see Chapter 23). Once in a body of water, mercury settles into the sediments and is converted to methyl mercury compounds, which readily enter the food chain (Figure 21–4). Methyl mercury compounds remain in the environment for a long time and are highly toxic to living organisms, including humans. Prolonged exposure to methyl mercury compounds causes kidney failure and mental retardation (in children). Mercury accumulation in the body also severely damages the nervous system.

*(Text continues on p. 474.)*

**Figure 21–4**
Mercury pollution is dangerously high in portions of the Everglades. (John Shaw/Tom Stack and Associates)

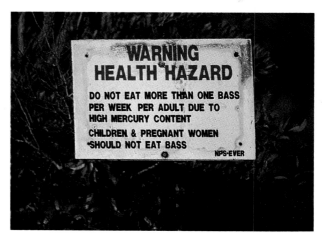

## Is It Safe To Drink the Water?

People are becoming increasingly, and understandably, worried about the safety of drinking water. More than 1,000 contaminants have been identified in public water supplies. Also, although municipal water supplies are monitored, what comes out of your tap at home is not, meaning that there is no way of knowing what contaminants are picked up during water transport and storage. People worry about their drinking water when it looks dirty, tastes bad, or smells funny, but the contaminants that you cannot see, taste, or smell are potentially more dangerous. Water is often discolored by iron or manganese, which also affects its taste and odor. However, iron and manganese are not known to have any bad effects on human health. The most common water pollutants that *are* dangerous are lead, radon, and nitrate, which don't affect the water's color or smell.

Some people think they're avoiding water contaminants by using bottled water for drinking and cooking. Although this may be true, there is no way to know for sure unless you have the bottled water tested. Some bottled waters contain the same contaminants as tap water.

We hear the worst news about our water supply from sensationally written news reports and from the water treatment industry. Both of these groups have good reason to use scare tactics—to sell newspapers and to sell water treatment supplies. Every person should be concerned about the quality of drinking water—concerned enough to become an informed consumer. But our concern should not become paranoia. Our drinking water is generally pure and safe enough to consume.

**Lead**   Millions of Americans, many of them children, have damaging levels of lead in their bodies. Extremely high amounts of lead in the body can result in convulsions, kidney disease, hallucinations, and other problems. Low-level lead poisoning generally doesn't cause such extreme symptoms (which makes it more difficult to detect), but it can cause irritability, anemia, and blood enzyme changes.

The three groups of people at greatest risk from lead poisoning are middle-aged men, pregnant women, and young children. Middle-aged men with high levels of lead are more likely to develop **hypertension,** or high blood pressure. High lead levels in pregnant women increase the risk of miscarriages, premature deliveries, and stillbirths. Children with even low levels of lead in their blood may suffer from a variety of mental and physical impairments, including partial hearing loss, hyperactivity, lowered IQ, and learning disabilities.

People used to think of lead poisoning as being confined to inner-city children who ate paint chips that contained lead. But lead lurks in many more places in the environment than simply paint. Although low amounts of lead originate in natural sources such as volcanoes and wind-blown dust, most of the lead contaminating our world can be traced to human activities. Anti-knock agents in gasoline are lead additives that are released into the atmosphere when the fuel is burned. Lead contaminates the soil, surface water, and groundwater when incinerator ash is dumped into ordinary sanitary landfills (see Chapter 23). It may be spewed into the atmosphere from old factories that lack air pollution control devices. We ingest additional amounts of lead—from pesticide and fertilizer residues on produce, from food cans that are soldered with lead, and even from certain types of dinnerware on which our food is served.

Tap water often contains high levels of lead that is usually from the corrosion of old lead water pipes, and lead solder in newer pipes, in buildings rather than from the municipal water supply. Lead piping is slowly being replaced, but in the meantime, it is relatively easy to reduce the amount of lead you are exposed to in water. Simply avoid using hot tap water for cooking or drinking, and let the water run in your tap for about 1 or 2 minutes each morning to flush out the lead that accumulated in the water during the night. To determine whether you have a problem with lead in your water, you can have a certified laboratory test.

You can take additional steps to decrease your exposure to lead and reduce its impact on your health. You can test other possible sources of lead pollution in your home, including the glaze on imported dishes and cookware. If you are in one of the high-risk groups, you should have the lead level of your blood evaluated when you have routine blood tests. To reduce the impact of lead on your health, make sure you eat a balanced diet that is low in saturated fat and contains adequate amounts of minerals and protein. Studies have shown that lead poisoning is less likely when you eat such a diet.

**Radon**   Radon is a naturally occurring radioactive gas. It is produced in the Earth's crust and increases the risk of lung cancer when it is inhaled over long periods of time (see Chapter 19). It is also possible for radon to enter groundwater. When you shower, wash dishes, or wash clothes with radon-contaminated water, the radon gets into the air in your home. In addition, there is concern about long-term exposure to radon in drinking water. Don't worry about testing for radon in your water until you first test for indoor levels of radon in the air, however. If that test indicates a high level of radon, and if your drinking water is groundwater (that is, if it comes from a well and not from a reservoir), then you should have your water tested for radon. Radon can be removed from groundwater by aerating the water and then venting the gas outside or by running the water through an activated charcoal filter.

**Nitrates**   Nitrate contamination comes mostly from agricultural sources (fertilizers and animal wastes), although some comes from septic tanks. Because nitrates leach readily through the soil, nitrate water pollution is mainly a problem in groundwater. If you live in a rural area, you should have your well water tested periodically for nitrates. The health effects of nitrates are discussed in Chapter 18.

**The Fluoride Debate**   The addition of small amounts of fluoride to most municipal drinking water has been practiced since the mid-1940s to help prevent tooth decay. Fluoride has also been added to many toothpastes since that time, for the same reason. This practice has been controversial, with opponents questioning the safety and effectiveness of fluoride and supporters saying it is completely safe and very effective in preventing decay.

Most dental health officials think fluoride is the main reason for the 50 percent decrease in tooth decay observed in children during the past 20 years. Recently, however, the role of fluoride in this impressive decrease has once again been questioned, because the benefits of improved nutrition and dental hygiene may have been partly responsible for the decrease in tooth decay. It is difficult to design a definitive study on the effectiveness of fluoride, because almost all children in the United States have fluoride in their drinking water and/or in their toothpaste. Finding a control group that has never used or come into contact with fluoride is therefore extremely difficult.

At the request of the Environmental Protection Agency, the U.S. National Academy of Sciences is currently conducting a review of fluoride. Once the data are fully evaluated, conclusions may warrant reconsidering the addition of fluoride to drinking water and toothpastes.

A large quantity of road salt (sodium chloride and calcium chloride) is applied in northern states during winter to treat and prevent icy, slippery road conditions. Besides damaging roads and corroding cars, these salts harm plants growing along roadsides. Precipitation washes road salt into waterways and groundwater, adversely affecting freshwater aquatic life and the quality of drinking water.

Coal mines expose sulfur-containing compounds such as iron pyrites ($FeS_2$) to water and air. This exposure results in the production of sulfuric acid ($H_2SO_4$), which then drains from the mine into waterways and contributes to the acid deposition problem in lakes and streams (see Chapter 20). Acid drainage from mines is most acute in states east of the Mississippi.

### Radioactive Substances

Radioactive materials get into water from several sources, including the mining and processing of radioactive minerals such as uranium and thorium. Many industries use radioactive substances; although nuclear power plants and the nuclear weapons industry make use of the largest amounts, medical and scientific research facilities also utilize them.

It is possible for radiation to inadvertently escape from any of these facilities, polluting the air, water, and soil. Accidents at nuclear power plants can release into the atmosphere large quantities of radiation, which eventually contaminates soil and water. Radiation from natural sources can also pollute groundwater. The health effects of exposure to radiation are discussed in Chapter 11.

### Thermal Pollution

A number of industries, including steam-generated electric power plants, use water to remove excess heat from their operations. Afterward, the heated water is allowed to cool a little before it is returned to waterways, but its temperature is still higher than it was originally (Figure 21–5). The result is that the waterway is warmed slightly.

A rise in temperature of a body of water has a number of chemical, physical, and biological effects. Chemical reactions, including decomposition

**Figure 21–5**

The emission of thermal pollution from an electric power plant affects the dissolved oxygen level as well as the temperature of the water downstream.

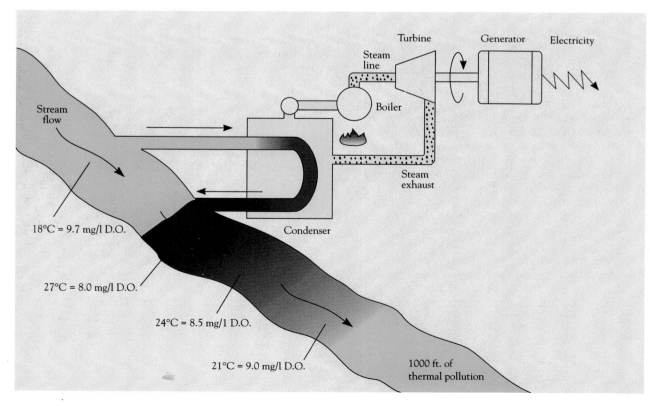

of wastes, occur faster, depleting the water of oxygen. Moreover, less oxygen dissolves in warm water than in cool water, and the amount of oxygen dissolved in water has important effects on aquatic life. There may be subtle changes in the activities and behavior of living organisms in thermally polluted water, because temperature affects reproductive cycles, digestion rates, and respiration rates. For example, at higher temperatures fish require more food to maintain body weight; they also typically have shorter life spans and smaller populations. In cases of extreme thermal pollution, fish and other aquatic organisms die.

## SOURCES OF WATER POLLUTION

Water pollutants are both natural and anthropogenic (from humans). For example, mercury from natural sources in the Earth's crust and oceans contaminates the biosphere to a far greater extent than does mercury emitted by human activities. Nitrate pollution has both natural and human sources—the nitrate that occurs in soil and the inorganic fertilizers that are added to it, respectively. Although natural sources of pollution are sometimes of local concern, pollution caused by human activities is generally more widespread. The four major sources of human-induced water pollution are industry, municipalities (that is, domestic activities), shipping, and agriculture.

### Point Source Pollution and Nonpoint Source Pollution

The sources of water pollution are classified into two types, point source pollution and nonpoint source pollution (Figure 21–6). **Point source pollution** is discharged into the environment through pipes, sewers, or ditches from specific sites such as factories or sewage treatment plants. **Nonpoint source pollution** is caused by land pollutants that enter bodies of water over large areas rather than at a single point. It includes agricultural runoff, mining wastes, urban wastes, and construction sediments. Soil erosion is a major cause of nonpoint source pollution.

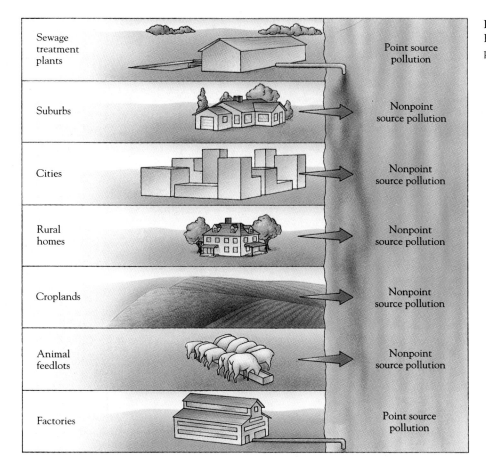

**Figure 21–6**
Examples of point and nonpoint source pollution.

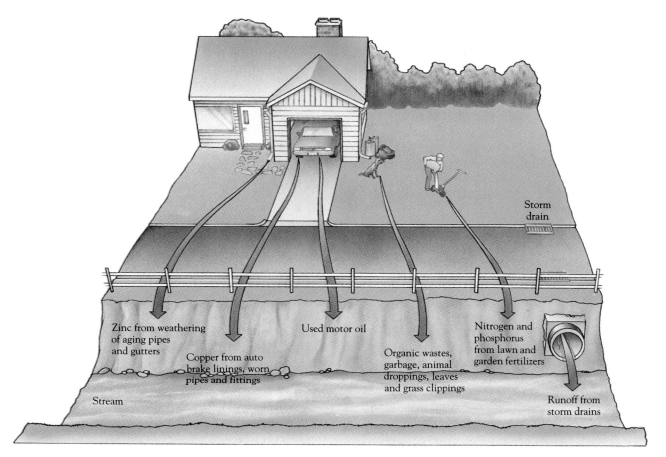

**Figure 21–7**
Everyday activities can result in polluted storm runoff water. The largest single pollutant is organic waste, which, when it decays, removes dissolved oxygen from water. Fertilizers cause excessive algal growth, which further depletes the water of oxygen, harming aquatic organisms. Other everyday pollutants include used motor oil (often poured into storm drains) and heavy metals. These pollutants are carried from storm drains on streets to streams and rivers.

## Industrial Wastes in Water

Different industries generate different types of pollutants. Paper mills, for example, produce toxic compounds and sludge (a slimy solid mixture), whereas food processing industries produce organic wastes that are readily decomposed but have a high biological oxygen demand.

Many industries in the United States treat their wastewater with advanced treatment methods. The electronics industry, for example, produces wastewater containing high levels of heavy metals such as copper, lead, and manganese, but uses special techniques—ion exchange and electrolytic recovery—to reclaim those heavy metals. Plates with commercial value are produced from the recovered metals—which would otherwise

have become a component of hazardous toxic sludge. Although U.S. industries do not usually dump very toxic wastes into water, disposal is still sometimes a problem (see Chapter 23).

## Municipal Waste Pollution

Although sewage is the main pollutant produced by cities and towns, municipal waste pollution also has a nonpoint source: storm runoff (Figure 21–7). The water quality of storm runoff is often worse than that of sewage. Storm runoff carries salt from roadways, untreated garbage, construction sediments, and traffic emissions (via rain that washes pollutants out of the air). It often contains such contaminants as asbestos, chlorides, copper, cyanides,

**Figure 21–8**
Water pollution in a Mediterranean harbor in Italy. Much of this pollution comes from shipping. (Visuals Unlimited/Frank M. Hanna)

hydrocarbons, lead, motor oil, organic wastes, phosphates, sulfuric acid, and zinc.

Municipal waste pollution from sewage is a greater problem in developing countries, many of which lack water treatment facilities, than in developed nations. Sewage from many densely populated cities in Asia, Latin America, and Africa is dumped directly into rivers or coastal harbors.

## Pollution from Shipping

Shipping introduces several types of pollution, including oil, plastic debris, and sewage, into oceans and rivers (Figure 21–8). Oil pollution is particularly acute in harbors and shipping lanes. Oil damages aquatic ecosystems in part because it quickly spreads over the surface of water, forming a film. It then coats marine birds, seals, and sea otters, destroying their natural insulation and their buoyancy, so that they can either die from loss of body heat or drown. When ingested, oil damages an animal's liver, kidneys, and lungs. Heavier, tar-like oil sinks, killing bottom-dwelling organisms such as shellfish.

## Water Pollution from Agriculture

Agriculture is a major source of nonpoint pollution throughout the world, and agricultural practices produce several types of pollutants that contribute to water pollution (Figure 21–9). Chemical pesticides that run off or leach into water are highly toxic and can adversely affect human health as well as the survival of aquatic organisms (see Chapter 22). Fertilizer runoff causes water enrichment. Animal wastes and plant residues in waterways produce high biological oxygen demand and a high suspended solids level as well as water enrichment.

**Figure 21–9**
Agriculture produces a number of water pollutants. (a) The runoff from fields where cattle graze (shown) or feedlots contributes animal wastes to water. (b) Liquid fertilizer is applied to newly seeded ground. Fertilizers and pesticides not only contaminate surface waters but also leach into groundwater. (a, USDA; b, Grant Heilman, from Grant Heilman Photography)

(a)

(b)

**Figure 21–10**
When irrigated land becomes salinized, its agricultural value is reduced or eliminated. (USDA/
Soil Conservation Service)

Soil erosion from fields and rangelands causes sediment pollution in waterways. In addition, some agricultural chemicals that are not very soluble in water (for example, certain pesticides) find their way into waterways by adhering to sediment particles. Thus, soil conservation methods not only conserve the soil but reduce water pollution.

**Irrigation and Salinization of the Soil**    Soils found in arid and semiarid regions often contain high concentrations of mineral salts. In these areas, the amount of water that drains into lower soil layers is minimal because the little precipitation that falls quickly evaporates, leaving behind the salt. In contrast, humid climates have enough precipitation to leach salts out of the soils and into waterways and groundwater.

Irrigation of agricultural fields often results in their becoming increasingly salty (see Chapters 1 and 13, and Figure 21–10). Also, when irrigated soil becomes waterlogged, salts may be carried by capillary movement from groundwater to the soil surface, where they are deposited as a crust of salt.

**Effects of Soil Salinization**    Most plants cannot obtain all the water they need from salty soil. The cause is a water balance problem that exists because water always moves from an area of higher concentration (of water) to an area of lower concentration.[1] Under normal conditions, the dissolved materials in plant cells give them a lower concentration of water than that in soil. As a result, water moves from the soil into plant roots. When soil water contains a large quantity of dissolved salts, however, its concentration of water may be lower than that in the plant cells; if so, water consequently moves *out* of the plant roots and into the salty soil (Figure 21–11).

Obviously, most plants cannot survive under these conditions. Plant species that thrive in saline soils have special adaptations that enable them to tolerate the high amount of salt. Most crops, unless they have been genetically selected to tolerate high salt, are not productive in saline soil (Figure 21–12).

---

[1] The concentration of water is determined by the amount of dissolved materials in it. For example, a solution containing 10 percent dissolved salt has 90 percent water. Pure water is 100 percent water. Because water obeys the law of diffusion, it moves from an area of higher concentration to an area of lower concentration. Water moves into an area that contains a higher amount of dissolved salt because that area's relative concentration of *water* is lower.

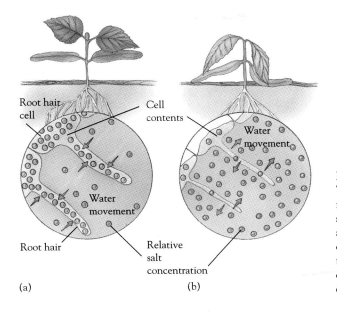

**Figure 21–11**
The effect of salinized soil on water absorption by the roots of plants. (a) Normally, the water concentration inside plant cells is lower than that in the soil, resulting in a net movement of water into root cells. (b) When soil contains a high amount of salt, its relative water concentration can be lower than the water concentration inside cells. This causes water to move *out* of roots into the soil, even when the soil is wet.

**Correcting Saline Soils** In principle, the way to remove excess salt from saline soils is to add enough water to leach away salt. Although this sounds like a straightforward process, it is extremely difficult, and in many cases impossible, to accomplish. For one thing, saline soils usually occur in arid or semi-arid lands where water is in short supply. Also, many soils don't have good drainage properties, so adding lots of water simply causes them to become waterlogged. Another factor that should not be overlooked is that even if the salt is flushed out of

the soil, it has to go somewhere. The excess salt is usually carried to groundwater or to rivers and streams, where it becomes a water pollutant.

## EUTROPHICATION: AN ENRICHMENT PROBLEM

Normal lakes that have minimal levels of nutrients are said to be unenriched, or **oligotrophic.** An oligotrophic lake has clear water and supports small populations of aquatic organisms. **Eutrophication** is the enrichment of water by nutrients; a lake that is enriched is said to be **eutrophic.** The water in a eutrophic lake is cloudy and usually resembles pea soup because of the presence of vast numbers of algae and cyanobacteria that are supported by the nutrients (Figure 21–13).

Although eutrophic lakes contain large populations of aquatic animals, different kinds of organisms predominate there than in oligotrophic lakes. For example, an unenriched lake may contain pike, sturgeon, and whitefish; all three are found in the deeper, colder part of the lake, where there is a higher concentration of dissolved oxygen. In eutrophic lakes, on the other hand, the deeper, colder levels of water are depleted of dissolved oxygen because of the high biological oxygen demand caused by decomposition on the lake floor. Therefore, fish such as pike, sturgeon, and whitefish die out and are replaced by warm-water fish, such as

**Figure 21–12**
Wheat (shown) withers and dies in saline soil. Most crops grown in salty water are less productive than those grown in freshwater. (USDA/Soil Conservation Service)

(a)

(b)

**Figure 21–13**
The effect of enrichment on a lake or pond. (a) Crater Lake in Oregon, like other oligo-trophic lakes, is low in nutrients. (Rich Buzzelli/Tom Stack and Associates) (b) Eutrophic lakes and ponds, such as this one in western New York state, are often covered with slimy, smelly mats of algae and cyanobacteria. (Visuals Unlimited/W. A. Banaszewski)

catfish and carp, that can tolerate lesser amounts of dissolved oxygen (Figure 21–14).

Over vast periods of time, oligotrophic lakes and slow-moving streams and rivers become eutrophic naturally. As natural eutrophication occurs, these bodies of water are slowly enriched and grow shallower from the immense number of dead organ-isms that have settled in the sediments over a long period. Gradually, plants such as water lilies and cattails take root in the nutrient-rich sediments and begin to fill the shallow waters, forming a swamp or marsh.

Eutrophication can be markedly accelerated by human activities. This fast, human-induced process

**Figure 21–14**
A comparison of the features of oligotrophic and eutrophic lakes.

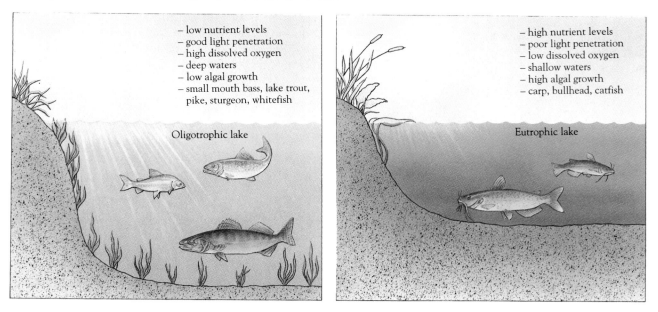

– low nutrient levels
– good light penetration
– high dissolved oxygen
– deep waters
– low algal growth
– small mouth bass, lake trout, pike, sturgeon, whitefish

Oligotrophic lake

– high nutrient levels
– poor light penetration
– low dissolved oxygen
– shallow waters
– high algal growth
– carp, bullhead, catfish

Eutrophic lake

## Envirobrief

### Sources of Phosphorus Pollution

Phosphorus pollution causes excessive algal growth, which robs the water of oxygen needed by fish and other aquatic life. In the process, it accelerates eutrophication.

Phosphorus gets into the water supply from several sources, including detergents that contain phosphates, fertilizers applied to fields and lawns, human and animal wastes, soil erosion, and phosphate mining.

Phosphates, which contain phosphorus, are added as a cleaning agent to detergents and household cleaners. Since the 1970s, phosphate detergents have been banned in many states, including Delaware, Georgia, Indiana, Maryland, Michigan, New York, Virginia, and Wisconsin. Other states limit the use of phosphates in detergents.

Although phosphate detergents have been banned in Virginia, the levels of phosphorus in streams in Fairfax County, Virginia, continue to increase. The rise has been attributed to the increased use of lawn fertilizers by homeowners in this suburban county near Washington, D.C.

Sources: *Consumer Reports*, February 1991; *Washington Post*, April 28, 1991.

## Envirobrief

### Using Wildlife To Detect Water Pollution

Scientists use a variety of wildlife, including shellfish, water birds, and fish, to detect and monitor water pollution.

The U.S. government uses mussels to monitor toxic compounds, both organic and inorganic, in the sediments of coastal waters. The mussels are harvested from the water, and their tissues are analyzed for the presence of pesticides, heavy metals, and industrial chemicals.

The Canadian Wildlife Service monitors herring gull eggs, which accumulate toxic pollutants, around the Great Lakes. Comparisons of different areas in the Great Lakes region are possible because individual herring gulls tend to stay in one place, accumulating toxins from that area only.

A small tropical fish monitors water pollution in the Stour River, in southern England, around the clock. Native to muddy waters in Nigeria, the elephant fish emits 300 to 500 pulses of electricity per minute to help it "see" in its environment. Interestingly, when water pollution is present, the fish emits more than 1,000 pulses per minute. Scientists keep the little fish in individual tanks, through which warmed river water is pumped. If more than half of the fish suddenly increase their electrical emissions (which are monitored by computer), scientists are called in to test the water.

Sources: *Science* 249, August 31, 1990; *Environment* 32:4, May 1990; *Washington Post*, January 30, 1991.

is usually called **artificial eutrophication** to distinguish it from natural eutrophication, and it results from the enrichment of water by inorganic plant nutrients—most commonly in sewage and fertilizer runoff.

### Controlling Eutrophication

Water, sunlight, carbonate (dissolved carbon dioxide), nitrogen, phosphorus, and certain other inorganic plant nutrients are the main requirements for algal growth, which is limited by the essential material that is in shortest supply (recall the law of the minimum, discussed in Chapter 4). Because phosphorus is often the limiting factor in freshwater lakes, the most effective way to slow artificial eutrophication is usually to limit the amount of phosphorus entering the aquatic system. For example, when treated sewage was no longer dumped in Lake Washington (see Chapter 2), the phosphorus level in the lake declined by about 75 percent and there

was a corresponding drop in algal growth. Nitrogen (as nitrates) has also been implicated in eutrophication. In Sweden, for example, coastal water eutrophication of the Kattegat, the narrow waterway between Denmark and Sweden, has been linked to excessive nitrogen runoff from agriculture.

## GROUNDWATER POLLUTION

More than 50 percent of the people in the United States obtain their drinking water from groundwater, which is also withdrawn for irrigation and industry (see Chapter 13). In recent years attention has been drawn to the quality of the nation's groundwater, which can become contaminated in a number of ways. Hazardous substances, including radioactive compounds, pesticides, fertilizers, organic compounds, and heavy metals, can seep into groundwater from municipal sanitary landfills as well as from hazardous landfills (see Chapter 23). Industry, mining, and agricultural operations also produce hazardous substances that, if disposed of improperly, can eventually contaminate groundwater.

Contamination of groundwater is a relatively recent environmental concern because people used to think that the soil, through which surface water must seep in order to become groundwater, filtered out any contaminants, thereby ensuring the purity of groundwater. This assumption proved false when the quality of groundwater began to be monitored and contaminants were discovered at certain sites.

Currently, most of the groundwater supplies in the United States are of good quality, although there are some local problems that have led to well closures and raised public health concerns. Cleanup of polluted groundwater is very costly, takes years, and in some cases is not technically feasible. Compounding the cleanup problem is the challenge of safely disposing of the toxic materials removed from groundwater—which, if they are not handled properly, could contaminate groundwater once again.

### Protection Through Prediction

Groundwater in Florida is typically close to the surface—a mere 10 feet underground in places. Florida also has a large agricultural industry, and pesticides are used heavily to control many pests that thrive in the state's hot, humid climate. These two factors have caused an acute groundwater contamination problem in certain places.

In 1991, soil scientists at the University of Florida came up with two computer programs, CHEMRANK and CMLS, to fight the problem. CHEMRANK ranks agricultural chemicals according to their mobility and persistence in the environment. CMLS (Chemical Movement in Layered Soils) shows how soil properties, chemical characteristics, and weather factors affect chemical movement and persistence. By referring to CHEMRANK and CMLS, farmers should be able to determine which pesticides are best suited to water and soil conditions on their property and avoid those that cause long-term damage.

**Figure 21–15**
The water supply for a town can be stored in a reservoir. The water is treated, often by chlorination, before use. After use, the quality of the water is fully or partially restored by sewage treatment.

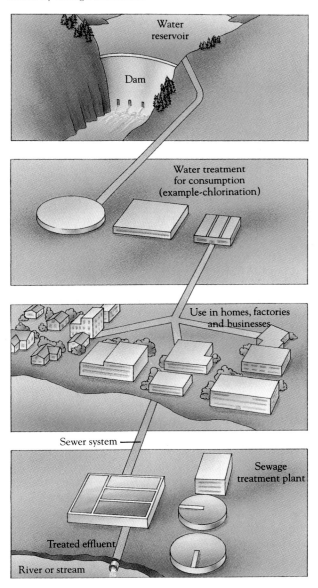

# IMPROVING WATER QUALITY

Water quality can be improved both by reducing the number of contaminants that go into the water supply and by cleaning up wastewater—water that already contains contaminants (Figure 21–15). Technology assists in both processes. In the United States, most municipal water supplies are treated before being used so that the water is safe to drink. For example, water is chlorinated to disinfect it, thereby preventing waterborne diseases. Also, sewage and other wastewater are usually treated to remove many contaminants before being discharged into rivers, lakes, and oceans.

Many governments have passed legislation aimed at controlling water pollution (see Focus On: Water Pollution Problems in Other Countries). Point source pollutants lend themselves to effective control more readily than nonpoint source pollutants. Also, as will be discussed in a later section, water pollution laws are generally difficult to monitor and enforce.

## Treating Wastewater

Wastewater, including sewage, usually undergoes several treatments at a sewage treatment plant to prevent environmental and public health problems. **Primary treatment** removes suspended and floating particles, such as sand and silt, by mechanical processes such as screening and gravitational settling (Figure 21–16). Primary treatment, however, does little to eliminate the inorganic and organic compounds that remain suspended in the wastewater.

**Secondary treatment** uses microorganisms to decompose the suspended organic material in wastewater. One of the several types of secondary treatment is trickling filters, in which wastewater trickles through aerated rock beds that contain bacteria and other microorganisms, which degrade (decompose) the organic material in the water. In another type of secondary treatment, water is aerated and circulated through bacteria-rich particles; the bacteria degrade suspended organic material. After several hours, the particles and microorganisms are allowed to settle out, forming **sewage sludge,** a slimy mixture of bacteria-laden solids. Water that has undergone primary and secondary treatment is clear and free of organic wastes such as sewage. Most wastewater treatment facilities in the United States perform both primary and secondary treatment (Figure 21–17).

Even after primary and secondary treatment, wastewater still contains pollutants; they include dissolved minerals, heavy metals, viruses, and synthetic organic compounds (Figure 21–18). Advanced wastewater treatment methods, also known as **tertiary treatment,** include a variety of biologi-

(*Text continues on p. 486.*)

**Figure 21–16**
Some municipalities use primary sewage treatment only.

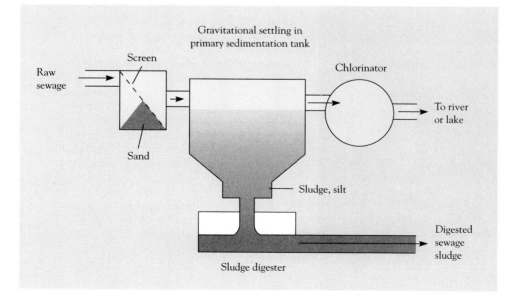

## Water Pollution Problems in Other Countries

Almost every nation in the world faces problems of water pollution. To give you an international perspective on water pollution, let's look at some specific issues in South America, Africa, Asia, and Europe.

**Rio de Janeiro, Brazil**   One of the most beautiful places on Earth is the Bay of Guanabara, the port of entry into Rio de Janeiro. Although it is beautiful from a distance, up close the bay is a disgusting, smelly pool of wastes. It suffers from the concentrated pollution of industrial and urban wastes and oil spills. Raw sewage from more than 6 million people is discharged directly into the water, and large numbers of chemical plants, shipyards, and refineries dump their untreated wastewater in the bay. Also, sediment from soil erosion has filled in much of the bay. Because pollution is generally viewed as an unavoidable aspect of industrialization in this rapidly developing country, the Brazilian government has shown little interest in cleaning it up.

The United Nations has sponsored the installation of hand-powered water pumps like this one built by a Kenyan women's group. (Rainer Drexel/Bilderberg)

Sailing on the Bay of Guanabara near Rio de Janiero. Although the water looks inviting from a distance, it is dangerously polluted. (Charlotte Kahler)

**Kwale, Kenya**   For many Africans, serious health problems are caused by surface water supplies that are contaminated with disease-causing organisms. A hand pump project sponsored by the World Bank and the United Nations Development Programme has solved the water safety problem for many Kenyans. For example, Kwale, Kenya, has been the site of cholera and diarrhea outbreaks in the past. Thanks to the installation of a village well with a hand pump, groundwater that is clean and healthful is now available to the inhabitants of Kwale.

**Ganges River, India**   The Ganges River symbolizes the spirituality and culture of the Indian people. It is also highly polluted. Little of the sewage and industrial waste produced by the 300 million people who live in the Ganges River basin is treated. Another major source of contamination is the 35,000 human bodies per year that are cremated in the open air in Varanasi, the holy city of the Hindus. Bodies that are incompletely burned are dumped into the Ganges River, where their decomposition adds to the biological oxygen demand of the river.

Bathing and washing clothes in the Ganges River is a common practice in India. The river is contaminated by raw sewage discharged directly into the river at a number of locations. (Mike Barlow/Dembinsky Photo Associates)

The Indian government has initiated an ambitious cleanup project that includes construction of water treatment plants in 27 large cities in the river basin. In addition, 32 electric crematoriums are being set up along the banks.

**Po River, Italy**　The Po River, which flows across northern Italy, empties into the Adriatic Sea. The Po is Italy's equivalent of the Mississippi River, and it is heavily polluted. Many cities, including Milan with 1.5 million residents, dump their untreated sewage into the Po, and Italian agriculture, which relies heavily on chemicals, is responsible for massive amounts of nonpoint source pollution.

Almost one-third of all Italians live in the Po River basin. The health of many Italians is threatened because the Po is the source of their drinking water. In addition, pollution from the Po has seriously jeopardized tourism and fishing in the Adriatic Sea. Although Italians recognize the problems of the Po and would like to do something about them, the improvement of water quality will be very difficult to implement because the river is under the jurisdiction of dozens of local and regional governments. The cleanup of the Po will require the implementation of a national plan over a period of several decades.

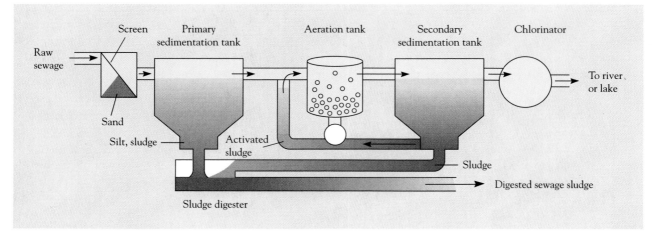

**Figure 21–17**
The steps involved in primary and secondary sewage treatment.

cal, chemical, and physical processes. Tertiary treatment must be employed to remove phosphorus and nitrogen, the nutrients most commonly associated with enrichment. Tertiary treatment can also be used to purify wastewater so that it can be reused in communities where water is scarce. Currently, tertiary treatment is not performed by most municipal water treatment plants in the United States.

**Disposal of Sewage Sludge**   A major problem associated with wastewater treatment is disposal of the

sewage sludge that is formed during primary and secondary treatment. Five possible ways to handle sewage sludge are anaerobic digestion, composting, incineration, disposal in a sanitary landfill, and ocean dumping. In anaerobic digestion, the sewage sludge is placed in large circular digesters and kept warm (about 95°F), which allows anaerobic bacteria to break down the organic material into gases such as methane and carbon dioxide; the methane can be cycled and burned to heat the digesters. After a few weeks of digestion, the sewage sludge

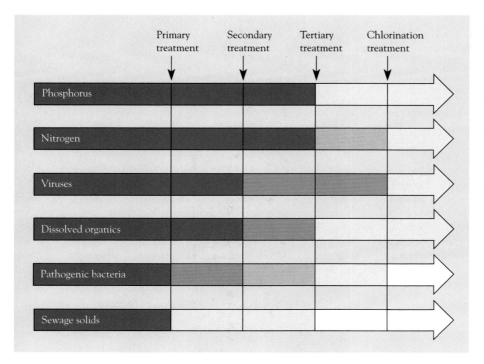

**Figure 21–18**
The effectiveness of primary, secondary, and tertiary sewage treatment in removing various water pollutants. Few municipalities offer tertiary treatment. Note how ineffective secondary treatment is in removing a number of contaminants, such as phosphorus and nitrogen. Also note that even after all treatments, many pollutants are still detectable.

resembles humus (see Chapter 14) and can be used as a soil conditioner.

Sewage sludge can also be dried and used as a fertilizer (Figure 21–19). It has the advantage of being rich in plant nutrients, although sometimes it contains too many heavy metals to be used commercially. (This happens when sewer systems mix industrial waste, which may contain toxic substances, with household waste.) Farmers are sometimes reluctant to use sewage sludge fertilizer out of concern that consumers won't purchase the food grown in it.

Although sewage sludge can be used to condition soil, it is generally treated as a solid waste. Dried sewage sludge is often incinerated, which contributes to air pollution, although sometimes the heat produced by this process is used constructively. Alternatively, sewage sludge can be disposed of in sanitary landfills. In the past, it was dumped in the ocean. Coastal cities such as New York traditionally dumped their sewage sludge into the ocean. However, in 1988 the U.S. Congress passed legislation—the Ocean Dumping Ban Act—that barred ocean dumping of sewage sludge and industrial waste as of the end of 1991. All cities stopped disposing of municipal sludge in the ocean on schedule except New York City, which continued to dump its sludge until June, 1992.

As landfill space becomes more costly, many cities are looking for alternative ways to handle

sewage sludge. Houston plans to use a special process to reduce its sewage sludge to sterile ash with approximately 5 percent the original volume, thus cutting the cost of disposal. Meanwhile, Texas is testing sludge ash as a paving material.

## Laws Controlling Water Pollution

Governments generally control point source pollution in one of two ways, either by imposing penalties on polluters (a common approach in the United States) or by taxing polluters to pay for the cleanup (common in Japan). Although most countries have passed laws to control water pollution, enforcement is difficult, even in developed countries. Typically, too few resources are allotted for enforcement. For example, in 1989 the 43 enforcement agents in the Office of Drinking Water of the Environmental Protection Agency (EPA) were expected to handle 80,000 complaints about drinking water safety; the agents were able to resolve 800 of those complaints in a year (see Meeting the Challenge: Using Citizen Watchdogs To Monitor Water Pollution).

The United States has attempted to control water pollution through legislation since the passage of the Refuse Act of 1899, which was intended to reduce the release of pollutants into navigable rivers. The three laws that have the most impact on water quality today are the Safe Drinking Water Act, the Clean Water Act (originally called the Water Pollution Control Act), and the Water Quality Act.

**Safe Drinking Water Act**  Prior to 1974, individual states set their own standards for drinking water, which of course varied a great deal from state to state. In 1974 the Safe Drinking Water Act was passed, which set uniform federal standards for drinking water in order to guarantee safe public water supplies throughout the United States. This law required that the EPA determine the **maximum containment level** (the maximum permissible amount) of any water pollutant that might adversely affect human health.

By 1986, 12 years after the passage of the Safe Drinking Water Act, the EPA had determined the maximum containment levels for only 26 pollutants; approximately 700 potential pollutants in municipal drinking water supplies remained to be characterized. The 1986 amendment to the Safe Drinking Water Act demanded, among other things, that the EPA increase its rate of evaluation of other water pollutants possibly requiring regula-

**Figure 21–19**
Sewage sludge that is produced from secondary treatment can be used as a fertilizer. Here the sewage sludge is being applied to strip-mined land that is being reclaimed. (Visuals Unlimited/R. F. Ashley)

## Using Citizen Watchdogs To Monitor Water Pollution

Lack of funds and insufficient staff prevent well-intentioned government agencies from effectively monitoring and enforcing environmental laws such as the Safe Drinking Water Act. For example, the San Francisco Bay Conservation and Development Commission is the regulatory agency charged with patrolling 1,000 miles of shoreline and 600 square miles of water to ascertain that no one is illegally polluting San Francisco Bay. In addition, it handles hundreds of cases arising from its monitoring activities. The agency, composed of just two people, is severely understaffed and unable to adequately protect the bay.

San Francisco Bay, like most other aquatic ecosystems, is endangered by the combined impact of many different pollution sources rather than from a single disaster, such as occurred when the *Exxon Valdez* spilled oil in Alaska (see Chapter 10). Thus, continual monitoring is necessary to ensure that many small polluters don't collectively do irreparable harm to the bay.

A growing number of private citizens have become actively involved in monitoring and enforcing environmental laws. Provisions in the Safe Drinking Water Act and other key environmental laws allow citizens to file suit when the government does not. Citizen action groups also pressure firms to clean up; for example, as a direct result of citizen action groups, the chemical company Monsanto reduced its toxic emissions by almost 40 percent, and Du Pont plans a 50 percent reduction by 1995.

In San Francisco Bay, some 300 citizen watchdogs called "Bay-Keepers" monitor the bay from kayaks, airplanes, and helicopters. Law students at local universities advise the Keepers on issues of litigation. The Bay-Keepers are modeled after the River-Keepers of the Hudson River, who organized in 1983 and were probably the first such group to monitor polluters. Similar Keeper groups have sprung up across the country. For example, Keepers monitor Long Island Sound, the Delaware River, and Puget Sound.

A Bay-keeper monitors a steel plant that had illegally dumped debris into San Francisco Bay in the past. (Felix Rigau Photography)

tion. As of January 1992, maximum containment levels for 62 pollutants were finalized.

**Clean Water and Water Quality Acts** The quality of surface waters in the United States is most affected by the Clean Water Act and the Water Quality Act of 1987. The Clean Water Act was originally passed as the Water Pollution Control Act of 1972; it was amended and renamed the Clean Water Act of 1977, and additional amendments were made in 1981 and 1987. The Clean Water Act is scheduled to be amended by Congress again in 1992. Under the provisions of these acts, the EPA is required to set up and monitor **national emission limitations**—the maximum permissible amounts of water pollutants that can be discharged into rivers, lakes, and oceans from sewage treatment plants, factories, and other point sources.

Overall, the Clean Water Act has been effective at improving the quality of water from point sources despite the relatively low fines it imposes on polluters. It is not hard to identify point sources, which must obtain permits from the National Pollutant Discharge Elimination System to discharge untreated wastewater (see Focus On: Pigeon River to find out about the failure of a point source polluter to comply with the act). The United States has improved its water quality in the past several decades, thereby demonstrating that the environment can recover rapidly once pollutants are eliminated.

In 1986, the EPA conducted a national water quality inventory, which showed that nonpoint source pollution was a major cause of water pollution. Although the Water Quality Act of 1987 had provisions to reduce nonpoint source pollution, it is much more difficult (and expensive) to control than point source pollution, and to date, U.S. environmental policies have failed to effectively address it. Nonpoint source pollution could be reduced by regulating land use, agricultural practices, and many other activities. The problem is that such regulation would require the interaction and cooperation of many government agencies, environmental organizations, and private citizens, and such coordination is enormously challenging.

**Laws That Protect Groundwater** A number of federal laws attempt to control groundwater pollution. The Safe Drinking Water Act contains provisions to protect underground aquifers that are important sources of drinking water. In addition, underground injection of wastes is regulated by the Safe Drinking Water Act in an effort to prevent groundwater contamination. The Resource, Conservation, and Recovery Act of 1976 (see Chapter 23) deals with the storage and disposal of hazardous wastes and helps prevent groundwater contamination. A number of miscellaneous laws related to pesticides, strip mining, and cleanup of abandoned hazardous waste sites also indirectly protect groundwater.

The many laws that directly or indirectly affect groundwater quality, which were passed at different times and for different reasons, provide disjointed and, at times, inconsistent protection of groundwater. The EPA makes an effort to coordinate all these laws, but groundwater contamination still occurs.

## WHAT ABOUT SOIL POLLUTION?

Soil pollution is important not only in its own right but because so many soil pollutants tend to get into water. Most of this chapter's information on water pollution applies to soil pollution as well.

Except for salt, the chief soil pollutants originate as agricultural chemicals, including fertilizers and pesticides. Until recently, it was assumed that most pesticides in current use had few long-term effects because tests had indicated that they evaporated, decomposed, or leached out of the soil in a relatively short time. However, in the late 1980s, soil scientists developed a new method of testing for contaminants and determined that many agricultural chemicals persist in the soil by seeping into tiny cracks, called micropores, and adhering to soil particles. It appears to soil scientists that the soil is behaving like a reservoir that stores contaminants and continuously releases them to surface water and groundwater, as well as topsoil, over a long period.

This finding is disturbing for several reasons. First, more than 907,000 metric tons (1 million tons) of chemicals are applied to U.S. soils each year, and we do not really know what their long-term effects are. Whereas there has been a great deal of research on air and water pollution, comparatively little has been done on soil pollution. Second, and perhaps even more unsettling, the only sure way to remove contaminants from soil is to dig up the soil from an entire field and incinerate it. This solution, besides being impractical, would kill all beneficial soil organisms.

Fortunately, there is a growing interest in alternative farming practices that reduce the need for large chemical applications (see Chapter 18). Alternative agriculture may help solve the dual problems of producing enough food *and* preventing environmental contamination.

## Pigeon River

The Pigeon River flows through the Great Smoky Mountains from North Carolina into Tennessee. From its source in the Pisgah National Forest, it flows, sparklingly clear, to the town of Canton, North Carolina. Canoeing, white-water rafting, and trout fishing are popular in this initial stretch of the Pigeon. In Canton, however, the river changes drastically. Canton's largest industry, a paper and pulp mill, discharges wastewater into the Pigeon River that turns the water brown and contaminates it with toxic chemicals.

When the Pigeon River flows into Hartford, Tennessee, the first town downstream from Canton, its waters are the color of cola and its smell is offensive. Except for sludge worms, little aquatic life is left in the river. Hartford has the nickname "Widowville" because so many men there have died of cancer. People in Hartford suspect that the upstream toxic discharges into the river are responsible, but it is very difficult to prove a causal relationship.

Although Tennessee has been trying to get North Carolina to make the mill clean up its act, all efforts so far have been unsuccessful. The EPA has also been ineffectual.

Why? The mill provides approximately 2,000 high-paying jobs in an area that is beset with high unemployment. The mill is antiquated—it has been operating since 1908—and mill officials say that modernization would clean up the water but would also cost jobs. In addition, the company has said that modernization would not be profitable and that it may close down entirely, laying off all its workers. If it does, because the economy of the region cannot absorb them, the employees of the mill will be faced with a hard choice: to either find low-paying, unskilled jobs in the area or move away from their homes. Thus, the situation at Pigeon River is a classic confrontation between economic concerns and the environment.

Although the mill at Canton has violated the Clean Water Act, North Carolina has always granted it a wastewater permit anyway, in light of its economic importance. The Pigeon River controversy is far from over. Legal battles, which have been going on for years, will determine the fate of the mill, the river, and the people who live on its banks. In the meantime, things remain pretty much the same as they have been for the past 85 years along the Pigeon River.

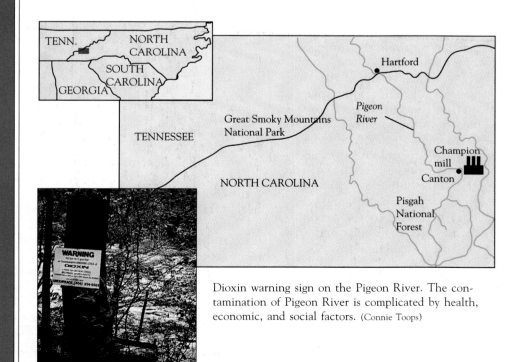

Dioxin warning sign on the Pigeon River. The contamination of Pigeon River is complicated by health, economic, and social factors. (Connie Toops)

# SUMMARY

**1.** The eight main types of water pollution are sediments, sewage, disease-causing agents, inorganic plant nutrients, organic compounds, inorganic chemicals, radioactive substances, and thermal pollution. Sediment pollution, which is caused primarily by soil erosion, increases water turbidity, thereby reducing the photosynthetic productivity of the water. Sewage contributes to enrichment (fertilization of water) and creates an oxygen demand as it is decomposed. Disease-causing agents are transmitted in sewage.

**2.** Inorganic plant nutrients such as nitrogen and phosphorus contribute to water enrichment. Many organic compounds are synthetic and do not decompose readily; some of these are quite toxic to living organisms. Inorganic chemicals include road salt, acids from mine drainage, and toxins such as mercury. Radioactive substances and heat can also pollute water.

**3.** Water pollutants come from both natural sources and human activities. Pollution that enters the water at specific sites, such as pipes from industrial or sewage treatment plants, is called point source pollution. Nonpoint source pollution, such as fertilizer runoff, comes from the land rather than from a single point of entry. The four human activities that contribute the most water pollution are industry, municipalities, shipping, and agriculture.

**4.** Soil salinization, a common problem in irrigated arid and semiarid regions, makes soil unfit for growing most crops. It is extremely difficult to remove excess salts from salinized soils.

**5.** Eutrophication, the enrichment of oligotrophic bodies of water by nutrients, results in high photosynthetic productivity, which supports overpopulation of algae. Eutrophic bodies of water tend to fill in rapidly as dead organisms settle to the bottom. Eutrophication also kills fish and causes a decline in water quality as large numbers of algae die and decompose rapidly.

**6.** Groundwater can become contaminated by hazardous substances that seep from landfills as well as from industry, mining, and agricultural operations. Currently, most of the groundwater supplies in the United States are of good quality, although there are some local problems. Because cleanup of polluted groundwater is very costly, takes years, and in some cases is not technically feasible, it is important to prevent groundwater contamination from occurring in the first place.

**7.** Water quality is improved both by reducing the number of contaminants that we put into water and by cleaning up polluted water. Wastewater treatment includes primary treatment (the physical settling of solid matter), secondary treatment (the biological degradation of organic wastes), and sometimes tertiary treatment (the removal of special contaminants such as organic chemicals, nitrogen, and phosphorus). One of the most pressing problems of wastewater treatment is disposal of the sewage sludge that results from secondary treatment.

**8.** Laws attempt to control water pollution. The Safe Drinking Water Act requires the EPA to establish maximum levels for water pollutants that might affect human health. The Clean Water Act and the Water Quality Act require the EPA to establish national emission limitations for wastewater that is discharged into U.S. surface waters. Legislation has been more effective in controlling point source pollution than in controlling nonpoint source pollution.

# DISCUSSION QUESTIONS

**1.** Explain why sewage may kill fish when it is added directly to a body of water.

**2.** Tell whether each of the following represents point source pollution or nonpoint source pollution: fertilizer runoff from farms, thermal pollution from a power plant, storm runoff from city streets, sewage from a ship, sediments from deforestation.

**3.** Explain why saline soils are physiologically dry for plants even though they may be physically wet.

**4.** The United States has a Clean Air Act and a Clean Water Act. Should we also have a Clean Soil Act? Present at least one argument for and one argument against such legislation.

**5.** Compare the potential pollution problems of groundwater and surface water used as sources of drinking water.

**6.** Imagine that you are heading a task force that is assigned the task of solving the Pigeon River pollution problem. What would you recommend? How would your recommendations affect the people in North Carolina? In Tennessee?

# SUGGESTED READINGS

Conniff, R. The transformation of a river—from "sewer" to suburbs in 20 years. *Smithsonian*, April 1990. The cleanup of the Connecticut River, which used to be shunned because it was so foul, has brought people flocking to its shores.

Fit to drink? *Consumer Reports*, January 1990, 27–43. Examines the major contaminants found in drinking water.

Holloway, M. Abyssal proposal. *Scientific American*, February 1992. Examines some of the ramifications of disposing sewage sludge in the ocean depths, which has recently been proposed by some scientists.

Maurits la Riviere, J. W. Threats to the world's water. *Scientific American*, September 1989. International water issues are caused by such factors as population growth and poor agricultural practices.

Okun, D. A. A water and sanitation strategy for the developing world. *Environment* 33:8, October 1991. Examines water quality and management in Asia, Africa, and Latin America.

Satchell, M. Fight for Pigeon River. *U.S. News & World Report*, 4 December 1989. The Pigeon River story represents a classic confrontation between jobs and the environment. (A short update of the story appears in the December 16, 1991, issue of *Business Week*.)

Stewart, D. Nothing goes to waste in Arcata's teaming marshes. *Smithsonian*, April 1990. How Arcata, California, turned its sewage into a natural resource.

# Chapter 22

*The salad vac is an environmentally safe method of controlling certain insect pests. Each vacuuming eliminates the need for one application of pesticide. (Richard Steven Street)*

# The Pesticide Dilemma

Pesticides have saved millions of lives by killing insects that carry disease and by increasing the amount of food we grow. Modern agriculture depends on pesticides to produce blemish-free fruits and vegetables at a reasonable cost to farmers (and therefore to consumers). However, pesticides also cause environmental and human health problems, and it appears that in many cases their harmful effects outweigh their benefits. Pesticides rarely affect the pest species alone, and the balance of nature, including predator-prey relationships, is upset. Certain pesticides concentrate at higher levels of the food chain. Humans who apply and work with pesticides are at risk from pesticide poisoning (short-term) and cancer (long-term), and people who eat traces of pesticide on food are concerned about its long-term effects.

## CONTROLLING PESTS

Picture a giant vacuum cleaner slowly moving over rows of strawberry or vegetable crops and sucking insects off the plants. Such a machine, given names such as "salad vac" and "bug vac," was invented as a substitute for the chemical poisons we call pesticides. Increasing public concern about pesticide residues on food has caused farmers to look seriously at other ways to control pests—even by vacuuming insects.

In this chapter we examine the types and uses of pesticides, their benefits and disadvantages. We also consider some alternatives to pesticides and the pesticide laws that are supposed to protect our health and the environment.

## WHAT IS A PESTICIDE?

Any organism that interferes in some way with human welfare or activities is called a **pest.** Some weeds, insects, rodents, bacteria, fungi, and other pest organisms compete with humans for food; other pests cause or spread disease. The definition of "pest" is subjective; a mosquito may be a pest to you, but it is not a pest to the bat or bird that eats it. People try to control pests, usually by reducing the size of the pest population. Toxic chemicals called **pesticides** are the most common way of doing this, particularly in agriculture.

Pesticides can be grouped by their target organisms, that is, by the pests they are supposed to eliminate. Thus, **insecticides** kill insects, **herbicides** kill plants, **fungicides** kill fungi, and **rodenticides** kill rodents such as rats and mice (Figure 22–1).

Agriculture is the sector that uses the most pesticides in the United States—approximately 74 percent of the estimated 0.43 million metric tons (0.48 million tons) used each year. U.S. farmers spend $4.1 billion each year on pesticide treatments for crops. Other pesticide users are government and industry (almost 13 percent of total pesticide use), households (almost 13 percent), and forests (less than 1 percent).

### The "Perfect" Pesticide

The ideal pesticide would be a **narrow-spectrum pesticide,** which would kill only the organism for which it was intended and not harm any other species. The perfect pesticide would also readily be degraded, or broken down, either by natural chemical decomposition or by biological organisms, into

**Figure 22–1**
Wide strips under the trees in this orchard lack grass and other plants that would compete with the trees for water and mineral nutrients. Herbicides, the chemicals that kill plants, are responsible for the dearth of vegetation. (USDA)

safe materials such as water, carbon dioxide, and oxygen. Finally, the ideal pesticide would stay exactly where it was put and would not move in the environment.

Unfortunately, there is no such thing as an ideal pesticide. Most pesticides are **broad-spectrum pesticides,** which kill a variety of organisms in addition to the target pest. Many pesticides either don't degrade readily or else break down into compounds that are equally as dangerous as, if not more dangerous than, the original pesticide. And pesticides move around a great deal throughout the environment.

### First-Generation and Second-Generation Pesticides

Before the 1940s, pesticides were of two main types, inorganic compounds (also called minerals) and organic compounds. Inorganic compounds that contain lead, mercury, and arsenic are extremely toxic to pests but are not used much today, in part because their chemical stability (they are not broken down by natural processes) allows them to persist and accumulate in soil and water. This accumulation poses a threat to humans and wildlife, which, like the target pests, are susceptible to inorganic compounds.

Plants provide humans with a number of natural organic compounds that are poisonous, particularly to insects. Such plant-derived pesticides are called **botanicals.** Botanicals include nicotine from tobacco, pyrethrum from chrysanthemum flowers, and rotenone from roots of the derris plant, all of which are used to kill insects (Figure 22–2). Botanicals are easily degraded by living organisms and, therefore, do not persist in the environment.

In the 1940s a large number of synthetic organic pesticides began to be produced. Earlier pesticides—both inorganic compounds and botanicals—are called **first-generation pesticides** to distinguish them from the vast array of synthetic poisons in use today, called **second-generation pesticides.** The insect-killing ability of **dichloro-diphenyl-trichloroethane (DDT),** the first of the second-generation pesticides, was recognized in 1939. Today there are thousands of pesticide products, made up of combinations of more than 1,000 different chemicals.

### The Major Kinds of Insecticides

The largest category of pesticides, the insecticides, are usually classified into groups based on chemical structure. Three of the most important kinds of second-generation insecticides are the chlorinated hydrocarbons, organophosphates, and carbamates

**Figure 22–2**
Botanicals are chemicals from plants that can be used as pesticides. Chrysanthemum flowers, shown here as they are harvested in Rwanda, are the source of the insecticide pyrethrum.
(Robert E. Ford/Terraphotographics)

**Table 22–1
Examples of Pesticides**

**Insecticides**
Botanicals
   Nicotine
   Pyrethrum
   Rotenone
Chlorinated hydrocarbons
   Aldrin
   BHC (benzene hexachloride)
   Chlordane
   DDT (dichloro-diphenyl-trichloroethane)
   Dieldrin
   Endrin
   Heptachlor
   Kepone
Organic phosphorus compounds
   Malathion
   Parathion
   Diazinon
Carbamates
   Carbaryl (Sevin)
   Carbofuran
   Propoxur (Baygon)
   Methylcarbamate (Zectran)

**Herbicides**
Selective herbicides
   Atrazine
   2,4-D (2,4-dichlorophenoxyacetic acid)
   2,4,5-T (2,4,5-trichlorophenoxyacetic acid)
   Picloram
   Silvex
Nonselective herbicide
   Paraquat

(Table 22–1). DDT is an example of a **chlorinated hydrocarbon,** an organic compound containing chlorine. After DDT's insecticidal properties were recognized, many more chlorinated hydrocarbons were synthesized as pesticides. Generally speaking, chlorinated hydrocarbons are broad-spectrum insecticides. They are slow to degrade and therefore persist in the environment (even inside living organisms) for many months or even years. They were widely used from the 1940s to the 1960s, but since then their use has largely been restricted, mainly because of problems associated with their persistence in the environment. Much of the public first became aware of the problems with pesticides in 1963, when Rachel Carson published her book *Silent Spring.*

**Organophosphates,** organic compounds that contain phosphorus, were developed during World War II as an outgrowth of German research on nerve gas. Organophosphates are more poisonous than other types of insecticides, and they are also toxic to animals other than insects. The toxicity of organophosphates in humans is comparable to that of some of our most dangerous poisons—arsenic, strychnine, and cyanide. Organophosphates do not persist in the environment as long as chlorinated hydrocarbons do, because organophosphates are usually degraded more easily by microorganisms. As a result, they have generally replaced the chlorinated hydrocarbons in large-scale uses such as agriculture.

The third group of insecticides, the **carbamates,** are relatively new on the pesticide scene. Carbamates are broad-spectrum insecticides, derived from carbamic acid, that are generally not as toxic to mammals as the organophosphates. Two common carbamates are carbaryl (trade name Sevin) and propoxur (trade name Baygon).

## The Major Kinds of Herbicides

Unwanted vegetation such as weeds in crops or lawns can be killed with herbicides (Table 22–1). Like insecticides, herbicides can be classified into groups based on chemical structure, but this method is cumbersome because there are at least 12 different chemical groups that are used as herbicides. It is more common to group herbicides according to how they act and what they kill. **Selective herbicides** kill only certain types of plants, whereas **nonselective herbicides** kill all vegetation. Selective herbicides can be further classified according to the types of plants they affect. **Broad-leaf herbicides** kill plants with broad leaves but do not kill grasses; **grass herbicides** kill grasses but are safe for most other plants.

Two common herbicides with similar structures are **2,4-**dichlorophenoxyacetic acid (2,4-D) and **2,4,5-**trichlorophenoxyacetic acid (2,4,5-T), both developed in the United States in the 1940s. These broad-leaf herbicides are similar in structure to a natural growth hormone in certain plants and therefore disrupt the plants' natural growth processes; they kill plants such as dandelions but don't harm grasses. You may recall that many of the world's important crops, including wheat, corn, and rice, are cereal grains, which are grasses (see Chapter 18). Both 2,4-D and 2,4,5-T can be used to kill weeds that compete with these crops, although 2,4,5-T is no longer used in this country (see Focus On: The Wartime Use of Herbicides).

### The Wartime Use of Herbicides

One of the controversial aspects of the Vietnam War was the defoliation program carried on by the United States in South Vietnam. From 1961 to 1971, the United States sprayed herbicides over large areas of South Vietnam to expose hiding places and destroy crops planted by the Vietcong and North Vietnamese troops. The three mixtures of herbicides used were designated Agent White, Agent Blue, and Agent Orange.

The negative impact of these herbicides on the environment is still being felt today. It is estimated that between 20 and 50 percent of the ecologically important mangrove forests of South Vietnam were destroyed. These forests were replaced with shrubs and may take decades to return. Approximately 30 percent of the nation's commercially valuable hardwood forests were killed and have been replaced by bamboo and weedy grasses. It is estimated that the crops destroyed during the ten-year spraying period could have fed 600,000 people per year for each of those ten years.

In addition to the ecological damage caused by the spraying of herbicides, there were negative effects on the native people and the U.S. servicemen stationed in Vietnam. Agent Orange, which contained a mixture of 2,4-D and 2,4,5-T, also had small amounts of dioxin—an extremely dangerous poison that is formed during manufacture of these herbicides. It is **teratogenic,** which means that it causes birth defects. During the period of herbicide spraying, the number of birth defects and stillbirths in Vietnam increased. It also appears that American service veterans who were exposed to high levels of Agent Orange have more health problems than other veterans. A Veterans Administra-

The spraying of herbicides over forested areas during the Vietnam War. (This photo was taken in 1966.) (UPI/Bettmann)

tion mortality study of marines who served in Vietnam found higher-than-normal levels of cancer, particularly of the lymph system and lungs.*

It is always easier to criticize bad policies after their effects are fully understood. In retrospect, the defoliation program in Vietnam was an ecological disaster that also harmed many people.

*Two recent studies, one of U.S. chemical workers (1991) and one of workers at a German herbicide plant (1992), confirm a strong link between prolonged exposure to high levels of dioxin and increased risk of cancer.

## BENEFITS OF PESTICIDES

Each day a war is waged as farmers, struggling to produce bountiful crops, battle insects and weeds. Similarly, health officials fight their own war against the ravages of human diseases transmitted by insects. One of the most effective weapons in the arsenals of farmers and health officials is the pesticide.

### Disease Control

Several devastating human diseases are transmitted by insects. Fleas and lice, for example, carry the microorganism that causes typhus in humans. Ma-

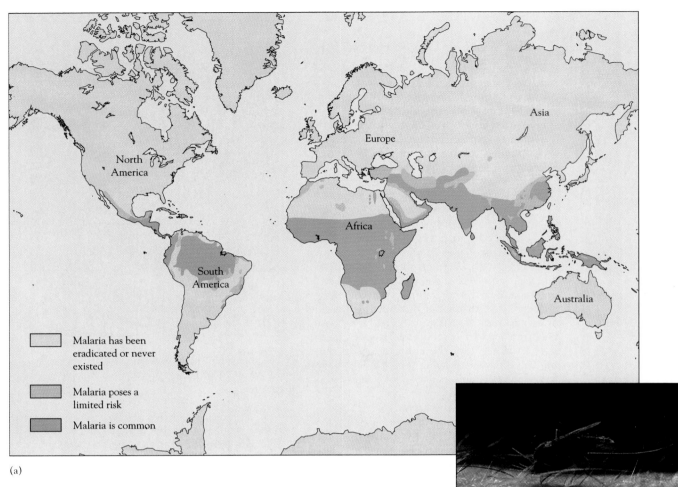

(a)

(b)

**Figure 22–3**

Worldwide, millions of lives have been saved by insecticides that were sprayed to control mosquitoes and other insect carriers of disease. (a) Where malaria occurs. (b) The *Anopheles* mosquito transmits the protozoon that causes malaria. (David M. Dennis/Tom Stack and Associates)

laria, which is caused by a protozoon, is transmitted to millions of humans each year by mosquitoes (Figure 22–3). Worldwide, approximately 100 million people currently suffer from malaria.

Pesticides, particularly DDT, have helped control the population of mosquitoes, thereby reducing the incidence of malaria. Consider Sri Lanka. In the early 1950s, more than 2 million cases of malaria were reported in Sri Lanka each year. When spraying of DDT was initiated to control mosquitoes, malaria cases dropped to almost zero. When DDT spraying was discontinued in 1964, malaria reappeared almost immediately; by 1968, its annual incidence had increased to greater than 1 million cases per year. Despite the negative effects of DDT on the environment and wildlife, the Sri Lankan

government decided to begin spraying DDT once again in 1968. Today, insecticides are still used in Sri Lanka to help control malaria, although DDT has been replaced by other, less persistent pesticides such as malathion.

## Crop Protection

Although exact assessments are difficult to make, it is widely estimated that almost half of the world's crops are eaten or destroyed by pests (Figure 22–4). Given our expanding population and world hunger, it is easy to see why control of agricultural pests is desirable.

Pesticides reduce the amount of a crop that is lost through competition with weeds, consumption

**Figure 22-4**
A tree in Burkina Faso, a country in Africa, is covered with locusts, which are also visible in the air. Infestations of locusts and other destructive insects can completely destroy crops. Insecticides help control these pests. (Robert E. Ford/ Terraphotographics)

by insects, and diseases caused by plant **pathogens** (microorganisms, such as fungi and bacteria, that cause disease). Although many insect species are beneficial from a human viewpoint (two examples are honeybees, which pollinate crops, and ladybugs, which prey on crop-eating insects), a large number are considered pests. Of these, about 200 species have the potential to cause large economic losses in agriculture. For example, the Colorado potato beetle is one of many insects that voraciously consume the leaves of the potato plant, reducing the plant's ability to produce large tubers for harvest.

Serious agricultural losses are minimized in the United States and other developed nations primarily by the heavy application of pesticides. Pesticide use is usually justified economically, in that farmers save an estimated $3 to $5 in crops for every $1 that they invest in pesticides. In developing countries where pesticides are not used in appreciable amounts, the losses due to agricultural pests can be considerable.

Why are agricultural pests found in such great numbers in our fields? Part of the reason is that agriculture is usually a **monoculture;** that is, only one type of plant is grown on a given piece of land. The cultivated field thus represents a very simple ecosystem. In contrast, natural ecosystems are extremely complex and contain many different species, including predators and parasites of the pest species, which control the pests' populations, and plant species that are not suitable as food for pests.

A monoculture reduces the dangers and accidents that might befall a pest as it searches for food. A Colorado potato beetle in a forest would have a hard time finding anything to eat, but a 500-acre potato field is like a big banquet table set just for the pest. It eats, prospers, and reproduces. In the absence of many natural predators and in the presence of plenty of food, the population thrives and grows, and more crops get damaged.

## PROBLEMS ASSOCIATED WITH PESTICIDE USAGE

Although pesticides have their benefits, they are accompanied by a number of problems. For one thing, many pest species develop a resistance to pesticides after repeated exposure to them. Also, pesticides affect numerous species in addition to the target pests, creating imbalances in the ecosystem (including agricultural fields) and posing a threat to human health. And, as mentioned earlier, the ability of some pesticides to resist degradation and readily move around in the environment creates even more problems for humans and wildlife.

### Development of Genetic Resistance

The prolonged use of a particular pesticide can cause a pest species to develop genetic resistance to the pesticide. **Genetic resistance** is any inherited characteristic that decreases the effect of a pesticide on a pest. The short generation times (the period between the birth of one generation and that of another) and large populations that are characteristic of most pests favor rapid evolution, which allows the pest to quickly adapt to the pesticides used against it. Thus, an insecticide that kills a large

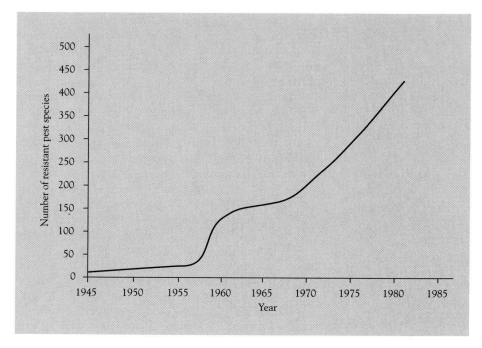

**Figure 22–5**
There has been a dramatic increase in the number of insect species exhibiting genetic resistance to insecticides, particularly since the 1950s.

portion of an insect population becomes less effective after prolonged use because the survivors and their offspring are genetically resistant. In the 50 years during which pesticides have been widely used, more than 400 species of insects have developed genetic resistance to certain pesticides (Figure 22–5).

For example, some insects that attack cotton have become so resistant to insecticide applications that chemical control is no longer effective (Table 22–2). The same is true of the soybean looper, an insect that eats soybean plants and has genetic resistance to all pesticides that are registered for use on soybeans. Because repeated appli-

**Table 22–2**
**Change in Pesticide Effectiveness Caused by Genetic Resistance**

| | Average Amounts of Pesticide Needed To Kill Two Insect Pests on Cotton* | | | |
| | Bollworm | | Tobacco Budworm | |
| Compound | 1960 | 1965 | 1960 | 1965 |
| --- | --- | --- | --- | --- |
| DDT | 0.03 | 1,000+ | 0.13 | 16.51 |
| Endrin | 0.01 | 0.13 | 0.06 | 12.94 |
| Carbaryl | 0.12 | 0.54 | 0.30 | 54.57 |

*In milligrams of pesticide per gram of caterpillars (larvae).
Source: Adkisson, P. L., et al., "Controlling Cotton's Insect Pests: A New System," *Science* 216, 1982.

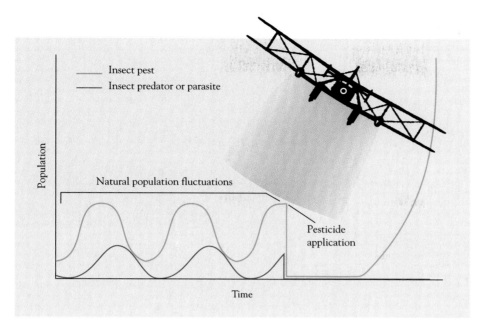

**Figure 22–6**
Natural population fluctuations are controlled by a variety of factors, including the presence of predators and parasites. When pesticides are applied, the predators and parasites of the pest species are also affected. The effect on predator/parasite populations disrupts the normal interactions between species and can cause a huge increase in the population of the pest species a short time after the pesticide is applied.

cations of insecticides to control malarial mosquitoes have resulted in their genetic resistance, insecticides are not as useful in the war against malaria as they were a few decades ago.

## Creation of Imbalances in the Ecosystem

One of the worst problems associated with pesticide usage is that pesticides affect species other than the pests for which they are intended. Beneficial insects such as honeybees and ladybugs are killed as effectively as insect pests. In a study of the effects of spraying the insecticide dieldrin to kill Japanese beetles, scientists found a large number of dead animals in the treated area, including a variety of birds, rabbits, ground squirrels, cats, and beneficial insects. (Use of dieldrin in the United States has since been canceled.)

Wildlife doesn't have to be killed to be negatively affected by pesticides. Quite often the stress of carrying pesticides in its body makes an organism more vulnerable to other diseases or stresses in its environment.

Because the natural enemies of pests—organisms that prey on the pests to survive—often starve or migrate in search of food after pesticide has been sprayed in an area, pesticides are indirectly responsible for a large reduction in the populations of natural enemies of pests. Pesticides also kill natural enemies directly, because predators consume a lot of the pesticide by consuming the pests. After a brief period, the pest population rebounds and gets larger than ever, partly because no natural predators are left to keep its numbers in check (Figure 22–6).

Despite a 33-fold increase in pesticide use in the United States since 1945, crop losses due to pests have not declined significantly (Table 22–3). Increasing resistance to pesticides in pests and the destruction of the natural enemies of pests provide a partial explanation. Changes in agricultural practices are also to blame; for example, crop rotation (see Chapter 14), a proven way of controlling cer-

**Table 22–3**
**Percentage of Crops Lost Annually to Pests in the United States**

| Period | Insects | Diseases | Weeds |
|---|---|---|---|
| 1989 | 13.0 | 12.0 | 12.0 |
| 1974 | 13.0 | 12.0 | 8.0 |
| 1951–1960 | 12.9 | 12.2 | 8.5 |
| 1942–1951 | 7.1 | 10.5 | 13.8 |

Source: Pimentel, D., et al., "Environmental and Economic Effects of Reducing Pesticide Use," *BioScience* 41:6, June 1991.

**Figure 22–7**
Pesticide use sometimes results in the creation of new pest species. (a) Red scale on green (unripened) citrus fruit. An infestation of red scale insects on lemons occurred after DDT was sprayed to control a different pest. Prior to DDT treatment, red scale was not a problem on citrus crops. (b) A comparison of red scale populations on DDT-treated and untreated trees. (a, William E. Ferguson)

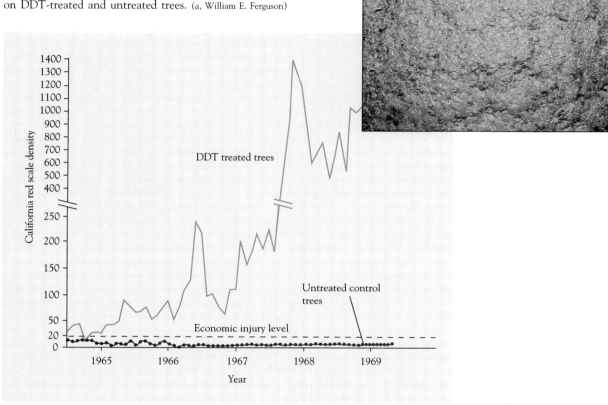

(a)

(b)

tain pests, is not practiced as much today as it was several decades ago.

**Creation of New Pests**   In some instances, the use of a pesticide has resulted in the creation of a pest problem that did not exist before. Creation of new pests (that is, turning nonpest organisms into pests) is possible because the natural predators, parasites, and competitors of a certain organism may be largely killed by a pesticide, allowing the organism's population to balloon. The use of DDT to control certain insect pests on lemon trees, for example, was documented as causing an outbreak of a scale insect (a sucking insect that attacks plants) that had not been a problem before spraying (Figure 22–7). In like manner, the European red mite became an important pest on apple trees in the north-

eastern United States after the introduction of pesticides.

**Persistence and Biological Magnification**

Certain problems of chlorinated hydrocarbon pesticide use were first demonstrated by the effects of DDT on many bird species. Falcons, pelicans, bald eagles, ospreys, and a number of other birds are very sensitive to traces of DDT in their tissues. A substantial body of scientific evidence indicates that one of the effects of DDT on these birds is that they lay eggs with extremely thin, fragile shells that usually break during incubation, causing the chicks' deaths. After DDT was banned in the United States, the reproductive success of many birds improved (Figure 22–8).

**Figure 22–8**
The effect of DDT on birds.
(a) Many birds, including the bald eagle, suffered reproductive failure after DDT accumulated in their tissues and interfered with their ability to produce strong eggshells. (b) A comparison of the number of successful bald eagle offspring with the level of DDT residues in their eggs. Note that reproductive success improved after DDT levels decreased.
(*a*, A. Carey/VIREO)

(a)

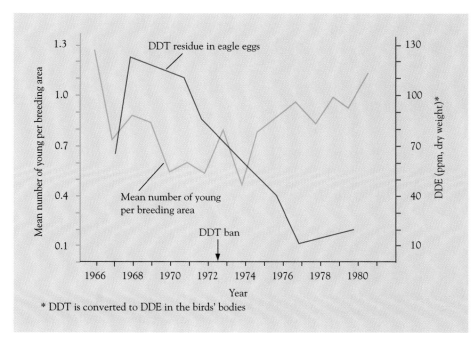

(b)

The impact of DDT on birds is the result of two characteristics of DDT, its persistence and its biological magnification. Some pesticides, particularly chlorinated hydrocarbons, are extremely stable and may take many years to be broken down into less toxic forms. The persistence of synthetic pesticides is a result of their novel chemical structures. Natural decomposers such as bacteria have not evolved ways to degrade them, so they accumulate in the environment and in the food chain.

When a pesticide is not metabolized or excreted by an organism, it simply gets stored, usually in fatty tissues. Organisms higher on the food chain tend to have greater concentrations of pesticide stored in their bodies than those lower on the food chain; this tendency is known as **biological magnification.**[1]

As an example of the concentrating characteristic of persistent pesticides, consider a hypothetical food chain: plant → insect → frog → hawk. When pesticide is sprayed on a plant, it is extremely dilute; we will assign it an arbitrary concentration of "1" per leaf. Each insect grazing on the plant consumes ten leaves, concentrating the pesticide in its tissues to a value of "10." (Assume

[1] Other toxic substances besides pesticides may exhibit biological magnification, including radioactive isotopes, heavy metals such as mercury, and industrial chemicals such as PCBs (see Chapters 11, 21, and 23).

DDT
in water

0.00005 ppm

DDT
in algae
and plants
0.04 ppm

DDT
in plant-
eating fish
0.2–1.2 ppm

DDT
in large
fish
1–2 ppm

DDT
in fish-
eating birds
3–76 ppm

**Figure 22–9**
The biological magnification of DDT in an aquatic ecosystem. Note how the
level of DDT concentrates in the tissues of various organisms as you move along
the food chain from producers to consumers. The green heron at the end of the
food chain has approximately one million times more DDT in its tissue than the
concentration of DDT in the water.

that these insects are genetically resistant to the
pesticide so that they stay alive.) A frog that eats
ten insects laced with pesticide ends up with a pesticide level of "100." The top carnivore in this example, a hawk, will have a pesticide value of
"1,000" if it eats ten contaminated frogs
(Figure 22–9). Although this example involves a
bird at the top of the food chain, it is important to
recognize that *all* top carnivores, from fish to humans, are at risk from biological magnification.

## Mobility in the Environment

Another problem associated with pesticides is that
they do not stay where they are applied, but tend to
move through the soil, water, and air, sometimes
long distances (Figure 22–10). For example, fish
can be affected by pesticides that were applied to
agricultural lands and then washed into rivers and
streams when it rained. If the pesticide level in its
aquatic ecosystem is high enough, the fish may be
killed. If the level is sublethal—that is, not enough
to kill the fish—the fish may still suffer from undesirable effects such as bone degeneration.

## Risks to Human Health

Wildlife are not the only organisms harmed by pesticides. Exposure to pesticides can also damage
human health. Pesticide poisoning caused by short-
term exposure to high levels of pesticides can result
in harm to organs and even death, whereas long-
term exposure to lower levels of pesticide can cause
cancer. There is also concern about the potential
effects of long-term exposure to the trace amounts
of pesticides on the food we eat.

**Short-Term Effects of Pesticides** Approximately
20,000 people are poisoned by pesticides in the
United States each year. Most of these are farm
workers or others whose occupations involve daily
contact with large quantities of pesticides
(Figure 22–11). A person with a mild case of pesticide poisoning may exhibit symptoms such as nausea, vomiting, and headaches. More serious cases,
particularly organophosphate poisonings, may result in permanent damage to the nervous system
and other body organs. Although the number is
low, people do die from overexposure to pesticides;

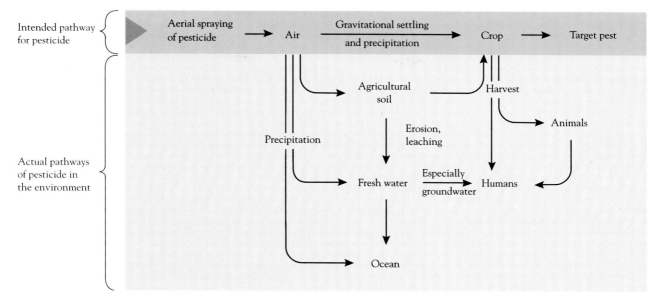

**Figure 22–10**
The actual pathways of pesticides in the environment, including living organisms.
(The *intended* pathway of pesticides is marked in orange.)

almost any pesticide can kill a human if the dose is large enough. The incidence of pesticide poisoning is highest in developing countries, in part because they often use dangerous pesticides that have been banned by developed nations. Also, pesticide users in developing nations often are not trained in the safe handling and storage of pesticides.

**Long-Term Effects of Pesticides**   Many studies of farm workers and others who are exposed to low levels of pesticides over many years show an association between cancer and long-term exposure to pesticides. A type of lymphoma (a cancer of the lymph system) has been associated with the herbicide 2,4-D, for example. Other pesticides have been linked to a variety of cancers, including leukemia and cancer of the brain, lungs, and testicles. In addition, the children of agricultural workers are at greater risk than other children for birth defects, particularly stunted limbs.

Humanity is conducting a worldwide experiment on the effects of long-term exposure to traces of pesticides, and whether we know it or not, we're *all* taking part in the experiment, because every day we consume minute amounts of pesticides on our food. Apples, grapes, lettuce, oranges, potatoes, tomatoes, dairy products, and beef are the common foods that are most likely to contain traces of pesticides. The tissues of the typical American contain

traces of DDT and certain other pesticides. So far there is no evidence that long-term exposure to traces of pesticides results in cancer or other health problems, but there is also no guarantee that the evidence will not emerge at a later date.

**Figure 22–11**
Migrant farm workers harvest tomatoes in Florida. Farm workers and others who have close daily contact with pesticides suffer a number of health problems. (E. R. Degginger)

## ALTERNATIVES TO PESTICIDES

Given the many problems associated with pesticides, it is clear that they are not a viable long-term solution to pest control. Fortunately, pesticides are not the only weapon in our arsenal. Alternative ways to control pests include cultivation methods, biological controls, genetic controls, hormones and pheromones, quarantine, and irradiation. A combination of these methods in agriculture, often including a limited use of pesticides, is known as **integrated pest management (IPM).** IPM may be the most effective way to control pests.

### Using Cultivation Methods To Control Pests

Sometimes agricultural practices can be altered in such a way that a pest is adversely affected or discouraged from causing damage. Although some practices (such as the insect vacuum mentioned at the beginning of the chapter) are quite new, a number of cultivation methods that discourage pests have been practiced for many years. The proper timing of planting, fertilizing, and irrigating promotes healthy, vigorous plants that are more able to resist pests because they are not being stressed by other environmental factors. A technique that has been used with success in alfalfa crops is **strip cutting,** in which only one segment of the crop is harvested at a time. The unharvested portion of the crop provides an undisturbed habitat for natural predators and parasites of the pest species. The rotation of crops also helps control pests. When corn is not planted in the same field for two years in a row, for example, the corn rootworm is effectively controlled.

### Biological Controls

**Biological controls** involve the use of naturally occurring disease organisms, parasites, or predators to control pests. As an example, suppose that an insect species is accidentally introduced into a country where it was not found previously, and becomes a pest. It might be possible to control this pest by going to its native country and identifying an organism there that is an exclusive predator or parasite of the pest species. That predator or parasite, if successfully introduced, may be able to lower the population of the pest species so that it is no longer a problem. Cottony-cushion scale and prickly pear provide two examples of successfully using one organism to control another.

### Envirobrief

**Using Heat Instead of Poison**

An entomologist who retired from the University of California at Los Angeles has invented an environmentally benign form of pest control that kills bugs by heating them past their level of tolerance. "Thermal pest eradication," a process now patented by Isothermics, Inc., of Anaheim, California, gets rid of drywood termites, cockroaches, and other damaging insects.

The infested structure is draped with tarpaulins; special heaters boost the indoor temperature to 66°C (150°F) for about 6 hours; and because the insects cannot sweat, they die, often bursting in the process. Inspired by an ancient method of using hot rocks to drive vermin from clothing, the heat treatment is an nonchemical alternative to insecticidal fumigation. It is being used by a few pest control companies in California and in Florida, where the drywood termite causes structural damage in the neighborhood of $119 million annually.

**Biological Control of Cottony-Cushion Scale**   The cottony-cushion scale is a small insect that sucks the sap from the branches and bark of many fruit trees, including citrus trees (Figure 22–12). It is native to Australia but was accidentally introduced to the United States in the late 1880s. Subsequently, an American entomologist (insect biologist) went to Australia and returned with several possible biological control agents. One, the vedalia beetle, was found to be very effective in controlling scale, which it eats voraciously (and exclusively). Within two years of its introduction, the vedalia beetle had eliminated the cottony-cushion scale from citrus orchards. Today both the cottony-cushion scale and the vedalia beetle are present in very low numbers, and the scale is not considered an economically important pest.

**Biological Control of Prickly Pear**   Prickly pear is a common name for several hundred species of cactus that are native to North and South America. A number of these plants were introduced to Australia in the 1800s to add variety to cultivated gardens. One prickly pear species, *Opuntia stricta,*

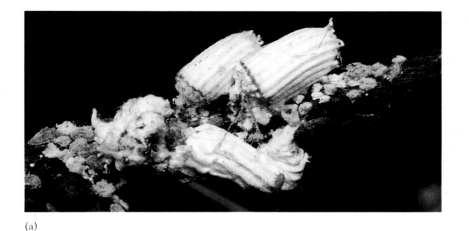

(a)

(b)

**Figure 22–12**
The biological control of cottony-cushion scale. (a) The cottony-cushion scale is an insect that attacks the stems and bark of a number of important crops. The white, ridged mass on top of each female contains up to 800 eggs. The males are reddish-brown in color. (b) A vedalia beetle larva feeds on cottony-cushion scale. Both adults and larvae of the vedalia beetle control the cottony-cushion scale. (*a* and *b*, Peter J. Bryant/Biological Photo Service)

spread rapidly over much of eastern Australia. In many areas it grew densely, covering the entire ground and eliminating native plants (Figure 22–13).

A number of insects were sent to Australia as possible biological controls, but only one, a moth, was effective. The female moth lays her eggs on the prickly pear, and the larvae that hatch from the eggs burrow into the cactus tissue and feed on it. Bacteria and fungi then invade the burrows and attack the plant. Within several years of its introduction in 1926, the moth had reduced the population of prickly pear to an occasional plant. Today the prickly pear is generally not a nuisance in Australia.

Although some examples of biological control are quite spectacular, finding an effective parasite or predator is very difficult. And just because a parasite or predator has been identified does not mean it can become successfully established in a new environment. Slight variations in environmental

**Figure 22–13**
The biological control of prickly pear. *Cactoblastis* caterpillars eating out the center of a prickly pear segment. Large areas of Australia were infested with prickly pear. This same area was devoid of prickly pear within a short time after the introduction of *Cactoblastis*. (G. R. Roberts)

factors such as temperature and moisture can alter the effectiveness of the biological control organism in its new habitat. Care must also be taken to ensure that the introduced control agent doesn't attack unintended hosts and become a pest itself.

Insects are not the only biological control agents. Bacteria that harm insect pests have also been used successfully as biological controls. *Bacillus popilliae*, which causes milky spore disease in insects, can be applied as a dust on the ground to control the grub stage of Japanese beetles. *Bacillus thuringiensis* (Bt), which produces a toxin that is poisonous to some insects when they eat it, is used against insects such as the cabbage looper, a green caterpillar that damages many vegetable crops, and the corn earworm (see Focus On: *Bt*, Its Potential and Problems). Viruses can also be introduced as biological control agents.

## Genetic Controls

Like biological control, genetic control of pests involves the use of living organisms. Instead of using another species to reduce the pest population, however, genetic control strategies suppress pests in other ways. One is to reduce the pest population by sterilizing some of its members; another is to breed crop plants and animals so that they can resist pests.

**The Sterile Male Technique**  Altering the genetic makeup of an insect pest is one type of genetic control. One of the most common methods of alteration is the **sterile male technique,** in which large numbers of males are sterilized, usually with radiation or chemicals (Figure 22–14).[2] They are then released into the wild, where they reduce the reproductive potential of the pest population by mating with normal females, who then lay eggs that never hatch; as a result, of course, the population of the next generation is much smaller.

One disadvantage of the sterile male technique is that it must be carried out continually to be effective. If sterilization is discontinued, the pest population rebounds to a high level in a few generations (which, you will recall, are very short). The procedure is also expensive, as it requires the rearing and sterilization of large numbers of insects in a laboratory or production facility. For example, during the

---

[2]Males are sterilized rather than females because the male insects mate several to many times, but the females mate only once. Thus, releasing a single sterile male may prevent successful reproduction by several females, whereas releasing a single sterile female would prevent successful reproduction by only that female.

## Envirobrief

### Killing Mosquitoes with Coconuts

Malaria, transmitted by mosquitoes, is a serious problem in developing countries. It kills hundreds of thousands of people each year. Millions more who don't die from it suffer regularly from a high fever that leaves them weakened and vulnerable to other diseases. In Peru, an extremely poor country, malaria has a devastating effect on the economy. It keeps workers from their jobs and children from school.

In the late 1980s, Peruvian researchers discovered that *Bacillus thuringiensis var israelensis* H-14 (Bti), a bacterium lethal to mosquito larvae, thrives in coconut milk, which is common throughout much of the developing world. Villages that cannot afford pesticides or don't want their environment contaminated by the chemicals can now use a simple method. A swab of bacteria is inserted through a hole drilled in a coconut. The hole is plugged with cotton. Three days later, the contents of two or three coconuts are poured into a local pond. The bacteria, harmless to humans and animals, kill mosquito larvae by eating through their stomachs after being ingested. Field tests have shown that a single Bti application will kill virtually all the mosquito larvae in a pond and stop further larval growth for up to 45 days.

1990 medfly outbreak in California, as many as 400 million sterile male medflies were released each week.

**Developing Resistant Crops**  Selective breeding has been used to develop many varieties of crops that are resistant to disease organisms or insects. Traditional breeding of crop plants typically involves identifying individual plants that are in an area where the pest is common but do not appear to be damaged by the pest. These individuals are then crossed with standard crop varieties in an effort to produce a pest-resistant version. It may take 10 to 20 years to develop a resistant crop variety, but the benefits are usually worth the time and expense.

Genetic engineering offers great promise in the field of breeding pest-resistant plants (see

## Bt, Its Potential and Problems

The soil bacterium *Bacillus thuringiensis*, or *Bt*, serves as a natural pesticide that is toxic to insects and yet environmentally "friendly." When eaten by insects, *Bt* produces spores that release a poison. It has been marketed since the 1950s but was not sold on a large scale until recently, mainly because there are many different varieties of the *Bt* bacterium, and each variety is toxic to only a small group of insects. For example, the *Bt* variety that works against corn borers would not be effective against Colorado potato beetles. As a result, *Bt* was not economically competitive against chemical pesticides, each of which could kill many different kinds of pests on many different crops.

The potential of *Bt*'s toxin as a natural pesticide has greatly increased because of improvements made by genetic engineers. For example, genetic engineers modified the gene coding for the toxin so that it affects a wider range of insect pests. Also, genetic engineers inserted the *Bt* gene that codes for the toxin into other bacteria. For example, they inserted the *Bt* gene into *Clavibacter xyli*, which normally grows inside the vascular tissue of corn. Regular *Clavibacter* has no effect on the corn plant in which it grows, but the geneti-

cally altered *Clavibacter* actually helps corn. Plants that contain the genetically altered bacterium receive a continuous supply of toxin, which provides a natural defense against insects that eat corn plants. Genetic engineers have also inserted the *Bt* gene into crop plants such as tomato and cotton; the genetically altered plants are more resistant to pests such as the tomato pinworm and the cotton bollworm.

The future of the *Bt* toxin as an effective substitute for chemical pesticides is not completely secure, however. Beginning in the late 1980s, several farmers began to notice that *Bt* was not working as well against the diamondback moth as it had in the past. All of the farmers who reported this reduction in effectiveness had used *Bt* frequently and in large amounts on their fields. It appears that certain insects, including the diamondback moth, have developed resistance to this natural toxin in much the same way that they develop resistance to chemical pesticides. If *Bt* continues to be used in greater and greater amounts, it is likely that more insect pests will develop genetic resistance to it, greatly reducing *Bt*'s potential as a natural pesticide.

Chapter 18). For example, a gene from the soil bacterium *Bacillus thuringiensis* (already discussed in the section on biological control) has been introduced into cotton plants. Caterpillars that eat cotton leaves from these genetically altered plants die or exhibit stunted growth.

Although selective breeding has resulted in many disease-resistant crops, it has not been an unqualified success. Fungi, bacteria, and other plant pathogens evolve rapidly. As a result, they can quickly adapt to the disease-resistant host plant, meaning that the new pathogen strains can cause disease in the formerly disease-resistant plant variety. Plant breeders, then, are in a continual race to keep one step ahead of plant pathogens.

## Pheromones and Hormones

**Pheromones** are natural substances produced by animals to stimulate a response in other members of the same species. Pheromones are commonly called

**Figure 22–14**
The screwworm is a serious insect pest that can kill cattle. Screwworm larvae (maggots), hatched from eggs laid by a female screwworm fly, invade the ear of a steer, creating a large wound. Such infestations can cause death in less than two weeks. (USDA)

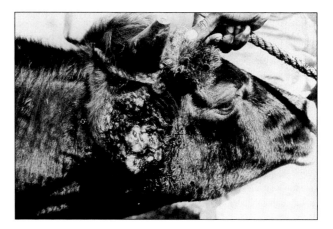

**Figure 22–15**
A Japanese beetle trap uses a sexual attractant to lure Japanese beetles, which fall into the bag and die. Such insect
traps draw large numbers of Japanese beetles. (Visuals Unlim-
ited/John D. Cunningham)

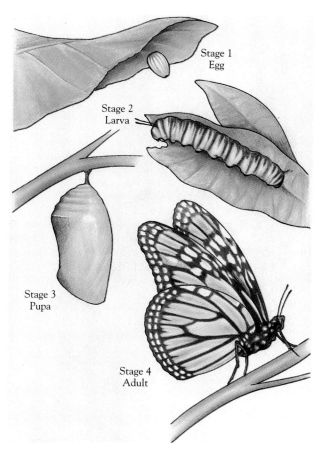

**Figure 22–16**
Different stages in the life cycle of an insect with complete
metamorphosis. The egg hatches into a larva (caterpillar),
which later becomes a pupa. Metamorphosis is complete
when the pupa becomes an adult insect.

sexual attractants because they are often produced
by an individual to attract members of the opposite
sex for mating. Each insect species produces its own
specific pheromone, so once the chemical structure
is known, it is possible to make use of pheromones
to control individual pest species. Pheromones
have been successfully used to lure insects such as
Japanese beetles to traps, where they are killed
(Figure 22–15). Alternatively, pheromones can be
released into the atmosphere to confuse insects so
that they cannot locate mates.

Insect **hormones** are natural chemicals produced by insects to regulate their own growth and
metamorphosis—the process by which an insect's
body changes in form from a larva to a pupa and
from a pupa to an adult (Figure 22–16). Specific
hormones must be present at certain times in the
life cycle of the insect; if they are present at the
wrong time, the insect develops abnormally and
dies. Many such hormones have been identified,
and synthetic hormones with similar structures

have been made. The possibility of using these substances to control pests is being actively pursued by
entomologists, but several concerns have been
raised. First, although it is assumed that humans
and wildlife are not affected by insect hormones,
we do not know this for certain. Second, each insect species produces its own specific hormones, but
many species have similar hormones. Using hormones to control insect pests might conceivably
harm beneficial insects as well; some evidence suggests that a few synthetic hormones affect several
insect species.

## Quarantine

Governments attempt to prevent the importation
of foreign pests and diseases by practicing **quarantine,** or restriction of the importation of exotic
plant and animal material that might harbor pests.

If a foreign pest is accidentally introduced, quarantine of the area where it is detected helps prevent its spread. For example, if a foreign pest is detected on a farm, the farmer may be required to destroy the entire crop.

Quarantine is an effective, although not foolproof, means of control. The U.S. Department of Agriculture has blocked the accidental importation of Mediterranean fruit flies (also called medflies) on more than 100 separate occasions. The few occasions when quarantine failed and these insects successfully passed into the United States have done millions of dollars' worth of crop damage and required the expenditure of additional millions to eradicate the pests. For example, when the medfly invaded California's crop-rich San Joaquin Valley in the early 1980s, farmers lost $75 million. (The adult medfly lays its eggs on 250 different fruits and vegetables; when the eggs hatch, the maggots feed on the fruits and vegetables and turn them into a disgusting mush.) Eradication efforts, which include the use of helicopters to spray the insecticide malathion over hundreds of square kilometers and the rearing and releasing of millions of sterile males to breed the medfly out of existence, are extremely costly.

## Integrated Pest Management

Many pests cannot be controlled effectively with a single technique; a combination of control methods is often more effective. Integrated pest management (IPM) combines the use of biological, cultural, and chemical controls. Nonchemical controls are used as much as possible, and pesticides are used sparingly and only when other methods fail (see Meeting the Challenge: Reducing Agricultural Pesticide Use by 50 Percent in the United States). Thus, IPM allows us to control pests with a minimum of environmental disturbance.

In order to be effective, IPM requires thorough knowledge of the life cycles of the pests and their hosts as well as all of their interactions. The timing of treatments is critical and is determined by carefully monitoring the concentration of pests (Figure 22–17). IPM also optimizes natural controls by utilizing agricultural techniques that discourage pests. IPM is an important part of the "new" agricultural methods known as alternative agriculture (discussed in Chapter 18).

Cotton, which is attacked by many insect pests, has responded well to integrated pest management. Cotton has the heaviest insecticide application of any crop: although only about 1 percent of agricul-

### Organic Wine

In 1991, the world's largest wine maker, E. & J. Gallo Winery, applied for organic certification on a 2,700-acre vineyard near Fresno, California. Gallo is one of several large U.S. wineries, mostly in California, that have started to convert to chemical-free production by using predatory wasps instead of pesticides, and compost and cover crops instead of fertilizers and herbicides. One reason for the conversion is that the vintners fear government reprisal if trace chemicals are found in their wines. There is a precedent for such measures: in the 1980s, the U.S. government banned the importation of 79 wines from France, Italy, and Spain for more than a year because they contained traces of procymidone, a fungicide used widely in European vineyards. Fortunately, grapes are one of the easiest crops to grow without human-made chemicals, and their skins make excellent compost after pressing.

tural land in the United States is used for this crop, cotton accounts for almost 50 percent of all the insecticides used in agriculture! By simple techniques such as planting a strip of alfalfa adjacent to the cotton field, the need for chemical pesticides is lessened. Lygus bugs, a significant pest of cotton,

**Figure 22–17**

Insect trap. An important aspect of integrated pest management is the careful and continual monitoring of insect pests, which provides an early warning of the buildup of pest populations. (Connie Toops)

move from the cotton field to the strip of alfalfa, which they prefer as a food.

### Irradiating Foods

It is possible to prevent insects and other pests from damaging harvested food without using pesticides. The food can be exposed to ionizing radiation (usually gamma rays from cobalt 60), which kills many microorganisms. Numerous countries, including Canada, much of western Europe, Japan, Russia, and Israel, extend the shelf lives of foods with irradiation. The Food and Drug Administration of the U.S. government approved this process for fruits and vegetables as well as certain meats in 1986, and the first irradiated food in the United States (strawberries) was sold in January 1992.

The irradiation of foods is somewhat controversial. Many consumers are concerned because they fear that irradiated food is radioactive. This is *not* true! In the same way that you do not become radioactive when your teeth are x-rayed, irradiated food is not radioactive. Nevertheless, most U.S. consumers refuse to purchase food that is labeled as having been irradiated.

Critics of irradiation are concerned because irradiation forms traces of certain chemicals, some of which have been demonstrated to be carcinogenic in laboratory animals. They also point out that we do not know the long-term effects of eating irradiated foods. Proponents of irradiation argue that some of these same trace chemicals are also produced by cooking processes such as frying and barbecuing. They further assert that more than 1,000 investigations on irradiation of foods, conducted worldwide, have demonstrated that it is safe. Furthermore, irradiation lessens the use of dangerous pesticides and food additives. Opponents remain unconvinced, and three states—Maine, New York, and New Jersey—have banned the sale of irradiated food.

## CONTROLLING PESTICIDE USE

Several laws have been passed to regulate pesticides in the interest of protecting human health and the well-being of the environment. The **Food, Drug, and Cosmetics Act (FDCA),** passed in 1938, recognized the need to regulate pesticides found in food but did not provide a means of regulation. The FDCA was made more effective in 1954 with the passage of the **Pesticide Chemicals Amendment** (also called the **Miller Amendment),** which required the establishment of acceptable and unacceptable levels of pesticides in food.

An updated Food, Drug, and Cosmetics Act, passed in 1958, contained an important section known as the **Delaney Clause,** which stated that no substance capable of causing cancer will be permitted in processed food. (Processed foods are foods that are prepared in some way—such as frozen, canned, dehydrated, or preserved—before being sold. The Delaney Clause recognized that pesticides tend to concentrate in condensed processed foods, such as tomato paste and applesauce.)

The Delaney Clause, although commendable for its intent, has several inconsistencies. First, it does not cover pesticides on raw foods such as fresh fruits and vegetables, milk, meats, fish, and poultry. Second, because the Environmental Protection Agency (EPA) lacks sufficient data on the cancer-causing risks of pesticides that have been used for a long time, the Delaney Clause has been applied only to pesticides that were registered after strict tests were put into effect in 1978. This situation has given rise to one of the paradoxes of the Delaney Clause: there have been cases in which a newer pesticide that posed minimal risk was banned because of the Delaney Clause; meanwhile, an older pesticide, which the newer one was to have replaced, is still used despite the fact that it is many times more dangerous than the new one.

For example, a new fungicide called fosetyl Al was determined to have weak but positive cancer-causing effects in test animals. The risk of developing cancer as a result of drinking beer made from hops treated with fosetyl Al was determined by the EPA to be $1 \times 10^{-8}$ (1 in 100 million or less), so the EPA forbade its use, citing the Delaney Clause. Fosetyl Al would have replaced an older fungicide called EBDC, which has a much greater estimated risk, $1 \times 10^{-5}$ (1 in 100,000 or less).[3]

The **Federal Insecticide, Fungicide, and Rodenticide Act (FIFRA)** was originally passed in 1947 to regulate the effectiveness of pesticides—that is, to prevent people from buying pesticides that didn't work. FIFRA has been amended over the years to require testing and registration of the active ingredients of pesticides. Also, any pesticide that does not meet the tolerance standards established by the FDCA must be denied registration by FIFRA.

---

[3]On March 2, 1992, the federal government banned or restricted the use of EBDC on certain crops, although EBDC applications are still permitted on hops.

# Reducing Agricultural Pesticide Use by 50 Percent in the United States

Pesticides have benefited farmers (by increasing agricultural productivity) and consumers (by lowering food prices). But pesticide use has had its price—not necessarily in economic terms, for it is difficult to assign monetary values to many of its effects—but in terms of health problems and damage to agricultural and natural ecosystems. Society is increasingly concerned about pesticide use. For example, Proposition 128 on the 1990 California ballot asserted that no pesticide known to cause cancer could be applied to foods. Although this proposition was defeated, the fact that it was even proposed and considered by voters indicates a high level of public concern over pesticides.

Is it feasible to ban all pesticides known to cause cancer in laboratory animals? Probably not—at least not now. In many cases, substitute pesticides either do not exist, are less effective, or are considerably more expensive. A pesticide ban would increase food prices, although estimates on the magnitude of that increase vary considerably from one study to another. A pesticide ban would also cause considerable economic hardship for certain growers, although other farmers might benefit. For example, some insects are more troublesome in certain areas than in others. A farmer growing a crop in an area where an insect was very harmful might not be able to afford the crop losses that would occur without the use of a banned pesticide. Growers in areas where the insect was less of a problem could then increase their production of that crop, benefiting financially from the first farmer's loss.

Since it is impractical to ban large numbers of pesticides right now, is there another way to provide greater protection to the environment and human health without reducing crop yields? Governments in Denmark and Sweden think so. They are implementing plans to reduce pesticide use by 50 percent during the 1990s. The Netherlands is developing a similar program. Strategies to reduce pesticide use include applying pesticide only when needed and using improved application equipment.

Too often pesticides are applied unnecessarily. Pesticide use can be decreased by continually monitoring pests so that pesticides are applied only when pests become a problem. For example, a 1991 study at Cornell University determined that a monitoring program might reduce the use of insecticides on cotton by 20 percent.

The use of aircraft to apply pesticide is extremely wasteful; 50 percent to 75 percent of the pesticide doesn't reach the target area, but instead drifts in air currents until it settles on soil or water. Pesticides applied on land with traditional methods also drift in air currents. Advances in the design of equipment for applying pesticides could reduce pesticide use considerably. For example, a recently developed rope-wick applicator reduces herbicide use on soybean fields by approximately 90 percent.

Pesticide use can also be reduced considerably through alternative pest control strategies. The widespread adoption of integrated pest management and other pesticide-reducing strategies makes it feasible for the United States to reduce pesticide use by 50 percent within a 5- to 10-year period.

Aerial spraying of insecticide on corn. Most pesticide applied by aircraft never reaches the target area. (Grant Heilman, Grant Heilman Photography)

**Table 22–4**
**Worst-Case Estimates of Risk of Cancer from Pesticide Residues on Food**

| Food | Cancer Risk* |
|---|---|
| Tomatoes | $8.75 \times 10^{-4\dagger}$ |
| Beef | $6.49 \times 10^{-4}$ |
| Potatoes | $5.21 \times 10^{-4}$ |
| Oranges | $3.76 \times 10^{-4}$ |
| Lettuce | $3.44 \times 10^{-4}$ |
| Apples | $3.23 \times 10^{-4}$ |
| Peaches | $3.23 \times 10^{-4}$ |
| Pork | $2.67 \times 10^{-4}$ |
| Wheat | $1.92 \times 10^{-4}$ |
| Soybeans | $1.28 \times 10^{-4}$ |

*Please note that these figures are worst-case estimates. Four assumptions were made in arriving at these values: (1) the entire U.S. crop (of tomatoes, for example) is treated (2) with *all* pesticides that are registered for use on that crop; (3) the pesticides are applied the maximum number of times (4) at the maximum rate, or amount, each time.

†As an example of how to interpret these figures, tomatoes are estimated to cause an average of 8.75 deaths from cancer for every 10,000 people.

From *Regulating Pesticides in Foods: The Delaney Paradox,* National Research Council, National Academy Press, Washington, D.C., 1987.

In 1972 the EPA was given the authority to regulate pesticide use under the terms of the FDCA and FIFRA. Since that time, the EPA has banned or restricted the use of many chlorinated hydrocarbons (Table 22–1). In 1972 the EPA banned DDT for almost all uses. Aldrin and dieldrin were outlawed in 1974 after more than 80 percent of all dairy products, fish, meat, poultry, and fruits were found to contain residues of these insecticides. The banning of chlordane and heptachlor occurred in 1975, and kepone was banned in 1976.

A two-year study by the National Research Council[4] concluded in 1987 that U.S. laws regarding pesticide residues in food are not adequate to protect the public from cancer-causing pesticides (Table 22–4). It included a number of recommendations that were made into law—an amended

[4]The National Research Council is a private, nonprofit society of distinguished scholars. It was organized by the National Academy of Sciences to advise the federal government on complex issues in science and technology.

FIFRA—in 1988. The new law requires re-registration of older pesticides before the end of the 1990s, which will subject them to the same toxicity tests that new pesticides must face. Also, when a pesticide's registration is canceled, the EPA no longer has to pay the manufacturer for unused supplies of the pesticide, a practice that used to severely strain the EPA's limited budget. Manufacturers will now have to pay to dispose of their unused stocks.

Not everyone is happy with the new law. Although it is stricter than previous legislation, it represents a compromise between agricultural interests, including pesticide manufacturers, and those opposed to all uses of pesticides. The new law does not address a very important issue, the contamination of groundwater by pesticides. It also fails to address the establishment of standards for pesticide residues on foods and the safety of farm workers who are exposed to high levels of pesticides.

A modification of the Delaney Clause was instituted by the EPA in 1988: the maximum amount of pesticide residue allowed on the processed food eaten during a lifetime must pose a "negligible" risk of no more than one case of cancer for every million people. Although this could eliminate the type of Delaney Clause paradox described earlier (provided the EPA follows through and cancels the registration of older, more dangerous pesticides), a federal appeals court in San Francisco unanimously struck down the modification in July 1992. The three-judge panel said that only Congress had the authority to change the Delaney Clause. Two bills that would do that are pending in Congress as we go to press.

## The Importation of Banned Pesticides

The fact that many dangerous pesticides are no longer being used in the United States is no guarantee that traces of those pesticides are not in our food. Although many pesticides have been restricted or banned in the United States, they are widely used in other parts of the world. Much of our food that is imported from other countries, including coffee, chocolate, bananas, and many other fruits, may contain traces of banned pesticides such as DDT, dieldrin, chlordane, and benzene hexachloride.

## The Export of Banned Pesticides

A number of American companies manufacture pesticides that have been banned from use in the United States and export them to developing countries, particularly in Asia, Africa, and Latin America. Other developed nations also export banned pesticides. The Food and Agricultural Organization (FAO) of the United Nations is attempting to help developing nations become more aware of dangerous pesticides. It has established a "red alert" list of more than 50 pesticides that have been banned in five or more countries. Many of these pesticides are the ones responsible for the more than 1 million cases of pesticide poisoning that occur worldwide, mostly in developing countries, each year. The FAO further requires that the manufacturers of these pesticides inform importing countries about why such pesticides were banned. The United States supports these international guidelines and exports banned pesticides only with the informed consent of the importing country.

### Changing Attitudes

Heavy pesticide use can be attributed partly to the consumer, who has come to expect perfect, unblemished produce. There is no question that pesticides help farmers grow crops that are more visually appealing, but it comes to this: would you rather buy apples that are smaller and have an occasional blemish or worm but are pesticide-free, or apples that are free from all imperfections but contain traces of pesticides? Until consumers change their attitudes and demand produce that is grown without pesticides, farmers will continue to use pesticides (Figure 22–18).

Many farmers are exploring alternatives to pesticides on their own because they come into contact with pesticides on a regular basis. Farmers are aware of the dangers and problems associated with pesticide use and are concerned for their own safety and that of their families. They don't want to inhale pesticide or let it settle on their skin when they are applying it. They don't want to drink well water or eat food with traces of pesticide anymore than the typical consumer does.

**Figure 22–18**

Pesticide-free produce in a supermarket. If consumers demand food that is grown in the absence of pesticides, then farmers will grow more pesticide-free crops. (Steve Feld)

### Pesticide Risk Assessment

Although the effects of long-term exposure to low levels of carcinogenic pesticides should be of concern to all informed consumers, panic, which is often fueled by sensational news reports, is not justified. It is important to keep a balanced perspective when considering pesticides. The threat of cancer from consuming pesticide residues on our food is quite small compared to the threat of cancer from smoking cigarettes or from overexposure to ultraviolet radiation from the sun.

However, it is difficult for consumers to make informed decisions about the risks of pesticide residues on food, because we have no way of knowing what kinds of pesticides have been used on the foods we eat. A requirement that all foods list such chemicals would go a long way toward helping us assess risks as well as discouraging heavy pesticide use.

## SUMMARY

**1.** Pesticides are toxic chemicals that are used to kill pests, including insects, weeds, fungi, and rodents. Most are broad-spectrum pesticides, which kill a variety of organisms besides the target organism.

**2.** Insecticides are classified into groups based on their chemical structure—such as chlorinated hydrocarbons, organophosphates, and carbamates. Herbicides may be classified as either nonselective or selective, with selective herbicides including broad-leaf herbicides and grass herbicides.

**3.** Pesticides provide important benefits to humans, including the prevention of diseases that are transmitted by insects, such as malaria. In addition, pesticides reduce crop losses from pests, thereby increasing agricultural productivity.

**4.** Some serious problems are associated with pesticide use. Humans may be poisoned by exposure to large amounts, whereas lower levels of many pesticides pose a long-term threat of cancer.

**5.** In addition to its negative effect on human health, pesticide use has caused environmental problems. These include the development of genetic resistance in the pest, adverse effects on nontarget species, the creation of new pests, and mobility in the environment. Persistence and biological magnification are other problems associated with some types of pesticides.

**6.** The need for pesticides in agriculture can be less-ened or even eliminated by using quarantine, biological controls, genetic controls, and pheromones and hormones. Cultivation techniques such as crop rotation are also effective in controlling pests. Integrated pest management in agriculture stresses biological controls and cultivation methods along with the judicious use of pesticides. Irradiation of food can be used to control pests after food has been harvested.

**7.** Pesticide registration and use are controlled by the Food, Drug, and Cosmetics Act (FDCA) and the Federal Insecticide, Fungicide, and Rodenticide Act (FIFRA). These acts have been updated and made more useful over the years but are still not completely effective. More legislation is pending before Congress.

## DISCUSSION QUESTIONS

**1.** What is the dilemma referred to in the title of this chapter?

**2.** Overall, do you think the benefits of pesticide use outweigh its disadvantages? Give at least two reasons for your answer.

**3.** Sometimes pesticide use can increase the damage done by pests. Explain.

**4.** Biological control is often much more successful on a small island than on a continent. Offer at least one reason why this might be the case.

**5.** It is more effective to use the sterile male technique when an insect population is small than when it is large. Explain.

**6.** Which of the following uses of pesticides do you think are most important? Which are least important? Explain your views.

(a) Keeping roadsides free of weeds
(b) Controlling malaria
(c) Controlling crop damage
(d) Producing blemish-free fruits and vegetables

## SUGGESTED READINGS

Borchelt, R. *Regulating Pesticides in Food: The Delaney Paradox*, National Research Council News Report. National Academy Press, Washington, D.C., 1987. Conflicting standards exist in the FDCA and FIFRA provisions that regulate pesticide residues in food.

Carey, J. R. Establishment of the Mediterranean fruit fly in California. *Science*, vol. 253, 20 September 1991. The author presents evidence that the medfly, one of the most destructive agricultural pests in the world, has become established in southern California.

Carson, R. L. *Silent Spring*. Houghton Mifflin, Boston, 1962. The classic book that first alerted the public to the environmental dangers of pesticide use.

Dover, M. J., and B. A. Croft. Pesticide resistance and public policy. *BioScience* 36:2, February 1986. Strategies for managing pesticide resist-ance. This issue also contains an article on the amounts of pesticide applied versus the amount that reaches the target pest.

Lambert, B., and M. Peferoen. Insecticidal promise of *Bacillus thuringiensis*. *BioScience* 42:2, February 1992. What is known (and not known) about the bacterium that is used to produce a large number of natural insecticides.

Marshall, E. Malaria research—What next? *Science*, vol. 247, 26 January 1990. Malaria, the most common infectious disease in the world, is on the increase. Possible ways to stop or control it are considered.

Pimentel, D., et al. Environmental and economic effects of reducing pesticide use. *BioScience* 41:6, June 1991. The authors present a convincing case that pesticide use can be reduced by 50 percent in the United States without a substantial drop in productivity.

Sonsino, S. Radiation meets the public's taste. *New Scientist*, vol. 113, 19 February 1987. Irradiation is a controversial process that destroys microorganisms in food, keeping it fresh longer.

Strobel, G. A. Biological control of weeds. *Scientific American*, July 1991. Explores environmentally safe alternatives to herbicides.

Weber, P. A place for pesticides? *World Watch* 5:3, May/June 1992. How some farmers are reducing their dependence on pesticides.

*Tons of garbage heaped on a barge sail past the Statue of Liberty on its way to Fresh Kills Landfill in Staten Island, New York, the largest municipal garbage dump in the world. (Louie Psihoyos/Matrix International, Inc.)*

# Solid and Hazardous Wastes

he continual increase in the amount of solid waste being generated is a major problem today, particularly because the number of sanitary landfills to receive the trash is declining. There is no single solution to the challenge of solid waste disposal; a combination of source reduction, recycling, composting, burning, and burying in sanitary landfills is currently the optimal way to manage solid waste. Source reduction is the most underutilized aspect of waste management. Industrial processes can be designed to reduce not only the volume of solid waste but the amount of hazardous materials in solid waste. Hazardous waste should be treated to eliminate or reduce its toxicity, and the small amount of hazardous waste that remains should be disposed of in hazardous waste landfills that are designed to protect the environment from contamination.

## THE SOLID WASTE PROBLEM

Every man, woman, and child in the United States produces an average of 1.6 kilograms (3.5 pounds) of solid waste per day, which corresponds to a total of 160 million metric tons (176 million tons) per year in the United States. And the problem worsens each year as the U.S. population increases and as per-capita consumption continues to increase (Figure 23–1).

The solid waste problem has been made abundantly clear by several highly publicized instances of garbage barges wandering from port to port and from country to country, trying to find someone willing to accept their cargo. In 1987, for example, a garbage barge from Islip, New York, was towed by the tugboat *Break of Dawn* to North Carolina (Figure 23–2). When North Carolina refused to accept the garbage, the *Break of Dawn* set off on a journey of many months. In total, six states and three countries rejected the waste, which was eventually returned to New York to be incinerated.

A more ominous example of our solid waste problem is the story of the *Khian Sea*, a Bahamian ship that was hired by the city of Philadelphia in 1986. It was to transport thousands of tons of incinerated ash to Panama to be buried under a road for a new tourist resort. Incinerated ash contains toxic chemicals that could have adversely affected the fragile wetlands through which the road was being built, so Panama rejected the waste. The *Khian Sea* spent the next two years wandering, attempting to give its cargo to countries on five different continents. The ship reappeared in 1988, empty of cargo and with no explanations. Whether the ash was illegally dumped in some foreign nation or at sea is unknown.

Waste is an unavoidable consequence of prosperous, high-technology, disposable economies; it is a problem not only in the United States but in other industrialized nations. Many products that have the potential to be repaired, reused, or recycled are simply thrown away. Others, including paper napkins and disposable diapers, are designed to be used once and then discarded. Packaging, which not only makes a product more attractive and therefore more likely to sell, but protects it and keeps it sanitary, also contributes to waste. Nobody likes to think about garbage, but the fact is that solid waste is a pressing concern of modern society— we keep creating it, and places to dispose of it safely are dwindling in number. In this chapter we exam-

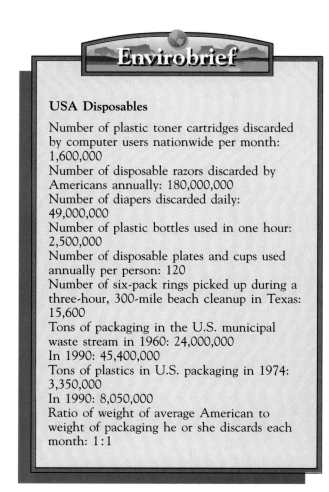

**Envirobrief**

**USA Disposables**

Number of plastic toner cartridges discarded by computer users nationwide per month: 1,600,000
Number of disposable razors discarded by Americans annually: 180,000,000
Number of diapers discarded daily: 49,000,000
Number of plastic bottles used in one hour: 2,500,000
Number of disposable plates and cups used annually per person: 120
Number of six-pack rings picked up during a three-hour, 300-mile beach cleanup in Texas: 15,600
Tons of packaging in the U.S. municipal waste stream in 1960: 24,000,000
In 1990: 45,400,000
Tons of plastics in U.S. packaging in 1974: 3,350,000
In 1990: 8,050,000
Ratio of weight of average American to weight of packaging he or she discards each month: 1:1

**Envirobrief**

**Junk Mail Blues**

Number of pieces of junk mail posted in the United States in 1990: 63.7 billion
Approximate number of acres of forest cut to create 63.7 billion pieces of mail: 74,000
Number of pounds of junk mail received each year by the average Los Angeles household: 170
Number of pieces of junk mail yielded by a forest the size of Manhattan: 10 billion
Number of pieces per year of junk mail never opened by U.S. recipients: 10 billion
Proportion of all discarded paper in the United States accounted for by junk mail: 5 percent

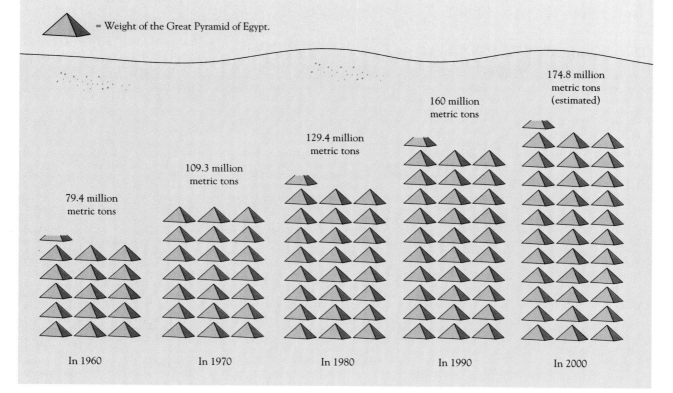

= Weight of the Great Pyramid of Egypt.

79.4 million metric tons

In 1960

109.3 million metric tons

In 1970

129.4 million metric tons

In 1980

160 million metric tons

In 1990

174.8 million metric tons (estimated)

In 2000

(a) One year's garbage in the United States.

**Figure 23–1**
As long as Americans continue to consume more and more of everything, huge amounts of solid waste will be produced. (a) The amount of municipal solid waste has been rising steadily in the United States. (b) How municipal solid waste generated by a typical American household increases over time.

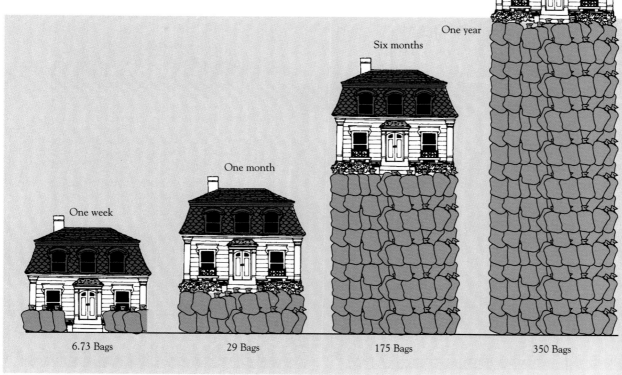

One week

6.73 Bags

One month

29 Bags

Six months

175 Bags

One year

350 Bags

(b) Garbage generated by average American household.

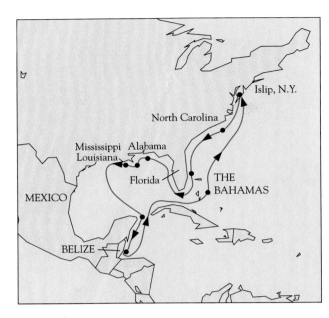

**Figure 23–2**
The tugboat *Break of Dawn* transported a load of garbage from New York to six states and three different countries before returning its unwanted cargo to where it originated. (© Jeffrey Cardinas/SYGMA)

ine the problems and the opportunities associated with the management of solid waste.

approximately 98.5 percent of solid waste in the United States is from nonmunicipal sources.

## TYPES OF SOLID WASTES

**Municipal solid waste,** or trash, is solid waste that is generated primarily in homes, although it also includes the waste from commercial and institutional facilities. Municipal solid waste is a heterogeneous mixture composed primarily of paper and paperboard, yard waste, glass, metals, plastics, food waste, and other materials such as rubber, leather, and textiles (Figure 23–3). The proportions of the major types of solid waste in this mixture change over time. Today's solid waste contains more paper and plastics than in the past, whereas the amounts of glass and steel have declined.

Municipal solid waste is a relatively small portion of all the solid waste produced. **Nonmunicipal solid waste,** which includes wastes from industry, agriculture, and mining, is produced in substantially larger amounts than municipal solid waste;

## DISPOSAL OF TRASH

Trash has traditionally been regarded as material that is no longer useful and that should be disposed of. There are four ways to get rid of trash: dump it, bury it, burn it, or compost it. Alternatively, we can dispose of trash by exporting it to another country (see Focus On: International Issues in Waste Management, page 526).

### Open Dumps

The old method of solid waste disposal was dumping. Open dumps were unsanitary, malodorous places in which disease-carrying vermin such as rats and flies proliferated. Methane gas was released into the surrounding air as microorganisms decomposed the garbage, and fires polluted the air with acrid smoke. Liquid that oozed and seeped through

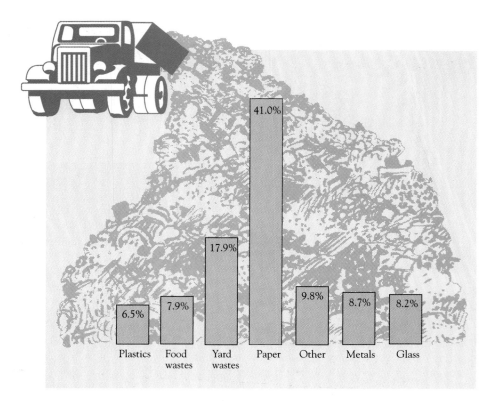

**Figure 23–3**
What's in our garbage? Municipal solid waste is composed of paper, yard waste, metal, glass, plastic, food scraps, and other materials such as rubber and leather.

the garbage heap ultimately found its way into the soil, surface water, and groundwater. Hazardous materials (Table 23–1) that were dissolved in this liquid often contaminated soil and water.

## Sanitary Landfills

Open dumps have largely been replaced by sanitary landfills, which receive approximately 80 percent of the solid waste generated in the United States today. Sanitary landfills differ from open dumps in that the trash is placed in a hole, compacted, and covered with a thin layer of soil every day (Figure 23–4). This process reduces the number of rats and other vermin usually associated with garbage, lessens the danger of fires, and decreases the amount of odor. If a sanitary landfill is operated in accordance with strict safety guidelines, it does not pollute local surface and groundwater. Newer landfills possess sophisticated systems to collect leachate (liquid that seeps, or leaches, through the trash) and gases that form during decomposition.

The location of an "ideal" sanitary landfill is based on a variety of factors, including the geology of the area and the proximity of nearby bodies of water and wetlands. The landfill should be far enough away from centers of dense population to be inoffensive but close enough so as not to create high transportation costs. In an ideal situation, safety is ensured by the presence of layers of compacted clay and plastic sheets at the bottom of the landfill, which prevent liquid wastes from leaching into groundwater (Figure 23–5). Nearby groundwater should be continually monitored for possible pollution, and precipitation runoff from the landfill should be collected and treated to remove any possible contaminants.

**Problems Associated with Sanitary Landfills**  Although the operation of sanitary landfills has improved over the years with the passage of stricter and stricter guidelines, very few landfills are ideal. Most sanitary landfills in operation today do not meet current legal standards.

One of the problems associated with sanitary landfills is the production of methane gas by microorganisms that decompose organic material anaerobically (in the absence of oxygen). This methane

**Table 23–1**
**Examples of Hazardous Waste**

| Hazardous Material | Possible Sources |
| --- | --- |
| Acids | Ash from power plants and incinerators; petroleum products |
| CFCs | Coolant in air conditioners and refrigerators |
| Cyanides | Metal refining; fumigants in ships, railway cars, and warehouses |
| Dioxins | Ash from incinerators (?) |
| Explosives | Old military installations |
| Heavy metals | Paints, pigments, batteries, ash from incinerators, sewage sludge with industrial waste, improper disposal in landfills |
|    Arsenic | Industrial processes, pesticides, additives to glass, paints |
|    Cadmium | Rechargeable batteries, incineration, paints, plastics (?) |
|    Lead | Lead-acid storage batteries, stains and paints; TV picture tubes and electronics discarded in landfills |
|    Mercury | Paints, household cleaners (disinfectants), industrial processes, medicines, seed fungicides |
| Infectious wastes | Hospitals, research labs |
| Nerve gas | Old military installations |
| Organic solvents | Industrial processes, household cleaners, leather, plastics, pet maintenance (soaps), adhesives, cosmetics |
| PCBs | Older appliances (built before 1980); electrical transformers and capacitors |
| Pesticides | Household products |
| Radioactive wastes | Nuclear power plants, nuclear medicine facilities, weapons factories |

may seep through the trash and accumulate in underground pockets, creating the possibility of an explosion. It is even possible for methane to seep into basements of nearby homes, creating an extremely dangerous situation. Some landfills collect the methane and burn it as a fuel. A landfill near Virginia Beach, Virginia, for example, uses methane to generate electricity.

Another problem is the contamination of surface water and groundwater by leachate from sanitary landfills without proper linings. Even household trash contains toxic chemicals that can be carried into groundwater. Pollutants in the leachate may include heavy metals, pesticides, and organic compounds.

Although we currently dispose of about 80 percent of our solid waste in landfills, they are not a long-term remedy for waste disposal because they are filling up. During the decade from 1978 to 1988, the number of landfills in operation decreased from 20,000 to 6,000 (most recent data available). It is likely that half the landfills currently in operation will be full in just a few short years. The situation is critical in some states, such as New Jersey and Florida, because it is anticipated that almost all of their landfills will be full in less than ten years (Figure 23–6).

Compounding the problem is the fact that not enough new sanitary landfills are being opened to replace the old ones. The reasons for this are many and complex. Many desirable sites have already been taken. Also, people living near a proposed site are usually adamantly opposed to a landfill near their homes. Recall from Chapter 11 that the opposition of people to the location of hazardous facilities near their homes is known as the **not in my backyard**, or NIMBY, response. This attitude is partly the result of past problems with sanitary landfills—everything from offensive odors to dangerous contamination of drinking water. It is also caused by the fear that property values will be lowered by a nearby facility.

**Closing a Sanitary Landfill** Considerable expense is involved in closing a sanitary landfill once it is

(Text continues on p. 526)

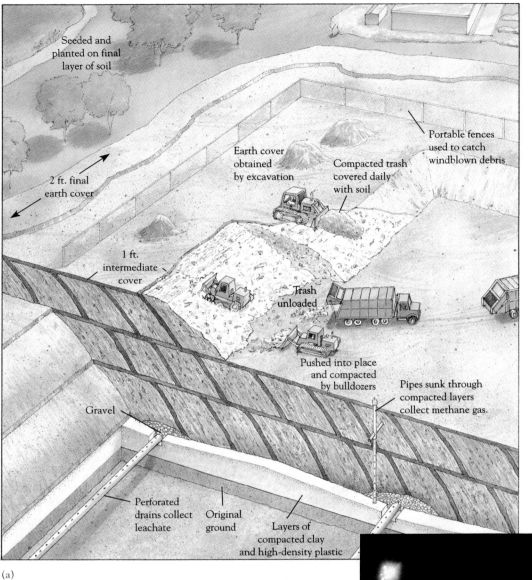

(a)

(b)

**Figure 23–4**
A sanitary landfill is much more than a hole in the ground.
(a) Solid waste is compacted into smaller sections, called cells,
and covered with soil. (b) After a landfill is full, it can be planted
and returned to a more or less natural condition. Alternatively, it
can be developed as a ski slope, baseball field, or other use.
Shown is an Atlanta softball field that was built over a landfill.
(Ron Sherman)

**Figure 23–5**
A modern sanitary landfill contains plastic liners (shown) to trap leachate and prevent it from draining any deeper into the soil and contaminating groundwater. (Terry Wild Studio)

**Figure 23–6**
We are running out of time. Many states have less than 5 years before all existing landfills will be full. Certain municipal areas in states that do not appear to have an immediate problem are also running out of landfill space.

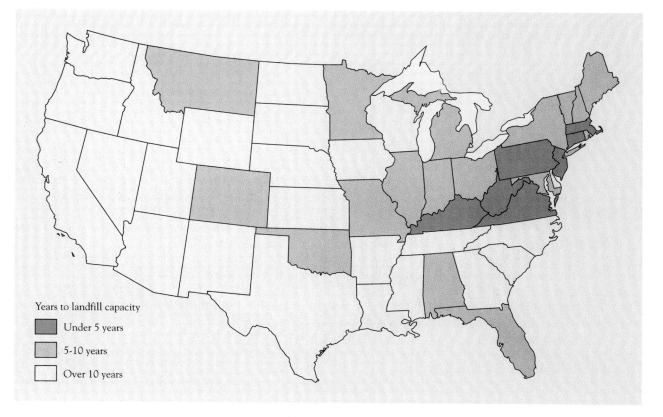

Years to landfill capacity

Under 5 years

5-10 years

Over 10 years

## International Issues in Waste Management

Although there are ways to reduce and dispose of waste in an environmentally sound manner, industrialized countries sometimes choose to send their waste to other countries. Some wastes are exported for legitimate recycling, but some are exported strictly for disposal.

The export of both solid and hazardous wastes by the United States, Japan, and the European Economic Community is one of the most controversial aspects of waste management today. Usually the recipients of this waste are developing nations in Africa, Central and South America, and the Pacific Rim of Asia. The governments, industries, and citizens of these countries are inexperienced and ill equipped to handle such materials. As a result, the wastes often create the same types of environmentally hazardous sites in developing nations that industrialized nations are trying to clean up at home.

Sometimes the governments of developing nations are not even informed when hazardous waste is delivered to their countries. In the 1980s, for example, an Italian business placed 8,000 drums of toxic waste contaminated with PCBs in Nigeria without the Nigerian government's permission. Naturally, Nigeria was outraged when it discovered

what had occurred. It pressured Italy to retrieve the waste, which it did—in another infamous waste ship, the *Karin B*. This ship tried unsuccessfully to dump its cargo in five different European countries before returning it to Italy.

The United Nations Environment Programme developed a treaty to restrict the international transport of hazardous wastes. Although many developing countries supported a total ban on the export of hazardous waste, the United States vigorously opposed such a ban. The final treaty, called the Basel Convention on the Control of Transboundary Movements of Hazardous Waste and Their Disposal, was signed in 1989. It allows countries to export hazardous waste only with prior consent of the country that is to receive the waste. Many developing nations refused to sign the treaty because it legitimizes a process they consider environmentally and morally abhorrent.

In the authors' opinion, the United States government has an opportunity and an obligation to become a world leader in waste management by reducing the volume of solid and hazardous waste and by safely disposing of unavoidable wastes *in the United States,* not in other nations. What is your opinion?

---

full. The danger of groundwater pollution and gas explosions remains for a long time, so appropriate monitoring must continue long after the landfill is closed. In addition, no homes or other buildings can be built on a closed sanitary landfill for many years.

Despite their numerous problems, sanitary landfills are currently our primary method of solid waste disposal. Alternative ways to deal with solid waste are increasingly being adopted, however. These include incineration and composting as well as methods of using and reducing trash. Regardless of what waste management practices are used, there will probably always be a need for sanitary landfills for unburnable and unusable solid waste.

**The Special Problem of Plastic**  The amount of plastic in our trash is growing faster than any other

component of municipal solid waste. More than half of this plastic is from packaging. Plastics are chemical **polymers** that are composed of chains of repeating carbon compounds. The properties of the many types of plastics—polypropylene, polyethylene, and polystyrene, to name a few—differ based on their chemical compositions.

One of the characteristics of most plastics is that they are chemically stable and do not readily break down, or decompose. This characteristic, which is essential in the packaging of certain products, such as food, causes long-term problems: most plastic debris will probably last for centuries.

In response to concerns about the volume of plastic waste, some places have actually banned the use of certain types of plastic, such as the polyvinyl chloride employed in packaging. Special plastics that have the ability to degrade or disintegrate have

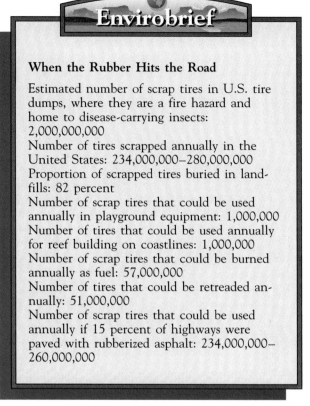

### When the Rubber Hits the Road

Estimated number of scrap tires in U.S. tire dumps, where they are a fire hazard and home to disease-carrying insects: 2,000,000,000
Number of tires scrapped annually in the United States: 234,000,000–280,000,000
Proportion of scrapped tires buried in landfills: 82 percent
Number of scrap tires that could be used annually in playground equipment: 1,000,000
Number of tires that could be used annually for reef building on coastlines: 1,000,000
Number of scrap tires that could be burned annually as fuel: 57,000,000
Number of tires that could be retreaded annually: 51,000,000
Number of scrap tires that could be used annually if 15 percent of highways were paved with rubberized asphalt: 234,000,000–260,000,000

been developed. Some of these, however, are **photodegradable;** they break down only upon exposure to sunlight, which means they won't break down in a sanitary landfill. Other plastics are **biodegradable** (decomposed by living organisms). Whether biodegradable plastics actually break down under the conditions found in a sanitary landfill is not yet clear, although preliminary studies indicate that they probably do not.

## Incineration

When solid waste is incinerated, two positive things are accomplished. First, the volume of solid waste is reduced by up to 90 percent; the ash that remains is, of course, much more compact than trash that has not been burned. Second, incineration produces heat that, if properly channeled, can produce steam to warm buildings or generate electricity. In 1991 the United States had 128 waste-to-energy incinerators, which burned approximately 15 percent of the nation's solid waste. By comparison, only 1 percent of U.S. solid waste was incinerated in 1970.

Some materials are best removed from solid waste before incineration occurs. For example, glass doesn't burn, and when it melts, it is difficult to remove from the incinerator. Although food wastes burn, their high moisture content often decreases the efficiency of incineration, so it is better to remove them before incineration. The best materials for incineration are paper, plastics, and rubber.

Paper is a good candidate for incineration because it burns readily and produces a great amount of heat. One potential complication is the presence in the ink and paper of toxic compounds, which might be emitted during incineration. Some types of paper release dioxins into the atmosphere when burned; dioxins are very toxic compounds that threaten human health (see Chapter 22).

Plastic produces a lot of heat when it is incinerated: a kilogram of plastic waste yields almost as much heat as a kilogram of fuel oil. As with paper, there is concern about some of the pollutants that might be emitted during the incineration of plastic. Polyvinyl chloride, a common component of many plastics, may release dioxins and other toxic compounds when incinerated. More research is needed to determine whether these or other hazardous substances are actually produced during the incineration of plastics.

One of the most difficult materials to manage is rubber. Discarded tires—some 279 million each year in the United States—are made of vulcanized rubber, which cannot be melted and reused. The number of products that can be made from old tires is limited, although some research in product development is being carried out. Disposal of tires in sanitary landfills is a real problem, because tires, being relatively large and light, have a tendency to move upward in the volume of trash; after a period of time, they can be found at the top of the trash heap. These tires are a fire hazard and collect rainwater, providing a good breeding place for mosquitoes. One of the best uses for old tires is incineration, because burning rubber produces much heat. In order to burn properly in municipal incinerators, tires must first be shredded.

**Types of Incinerators**   The three types of incinerators are mass burn, modular, and refuse-derived fuel. **Mass burn incinerators** are large furnaces that burn *all* solid waste except for unburnable items such as refrigerators. Most mass burn incinerators are large and are designed to recover the energy produced from combustion (Figure 23–7). **Modular incinerators** are smaller incinerators that burn all solid waste. They are assembled at factories and so are less expensive to build. In **refuse-derived fuel**

**incinerators,** only the combustible portion of solid waste is burned. First, noncombustible materials such as glass and metals are removed by machine or by hand. The remaining solid waste, including plastic and paper, is shredded or shaped into pellets and burned.

**Problems Associated with Incineration**   The possible production of toxic air pollutants is the main reason people oppose incineration in their proximity. Incinerators can pollute the air with carbon monoxide, particulates, heavy metals, and other toxic materials unless expensive air pollution control devices are used. Such devices include **lime scrubbers,** towers in which a chemical spray neutralizes acidic gases, and **electrostatic precipitators,** which give ash a positive electrical charge so that it adheres to negatively charged plates rather than going out the chimney. (See Chapters 10 and 19 for additional information on these air pollution control devices.)

**Figure 23–7**
A modern incinerator has pollution control devices such as lime scrubbers and electrostatic precipitators to trap dangerous and dirty emissions.

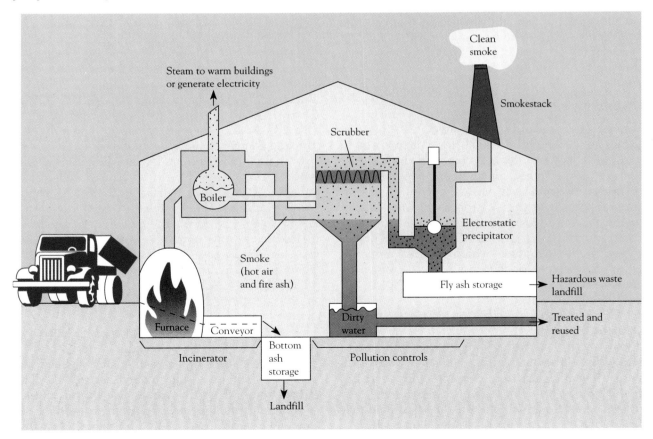

Incinerators produce large quantities of ash that must be disposed of properly. Two kinds of ash are produced, bottom ash and fly ash. **Bottom ash** (also known as **slag**) is the residual ash left at the bottom of the incinerator when combustion is completed. **Fly ash** is the ash from the flue (chimney) that is trapped by electrostatic precipitators. Fly ash usually contains more toxic materials, including heavy metals and possibly dioxins, than bottom ash.

Currently, both types of incinerator ash are best disposed of in special sanitary landfills or hazardous waste sites. What happens to the toxic materials in incinerator ash when it is placed in an average sanitary landfill, as is usually the case, is unknown, but there are concerns that the toxic materials could contaminate groundwater. Scientists are working on other ways to dispose of or use incinerator ash. A team from the State University of New York has mixed toxic ash with cement to make concrete blocks for buildings. It appears from preliminary data that this process locks up the hazardous materials so that they cannot contaminate the environment.

Site selection for incinerators, like that for sanitary landfills, is controversial. People may recognize the need for an incinerator, but they don't want it near their homes (Figure 23–8). Another drawback of incinerators is their high cost. Prices have been escalating because costly pollution control devices are now required.

## Composting

Yard waste—grass clippings, branches, and leaves—is a large component of municipal solid waste. As space in sanitary landfills becomes more limited, other ways to dispose of yard waste are being developed and implemented. One of the best ways is to convert organic wastes into soil conditioners such as compost or mulch (see You Can Make a Difference: Practicing Environmental Principles, in Chapter 14). Food scraps, sewage sludge, and agricultural manure are other forms of solid waste that can be used to make compost. Compost and mulch can be used in public parks and playgrounds, for landscaping, or as part of the daily soil cover at sanitary landfills; or they can be sold to gardeners.

Composting as a way to manage solid waste first became popular in Europe. A number of municipalities in the United States have composting facilities as part of their comprehensive solid waste management plans (Figure 23–9). At least ten

**Figure 23–8**
An environmental protest against a solid waste incinerator in East Liverpool, Ohio, on May 17, 1992. The NIMBY ("*Not In My Backyard*") response is often seen when municipal planners try to find locations for incinerators or sanitary landfills. Part of the reason for this is that many communities are not brought into the decision-making process enough. (Visuals Unlimited/Bill Beatty)

**Figure 23–9**
Leaves, twigs, and other yard wastes are collected and delivered to the Ann Arbor Municipal Recycling Center in Michigan. Composting is a great way to dispose of yard waste without filling up valuable landfill space. (The plastic trash bags will be screened out later.) (Larime Photographic/ Dembinsky Photo Associates)

states have banned yard wastes from sanitary land-fills to help alleviate the solid waste crisis. This trend is likely to continue, making composting even more desirable.

## REDUCING THE VOLUME OF TRASH

Every time an item is recycled or reused, the volume of solid waste is reduced (see Meeting the Challenge: Recycling Automobiles in Germany). We already discussed recycling, reuse, and demate-rialization in Chapter 15, in the context of resource conservation. Now we examine the impact of these practices on solid waste.

### An Industrial Ecosystem

The J. R. Simplot Company supplies more than half of the french fries sold by McDon-ald's each year. According to a *Newsweek* re-port in 1990, Jack Richard Simplot, the chief executive of the firm, has turned waste into money with an extraordinary waste reduction system.

Because french fries are a quality-controlled product, more than 50 percent of Simplot's potatoes normally used to end up as waste, until he learned that the peelings could be mixed with grain to feed cattle. His potato residues now feed more than 150,000 animals. But that's only the beginning. Wastewater from potato processing is used for irrigation. Cattle manure is anaerobically di-gested to create methane for power plants. Power is also derived from ethanol made from potatoes. The sludge from the ethanol process is converted to low-cost fish food for a commercial fish breeding operation. The water from the fish tanks is filtered through tubs filled with water hyacinths. The flowers process nitrogen from the fish waste so that the water can be reused. The hyacinths are fed to the cattle. And after the fish are har-vested, their offal is mixed back into food for more fish.

## Recycling

It is possible to recycle many materials found in solid waste. Recycling is preferred over incineration and landfill disposal because it conserves our natu-ral resources and is more environmentally benign. Recycling also has a positive effect on the economy by creating jobs.

Solid waste is a mixture of many different ma-terials that must be separated from one another be-fore recycling can be accomplished. It is easy to separate materials such as glass bottles and newspa-pers, but the separation of materials in items with complex compositions is difficult. Some food con-tainers are composed of thin layers of metal foil, plastic, and paper, for example. Trying to separate these layers is a daunting prospect, to say the least.

**Figure 23–10**
Recycling is becoming a way of life in many communities across the United States. In Howard County, Maryland, paper (in paper bags), plastic containers, glass containers, and metal cans are picked up for recycling once each week. (Dennis Drenner)

Many areas have started recycling programs. As of 1991, 28 states had laws that mandated recycling and waste reduction. Some 1,500 communities in 35 states have adopted a low-technology approach. They rely on their citizens to sort their solid waste and leave it in special bins at curbside for pickup (Figure 23–10). Other communities collect trash as a mixture and take it to special resource recovery facilities (Figure 23–11), where it is either hand-sorted or separated using a variety of technologies, including magnets, screens, and conveyor belts.

The most successful large-scale recycling program in the United States is in Seattle, Washington. By 1989 Seattle had reduced its solid waste by 37 percent, largely through recycling. The city's ultimate goal is to reduce its solid waste by 60 percent by 1998.

Most people think that recycling involves merely separating certain materials from the solid waste stream, but that is only the first step in recycling. In order for recycling to work, there must be a market for the recycled goods, and the recycled products must be used in preference to virgin products.

**Paper** Americans recycle less than 30 percent of their paper. The rest ends up in landfills or incinerators. This is a dismal record compared to most other developed countries, which recycle up to twice as much waste paper as the United States does. Part of the reason paper is not recycled more in the United States is that many of our paper mills are old and are not equipped to process waste paper. The number of mills that can handle waste paper is slowly growing; in 1989 just over 25 percent were able to use waste paper.

In addition to a slow increase in paper recycling in the United States, there is a growing demand for U.S. waste paper in other countries. Mexico, for example, imports a large quantity of waste paper and cardboard from the United States. Used paper is also in great demand in China, Taiwan, and Korea.

As mentioned earlier, paper recycling does not work unless there is a market for recycled paper products. Sometimes the demand can be created by legislation. In Toronto, Canada, for example, the city council passed a law, effective in 1991, that daily newspapers must contain at least 50 percent

**Figure 23–11**
Resource recovery. (a) Plastic containers can be separated from other garbage and recycled. (Larime Photographic / Dembinsky Photo Associates) (b) When shredded, plastic can be used to make polyester carpet and other products. (P. Degginger / H. Armstrong Roberts)

(b)

(a)

# Recycling Automobiles in Germany

The Leer Volkswagen plant in Germany is a research and development laboratory that is concerned with cars—not producing them, but disassembling and recycling them. Many auto manufacturers, both in Germany and in other countries, are beginning to develop recycling strategies, because scrapped cars contribute to the growing solid waste problem (for example, Germans discard approximately 2 million cars each year). Although about 75 percent of a scrapped car is easily recyclable as scrap metal and secondhand parts, the remaining 25 percent is not easy to recycle and usually ends up in sanitary landfills.

Volkswagen is one of several auto manufacturers in Germany that have announced that they will take back the models being produced now when they are ready to be disposed of (in about ten years for most cars) and recycle as much of them as is possible. In the Leer plant, Volkswagen is researching ways to recycle or use old auto parts, a complex problem because automobiles typically contain about 600 different materials—glasses, metals, plastics, fabrics, rubber, leather, and so on. Economics is an important aspect of the problem, as the company wants to make money or at least break even in its recycling endeavor.

Workers at the Leer plant begin disassembling a used car by draining all fluids (such as transmission fluid, oil, and brake fluid) and recycling them or processing them for disposal. Large components, such as the engine, tires, and battery, are then removed and either sold as used parts or further disassembled for their materials; for example, catalytic converters are disassembled because they contain valuable amounts of platinum and rhodium.

Plastic, such as that used for interior molding and bumpers, is removed next. Volkswagen currently shreds old plastic bumpers and uses them to make new bumpers. The company would like to eventually recycle all plastic parts, but that is a bigger challenge than it might seem: there are currently no industry standards for plastic parts, and as a result, the kinds and amounts of plastic from which cars are made vary a great deal.

It may never be feasible to recycle all parts of a used automobile. For example, as yet there is no use for glass with heat strips embedded in it (the rear window). Other parts—steering wheels, for example—have such complex compositions that it is simply too expensive to separate them into their component materials.

Auto manufacturers in the United States are beginning to consider the challenge of recycling old cars, although none is currently doing what Volkswagen is doing. Legislators are watching German efforts as well, perhaps with the idea of someday creating legislation that would require U.S. manufacturers to accept scrapped autos from consumers after they have outlived their usefulness.

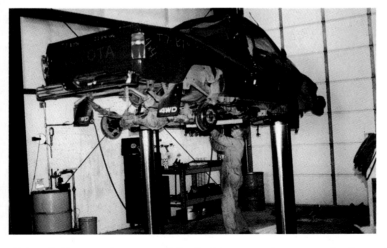

Although Germany has assumed the lead in attempting to recycle almost all components of old automobiles, the United States currently recycles portions of over 11 million vehicles each year. The automobiles are disassembled and some of their parts are sold to auto body shops, to new and used car dealers, and to repair shops. In addition, about 37 percent of the iron and steel scrap reprocessed in the United States comes from old automobiles. (Automotive Dismantlers and Recyclers Association)

recycled fiber or the publishers will not be allowed to have vending boxes on city streets. Maryland has taken a different approach. It taxes newspapers that do not contain at least 12 percent recycled fiber, at $10 per ton.

**Glass** Glass is another component of solid waste that is appropriate for recycling. Recycled glass costs less than glass made from virgin materials. Glass food and beverage containers are crushed to form **cullet,** which can be melted and used by glass manufacturers to make new products without any special adaptations in their factories. Although cullet is much more valuable when glass containers of different colors are separated before being crushed, cullet made from a mixture of colors can be used to make glassphalt, a composite of glass and asphalt that makes an attractive roadway.

**Aluminum** The recycling of aluminum (see Chapter 15) is one of the best success stories in U.S. recycling, largely because of economic factors. Making a new aluminum can from a recycled one requires a fraction of the energy it would take to make a new can from raw metal. Because energy costs are high, there is strong economic incentive to recycle aluminum. Approximately 55 percent of all aluminum cans in the United States were recycled in 1988.

**Metals Other Than Aluminum** Other recyclable metals include lead, gold, iron and steel, silver, and zinc. One of the obstacles to recycling metal products discarded in municipal solid waste is that their metallic compositions are often unknown. It is also difficult to extract metal from products, such as stoves, that contain other materials besides metal (plastic, rubber, or glass, for instance). In contrast, any waste metal produced at factories is recycled easily because its composition is known.

The economy has a large influence on whether metal is recycled or discarded; greater recycling occurs when the economy is strong than when there is a recession. Thus, although the supply of metal waste is fairly constant, the amount of recycling varies from year to year.

One exception to this generalization is steel. Before the 1970s, almost all steel was produced in mills that processed raw ores. Starting in the 1970s and continuing to the present, however, "minimills" that produce steel products completely from scrap became increasingly important. Minimills have experienced fewer economic difficulties than traditional steel mills in recent years.

**Plastic** Overall, less than 1 percent of plastic was recycled in 1988. However, public pressure to recycle plastic has been so strong that the demand for recycled plastic is currently greater than the supply. As a result, more plastic is being recycled each year.

Despite the limited amount of plastic recycling, it offers great promise of both environmental and economic benefits. Recycled plastic costs about two-thirds as much as virgin plastic. Certain types of scrap plastic sell for approximately 6 cents per pound. In comparison, scrap paper sells for only 1.5 cents per pound. Increasingly, local and state governments are supporting or requiring the recycling of plastic.

Polyethylene terphthalate (PET), the plastic used in soda bottles, is recycled more than any other plastic. Industry sources declare a 12 percent recycling rate for PET, which is incorporated into such diverse products as carpeting, automobile parts, and tennis ball felt.

Polystyrene (one form of which is Styrofoam) is an example of a plastic with great recycling potential. Cups, silverware, and packaging material that are made of polystyrene can be recycled into a variety of products, from coat hangers to flower pots to foam insulation to toys. Because approximately 2.3 billion kilograms (5 billion pounds) of polystyrene are produced each year, large-scale recycling would make a major dent in the amount of polystyrene that ends up in landfills. (By comparison, 0.7 billion kilograms [1.5 billion pounds] of PET is produced each year.)

One of the challenges associated with recycling plastic is that there are many different kinds. Forty-six different plastics are common in consumer products, and many products contain multiple kinds of plastic. A plastic ketchup bottle, for example, may have up to six different layers of plastic bonded together. In order to effectively recycle high-quality plastic, the different types must be sorted or separated. If two or more resins are recycled together, the resultant plastic is of lower quality.

Low-quality plastic mixtures are used to make a construction material similar to wood (sometimes called "plastic lumber") that is particularly useful for outside products, such as fence posts and park benches, because of its durability. One of the more interesting examples of waste plastic recycling is found in Cologne, Germany, where a plastic noise barrier was constructed along a four-lane highway.

## Reuse

Years ago, refillable bottles were used a great deal in the United States (see Chapter 15). Today their use is rare, although about ten states still have them (Figure 23–12). In order for a glass bottle to be reused, it must be considerably thicker (and therefore heavier) than one-use bottles. Because of the increased weight, transportation costs are higher. In the past, reuse of glass bottles worked because there were many small bottlers scattered across the United States, helping to minimize transportation costs. Today there are approximately one-tenth as many bottlers. Because of this centralization, it is economically difficult to go back to the days of refillable bottles; the price of beverages would have to increase to absorb increased transportation costs.

Although the quantity of reusable glass bottles in the United States has declined, certain countries still reuse glass to a large extent. In Japan, almost all beer and sake bottles are reused as many as 20 times; bottles in Ecuador may remain in use for 10 years or longer. European countries such as Germany and Switzerland have passed legislation that promotes the refilling of beverage containers.

**Can Refillable Bottles Make a Comeback in the United States?**   Some beverage producers have recently reversed the trend toward disposable bottles. A number of dairies have started delivering milk directly to consumers in bottles made of acrylic plastic that can be reused 50 to 100 times before being disposed of. A few breweries have also switched from disposable to refillable bottles.

Despite these examples of a shift toward refillables, however, it is unlikely that the United States will switch to refillable bottles on a large scale unless legislators and consumers take action. If a federal deposit law (a national "bottle bill") were passed, for example, consumers would return bottles to get the small deposit—usually a nickel or a dime per bottle. Other legislation, such as a requirement that bottlers sell a certain percentage of

**Figure 23–12**
Only a few states have deposit or redemption systems for glass. Recycling glass bottles is much more common than reuse.

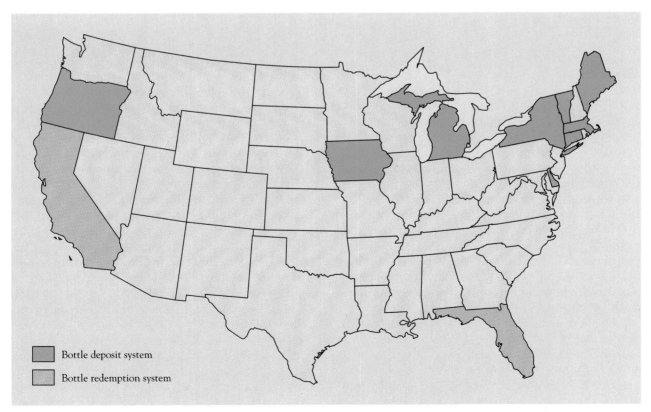

their product in refillable bottles, would also encourage reuse.

## Source Reduction

Industries can often design and manufacture their products in ways that decrease not only the volume of solid waste but the amount of hazardous materials in the solid waste that remains. Such a strategy is known as **source reduction,** and it is the most underutilized aspect of waste management. Source reduction is accomplished in a number of different ways. Innovations and product modifications can reduce the waste produced after a product has been used by the consumer. Dry-cell batteries, for example, contain much less toxic mercury today than they did in the early 1980s. The 35 percent reduction in the weight of aluminum cans since the 1970s provides another example of source reduction.

Many industries are assessing ways to reduce the amounts of hazardous materials they use in

**Focus On**

## PCBs

**Polychlorinated biphenyls (PCBs)** are a group of about 70 industrial chemicals composed of carbon, hydrogen, and chlorine. They are clear or light yellow liquids with an oily consistency. PCBs were manufactured in the United States between 1929 and 1979 for a wide variety of uses. They have been employed as cooling fluids in electrical transformers, electrical capacitors, vacuum pumps, and gas-transmission turbines. PCBs were also used in hydraulic fluids, fire retardants, adhesives, lubricants, pesticide extenders, inks, and other materials.

The first indication that PCBs were dangerous was in 1968, when hundreds of Japanese who ate rice bran oil that had accidentally been contaminated with PCBs experienced a number of serious effects, including liver and kidney damage. A similar mass poisoning, also attributed to PCB-contaminated rice oil, occurred in Taiwan in 1979. Since then, toxicity tests conducted on animals indicate that PCBs harm the skin, eyes, reproductive capacity, and gastrointestinal system. They may also be carcinogenic.

Some of the properties of PCBs that make them so useful in industry also make them dangerous to living things and difficult to eliminate from the environment. PCBs are chemically stable and resist chemical and biological degradation. Like DDT, PCBs accumulate in fatty tissues and are subject to biological magnification in the food chain (see Chapter 22).

Prior to their banning by the EPA in the 1970s, large quantities of PCBs were dumped into landfills, sewers, and fields. Such improper disposal is one of the reasons PCBs are still a threat today. Also, when sealed electrical transformers and capacitors leak or catch fire, PCB contamination of the environment occurs.

One of the most effective ways to destroy PCBs is by high-temperature incineration. However, incineration is not practical for the removal of PCBs that have leached into the soil and water because, among other difficulties, the cost of incinerating large quantities of soil is prohibitively high. Another way to remove PCBs from soil and water is to extract them by using solvents. This method is undesirable for two reasons: first, the solvents themselves are hazardous chemicals, and second, extraction is also costly because wells must be dug to collect the chemically contaminated groundwater in order to decontaminate it.

More recently, researchers have discovered a number of bacteria that can degrade PCBs at a fraction of the cost of incineration (see Focus On: The Biological Treatment of Hazardous Wastes). One of the limitations of this method is that if PCB-eating bacteria are sprayed on the surface of the soil, they cannot decompose the PCBs that have already leached deep into the soil. Although these microorganisms show promise as a help in removing PCBs from the environment, additional research will have to be conducted to make the biological degradation of PCBs practical.

their daily plant operations. For example, chlorinated solvents are widely used in electronics, dry cleaning, foam insulation, and industrial cleaning. (Some of the dangers to the environment of chlorinated solvents are considered in Chapter 20.) It is sometimes possible to accomplish source reduction by substituting a less hazardous material, but substantial source reduction of chlorinated solvents can be realized simply by reducing solvent emissions. Most chlorinated solvent pollution gets into the environment by evaporation during industrial processes. Installing solvent-saving devices not only benefits the environment but provides economic gains, because smaller amounts of chlorinated solvents must be purchased.

**Dematerialization,** the progressive decrease in the size and weight of a product as a result of technological improvements, was discussed in Chapter 15. *Dematerialization results in source reduction only if the new product is as durable as the one it replaced.* If smaller, lighter products have shorter life spans (so that they have to be thrown out and replaced more often), source reduction is not accomplished.

Advertising campaigns aimed at persuading consumers to use products that generate less solid waste have only recently begun to be utilized. For example, Procter & Gamble, the manufacturer of Downy fabric softener, advertises a collapsible, biodegradable container of concentrated fabric softener that can be used to refill the original, rigid polyethylene container. If consumers bought this sort of product on a large scale, they would have a significant impact on the quantity of municipal solid waste.

## HAZARDOUS WASTE

Beginning in 1977 with the discovery that toxic wastes from an abandoned chemical dump had contaminated homes—and possibly people—in Love Canal, New York, toxic waste has held national attention (Figure 23–13). The Love Canal tragedy created an immediate concern about the hazardous wastes in thousands of old landfills, dumps, and junkyards across the United States, and that concern has been with us in one form or another ever since. In the summer of 1988, for example, hazardous wastes made headlines again when medical wastes that had been illegally dumped in the ocean washed onto beaches from Maine to Florida (Figure

**Figure 23–13**

Hazardous waste in the news. (a) An elderly man who refuses to leave his home mows his yard next to the fenced-in, abandoned homes at Love Canal, New York. In 1978, Love Canal had the dubious distinction of being the first area ever declared a national emergency disaster area because of toxic waste. A local industry, Hooker Chemical Company, disposed of its chemical waste in Love Canal. When the site was filled, Hooker added topsoil and donated the land to the local Board of Education. A school and houses were built on the site, which began oozing toxic waste a number of years later. Over 300 chemicals, many of them carcinogenic, have been identified in Love Canal's toxic waste. (UPI/Bettmann) (b) Medical debris that was illegally disposed in the Atlantic Ocean washed ashore on many beaches in 1987 and 1988. (Visuals Unlimited/Peter Ziminski)

(a)

(b)

23–13). Some of this refuse, including used syringes and vials of blood contaminated with the AIDS virus, was extremely hazardous.

Other industrialized countries have the same problems with toxic waste management that Americans do. How shall we deal with the bewildering array of hazardous waste that is produced by mining, industrial processes, incinerators, military activities, and thousands of small businesses in ever-increasing amounts? How do we clean up the hazardous materials that are already contaminating our world?

## Types of Hazardous Waste

Any discarded solid, liquid, or gas that threatens human health or the environment is known as **hazardous** or **toxic waste.** Hazardous waste, which accounts for about 1 percent of the solid waste stream in the United States, includes chemicals that are dangerously reactive, corrosive, explosive, or poisonous.

More than 700,000 different chemicals are known to exist. How many are hazardous is unknown, because most have never been tested for toxicity, but without a doubt, hazardous substances number in the thousands. They include a variety of acids, dioxins, abandoned explosives, heavy metals, infectious wastes, nerve gas, organic solvents, **poly-chlorinated biphenyls (PCBs)**, pesticides, and radioactive substances (see Table 23–1 and Focus On: PCBs on page 535). Many of these substances have been discussed in this book, particularly in Chapters 11, 19, 21, and 22, which discuss radioactive wastes, air pollution, water and soil pollution, and pesticides.

## MANAGEMENT OF HAZARDOUS WASTE

Humans have the technology to manage toxic wastes in an environmentally responsible way, but it is extremely expensive. Although great strides have been made in educating the public about the problems posed by hazardous wastes, we have only begun to address most of the issues of hazardous waste disposal. No country currently has an effective hazardous waste management program, but several European countries have led the way by producing smaller amounts of hazardous waste and by using fewer hazardous substances.

## Current Management Policies

Currently, two federal laws dictate how hazardous wastes should be managed: (1) the Resource Conservation and Recovery Act, which is concerned with managing hazardous wastes that are now being produced; and (2) the Superfund Act, which provides for the cleanup of abandoned and inactive hazardous waste sites.

The Resource Conservation and Recovery Act was passed in 1976 and amended in 1984. It is up for reauthorization in 1992 (as we go to press). Among other things, it instructed the Environmental Protection Agency (EPA) to identify which wastes were hazardous and to provide guidelines and standards to states so that they could initiate hazardous waste management programs. The act also provided for the elimination, by 1990, of disposal on land as a way of handling hazardous waste. The deadline was not met, but the management of toxic waste is gradually improving. Certainly, the situation is better than it was 20 years ago.

In 1980 the Comprehensive Environmental Response, Compensation, and Liability Act (CERCLA), commonly known as the Superfund Act, established a program to tackle the huge challenge of cleaning up abandoned and illegal toxic waste sites across the country. At many of these sites, hazardous chemicals have migrated deep into the soil and polluted groundwater. The greatest threat to human health from toxic waste sites comes from drinking water laced with such contaminants.

The Superfund program got off to a slow start, in part because the EPA, which was authorized to administer it, experienced severe budget cuts. An amendment passed in 1986 provided the EPA with more money and instructions to work faster. As of January 1992, Congress had appropriated $9 billion to implement the Superfund Act, and several additional billions of dollars have been authorized.

## Cleaning Up Existing Toxic Wastes

The federal government estimates that the United States has more than 400,000 hazardous waste sites with leaking chemical storage tanks and drums (both above and below ground), pesticide dumps, and piles of mining wastes. This figure does not include the hundreds or thousands of toxic waste sites at military bases and nuclear weapons facilities.

As of October 1991, 34,618 sites were in the CERCLA inventory, which means that they have been identified by the EPA as possibly qualifying for cleanup. These sites are not identified according to any particular criteria; some are dumps that local or state officials have known about for years, and others are identified by concerned citizens. The sites in the inventory are evaluated and ranked to identify those with extremely serious hazards. The ranking system uses data from preliminary assessments, site inspections, and expanded site inspections, which include contamination tests of soil and groundwater and sampling of toxic waste.

The sites that pose the greatest threat to public health and the environment are placed on the **Superfund National Priorities List,** which means that the federal government will assist in their cleanup (Figure 23–14). As of August 1992, 1,176 sites were on the National Priorities List, and the number is sure to grow as other sites are evaluated. One-fifth of these sites are open dumps or sanitary landfills that operated before state and federal regulations were imposed and received trash that would

be considered toxic waste today. The five states with the greatest number of sites on the priorities list are New Jersey (108 sites as of February 1992), Pennsylvania (100 sites), California (95 sites), New York (84 sites), and Michigan (77 sites).

One reason for the urgency about cleaning up the sites on the National Priorities List is their locations. Most were originally in rural areas on the outskirts of cities. With the growth of cities and their suburbs, however, many of the dumps are now surrounded by residential developments.

Because the federal government cannot assume major responsibility for cleaning up every old dumping ground in the United States, cleanup costs for each site are shared by the current landowner, prior owners, anyone who has dumped wastes on the site, and anyone who has transported wastes to the site. For some sites, as many as several hundred different parties are considered liable for cleanup costs. Despite the urgency of cleaning up sites on the National Priorities List, it will take

**Figure 23–14**

Cleaning up hazardous waste. (a) Toxic waste in deteriorating drums at a site near Washington, D.C. The metal drums in which much of the waste is stored have corroded and started to leak. Old toxic waste dumps are commonplace around the United States. (USDA/Soil Conservation Service) (b) Cleanup of a hazardous waste site. Removal and destruction of the waste are complicated by the fact that usually nobody knows what chemicals are present. (Gary Milburn/Tom Stack and Associates)

(b)

(a)

### The Biological Treatment of Hazardous Wastes

Until recently, companies wanting to clean up soil contaminated with hazardous wastes had few options: they could dig up the soil and deliver it to a hazardous waste landfill, or they could dig up the soil and incinerate it to break down the poisons with intense heat. Because both processes are prohibitively expensive, innovative approaches for dealing with hazardous wastes are being developed. In bioremediation, the contaminated site is exposed to an army of microorganisms, which gobble up the poison and leave behind harmless substances such as carbon dioxide and chlorides. To date, more than 1,000 different species of bacteria and fungi are being used to clean up various forms of pollution.

Bioremediation was used in the early 1990s to clean up an Iowa site that had been contaminated by a poisonous wood preservative called pentachlorophenol (penta for short). A bacterium called *Flavobacterium*, which can break down the penta molecule, was brought in to clean up the site.

During bioremediation, conditions at the hazardous waste site are modified so that the bacterium will thrive in large enough numbers to be effective. The site in Iowa was small, so engineers mixed the contaminated soil with sand (to make it more porous) and then injected *Flavobacterium* into the soil. They pumped air through the soil (to increase its oxygen level) and added a few soil nutrients such as phosphorus; both oxygen and phosphorus are required by *Flavobacterium*. Engineers also created a drainage system at the bottom of the pit to pipe any penta-laden water that leached through the soil back to the surface for another go-around with the bacteria.

Bioremediation takes a little longer to work than traditional hazardous waste disposal methods. One year after *Flavobacterium* was introduced at the Iowa site, two-thirds of the penta was gone. It took the bacteria between two and three years to degrade all of the penta.

How are toxin-eating microbes such as *Flavobacterium* discovered? The penta-eating bacterium was found by a University of Idaho microbiologist, Ronald Crawford, who was investigating the effects of penta on stream ecosystems. Crawford found some streams where penta disappeared shortly after it was added. He collected bacteria from these streams and brought them into his lab, where he fed them a steady diet of penta. Eventually, he identified *Flavobacterium* as the champion penta-eater.

---

many years to complete the job. Civil litigation and legal infighting about who should pay what will prolong the cleanup.

As of January 1992, about 35 hazardous waste sites had been cleaned up enough to be deleted from the National Priorities List; many other sites have been partially corrected. A variety of methods are employed to clean up a hazardous waste site; some are innovative technologies such as **bioremediation,** which uses microorganisms to break down hazardous wastes (see Focus On: The Biological Treatment of Hazardous Wastes).

### Managing the Toxic Wastes We're Producing Now

Many people think, incorrectly, that the establishment of the Superfund has eliminated the problem of toxic waste. The Superfund deals only with hazardous waste produced in the past; it does nothing to eliminate the large amount of toxic waste that continues to be produced even as you read this page.

There are three ways to manage hazardous waste: (1) source reduction, (2) conversion to less hazardous materials, and (3) long-term storage. As with municipal solid waste, the most effective of the three is source reduction—that is, recycling hazardous materials and substituting less hazardous or nonhazardous materials for hazardous ones in industrial processes. No matter how efficient source reduction becomes, however, it will never entirely eliminate hazardous waste.

The second best way to deal with hazardous waste is to reduce its toxicity. This can be accomplished by chemical, physical, or biological means, depending on the nature of the hazardous waste. One way to detoxify organic compounds is by high-

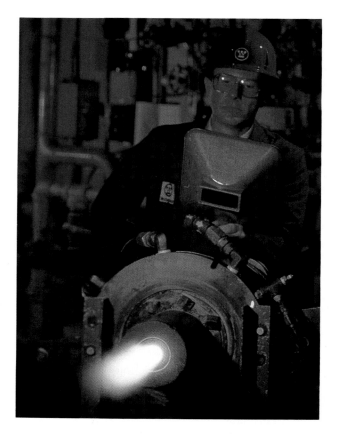

**Figure 23–15**
The plasma torch produces a flame that is hotter than 10,000°C. At this temperature, toxic molecules such as PCBs and dioxins are broken down into nonhazardous or less hazardous gases such as hydrogen and nitrogen. (Courtesy of Westinghouse Electric Corporation)

temperature incineration. The high heat of combustion reduces these dangerous compounds, such as pesticides, PCBs, and organic solvents, into safe products such as water and carbon dioxide. The incineration ash is hazardous, however, and must be disposed of in a landfill designed specifically for hazardous materials. New methods to reduce the toxicity of hazardous wastes are also being developed—such as the plasma torch, which produces such high temperatures (greater than 10,000°C) that hazardous waste is almost completely converted to harmless gases (Figure 23–15).

Hazardous waste that is produced in spite of source reduction and that is not completely detoxified must be placed in long-term storage. Landfills that are equipped to store hazardous substances have many special features, including relatively impermeable layers of compacted clay and double plastic liners at the bottom of the landfill to prevent leaching of hazardous substances into surface water and groundwater. Liquid that percolates through a hazardous waste landfill is collected and treated to remove contaminants. The entire facility and nearby groundwater deposits are carefully monitored to make sure there is no leaking. Only solid chemicals (*not liquids*) that have been treated to detoxify them as much as possible are accepted at hazardous waste landfills. These chemicals are placed in sealed barrels before being stored in the hazardous waste landfill.

Very few facilities are certified to handle toxic wastes (Figure 23–16). In 1992 there were only 21

**Figure 23–16**
A landfill near Detroit, Michigan, is designed to accommodate hazardous waste. The bottom of the landfill is several feet of clay covered by 3 plastic liners (shown). Barrels of hazardous waste are then deposited and covered with soil.
(Dennis Barnes)

commercial hazardous waste landfills in the United States. As a result, most of our hazardous waste is still placed in unlined sanitary landfills, burned in incinerators that lack the required pollution control devices, or discharged into sewers.

## FUTURE MANAGEMENT OF SOLID AND HAZARDOUS WASTES

The United States must find a way to deal safely with its growing solid waste. Source reduction has the greatest promise as a means of reducing the amount and toxicity of solid waste in the future. Increased recycling and reuse will also diminish the amount of solid waste that must be disposed of in incinerators and landfills.

The most effective way to deal with solid waste is with a combination of techniques. In **integrated waste management,** a variety of options—such as source reduction, recycling, and composting—are incorporated into an overall waste management plan (Figure 23–17). Even on a large scale, recycling and source reduction will not entirely eliminate the need for disposal facilities such as incinerators and landfills. They will, however, substantially *reduce* the amount of trash requiring disposal in incinerators and landfills.

### Changing Attitudes

In the United States we have become accustomed to the convenience of a throwaway society. The products we purchase rarely, if ever, have disposal expenses built into their prices, so we are not aware

**Figure 23–17**
Integrated waste management.

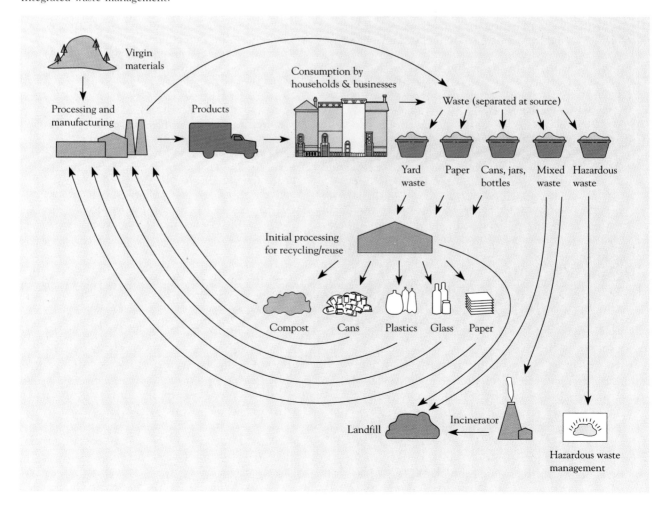

## The Environmental Costs of Diapers

Should babies wear disposable diapers or cloth diapers? New parents must choose whether to use about 6,000 disposables or 50 cloth diapers between a child's birth and his or her toilet training. Until the late 1980s, convenience and cost were the only issues parents considered in choosing diapers: disposables are more convenient, and cloth diapers are less expensive. More recently, the environmental effects of diapers have become a consideration for many parents.

Approximately 16 billion disposable diapers are produced in the United States each year, making up more than 85 percent of the diaper market. Disposables ultimately end up in sanitary landfills; they account for almost 2 percent of the total municipal solid waste produced in this country. In the early 1990s, as the nation's solid waste problem became more acute, parents were urged by certain environmental groups to convert to cloth diapers to help "save the environment." Several states with particularly acute sanitary landfill shortages contemplated an outright ban on disposables in favor of cloth diapers, which contribute only about one-hundredth as much solid waste as disposables do.

It turns out that the diaper issue is not as clear-cut as originally thought. Analysis of the entire life cycles of both disposable and cloth diapers reveals that both types have environmental costs. Cloth diapers begin their lives as cotton growing in a field, and during this stage the cotton requires pesticides, which contribute to air, soil, and water pollution. Cotton plants also have the potential to erode the soil, and some cotton fields are irrigated, which can contribute to groundwater depletion and soil salinization.

Energy is required to manufacture cotton diapers and to transport them to stores or homes, and the trucks and cars that transport them produce air pollution. After being used, cloth diapers are washed and reused. Washing cotton diapers requires lots of water, and detergents, bleaches, and babies' wastes produce water pollution. Energy is required to run the washing machine and dryer.

Disposable diapers begin their lives as trees. The harvesting of the trees has potential to cause soil erosion and contribute to deforestation. The production of paper from trees requires many chemicals that can pollute both air and water. Disposable diapers also contain plastic, which is made from petroleum that must be extracted from the Earth and refined, producing both air and water pollution. As with cloth diapers, energy is required to manufacture and transport disposable diapers, and the vehicles that do the transporting pollute the air. Disposable diapers usually end up in sanitary landfills, which are filling up rapidly.

Comparing the environmental effects of cloth diapers and disposable diapers is a bit like comparing apples and oranges, but it is clear that both types have environmental costs. Neither cloth nor disposable diapers has a clear-cut environmental advantage. The August 1991 issue of *Consumer Reports* offered perhaps the best suggestion to parents: if you live in an area where landfill space is becoming a problem (such as the Northeast and Midwest), try using cloth diapers; if you live in an area where water is in short supply (such as in the Southwest), try using disposable diapers.

---

of their *true* cost to our communities. And many municipalities charge all citizens equally for trash collection, regardless of whether they generate a lot or a little trash.

If all other factors are equal, most Americans do not select a particular product because it produces less waste than another. Most of us like the neatness, squeezability, and unbreakability of plastic ketchup bottles, and we prefer the convenience of disposable diapers over cloth diapers (see Focus On: The Environmental Costs of Diapers). We value attractive packaging. In contrast, Europeans are generally more willing than Americans to accept new products or changes in packaging that produce less solid waste. Europeans' recycling technologies are also much more advanced.

One of the more effective ways to help consumers become aware of the amount of garbage they generate is by charging them for trash collection based on quantity. When a town switches from a flat annual fee for garbage collection to fees based on the number of trash bags or cans used, customers substantially reduce their solid waste. The same people are also encouraged to recycle if the materials they separate out from their solid waste are picked up at curbside free of charge.

## Locating Future Waste-Handling Facilities

Even if we reduce municipal solid waste to near zero through source reduction and recycling, there will still be a need for sanitary landfills, if only to contain the residues produced during recycling and incineration. In addition, facilities that handle hazardous wastes will always be needed, even though source reduction decreases the amounts of such materials being produced.

The selection of a site does not have to be a political bombshell if communities are fully informed and are allowed to participate throughout the selection process. In the past, communities typically found out that they were being considered for such a site only when the waste disposal company applied for a building permit in their area. At such a late date, public opposition is usually strong (and successful).

More states and communities should follow the example of the state of Washington. During the late 1980s the need for a hazardous waste disposal facility in Washington became apparent. From the start of its search for a site, the private waste disposal company ECOS let everyone know what it was planning. First the entire state area was evaluated, and environmentally sensitive areas (such as wetlands), areas with historical or archaeological importance, and inaccessible areas were eliminated from consideration. ECOS then requested community participation through a series of statewide press releases. Meetings were organized in nine of the communities that expressed interest in hosting such a facility. Citizen advisory committees then formed to negotiate for strict environmental precautions and economic benefits for the community. Two towns voted in favor of locating such a site in their vicinity, and in 1989 the community of Lind was selected. Only then did the waste disposal company apply for a permit.

## SUMMARY

**1.** One of the most urgent problems of industrialized nations is the disposal of solid and hazardous wastes, which increase each year. Traditionally, solid waste has been disposed of in open dumps and, more recently, in sanitary landfills.

**2.** Sanitary landfills are less likely to harbor disease-carrying vermin than open dumps. However, like dumps, most landfills have the potential to contaminate soil, surface water, and groundwater. New landfills are not numerous enough to replace the older ones as they fill up, because most people oppose the establishment of sanitary landfills near their homes.

**3.** Incineration reduces solid waste to a fraction of its original volume, and the heat produced during incineration can be used to warm buildings or generate electricity. One of the drawbacks of incineration is the great expense of installing pollution control devices on the incinerators. These controls make the gaseous emissions from incinerators safe (as far as we know) but make the ash that remains behind more toxic. Another problem associated with incineration is finding appropriate sites, as most people oppose putting incinerators near residential areas.

**4.** Composting is increasingly being used to reduce the amount of organic waste, particularly yard waste, in the solid waste stream.

**5.** Recycling, reuse, and source reduction help decrease the volume of trash. Many communities are recycling paper, glass, metals, and plastic. Reuse of refillable bottles is not as widespread in the United States as it was a few years ago. Other countries reuse containers to a greater extent than the United States in order to conserve resources as well as reduce solid waste.

**6.** Source reduction, which has great potential to reduce the volume of trash, can be accomplished in a variety of ways; these include substituting raw materials that introduce less hazardous or solid waste during the manufacturing process and recycling hazardous and solid waste at the plants where it is generated. Consumers can also practice source reduction by purchasing durable products that are designed to generate less solid waste.

**7.** Hazardous wastes are solids, liquids, and gases that pose a real or potential threat to the environment or to human health. In the past, hazardous waste was dumped, along with other solid waste, in open dumps and regular landfills. As a result, we are faced with the daunting prospect of cleaning up old toxic waste dumps and the soil and water they have polluted.

**8.** Hazardous wastes being produced today are supposed to be disposed of (1) in special landfills designed to minimize the risk of environmental contamination or (2) by high-temperature incineration. Unfortunately, very few sites are approved for the disposal of hazardous wastes. Source reduction is the best way to reduce hazardous wastes.

**9.** There is no single solution to the problem of solid and hazardous wastes. The best approach is to use integrated waste management, which is a combination of techniques including source reduction, recycling, reuse, and composting, to reduce the amount of waste to be discarded. Combustible materials that remain can then be incinerated to further reduce the volume of trash. Incinerators must be equipped with pollution control devices to ensure that the pollutants are not simply being transferred from one medium to another. The remaining solid waste, a small fraction of the original solid waste, should be disposed of in an environmentally safe sanitary landfill.

## DISCUSSION QUESTIONS

**1.** List what you think are the best ways to treat the following types of waste, and explain the benefits of the processes you recommend: paper, glass, metals, food wastes, yard wastes, PCBs.

**2.** Keep an inventory of the plastic materials you discard during one week. Include all forms of packaging (for instance, meat trays and milk cartons) as well as disposable items such as Styrofoam cups.

**3.** How do industries such as Goodwill, which accepts donations of clothing, appliances, and furniture for resale, affect the volume of solid waste?

**4.** It could be argued that a business that collects and sells its waste paper isn't really recycling unless it also buys products made from recycled paper. Explain.

**5.** Suppose hazardous chemicals were suspected to be leaking from an old dump near your home. Outline the steps you would take to (1) have the site evaluated to determine if there is a danger, and (2) mobilize the local community to get the site cleaned up.

**6.** Why must a sanitary landfill always be included in any integrated waste management plan?

**7.** The Organization for African Unity has vigorously opposed the export of hazardous wastes from industrialized countries to developing nations. They call this practice "toxic terrorism." Explain.

**8.** The United States is testing the feasibility of incinerating hazardous waste at sea. Why do you think the government would take the trouble to transport toxic waste to sea for disposal? Give at least two advantages and two disadvantages of this type of disposal.

## SUGGESTED READINGS

Chollar, S. The poison eaters. *Discover*, April 1990. How microorganisms help clean up toxic waste sites.

Frosch, R. A., and N. E. Gallopoulos. Strategies for manufacturing. *Scientific American*, September 1989. How industry can operate competitively without producing as much waste or pollution as before and without exhausting resources.

Gillis, A. M. Shrinking the trash heap. *BioScience* 42:2, February 1992. The potential of composting to reduce the amount of municipal solid waste.

Kunreuther, H., and R. Patrick. Managing the risks of hazardous waste. *Environment* 33:3, April 1991. Management of hazardous waste is complicated by the fact that the public perception of risk from hazardous materials differs from the views of experts.

Levenson, H. Wasting away. *Environment* 32:2, March 1990. Policies and economic incentives that would reduce the quantity and toxicity of municipal solid waste.

Rathje, W. L. Once and future landfills. *National Geographic*, May 1991. All you ever wanted to know about sanitary landfills.

Wolkomir, R. I learned that it just keeps getting deeper. *Smithsonian*, April 1990. The experiences of a reporter who worked for a week at his local landfill.

Yazdani, A. Source reduction research partnership: A unique joint venture. *Environment* 31:9, November 1989. How the Environmental Defense Fund and the Metropolitan Water District of Southern California formed a partnership to reduce the industrial use of chlorinated solvents.

Young, J. E. Tossing the throwaway habit. *World Watch*, May–June 1991. Source reduction, reuse, and recycling could cut down on waste, eliminating the need for incinerators and reducing the need for landfills.

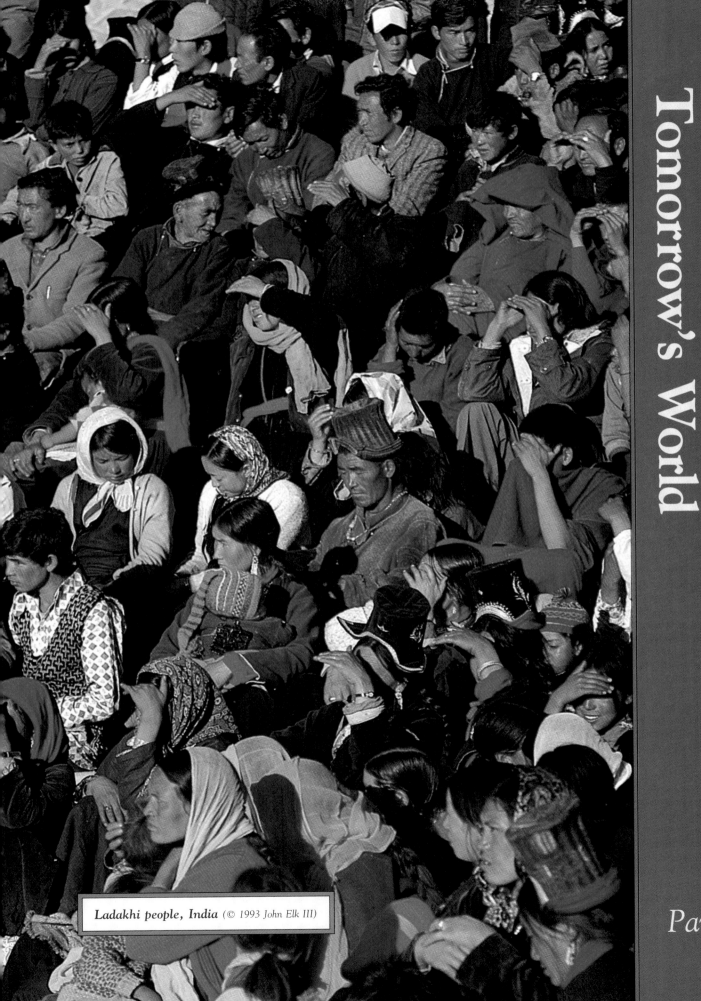

Tomorrow's World

Ladakhi people, India (© 1993 John Elk III)

Part 7

# Jerry Greenfield

## Ben & Jerry's Ice Cream: The Business of Social Responsibility

*Jerry Greenfield received his undergraduate degree in biology from Oberlin College, and worked as a lab assistant for a time after graduating. In 1978, upon completing a $5 correspondence course in ice-cream-making from Penn State University, he and his childhood friend, Ben Cohen, opened their first scoop shop in a renovated gas station in Burlington, Vermont. Today, Ben & Jerry's Homemade Ice Cream, Inc. is a $97 million business. But both founders have always emphasized that social responsibility must go hand-in-hand with profit-making. The company uses 7.5 percent of its annual pre-tax profits to fund community-oriented projects, and their extensive environmental program has this goal: "To be 100 percent involved in responsibly managing our solid waste stream, conserving energy and resources, exploring sustainable, renewable energy sources, and developing environmentally beneficial community outreach programs."*

**The success of Ben & Jerry's seems proof that an environmentally responsible business is not only possible, but profitable. How have you accomplished this? How are your priorities different from those of traditional companies?**

We believe that a business can be profitable and still be a caring company, whether it's how you treat employees or how you treat the environment. Traditionally, businesses have thought that if they try to put money, energy, time, or resources into social concerns or environmental concerns, that would weaken their focus and ability to use those same resources for making money. We're trying to show that you can do both, that making a profit and caring for the environment are actually complementary goals. Our whole business is based on the idea that if we support the community, respond to its social and environmental concerns, the community will return that support.

**Do you have a different, or nontraditional, definition of "profit"?**

We have what we call a two-part bottom line. We look at how much money is left over at the end of the year, the *financial* bottom line, and also how much we contributed to the community, our *social* bottom line, which takes into account environmental issues, as well. We evaluate our company on both of those components. Business tends to value things that can be measured. Traditionally, the easiest thing to measure has been profitability, so that's been the focus. Until you make a conscious decision to measure other things, they won't seem very important. Business must therefore figure out a way to measure its social or environmental return. That's a very difficult thing to do because it's not nearly as clear-cut as financial return.

**How does Ben & Jerry's do that?**

Our annual report includes both a financial and a social report. We've used independent social auditors, who report on the progress of our social and environmental programs. We're now moving toward establishing some sort of internal reporting system, however. We will set up goals and objectives and evaluate our own performance.

Ben & Jerry's, since its inception, has been concerned with many kinds of social issues, but did not always have the extensive environmental program it has today. How was the program established, and what are the "Green Teams"?

There was one person in the company, Gail Mayville, who really pushed environmental concerns to the top of our agenda. She was an administrative assistant at the time. She kept urging us to develop an environmental program, and eventually she implemented it herself in 1988. She is now our office manager, and makes sure that our daily operations are as environmentally sound as possible. She is constantly looking for new ways to recycle and use recycled material.

The Green Teams are volunteers from within the company who have taken it upon themselves to identify environmental issues and to improve the company's environmental performance. They examine all sorts of things from the kind of inks we use in the art room to our recycling policy, packaging, and energy usage.

**Was it a difficult transition to embark on this new program?**

It was a change of thinking for everyone. We can no longer think only about how to obtain the greatest return for the money, or even how to produce the highest quality product, but must also consider the environmental impact of our actions. It adds another layer of complexity to our business, but it's the right thing to do. And since the people in this company have a natural inclination toward environmental responsibility, the adjustment was not very difficult for us.

**Many companies would argue that it wouldn't be cost-effective to introduce new environmental policies and change their procedures.**

What they're really saying is that it wouldn't be cost-effective in the *short run*, which is a very narrow viewpoint. If you look at business and people as part of a global community, however, and as inhabitants of an Earth that we all depend upon, it doesn't make any sense to be trashing our world. The trick is to find policies that are profitable, or that cost little or nothing, and are good for the environment, as well. Then there will be times when, in the short run, environmentally responsible policies *will* cost more. Ultimately you have to believe that, in the long term, a healthy world will benefit everyone, including business.

**Do you think that environmental programs like yours could successfully be adopted by other types of companies?**

Yes, I do. Some businesses are, of course, based on processes that degrade the environment. The paper industry, for example, would have to make some major practical and philosophical changes. But in business, and in life in general, nothing is really pure. Everything is in shades of gray. So it's largely a matter of everyone improving their own activities as best they can and becoming involved in the restorative process. I think that's where a big change might be: implementing environmentally restorative processes that will repair the Earth instead of destroying it. It's also an area in which new businesses and jobs can be created, so that the economy could grow while the environment is being restored at the same time.

**What are some of the ways in which you have formed alliances with other socially responsible businesses, creating the support system that enables you to accomplish so many of your goals? Are there many such businesses in this country?**

There is a loose association of like-minded businesses called Social Ventures Network based in San Francisco. In addition to business people, members include investors, futurists, and other far-sighted people, all dedicated to the development of a just and sustainable society. We meet several times during the year to exchange information, and try to bring attention to new ventures that advocate these values. Out of Social Ventures Network has grown BSR, Businesses for Social Responsibility, a Washington, D.C., association formed in 1992 to provide an alternative to the National Chamber of Commerce.

There are a lot of small and mid-sized companies that are not as high-profile as Ben & Jerry's, but are really in the forefront of environmental awareness. They're part of a growing wave of businesses experimenting with new, cutting-edge environmental policies and practices.

# Chapter 24

The choices we make today will determine whether future generations will inherit a world of prosperity or decline. (© 1993 Lawrence Migdale)

# Tomorrow's World

**Y**ou have been presented with a broad overview of the environment, including both the principles necessary to understand how ecosystems operate and a detailed review of the individual issues that are having an important influence on today's world. Now we wish to speak directly to you, not only as students but also as citizens. Our world is in very serious trouble today, presenting us with an enormous challenge. What follows represents where we stand, our opinions on what must be done to meet today's critical environmental challenge. There are reasonable people who hold other views, but we believe that we would not be doing our job as responsible teachers if we did not lay out our views for you. Read, think, discuss, and come to your own conclusions. And then, if our world is to have any future—act.

# THE PROBLEMS WE FACE

With a few short years separating us from the end of the century, most Americans continue to act as if only the immediate future mattered. We tend to view changes in the environment—the water, soil, plants, and animals on which we base all of our activities and every bit of our prosperity—as if they would simply always be there. Major trends "happen" to us, as the weather does, and we continue to hope that the future will somehow take care of itself. Meanwhile, our rapidly growing numbers are relentlessly, and in some cases permanently, exacting a toll on the ability of the Earth to support us.

When the century began, neither human numbers nor technology had the power to radically alter planetary systems. As the century closes, not only do vastly increased human numbers and their activities have that power, but major unintended changes are occurring in the atmosphere, in soils, in waters, among plants and animals, and in the relationships among all of these. The rate of change is outstripping the ability of scientific disciplines and our current capabilities to assess and advise. It is frustrating the attempts of political and economic institutions, which evolved in a different, more fragmented world, to adapt and cope. [Bruntland report, World Commission on Environment and Development, 1987, p. 22 by permission of Oxford University Press]

We live in a world in which the rapidly growing human population is consuming, co-opting, or wasting a major portion of total biological productivity. Related to this development is the fact that up to one fourth of all the species of plants, animals, fungi, and microorganisms are likely to disappear forever over the next few decades. In addition, most scientists who have analyzed the problem have concluded that the Earth's temperature is already climbing steadily because of the progressive increase in carbon dioxide, methane, nitrogen oxides, and CFCs in the atmosphere. The concentration of ozone in the upper atmosphere has apparently fallen sufficiently to account for already at least a 20 percent increase in the incidence of skin cancer at middle latitudes, and the problem continues to worsen. Our topsoil is being lost and our waters polluted at unprecedented rates. For one out of three people in the developing world, the only water available to drink is unsafe and possibly deadly. Meanwhile, we are rapidly destroying what is left of our forests, and deserts are spreading in many regions. All of these trends constitute a clear call to action.

In this textbook we have explored these problems individually. Now we shall examine where our world stands today, and discuss some ways in which we might appropriately address the future.

# ISSUE #1: POPULATION PRESSURES

In Chapter 9 we reviewed the growth of the world population from 2.5 billion in 1950 to more than 5.4 billion in 1992; it is continuing to increase, at a rate of nearly 100 million additional people each year. While this astonishingly rapid growth has been taking place, the proportion of people living in the industrialized countries of the world has dropped from an estimated 33 percent, or one out of every three people in the world, in 1950 to an estimated 20 percent, or one out of five, by the end of the 1990s. This proportion is projected to fall to about one sixth by the year 2020. In other words, the proportion of the world's people living in industrialized countries will have dropped in a period of 70 years—the space of an average human lifetime—from one out of three to one out of six.

## Achieving Stability

The rate of world population growth peaked at 1.9 percent in 1970 and fell to about 1.7 percent by the early 1990s, but because of the greater and greater numbers of people in the base population, increasing figures are still being added each year. The World Bank has recently calculated that world population could stabilize at a level of approximately 10.4 billion people by the end of the next century; United Nations projections for a stable population range from 9 to 14 billion people. In order to be able to restrict the global population even to one of these enormous totals, however, there needs to be sustained worldwide attention to family planning. Our population will not automatically reach a level of 9 billion or 14 billion people and then stop growing. If we continue to pay consistent attention to this problem and devote the resources that are necessary to bring about success, we can make it happen. If we do not, we simply will not achieve stability.

In addition, it is important to note that the difference between 9 and 14 billion—the range of the estimates—is equal to the *entire* world population in the early 1990s, a population that, as we have seen in this text, is *already* high enough to be consuming the world's resources at an unsustainable rate! This realization reinforces the need for sustained attention by all governments to the need for family planning throughout the world; it is important both in industrialized countries, which consume most of the world's resources even though they constitute a rapidly shrinking minority of the world's population, and in developing countries.

(Robert Caputo, © 1988 National Geographic Society)

(Pamela Spaulding, © 1988 National Geographic Society)

## Urbanization

At the beginning of the Industrial Revolution, approximately in 1800, only 3 percent of the world's people were living in cities; fully 97 percent were rural, living on farms or in small towns. In the two centuries since then, population distribution has changed radically—toward the cities. More people will live in Mexico City by the end of the 1990s than were living in all the cities of the world combined 200 years earlier. This is a staggering difference in the way people live. Almost half of the world population is in cities now, with a very high proportion of the people who are being added to the population also being city dwellers. In the less developed world, this tends to lead to increasing exploitation of those who live in rural areas and produce the food, wood, and other commodities on which city dwellers depend. Even in industrialized countries such as the United States, urbanization tends to make us collectively less able to understand and appreciate biological productivity, on which our common future depends. The great majority of all world population growth over the next few decades will take place in the cities, with all of the problems that implies.

(Tim Holt/Photo Researchers, Inc.)

### Garbage by the Pound

An increasing number of cities will reduce their solid waste in the future by charging residents for the amount of garbage they throw away. By 1992, over 100 U.S. communities were charging for trash collection based on volume. If you leave two containers on the street, your bill is higher than someone who leaves only one. This system rewards people who use less packaging and compost their kitchen scraps. It penalizes those who don't. Eventually, many cities will go even further and charge by weight, using trucks with automated weighing and loading arms. Bar-coded trash containers will be hoisted, weighed, and scanned with a laser. An onboard computer will automatically issue an invoice.

## ISSUE #2: RICH VERSUS POOR

In today's world, the distribution of resources is very unequal. Those who live in industrialized countries, a rapidly shrinking 23 percent of the global population, control about 85 percent of the world's finances, as measured by summing gross domestic products. The per-capita income in industrialized countries in 1992 was about $17,900; the per-capita income of the other 77 percent of the people in the world was less than a twentieth of that amount, about $810. In general, developing countries are using up their own resources, which otherwise might have been renewable, much more rapidly than are the citizens of industrialized countries who often consume other people's resources. As a result, the disparity in wealth and access to resources between industrialized and developing nations only grows greater with time.

The difference between rich and poor is evident at every level of consumption. For example,

### The Challenge Is Now

The 1990s and the two decades that follow will be of fundamental importance to the pattern of growth of human populations, and thus to their impact on global ecology. During the 1990s alone, nearly a billion people will be added to the world population, with approximately a billion more being added during each of the first two decades of the 21st century. About 95 percent of the increase will occur in developing countries. From every point of view, these three decades are likely to be the most stressful that the world has ever experienced. We need to take action immediately, knowing that our best chance to affect the future is now, not at some distant time. With the addition of nearly 3 billion people to be added, in the relatively near future, to the 4.2 billion who now live in developing countries, we have in place the recipe for a global disaster of unprecedented dimensions.

*During the 1990s alone, nearly a billion people will be added to the world population.*

(E. R. Degginger)

the 77 percent of the world's people who live in developing countries utilize only about 20 percent of the world's industrial energy. Whereas the United States, with 4.5 percent of the world's population, produces about one fourth of the world's carbon dioxide emissions, China, with 1.2 billion people and the world's largest proven coal deposits, produces only about one fifth as much. In addition, 1.5 billion of the 4.2 billion people living in developing countries depend on firewood as their principal source of fuel, consuming a very high proportion of the trees and shrubs in their vicinity with an intensity that continually increases. About one third of these people lack access to dependable supplies of fresh water, a condition that condemns mainly women and girls to a lifetime of carrying wood and water to smoky, unhealthful dwellings, where their days are spent tending cooking fires, without any hope of improving their lot. For all the raw materials that contribute to living standards, such as iron, nickel, and aluminum, the three-fourths of the world's population who live in developing countries typically consume less than a quarter of these materials.

## Poverty

The developing world experiences far greater poverty than most people living in industrialized countries can appreciate. According to the World Bank, about 1.2 billion of the citizens of developing countries live in absolute poverty, with incomes of less than $370 per year. These people are frequently unable to obtain the necessities of life—food, shelter, and clothing. About half of them are seriously malnourished, receiving less than 80 percent of the United Nations–recommended minimum caloric intake; consequently, their bodies are not being maintained, and their brains cannot develop properly. In addition, UNICEF has estimated that about 13 million babies per year, or more than 35,000 each day, starve to death or die of diseases related to starvation. Against such a background, it is incredible that many authorities assert that the world is functioning well and that we are fortunate to have avoided the widespread starvation that Paul Ehrlich[1] and other "ecological extremists" in the 1960s predicted would occur by now. How could

[1] Paul Ehrlich is Bing Professor of Population Studies at Stanford University, and an internationally recognized expert and author on population biology. (See interview with Paul Ehrlich in Part III.)

(Thomas S. England/Photo Researchers, Inc.)

"worldwide disaster" be defined if we are not experiencing one now?

One of the greatest barriers to equalizing the gap between industrialized and developing nations is the lack of trained professionals in the latter group. Only about 6 percent of the world's scientists and engineers live in these countries. Considering that a majority of those scientists and engineers live in only four nations (China, Brazil, Mexico, and India), you can see that for most developing countries, trained technical personnel are virtually lacking. Therefore, how are they to decide, based on their own knowledge, whether to join international agreements concerning the environment or how best to manage their own natural resources?

Developing countries are home to at least 80 percent of the world's biodiversity—the plants, animals, fungi, and microorganisms on which we all depend. How will they use, manage, and conserve these precious resources? Addressing the inequality in scientific and technical personnel between industrialized and developing countries must become one of the most pressing items on the agenda of tomorrow's world.

## International Debt

In industrialized nations such as the United States, we tend to think of tropical countries as competitors, potential markets, debtors, drug producers, and sources of illegal immigrants—politically unstable, often corrupt, and, in general, a problem. What we do not spend much time thinking about is the common stake we have in the management of the global ecosystem, or the ways in which we depend on one another. We like to think of the $1.2 trillion debt owed by developing countries as something that arose because of the misjudgment of opportunistic bureaucrats and businessmen in the developing nations, or of corrupt politicians, or both. We rarely stop to remember that the people who live in such nations have access to no more than one sixth of the world's wealth, or that the operation of the global economic system is directly contributing to the destruction of their resources and thus to famine, misuse of potentially sustainable natural resources, and political and economic instability throughout the world.

Even with the relatively good prospects that are being enjoyed by many developing countries today, the *international debt* has continued to grow. World Bank officials have stated very pointedly that the growth of this debt means "the relapse into poverty of large sections of the population" of the developing world. During the 1980s, the debt burden (that is, payment of interest) compressed personal incomes by about one seventh in middle-income developing countries and by one fourth in poorer countries. Increasing social and economic problems appeared, and world economic growth slowed appreciably. The U.S. trade deficit has increased whenever the ability of debtor countries to accept exports has worsened, and inflation is a serious problem in many nations.

It is in the interest of wealthy industrialized nations to attempt to reverse the net flow of resources from poor to wealthy nations, and thus to construct political, economic, and ecological stability for the future. Direct development assistance, especially assistance targeted to the very poor, is of fundamental importance in this process.

---

*I*t *is in the interest of wealthy industrialized nations to attempt to reverse the net flow of resources from poor to wealthy nations.*

---

## Immigration

Another important way in which we are linked to the developing world is revealed by the massive immigration of poor people from the tropics and subtropics into the industrialized nations of the temperate zone. The U.S. Immigration and Naturalization Service estimates that approximately 1.8 million aliens were apprehended at the Mexican border alone in 1986, suggesting that perhaps twice that many may have entered successfully. The U.S. Census Bureau estimates that net immigration now accounts for 28 percent of population growth in the United States and will account for *all* growth by the 2030s if present trends continue.

The same pattern of fleeing poverty can be seen worldwide. The United Nations has estimated that more than 10 million Africans have left their homes in search of food, often crossing national borders. The numbers of people seeking to enter the United States and other industrialized countries are only likely to increase greatly in the future. This pattern is the direct result of mounting populations and economic pressures in the developing countries. If we continue to ignore the driving forces that underlie immigration, we will never succeed in decreasing the numbers.

## Political Instability

Finally, the most immediate danger to us of problems in the developing world concerns political instability. Industrialized nations can no longer solve the problems of these countries by direct intervention or, at the other extreme, by pretending that there is no problem. On the other hand, we have a direct interest in promoting stability throughout the world, because the global economy is interconnected in a very complex way. The United States, the richest nation on Earth and therefore the one with the most to lose from global instability, should logically invest more in international development than any other nation, but it invests the *least* per capita.

Here ecology and politics meet head on. True stability in developing countries will involve the incorporation of poor people into the economies of their regions, as well as the management of natural resources so that they will continue to be productive in the long term. It can be achieved in no other way, and it cannot be achieved without the active cooperation of industrialized nations.

## ISSUE #3: OUR MANAGEMENT OF THE BIOSPHERE

The rapidly growing global human population is a dominant ecological force without precedent in the world's history. The bright lights of densely-populated urban centers visible almost everywhere on Earth in nighttime satellite pictures (see Earth at Night, page 566), the large-scale changes we are causing in the composition of the atmosphere, the die-off of trees throughout much of the world because of industrial pollution, the very fact that there is no longer a square inch anywhere on Earth where chemical pollutants do not fall—all of these relationships signal our huge and growing impact on the productive capacity of the Earth.

(Visuals Unlimited/Science VU)

## Productivity

Peter Vitousek and his colleagues at Stanford University calculated that, in the middle 1980s, the proportion of total net photosynthetic productivity on land that humans used directly, wasted, or diverted was about 40 percent of the total amount available! This is an alarming figure in view of the extremely rapid growth of the human population and the universal desire for improved living standards. There is, in fact, no plan for obtaining access to a greater proportion of the world's productivity, which must therefore come under ever-increasing strain as the human population grows. No change in law or behavior can increase the amount of light coming from the sun. We can address the problem of consumption and resource management now, or be forced to address an even more desperate problem later.

## Global Warming

What are some of the consequences of the pressure we are exerting on the global environment? You have encountered many of them in this text. One of the most widely discussed is the rapid enhancement of the greenhouse effect. Both industrial and developing countries contribute to major increases in carbon dioxide in the atmosphere, as well as to the increasing amounts of nitrogen oxides, ozone, methane, and CFCs, which also trap radiant energy from the sun. The changing trend in atmospheric composition has led most modelers to conclude that there will be a rise of 2°C–5°C in global temperatures by the middle of the next century.

News reports often trivialize the effects of such a rise, presenting them as though we could cope with them by building dikes, turning up our air conditioners, changing our farming system, and the like. Such simplistic characterizations do not take into account the global effects of warming. Simply saying "Well, we got through the Dust Bowl all right, didn't we?" hides the far more serious nature of this problem, which is that a 5°C change is approximately equal to the change from the height of the last expansion of glacial ice at the end of the Pleistocene to the temperature today—and we are talking about this change occurring in less than a century. Obviously, climatic change of that magnitude will significantly alter patterns of rainfall and temperature, which will have a very major and unpredictable impact worldwide. Furthermore, unless actions are taken soon, the temperature will not stop there—it will simply go on rising. Imagine putting a globe in a microwave and leaving the switch turned to "ON."

(G. R. Roberts)

## Desertification

Overuse of agricultural lands has caused their rapid deterioration worldwide. It is calculated that about 10 percent of the world's land surface has already been desertified (that is, its topsoil has been lost) and that an additional 25 percent is at risk. It is difficult to imagine a more frightening statistic than that an estimated 20 percent of the topsoil from the world's arable lands was lost while the global population was increasing from 2.5 billion in 1950 to more than 5.4 billion in 1992. Topsoil is now being lost at a rate of about 24 billion metric tons per year. Six to seven million hectares of agricultural land are lost every year, and an additional 1.5 million hectares to salinization from overirrigation.

(Robert E. Ford / Terraphotographics)

## A Hungry World

Grain production per capita has largely kept pace with human population growth over the past 40 years. In Europe and the United States, we tend to overproduce and to be concerned with the expense of subsidies for agricultural productivity. We largely ignore the fact that much of our agricultural productivity is taking place at high cost in energy, and sometimes as the result of expending nonrenewable supplies of groundwater. Our overproduction, in turn, tends to cause us to fail to acknowledge the inadequate productivity in much of the world and the serious need for additional food associated with rapidly growing populations in developing countries.

In addition, and for similar reasons, we tend to be tentative about applying the fruits of biotechnology to crop improvement. The application of genetic technology and other advances must play a critical future role in growing crops on marginal lands, something that is absolutely necessary if the world's people are to be fed adequately. As the human population continues to grow, it is obvious that the world's food production systems will be stretched to their very limits, a frightening prospect in light of current rates of topsoil loss. It is essential that we use the tools of biotechnology to attack this problem vigorously.

## ISSUE #4: CUTTING THE WORLD'S FORESTS

The world's forests are being cut, burned, or seriously disturbed at a frightening rate. Tropical lowland moist forests, or rain forests—biologically the richest areas on Earth—have already been reduced to half their original area. What remains in Asia, Africa, and Latin America collectively amounts to less than two thirds of the area of the United States and is being destroyed so quickly that most will be gone within 30 years.

We know very little about replacing most tropical forests with productive agriculture and forestry. For many tropical soils, the combination of inefficient or short-term exploitation with disorganized logging and clearing (often by burning) results in irreversible destruction of their potential productivity. Methods of forest clearing that were suitable when population levels were lower and forests had time to recover from temporary disturbances simply do not work any longer: they convert a potentially

(Dr. Nigel Smith, © 1993 Earth Scenes)

calculations, susceptible natural resources will continue to be consumed, driven by short-term economics, and often lost permanently.

The second reason that the world's forests are being lost is the pressures created by rapid population growth and widespread poverty, often leading to unwise or unplanned use of resources. In many developing countries, the forests have traditionally served as a "safety valve" for the poor, who, by consuming small tracts of forest on a one-time basis and moving on, find a source of food, shelter, and clothing for themselves and their families. But now the numbers of people in developing countries are too great for their forests to support—and in three decades nearly 3 billion more people will be living there.

To give but one example of the impact of relentless population pressure on the forests, consider firewood. Roughly 1.5 billion people in developing countries depend upon firewood as their major source of fuel. The demand greatly exceeds the supply. In India, for example, where forests can sustain an annual harvest of only 39 million metric tons of wood, the annual demand is for 133 million metric tons. As the price of fuel wood rises sharply, its use is increasingly being restricted to relatively affluent people, with the poor in many regions losing the ability to cook their food. As a direct result, diseases that were once thought to be under control (the microbes that cause them are killed by heat) are again spreading rapidly in northern India, and prospects for the future there seem grim.

sustainable resource (forests growing in infertile soil) into an unsustainable one. Clearing the woods and prairies of Eurasia and North America traditionally led to the establishment of productive farms, for the soil was rich; in contrast, clearing the forests of tropical Africa or Latin America often creates wastelands. The relative infertility of many tropical soils, their thin and easily disturbed surface layer of organic matter, and the high temperature and precipitation levels of tropical regions often combine to make the attainment of sustainable agriculture or forestry systems extremely difficult.

The world's forests are being lost for two principal reasons. First, they are being converted to cash. Like natural resources all over the world, they are being extracted and sold, and the resulting funds invested. Old-growth forests in Oregon, Washington, Alaska, and Siberia are being lost for the same reasons, with the cutting actually being subsidized by the central government in each case. Until concepts of natural productivity and the central role of sustainability are built more securely into economic

(Robert E. Ford/Terraphotographics)

*Roughly 1.5 billion people in developing countries depend upon firewood as their major source of fuel.*

Tropical forests are rapidly being cleared and destroyed not only because of the needs of the people who live in or near them, but also because of the demands of the global economy. Many products—foods, such as beef, bananas, coffee, and tea; medicines; and hardwoods, for example—come to the industrialized world from the tropics. As timber is harvested or trees are cleared for other reasons, however, only a very limited amount of replanting is taking place. It is estimated that only one tree is planted for every ten cut in the Latin American tropics, and in Africa the proportion is much lower. The industrial-world consumption of tropical hardwoods has risen more than 15 times since 1950. Japan accounts for about three fifths of this consumption, while reducing the production of its own forests by half during the same period. Few nations have developed forestry plans for themselves, and there is almost no international coordination of forestry policies.

(Chip Clark)

## ISSUE #5: LOSS OF BIODIVERSITY

The most serious long-term global problem resulting from deforestation will be the loss of a large portion of the Earth's biological diversity within a few decades. Since most tropical species have highly specific ecological requirements, many are unable to exist under the radically different conditions that arise as a tropical forest is disturbed.

Over the next 30 years we can expect the rate of extinction to average dozens of species a day. How big a loss is this? Unfortunately, we still have very limited knowledge about the world's biological diversity. The world has but one living library; people have read very few of its books and don't even have a complete catalogue of the volumes it contains—but the library is being burned unread. We have given scientific names to only about 1.4 million kinds of organisms. Although we know flowering plants, vertebrate animals, butterflies, and a few other groups reasonably well, we have only a small amount of information about most of the other living inhabitants of this planet, so we are ignorant of much of what we are losing.

The loss is driven by the operation, throughout the world, of an outmoded economic system that fails to place value on what cannot immediately be turned into cash. National economic planners have not taken into account the very real costs of irreversible loss of productivity, and as a result the world continues its routine business, dealing with natural resources as if somehow they will be renewed. To act in this way is a tragic error for which our descendants should not forgive us.

### Why the Loss Is Important

Although to the well informed the point of view seems incredible, some economists and government officials argue that today's wave of extinction is nothing special because "extinction is normal." In fact, between the great extinction at the end of the Cretaceous Period (65 million years ago), when the dinosaurs disappeared from the Earth forever, and the recent sharp rise, the rate of extinction remained at about one thousandth to one ten thousandth of the present level. Industrial nations have a great deal to lose as more and more species be-

(Doug Wechsler)

come extinct, because much of our prosperity is grounded in biodiversity, and much of what we hope to achieve in biotechnology ultimately has its roots in an understanding of biodiversity coupled with the ability to manage it well.

Plants, for example, are natural biochemical factories. Oral contraceptives for many years were produced from a species of Mexican yam; muscle relaxants used in surgery worldwide come from an Amazonian vine; a drug used to treat Hodgkin's disease comes from the rosy periwinkle, a native of Madagascar; and the gene pool of commercial corn (an annual) has recently been enriched by the discovery, in a small forest clearing in the mountains of Jalisco, Mexico, of a wild perennial relative. Much of plant diversity is linked to the variety of chemical defenses they have evolved to deter herbivores, and many of these chemical compounds are of great value to us. In addition to their use as foods and medicines, many plants have extraordinary possibilities as sources of oils, waxes, fibers, and other commodities of interest to modern industrial societies.

Genetic engineering provides the exciting future possibility of transferring individual genes from donor organisms (which may themselves be of no economic interest whatsoever) into crop plants. As molecular techniques become more sophisticated, we shall come to depend on genetic diversity even more heavily than we do now. Because every species of plant or animal has tens of thousands of individual genes, these organisms constitute enormous living gene libraries where genetic engineers

can prospect for genes to improve crops, protect them from insects and disease, and enable them to grow under more marginal conditions. Such improvements could dramatically boost efforts to make small farms in the tropics sustainable, and to increase the productivity of large ones. Efforts to give tropical countries a financial stake in the use of biodiversity as a source of unique genes for genetic engineering, such as the biodiversity convention adopted by most of the world's nations at the 1992 Earth summit in Brazil, represent very important steps toward preserving biodiversity. They create an economic incentive for preservation.

## ISSUE #6: WORKING TOGETHER

One of the key steps in any rational approach to world problems is to promote *internationalism*—a widespread understanding that all of our human problems are interconnected. From the vantage point of a country such as the United States, attaining this objective is more difficult than it sounds; like most people everywhere, we tend to operate as if we were alone.

### Narrow Perspectives

Americans are growing more isolated from the rest of the world just when their dependency is becoming critical. We worry a great deal about Japan and Korea as economic competitors, but how many

## Duty of Care

Two important new concepts of corporate responsibility for the environment are being applied with growing frequency in Europe and may soon make their way to the United States. The first is a simple but powerful term: "duty of care." It implies that corporations have a duty to protect the environment, whether the law requires it or not. Duty of care probably originated in England's Environmental Protection Act, which refers to manufacturers' "duty of care" to safely dispose of hazardous waste. The second concept is known as the "precautionary principle." Written into international laws regarding the dumping of toxic waste at sea, it infers that a lack of scientific evidence is no excuse to avoid protecting the environment if meaningful evidence of damage exists. If you wait for scientists to agree, it may be too late.

Americans study Japanese language, history, or culture? It is ironic that the wealthiest nation on Earth would be turning its back on the rest of the world just as most nations are opening up. We have by far the most to lose; anything we can do to foster an international outlook will be of fundamental importance to our common future.

As an example of what might be done, consider the possibilities for improving relationships between North America and Latin America. As industrialized countries, the United States and Canada comprised about 283 million people in 1992, their populations growing relatively slowly at 0.8 percent per year, with an average per-capita income of just over $21,500. At the same time, the developing countries of Latin America comprised about 453 million people, their populations growing rapidly at 2.1 percent per year, with a per-capita income of just $2,170. By early in the next century there will be twice as many people in Latin America as in the United States and Canada. What kinds of understandings and relationships shall we forge to expedite our common task of managing the lands and waters of the Western Hemisphere for our common benefit?

It is very much in the immediate and long-term interest of the United States and Canada to encourage the cooperation that is growing among the nations of Latin America, and their efforts to help themselves. This relationship was explicitly acknowledged in the early 1990s by discussions about the formation of a North American Free Trade Agreement, which would involve Canada, Mexico, and the United States. Positive political and economic developments such as this agreement should be encouraged, and development funds should be made available in any case, to promote world stability. When the United States deals with its southern neighbors, the issues emphasized tend to be confrontational ones, including drug trafficking, illegal immigration, debt repayment, protective trade practices, and national policies toward repression of political conflict. Why could we not begin to move the agenda toward cooperative initiatives in science, technology, and the humanities; toward sustainable development and conservation; and toward mutual cultural understanding? Deep and mutually considerate exchanges of this sort would prepare us much better for the increasingly intense interactions that are certain to characterize our relationships in the next century.

## The Role of Women

The roles of women worldwide are a matter for the most serious consideration. Although substantial numbers of women have entered the work forces of industrialized countries, their opportunities are still much more limited than those available to men. Worldwide, no single factor will influence the rate of population growth more than assigning a higher status to women. Family planning is very much in the interest of individual women's health and welfare, but access to education and contraceptives is becoming increasingly difficult. Family planning contributions to developing countries from both Europe and the United States have dropped sharply over the past 15 years. As the populations of developing countries have increased dramatically, assistance for dealing with the problem has decreased.

As Lester Brown[2] has pointed out so well, women hold a paradoxical place in many societies. As part of their traditional duties as mothers and wives, they are expected to bear the whole respon-

[2]Lester Brown is founder of the Worldwatch Institute, a research organization providing information on global environmental issues to concerned citizens and policy-makers. (See interview with Lester Brown in Part I.)

(Larry C. Price)

sibility for child care; at the same time, they are often expected to contribute significantly to the family income by direct labor. In many developing countries women have few rights and little legal ability to protect their property, their rights to their children, their income, or anything else. Improving the status of women is a crucial aspect of development. As Lester Brown puts it, "Until female education is widespread—until women gain at least partial control over the resources that shape their economic lives—high fertility, poverty, and environmental degradation will persist in many regions."

Of even more fundamental importance is the simple fact that the human race will never be able to accomplish its best, and deal most effectively with the planet that sustains us all, as long as we discriminate based on gender, race, or any other arbitrary characteristic. We need all of the abilities that each of us can bring to bear for the common benefit, and this relationship alone, not even considering simply social justice or morality, ought to be sufficient to make each one of us avoid discrimination in all of its forms.

## The Poor

One out of four people in the world lives in abject poverty. From day to day these people have no idea where their next meal is coming from, whether their children will be properly clothed, or whether they will find adequate shelter. Rich nations tend to view poor ones as if they should somehow pull themselves together and do a better job of feeding their people, while the rich nations are willing to provide only a limited amount of help to them. Nor do people typically work together within developing countries. The more affluent classes, and those centered in the cities, tend to dominate the rural poor; in general, laws and customs coincide to ensure the continuation of such domination. The outcome of this negligence of the poor is that their ranks are swelling rapidly.

What is wrong with such attitudes? First of all, they are morally indefensible. Faced with the facts, most people would find unacceptable the deaths of 40,000 babies each day, most of which could have been avoided by access to adequate supplies of food or simple medical techniques and supplies. There is no justification for the bigoted view that these deaths prevent the world from becoming overpopulated; the world is already overpopulated, and for us to allow so many to starve, to go hungry, and to live in absolute poverty (and we do allow it, whether we choose to look at the problem squarely or not) is to threaten the future of the global ecosystem that sustains us all. Everyone must have a reasonable share of the Earth's productivity, in the face of our current numbers and their projections, or our civilization will simply come unraveled.

## WHAT SHOULD BE DONE: AN AGENDA FOR TODAY

The six problems just discussed carry an urgent message to us all, that the explosive and unevenly distributed growth of an unprecedented human population is putting unsupportable strains on the global ecosystem. The long period of transition from a hunter-gatherer society, in which widely dispersed humans were one of millions of species of organisms, to a modern industrial society, in which humans consume, co-opt, or waste 40 percent of the total global productivity, has now reached a point where our traditional modes of operation seem increasingly unlikely to lead us to stability and prosperity. In the light of this grave threat to

(Louie Psihoyos/Matrix International, Inc.)

our common future, we must weigh our options carefully as we seek ways to avoid poisoning the air we breathe and the water we drink, to avoid exterminating the organisms that we will need to build a stable life for our children, and to collectively and sustainably use the biosphere. There are a number of critical areas where significant progress can be made as we approach the year 2000, and we shall now discuss some of the concerns that we consider to be top priorities for global attention.

**1. Population stability** must be attained throughout the world, and the industrialized countries must assist others in carrying out their plans in this respect. It is especially important for industrialized countries to attain stable population levels, since their people consume such a disproportionately large share of what the world is capable of producing. There is no hope for a peaceful world without overall population stability, and no hope for regional economic sustainability without regional population stability. Thus, it is difficult to understand why international assistance for family planning is decreasing, or why the United States would choose not to make available to other countries, on their request, the same contraceptive means that are fully available domestically.

**2. Women's rights** need to find fuller expression everywhere. Ensuring that all women have access to the full range of human opportunities would address environmental problems in two significant ways. First, it would greatly accelerate our progress toward global population stability. Second, because we will need *all* of the talent available in order to build a sustainable society, we cannot afford to limit any individual's potential for contributing to this effort.

**3. New energy sources** must be sought. The greenhouse effect associated with increasing concentrations of carbon dioxide and other gases in the atmosphere ought to stimulate research into alternative energy sources. Even if we officially choose to believe that the supplies of fossil fuels such as coal and oil are infinite, burning them in the quantities we do now is certainly not an environmentally sound practice. To make major changes

(Courtesy of U.S. Windpower, Livermore, CA)

in our sources of energy will require so high a level of readjustment that the effort must be begun well before it is to have a significant impact. Renewable energy sources such as solar/hydrogen power, and biomass production should be explored while there is still time to accomplish changes.

(Photo by Gene Elle Calvin. Courtesy of Melvin Calvin, Lawrence Berkeley Laboratory)

**4. A comprehensive energy plan** must be implemented for the world. At present, the three fourths of the world's people who live in developing countries use about 20 percent of the world's industrial energy. Their numbers are growing rapidly, however, and even without any increase in their standard of living, these countries will use much larger quantities of energy in the future. If and when they industrialize, and if there has been a global consensus on limiting carbon dioxide in the atmosphere, we who live in the industrial nations of the world will find ourselves insisting that developing countries not burn coal, or that they take steps to remove gases from coal smoke that are far more rigorous and expensive than any we have used in our own countries. Such a strategy seem unlikely to prevail—unless we all share in paying for it. The key is for those of us who live in the industrialized nations to realize that the implementation of a comprehensive energy plan for the developing world is a necessary ingredient for our own future security as well as for global stability. It must involve reducing our own profligate use of energy,

which certainly cannot be sustained—much less used as a model for the rest of the world.

**5. Regional cooperation** will be necessary to solve many pollution and conservation problems. Acid precipitation, for example, is a problem that must be solved regionally. A country, such as the United Kingdom or the United States, that "exports" acid precipitation to other countries by erecting tall smokestacks with inadequate scrubbers to remove sulfates saves money, but it causes other countries to incur substantial costs and may cause them to adopt highly negative and ultimately dangerous attitudes toward us. When the problem is viewed regionally, there is an overall economic gain to ending the pollution. The Earth summit that took place in Brazil in 1992 was a significant step in the process of considering common problems in an appropriate governmental, collective context. Such efforts, which should be supported strongly, have begun to form a framework for the actions that will be necessary to achieve a secure future.

(Reuters/Bettmann Archives)

**6. Soil and water** must be better conserved. Plans for the sustainable use of the soil and water of all regions of the world must be developed, with provisions for sustainable forms of agriculture and forestry. Underdeveloped countries can achieve stability only if their best lands—those capable of sustainable productivity—are developed properly and if appropriate land-use schemes are implemented. Many experts agree that all of the agricultural and forestry needs of the poor people who live in the tropics could certainly be met by proper development of lands that have already been deforested. It is a profound human tragedy that we are not accomplishing this development. Instead, the rural poor are ignored, left to "mine" undisturbed forests on a one-time basis, and so convert potentially renewable resources into nonrenewable wasteland.

**7. Biodiversity** must be truly protected throughout the world. We live in a situation where up to one fourth, or more, of the world's species of plants, animals, fungi, and microorganisms are likely to become extinct over the next several decades.

(Stouffer Enterprises, Inc. © 1993 Animals Animals)

Once they are gone, they will be gone forever; their disappearance is the crime for which our descendants will be least likely to forgive us. Our entire sustainable use of the world's resources depends on the wise management of biodiversity, because we obtain individual products from individual kinds of organisms, and we depend on the management of communities of organisms, regardless of how poorly we understand them, for the global preservation of soil, water, and air.

**8. Biotechnology** provides us with remarkable new opportunities for the improvement of agriculture and forestry systems, and should be utilized in the development of improved crops throughout the world. We simply cannot afford our current inefficient use of biological resources, or the hunger that results from insufficient supplies of food in many regions. A global effort should be made to utilize the tools that are available to us for the improvement of many traditional tropical crops, such as manioc and yams, and for the development of additional ones that can be grown in areas not now under cultivation. Efforts to cause alarm about the use of biotechnology only tie our hands and contribute directly to human suffering.

### Telecommuting: Good for the Planet

Telecommuting is the term that describes working at home by using computers, modems, and fax machines linked through telephone lines to an office in another location. It is a growing trend. In the future, it will be common for people to work at home most of the time and travel to office only when they need to meet with fellow workers.

In 1990, the environmental consulting firm Arthur D. Little studied the potential impact in the United States if telecommunications (including videoconferencing) were substituted for transportation 10 to 20 percent of the time in working, shopping, business meetings, and the movement of electronic data. The benefits: $23 billion saved (including $610 million through reduced airborne pollution, $1.5 billion through reduced energy expenditures, and $15.38 billion through increased productivity), 1.8 million tons of pollution avoided, 3.5 million gallons of gasoline not burned, and 3.1 billion hours of personal time gained.

**9. Individual values** could use improvement, particularly among those making economic decisions. Often, businesspeople acquire assets and manage them guided strictly by economic considerations, without thinking about the serious environmental consequences of their actions. Many of these consequences have been discussed in this book. New methods of conducting business that internalize environmental costs are emerging, as are new work patterns such as telecommuting, in which travel is minimized.

## WHAT KIND OF WORLD DO WE WANT?

Perhaps the most important single lesson to have learned from this textbook is that those of us who live in industrialized countries are the core of the problem facing the global ecosystem today. We number less than one quarter of the world's people, and our activities alone are more than sufficient to create global instability. For example, the United States, with 4.7 percent of the world's population,

(NASA)

generates about 21 percent of the world's carbon dioxide. Similar relationships can be demonstrated in almost any area of resource consumption, indicating clearly that the industrialized countries of the world must act forcefully to reduce their levels of consumption if we are all going to be able to attain stability.

As we strive aggressively to increase our high standards of living in industrialized countries, from a level that would be considered utopian by most people on Earth, we drain resources from the entire globe and thus contribute willy-nilly to the creation of a future in which neither we nor our children nor our grandchildren will be able to live in anything like the affluence or, more importantly, opportunity that we experience now.

The citizens of industrialized nations often seem to assume implicitly that overall prosperity can be created as a result of science and technology, and that primary productivity will take care of itself. If we are to improve the world situation, however, we must find ways to support the farmers and others who produce the crops on which we ultimately depend, and to involve these citizens in the affairs of the countries where they live.

To put it concisely, we must radically change our view of the world and adopt new ways of thinking, or we will perish together. We must learn to understand, respect, and work with one another, regardless of the differences that exist between us. The most heartening aspect of the situation that we confront is that people, given the motivation, do have the ability to make substantial changes.

At the deepest level, the most critical environmental problems, from which all others arise, are our own attitudes and values. We are totally out of

touch and out of balance with the world, and until we reconnect and readjust in some significant way, all solutions will be stopgap ones. As a society we don't feel part of the global ecosystem; we feel separate, above it, and therefore in a position to consume and abuse without thought of consequence. This book has been about consequences. In the last 30 years and especially in the last decade, we have come to recognize the nature of the impact of human activity on the biosphere. We now understand this impact enough to know that we *cannot* continue to act as we have been acting and expect any sort of viable future for our species. If all we do as a result of this new knowledge is to make some shifts in consumer choices and write a few letters, it won't be nearly enough.

*T*he most critical environmental problems, from which all others arise, are our own attitudes and values.

Your generation must become the next pioneers. A pioneer is one who ventures into unexplored territory, a process that is simultaneously terrifying and profoundly exciting. The unexplored territory in this case is the development of a truly different way for humans to exist in the world. No models exist for this kind of change. You must forge a new revolution, akin in scope and effect to the Agricultural Revolution or the Industrial Revolution, yet totally different because it must be deliberate. You must help create the political will for it with your numbers and your commitment. You must create the economic power with your thoughtful decisions as both consumers and leaders. You must create social change with your acceptance and respect of the differences between peoples.

This change will require reconnecting to the natural world. That means, at a personal level, taking opportunities to be in the wilderness (even if it is a city park or a backyard garden), to listen to the wind, and to look at the exquisite variety of plants and rocks and insects with which we coexist. Humans evolved in nature. Our immensely complex and multidimensional brains developed precisely because we were able to interact with growing things, weather patterns, and other animals. The world we have created now screens us from all that. The sophisticated devices we imagined and manufactured—things such as televisions, computers,

(Visuals Unlimited/John Gerlach)

to mean success, at tremendous cost to the planet. Such ideas are deeply imbedded and extremely compelling, and will be difficult to change. You may need to throw off some of the myths of Western culture, such as the belief that "faster" and "more" inevitably mean "better." You will at least be called on to examine those myths very deeply and very thoughtfully, and to decide what they mean to you.

It will require reinventing economic constructs, building the cost of environmental impact and damage into our accounting systems, causing market forces to work in favor of environmental protection. Business activities must involve developing cooperative partnerships with people all over the world and making decisions based on long-term benefits to the environment.

This is an overwhelming responsibility. The choices we make now will have a greater impact on the future than those that any generation has had before, whether these choices lead us further into or out of ecological collapse. Even choosing to do nothing will have profound consequences for the future. At the same time, it is an incredible opportunity. This is a time in history when the best of human qualities—vision, courage, imagination, and concern—will play a critical role in establishing the contours of tomorrow's world.

and automobiles—have come to define our world. One of your challenges will be to use technology as a tool but *not* to let it define your interaction with the world.

The new revolution will involve revaluing ourselves and our lives according to a different set of ideals. Wealth and material possessions have come

(Joanna Pinneo)

EARTH at NIGHT

This composite image made from U.S. weather satellite photographs is a dramatic depiction of the human presence on Earth. Clearly visible are the lights of urban centers, rural fires, gas flares from oil wells, the aurora borealis, and more. Africa, Europe, Asia, and Australasia appear on the next two pages.

CREDITS: United States Air Force / DMSP Archives, National Snow & Ice Data Center, NOAA, University of Colorado / Text & image © 1985 W.T. Sullivan, III, University of Washington, Seattle Poster design & image processing by Kerry Meyer / Poster © 1986 Hansen Planetarium, Salt Lake City, Utah, U.S.A.

## IDENTIFICATION CHART

G = natural gas burn-off
F = agricultural fires

### NORTH AMERICA

1. G (Prudhoe Bay)
2. Fairbanks
3. Anchorage + G
4. Edmonton
5. Winnipeg
6. Calgary
7. Medicine Hat
8. Vancouver
9. Seattle
10. Portland
11. Interstate Highway 5
12. San Francisco-Oakland
13. Los Angeles
14. Honolulu
15. San Diego
16. Phoenix
17. Tucson
18. El Paso
19. Chihuahua
20. Dallas-Ft. Worth
21. Torreon
22. Guadalajara
23. Monterrey
24. Mexico City
25. Puebla
26. G (Tabasco)
27. Tampa
28. San Antonio
29. Houston
30. New Orleans
31. Miami
32. Jacksonville
33. Atlanta
34. Kansas City
35. St. Louis
36. Norfolk
37. Washington D.C.-Baltimore
38. Philadelphia-Newark-New York City-Long Island
39. Boston-Providence
40. Montréal
41. Québec
42. Aurora borealis
43. Pittsburgh
44. Buffalo
45. Toronto
46. Cleveland
47. Detroit
48. Chicago-Milwaukee
49. Minneapolis-St. Paul
50. Denver
51. Salt Lake City

### CENTRAL & SOUTH AMERICA

52. San José
53. Panamá
54. Bogotá + G
55. Quito + G
56. Lima
57. Santiago
58. Rosario
59. Buenos Aires
60. Montevideo
61. Pôrto Alegre
62. São Paulo
63. Rio de Janeiro
64. Brasília
65. Salvador
66. Recife
67. Manaus
68. Caracas + G
69. Puerto Rico
70. Kingston
71. Havana

### NEW ZEALAND

72. Christchurch
73. Wellington
74. Auckland

A tapestry of city lights and rural fires announces our presence on this planet. In contrast to daytime images, where only natural features are easily visible, the activities of humankind at night are readily traced in this mosaic of images from U.S. Air Force weather satellites. Much of the light leakage to space corresponds to street and building lights in urbanized regions, especially in Europe, North America, and eastern Asia. Also nicely etched are transportation features such as the Trans-Siberian railroad, the main railroad through central China, the spoke pattern centered on the hub of Moscow, and Interstate Highway 5 along the western coast of the United States. The delimiting effects of geographical features such as the Nile River, the Sahara Desert, the Himalayas, and the Australian Outback are also apparent. In the tropics, the major sources of light are controlled fires—the result of grassland burning, slash-and-burn agriculture, and clearing of forests. The frequency of these fires depends on season, but in the present image they are prominent throughout the highlands of Southeast Asia, the sub-Saharan savannas, and East Africa. Other lights arise from huge burn-offs of natural gas associated with oil wells. Gas flares show clearly in Indonesia, the Tashkent region of the Soviet Union, Siberia, the Middle East, North and West Africa, and northwestern South America. In the Sea of Japan, the large blotch of light emanates from a fishing fleet that hangs multitudes of lights on its boats in order to lure squid and saury to the surface. The only natural source of light is the aurora over Greenland. Aurorae (Northern Lights) occur when high-energy particles from the sun enter polar magnetic regions, creating currents that cause the upper atmosphere to glow like a gigantic fluorescent tube.

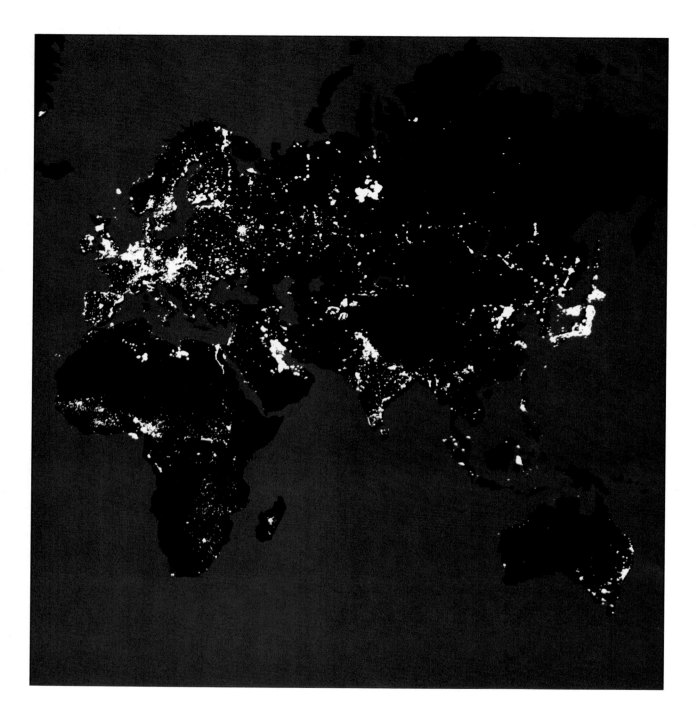

The image is replete with lessons in geography, economics, anthropology, and environmental science. And for astronomers, who find themselves limited to increasingly remote and expensive sites, it also illustrates their constant battle against the "light pollution" that damages observations of faint stars and galaxies. But is there not a wider loss? The image testifies that hundreds of millions of people today have no dark sky and are thus denied the nighttime universe. No longer do they know the exquisite thrill of a meteor shooting across the sky, nor the humility brought on by the resplendence of two thousand stars wreathed by the Milky Way. At a time when the very survival of our species depends on finding a common vision, we have wrapped Earth in a glowing fog.

# IDENTIFICATION CHART

G = natural gas burn-off
F = agricultural fires

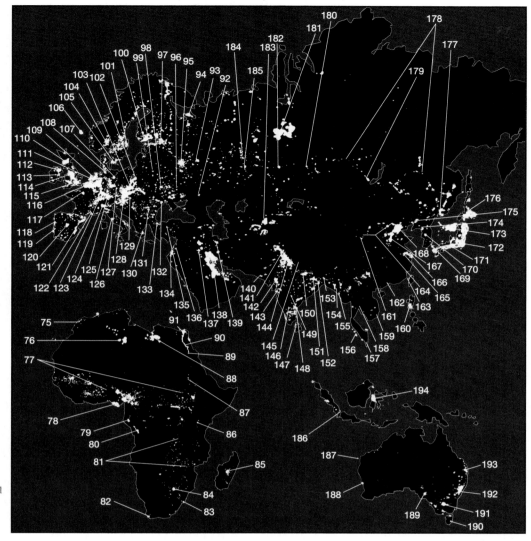

## AFRICA

75 Casablanca
76 G (Algeria)
77 F (sub-Saharan savanna)
78 G (Nigeria)
79 G (Gabon)
80 Kinshasa
81 F (Southeast Africa)
82 Capetown
83 Durban
84 Johannesburg
85 F (Malagasy [Madagascar])
86 Nairobi
87 Khartoum
88 G (Libya)
89 Aswan
90 Nile River
91 Cairo + Nile River delta

## EUROPE

92 Volgograd
93 Gor'kiy
94 Arkhangel'sk
95 Moscow
96 Khar'kov
97 Leningrad
98 Kiev
99 Minsk
100 Helsinki
101 Stockholm
102 Warsaw
103 Cracow
104 Oslo
105 Copenhagen
106 G (North Sea)
107 Berlin
108 Hamburg
109 Essen-Cologne
110 Glasgow-Edinburgh
111 The Netherlands
112 Northern Ireland
113 Dublin
114 Manchester-Birmingham-Liverpool
115 London
116 Belgium
117 Paris
118 Porto
119 Lisbon
120 Madrid
121 Barcelona
122 Marseille
123 Zurich
124 Milan
125 Munich
126 Rome
127 Prague
128 Vienna
129 Budapest
130 Athens
131 Bucharest
132 Odessa

## ASIA

133 Istanbul
134 Israel
135 Ankara
136 Damascus
137 Baghdad
138 G (Saudi Arabia & Persian Gulf States)
139 Teheran
140 Karachi
141 Lahore
142 Ahmadabad
143 Bombay
144 Delhi
145 Himalaya Mountains
146 Bangalore
147 Colombo
148 Madras
149 Hyderabad
150 Allahabad
151 Patna
152 Calcutta + Ganges River delta
153 Dacca
154 F (SE Asian highlands)
155 Rangoon
156 Kuala Lumpur
157 Singapore
158 Bangkok
159 Ho Chi Minh City (Saigon)
160 Manila
161 Lanzhou (Lanchow)
162 Hong Kong
163 Taiwan
164 Jing-Bao/Bao-Lan Railroads + Yellow River
165 Nanjing (Nanking)
166 Shanghai + Yangtze River delta
167 Taiyüan
168 Beijing (Peking)-Tianjin (Tientsin)
169 Pusan
170 Seoul
171 Osaka-Kobe-Kyoto
172 Tokyo-Yokohama
173 Japanese squid fleet
174 Pyongyang
175 Shenyang
176 Sapporo
177 Vladivostok
178 Trans-Siberian Railroad
179 Irkutsk
180 Novosibirsk
181 G (Surgut, Siberia)
182 Omsk
183 Tashkent + G
184 Sverdlovsk
185 Chelyabinsk

## AUSTRALASIA

186 Jakarta
187 Northwest Cape
188 Perth
189 Adelaide
190 Hobart
191 Melbourne
192 Sydney
193 Brisbane
194 G (Sulawesi, Indonesia)

CAUTIONARY NOTES: There are several aspects of this image that the discriminating viewer should keep in mind. First, it is based on a mosaic of about forty individual photographs, each of which has its own distortions. These photos have been approximately reduced to a common scale, but the final image corresponds to a Mercator projection only to about 5% accuracy. The individual photos were also taken with a variety of exposures and under varying moonlight. Several additional processing steps then led to this poster, with the result that quantitative light intensities in different regions can only roughly be judged. The photos in the mosaic were taken at various times and seasons over the period 1974-84; this affects in particular the occurrence of tropical fires, which are highly seasonal. The aurora included in the image is an artist's rendering based on satellite photography. Its positioning is only meant to be suggestive—aurorae in fact usually occur in a ring centered on the north or south magnetic pole. Finally, although most dark regions are truly lacking in light, sections of some photos were clouded out, and suitable photos were not available for a few regions such as many remote islands and portions of South China and southwestern Africa.

CREDITS: United States Air Force / DMSP Archives, National Snow & Ice Data Center, NOAA, University of Colorado / Text & image © 1985 W.T. Sullivan, III, University of Washington, Seattle
Poster design & image processing by Kerry Meyer / Poster © 1986 Hansen Planetarium, Salt Lake City, Utah, U.S.A.

**abiotic** Nonliving. Compare *biotic*.

**acid** A substance that releases hydrogen ions (protons) in water. Acids have a sour taste and turn blue litmus paper red. Compare *base*.

**acid deposition** A type of air pollution that includes acid which falls from the atmosphere to the Earth as precipitation or as dry acidic particles. See *acid precipitation*.

**acid mine drainage** Pollution caused when acids and other toxic compounds wash from mines into nearby lakes and streams.

**acid precipitation** Precipitation that is acidic as a result of both sulfur and nitrogen oxides forming acids when they react with water in the Earth's atmosphere; partially due to the combustion of coal; includes acid rain, acid snow, and acid fog.

**active solar heating** A heating method in which a series of collection devices mounted on a roof or in a field are used to absorb solar energy for space heating or heating water. Pumps or fans distribute the collected heat. Compare *passive solar heating*.

**adaptive radiation** The evolution of a large number of related species from an unspecialized ancestral organism.

**aerobic respiration** The process by which cells utilize oxygen to break down organic molecules into waste products (such as carbon dioxide and water) with the release of energy that can be used for biological work.

**age structure** The percentage of a population, including the number of people of each sex, at different ages. Age structure diagrams represent the number of males and females at each age in a population.

**A-horizon** The topsoil; located just beneath the O-horizon of the soil. The A-horizon is rich in various kinds of decomposing organic matter.

**air pollution** Various chemicals (gases, liquids, or solids) present in high enough levels in the atmosphere to harm humans, other animals, plants, or materials. Excess noise and heat are also considered air pollution.

**alcohol fuels** A liquid fuel substitute such as methanol or ethanol that may eventually replace gasoline.

**algae** Single-celled or simple multicellular photosynthetic organisms; important producers in aquatic ecosystems.

**allelopathy** An adaptation in which toxic substances secreted by roots or shed leaves inhibit the establishment of competing plants nearby.

**alpine tundra** A distinctive ecosystem located in the higher elevations of mountains, above the tree line.

**alternative agriculture** Agricultural methods that rely on beneficial biological processes and environmentally friendly chemicals rather than conventional agricultural techniques. Also called *sustainable* or *low-input agriculture*. Compare *high-input agriculture*.

**altitude** The height of a thing above sea level. Compare *latitude*.

**amino acids** Organic compounds that are linked together to form proteins. See *essential amino acid*.

**ammonification** The conversion of nitrogen-containing organic compounds to ammonia ($NH_3$) by certain bacteria (ammonifying bacteria) in the soil; part of the nitrogen cycle.

**annual** A plant that grows, reproduces, and dies in one growing season. Compare *biennial* and *perennial*.

**anthracite** The highest grade of coal; has the highest heat content and burns the cleanest of any grade of coal. Also known as *hard coal*. Compare *bituminous coal* and *lignite*.

**anticline** An upward folding of rock layers, or strata.

**antioxidant** Food additive that prevents oxidation (breakdown) of food molecules.

**aquaculture** The rearing of aquatic organisms (fish and shellfish), either freshwater or marine, for human consumption. See *mariculture*.

**aquatic** Pertaining to the water. Compare *terrestrial*.

**aquifer** Porous layer of underground rock in which groundwater is stored. See *confined aquifer* and *unconfined aquifer*.

**aquifer depletion** The removal by humans of more groundwater than can be recharged by precipitation or melting snow.

**arid land** A fragile ecosystem in which lack of precipitation limits plant growth. Arid lands are found in both temperate and tropical regions. Also called *desert*.

**artesian aquifer** See *confined aquifer*.

**artificial eutrophication** Overnourishment of an aquatic ecosystem by nutrients such as nitrates and phosphates. In artificial eutrophication the pace of eutrophication is accelerated quite rapidly due to human activities such as agriculture and discharges from sewage treatment plants.

Also called *cultural eutrophication*. See *eutrophication*.

**assimilation**  The conversion of inorganic nitrogen (nitrate, $NO_3^-$, or ammonia, $NH_3$) to the organic molecules of living things; part of the nitrogen cycle.

**atmosphere**  The gaseous envelope surrounding Earth.

**atom**  The smallest quantity of an element that can retain the chemical properties of that element; composed of protons, neutrons, and electrons.

**atomic mass**  A number that represents the sum of the number of protons and neutrons in the nucleus of an atom. The atomic mass represents the relative mass of an atom. Compare *atomic number*.

**atomic number**  A number that represents the number of protons in the nucleus of an atom. Each element has its own characteristic atomic number. Compare *atomic mass*.

**autotroph**  See *producer*.

**background extinction**  The continuous, low-level extinction of species that has occurred throughout much of the history of life. Compare with *mass extinction*.

**bacteria**  Unicellular prokaryotic microorganisms. Most bacteria are decomposers, but some are autotrophs and some are parasites.

**base**  A compound that releases hydroxyl ions ($OH^-$) when dissolved in water. A base turns red litmus paper blue. Compare *acid*.

**benthos**  Bottom-dwelling sea creatures that fix themselves to one spot, burrow into the sand, or simply walk about on the ocean floor.

**B-horizon**  The light-colored, partially weathered soil layer underneath the A-horizon; subsoil. The B-horizon contains much less organic material than the A-horizon.

**biennial**  A plant that requires two years to complete its life cycle. Compare *annual* and *perennial*.

**bioaccumulation**  See *biological magnification*.

**biodegradable**  A chemical pollutant decomposed (broken down) by living organisms or other natural processes. Compare *nondegradable*.

**biodiversity**  See *biological diversity*.

**biogas**  A clean fuel, usually composed of a mixture of gases, whose combustion produces fewer pollutants than either coal or biomass. Biogas is produced from plant matter of one form or another.

**biogas digester**  A vat that uses bacteria to break down household wastes, including sewage. The gas produced by the bacteria is burned as a fuel.

**biogeochemical cycle**  Process by which matter cycles from the living world to the nonliving physical environment and back again. Examples of biogeochemical cycles include the carbon cycle, the nitrogen cycle, and the phosphorus cycle.

**biological amplification**  See *biological magnification*.

**biological control**  A method of pest control that involves the use of naturally occurring disease organisms, parasites, or predators to control pests. Also called *biological pest control*.

**biological diversity**  Number and variety of living organisms; includes genetic diversity, species diversity, and ecological diversity. Also called *biodiversity*.

**biological magnification**  The increased concentration of toxic chemicals such as PCBs, heavy metals, and pesticides in the tissues of organisms higher in the food chain. Also called *bioaccumulation* and *biological amplification*.

**biological oxygen demand (BOD)**  The amount of oxygen needed by microorganisms to decompose (by aerobic respiration) the organic material in a given volume of water.

**biological pest control**  See *biological control*.

**biomass**  (1) A quantitative estimate of the total mass, or amount, of living plant or animal material. Often expressed as the dry weight of all the organic material that comprises living organisms in a particular ecosystem. (2) Plant and animal materials used as fuel.

**biome**  A large, relatively distinct ecosystem characterized by a similar climate, soil, plants, and animals, regardless of where it occurs on Earth.

**bioremediation**  A method employed to clean up a hazardous waste site that uses microorganisms to break down the toxic pollutants.

**biosphere**  All of the Earth's living organisms. Includes all the communities on Earth. Compare *ecosphere*.

**biotechnology**  See *genetic engineering*.

**biotic**  Living. Compare *abiotic*.

**biotic potential**  The maximum rate at which a population could increase under ideal conditions.

**birth rate**  The number of births per 1000 people per year. Also called *crude birth rate* and *natality*.

**bituminous coal**  The most common form of coal; produces a high amount of heat and is used extensively by electric power plants. Also called *soft coal*. Compare *anthracite* and *lignite*.

**bloom**  Large algal populations caused by the sudden presence of large amounts of essential nutrients (such as nitrates and phosphates) in surface waters.

**boreal forest**  See *taiga*.

**botanical**  A plant-derived chemical used as a pesticide.

**bottom ash**  The residual ash left at the bottom of an incinerator when combustion is completed. Also called *slag*. Compare *fly ash*.

**breeder nuclear fission**  A type of nuclear fission in which nonfissionable U-238 is converted to fissionable P-239. Thus, a breeder nuclear reactor produces more fuel than it consumes.

**broad leaf herbicide**  A herbicide that kills plants with broad leaves but does not kill grasses (such as corn, wheat, and rice).

**calorie**  A unit of heat energy; the amount of heat required to raise 1 g of water 1°C.

**cancer**  A malignant tumor anywhere in the body. Cancers tend to spread throughout the body.

**carbamates**  A class of broad-spectrum pesticides that are derived from carbamic acid.

**carbohydrate** An organic compound containing carbon, hydrogen, and oxygen in the ratio of 1C:2H:1O. Carbohydrates include sugars and starches, molecules metabolized readily by the body as a source of energy.

**carbon cycle** The worldwide circulation of carbon from the abiotic environment into living things and back into the abiotic environment.

**carcinogen** Any substance that causes cancer or accelerates its development.

**carnivore** An animal that feeds on other animals; flesh-eater. Compare *herbivore* and *omnivore*. See also *secondary consumer*.

**carrying capacity** The maximum number of individuals of a given species that a particular habitat can support sustainably (long-term).

**cell** The basic structural and functional unit of life. Simple organisms are composed of single cells, whereas complex organisms are composed of many cells.

**cell respiration** A process in which the energy of organic molecules is released within cells of plants, animals, or other living organisms. Aerobic respiration is a type of cell respiration.

**central-receiver energy generation** See *solar power tower*.

**CFCs** See *chlorofluorocarbons*.

**chain reaction** A reaction maintained because it forms as products the very materials that are used as reactants in the reaction. For example, a chain reaction occurs during nuclear fission when neutrons collide with other U-235 atoms and those atoms are split, releasing more neutrons, which then collide with additional U-235 atoms.

**chaparral** A biome with a Mediterranean climate (mild, moist winters and hot, dry summers). Chaparral vegetation is characterized by small-leaved evergreen shrubs and small trees.

**chlorinated hydrocarbon** A synthetic organic compound that contains chlorine and is used as a pesticide (for example, DDT) or an industrial compound (for example, PCBs).

**chlorofluorocarbons** Human-made organic compounds composed of carbon, chlorine, and fluorine that have a number of applications, including refrigerants and starting materials for certain plastics. In the atmosphere chlorofluorocarbons break down stratospheric ozone. Also called *CFCs*.

**chlorophyll** A green pigment that absorbs radiant energy for photosynthesis.

**C-horizon** The largely unweathered layer in the soil located beneath the B-horizon. The C-horizon borders solid parent rock.

**circumpolar vortex** A mass of cold air that circulates around the southern polar region, in effect isolating it from the warmer air in the rest of the world.

**clay** The smallest inorganic soil particles. Compare *silt* and *sand*.

**clean-coal technologies** New methods for burning coal that do not contaminate the atmosphere with sulfur oxides and produce significantly fewer nitrogen oxides.

**clearcutting** A forest management technique that involves the removal of all trees from an area at a single time. Compare *seed-tree cutting*, *selective cutting*, *shelterwood cutting*, and *whole-tree harvesting*.

**climate** The average weather conditions that occur in a place over a period of years. Includes temperature and precipitation. Compare *weather*.

**climax** See *climax community*.

**climax community** The relatively stable stage at the end of an ecological succession. The climax community remains more or less unchanged as long and climate and other environmental factors do not change. See *succession*.

**coal** A black combustible solid found in the Earth's crust. Formed from the remains of ancient plants that lived millions of years ago. Used as a fuel. See *fossil fuel*.

**coal gasification** The technique of producing combustible gases (such as carbon monoxide and hydrogen) from coal.

**coal liquefaction** The process by which coal is used to produce a nonalcohol liquid fuel.

**coal spoils** Land badly disturbed and damaged as a result of coal mining.

**coastal wetland** Marshes, bays, tidal flats, and swamps that are found along a coastline. See *salt marsh* and *wetland*.

**cochlea** The part of the ear that perceives sound.

**coevolution** The interdependent evolution of two or more species that occurs as a result of their interactions over a long period of time. Flowering plants and their animal pollinators are an example of coevolution because each has profoundly affected the other's characteristics.

**cogeneration** A nontraditional energy technology that involves recycling waste heat As a result, both electricity and steam or hot water are produced from the same fuel.

**coloring agent** Natural or synthetic food additives used to make food visually appealing.

**combustion** The process of burning by which organic molecules are rapidly oxidized, converting them into carbon dioxide and water with an accompanying release of heat and light.

**command and control** Pollution control laws that work by setting pollution ceilings. Examples include the Clean Water and Clean Air Acts.

**commensalism** A type of symbiosis in which one organism benefits and the other one is neither harmed nor helped. See *symbiosis*. Compare *mutualism* and *parasitism*.

**commercial extinction** Depletion of the population of a commercially important species to the point that it is unprofitable to harvest.

**commercial harvest** The harvest of commercially important living organisms from the wild. Examples include the commercial harvest of parrots (for the pet trade) and cacti (for houseplants).

**commercial hunting**  Killing commercially important animals for profit (from the sale of their furs, meat, and so on). Compare *sport hunting* and *subsistence hunting*.

**community**  An association of organisms of different species living together in a defined habitat with some degree of mutual interdependence. Compare *ecosystem*.

**competition**  The interaction among organisms that vie for the same resources in an ecosystem (such as food, living space, or other resources).

**competitive exclusion**  The concept that no two species with identical living requirements can occupy the same ecological niche indefinitely. Eventually, one species will be excluded by the other as a result of interspecific (between-species) competition for a resource in limited supply.

**compost**  A natural soil and humus mixture that improves soil fertility and soil structure.

**confined aquifer**  A groundwater storage area trapped between two impermeable layers of rock. Also called *artesian aquifer*. Compare *unconfined aquifer*.

**conifer**  Any of a group of woody trees or shrubs (gymnosperms) that bears needlelike leaves and seeds in cones.

**conservation tillage**  A method of cultivation in which residues from previous crops are left in the soil, partially covering it and helping to hold it in place until the newly planted seeds are established. See *reduced-tillage* and *no-tillage*. Compare *conventional tillage*.

**conservationist**  A person who supports the conservation of natural resources.

**consumer**  Organism that cannot synthesize its own food from inorganic materials and therefore must use the bodies of other organisms as a source of energy and body-building materials. Also called *heterotroph*. Compare *producer*.

**consumption overpopulation**  Circumstance in which each individual in a population consumes too large a share of resources—that is, much more than is needed to survive. Consumption overpopulation results in pollution, environmental degradation, and resource depletion. See *overpopulation*. Compare *people overpopulation*.

**containment building**  A safety feature of nuclear power plants that provides an additional line of defense against any accidental leak of radiation.

**continental shelf**  The submerged, relatively flat ocean bottom that surrounds continents. The continental shelf extends out into the ocean to the point where the ocean floor begins a steep descent.

**contour plowing**  Plowing that matches the natural contour of the land—that is, the furrows run around rather than up and down a hill.

**contraceptive**  A device or drug used to intentionally prevent pregnancy.

**control**  An essential part of every scientific experiment in which the experimental variable remains con-stant. The control provides a standard of comparison in order to verify the results of an experiment.

**conventional tillage**  The traditional method of cultivation in which the soil is broken up by plowing before seeds are planted. Compare *conservation tillage*.

**cooling tower**  Part of an electric power generating plant within which heated water is cooled.

**Coriolis effect**  The tendency of moving air or water to be deflected from its path to the right in the Northern Hemisphere and to the left in the Southern Hemisphere. Caused by the direction of the Earth's rotation.

**crop rotation**  The planting of different crops in the same field over a period of years. Crop rotation reduces mineral depletion of the soil because the mineral requirements of each crop vary.

**crude birth rate**  See *birth rate*.

**crude death rate**  See *death rate*.

**crude oil**  See *petroleum*.

**cullet**  Crushed glass food and beverage containers that are melted to make new products.

**cultural eutrophication**  See *artificial eutrophication*.

**death rate**  The number of deaths per 1000 people per year. Also called *crude death rate* and *mortality*.

**debt-for-nature swap**  The cancellation of part of a country's foreign debt in exchange for their agreement to protect certain land (or other resources) from detrimental development.

**decibel (dB)**  A numerical scale that expresses the relative loudness of sound.

**decibel-A (dBA)**  A modified decibel scale that takes into account high-pitched sounds to which the human ear is more sensitive.

**decommission**  The dismantling of an old nuclear power plant after it closes. Compare *entombment*.

**decomposer**  A heterotroph that breaks down organic material and uses the decomposition products to supply it with energy. Decomposers are microorganisms of decay. Also called *saprophyte*. Compare *detritivore*.

**deductive reasoning**  Reasoning that operates from generalities to specifics and can make relationships among data more apparent. Compare *inductive reasoning*.

**deep ecology**  The conviction that all creatures have the right to exist and that humans should not cause the extinction of other living things.

**deforestation**  The removal of forest without adequate replanting.

**delta**  A deposit of sand or soil at the mouth of a river.

**dematerialization**  The decrease in the size and weight of a product as a result of technological improvements that occur over time.

**demographics**  The branch of sociology that deals with population statistics (such as density and distribution).

**denitrification**  The conversion of nitrate ($NO_3^-$) to nitrogen gas (N$_2$) by certain bacteria (denitrifying bacteria) in the soil; part of the nitrogen cycle.

**depletion time**  An estimate of the time remaining before 80 percent of the known reserves of a particular nonrenewable resource are expended (at current rates of extraction and use).

**derelict land**  Land area degraded by mining.

**desalination**  See *desalinization.*

**desalinization**  The removal of salt from ocean or brackish (somewhat salty) water. Also called *desalination.*

**desert**  See *arid land.*

**desertification**  Degradation of once-fertile rangeland (or tropical dry forest) into nonproductive desert. Caused partly by soil erosion, deforestation, and overgrazing.

**detritivore**  An organism (such as an earthworm or crab) that consumes fragments of dead organisms. Also called *detritus feeder.* Compare *decomposer.*

**detritus feeder**  See *detritivore.*

**deuterium**  An isotope of hydrogen that contains one proton and one neutron per atom. Compare *tritium.*

**developed country**  A country industrialized and characterized by a low fertility rate, low infant mortality rate, and high per capita income. Developed countries include the United States, Canada, Japan, and European countries. Also called *highly developed country.* Compare *developing country.*

**developing country**  A country not highly industrialized and characterized by a high fertility rate, high infant mortality rate, and low per capita income. Most developing countries are located in Africa, Asia, and Latin America, and fall into two subcategories: *moderately developed* and *less developed.* Compare *developed country.*

**disease**  A departure from the body's normal healthy state as a result of infectious organisms, environmental stresses, or some inherent weakness.

**distillation**  A heat-dependent process used to purify or separate complex mixtures. Saltwater or brackish water may be distilled to remove the salt from the water. Compare *reverse osmosis.*

**DNA**  Deoxyribonucleic acid. Present in a cell's chromosomes, DNA contains the genetic information for all living organisms.

**domesticated**  Adapted to humans. Describes plants and animals which, during their association with humans, have become so altered from their original ancestors that it is doubtful they could survive and compete successfully in the wild.

**doubling time**  The amount of time it takes for a population to double in size, assuming that its current rate of increase doesn't change.

**drainage basin**  The area of land drained by a river system. Also called *watershed.*

**drip irrigation**  See *microirrigation.*

**dry deposition**  A form of acid deposition in which dry, sulfate-containing particles settle out of the air. Compare *wet deposition.*

**dust bowl**  A semiarid region that has become desertlike as a result of extended drought and severe dust storms.

**dust dome**  A dome of heated air that surrounds an urban area and contains a lot of air pollution. See *urban heat island.*

**ecological diversity**  Biological diversity that encompasses the variety among ecosystems—forests, grasslands, deserts, lakes, estuaries, and oceans, for example. Compare with *genetic diversity* and *species diversity.*

**ecological niche**  See *niche.*

**ecological pyramid**  A graphic representation of the relative energy value at each trophic level. See *pyramid of biomass, pyramid of energy,* and *pyramid of numbers.*

**ecology**  A discipline of biology that studies the interrelationships between living organisms and their environment.

**economics**  The study of how people (individuals, businesses, or countries) use their limited economic resources to fulfill their needs and wants. Economics encompasses the production, consumption, and distribution of goods.

**ecosphere**  The interactions among and between all the Earth's living organisms and the air (atmosphere), land (lithosphere), and water (hydrosphere) that they occupy. Compare *biosphere.*

**ecosystem**  The interacting system that encompasses a community and its nonliving, physical environment. Compare *community.*

**electrostatic precipitator**  An air pollution control device that gives ash a positive electrical charge so that it adheres to negatively charged plates.

**emigration**  A type of migration in which individuals leave a population and thus decrease its size. Compare *immigration.*

**emission charge**  A government policy that controls pollution by charging the polluter for each given unit of emissions; that is, by establishing a tax on pollution.

**emission reduction credit (ERC)**  A waste-discharge permit which can be bought and sold by companies that produce emissions.

**endangered species**  A species whose numbers are so severely reduced that it is in imminent danger of becoming extinct. Compare *threatened species.*

**energy**  The capacity or ability to do work.

**energy conservation**  Saving energy by reducing energy use and waste. Carpooling to work or school is an example of energy conservation. Compare *energy efficiency.*

**energy efficiency**  Using less energy to accomplish the same task. Purchasing an appliance such as a refrigerator that uses less energy to keep food cold is an example of energy efficiency. Compare *energy conservation.*

**energy flow**  The passage of energy in a one-way direction through an ecosystem.

**energy intensity**  A country's or region's total energy consumption divided by its gross natural product.

**enrichment**  The process by which uranium ore is refined after mining to increase the concentration of fissionable U-235.

**entombment**  An option after the closing of an old nuclear power plant in which the entire power plant is permanently encased in concrete. Compare *decommission*.

**entropy**  A measure of the randomness or disorder of a system.

**environment**  All the external conditions, both abiotic and biotic, that affect an organism or group of organisms.

**environmental impact statement (EIS)**  A statement that accompanies federal recommendations or proposals and is supposed to help federal officials and the public make informed decisions. Required by the National Environmental Policy Act of 1970.

**environmental resistance**  Limits set by the environment that prevent organisms from reproducing indefinitely at their biotic potential.

**environmental science**  The interdisciplinary study of how humanity affects other living organisms and the nonliving physical environment.

**environmentalist**  A person who works to solve environmental problems such as overpopulation, pollution of the Earth's air, water, and soil, and depletion of natural resources.

**epiphytes**  A small organism that grows on another organism but is not parasitic on it. Small plants that live attached to the bark of a tree's branches are epiphytes.

**essential amino acid**  Any of the eight amino acids that must be obtained in the diet because humans cannot synthesize them from simpler materials.

**estuary**  A coastal body of water that connects to oceans, in which fresh water from the land mixes with saltwater from the oceans.

**ethanol**  A colorless, flammable liquid, $C_2H_5OH$. Also called *ethyl alcohol*.

**ethyl alcohol**  See *ethanol*.

**eutrophic lake**  A lake enriched with nutrients such as nitrate and phosphate and consequently choked with plant or algal organisms. Water in a eutrophic lake contains very little dissolved oxygen. Compare *oligotrophic lake*.

**eutrophication**  The enrichment of a lake or pond by nutrients. Eutrophication that occurs naturally is a very slow process in which the lake gradually fills in and converts to a marsh, eventually disappearing. See *artificial eutrophication*.

**evaporation**  The conversion of water from a liquid to a vapor. Also called *vaporization*.

**evaporite deposit**  A massive salt deposit that forms when a body of water dries up.

**evolution**  The cumulative changes in the characteristics of populations or organisms that occur during successive generations in a line of descent. Evolu-tion explains the origin of all the organisms that exist today or have ever existed.

**ex situ conservation**  Conservation efforts that involve conserving biological diversity in human-controlled settings. Compare *in situ conservation*.

**exosphere**  The outermost layer of the atmosphere, bordered by the thermosphere and interplanetary space.

**exponential growth**  Growth that occurs at a constant rate of increase over a period of time. When the increase in number versus time is plotted on a graph, exponential growth produces a characteristic J-shaped curve.

**extinction**  The elimination of a species from Earth; occurs when the last individual member of a species dies.

**facultative parasite**  An organism that is normally saprophytic but, given the opportunity, becomes parasitic. Compare *obligate parasite*.

**fall turnover**  A mixing of the lake waters in temperate lakes, caused by falling temperatures in autumn. Compare *spring turnover*.

**family planning**  Providing the services, including information about birth control methods, to help people have the number of children they want.

**famine**  Widespread starvation caused by a drastic shortage of food. Famine is caused by crop failures that are brought on by drought, war, flood, or some other catastrophic event.

**fecal coliform test**  A water quality test for the presence of *E. coli* (fecal bacteria common in the intestinal tracts of people and animals). The presence of fecal bacteria in a water supply indicates a chance that pathogenic organisms may be present as well.

**first law of thermodynamics**  Energy cannot be created or destroyed, although it can be transformed from one form to another. Compare *second law of thermodynamics*.

**fission**  A nuclear reaction in which large atoms of certain elements are each split into two smaller atoms with the release of a large amount of energy. Compare *fusion*.

**flood plain**  The area bordering a river that is subject to flooding.

**fluidized-bed combustion**  A clean-coal technology in which crushed coal is mixed with particles of limestone in a strong air current during combustion. The limestone neutralizes the acidic sulfur compounds produced during combustion.

**fly ash**  The ash from the flue (chimney) that is trapped by electrostatic precipitators. Compare *bottom ash*.

**flyway**  An established route that ducks, geese, and shorebirds follow during their annual migrations.

**food additive**  A chemical added to food because it enhances the taste, color, or texture of the food, improves its nutrient value, reduces spoilage, prolongs shelf life, or helps maintain its consistency.

**food chain**  The successive series of organisms through

which energy flows in an ecosystem. Each organism in the series eats or decomposes the preceding organism in the chain.

**food web** A complex interconnection of all the food chains in an ecosystem.

**forest decline** A gradual deterioration (and often death) of many trees in a forest. The cause of forest decline is unclear at the present time, and it may involve a combination of factors.

**fossil fuel** Combustible deposits in the Earth's crust. Fossil fuels are composed of the remnants of prehistoric organisms that existed millions of years ago. Examples include oil, natural gas, and coal.

**fundamental niche** The potential ecological niche that an organism could have if there were no competition from other species. See *niche*. Compare *realized niche*.

**fungicide** A toxic chemical that kills fungi.

**fusion** A nuclear reaction in which two smaller atoms are combined to make one larger atom with the release of a large amount of energy. Compare *fission*.

**gas hydrates** Reserves of ice-encrusted natural gas that are located in porous rock deep in the Earth's crust.

**genetic diversity** Biological diversity that encompasses the genetic variety among individuals within a single species. Compare *ecological diversity* and *species diversity*.

**genetic engineering** The ability to take a specific gene from one cell and place it into another cell where it is expressed. Also called *biotechnology*.

**genetic resistance** An inherited characteristic that decreases the effect of a pesticide on a pest. Over time, the repeated exposure of a pest population to a pesticide causes an increase in the number of individuals that can tolerate the pesticide.

**geothermal energy** Heat produced deep in the Earth from the natural decay of radioactive elements.

**global commons** Those resources of our environment that are available to everyone but for which no single individual has responsibility.

**green revolution** The period of time during the 20th century when plant scientists developed genetically uniform, high-yielding varieties of important food crops such as rice and wheat.

**greenhouse effect** The global warming of our atmosphere produced by the buildup of carbon dioxide and other greenhouse gases, which trap the sun's radiation in much the same way that glass does in a greenhouse. Greenhouse gases allow the sun's energy to penetrate to the Earth's surface but do not allow as much of it to escape as heat.

**gross primary productivity** The rate at which energy accumulates in an ecosystem (as biomass) during photosynthesis. Compare *net primary productivity*.

**groundwater** The supply of fresh water under the Earth's surface. Groundwater is stored in porous layers of underground rock called aquifers. Compare *surface water*.

**growth rate** The rate of change of a population's size; calculated by subtracting the death rate from the birth rate (in populations with little or no migration).

**gyre** A circular, prevailing wind that generates circular ocean currents.

**habitat** The natural environment or place where an organism, population, or species lives.

**hair cell** One of a group of cells inside the cochlea of the ear that detects differences in pressure caused by sound waves.

**hard coal** See *anthracite*.

**hazardous waste** Any discarded chemical (solid, liquid, or gas) that threatens human health or the environment. Also called *toxic waste*.

**herbicide** A toxic chemical that kills plants.

**herbivore** An animal that feeds on plants or algae. Compare *carnivore* and *omnivore*. Also see *primary consumer*.

**heterocysts** Oxygen-excluding cells of cyanobacteria that fix nitrogen.

**heterotroph** See *consumer*.

**high-grade ore** An ore that contains relatively large amounts or a particular mineral. Compare *low-grade ore*.

**high-input agriculture** Agriculture that relies on large inputs of energy in the form of fossil fuels. Also called *industrialized agriculture*. Compare *alternative agriculture*.

**high-level radioactive waste** Any radioactive solid, liquid, or gas that initially gives off large amounts of ionizing radiation. Compare *low-level radioactive waste*.

**highly developed country** See *developed country*.

**host** The organism in a parasitic relationship that nourishes a parasite. See *parasitism* and *parasite*.

**humus** Black or dark brown decomposed organic material.

**hydrocarbons** A diverse group of organic compounds that contain only hydrogen and carbon.

**hydrogen bond** A bond between water molecules, formed when the negative (oxygen) end of one water molecule is attracted to the positive (hydrogen) end of another water molecule. Hydrogen bonding is the basis for a number of water's physical properties.

**hydrologic cycle** The water cycle, which includes evaporation, precipitation, and flow to the seas. The hydrologic cycle supplies terrestrial organisms with a continual supply of fresh water.

**hydropower** The energy of flowing or falling water used to generate electricity.

**hydrosphere** The Earth's supply of water (both liquid and frozen, fresh and salty).

**hypertension** High blood pressure.

**hypothesis** An educated guess that might be true and is testable by observation and experimentation. Compare *theory*.

**immigration**  A type of migration in which individuals enter a population and thus increase the population size. Compare *emigration*.

**indicator species**  A species that indicates particular conditions. For example, a large population of *Euglena* is indicative of polluted waters.

**inductive reasoning**  Reasoning that uses specific examples to draw a general conclusion or discover a general principle. Compare *deductive reasoning*.

**industrial smog**  The traditional, London-type smoke pollution, which consists principally of sulfur oxides and particulate matter. Compare *photochemical smog*.

**industrialized agriculture**  See *high-input agriculture*.

**infant mortality rate**  The number of infant deaths per 1000 live births. (An infant is a child in its first year of life.)

**infectious disease**  A disease caused by a microorganism (such as a bacterium or fungus) or infectious agent (such as a virus). Infectious diseases can be transmitted from one individual to another.

**infrared radiation**  Electromagnetic radiation with wavelength longer than visible light but shorter than microwaves. Humans perceive infrared radiation as invisible waves of heat energy.

**inorganic chemical**  A chemical that does not contain carbon and is not associated with life. Inorganic chemicals that are pollutants include mercury compounds, road salt, and acid drainage from mines.

**inorganic fertilizer**  Plant nutrients (especially nitrates, phosphates, and potassium) that are manufactured commercially.

**inorganic plant nutrient**  A nutrient such as phosphate or nitrate that stimulates plant or algal growth. Excessive amounts of inorganic plant nutrients, which may come from animal wastes and plant residues as well as fertilizer runoff, can cause both soil and water pollution.

**insecticide**  A toxic chemical that kills insects.

**in situ conservation**  Conservation efforts that concentrate on preserving biological diversity in the wild. Compare *ex situ conservation*.

**integrated pest management (IPM)**  A combination of pest control methods (biological, chemical, and cultivation) that, if used in the proper order and at the proper times, keep the size of a pest population low enough that it does not cause substantial economic loss.

**integrated waste management**  A combination of the best waste management techniques into a consolidated program to deal effectively with solid waste.

**intertidal zone**  The shoreline area between the low tide mark and the high tide mark.

**ionizing radiation**  Radiation that contains enough energy to eject electrons from atoms, forming positively charged ions. Ionizing radiation can damage living tissue.

**isotope**  An alternate form of the same element that has a different atomic mass. That is, an isotope has a different number of neutrons but the same number of protons and electrons.

**kinetic energy**  The energy of a body that results from its motion. Compare *potential energy*.

**kwashiorkor**  Malnutrition, most common in infants and in young children, that results from protein deficiency. Compare *marasmus*.

**laterization**  A soil process that produces a rock-hard soil. Laterization occurs in certain tropical areas that have very infertile soils with little organic matter.

**latitude**  The distance, measured in degrees north or south, from the equator. Compare *altitude*.

**law of the minimum**  The concept that the growth of each organism is limited by whatever essential factor is in shortest supply (or is present in harmful excess).

**leaching**  The process by which dissolved materials (nutrients or contaminants) are washed away or carried with water down through the various layers of the soil.

**less developed country**  A developing country with a low level of industrialization, a very high fertility rate, a very high infant mortality rate, and a very low per capita income (relative to highly developed countries). Compare *moderately developed country* and *highly developed country*.

**lignite**  A grade of coal brown or brown-black in color that has a soft, woody texture (softer than bituminous coal). Compare *anthracite* and *bituminous coal*.

**lime scrubber**  An air pollution control device in which a chemical spray neutralizes acidic gases.

**limiting factor**  Whatever environmental variable tends to restrict the growth, distribution, or abundance of a particular population.

**limnetic zone**  The open-water area away from the shore of a lake or pond that extends down as far as sunlight penetrates.

**lipid**  A diverse group of organic molecules that are metabolized by cell respiration to provide the body with a high level of energy. Lipids are commonly called *fats* and *oils*.

**liquefied petroleum gas**  A mixture of liquefied propane and butane. Liquefied petroleum gas is stored in pressurized tanks.

**lithosphere**  The soil and rock of Earth's crust.

**littoral zone**  The shallow-water area along the shore of a lake or pond.

**loam**  A soil that has approximately equal portions of sand, silt, and clay. A loamy soil that also contains organic material makes an excellent agricultural soil.

**low-grade ore**  An ore that contains relatively small amounts of a particular mineral. Compare *high-grade ore*.

**low-input agriculture**   See *alternative agriculture*.

**low-level radioactive waste**   Any radioactive solid, liquid, or gas that gives off small amounts of ionizing radiation. Compare *high-level radioactive waste*.

**malnutrition**   A condition caused when a person does not receive enough specific essential nutrients in the diet. See *marasmus* and *kwashiorkor*. Compare *overnutrition* and *undernutrition*.

**manganese nodule**   A small (potato-sized) rock that contains manganese and other minerals. Manganese nodules are common in the ocean floor.

**marasmus**   A condition of progressive emaciation that is especially common in children and is caused by a diet low in both total calories and protein. Compare *kwashiorkor*.

**marginal cost of pollution**   The cost in environmental quality of a unit of pollution emitted into the environment.

**marginal cost of pollution abatement**   The cost to dispose of a unit of pollution in a nonpolluting way.

**mariculture**   The rearing of marine organisms (fish and shellfish) for human consumption. See *aquaculture*.

**mass extinction**   The extinction of numerous species during a relatively short period of geological time. Compare *background extinction*.

**maximum containment level**   The maximum permissible amount (by law) of a water pollutant that might adversely affect human health.

**meltdown**   The melting of a nuclear reactor's core (the metal encasing the uranium fuel). A meltdown would cause the release of a substantial amount of radiation into the environment.

**mesosphere**   The layer of the atmosphere between the stratosphere and the thermosphere. It is characterized by the lowest atmospheric temperatures.

**metal**   An element that is malleable, lustrous, and a good conductor of heat and electricity. See *mineral*. Compare *nonmetal*.

**methane**   The simplest hydrocarbon, $CH_4$, which is an odorless, colorless, flammable gas.

**methanol**   A colorless, flammable liquid, $CH_3OH$. Also called *methyl alcohol*.

**methyl alcohol**   See *methanol*.

**microclimate**   Local variations in climate produced by differences in elevation, in the steepness and direction of slopes, and in exposure to prevailing winds.

**microirrigation**   A type of irrigation that conserves water. In microirrigation pipes with tiny holes bored into them convey water directly to individual plants. Also called *drip* or *trickle irrigation*.

**mineral**   An element, inorganic compound, or mixture that occurs naturally in the Earth's crust. See *metal* and *nonmetal*.

**mineral reserve**   A mineral deposit that has been identified and is currently profitable to extract. Compare *mineral resource*.

**mineral resource**   Any undiscovered mineral deposits or a known deposit of low-grade ore that is currently unprofitable to extract; mineral resources are potential resources that may be profitable to extract in the future. Compare *mineral reserve*.

**moderately developed country**   A developing country with a medium level of industrialization, a high fertility rate, a high infant mortality rate, and a low per capita income (relative to highly developed countries). Compare *less developed country* and *highly developed country*.

**monoculture**   The cultivation of only one type of plant over a large area.

**mortality**   See *death rate*.

**mulch**   Material placed on the surface of soil around the bases of plants. A mulch helps to maintain soil moisture and reduce soil erosion. Organic mulches have the additional advantage of decomposing over time, thereby enriching the soil.

**municipal solid waste**   Solid waste generated primarily in homes and businesses; commonly called trash. Compare *nonmunicipal solid waste*.

**municipal solid waste composting**   The large-scale composting of the entire organic portion of a community's trash.

**mutation**   A change in the DNA (a gene) of an organism. A mutation in reproductive cells may be passed on to the next generation, where it may result in birth defects or genetic disease.

**mutualism**   A symbiotic relationship in which both partners benefit from the association. See *symbiosis*. Compare *parasitism* and *commensalism*.

**mycorrhizae**   A symbiotic association between a fungus and the roots of a plant. Most plants form mycorrhizal associations with fungi. This mutualistic relationship enables plants to absorb adequate amounts of essential minerals from the soil.

**narrow-spectrum pesticide**   An "ideal" pesticide that only kills the organism for which it is intended and does not harm any other species. Compare *wide-spectrum pesticide*.

**natality**   See *birth rate*.

**national emission limitations**   The maximum permissible amount (by law) of a particular pollutant that can be discharged into the nation's rivers, lakes, and oceans from point sources.

**natural resources**   Goods and services—coal, fresh water, clean air, arable land, and wildlife, for example—that are supplied by the environment.

**nekton**   Relatively strong-swimming aquatic organisms such as fish and turtles. Compare *plankton*.

**neo-Malthusians**   Economists who hold that developmental efforts are hampered by a rapidly expanding population.

**neritic province**   Open ocean from the shoreline to a depth of 200 meters.

**net primary productivity**   Energy that remains in an

ecosystem (as biomass) after cell respiration has occurred. Compare *gross primary productivity*.

**niche**  The totality of an organism's adaptations, its use of resources, and the life style to which it is fitted. The niche describes how an organism utilizes materials in its environment as well as how it interacts with other organisms. Also called *ecological niche*.

**nitrification**  The conversion of ammonia ($NH_3$) to nitrate ($NO_3^-$) by certain bacteria (nitrifying bacteria) in the soil; part of the nitrogen cycle.

**nitrogen cycle**  The worldwide circulation of nitrogen from the abiotic environment into living things and back into the abiotic environment.

**nitrogen fixation**  The conversion of atmospheric nitrogen ($N_2$) to ammonia ($NH_3$) by nitrogen-fixing bacteria and cyanobacteria; part of the nitrogen cycle.

**noise**  A loud or disagreeable sound.

**nomadic herding**  A type of subsistence agriculture in which livestock is supported on land too arid for successful crop growth. Nomadic herding is land intensive because the herders must continually move the animals in order to provide them with enough forage.

**nondegradable**  A chemical pollutant (such as the toxic elements mercury and lead) that cannot be decomposed (broken down) by living organisms or other natural processes. Compare *biodegradable*.

**nonmetal**  A mineral that is nonmalleable, nonlustrous, and a poor conductor of heat and electricity. See *mineral*. Compare *metal*.

**nonmunicipal solid waste**  Waste generated by industry, agriculture, and mining. Compare *municipal solid waste*.

**non-point-source pollution**  Pollutants that enter bodies of water over large areas rather than being concentrated at a single point of entry. Examples include agricultural fertilizer runoff and sediments from construction. Compare *point-source pollution*.

**nonrenewable resources**  Natural resources that are present in limited supplies and are depleted by use; include minerals such as copper and tin and fossil fuels such as oil and natural gas. Compare *renewable resources*.

**no-tillage**  A method of conservation tillage that leaves both the surface and subsurface soil undisturbed. Special machines punch holes in the soil for seeds. See *conservation tillage*. Compare *conventional tillage* and *reduced-tillage*.

**nuclear autumn**  A moderate cooling of the global climate caused by dust and smoke hurled into the stratosphere after a nuclear war. The envisioned effects of a nuclear autumn are less severe than those of a nuclear winter. Compare *nuclear winter*.

**nuclear energy**  The energy released from the nucleus of an atom in a nuclear reaction (fission or fusion) or during radioactive decay.

**nuclear winter**  A catastrophic cooling of the global climate caused by dust and smoke hurled into the

stratosphere after a nuclear war. Compare *nuclear autumn*.

**obligate parasite**  An organism that can only exist as a parasite. Compare *facultative parasite*.

**oceanic province**  That part of the open ocean that is deeper than 200 meters and comprises most of the ocean.

**ocean temperature gradient**  The differences in temperature at various ocean depths.

**O-horizon**  The uppermost layer of certain soils, composed of dead leaves and other organic matter.

**oil shales**  Sedimentary "oily rocks" that must be crushed, heated to high temperatures, and refined after they are mined in order to yield oil.

**oligotrophic lake**  A deep, clear lake that has minimal nutrients. Water in an oligotrophic lake contains a high level of dissolved oxygen. Compare *eutrophic lake*.

**omnivore**  An animal that eats a variety of plant and animal material. Compare *herbivore* and *carnivore*.

**open-pit mining**  A type of surface mining in which mineral resources are dug out of the ground, leaving a large hole. Compare *strip mining*. See *surface mining*.

**optimum amount of pollution**  The amount of pollution that is economically most desirable. It is determined by plotting two curves, the marginal cost of pollution and the marginal cost of pollution abatement; the point where the two curves meet is the optimum amount of pollution from an economic standpoint.

**ore**  Rock that contains a large enough concentration of a particular mineral that it can be profitably mined and extracted.

**organic agriculture**  Growing crops and livestock without the use of synthetic pesticides or inorganic fertilizers. Organic agriculture makes use of natural organic fertilizers (such as manure and compost) and chemical-free methods of pest control.

**organic compound**  A compound that contains the element carbon and is either naturally occurring (in living things) or synthetic (manufactured by humans). Many synthetic organic compounds persist in the environment for an extended period of time, and some are toxic to living organisms.

**organophosphate**  A synthetic organic compound that contains phosphorus and is very toxic; used as an insecticide.

**overburden**  Overlying layers of soil and rock over mineral deposits. The overburden is removed during surface mining.

**overgrazing**  The destruction of an area's vegetation that occurs when too many animals graze on the vegetation, consuming so much of it that it does not recover.

**overnutrition**  A condition caused by eating food in excess of that required to maintain a healthy body. Compare *malnutrition* and *undernutrition*.

**overpopulation** A situation in which a country or geographical area has more people than its resource base can support without damaging the environment. See *people overpopulation* and *consumption overpopulation*.

**oxide** A compound in which oxygen is chemically combined with some other element.

**ozone** A blue gas, $O_3$, that has a distinctive odor. Ozone is a human-made pollutant in one part of the atmosphere (the troposphere) but a natural and essential component in another (the stratosphere).

**parasite** Any organism that obtains nourishment from the living tissue of another organism (the host). See *parasitism* and *host*.

**parasitism** A symbiotic relationship in which one member (the parasite) benefits and the other (the host) is adversely affected. See *symbiosis*. Compare *mutualism* and *commensalism*.

**particulate matter** Solid particles and liquid droplets suspended in the atmosphere.

**parts per billion** The number of parts of a particular substance found in one billion parts of air, water, or some other material. Abbreviated ppb.

**parts per million** The number of parts of a particular substance found in one million parts of air, water, or some other material. Abbreviated ppm.

**passive solar heating** A heating method that uses the sun's energy to heat buildings or water without requiring mechanical devices (pumps or fans) to distribute the collected heat. Compare *active solar heating*.

**pathogen** An agent (usually a microorganism) that causes disease.

**people overpopulation** The situation in which there are too many people in a given geographical area. Even if those people use few resources per person (the minimum amount they need to survive), people overpopulation results in pollution, environmental degradation, and resource depletion. See *overpopulation*. Compare *consumption overpopulation*.

**perennial** A plant that lives for more than two years. Compare *annual* and *biennial*.

**permafrost** Permanently frozen subsoil characteristic of frigid areas such as the tundra.

**persistence** A characteristic of certain chemicals that are extremely stable and may take many years to be broken down into simpler forms by natural processes. Certain pesticides, for example, exhibit persistence and remain unaltered in the environment for years.

**pest** Any organism that interferes in some way with human welfare or activities.

**pesticide** Any toxic chemical used to kill pests. See *fungicide, herbicide, insecticide,* and *rodenticide*.

**petrochemicals** Chemicals obtained from crude oil that are used in the production of such diverse products as fertilizers, plastics, paints, pesticides, medicines, and synthetic fibers.

**petroleum** A thick, yellow to black, flammable liquid hydrocarbon mixture found in the Earth's crust. When petroleum is refined, the mixture is separated into a number of hydrocarbon compounds, including gasoline, kerosene, fuel oil, lubricating oils, paraffin, and asphalt. Also called *crude oil*.

**pH** A number from 0 to 14 that indicates the degree of acidity or alkalinity of a substance.

**pheromone** A substance secreted by one organism into the environment that influences the development or behavior of other members of the same species.

**photochemical smog** A brownish-orange haze formed by complex chemical reactions involving sunlight, nitrogen oxides, and hydrocarbons. Some of the pollutants in photochemical smog include peroxyacyl nitrates (PANs), ozone, and aldehydes. Compare *industrial smog*.

**photodegradable** Breaking down upon exposure to sunlight.

**photosynthesis** The biological process that captures light energy and transforms it into the chemical energy of organic molecules (such as glucose), which are manufactured from carbon dioxide and water. Photosynthesis is performed by plants, algae, and several kinds of bacteria.

**photovoltaic (PV) solar cell** A wafer or thin-film device that generates electricity when solar energy is absorbed.

**phytoplankton** Microscopic floating algae that are the base of most aquatic food chains. See *plankton*. Compare *zooplankton*.

**pioneer community** The first organisms (such as lichens or mosses) to colonize an area and begin the first stage of ecological succession. See *succession*.

**plankton** Small or microscopic aquatic organisms that are relatively feeble swimmers and thus, for the most part, are carried about by currents and waves. Composed of phytoplankton and zooplankton. Compare *nekton*.

**plate tectonics** The theory that explains how the Earth's crustal plates move and interact at their boundaries.

**point-source pollution** Water pollution that can be traced to a specific spot (such as a factory or sewage treatment plant) because it is discharged into the environment through pipes, sewers, or ditches. Compare *non-point-source pollution*.

**polar easterly** A prevailing wind that blows from the northeast near the North Pole or from the southeast near the South Pole.

**pollution** An unwanted change in the atmosphere, water, or soil that can harm humans or other living organisms.

**polychlorinated biphenyls (PCBs)** A group of toxic, oily, synthetic industrial chemicals. PCBs are chlorinated hydrocarbons (composed of carbon, hydrogen, and chlorine) that persist in the environment and exhibit biological magnification in food chains.

**polymer** A compound composed of repeating subunits.

**population**  A group of organisms of the same species that live in the same geographical area at the same time.

**potential energy**  Stored energy that is the result of the relative position of matter instead of its motion. Compare *kinetic energy*.

**preservative**  A chemical added to food to retard the growth of bacteria and fungi that cause food spoilage.

**prevailing wind**  A major surface wind that blows more or less continually.

**primary air pollutant**  A harmful chemical that enters directly into the atmosphere either from human activities or natural processes (such as volcanic eruptions). Compare *secondary air pollutant*.

**primary consumer**  A consumer that eats producers. Also called *herbivore*. Compare *secondary consumer*.

**primary succession**  An ecological succession that occurs on land that has not previously been inhabited by plants; no soil is present initially. Compare *secondary succession*.

**primary treatment**  Treating wastewater by removing suspended and floating particles (such as sand and silt) by mechanical processes (such as screens and physical settling). Compare *secondary treatment* and *tertiary treatment*.

**producer**  An organism (such as a chlorophyll-containing plant) that manufactures complex organic molecules from simple inorganic substances. In most ecosystems, producers are photosynthetic organisms. Also called *autotroph*. Compare *consumer*.

**profundal zone**  The deepest zone of a large lake.

**pronatalists**  Those who are in favor of population growth.

**protein**  A large, complex organic molecule composed of amino acid subunits; proteins are the principal structural components of cells.

**pyramid of biomass**  An ecological pyramid that illustrates the total biomass (for example, the total dry weight of all living organisms in a community) at each successive trophic level. See *ecological pyramid*. Compare *pyramid of energy* and *pyramid of numbers*.

**pyramid of energy**  An ecological pyramid that shows the energy flow through each trophic level of an ecosystem. See *ecological pyramid*. Compare *pyramid of biomass* and *pyramid of numbers*.

**pyramid of numbers**  An ecological pyramid that shows the number of organisms at each successive trophic level in a given ecosystem. See *ecological pyramid*. Compare *pyramid of biomass* and *pyramid of energy*.

**quarantine**  Practice in which the importation of exotic plant and animal material that might be harboring pests is restricted.

**radiation**  The emission of fast-moving particles or rays of energy from the nuclei of radioactive atoms.

**radioactive**  Atoms of unstable isotopes that spontaneously emit radiation.

**radiation decay**  The process in which a radioactive element emits radiation and, as a result, its nucleus changes into the nucleus of a different element.

**radioactive half-life**  The period of time required for one-half of a radioactive substance to change into a different material.

**radioactive isotopes**  Unstable isotopes that spontaneously emit radiation.

**radon**  A colorless, tasteless, radioactive gas produced during the radioactive decay of uranium in the Earth's crust.

**rain shadow**  An area on the downwind side of a mountain range with very little precipitation. Deserts often occur in rain shadows.

**range**  The area of the Earth in which a particular species occurs.

**realized niche**  The life style that an organism actually pursues, including the resources that it actually utilizes. An organism's realized niche is narrower than its fundamental niche because of competition from other species. See *niche*. Compare *fundamental niche*.

**recycling**  Conservation of the resources in used items by converting them into new products. For example, used aluminum cans are recycled by collecting, remelting, and reprocessing them into new cans. Compare *reuse*.

**reduced-tillage**  A method of conservation tillage in which the subsurface soil is tilled without disturbing the topsoil. See *conservation tillage*. Compare *conventional tillage* and *no-tillage*.

**renewable resources**  Resources that are replaced by natural processes and can be used forever, provided they are not overexploited in the short term. Examples include fresh water in lakes and rivers, fertile soil, and trees in forests. Compare *nonrenewable resources*.

**replacement-level fertility**  The number of children a couple must produce in order to "replace" themselves. The average number is greater than 2 because some children die before reaching reproductive age.

**reuse**  Conservation of the resources in used items by using them over and over again. For example, glass bottles can be collected, washed, and refilled again. Compare *recycling*.

**reverse osmosis**  A desalinization process that involves forcing saltwater through a membrane permeable to water but not to salt. Compare *distillation*.

**risk assessment**  The process of estimating the harmful effects on human health or the environment of exposure to a particular danger. Risk estimates are most useful when they are compared with one another.

**rodenticide**  A toxic chemical that kills rodents.

**runoff**  The movement of fresh water from precipitation and snowmelt to rivers, lakes, wetlands, and ultimately, the ocean.

**salinity**  The concentration of dissolved salts (such as sodium chloride) in a body of water.

**salinity gradient**  The difference in salt concentrations that occurs at different depths in the ocean and at different locations in estuaries.

**salt marsh** A wetland dominated by grasses in which the salinity fluctuates between that of seawater and fresh water. Salt marshes are usually located in estuaries.

**saltwater intrusion** The movement of seawater into a freshwater aquifer located near the coast; caused by groundwater depletion.

**sand** Inorganic soil particles that are larger than clay or silt. Compare *clay* and *silt*.

**sanitary landfill** The disposal of solid waste by burying it under a shallow layer of soil.

**saprophyte** See *decomposer*.

**savanna** A tropical grassland with widely scattered trees or clumps of trees; found in areas of low rainfall or seasonal rainfall with prolonged dry periods.

**scientific method** The way a scientist approaches a problem (by formulating a hypothesis and then testing it by means of an experiment).

**sclerophyllous leaf** A hard, small, leathery leaf that resists water loss; characteristic of perennial plants adapted to extremely dry habitats.

**scrubbers** Desulfurization systems that are used in smokestacks to decrease the amount of sulfur released in the air by 90 percent or more.

**second law of thermodynamics** When energy is converted from one form to another, some of it is degraded into a lower-quality, less useful form. Thus, with each successive energy transformation, less energy is available to do work. Compare *first law of thermodynamics*.

**secondary air pollutant** A harmful chemical that forms in the atmosphere when a primary air pollutant reacts chemically with other air pollutants or natural components of the atmosphere. Compare *primary air pollutant*.

**secondary consumer** An organism that consumes primary consumers. Also called *carnivore*. Compare *primary consumer*.

**secondary succession** An ecological succession that takes place after some disturbance destroys the existing vegetation; soil is already present. Compare *primary succession*.

**secondary treatment** Treating wastewater biologically, by using microorganisms to decompose the suspended organic material; occurs after primary treatment. Compare *primary treatment* and *tertiary treatment*.

**sediment pollution** Soil particles that enter the water as a result of erosion.

**sedimentation** (1) Letting solids settle out of wastewater by gravity during primary treatment. (2) The process in which eroded particles are transported by water and deposited as sediment on river deltas and the sea floor. If exposed to sufficient heat and pressure, sediments can solidify into sedimentary rock.

**seed-tree cutting** A forest management technique in which almost all trees are harvested from an area in a single cutting, but a few desirable trees are left behind to provide seeds for the regeneration of the forest. Compare *clearcutting, selective cutting, shelterwood cutting,* and *whole-tree harvesting*.

**selective cutting** A forest management technique in which mature trees are cut individually or in small clusters while the rest of the forest remains intact so that the forest can regenerate quickly (and naturally). Compare *clearcutting, seed-tree cutting, shelterwood cutting,* and *whole-tree harvesting*.

**semi-arid land** Land that receives more precipitation than a desert but is subject to frequent and prolonged droughts.

**sewage** Wastewater carried off by drains or sewers.

**sewage sludge** A slimy mixture of bacteria-laden solids that settles out from sewage wastewater during primary and secondary treatments.

**shelter belt** A row of trees planted as a windbreak to reduce soil erosion of agricultural land.

**shelterwood cutting** A forest management technique in which all mature trees in an area are harvested in a series of partial cuttings over a period of time. (Typically two or three harvests are made over a decade.) Compare *clearcutting, seed-tree cutting, selective cutting,* and *whole-tree harvesting*.

**shifting agriculture** Agriculture that involves clearing a small patch of tropical land to plant crops. Typically, the soil is depleted of nutrients within a few years and the plot must be abandoned. See *slash-and-burn agriculture*.

**sick building syndrome** Eye irritations, nausea, headaches, respiratory infections, depression, and fatigue caused by the presence of air pollution inside office buildings.

**silt** Medium-sized inorganic soil particles. Compare *clay* and *sand*.

**slag** See *bottom ash*.

**slash-and-burn agriculture** A type of shifting agriculture in which the forest is cut down, allowed to dry, and burned; the crops that are planted immediately afterwards thrive because the ashes provide nutrients. In a few years, however, the soil is depleted and the land must be abandoned. See *shifting agriculture*.

**smelting** Process in which ore is melted at high temperatures to help separate impurities from the molten metal.

**smog** Air pollution caused by a variety of pollutants. See *industrial smog* and *photochemical smog*.

**soft coal** See *bituminous coal*.

**soil** The uppermost layer of the Earth's crust, which supports terrestrial plants, animals, and microorganisms. Soil is a complex mixture of inorganic minerals (from the parent rock), organic material, water, air, and living organisms.

**soil erosion** The wearing away or removal of soil from the land; caused by wind and flowing water. Although soil erosion occurs naturally from precipitation and runoff, human activities (such as clearing land) accelerate it.

**soil horizons**    The horizontal layers into which many soils are organized.

**soil profile**    A section through the soil from the surface to the parent rock that reveals the horizons.

**soil salinization**    A process in which salt accumulates in the soil.

**solar energy**    Energy from the sun. Solar energy includes both direct solar radiation and indirect solar energy (such as wind, hydropower, and biomass).

**solar pond**    A technique to harvest the sun's energy by using a pond of water to collect solar energy.

**solar power tower**    A solar thermal technique in which a liquid is heated (using solar energy) to produce steam used to generate electricity. The solar energy is initially directed by a number of mirrors onto a central receiving area at the top of a tall tower. Also called *central-receiver energy generation.*

**solar thermal electric generation**    A means of producing electricity in which the sun's energy is directed by mirrors onto a fluid-filled pipe; the heated fluid is used to generate electricity.

**sound**    Vibrations in the air (or some other medium) that reach the ears and stimulate a sensation of hearing.

**source reduction**    An aspect of waste management in which products are designed and manufactured in ways that decrease not only the volume of solid waste but the amount of hazardous materials in the solid waste that remains.

**species**    A group of similar organisms that are able to interbreed with one another but unable to interbreed with other sorts of organisms.

**species diversity**    Biological diversity that encompasses the number of different species in an area. Compare *genetic diversity* and *ecological diversity.*

**spoil bank**    A hill of loose rock created when the overburden from a new trench is put into the old (already excavated) trench during strip mining.

**sport hunting**    Killing animals for recreation. Compare *commercial hunting* and *subsistence hunting.*

**spring turnover**    A mixing of the lake waters in temperate lakes that occurs in spring as ice melts and the surface water reaches 4°C, its temperature of greatest density. Compare *fall turnover.*

**stable runoff**    The share of runoff from precipitation available throughout the year. Most geographical areas have a heavy runoff during a few months (the spring months, for example) when precipitation and snowmelt are highest. Stable runoff is the amount that can be depended on every month.

**standing crop**    The current plant biomass.

**steppe**    A semi-arid temperate grassland that receives less precipitation than moister grasslands but more precipitation than deserts.

**sterile male technique**    A method of insect control that involves rearing, sterilizing, and releasing large numbers of males of the pest species.

**strata**    Layers of rock.

**stratosphere**    The layer of the atmosphere between the troposphere and the mesosphere. It contains a thin ozone layer that protects life by filtering out much of the sun's ultraviolet radiation.

**strip cropping**    A type of contour plowing that produces alternating strips of different crops that are planted along the natural contours of the land.

**strip cutting**    An agricultural harvesting technique that involves harvesting only one segment of the crop at a time.

**strip mining**    A type of surface mining in which a large trench is dug to extract the minerals. Compare *open-pit mining.* See *surface mining.*

**structural traps**    Underground geological structures that tend to trap oil or natural gas if it is present.

**subsidence**    The sinking or settling of land caused by aquifer depletion (as groundwater supplies are removed).

**subsistence agriculture**    The production of enough food to feed oneself and one's family with little left over to sell or reserve for bad times.

**subsistence hunting**    Killing wild animals for food and furs and other products needed for survival. Compare *commercial hunting* and *sport hunting.*

**subsurface mining**    The extraction of mineral and energy resources from deep underground deposits. Compare *surface mining.*

**succession**    The sequence of changes in a plant community over time. Includes the changes that occur from the initial colonization of the area to the climax community. See *pioneer community* and *climax community.*

**sulfide**    A compound in which an element is combined chemically with sulfur.

**surface mining**    The extraction of mineral and energy resources near the Earth's surface by first removing the soil, subsoil, and overlying rock strata. See *strip mining* and *open-pit mining.* Compare *subsurface mining.*

**surface water**    Fresh water found on the Earth's surface in streams and rivers, lakes, ponds, reservoirs, and wetlands. Compare *groundwater.*

**sustainable agriculture**    See *alternative agriculture.*

**sustainable manufacturing**    A manufacturing system based on minimizing waste by industry. Sustainable manufacturing involves such practices as recycling, reuse, and source reduction.

**symbionts**    The partners of a symbiotic relationship.

**symbiosis**    An intimate relationship between two or more organisms of different species. See *commensalism, mutualism,* and *parasitism.*

**synfuel**    A liquid or gaseous fuel synthesized from coal and other sources and used in place of oil or natural gas. Also called *synthetic fuel.*

**synthetic fuel**    See *synfuel.*

**taiga**    A region of coniferous forests (such as pine, spruce, and fir) in the northern hemisphere. The taiga is located just south of the tundra and stretches across both North America and Eurasia. Also called *boreal forest.*

**tailings**  Piles of loose rock produced when a mineral such as uranium is mined and processed (extracted and purified from the ore).

**tar sand**  An underground sand deposit so heavily permeated with tar or heavy oil that it doesn't move. The oil can be separated from the sand by heating.

**temperate deciduous forest**  A forest biome that occurs in temperate areas where annual precipitation ranges from about 75 cm to 125 cm.

**temperate grassland**  A grassland characterized by hot summers, cold winters, and less rainfall than is found in a temperate deciduous forest biome.

**temperate rain forest**  A coniferous biome characterized by cool weather, dense fog, and high precipitation. Found on the north Pacific coast of North America.

**terracing**  A soil conservation method that involves building dikes on hilly terrain in order to produce level terraced areas for agriculture.

**terrestrial**  Pertaining to the land. Compare *aquatic*.

**tertiary treatment**  Advanced wastewater treatment methods that occur after primary and secondary treatments and include a variety of biological, chemical, and physical processes. Compare *primary treatment* and *secondary treatment*.

**theory**  A widely accepted idea supported by a large body of observations and experiments. Compare *hypothesis*.

**thermal inversion**  A layer of cold air temporarily trapped near the ground by a warmer, upper layer. If a thermal inversion persists, air pollutants may build up to harmful or even dangerous levels.

**thermal pollution**  Water pollution that occurs when heated water produced during many industrial processes is released into waterways.

**thermal stratification**  The marked layering (separation into warm and cold layers) of temperate lakes during the summer. See *thermocline*.

**thermocline**  A marked and abrupt temperate transition in temperate lakes between warm surface water and cold deeper water. See *thermal stratification*.

**thermodynamics**  The branch of physics that deals with energy and its various forms and transformations.

**thermosphere**  The layer of the atmosphere between the mesosphere and the exosphere. Temperatures are very high due to the absorption of x-rays and short-wave ultraviolet radiation.

**threatened species**  A species in which the population is low enough for it to be at risk of becoming extinct, but not low enough that it is in imminent danger of extinction. Compare *endangered species*.

**topography**  A region's surface features (such as the presence or absence of mountains and valleys).

**total fertility rate**  The average number of children born to a woman during her lifetime.

**total resources**  The combination of a mineral's reserves and resources. See *mineral reserve* and *mineral resource*. Also called *world reserve base*.

**toxic waste**  See *hazardous waste*.

**toxicology**  The science of poisons, including their effects and any antidotes.

**trade wind**  A prevailing tropical wind that blows from the northeast (in the Northern Hemisphere) or from the southeast (in the Southern Hemisphere).

**transpiration**  The evaporation of water vapor from plants.

**trickle irrigation**  See *microirrigation*.

**tritium**  An isotope of hydrogen that contains one proton and two neutrons per atom. Compare *deuterium*.

**trophic level**  Each level in a food chain. All producers belong to the first trophic level, all herbivores belong to the second trophic level, and so on.

**tropical dry forest**  A tropical forest where enough precipitation falls to support trees, but not enough to support the lush vegetation of a tropical rain forest. Many tropical dry forests occur in areas with pronounced rainy and dry seasons.

**tropical rain forest**  A lush, species-rich forest biome that occurs in tropical areas where the climate is very moist throughout the year. Tropical rain forests are also characterized by old, infertile soils.

**troposphere**  The atmosphere from the Earth's surface to the stratosphere. It is characterized by the presence of clouds, turbulent winds, and decreasing temperature with increasing altitude.

**tundra**  The treeless biome in the far north that consists of boggy plains covered by lichens and small plants such as mosses. The tundra is characterized by harsh, very cold winters and extremely short summers.

**unconfined aquifer**  A groundwater storage area located above a layer of impermeable rock. Water in an unconfined aquifer is replaced by surface water that drains down from directly above it. Compare *confined aquifer*.

**undernutrition**  A condition caused when a person receives fewer calories in the diet than are needed; an undernourished person eats insufficient food to maintain a healthy body. Compare *overnutrition* and *malnutrition*.

**urban growth**  The rate at which a city's population grows.

**urban heat island**  Local heat buildup in an area of high population density. See *dust dome*.

**urbanization**  The increasing convergence of people from rural areas into cities.

**vaporization**  See *evaporation*.

**vector**  A living organism that transmits a pathogen from one organism to another.

**vitamin**  A complex organic molecule required in very small quantities for the normal metabolic functioning of living cells.

**waste-discharge permit**  A government policy that controls pollution by issuing permits allowing the holder to pollute a given amount. Holders are not allowed to produce more emission than the permit allows.

**water table**  The uppermost level of an unconfined

aquifer, below which the ground is saturated with water.

**watershed**   See *drainage basin.*

**weather**   The general condition of the atmosphere (temperature, moisture, cloudiness) at a particular time and place. Compare *climate.*

**weathering process**   A chemical or physical process that helps form soil from rock; during weathering, the rock is gradually broken down into smaller and smaller particles.

**westerly**   A prevailing wind that blows in the mid-latitudes from the southwest (in the Northern Hemisphere) or from the northwest (in the Southern Hemisphere).

**Western diseases**   A group of noninfectious diseases that are generally more commonplace in industrialized countries. Include obesity and heart disease.

**wet deposition**   A form of acid deposition in which acid falls to the Earth as precipitation. Compare *dry deposition.*

**wetland**   Land that is transitional between aquatic and terrestrial ecosystems and is covered with water for at least part of the year.

**whole-tree harvesting**   A forest management technique in which machines clearcut an area by harvesting each entire tree (including roots and branches) and cutting it into small chips. The chips are processed for paper products or fuel. Compare *clearcutting,*

*seed-tree cutting, selective cutting,* and *shelterwood cutting.*

**wide-spectrum pesticide**   A pesticide that kills a variety of organisms in addition to the pest against which it is used. Most pesticides are wide-spectrum pesticides. Compare *narrow-spectrum pesticide.*

**wilderness**   Any area that has not been greatly disturbed by human activities and that humans may visit but do not inhabit.

**wildlife management**   Efforts to handle wildlife populations and their habitats in order to assure their sustained welfare.

**wind**   Surface air currents that are caused by the solar warming of air.

**wind farm**   An array of wind turbines for utilizing wind energy by capturing it and converting it to electricity.

**world reserve base**   See *total resources.*

**zero population growth**   When the birth rate equals the death rate. A population with zero population growth remains the same size.

**zooplankton**   The nonphotosynthetic organisms that are part of the plankton. See *plankton.* Compare *phytoplankton.*

**zooxanthellae**   Algae that live inside coral animals and have a mutualistic relationship with them.

# Review of Basic Chemistry

## ELEMENTS

All matter, living and nonliving, is composed of chemical **elements,** substances that cannot be broken down into simpler substances by chemical reactions. There are 92 naturally occurring elements, ranging from hydrogen (the lightest) to uranium (the heaviest). In addition to the naturally occurring elements, about 17 elements heavier than uranium have been made in laboratories by bombarding elements with subatomic particles.

Instead of writing out the name of each element, chemists use a system of abbreviations called **chemical symbols**—usually the first one or two letters of the English or Latin name of the element. For example, O is the symbol for oxygen, C for carbon, Cl for chlorine, and Na for sodium (its Latin name is *natrium*).

### Atoms

The atom is the smallest subdivision of an element that retains the characteristic chemical properties of that element. Atoms are almost unimaginably small, much smaller than the tiniest particle visible under a light microscope.

An atom is composed of smaller components called subatomic particles—protons, neutrons, and electrons. **Protons** have a positive electrical charge; **neutrons** are uncharged particles with about the same mass as protons. Protons and neutrons make up almost all the mass of an atom and are concentrated in the atomic nucleus. **Electrons** have a negative electrical charge and an extremely small mass (only about 1/1800 of the mass of a proton). The electrons spin in the space surrounding the atomic nucleus.

Each kind of element has a fixed number of protons in the atomic nucleus. This number, called the **atomic number,** determines the chemical identity of the atom. The total number of protons plus neutrons in the atomic nucleus is termed the **atomic mass.** For example, the element oxygen has eight protons and eight neutrons in the nucleus; it therefore has an atomic number of 8 and an atomic mass of 16.

When an atom is uncombined, it contains the same number of electrons as protons. Some kinds of chemical combinations and certain other circumstances change the number of electrons, but chemical reactions do not affect anything in the atomic nucleus. Because electrons and protons have equal but opposite charges, an uncombined atom is electrically neutral.

The way electrons are arranged around an atomic nucleus is referred to as the atom's **electronic configuration.** Knowing the locations of electrons enables chemists to predict how atoms can combine to form different types of chemical compounds.

An atom may have several **energy levels,** or **electron shells,** where electrons are located. The lowest energy level is the one closest to the nucleus. Only two electrons can occupy this energy level. The second energy level can accommodate a maximum of eight electrons. Although the third and outer shells can each contain more than eight electrons, they are most stable when only eight are present. We may consider the first shell complete when it contains two electrons and every other shell complete when it contains eight electrons.

The energy levels correspond roughly to physical locations of electrons, called **orbitals.** There may be several orbitals within a given energy level. Electrons are thought to whirl around the nucleus in an unpredictable manner, now close to it, now farther away. Orbitals represent the places where electrons are most probably found.

### Isotopes

**Isotopes** are atoms of the same element that contain the same number of protons but different numbers of neutrons. Isotopes, therefore, have different atomic mass numbers. The three isotopes of hydrogen contain zero, one, and two neutrons, respectively. Elements usually occur in nature as mixtures of isotopes.

All isotopes of a given element have essentially the same chemical characteristics. Some isotopes with excess neutrons are unstable and tend to break down, or decay, to a more stable isotope (usually of a different element). Such isotopes are termed **radioisotopes,** since they emit high-energy radiation when they decay.

## Molecules

Two or more atoms may combine chemically to form a **molecule.** When two atoms of oxygen combine, for example, a molecule of oxygen is formed. Different kinds of atoms can combine to form **chemical compounds.** A chemical compound is a substance that consists of two or more different elements combined in a fixed ratio. Water is a chemical compound in which each molecule consists of two atoms of hydrogen combined with one atom of oxygen.

## Chemical Bonds

The chemical properties of an element are determined primarily by the number and arrangement of electrons in the *outermost* energy level (electron shell). In a few elements, called the **noble gases,** the outermost shell is filled. These elements are chemically inert, meaning that they will not readily combine with other elements. The electrons in the outermost energy level of an atom are referred to as **valence electrons.** The valence electrons are chiefly responsible for the chemical activity of an atom. When the outer shell of an atom contains fewer than eight electrons, the atom tends to lose, gain, or share electrons to achieve an outer shell of eight. (The exceptions are zero or two electrons in the case of the lightest elements, hydrogen and helium.)

The elements in a compound are always present in a certain proportion. This reflects the fact that atoms are attached to each other by chemical bonds in a precise way to form a compound. A **chemical bond** is the attractive force that holds two atoms together. Each bond represents a certain amount of potential chemical energy. The atoms of each element form a specific number of bonds with the atoms of other elements—a number dictated by the number of valence electrons.

## Ions

Some atoms have the ability to gain or lose electrons. Because the number of protons in the nucleus remains unchanged, the loss or gain of electrons produces an atom with a net positive or negative charge. Such electrically charged atoms are termed **ions.**

## CHEMICAL FORMULAS

A **chemical formula** is a shorthand method for describing the chemical composition of a molecule. Chemical symbols are used to indicate the types of atoms in the molecule, and subscript numbers are used to indicate the number of each atom present. The chemical formula for molecular oxygen, $O_2$, tells us that each molecule consists of two atoms of oxygen. This formula distinguishes it from another form of oxygen, ozone, which has three oxygen atoms and is written $O_3$. The chemical formula for water, $H_2O$, indicates that each molecule consists of two atoms of hydrogen and one atom of oxygen. Note that when a single atom of one type is present it is not necessary to write 1; it is not necessary to write $H_2O_1$.

## CHEMICAL EQUATIONS

The chemical reactions that occur between atoms and molecules—for example, between methane and oxygen—can be described on paper by means of **chemical equations.**

$$CH_4 + 2 O_2 \longrightarrow CO_2 + 2 H_2O$$
Methane    Oxygen         Carbon     Water
                          dioxide

Methane is broken down in this reaction.

In a chemical reaction, the **reactants** (the substances that participate in the reaction) are written on the left side of the equation and the **products** (the substances formed by the reaction) are written on the right side. The arrow means *yields* and indicates the direction in which the reaction tends to proceed. The number preceding a chemical symbol or formula indicates the number of atoms or molecules reacting. Thus, $2 O_2$ means two molecules of oxygen. The absence of a number indicates that only one atom or molecule is present. Thus, the equation can be translated into ordinary language as, "One molecule of methane reacts with two molecules of oxygen to yield one molecule of carbon dioxide and two molecules of water."

## ACIDS AND BASES

An **acid** is a compound that ionizes in solution to yield hydrogen ions ($H^+$)—that is, protons—and a negatively charged ion. Acids turn blue litmus paper red and have a sour taste. Hydrochloric acid (HCl) and sulfuric acid ($H_2SO_4$) are examples of acids.

$$HCl \xrightarrow{\text{in water}} H^+ + Cl^-$$

Hydrochloric       Hydrogen   Chloride
acid                ion        ion

The strength of an acid depends on the degree to which it ionizes in water, releasing hydrogen ions. Thus, HCl is a very strong acid because most of its molecules dissociate, producing hydrogen and chloride ions.

Most bases are substances that yield a hydroxide ion ($OH^-$) and a positively charged ion when dissolved in water. Bases turn red litmus paper blue and feel slippery to the touch. Sodium hydroxide (NaOH) and aqueous ammonia ($NH_4OH$) are examples of bases.

$$NaOH \xrightarrow{\text{in water}} Na^+ + OH^-$$

Sodium        Sodium   Hydroxide
hydroxide      ion        ion

Bases react with hydrogen ions and remove them from solution.

## pH

Since the concentration of hydrogen or hydroxide ions is usually small, it is convenient to express the degree of acidity or alkalinity in a solution in terms of **pH,** formally defined as the negative logarithm of the hydrogen ion concentration. The pH scale is logarithmic, extending from 0, the pH of a very strong acid, to 14, the pH of a very strong base. The pH of pure water is 7, neither acidic nor alkaline (basic), but neutral. Even though water does ionize slightly, the concentrations of $H^+$ ions and $OH^-$ ions are exactly equal; each of them has a concentration of $10^{-7}$, which is why we say that water has a pH of 7. Solutions with a pH of *less* than 7 are acidic and contain more $H^+$ ions than $OH^-$ ions. Solutions with a pH greater than 7 are alkaline and contain more $OH^-$ ions than $H^+$ ions.

Because the scale is logarithmic (to base 10), a solution with a pH of 6 has a hydrogen ion concentration that is ten times greater than a solution with a pH of 7, and is much more acidic. A pH of 5 represents another tenfold increase. Therefore, a solution with a pH of 4 is $10 \times 10$ or 100 times more acidic than a solution with a pH of 6.

The contents of most animal and plant cells are neither strongly acidic nor alkaline but are an essentially neutral mixture of acidic and basic substances. Most life cannot exist if the pH of the cell changes very much.

# How to Make a Difference

Individuals and groups all around the world are beginning to attack complex environmental problems. Youths from different countries and cultures are working together to clean up the Mediterranean Sea. Women in India are protecting trees from lumberjacks and planting more trees. Political parties that address environmental issues are gaining increasing political clout throughout Europe. Grassroots environmental groups are even organizing in the former Soviet Union to stop development projects that are environmentally unsound.

In this book we have explored some of the complexities of our environment, and we have examined many problems that are the direct result of human actions. Some environmental problems seem almost impossible to address, yet the quality of life and the kind of world we want this place to be, not only for our children and grandchildren but for ourselves, depends largely on the actions we take today.

## WHAT YOU CAN DO AS AN INDIVIDUAL

Throughout this text we have highlighted things you can do as an individual to conserve energy and other natural resources and reduce waste. Although individual savings are small, they represent substantial conservation when combined with other individual contributions. Knowing what you now know about the environment, you should try to break personal habits that harm the environment.

Because you are taking a course in environmental science, you should be well informed on most environmental issues. Make an effort to stay informed and share your knowledge with others. You can help shape environmental policies of the future, but you cannot be effective if you don't know what you're talking about.

Most important, should you be particularly concerned about one or more environmental issues, don't just fret over them. . . . **get committed.** Join appropriate environmental groups whose collective power carries more weight than single voices. Become politically involved on local, state, and national levels. Help to elect leaders who support a sustainable Earth society and influence those leaders in office to support the environmental causes you advocate. And be persistent.

In summary, there are three things you can do to help preserve and improve the environment: (1) Lower your consumption and reduce waste so that your life style reflects the conservation of natural resources; (2) stay informed on environmental topics and share your knowledge; (3) get involved in environmental issues.

## ENVIRONMENTAL ORGANIZATIONS YOU CAN JOIN

The following is an incomplete listing of environmental organizations. Some are grassroots groups that depend on their constituents—people like yourself—for funding and momentum. Others are private, not-for-profit organizations that are funded by foundations or other groups. A few are trade associations or professional societies.

Something almost all these groups have in common is that they offer internships for college-level students. Some groups have organized internship programs, while others have internships available only occasionally. Most offer paid internships, but a few are too small to be able to afford any stipend. You can generally arrange to receive college credit for paid or unpaid internships. When this list was being compiled, a number of the organizations were actively seeking interns, but most environmental organizations do not advertise such positions. Therefore, you must contact the organizations you are most interested in to see if they are currently offering anything of interest to you.

**Air Pollution Control Association,** P.O. Box 2861, Pittsburgh, PA 15230
An international professional organization for people involved in air pollution control.

**Alliance for Environmental Education,** 2111 Wilson Boulevard, Arlington, VA 22201
Works with educational organizations to define environmental issues.

**American Association for World Health,** 2001 S Street NW, Washington, DC 20009
A grassroots organization concerned with national health problems.

**American Forestry Association,** 1516 P Street NW, Washington, DC 20005
Emphasizes conservation of soil and forests, but also involved in air and water pollution issues. Their Global ReLeaf Program addresses the issue of global climate change.

**American Lung Association,** 1740 Broadway, New York, NY 10019
Educates schools and the general community on the health effects of indoor and outdoor air pollution.

**American Rivers, Inc.,** 801 Pennsylvania Avenue SE, Washington, DC 20003
Focuses on the preservation of the nation's rivers and landscapes.

**American Water Resources Association,** 5410 Grosvenor Lane, Bethesda, MD 20814
A multidisciplinary association composed of members interested in many different aspects of water resources.

**Center for Acid Rain and Clean Air Policy Analysis,** 444 N. Capitol St., Washington, DC 20001
A policy-oriented lobbying group established in 1985 by several state governors.

**Center for Environment, Commerce, and Energy,** 733 6th Street SE, Washington, DC 20003
The main mission of this organization is to increase minority participation in environmental issues. They coordinate a minority environmental internship program in which they place black students in internships with a variety of different environmental organizations.

**Center for Marine Conservation** (formerly Center for Environmental Education), 1725 DeSales NW, Washington, DC 20009
Educates the public on topics such as marine pollution issues and marine wildlife sanctuaries.

**Center for Population Options,** 1012 14th Street NW, Washington, DC 20005
A not-for-profit organization that directs its efforts at the prevention of early childbirth, both in the United States and in other countries.

**Center for Science in the Public Interest,** 1501 16th Street NW, Washington, DC 20036
Concerned with sustainable agriculture, alcohol, food, and nutrition.

**Citizens' Clearinghouse for Hazardous Waste,** P.O. Box 6806, Falls Church, VA 22040
Assists citizens' groups in the fight against toxic polluters and for environmental justice.

**Clean Water Action Project,** 1320 18th Street NW, Washington, DC 20036
Lobbies for stricter water pollution controls and safe drinking water.

**Concern, Inc.,** 1794 Columbia Road, NW, Washington, DC 20009
Focuses on environmental education and sends materials to schools and public and private organizations.

**Congress Watch,** 215 Pennsylvania Avenue SE, Washington, DC 20003
Lobbying group (affiliated with The Public Citizen) concerned with a wide range of issues, both environmental and nonenvironmental. Current environmental concern focuses on pesticides, nuclear energy, and alternative energy sources.

**Conservation Foundation,** 1250 24th Street NW, Washington, DC 20037
Conservation education and the ecological impact of foreign aid are some of the concerns emphasized by this group.

**Conservation International Foundation,** 1015 18th Street NW, Washington, DC 20036
This not-for-profit group, which originally splintered from the Nature Conservancy, provides funds and technical support for a variety of projects related to animal and plant conservation.

**Consumer Federation of America,** 1424 16th Street NW, Washington, DC 20036
An advocacy group composed of several hundred consumer organizations that lobbies on a broad range of energy and indoor air pollution issues.

**Cousteau Society,** 930 West 21st Street, Norfolk, VA 23517
Protection and preservation of the oceans for future generations. Although internships are not available, student volunteers are welcome in their Nor-

folk office. Students might wish to participate in Project Ocean Search, in which people from all walks of life spend 2 weeks studying an island ecosystem. Limited scholarships are available.

**Defenders of Wildlife,** 1244 19th Street NW, Washington, DC 20036
Works to preserve, enhance, and protect the diversity of wildlife and the habitats critical for their survival.

**Ducks Unlimited,** #1 Waterfowl Way, Long Grove, IL 60047
Restores and protects wetland habitat for migrating waterfowl. Currently protects over 5.3 million acres in Canada, Mexico, and the United States.

**Environmental Action, Inc.,** 1525 New Hampshire Avenue NW, Washington, DC 20036
Environmental advocacy group that was founded by the originators of Earth Day 1970. Recent concerns include solid and hazardous waste.

**Environmental Defense Fund, Inc.,** 257 Park Avenue S., 16th Floor, New York, NY 10010. Other offices are in Washington, DC; Oakland, CA; Boulder, CO; Raleigh, NC; Austin, TX; and Richmond, VA.
A national not-for-profit organization that uses science, economics, and law to develop economically viable solutions to today's environmental problems. Student internships are offered at all offices.

**Environmental Law Institute,** 1616 P Street NW, 2nd Floor, Washington, DC 20036
A not-for-profit grassroots organization involved in finding legal solutions to environmental problems.

**Friends of the Earth, Inc.,** 218 D Street SE, Washington, DC 20003; also 4512 University Way NE, Seattle, WA 98105
A global advocacy group that works on a wide range of issues at local, national, and international levels. Current projects include ozone protection, tropical deforestation, global warming, and oil spill legislation. Internships are offered from time to time in the Washington, DC, and Seattle offices.

**Greenpeace USA, Inc.,** 1432 U Street NW, Suite 201-A, Washington, DC 20009
An international organization that currently focuses on ocean ecology, hazardous wastes, disarmament, and atmospheric pollution issues.

**International Planned Parenthood Federation,** 902 Broadway, New York, NY 10010
Headquartered in London, this organization helps governments in over 120 different countries establish family planning services.

**Izaak Walton League of America,** 1401 Wilson Boulevard, Level B, Arlington, VA 22209
The conservation of wildlife, renewable natural resources, and water quality are the concerns of this group.

**League of Conservation Voters,** 1150 Connecticut Avenue NW, Suite 201, Washington, DC 20036
A nonpartisan political action group that works to elect pro-environmental candidates to Congress. They also publish a scorecard that rates members of Congress on their environmental votes.

**National Audubon Society,** 950 Third Avenue, New York, NY 10022. Also 9 regional offices and over 500 local chapters.
Interested in a wide range of environmental issues. Currently runs over 80 wildlife sanctuaries.

**National Parks and Conservation Association,** 1015 31st Street NW, Washington, DC 20007
Acquisition and protection of national parks. Active in environmental issues as they relate to the national parks.

**National Resources Defense Council,** 40 West 20th Street, New York, NY 10011. Offices in Washington, DC; Los Angeles; San Francisco; and Hawaii.
Organizing and litigation on a number of environmental issues, including atmospheric problems (ozone, global warming, clean air), nuclear weapons, and pesticides.

**National Solid Wastes Management Association,** 1730 Rhode Island Avenue NW, Washington, DC 20036
A trade association that works with federal and state governments and provides educational services on solid-waste management and resource recovery.

**National Wildlife Federation,** 1400 16th Street NW, Washington, DC 20036
The largest private conservation organization in the world. Conservation education is its primary mission. Its conservation internship program is primarily intended for recent college graduates, but college seniors sometimes qualify.

**Natural Resources Council of America,** 1015 31st Street NW, Washington, DC 20007
A consortium that serves over 75 different environmental groups.

**The Nature Conservancy,** 1815 North Lynn Street, Arlington, VA 22209
Global preservation of natural diversity. This group finds, protects, and maintains the best examples of

communities, ecosystems, and endangered species in the natural world.

**New Alchemy Institute,** 237 Hatchville Road, East Falmouth, MA 02536
This group concentrates on research and education with an emphasis on agricultural methods (including IPM), passive solar energy, and energy conservation.

**Physicians for Social Responsibility,** 1000 16th Street NW, Washington, DC 20036
Focuses on the effects of nuclear war and nuclear weapons on human health.

**Planet/Drum Foundation,** P.O. Box 31251, San Francisco, CA 94131
Ecological education with a focus on urban sustainability.

**Planned Parenthood Federation of America,** 810 Seventh Avenue, New York, NY 10019
Addresses the family planning needs of nearly 8 million men and women around the world each year.

**Population Crisis Committee,** 1120 19th Street NW, Washington, DC 20036
International voluntary family planning is the focus of this not-for-profit research group.

**The Population Institute,** 110 Maryland Avenue NE, Washington, DC 20002
Education, lobbying, and public policy on population growth, particularly in developing nations.

**Population Reference Bureau,** 777 14th Street NW, Washington, DC 20005
A private, not-for-profit educational organization that disseminates demographic and population information. PRB is a clearinghouse on United States and international population matters. Internships are only available for recent college graduates.

**Population Resource Center,** 500 East 66nd Street, New York, NY 10021; second office in Washington, DC.
Provides seminars and briefings on population issues for policy makers. Past focus has been on the United States, but recently they have done projects in Africa and Latin America.

**The Public Citizen,** 2000 P Street NW, Washington, DC 20036
Political action group founded by Ralph Nader. Composed of several sister organizations with different missions, including Congress Watch, Critical Mass (energy project), and the Health Research Group.

**Public Interest Research Group,** 215 Pennsylvania Avenue SE, Washington, DC 20003
An environmental and consumer advocacy and research group.

**Renew America,** 1400 16th Street NW, Washington, DC 20036
A not-for-profit organization that focuses on educational and information networking on national and global environmental issues.

**Resources for the Future,** 1616 P Street NW, Washington, DC 20036
Concerned with the quality of the environment and the conservation of natural resources.

**Scientists' Institute for Public Information,** 355 Lexington Avenue, New York, NY 10017
Strives to improve the understanding of science by the general public through working with the media. One of its departments focuses on the environment and the media.

**Sierra Club,** 730 Polk Street, San Francisco, CA 94109; 408 C Street NE, Washington, DC 20002
Interested in conserving the natural environment by influencing public policy decisions. Sierra Club members explore, enjoy, and protect the wild places of the Earth. Internships are offered in the San Francisco and Washington, DC offices.

**Smithsonian Institution,** 1000 Jefferson Drive SW, Washington, DC 20560
Sponsors a wide variety of environmental education and research programs.

**Student Conservation Association, Inc.,** P.O. Box 550, Charlestown, NH 03603
Resource assistant program places college students in internship positions at such places as national parks. Opportunities are available throughout the year.

**Union of Concerned Scientists,** 1616 P Street NW, Washington, DC 20036
Nuclear power safety and arms control are the main focuses of this group.

**Water Pollution Control Federation,** 601 Wythe Street, Alexandria, VA 22314
Focuses primarily on wastewater treatment and water quality. Although internships are not available, scholarships are sometimes offered to both undergraduate and graduate students.

**The Wilderness Society,** 900 17th Street NW, Washington, DC 20006
Wilderness, parks, and public lands are the focuses of this group.

**Wildlife Habitat Enhancement Council,** 1010 Wayne Avenue, Silver Spring, MD 20910
A not-for-profit consortium of corporations and conservation groups that makes recommendations on how to enhance corporate-held lands (such as rights-of-way under power transmission lines) for wildlife. Internships are offered for recent college graduates and possibly undergraduates.

**Wildlife Management Institute,** 1101 14th Street NW, Washington, DC 20005
A scientific and educational organization that works to improve the professional management and wise use of wildlife and other natural resources. Internships are offered only in rare instances.

**Wildlife Protectorate,** P.O. Box 19608, Alexandria, VA 22320
The protection of rare and endangered bird species is accomplished through the purchase of land held in trust. This newcomer group on the conservation scene has plans to obtain properties in Central America and the Caribbean as well as in the United States. Internships are currently not available, but student tours (primarily for grade school students) are conducted at one of their properties.

**Wildlife Society,** 5410 Grosvenor Lane, Bethesda, MD 20814
A professional society for the resource management field.

**World Environment Center,** 419 Park Avenue S., New York, NY 10016
A liaison between government and industry leaders to resolve environmental problems caused by industry.

**World Resources Institute,** 1709 New York Avenue NW, Washington, DC 20006
A policy research center that focuses on the environment and development.

**World Wildlife Fund,** 1250 24th Street NW, Washington, DC 20036
Finds ways to save endangered species, including the acquisition of wildlife habitat. Internships are offered from time to time, primarily for college graduates.

**Worldwide Network,** 1331 H Street NW, Suite 903, Washington, DC 20005
An international network for women involved in environmental issues, from women in rural villages to national leaders. Closely affiliated with the United Nations Environment Program (UNEP).

**Zero Population Growth,** 1400 16th Street NW, Washington, DC 20036

The largest nonprofit membership group concerned with the human population's impact on resources and the environment.

## GET POLITICALLY INVOLVED

It is important to let your elected officials know about your concerns. Writing to your Senator or Congressional Representative is a very effective way to help influence the formation of policies and laws affecting environmental issues. It also helps to stay in touch when laws that have been passed are being implemented.

### How to Write to Elected Officials

When addressing a letter, you should use the appropriate heading. If you are writing to the President of the United States, address your letter to: The President, The White House, 1600 Pennsylvania Avenue NW, Washington, DC 20500, Dear Mr. President. A letter to a Senator should be addressed to The Honorable _____, Senate Office Building, Washington, DC 20510, Dear Senator _____. When writing to your representative, address the letter to The Honorable _____, House Office Building, Washington, DC 20515, Dear Representative _____.

**The Do's.** Letters are more effective if you keep them short (1 page maximum) and to the point. Talk about the item of concern as specifically as possible: how it impacts you and others, what your position is on the matter, and why. If possible, make a specific request of the official; that is, let them know what you would like them to do. If you are writing about a particular bill, it would be helpful to know the number or name of the bill. Also, make sure you include your name and return address in case they wish to respond.

**The Don'ts.** Letters are more effective if you avoid being contentious or rude. Also, don't come across as high and mighty ("as a taxpayer who put you in office . . ."). Don't introduce more than one issue or allow the letter to ramble on and on.

### Federal and International Agencies You Can Contact

The following agencies can be contacted for information about specific environmental bills, laws, and issues.

**Agency for International Development**
320 21st Street NW, Washington, DC 20523

**Bureau of the Census**
U.S. Department of Commerce, Washington, DC 20233

**Bureau of Land Management**
U.S. Department of Interior, Interior Building, Room 5600, Washington, DC 20240

**Bureau of Mines**
U.S. Department of Interior, 2401 E Street NW, Washington, DC 20241

**Bureau of Reclamation**
U.S. Department of Interior, Interior Building, Room 7654, Washington, DC 20240

**Congressional Research Service, Library of Congress**
101 Independence Avenue SE, Washington, DC 20540

**Council on Environmental Quality**
722 Jackson Place NW, Washington, DC 20006

**Department of Agriculture**
14th and Independence Avenue SW, Washington, DC 20250

**Department of Energy**
Forrestal Building, 1000 Independence Avenue SW, Washington, DC 20585

**Department of the Interior**
18th and C Streets NW, Washington, DC 20240

**Environmental Protection Agency**
401 M Street SW, Washington, DC 20460

**Federal Energy Regulatory Commission**
825 North Capitol Street NE, Washington, DC 20426

**Fish and Wildlife Service**
U.S. Department of Interior, Washington, DC 20240

**Food and Drug Administration**
U.S. Department of Health and Human Services, 5600 Fishers Lane, Rockville, MD 20852

**Forest Service**
U.S. Department of Agriculture, P.O. Box 96090, Washington, DC 20090

**Geological Survey**
U.S. Department of Interior, 12201 Sunrise Valley Drive, Reston, VA 22092

**Government Printing Office**
941 North Capitol Street NE, Washington, DC 20402

**National Academy of Sciences**
2101 Constitution Avenue NW, Washington, DC 20418

**National Oceanic and Atmospheric Administration**
6010 Executive Boulevard, Rockville, MD 20852

**National Park Service**
U.S. Department of Interior, P.O. Box 37127, Washington, DC 20013-7127

**National Response Center for Water Pollution**
2100 2nd Street SW, Washington, DC 20593

**National Science Foundation**
1800 G Street NW, Washington, DC 20550

**Nuclear Regulatory Commission**
1717 H Street NW, Washington, DC 20555

**Occupational Safety and Health Administration**
U.S. Department of Labor, 200 Constitution Avenue NW, Washington, DC 20210

**Soil Conservation Service**
U.S. Department of Agriculture, P.O. Box 2890, Washington, DC 20013

**United Nations Environment Program**
P.O. Box 30552, Nairobi, Kenya; or, 2 United Nations Plaza, Room 803, New York, NY 10017

# Green Collar Professions

Careers relating to the environment are varied and their number is ever-increasing. The following list of "green collar" jobs, organized into three general categories, is intended only as a broad sample. It should be noted that many jobs relating to the environment are by nature cross-disciplinary. With a creative approach, you can contribute to the welfare of our environment through almost any career path.

## ENVIRONMENTAL PROTECTION (SOLID AND HAZARDOUS WASTE MANAGEMENT, POLLUTION CONTROL)

AIR QUALITY ENGINEER (i.e., analyzing and controlling air pollution)

ATMOSPHERIC SCIENTIST (i.e., measuring the chemical composition of the atmosphere)

CHEMICAL ENGINEER (i.e., designing systems for chemical waste disposal)

CHEMIST (i.e., testing toxins and their interactions for possible environmental effects)

ELECTRICAL ENGINEER (i.e., designing energy-efficient power sources)

EMERGENCY RESPONSE SPECIALIST (i.e., managing the emergency cleanup of chemical fires or spills)

ENVIRONMENTAL ENGINEER (i.e., maintaining public water resources)

ENVIRONMENTAL HEALTH SCIENTIST (i.e., monitoring the health of ecosystems)

ENVIRONMENTAL PROTECTION SPECIALIST (i.e., reviewing government contracts for compliance with EPA regulations)

HAZARDOUS WASTE MANAGER (i.e., transporting and disposing of hazardous materials)

HEALTH PHYSICIST (i.e., developing protective measures for radiation exposure)

INDUSTRIAL HYGIENIST (i.e., establishing procedures to minimize worker health risks associated with hazardous materials)

MECHANICAL ENGINEER (i.e., designing machinery to comply with environmental regulations)

NOISE-CONTROL SPECIALIST (i.e., developing methods of neutralizing noise at heavy-industry manufacturing sites)

NUCLEAR ENGINEER (i.e., improving design safety of nuclear power plants; disposing of nuclear waste)

RISK MANAGER (i.e., assessing potential environmental risks of proposed industrial procedures, chiefly for insurance purposes)

SAFETY ENGINEER (i.e., designing environmentally safe manufactured products)

SOLID WASTE MANAGER (i.e., managing the distribution, transportation, and disposal of solid wastes)

TOXICOLOGIST (i.e., determining the link between human illness or disease and exposure to toxic substances)

WATER QUALITY TECHNOLOGIST (i.e., testing for contaminants at water treatment plants)

## NATURAL RESOURCE MANAGEMENT (LAND AND WATER CONSERVATION, FISHERY AND WILDLIFE MANAGEMENT, FORESTRY, PARKS, AND OUTDOOR RECREATION)

AGRICULTURAL ENGINEER (i.e., designing agricultural systems for soil and water conservation)

AGRONOMIST (i.e., developing alternatives to chemical fertilizers and pesticides for use in food crop cultivation)

BIOSTATISTICIAN (i.e., measuring the statistical relation between injury or disease and exposure to pollution or toxins)

BOTANIST (i.e., identifying new plant species; using genetic engineering to develop new, commercially valuable plant breeds)

ECOLOGIST (i.e., studying how natural and human-altered ecosystems work)

FISHERY BIOLOGIST (i.e., studying the effects of changing environmental conditions on the survival and growth of fish)

FISHERY MANAGER (i.e., managing fish hatcheries)

FORESTER/PARK RANGER (i.e., planting and maintaining populations of tree species)

GEOLOGICAL ENGINEER (i.e., designing environmentally safe sanitary landfills; selecting hazardous waste disposal sites)

HYDROLOGIST/GEOLOGIST (i.e., managing water resources for agricultural use)

LANDSCAPE ARCHITECT (i.e., designing parks, recreational, and other public facilities that preserve to restore natural ecosystems)

METEOROLOGIST (i.e., monitoring the effects of atmospheric pollutants on weather patterns)

MICROBIOLOGIST (i.e., studying the effects of toxins on microorganisms as part of natural food chains; discovering microorganisms to be used in bioremediation)

MINING ENGINEER (i.e., designing mining tunnels to minimize dangerous methane gas leaks and groundwater contamination)

OCEANOGRAPHER (i.e., testing ocean water for the presence of sludge and industrial wastes)

PETROLEUM ENGINEER (i.e., ensuring that oil well sites are returned to their original condition after drilling is completed)

SOIL SCIENTIST (i.e., testing soil for toxins leaked from landfill sites)

WILDLIFE BIOLOGIST (i.e., researching wildlife populations)

WILDLIFE MANAGER (i.e., restoring or preserving wildlife populations and habitats)

ZOOLOGIST (i.e., studying the habitat requirements of animal species to prevent their extinction)

## COMMUNICATIONS AND PUBLIC AFFAIRS

ARCHITECT (i.e., designing energy-efficient buildings)

CIVIL ENGINEER (i.e., planning public works projects, including highway and sewage restoration)

COMPUTER SPECIALIST (i.e., developing and operating computer systems that monitor or simulate environmental problems)

DEMOGRAPHER (i.e., researching statistics on human population growth rates to assist in efforts to stabilize population growth)

EDUCATOR (i.e., teaching environmental studies at the undergraduate or graduate level or conducting employee training programs)

ENVIRONMENTAL LAWYER (i.e., specializing in the interpretation of new environmental legislation, regulations, and enforcement procedures)

JOURNALIST (i.e., reporting on environmental issues)

INTERPRETIVE NATURALIST (i.e., conducting educational tours in natural settings)

OCCUPATIONAL/ENVIRONMENTAL PHYSICIAN (i.e., treating patients exposed to radioactive or toxic materials)

TECHNICAL WRITER (i.e., preparing environmental impact statements)

URBAN/COMMUNITY PLANNER (i.e., designing residential areas to preserve natural ecosystems)

# Units of Measure: Some Useful Conversions

## Some Common Prefixes

| Prefix | Meaning | Example | |
|--------|---------|---------|---|
| giga | 1,000,000,000 | 1 gigaton | = 1,000,000,000 tons |
| mega | 1,000,000 | 1 megawatt | = 1,000,000 watts |
| kilo | 1000 | 1 kilojoule | = 1000 joules |
| centi | 0.01 | 1 centimeter | = 0.01 meter |
| milli | 0.001 | 1 milliliter | = 0.001 liter |

## Length: Standard Unit = Meter

1 meter = 39.37 in
1 inch = 2.54 cm
1 mile = 1.609 km

## Area: Standard Unit = Square meter ($m^2$)

1 hectare = 10,000 square meters
         = 2.471 acres
1 acre = 0.405 hectare

## Volume: Standard Unit = Cubic meter ($m^3$)

1 liter = 1000 $cm^3$ = 1.057 qt (U.S.)
1 gallon (U.S.) = 3.785 L

## Mass: Standard Unit = Kilogram

1 kilogram = 2.205 lb
1 metric ton = 1.103 ton
1 pound = 453.6 g

## Energy: Standard Unit = Joule

1 joule = 0.24 cal
1 calorie = 4.184 J
1 British thermal unit = 252 cal

## Electrical Power: Standard Unit = Watt

1 watt = 1 J/second

## Pressure: Standard Unit = Pascal

1 bar = $10^5$ Pa
1 atm = 1.01 bar = $1.01 \times 10^5$ Pa

## Temperature: Standard Unit = Centigrade

°C = (°F − 32) × 5/9
°F = °C × 9/5 + 32
1°C = 1.8°F

**Table 2–1**

Excerpted with permission from Table 2 of B. N. Ames: "Ranking Possible Carcinogenic Hazards," published 4/17/87. *Science* Vol. 236, p. 271. Copyright 1987 by the AAAS.

**Table 2–1**

Excerpted with permission from Table 2 of R. Wilson: "Risk Assessment and Comparisons: An Introduction," published 4/17/87. *Science* Vol. 236, p. 267. Copyright 1987 by the AAAS.

**Figure 2–5**

Adapted from Figs. 1–4 and 1–5 of W. T. Edmondson, *The Uses of Ecology: Lake Washington and Beyond*, University of Washington Press, 1991.

**Figure 2–6**

Adapted from Fig. 1–4 of W. T. Edmondson, *The Uses of Ecology: Lake Washington and Beyond*, University of Washington Press, 1991.

**Figure 6–2**

From BSCS: Biological Science: An Ecological Approach. © 1987 by BSCS. Kendall/Hunt Publishing Company, Dubuque, Iowa.

**Figure 7–1**

Michael Parkin, *Economics*, © 1990 by Addison-Wesley Publishing Company.

**Table 7–2**

Worldwatch Institute.

**Chapter 7, Figure in Focus On: Natural Resources, the Environment, and the National Income Accounts, p. 121.**

Jonathan Levin, "The Economy and the Environment: Revising the National Accounts": *IMF Survey*, 4 June 1990. (Washington: International Monetary Fund, 1990.)

**Figure 10–22**

Natural Resources Defense Council.

**Figure 10–23**

Reprinted by permission of THE WALL STREET JOURNAL, © 1991 Dow Jones & Company, Inc. All Rights Reserved Worldwide.

**Chapter 11, Figure in Focus On: The Effects of Radiation on Living Organisms, p. 220.**

Reprinted with permission from the National Council on Radiation Protection and Measurements. (From NCRP Report No. 93, 1987)

**Chapter 11, Figure in Focus On: The Effects of Nuclear War, p. 228.**

Adapted from "The Climatic Effects of Nuclear War," by Richard Turco et al. Copyright August 1984 by Scientific American, Inc. All rights reserved.

**Table 12–1**

Investor Responsibility Research Center.

**Figure 12–21**

Adapted from "Strategies for Energy Use," by John H. Gibbons, Peter D. Blair, and Holly L. Gwin. Copyright September 1989 by Scientific American, Inc. All rights reserved.

**Table 13–4**

Reprinted with permission from *The Mono Basin Ecosystem: Effects of Changing Lake Level*, © 1987 by the National Academy of Sciences. Published by National Academy Press, Washington, D.C.

**Figure 13–17a**

Mohamed T. El-Ashry, Jan van Schilfgaarde, and Susan Schiffman, "Salinity Pollution from Irrigated Agriculture," Journal of Soil and Water Conservation, Vol. 40. No. 1 (Jan.–Feb. 1985) p. 49.

**Figure 13–18**

Ernest Engelbert, Ann Foley Scheuring, WATER SCARCITY: IMPACTS ON WESTERN AGRICULTURE. Copyright © 1984 The Regents of the University of California.

**Figure 14–9a**

Redrawn by permission from Thompson and Troeh, *Soils and Soil Fertility*, McGraw-Hill, Inc., 1978.

**Figure 15–4b**

Redrawn by permission from p. 937 of *Chemistry* by Radel and Navidi; Copyright © 1990 by West Publishing Company. All rights reserved.

**Figure 18–1**

Data on world food production and food production per person for 1980, 1985, 1986, 1987, 1988, and 1989 from FAO Production Yearbook, Vol. 43, 1989.

**Table 18–2**

*Environment*, 24:9, pp. 6–11, 39–44, 1982. Reprinted with permission of the Helen Dwight Reid Educational Foundation. Published by Heldref Publications, 1319 18th St., N.W., Washington, D.C.

**Figure 18–6**

Worldwatch Institute.

**Chapter 19, Figure in Focus On: Smoking, p. 436.**

Worldwatch Institute.

**Chapter 19, quote on p. 423.**

Ulf Merbold.

**Figure 20–6**

Carbon Dioxide Information Analysis Center, Oak Ridge National Laboratory.

**Figure 20–10**

Climate Changes and Society, by W. W. Kellogg and R. Schware, 1981.

**Table 22–2**

Excerpted with permission from P. L. Adkisson: "Controlling Cotton's Insect Pests: A New System," published 4/2/82. *Science* Vol. 82, p. 216. Copyright 1982 by the AAAS.

**Table 22–3**

Pimental, David, "Environmental and Economic Effects of Reducing Pesticide Use," *BioScience* 41:6, June 1991. © 1991 American Institute of Biological Sciences.

**Figure 22–3b**

Reproduced, by permission, from International travel and health. Vaccination requirements and health advice: situation as of 1 January 1992. Geneva, World Health Organization, 1992.

**Table 22–4**

Excerpted with permission from *Regulating Pesticides in Food: The Delaney Paradox,* © 1987 by the National Academy of Sciences. Published by National Academy Press, Washington, D.C.

**Figure 22–7b**

Paul Debach, *Biological Control by Natural Enemies,* p. 4. London: Cambridge University Press, 1974.

**Figure 22–8b**

Redrawn with permission from J. W. Grier: "Ban of DDT and Subsequent Recovery of Reproduction in Bald Eagles," published 12/17/82. *Science* Vol. 218, p. 1232. Copyright 1982 by the AAAS.

**Figure 23–6**

National Solid Waste Management Association.

**Figure 23–12**

U.S. Congress, Office of Technology Assessment, *Facing America's Trash,* OTA-O-424 (Springfield, VA: National Technical Information Service, October 1989).

**Chapter 23, Figure in Focus on: The Environmental Costs of Diapers, p. 542.**

Copyright 1991 by Consumers Union of the United States, Inc., Yonkers, N.Y. 10703. Excerpted by permission from CONSUMER REPORTS, August 1991.

**The following organizations provided information used throughout the text:**

American Institute of Biological Sciences

Department of Agriculture

Department of Energy

Department of Interior

Environmental Protection Association

Los Alamos National Laboratory

National Park Service

Population Reference Bureau, Inc.

U.S. Geological Survey

U.S. Agency for International Development

United Nations

World Health Organization, Geneva

*Note:* Page numbers followed by f refer to footnotes.